Lecture Notes in Computer Science 999
Edited by G. Goos, J. Hartmanis and J. van Leeuwen

Springer
Berlin
Heidelberg
New York
Barcelona
Budapest
Hong Kong
London
Milan
Paris
Santa Clara
Singapore
Tokyo

Panos Antsaklis Wolf Kohn
Anil Nerode Shankar Sastry (Eds.)

Hybrid Systems II

Springer

Series Editors

Gerhard Goos, Karlsruhe University, Germany

Juris Hartmanis, Cornell University, NY, USA

Jan van Leeuwen, Utrecht University, The Netherlands

Volume Editors

Pano Antsaklis
Department of Electrical Engineering, The University of Notre Dame
Notre Dame, IN 46556, USA

Wolf Kohn
Sagent Corporation
3150 Richards Road SE, Bellevue, WA 98005, USA

Anil Nerode
Mathematical Sciences Institute, Cornell University
Ithaca, NY 14853, USA

Shankar Sastry
Department of Electrical Engineering and Computer Sciences
University of California, Berleley, CA 94720, USA

Cataloging-in-Publication data applied for

Die Deutsche Bibliothek - CIP-Einheitsaufnahme

Hybrid systems. - Berlin ; Heidelberg ; New York ; Barcelona ; Budapest ; Hong Kong ; London ; Milan ; Paris ; Tokyo : Springer.
Literaturangaben
2. Panos Antsaklis ... (ed.). - 1995
(Lecture notes in computer science ; Vol. 999)
ISBN 3-540-60472-3
NE: Antsaklis, Panos [Hrsg.]; GT

CR Subject Classification (1991): C.1.m, C.3, D.2.1, F.3.1, F.1-2

ISBN 3-540-60472-3 Springer-Verlag Berlin Heidelberg New York

Typesetting: Camera-ready by author
SPIN 10485820 06/3142 - 5 4 3 2 1 0 Printed on acid-free paper

Preface

Hybrid Systems are models for networks of digital and continuous devices, in which digital control programs sense and supervise continuous and discrete plants governed by differential or difference equations. Modern industrial society is filled with hybrid systems used for such varied purposes as aircraft control, computer synchronization, manufacturing, communication networks, traffic control, industrial process control, etc. Hybrid systems also provide the basic framework and methodology for the synthesis and analysis of autonomous and intelligent systems. Examples in this area include medical informatics systems, highway control and routing systems, robotics, and database management and retrieval systems.

In addition, hybrid systems theory provides the backbone for the formulation and implementation of learning control policies. In such policies, the control acquires knowledge (discrete data) to improve the behavior of the system as it evolves in time.

Hybrid Systems has become a distinctive area of study due to opportunities to improve on traditional control and estimation technologies by providing computationally effective methodologies for the implementation of digital programs that design or modify the control law in response to sensor detected events, or as a result of learning.

The areas of science and engineering that can be brought to bear on the issue of hybrid control are numerous. These include mathematical disciplines such as functional analysis, variational calculus, ordinary differential equations, linear partial differential equations, Lie algebras, differential geometry, dynamical systems; operations research disciplines such as linear and integer and non-smooth mathematical programming; engineering disciplines such as linear and optimal and intelligent control, stochastic processes, stochastic approximation, discrete event simulation; computer science disciplines such as distributed and agent-based systems, automata theory, and program validation and verification; and branches of mathematical logic and applied logic such as logic programming. We are gradually gaining an understanding of the subtle interplay of mathematical and physical disciplines involved in hybrid systems. The investigation of hybrid systems is creating a new and fascinating discipline bridging mathematics, control engineering, and computer science.

The first workshop on Hybrid Systems was held at the Mathematical Sciences Institute (MSI) of Cornell University June 10-12, 1991. The second workshop was held at the Technical University in Lyngby, Denmark, October 19-21,1992 and inspired the volume *Hybrid Systems* (Springer Lecture Notes in Computer Science 736, Grossman, Nerode, Rischel, Ravn, eds., 1993). The third workshop was held at MSI on October 28-30, 1994 and resulted in this volume, which consists of fully refereed papers.

The Mathematical Sciences Institute of Cornell University, a U.S. Army Research Office Center of Excellence, was the sponsor. We thank Prof. Victor Marek of the University of Kentucky for his coordination of this volume. We thank the anonymous referees for their conscientious efforts. We thank Wilson Kone, Diana Drake, and Valerie Kaines of MSI for their tireless work. Finally, we are very grateful for the consistent support given by Dr. Jagdish Chandra and the Army Research Office Mathematics and Computer Science program for development of this area from its inception.

June 1995

Panos Antsaklis, The University of Notre Dame, Notre Dame, Indiana
Wolf Kohn, Sagent Corporation, Bellevue, Washington
Anil Nerode, Cornell University, Ithaca, New York
Shankar Sastry, The University of California at Berkeley, Berkeley, California

Table of Contents

Symbolic Controller Synthesis for Discrete and Timed Systems 1
E. Asarin, O. Maler and A. Pnueli

A Calculus of Stochastic Systems for the Specification, Simulation, and Hidden State Estimation of Hybrid Stochastic/Non-stochastic Systems 21
A. Benveniste, B.C. Levy, E. Fabre and P. Le Guernic

Condensation of Information from Signals for Process Modeling and Control 45
J.D. Birdwell and B.C. Moore

On the Automatic Verification of Systems with Continuous Variables and Unbounded Discrete Data Structures 64
A. Bouajjani, R. Echahed and R. Robbana

On Dynamically Consistent Hybrid Systems 86
P.E. Caines and Y.-J. Wei

A Self-learning Neuro-Fuzzy System 106
N. DeClaris and M.-C. Su

Viable Control of Hybrid Systems 128
A. Deshpande and P. Varaiya

Modeling and Stability Issues in Hybrid Systems 148
M. Doğruel and U. Özgüner

Hierarchical Hybrid Control: a Case Study 166
D.N. Godbole, J. Lygeros and S. Sastry

Hybrid Systems and Quantum Automata: Preliminary Announcement ... 191
R.L. Grossman and M. Sweedler

Planar Hybrid Systems 202
J. Guckenheimer and S. Johnson

Programming in Hybrid Constraint Languages 226
V. Gupta, R. Jagadeesan, V. Saraswat and D.G. Bobrow

A Note on Abstract Interpretation Strategies for Hybrid Automata 252
T. A. Henzinger and P.-H. Ho

HYTECH: The Cornell HYbrid TECHnology Tool 265
T. A. Henzinger and P.-H. Ho

Hybrid Systems as Finsler Manifolds: Finite State Control as Approximation to Connections 294
W. Kohn, A. Nerode and J.B Remmel

Constructing Hybrid Control Systems from Robust Linear Control Agents 322
M. Lemmon, C. Bett, P. Szymanski and P. Antsaklis

Controllers as Fixed Points of Set-Valued Operators 344
A. Nerode, J.B. Remmel and A. Yakhnis

Verification of Hybrid Systems Using Abstractions 359
A. Puri and P. Varaiya

Control of Continuous Plants by Symbolic Output Feedback 370
J. Raisch

Hybrid Control of a Robot - a Case Study 391
A. P Ravn, H. Rischel, M. Holdgaard, T.J. Eriksen, F. Conrad and T. O Andersen

Verifying Time-bounded Properties for ELECTRE Reactive Programs with Stopwatch Automata 405
O. Roux, V. Rusu

Inductive Modeling: A Framework Marrying Systems Theory and Non-monotonic Reasoning 417
H.S. Sarjoughian and B. Zeigler

Semantics and Verification of Hierarchical CRP Programs 436
R.K. Shyamasundar and S. Ramesh

Interface and Controller Design for Hybrid Control Systems 462
J.A. Stiver, P. J. Antsaklis and M.D. Lemmon

Hybrid Objects 493
M. Tittus and B. Egardt

Modeling of Hybrid Systems Based on Extended Coloured Petri Nets 509
Y.Y. Yang, D.A. Linkens and S.P. Banks

DEVS Framework for Modeling, Simulation, Analysis, and Design of Hybrid Systems 529
B.P. Zeigler, H.S. Song, T.G Kim, H. Praehofer

Synthesis of Hybrid Constraint-Based Controllers 552
Y. Zhang and A.K. Mackworth

Author Index 569

Symbolic Controller Synthesis for Discrete and Timed Systems*

Eugene Asarin[1] Oded Maler[2] Amir Pnueli[3]

[1] Institute for Information Transmission Problems, 19 Ermolovoy st., Moscow, Russia, asarin@ippi.msk.su

[2] SPECTRE – VERIMAG, Miniparc-ZIRST, 38330 Montbonnot, France, Oded.Maler@imag.fr

[3] Dept. of Computer Science, Weizmann Inst. Rehovot 76100, Israel, amir@wisdom.weizmann.ac.il

Abstract. This paper presents algorithms for the *symbolic* synthesis of discrete and real-time controllers. At the semantic level the controller is synthesized by finding a winning strategy for certain games defined by automata or by timed-automata. The algorithms for finding such strategies need, this way or another, to search the state-space of the system which grows exponentially with the number of components. Symbolic methods allow such a search to be conducted without necessarily enumerating the state-space. This is achieved by representing sets of states using formulae (syntactic objects) over state variables. Although in the worst case such methods are as bad as enumerative ones, many huge practical problems can be treated by fine-tuned symbolic methods. In this paper the scope of these methods is extended from analysis to synthesis and from purely discrete systems to real-time systems.

We believe that these results will pave the way for the application of program synthesis techniques to the construction of real-time embedded systems from their specifications and to a solution of other related design problems associated with real-time systems in general and asynchronous circuits in particular.

1 Introduction

Apart from the different underlying state-spaces and time domains, perhaps the largest difference between control theory and computer science[4] lies in the relative weight of *analysis* and *synthesis* methods. Analysis can be roughly stated as: *Will this happen?* while synthesis as: *How can we make this happen?*

* This research was supported in part by the European Community projects BRA-REACT(6021), HYBRID EC-US-043 and INTAS-94-697 as well as by Research Grant #93-012-884 of Russian Foundation of Fundamental Research. VERIMAG is a joint laboratory of CNRS, INPG, UJF and VERILOG SA. SPECTRE is a project of INRIA.

[4] We take computer science to denote here the community of those who want to reason formally about the behavior of computers.

There is nothing inherent in the nature of the discrete or the continuous that makes control people more centered around synthesis and informaticians around analysis (also called "verification"). The study of "passive" analysis of continuous systems is simply not called "control" but rather the theory of dynamical systems, differential equations, etc. On the other hand, the attempts to build automatic program synthesis (or "derivation") methods, although numerous, have mostly been considered as sci-fi dreams, given the common belief that analysis/verification is already hard enough. One line of research in program synthesis concentrated on non-reactive programs, i.e., systems that operate in a "static" environment, e.g., [MWa80]. In this framework a program is derived as a constructive proof of an existential statement which constitutes the specification. In spite of the absence of external disturbances, this problem is much harder than the one we consider because no a-priori program structure is assumed.

For reactive systems (systems that maintain an ongoing interaction with a dynamic environment) the synthesis problem has been posed as early as 1957 by Church [Chu63] in the context of digital circuits. Church's problem was solved by Büchi and Landweber [BL69], (a readable exposition of their result appeared in [TB73]). More modern efforts toward automatic synthesis have been made by various authors such as [EC82], [MWo84], [PR89-a], [PR89-b], [ALW89] or [WD91], but apart from some impressive theoretical results (in particular, complexity bounds) the work on synthesis remained marginal compared to the vast literature on verification and, as far as we know, has not been transferred from academia to industry. Interestingly, a large body of work dealing with discrete synthesis came from outsiders to computer science, namely the DEDS model of Ramadge and Wonham [RW89] in which the controller can inhibit certain transitions of an automaton in order to achieve some behavioral specifications.

Meanwhile there have been some breakthroughs in the verification area. The analysis problem, although intractable in terms of worst-case asymptotic complexity, became feasible for industrial size problems, especially in hardware. This success is due to the use of *symbolic* methods [BCM+93], [McM93] that do not transform the description of the system into an enormous "flat" automaton but rather represent the transition relation as a formula over the state variables. Given such a formula $\mathcal{T}$ and a formula P describing some subset F of the state-space one can calculate a new formula P' characterizing the set F' of successors (or predecessors) of F. The goal of this paper is to discuss the transfer of this technology from analysis to synthesis and from discrete to real-time systems. The only work along similar lines we are aware of is that of Hoffmann and Wong-Toi [HW92-a], [HW92-b], [BHG+93] who introduce symbolic methods into the the Ramadge-Wonham model. Although this paper presents no new theoretical result, we hope that it will contribute to a better understanding of the nature of real-time control and its associated computational problems.

The rest of the paper is organized as follows. In section 2 we set the stage for the discrete case and give semantic and symbolic characterizations of winning strategies. The derivation of a discrete scheduler in section 3 illustrates the idea. Timed automata and real-time games are described in section 4 along with a solution of the symbolic synthesis problem. In section 5 we demonstrate a solution of a real-time version of the scheduler problem. Finally we mention application and implementation issues.

2 Discrete Systems

2.1 The Core Idea behind Discrete Synthesis

The most fruitful approach to discrete program synthesis is to view the ongoing interaction between the system one wants to design and the environment[5] in which it is supposed to operate as some variant of the von Neumann-Morgenstern discrete games. A strategy for a given game is a rule that tells the controller how to choose between several possible actions in any game position. A strategy is good if the controller, by following these rules, always wins (according to a given definition of winning) no matter what the environment does.

The existence (and extraction) of a strategy for finite games is done using the max-min principle of [NM44], disguised sometimes as searching AND-OR trees or as the elimination of an alternating pair of the logical quantifiers $\exists$ and $\forall$. This principle is illustrated using the game at the left part of figure 1. In this game the controller starts from position 0 and can choose between the two actions a_1 and a_2. Then the environment can choose between b_1 and b_2. The winning condition is specified via some subset F of $\{1,2,3,4\}$. A run of the game is winning if it ends up in an element of F. Suppose $F = \{1,4\}$ – in this case the first player has no winning strategy at state 0 because if it chooses a_1, the adversary can take b_2 and reach state 2. If it chooses a_2 the adversary can reach state 3 by taking b_1. Hence, 0 is not a winning position. If, on the other hand, we consider a game with the same transition structure but with $F = \{1,2\}$ then there is a winning strategy as the controller can by making a_1 "force" the environment into F.

The convention that each player plays in its own turn can be replaced by having a simultaneous move. In this case there are no two types of states and the transition between states is made by a *joint action* of the two players as in the left of figure 1. *This is the convention that we adopt in this paper*, and hence the notions of "first" and "second" player bear no ordinal meaning.[6] For games on continuous time the notion of turns becomes anyway meaningless.

Some readers may wonder whether such trivialities deserve a relatively long verbal exposition. We consider this to be the *essence* of any synthesis algorithm, a fact which is sometimes obscured by fancy complicated constructs. The mathematical formulation of this notion for a game with a state-space Q is via an operator $\pi : 2^Q \rightarrow 2^Q$ assigning for every $F \subseteq Q$ the set $\pi(F)$ denoting its *controllable predecessors*, that is, the set of states from which the controller can force

[5] As one control theoretician once said: "you CS people call *environment* everything that lies outside the computer".

[6] It can be shown that every von Neumann-Morgenstern game can be converted into an equivalent simultaneous action game.

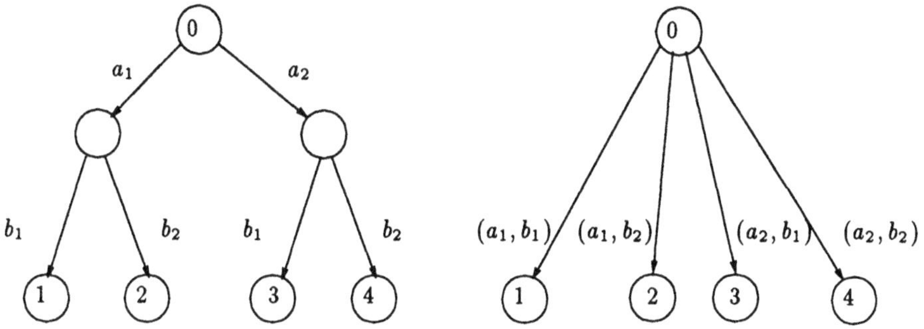

Fig. 1. A simple game.

its adversary into F. In the example above $\{0\} \notin \pi(\{1,4\})$ and $\{0\} \in \pi(\{1,2\})$. Calculating this operator, together with some set-theoretical operations constitutes the core of any synthesis algorithm.

2.2 Semantic Version

Definition 1 (Game Automaton). *A game automaton is $\mathcal{A} = (Q, \delta, q_0, F)$ where $Q = Q_x \times Q_y$ is a set of states, $\delta : Q \rightarrow 2^Q$ is the transition relation, $q_0 \in Q$ is the initial state of the game and $F \subseteq Q$ is a set of accepting states. We assume that δ admits a decomposition into $\delta_x : Q \rightarrow 2^{Q_x}$ and $\delta_y : Q \rightarrow 2^{Q_y}$ such that for every $q \in Q$, $\delta(q) = \delta_x(q) \times \delta_y(q)$.*

One can see that the game automaton is a product of two automata communicating via their states, that is, the transition made by the X-automaton may depend on the current state of the Y-automaton and vice versa.

A *run* of the game is any infinite sequence $\xi = q[1], q[2], \ldots$ such that $q[1] = q_0$ and for every i, $q[i+1] \in \delta(q[i])$. The first player wins a run of the game if the run always stays in F. Otherwise it loses the run.

Remark 1: We have considered one of the several possible definitions of winning, namely the $\Box$-condition (also known as "safety", or "closed" game). One can imagine the "dual" game as viewed from the second player's perspective – this player wins a run of the game if it succeeds to push the run at least once into $Q - F$. These are called $\Diamond$-games ("eventuality", "open"). Since the emphasis in this paper is on real-time and on symbolic methods, we will not refer here to more complicated winning conditions – see [Tho94] for a recent survey.

A strategy for the first player is a transition relation $\delta^* : Q \rightarrow 2^{Q_x}$ such that $\delta^*(q) \subseteq \delta(q)$ for every $q \in Q$. This can be viewed as a rule telling the controller which actions he can take while being at any $q \in Q$. A strategy can be deterministic ($|\delta(q)| = 1$) but it need not be so – in case we need later to implement a strategy we can choose arbitrarily some $q' \in \delta(q)$.

If we plug δ_x^* instead of δ_x we obtain a more restricted automaton $\mathcal{A}^*$ with $\delta^* \subseteq \delta$. A strategy δ^* is a winning one if all the runs of $\mathcal{A}^*$ are winning.

The search for a strategy is performed via the determination of the set F^* of winning states. This is done iteratively starting with $F_0 = F$. Every iteration we create F_{i+1} by removing from F_i all the states from which the first player *cannot* force the game into F_i. It is not hard to see that a state q belongs to F_i if, starting from q, the controller can stay in F for at least i steps. This procedure converges to the set of F^* of winning states. If $q_0 \in F^*$ the controller has a winning strategy. Formally:

Definition 2 (Controllable Predecessors). *Let $\mathcal{A} = (Q, \delta, q_0, F)$ be a game automaton. We define a function $\pi : 2^Q \mapsto 2^Q$ as*

$$\pi(P) = \{q : \exists q'_x \in \delta_x(q) \forall q'_y \in \delta_y(q) (q'_x, q'_y) \in P\} \tag{1}$$

The algorithm for calculating the winning states works as follows:

Algorithm 1 (Synthesis for Discrete $\Box$-Games).

$$\begin{array}{l} F_0 := F \\ \textbf{for } i = 0, 1, \ldots, \textbf{ repeat} \\ \qquad F_{i+1} := F_i \cap \pi(F_i) \\ \textbf{until } F_{i+1} = F_i \end{array}$$

The algorithm is illustrated in figure 2. The strategy for every $q \in Q$ is extracted as follows:

$$\delta^*(q) = \begin{cases} \emptyset & \text{if } q \notin F^* \\ \{q'_x : \forall q'_y \in \delta_y(q) (q'_x, q'_y) \in F^*\} & \text{otherwise} \end{cases}$$

Note that in the first case ($q \notin F^*$) we do not care because anyway we cannot win after arriving to such a state. This might be different had we been interested in more quantitative notions of winning, e.g., expected probability.
Remark 2: We extract the strategy *after* we have completed the calculation of the winning states. We could have done it incrementally by removing losing transitions after each iteration. In games with winning condition other than $\Box$, it might be better (and sometimes even necessary) to calculate δ^* iteratively along with the calculation of F^*.

This is all that has to be said about safety games. More complex games require more complicated iterative procedures and sometimes they cannot be won by simple strategies (those that depend only on q) but rather need some more information on the history of the game (as a continuous controller needs sometime to know the sign of the derivative of the state-variable). We refer the reader again to the survey [Tho94] as well as to [NYY92] and [TW94a].

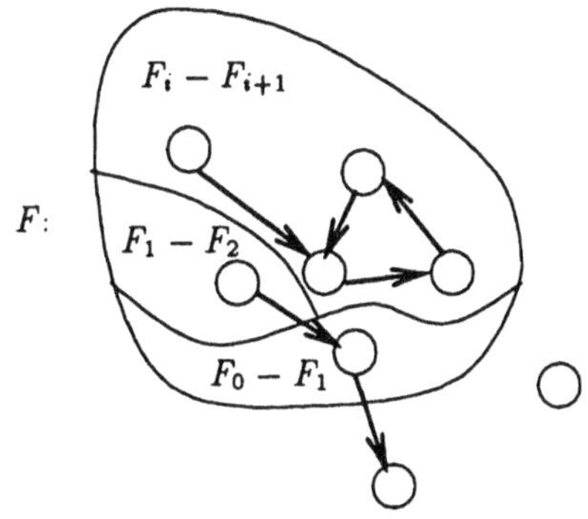

Fig. 2. An illustration of algorithm 1.

2.3 Symbolic Version

So far we have considered the state-space of $\mathcal{A}$ to be an amorphous set Q without any structure. Large complex systems, however, are usually composed of smaller sub-systems, and the global state-space is the Cartesian product of the local ones (and so is the transition relation). Instead of creating this huge automaton and applying analysis or synthesis algorithms to it, symbolic methods keep the transition relation in a syntactic form and operate on it.

Suppose a system is defined using a set $X = \{x_1, \ldots, x_k\}$ of Boolean variables. Hence $Q = \{0,1\}^k$ and every subset F of Q can be described by one (or more) Boolean formula P over X. A transition relation $R \subseteq \{0,1\}^k \times \{0,1\}^k$ can be written as a formula over a set $X \cup X'$ of variables where each x_i' represents the value of x_i in the "next" state.

A symbolic analysis method consists of a class $\mathcal{F}$ of syntactic objects (formulae) covering all the subsets of Q, such that for every set-theoretic operation on 2^Q there is a semantics-preserving operation on $\mathcal{F}$. In particular one needs a class of objects in which there are syntactic operations that correspond to: 1) set-theoretic operations, 2) the predecessor operator π, and 3) equality testing. This is all is needed in order to perform algorithm 1. Certain classes of syntactic objects, such as ordered BDDs [Bry86] have a canonicity property, namely one-to-one mapping between syntactic objects and the sets they denote – this makes equality testing trivial. In our exposition we will not insist on a particular representation (which is an implementation question) but rather use arbitrary Boolean formulae. Note that elimination of quantifiers is a simple syntactic operation for Boolean formulae:

$$\exists x_i\, P(x_1, \ldots, x_i, \ldots, x_k) = P(x_1 \ldots, 0, \ldots, x_k) \vee P(x_1, \ldots, 1, \ldots x_k)$$

and

$$\forall x_i\, P(x_1, \ldots, x_i, \ldots, x_k) = P(x_1 \ldots, 0, \ldots, x_k) \wedge P(x_1, \ldots, 1, \ldots x_k)$$

In order to adapt the symbolic method for synthesis we must first introduce some notion of interaction between automata and "ownership" of state-variables. Intuitively every process has its own set of local variables which cannot be changed by other processes. On the other hand, it may "read" the state of other processes and base its decision concerning which transition to make on the values of these variables. Syntactically, if the corresponding sets of variables are X and Y, the respective transition formulae of the two automata can be written as $\mathcal{T}_X(X,Y,X')$ and $\mathcal{T}_Y(Y,X,Y')$. When we compose them together we obtain a closed system where both X and Y are local variables and the transition formula is $\mathcal{T}(X,Y,X',Y') = \mathcal{T}_X(X,Y,X') \wedge \mathcal{T}_Y(Y,X,Y')$.

In a control setting we let the variables X denote the controllable variables, namely the variables owned by the controller and which he can change (or refrain from changing). The set Y consists of environmental variables which the controller cannot influence directly (however, since $\mathcal{T}_Y$ depends on X, the controller can influence them indirectly, which is essentially what control is all about).

We use quantifiers of the form $\exists X$ or $\forall X$ as an abbreviation for $\exists x_1 \exists x_2 \ldots$, etc., and assume without loss of generality that the transition relation is complete. Given $\mathcal{T}_X$, $\mathcal{T}_Y$ and a set F expressed by a formula $P(X,Y)$, the syntactic predecessor operator id defined as follows:

$$\pi(P)(X,Y) = \exists X' \left[\mathcal{T}_X(X,Y,X') \wedge \forall Y' \left(\mathcal{T}_Y(Y,X,Y') \Rightarrow P(X',Y')\right)\right] \quad (2)$$

To rephrase it verbally, the immediate controllable predecessor of P are all the states from which there is an X "action" such that for every Y action the resulting state satisfies P.

Having all the other ingredients of a symbolic method we can plug it into algorithm 1 and converge to a formula P^* characterizing all the winning states. The controller is derived from $\mathcal{T}$ and P^* as follows:

$$\mathcal{T}_X^*(X,Y,X') = P^*(X,Y) \wedge \forall Y' \left(\mathcal{T}_Y(Y,X,Y') \Rightarrow P^*(X',Y')\right) \quad (3)$$

The transition relation expressed by $\mathcal{T}_X^*$ is a subset of the one expressed by $\mathcal{T}_X$ and is obtained by restricting the possible values for X'. The transition relation of the whole system after the controller is synthesized is expressed by $\mathcal{T}^*(X,Y,X',Y') = \mathcal{T}_X^*(X,Y,X') \wedge \mathcal{T}_Y(Y,X,Y')$. All its runs are winning for the first player.

This is all the story. The question whether there exists an efficient implementation scheme allowing large-scale synthesis is an empirical open question. Some positive responses were reported in [BHG+93].

3 An Example: a Discrete Scheduler

Suppose we have two identical processes with the corresponding state variables S_1 and S_2. Each of them can be either in I (idle) or W (waiting). A process can be at $S_i = I$ as long as it wishes and can generate a request (move to $S_i = W$) only if at least 3 time units have passed since the previous request. It can move from $S_i = W$ to $S_i = I$ whenever the scheduler gives him permission by letting the variable $G_i = 1$ (we assume here that the service is immediate). For modeling this behavior we have for each i a variable C_i ranging over $\{0,1,2,3\}$ measuring the number of steps since the previous request. The system is depicted in figure 3 and the product of S_i and C_i is the automaton in figure 4. The whole system consists of a product $S_1 \circ C_1 \circ S_2 \circ C_2 \circ G$ where G is the automaton for the scheduler that we want to synthesize. It is a 2-variable (4-state) automaton that decides the values of G_1 and G_2. We have not drawn the whole system (the formulae are sufficiently large).

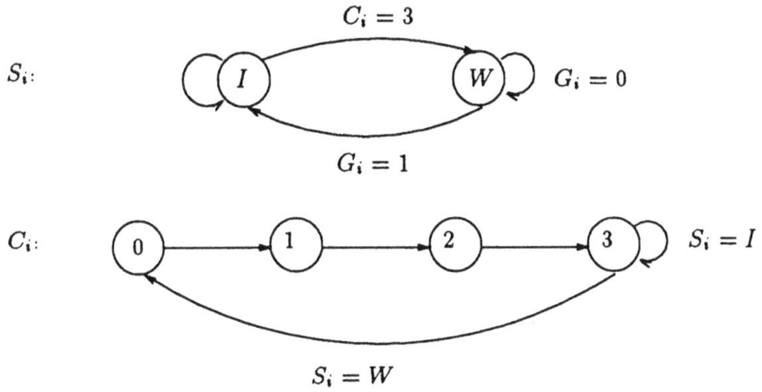

Fig. 3. The automata for process $i \in \{1,2\}$, responsible for the variables S_i and C_i. Unlabeled transitions can be made unconditionally.

The transition formula $\mathcal{T}_y$ appears below. Since we start with the most liberal controller that allows everything, we have $\mathcal{T}_x$ = **true** and $\mathcal{T} = \mathcal{T}_y$ does not mention the variables G_1' and G_2'.

$$\mathcal{T}_y(S_1, C_1, S_2, C_2, G_1, G_2, S_1', C_1', S_2', C_2') =$$

$$\begin{aligned}
&(S_1 = I \wedge (S_1' = I \vee (C_1 = 3 \wedge S_1' = W))\vee \\
&S_1 = W \wedge (G_1 = 1 \wedge S_1' = I \vee G_1 = 0 \wedge S_1' = W))\wedge \\
&(C_1 = 0 \wedge C_1' = 1 \vee C_1 = 1 \wedge C_1' = 2 \vee C_1 = 2 \wedge C_1' = 3\vee \\
&C_1 = 3 \wedge (S_1 = I \wedge C_1' = 3 \vee S_1 = W \wedge C_1' = 0))\wedge \\
&(S_2 = I \wedge (S_2' = I \vee (C_2 = 3 \wedge S_2' = W))\vee \\
&S_2 = W \wedge (G_2 = 1 \wedge S_2' = I \vee G_2 = 0 \wedge S_2' = W))\wedge \\
&(C_2 = 0 \wedge C_2' = 1 \vee C_2 = 1 \wedge C_2' = 2 \vee C_2 = 2 \wedge C_2' = 3\vee \\
&C_2 = 3 \wedge (S_2 = I \wedge C_2' = 3 \vee S_2 = W \wedge C_2' = 0))
\end{aligned}$$

The performance specifications (winning conditions) are that no process will wait in W more than 2 time units, and that mutual exclusion is satisfied, i.e., that at least one of G_1, G_2 is zero. This can be expressed by the formula

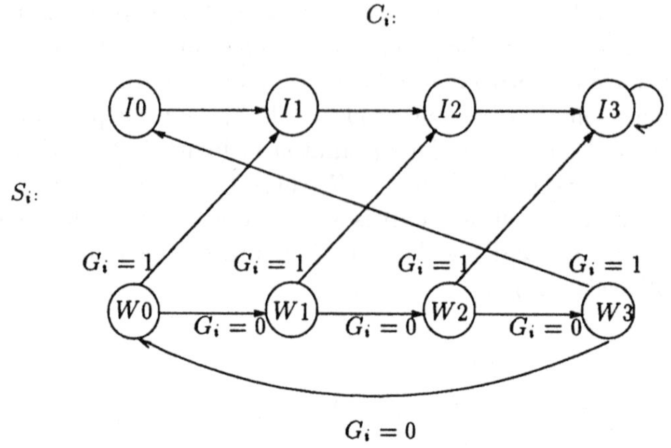

Fig. 4. The automaton $S_i \circ C_i$ for $i \in \{1,2\}$. Note that the only "external" variable this automaton refers to is G_i.

$$P(S_1, C_1, S_2, C_2, G_1, G_2) =$$
$$\neg(S_1 = W \wedge C_1 = 2 \vee S_2 = W \wedge C_2 = W \vee G_1 = 1 \wedge G_2 = 1)$$

By performing the algorithm we obtain the following sequence of formulae:

$$\begin{aligned} P_0 &= \quad \neg(S_1 = W \wedge C_1 = 2 \vee S_2 = W \wedge C_2 = W \vee G_1 = 1 \wedge G_2 = 1) \\ P_1 &= P_0 \wedge \neg(G_1 = 1 \wedge G_2 = 1) \\ P_2 &= P_1 \wedge \neg(S_1 = W \wedge C_1 = 1 \wedge G_1 = 0 \vee S_2 = W \wedge C_2 = 1 \wedge G_2 = 0) \\ P_3 &= P_2 \wedge \neg(S_1 = W \wedge C_1 = 0 \wedge G_1 = 0 \wedge S_2 = W \wedge C_2 = 0 \wedge G_2 = 0) \end{aligned}$$

The first iteration excludes violation of mutual exclusion. The second iteration excludes the case when for some i, $C_i = 1$ and $G_i = 0$ – in this case the next step will take us to a bad state $C_i = 2 \wedge S_i = W$. Finally the third iteration excludes the states where the two processes have moved to W and the scheduler has not allocated the resource to at least one of them. The resulting controller, appearing in figure 5, is specificed by the formula $\mathcal{T}_X^*$:

$$\mathcal{T}_x(G_1, G_2, S_1, C_1, S_2, C_2, G_1', G_2') =$$

$$\begin{aligned} &((S_1 = W \wedge C_1 = 0) \Rightarrow G_1' = 1) \wedge \\ &((S_2 = W \wedge C_2 = 0) \Rightarrow G_2' = 1) \wedge \\ &(G_1 = 0 \wedge S_1 = W \wedge C_1 = 3 \wedge \\ &G_2 = 0 \wedge S_2 = W \wedge C_2 = 3) \Rightarrow (G_1' = 1 \vee G_2' = 1) \end{aligned}$$

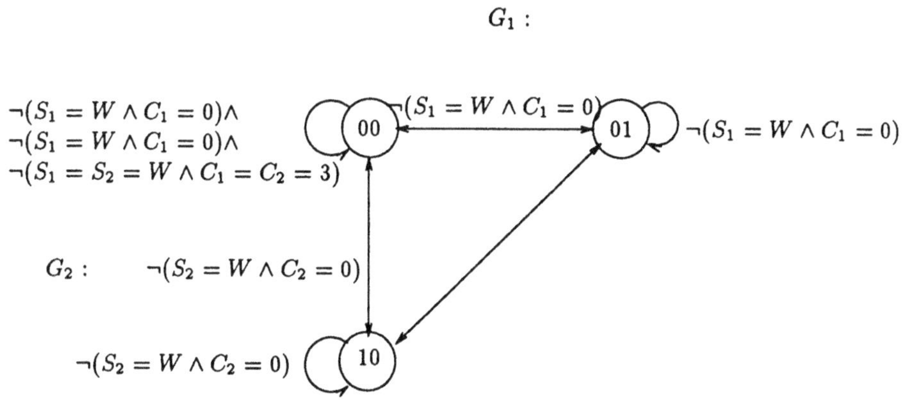

Fig. 5. The resulting scheduler obtained by restricting the complete 4-state automaton.

4 Timed Systems

4.1 Real-Time Games

In real-time games the outcome of the players' actions depend also on their timing as performing the same action "now" or "later" might have completely different consequences. For such games we take the model of *timed automata* [AD94], in which automata are equipped with auxiliary continuous variables called *clocks* which grow uniformly when the automaton is in some state. The clocks interact with the transitions by participating in pre-conditions (guards) for certain transitions and they are possibly reset when some transitions are taken.

In this continuous-time setting, a player might choose at a given moment to wait some time t and *then* take a transition. Unlike purely-discrete games, he should consider not only what the adversary can do *after* this action but also the possibility that the latter will *not wait* for t time, and perform an action at some $t' < t$. Thus the two-person game becomes a three-player game in which Time can interfere in favor of both other players.

While synthesizing a controller for timed automata one should be careful not letting any of the players win by "Zenonism", that is, by preventing the time from progressing as does the Tortoise in its race against Achilles.

4.2 Semantic Version

For the sake of readers not familiar with timed automata we start with an informal illustration of the behavior of these creatures. Consider the timed automaton of figure 6. It has two states and two clocks z_1 and z_2. Suppose it starts operating in the configuration $(q_1, 0, 0)$ (the two last coordinates denote the values of the clocks). Then it can stay at q_1 as long as the staying condition for q_1 is true, namely $z_1 < 2$. Meanwhile the values of the clocks grow and the set of all configurations reachable from $(q_1, 0, 0)$ without leaving q_1 is $\{(q_1, t, t) : 0 \leq t \leq 2\}$. However, after one second, the condition $z_1 \geq 1$ (the guard of the transition from q_1 to q_2) is satisfied and the automaton can move to q_2 while setting z_2 to 0. Hence the additional reachable configurations are $\{(q_2, t, 0) : 1 \leq t \leq 2\}$. Having entered q_2 in one of these configurations, the automaton can either stay there as long as $z_1 < 5 \wedge z_2 < 3$ or can unconditionally move to $(q_1, 0, 0)$, etc.

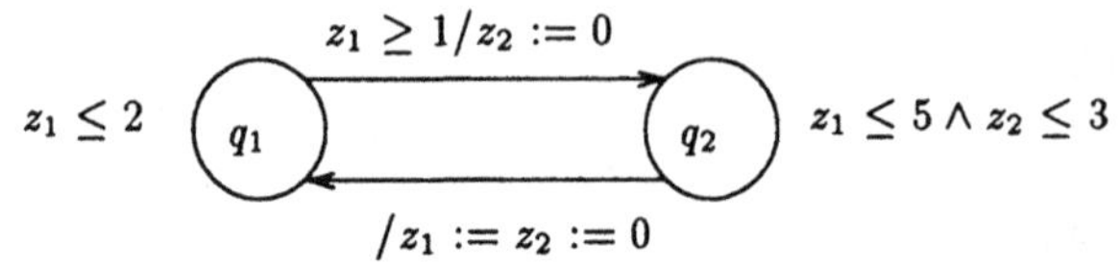

Fig. 6. A timed automaton.

Since the state-space of timed automata contains real-variables, we have an infinite-state automaton and a purely semantic approach where all the states are enumerated is impossible. We will use notation such as $T_{qq'}$ to denote the set of values in the clock space such that a transition from q to $q' \neq q$ is possible ("guards"). Similarly, T_{qq} denotes the set of clock values for which the automaton can stay in q ("staying conditions"). In timed automata such sets are restricted to be k-polyhedral subsets of $(\mathbb{R}^+)^d$, that is, the class of sets obtainable by set-theoretic operations from halfspaces of the form $\{v_1, \ldots, v_d : v_i \leq c\}$, $\{v_1, \ldots, v_d : v_i < c\}$, $\{v_1, \ldots, v_d : v_i - v_j \leq c\}$ or $\{v_1, \ldots, v_d : v_i - v_j < c\}$ for some integer $c \in \{0, \ldots, k\}$ for some positive integer k. These sets constitute the finite *region graph* [AD94] whose properties are the basis for all analysis methods for timed automata. Since we model interaction between two automata, the guards and staying conditions of one automaton may depend, in addition, on the state of the other automaton and thus can be a union of sets of the form $\{q_i\} \times L_i$ with $L_i \subseteq (\mathbb{R}^+)^d$. We will call such sets k-polyhedral as well.

A function $f : \mathbb{R}^d \to \mathbb{R}^d$ is a reset function if its sets some of its arguments to 0 and leaves the others intact. We will use $R_{qq'}$ the denote the reset function associated with every pair of states. Without loss of generality we assume that there is only one transition associated with every ordered pair of states. Finally, for $z \in (\mathbb{R}^+)^d$ we use $z + t$ to denote $z + t \cdot \mathbf{1}$ where $\mathbf{1} = (1, 1, \ldots, 1)$ is a d-dimensional unit vector.

Definition 3 (Timed Game Automaton). *A timed game automaton is $\mathcal{A} = (Q, Z, \delta_x, \delta_y, q_0, F)$ such that*

- *$Q = Q_x \times Q_y$ is a discrete set,*
- *$Z = Z_x \times Z_y = (\mathbb{R}^+)^d$ is the clock space (and $Q \times Z$ is called the configuration space),*

- $\delta_x : Q \times Z \times I\!R^+ \rightarrow 2^{Q_x \times Z_x}$ *and*
- $\delta_y : Q \times Z \times I\!R^+ \rightarrow 2^{Q_y \times Z_y}$ *are the transition relations for the two players.*
- $q_0 \in Q$, *and*
- $F \subseteq Q \times Z$ *is a set of accepting configurations.*

It is required that δ_x *and* δ_y *admit the following decomposition: For every* $q_x, q'_x \in Q_x$ *and* $q_y, q'_y \in Q_y$, *let* $T_{q_x q'_x} \subseteq Q_y \times Z$ *and* $T_{q_y q'_y} \subseteq Q_x \times Z$ *be k-polyhedral sets let* $R_{q_x q'_x} : Z_x \rightarrow Z_x$ *and* $R_{q_y q'_y} : Z_y \rightarrow Z_y$ *be reset functions. Then for every* $(q_x, z_x) \in Q_x \times Z_x$, $(q_y, z_y) \in Q_y \times Z_y$:

$$\begin{aligned}\delta_x((q_x, z_x, q_y, z_y), t) = \{(q'_x, z'_x) : &\forall t' \in [0, t)\, (q_y, z + t') \in T_{q_x q_x} \wedge \\ &(q_y, z + t) \in T_{q_x q'_x} \wedge z'_x = R_{q_x q'_x}(z + t)\}\end{aligned}$$

$$\begin{aligned}\delta_y((q_y, z_y, q_x, z_x), t) = \{(q'_y, z'_y) : &\forall t' \in [0, t)\, (q_x, z + t') \in T_{q_y q_y} \wedge \\ &(q_x, z + t) \in T_{q_y q'_y} \wedge z'_y = R_{q_y q'_y}(z + t)\}\end{aligned}$$

The meaning of $\delta_x((q_x, z_x, q_y, z_y), t)$ is the set of $Q_x \times Z_x$ configuration the first player can reach by waiting t time, and then making *at most* one transition – given that the other player has done nothing meanwhile. The meaning of δ_y is symmetric. This allows us to define the predecessors operator rather simply:

Definition 4 (Timed Controllable Predecessors). *For a given timed game automaton* $\mathcal{A} = (Q, Z, \delta_x, \delta_y, q_0, F)$ *we define a function* $\pi : 2^{Q \times Z} \mapsto 2^{Q \times Z}$ *as*

$$\begin{aligned}\pi(F) = \{&(q_x, z_x, q_y, z_y) : \\ &\exists t \geq 0 \quad (q'_x, z'_x) \in \delta_x((q_x, z_x, q_y, z_y), t) \quad \wedge \quad (q'_x, z'_x, q_y, z_y + t) \in F \wedge \\ &\quad \forall t' < t\ ((q'_y, z'_y) \in \delta_y((q_y, z_y, q_x, z_x), t') \Rightarrow (q_x, z_x + t', q_y, z_y) \in F) \wedge \\ &\qquad\qquad ((q'_y, z'_y) \in \delta_y((q_y, z_y, q_x, z_x), t) \ \Rightarrow \ (q'_x, z'_x + t, q_y, z_y) \in F) \wedge \\ &\quad t < \infty \Rightarrow q'_x \neq q_x\}\end{aligned}$$

Verbally this means that from the configuration (q_x, z_x, q_y, z_y) the controller can force the game to stay in F by taking a transition after waiting t such that whatever transition the environment can take during the interval $[0, t)$ will not steer the game out of F. This is the essence of the definition (lines 2 and 3). Line 4 takes care of the special case where $t' = t$ and the two players make their transition simultaneously. The last line makes sure that the first player will not play Zenonist tricks, i.e., will try to prevent the progress of time without taking any transition. He is allowed to refrain from action only if it chooses $t = \infty$. We assume, initially, that δ_x is strongly non-Zeno, i.e., there is minimal period of time d such that every two discrete transitions must be separated by an interval of at least d. This condition can be relaxed into a condition on cycles, but, as observed in the context of asynchronous circuits [MP95], you really do not need to interleave two discrete transitions in zero time.

An important fact about this operator (first stated explicitly in [MPS95], but really follows immediately form region-graph properties): *The class of k-polyhedral sets is closed under* π. This means that algorithm 1, when initiated with a k-polyhedral set F, is guaranteed to converge to a fixed point F^* as the number of k-polyhedral sets is finite. The strategy δ_x^*, which is a restriction of δ_x, can be obtained by restricting the sets $T_{q_x q_x'}$ as follows:

$$T^*_{q_x q_x'} = T_{q_x q_x'} \cap \{(q_y, z_x, z_y) : \delta_x((q_x, z_x, q_y, z_y), 0) \subseteq F^*\}$$

This approach to real-time synthesis has been first presented in [MPS95]. Alternative approaches, e.g., [OW90], [BW93] are based on a discrete time model. Wong-Toi and Hoffmann [WH92] use timed automata, but then they discretize the system into a an untimed automaton (essentially the region graph) and synthesize the controller using discrete symbolic methods.

4.3 Symbolic Version

For timed automata we need syntactic objects to represent subsets of the binary hypercube (discrete sets of states) as well as k-polyhedral subsets of the Euclidean space (which cannot be enumerated anyway). Systems of *linear inequalities* and sets of *vertices* are among the syntactic objects used to represent such sets. A very useful representation is based on the difference matrices of Dill [Dil89]. This representation is employed in the timed automata analysis tool KRONOS, developed at VERIMAG [DOY94]. This representation does not have a canonicity property (unless the set is convex). Following our presentation of the discrete case, we will not commit ourselves to this or that representation formalism but rather use arbitrary linear inequalities. Some of the ideas in this section are adapted from the symbolic analysis methods for timed automata [HNSY94], [ACD93].

We will use discrete sets of variables as before X and Y, whose sets of valuation constitute the sets Q_x and Q_y, and augment them with two sets of clock variables C_X and C_Y ranging over the non-negative reals and whose valuations are the elements of Z_x and Z_y respectively. All variable will have primed versions, X', C_x', Y' and C_y' to represent transition formulae.

The game automaton is described by the formulae $T_x(X, C_x, Y, C_y, X', C_x')$ and $T_y(Y, C_y, X, C_x, Y', C_y')$. Such formulae specify the instantaneous transition relation, namely the transitions that can be made in zero time. These formulae should also capture the "idle" transition and thus they contain a conjunction with $X = X' \Rightarrow C_x = C_x'$ and $Y = Y' \Rightarrow C_y = C_y'$ respectively. They are the syntactic equivalents of the $T_{qq'}$ of the semantic version.

The formulae $T_x(X, C_x, Y, C_y, X', C_x', C_y', t)$ and $T_y(Y, C_y, X, C_x, Y', C_y', C_x', t)$ which indicate what the two automata can make by waiting t and doing at most one transition, are constructed from T_x and T_y. They are the analogues of δ_x and δ_y. Note that t is a variable in these formulae:

$$\begin{aligned}
&\mathcal{T}_x(X, C_x, Y, C_y, X', C'_x, C'_y, t) = \\
&\forall t' < t \;\; T_x(X, C_x + t, Y, C_y + t, X, C_x + t) \wedge \\
&\qquad\qquad T_x(X, C_x + t, Y, C_y + t, X', C'_x)
\end{aligned}$$

$$\begin{aligned}
&\mathcal{T}_y(Y, C_y, X, C_x, Y', C'_y, C'_x, t) = \\
&\forall t' < t \;\; T_y(Y, C_y + t, X, C_x + t, Y, C_y + t) \wedge \\
&\qquad\qquad T_y(Y, C_y + t, X, C_x + t, Y', C'_y)
\end{aligned}$$

For a formula $P(X, C_x, Y, C_y)$ denoting a set of configurations we define a predecessor formula as:

$$\pi(P)(X, C_x, Y, C_y) =$$

$$\begin{pmatrix}
\exists t \geq 0 \, \exists X' \neq X \, \exists C'_x \;\; \mathcal{T}_x(X, C_x, Y, C_y, X', C'_x, t) \wedge \\
\forall t' \leq t \, \forall Y' \forall C'_y \;\; (\mathcal{T}_y(Y, C_y, X, C_x, Y', C'_y, t') \Rightarrow \\
((t = t' \wedge P(X', C'_x, Y', C'_y)) \vee (t < t' \wedge P(X, C_x + t, Y', C'_y)))
\end{pmatrix} \vee \quad (4)$$

$$\begin{pmatrix}
\forall t \geq 0 \;\; \mathcal{T}_x(X, C_x, Y, C_y, X, C_x + t, t) \wedge \\
\forall t' \leq t \, \forall Y' \forall C'_y \;\; (\mathcal{T}_y(Y, C_y, X, C_x, Y', C'_y, t') \Rightarrow P(X, C_x + t, Y', C'_y)
\end{pmatrix}$$

As before, we can apply algorithm 1 and converge to a formula P^*, from which we derive the controller as a restricted transition formula T_x^*. Note that this formula returns revised guards and staying conditions, something which is very useful for many design problems beside synthesis.

$$T_x^*(X, C_x, Y, C_y, X', C'_x) =$$

$$\begin{aligned}
&\exists t \geq 0 \, \mathcal{T}_x(X, C_x, Y, C_y, X', C'_x, t) \wedge \qquad (5) \\
&\;\; \forall t' < t \, \forall Y' \forall C'_y \;\; (\mathcal{T}_y(X, C_x, Y, C_y, Y', C'_y, t') \Rightarrow P^*(X', C'_x, Y', C'_y)) \wedge \\
&\;\; (t > 0 \wedge X = X') \vee (t = 0 \wedge X \neq X')
\end{aligned}$$

This concludes the symbolic synthesis for timed systems.

Remark 3: Logically speaking, the language we use here is some decidable fragment of the first-order logic over the reals with constants, addition and order.

5 An Example: a Real-Time Scheduler

We take again two processes and a scheduler. Each process (see figure 7) has one discrete variable S_i which can be in one of three states, Idle, Waiting and Busy. It has two clocks: c_i, which is used to enforce minimal inter-arrival time of 3, and d_i, which measures service time – we assume that every process must spend one time unit at B before coming back to I.

The scheduler is as in the discrete example, however we do not start with the most general scheduler but rather enforce a strong non-Zeno condition using

S_i:

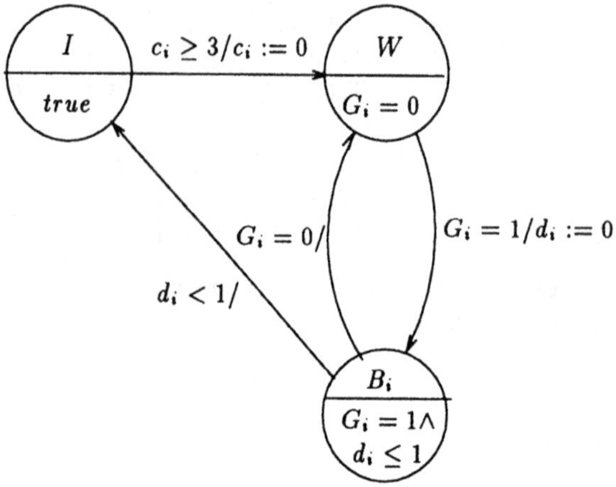

Fig. 7. A timed automaton for S_i, c_i and d_i, $i \in \{1, 2\}$.

the variable z. The scheduler has to spend at least 0.5 time at every state (see figure 8). For simplicity we exclude the bad state $G_1 = 1 \wedge G_2 = 1$ from the initial scheduler. The corresponding transition formulae are:

$$T_y(S_1, c_1, d_1, S_2, c_2, d_2, G_1, G_2, z, S_1', c_1', d_1', S_2', c_2', d_2') =$$

$$\begin{array}{l}
(S_1 = I \wedge S_1' = I \wedge c_1' = c_1 \wedge d_1' = d_1 \vee \\
\;S_1 = I \wedge c_1 \geq 3 \wedge S_1' = W \wedge c_1' = 0 \wedge d_1' = d_1 \vee \\
\;S_1 = W \wedge G_1 = 0 \wedge S_1' = W \wedge c_1' = c_1 \wedge d_1' = d_1 \vee \\
\;S_1 = W \wedge G_1 = 1 \wedge S_1' = B \wedge c_1' = c_1 \wedge d_1' = 0 \vee \\
\;S_1 = B \wedge G_1 = 1 \wedge d_1 < 1 \wedge S_1' = B \wedge c_1' = c_1 \wedge d_1' = d_1 \vee \\
\;S_1 = B \wedge G_1 = 0 \wedge S_1' = W \wedge c_1' = c_1 \wedge d_1' = d_1 \vee \\
\;S_1 = B \wedge d_1 = 1 \wedge S_1' = I \wedge c_1' = c_1 \wedge d_1' = d_1) \wedge \\
(S_2 = I \wedge S_2' = I \wedge c_2' = c_2 \wedge d_2' = d_2 \vee \\
\;S_2 = I \wedge c_2 \geq 3 \wedge S_2' = W \wedge c_2' = 0 \wedge d_2' = d_2 \vee \\
\;S_2 = W \wedge G_2 = 0 \wedge S_2' = W \wedge c_2' = c_2 \wedge d_2' = d_2 \vee \\
\;S_2 = W \wedge G_2 = 1 \wedge S_2' = B \wedge c_2' = c_2 \wedge d_2' = 0 \vee \\
\;S_2 = B \wedge G_2 = 1 \wedge d_2 < 1 \wedge S_2' = B \wedge c_2' = c_2 \wedge d_2' = d_2 \vee \\
\;S_2 = B \wedge G_2 = 0 \wedge S_2' = W \wedge c_2' = c_2 \wedge d_2' = d_2 \vee \\
\;S_2 = B \wedge d_2 = 1 \wedge S_2' = I \wedge c_2' = c_2 \wedge d_2' = d_2)
\end{array}$$

$$T_x(G_1, G_2, z, S_1, c_1, d_1, S_2, c_2, d_2, G_1', G_2', z') =$$

$$\begin{array}{l}
(G_1 = 0 \vee G_2 = 0) \wedge (G_1' = 0 \vee G_2' = 0) \wedge \\
\;G_1' = G_1 \wedge G_2' = G_2 \wedge z' = z \vee \\
(G_1' \neq G_1 \vee G_2' \neq G_2) \wedge z \geq 0.5 \wedge z' = 0
\end{array}$$

The winning condition is

$$P(S_1, c_1, d_1, S_2, c_2, d_2, G_1, G_2, z) = \neg(S_1 \neq I \wedge c_1 > 3 \vee S_2 \neq I \wedge c_2 > 3)$$

The iteration for the winning states goes as follows:

$$\begin{aligned}
P_0 =& \quad \neg(S_1 \neq I \wedge c_1 > 3 \vee S_2 \neq I \wedge c_2 > 3) \\
P_1 = P_0 \wedge& \neg(S_1 = B \wedge z < 0.5 \wedge c_1 - d_1 > 2 \wedge c_1 - z > 2.5 \vee \\
& S_2 = B \wedge z < 0.5 \wedge c_2 - d_2 > 2 \wedge c_2 - z > 2.5 \vee \\
& S_1 = W_1 \wedge z < 0.5 \wedge c_1 - z > 2.5 \vee \\
& S_2 = W_2 \wedge z < 0.5 \wedge c_2 - z > 2.5) \\
P_2 = P_1 \wedge& \neg(S_1 = W \wedge c_1 > 2.5 \vee \\
& S_2 = W \wedge c_2 > 2.5 \vee \\
& S_1 = B \wedge c_1 - d_1 > 2 \wedge c_1 > 2.5 \vee \\
& S_2 = B \wedge c_2 - d_2 > 2 \wedge c_2 > 2.5) \\
P_3 = P_2 \wedge& \neg(S_1 = W \wedge z < 0.5 \wedge c_1 - z > 2 \vee \\
& S_2 = W \wedge z < 0.5 \wedge c_2 - z > 2 \vee \\
& S_1 = B \wedge z < 0.5 \wedge c_1 - d_1 > 2 \wedge c_1 - z > 2 \vee \\
& S_2 = B \wedge z < 0.5 \wedge c_2 - d_2 > 2 \wedge c_2 - z > 2) \\
P_4 = P_3 \wedge& \neg(S_1 = W \wedge c_1 > 2 \vee \\
& S_2 = W \wedge c_2 > 2 \vee \\
& S_1 = B \wedge c_1 - d_1 > 2 \vee \\
& S_2 = B \wedge c_2 - d_2 > 2) \\
P_5 = P_4 \wedge& \neg(S_1 = W \wedge z < 0.5 \wedge c_1 - z > 1.5 \vee \\
& S_2 = W \wedge z < 0.5 \wedge c_2 - z > 1.5) \\
P_6 = P_5 \wedge& \neg(S_1 = S_2 = W \wedge c_1 > 1.5 \wedge c_2 > 1.5 \vee \\
& S_1 = W \wedge S_2 = B \wedge c_2 > 1.5 \wedge c_1 - d_2 > 1 \vee \\
& S_1 = B \wedge S_2 = W \wedge c_1 > 1.5 \wedge c_2 - d_1 > 1) \\
P_7 = P_6 \wedge& \neg(S_1 = S_2 = W \wedge z < 0.5 \wedge c_1 - z > 1 \wedge c_2 - z > 1 \vee \\
& S_1 = W \wedge S_2 = B \wedge z < 0.5 \wedge c_2 - z > 1 \wedge c_1 - d_2 > 1 \vee \\
& S_1 = B \wedge S_2 = W \wedge z < 0.5 \wedge c_1 - z > 1 \wedge c_2 - d_1 > 1) \\
P_8 = P_7 \wedge& \neg(S_1 = S_2 = W \wedge c_1 > 1 \wedge c_2 > 1) \\
P_9 = P_8 \wedge& \neg(S_1 = S_2 = W \wedge z < 0.5 \wedge c_1 - z > 0.5 \wedge c_2 - z > 0.5)
\end{aligned}$$

The first 5 iterations deal with failures not related to interaction of the two processes. At each step a part of "size" 0.5 is subtracted from the set of winning configurations (this is due to anti-Zeno constant 0.5 of the initial controller). The condition $\neg P_4$ means that it is too late to serve a process. The condition P_5 and the first line of P_6 exclude a situation when we are unable to start a service at the due time because of switching delays. The last 3 iterations correspond to failures related to interaction of the processes. This is done in steps of "size" 0.5. The synthesized controller is expressed by the formula:[7]

[7] We leave the drawing of the controller to the reader.

$T_x^*(G_1, G_2, z, S_1, c_1, d_1, S_2, c_2, d_2, G_1', G_2', z') =$

$P^*(S_1, c_1, d_1, S_2, c_2, d_2, G_1, G_2, z) \wedge$
$T_x(G_1, G_2, z, S_1, c_1, d_1, S_2, c_2, d_2, G_1', G_2', z') \wedge$
$(S_1 \neq I \wedge S_2 \neq I \wedge c_1 > 0.5 \wedge c_2 > 0.5 \wedge (G_1 = 1 \dot{\vee} G_2 = 1)) \Rightarrow (G_1' = 1 \vee G_2' = 1) \wedge$
$(S_1 = B \wedge S_2 = W \wedge c_1 > 1 \wedge c_2 > 1) \Rightarrow G_1' = 1 \wedge$
$(S_1 = W \wedge S_2 = B \wedge c_1 > 1 \wedge c_2 > 1) \Rightarrow G_2' = 1 \wedge$
$(S_1 = S_2 = W \wedge c_1 = c_2 = 1) \Rightarrow (G_1' = 1 \vee G_2' = 1) \wedge$
$(S_1 = S_2 = W \wedge c_1 > 1 \wedge c_2 = 1) \Rightarrow G_1' = 1 \wedge$
$(S_1 = W = S_2 = W \wedge c_1 = 1 \wedge c_2 > 1) \Rightarrow G_2' = 1 \wedge$
$(S_1 = B \wedge C_1 > 1.5) \Rightarrow G_1' = 1 \wedge$
$(S_2 = B \wedge C_2 > 1.5) \Rightarrow G_2' = 1 \wedge$
$(S_1 = W \wedge G_2 = 1 \wedge c_1 > 1.5) \Rightarrow (G_1' = 1 \vee G_2' = 1) \wedge$
$(S_2 = W \wedge G_1 = 1 \wedge c_2 > 1.5) \Rightarrow (G_1' = 1 \vee G_2' = 1) \wedge$
$(S_1 = W \wedge G_1 = G_2 = 0 \wedge c_1 > 1.5) \Rightarrow G_2' = 0 \wedge$
$(S_2 = W \wedge G_1 = G_2 = 0 \wedge c_2 > 1.5) \Rightarrow G_1' = 0 \wedge$
$(S_1 \neq I \wedge c_1 \geq 2) \Rightarrow G_1' = 1 \wedge$
$(S_2 \neq I \wedge c_2 \geq 2) \Rightarrow G_2' = 1$

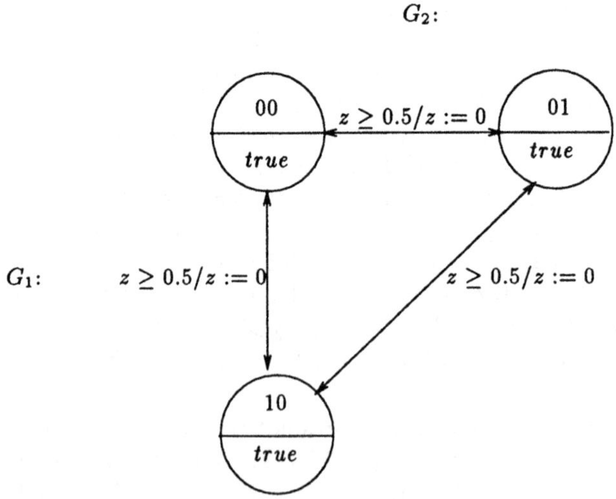

Fig. 8. The initial scheduler.

6 Discussion

We have demonstrated how the synthesis problem for real-time systems can be formulated and solved symbolically. As one potential application let us mention the delay analysis of asynchronous circuits. It has been shown [MP95] that every such circuit can be modeled as a timed automaton such that every wire requires one Boolean variable and one clock variable. The guards and staying conditions for the automaton are derived from the delay characteristics of the gates and from constraints on the variability of the input signals. Using the method described in this paper, we can formulate a game by considering the inputs as environmental variables (Y) and the outputs of the gates as controllable variables. By solving the synthesis problem for some winning condition (specification), we obtain the minimal timing requirements the gates need to satisfy in order to meet the specification. If we switch the roles and treat the gate variables as fixed and the inputs as controllable, we solve the opposite problem: what is the largest class of input signals against which the circuit will behave properly.

Without a convincing implementation, all this remains, of course, wishful thinking. For the discrete part of the system, there is no a-priori reason to believe that the practical hardness of synthesis will be much bigger than that of verification. The main challenge is to find more efficient data-structures and algorithms for treating k-polyhedral sets, and combining them with the discrete ones.

Acknowledgments This work grew out of discussions with J. Sifakis. We thank A. Bouajjani and Y. Lakhneche for commenting on previous drafts of the paper and A. Nerode for his help in stopping Time.

References

[AD94] R. Alur and D.L. Dill, A Theory of Timed Automata, *Theoretical Computer Science* 126, 183–235, 1994.

[ACD93] R. Alur, C. Courcoubetis, and D.L. Dill, Model Checking in Dense Real Time, Information and Computation 104, 2–34, 1993.

[ALW89] M. Abadi, L. Lamport, and P. Wolper, Realizable and Unrealizable Concurrent Program Specifications. In *Proc. 16th ICALP*, volume 372 of Lect. Notes in Comp. Sci., pages 1–17. Springer-Verlag, 1989.

[BHG+93] S. Balemi, G.J. Hoffmann, P. Gyugyi, H. Wong-Toi and G.F. Franklin, Supervisory Control of a Rapid Thermal Multiprocessor, *IEEE Trans. on Automatic Control* 38, 1040–1059, 1993.

[Bry86] R.E. Bryant, Graph-based Algorithms for Boolean Function Manipulation, *IEEE Trans. on Computers* C-35, 677-691, 1986.

[BCM+93] J.R. Burch, E.M. Clarke, K.L. McMillan, D.L. Dill, and L.J. Hwang, Symbolic Model-Checking: 10^{20} States and Beyond, *Proc. LICS'90*, Philadelphia, 1990.

[BW93] B.A. Brandin and W.M. Wonham, Supervisory Control of Timed Discrete-event Systems, *IEEE Transactions on Automatic Control*, 39, 329-342, 1994.

[BL69] J.R. Büchi and L.H. Landweber, Solving Sequential Conditions by Finite-state Operators, *Trans. of the AMS* 138, 295-311, 1969.

[Chu63] A. Church, Logic, Arithmetic and Automata, in *Proc. of the Int. Cong. of Mathematicians 1962*, 23-35, 1963.

[DOY94] C. Daws, A. Olivero and S. Yovine, Verifying ET-LOTOS Programs with KRONOS, *Proc. FORTE'94*, Bern, 1994.

[Dil89] D.L. Dill, Timing Assumptions and Verification of Finite-State Concurrent Systems, in J. Sifakis (Ed.), *Automatic Verification Methods for Finite State Systems*, volume 407 of Lect. Notes in Comp. Sci., Springer, 1989.

[EC82] E.A. Emerson and E.M. Clarke, Using Branching Time Temporal Logic to Synthesize Synchronization Skeletons, *Science of Computer Programming* 2, 241-266, 1982.

[HNSY94] T. Henzinger, X. Nicollin, J. Sifakis, and S. Yovine, Symbolic Model-checking for Real-time Systems, *Information and Computation* 111, 193–244, 1994.

[HW92-a] G. Hoffmann and H. Wong-Toi, Symbolic synthesis of supervisory controllers, *Proc. of the 1992 American Control Conference*, 2789-2793, 1992.

[HW92-b] G. Hoffmann and H. Wong-Toi, Symbolic Supervisory Synthesis for the Animal Maze, *Proc. of Workshop on Discrete Event Systems*, 189-197, Birkhauser Verlag, 1992.

[MPS95] O. Maler, A. Pnueli and J. Sifakis, On the Synthesis of Discrete Controllers for Timed Systems, In E.W. Mayr and C. Puech (Eds.), Proc. STACS '95, volume 900 of Lect. Notes in Comp. Sci., 229-242, Springer-Verlag, 1995.

[MP95] O. Maler and A. Pnueli, Timing Analysis of Asynchronous Circuits using Timed Automata, *Proc. Charme'95*, to appear, 1995.

[MWa80] Z. Manna and R.J. Waldinger, A Deductive Approach to Program Synthesis, *ACM Trans. of Prog. Lang. and Sys.* 2, 90-121, 1980.

[MWo84] Z. Manna and P. Wolper, Synthesis of Communication Processes from Temporal Logic Specifications, *ACM Trans. of Prog. Lang. and Sys.* 6, 68-93, 1984.

[McM93] K.L. McMillan, *Symbolic Model-Checking: an Approach to the State-Explosion problem*, Kluwer, 1993.

[NYY92] A. Nerode, A. Yakhnis and V. Yakhnis, Concurrent Programs as Strategies in Games, in Y. Moschovakis (Ed.), *Logic From Computer Science*, Springer, 1992.

[NM44] J. von Neumann and O. Morgenstern, *Theory of Games and Economic Behavior*, Princeton University Press, 1944.

[OW90] J.S. Ostroff and W.M. Wonham, A Framework for Real-time Discrete Event Control, *IEEE Trans. on Automatic Control* 35, 386-397, 1990.

[PR89-a] A. Pnueli and R. Rosner. On the Synthesis of a Reactive Module, In *Proc. 16th ACM Symp. Princ. of Prog. Lang.*, pages 179–190, 1989.

[PR89-b] A. Pnueli and R. Rosner. On the Synthesis of an Asynchronous Reactive Module, In *Proc. 16th ICALP*, volume 372 of Lect. Notes in Comp. Sci., 653–671, 1989.

[RW89] P.J. Ramadge and W.M. Wonham, The Control of Discrete Event Systems, *Proc. of the IEEE* 77, 81-98, 1989.

[TW94a] J.G. Thistle and W.M. Wonham, Control of Infinite Behavior of Finite Automata, *SIAM J. of Control and Optimization* 32, 1075-1097, 1994.

[Tho94] W. Thomas, On the Synthesis of Strategies in Infinite Games, In E.W. Mayr and C. Puech (Eds.), *Proc. STACS '95*, volume 900 of Lect. Notes in Comp. Sci., 1-13, Springer-Verlag, 1995.

[TB73] B.A. Trakhtenbrot and Y.M. Barzdin, *Finite Automata: Behavior and Synthesis*, North-Holland, Amsterdam, 1973.

[WD91] H. Wong-Toi and D.L. Dill, Synthesizing Processes and Schedulers from Temporal Specifications, in E.M. Clarke and R.P. Kurshan (Eds.), *Computer-Aided Verification '90*, DIMACS Series, AMS, 177-186, 1991.

[WH92] H. Wong-Toi and G. Hoffmann, The Control of Dense Real-Time Discrete Event Systems, Technical report STAN-CS-92-1411, Stanford University, 1992.

A Calculus of Stochastic Systems

for the Specification, Simulation, and Hidden State Estimation of Hybrid Stochastic/Non-stochastic Systems

Albert Benveniste[1], Bernard C. Levy[2], Eric Fabre[1], Paul Le Guernic[1]

[1] IRISA-INRIA, Campus Universitaire de Beaulieu, 35042 Rennes Cedex, France, name@irisa.fr

[2] Dept. of of Electrical and Computer Engineering, Univ. of California, Davis, CA 95616, USA, levy@ece.ucdavis.edu

Abstract. In this paper, we consider *mixed systems* containing both stochastic and non-stochastic[3] components. To compose such systems, we introduce a general combinator which allows the specification of an arbitrary mixed system in terms of elementary components of only two types. Thus, systems are obtained hierarchically, by composing subsystems, where each subsystem can be viewed as an "increment" in the decomposition of the full system. The resulting mixed stochastic system specifications are generally not "executable", since they do not necessarily permit the incremental simulation of the system variables. Such a simulation requires compiling the dependency relations existing between the system variables. Another issue involves finding the most likely internal states of a stochastic system from a set of observations. We provide a small set of primitives for transforming mixed systems, which allows the solution of the two problems of incremental simulation and estimation of stochastic systems within a common framework. The complete model is called CSS (*a Calculus of Stochastic Systems*), and is implemented by the SIG language, derived from the SIGNAL synchronous language. Our results are applicable to pattern recognition problems formulated in terms of Markov random fields or hidden Markov models (HMMs), and to the automatic generation of diagnostic systems for industrial plants starting from their risk analysis. A full version of this paper will appear in *Theoretical Computer Science*, and is available [1].

Keywords: stochastic systems, mixed systems, belief functions, communicating processes, simulation, estimation.

1 Introduction and motivation

This paper proposes a general framework for the specification and use of probabilistic models in applications of large computational complexity. To serve as reference in our

[3] Throughout this paper, we use the word "non-stochastic" to refer to systems which have no random part. In control science or statistics, such systems would be called "deterministic" as opposed to "stochastic"; however this name would be misleading in computer science, where "deterministic" vs. "nondeterministic" has a totally different meaning. This is why we decided to use the word "non-stochastic" here.

subsequent discussion, we now describe several real applications which either employ, or could benefit from the use of probabilistic methods.

- *Queuing networks, performance evaluation, and risk analysis* typically require a number of tools for the specification and simulation of systems, and to compute statistics of interest. The modelling and simulation tasks usually require modular models, which are often variations of stochastic Petri nets [2]. The computation of statistics relies on the underlying Markov chain associated to the Petri net specification.
- *Pattern recognition applications*, depending on whether they focus on one-dimensional signals, such as for speech recognition, or multidimensional ones, as in image analysis and understanding, frequently rely on hidden Markov models (HMMs) [3] or Markov random fields [4, 5]. Both classes of models have proved quite successful in their respective application areas. In particular, the best speech recognition systems currently available are based on HMMs. The nonintrusive appliance load monitoring problem described recently in [6] represents another interesting pattern recognition problem, where one seeks to determine which appliances switch on and off in an individual household, based on measurements of the total load power. In this context, appliances can be modeled in terms of communicating stochastic automata.
- *Model based monitoring and diagnostics procedures for complex systems* rely often on a blend of statistical approaches [7] for numerical systems, and symbolic techniques of artificial intelligence for systems of a combinatorial nature. However, somewhat surprisingly, while models play a significant role in the development of monitoring schemes, risk analysis considerations are usually not included. Risk analysis is mainly used to assess the safety margins of designs, but does not seem to enter the synthesis of on-line monitoring and diagnostics systems, even though such an inclusion would be highly beneficial.

Such applications require the following functions:

- *System specification* is a first issue for complex systems. Because most of the applications we have described, such as load appliance monitoring, or the monitoring and diagnostics of large-scale systems, involve a mixture of random and nonrandom phenomena, a *mixed* stochastic/non-stochastic form of modelling is desirable. Several other key features that would need to be included are modularity, i.e. the ability to specify large subsystems from small interacting modules, ease of modification, and the possibility to reuse subsystems in new applications.
- The ability to *simulate* systems, as well as evaluate statistics of interest is also a necessity. Again, modularity would be desirable in this context, although it may be less critical than for system specification. As for simulation, an important challenge is the *fast simulation of rare events* of interest, such as for fault-tolerance applications.
- Pattern recognition and diagnostics applications require the *estimation of hidden quantities of interest*, such as spoken words in speech recognition, appliance loads

for the nonintrusive appliance load monitoring application, or the origin and assessment of faults in failure diagnostics. Modularity would again be welcome in this context.

There exists a vast literature on the application of statistics and probability to the modelling, estimation, identification [8], and diagnostics [7] of dynamical systems. Unfortunately, modularity issues are almost never addressed by either statisticians or control engineers, and as a consequence, probabilistic and statistical techniques are used only rarely in the analysis of large scale systems (except in the area of performance evaluation, see below).

Stochastic Petri net models [2, 9] are often used to specify stochastic systems, in applications such as queing networks with synchronization, or fault-tolerance studies. They are commonly employed to evaluate statistics of interest in performance evaluation. However, such computations rely on the underlying Markov chain of the Petri net model, so that Petri nets by themselves do not simplify the computation of statistics. In [10, 11], however, particular structures of the transition matrix associated to certain Markov chains are used to decompose the statistical analysis of the system under consideration. Clearly, many real applications have been tackled by employing approaches developed within the Petri net community, and software products are available.

In a different area, *probabilistic communicating process algebras* and related logics have been studied in theoretical computer science [12, 13, 14]. The common approach to such studies consists in enriching with probability available models of communicating process algebras, such as CCS, CSP, etc., and related kinds of temporal logics [15, 16, 17]. Expressive power and system equivalence are analyzed, as well as the decidability of related logics. These approaches benefit from the fundamental advances achieved by this community to handle modularity, communication, and interaction between processes. However, to our knowledge, no real application has been reported based on such approaches, and no service is really provided beyond modelling.

This paper proposes a new and flexible form of calculus, called CSS, for the *specification, simulation, and hidden state estimation of mixed stochastic/non-stochastic systems*. The model of mixed stochastic/non-stochastic systems that we employ is introduced in Section 2.2. Mixed stochastic/non-stochastic systems interact via a single combinator that we call the *composition* and denote by " | ". The combination of mixed systems with | yields again mixed systems, and | is both associative and commutative. When applied to purely non-stochastic systems, the composition operator | behaves like the conjunction of systems of relations in mathematics. The shared variables of the two systems provide the only mechanism for system interaction. On the other hand, when two purely probabilistic systems with no shared variables are combined, we obtain two statistically independent systems. Also, combining a purely stochastic system with a purely non-stochastic one, viewed as a constraint, gives the conditional distribution of the original stochastic system, given that the constraint is satisfied; this provides a very simple mechanism to specify conditional distributions. The combination of purely non-stochastic and stochastic building blocks with | allows the specification of arbitrary mixed systems. A concrete syntax based on the SIG minilanguage is provided to implement the operations of CSS. Note that we restrict our attention here to systems with only a finite number of variables. Dynamical systems, i.e., systems defined over infinite index sets, have been examined in [18], and their study raises a number of technical issues that will be tackled elsewhere. Also, throughout the paper, only *finitely valued* variables are considered. Although our results hold in more general situations, such as for the case of linear Gaussian systems which is examined in detail in [19], a precise description of such cases will not be attempted here.

Section 3 examines the *simulation* of mixed stochastic/non-stochastic systems. Simulation is operational in nature. In contrast, the system specification provided by CSS is nonoperational, since it relies on relations. We are therefore confronted with the issue of converting a system specification into a simulation. Since many of the applications we have in mind are of a real time nature, we would like to perform simulations *incrementally*, in order to ensure their efficiency. For instance, Markov chains or stochastic automata are naturally simulated by using the Kolmogorov chain rule, so that states are generated incrementally. The Bayes rule $\mathrm{p}(x, y) = \mathrm{p}(y|x)\mathrm{p}(x)$ provides a way to simulate incrementally the random variables (X, Y) with joint distribution $\mathrm{p}(x, y)$. We only need to draw X according to the distribution $\mathrm{p}(x)$ and then, for X given, draw Y based on the conditional distribution $\mathrm{p}(y|X)$. We generalize the notions of marginal and conditional distributions to mixed systems, and use them to extend Bayes rule to these systems. The primitives implementing the marginal and conditional are introduced in SIG and are used to derive graph transformation rules which can be used to convert a compound system to an equivalent form which admits an incremental simulation.

The maximum likelihood (ML) *estimation* of mixed stochastic/non-stochastic systems is considered in [1], we discuss it here briefly. Consider a triple (X, Y, Z) of random variables, where Z is observed, and the two unknown random variables X and Y admit the conditional distribution $\mathrm{p}(x, y|z)$. The ML estimate, also known as the maximum a posteriori (MAP) estimate, of (X, Y) given Z is given by $(\widehat{x}, \widehat{y}) = \arg\max_{x,y} \mathrm{p}(x, y|z)$. To find these estimates incrementally, we can first compute the so-called "generalized likelihood" function $\mathrm{p}_{\mathcal{L}}(x|z) = \max_y \mathrm{p}(x, y|z)$. Next, compute the conditional likelihood $\mathrm{p}_{\mathcal{L}}(y|x, z) = \mathrm{p}(x, y|z)/\mathrm{p}_{\mathcal{L}}(x|z)$ of Y given X and Z, so that the following factorization holds: $\mathrm{p}(x, y) = \mathrm{p}_{\mathcal{L}}(y|x)\mathrm{p}_{\mathcal{L}}(x)$. Then, the desired ML estimates can be generated sequentially from $\widehat{x} = \arg\max_x \mathrm{p}_{\mathcal{L}}(x|z)$ and $\widehat{y} = \arg\max_y \mathrm{p}_{\mathcal{L}}(y|\widehat{x}, z)$. This incremental estimation procedure, which is called the Viterbi alborithm in the HMM literature [3, 20], just corresponds to a simple case of dynamic programming. We extend the notions of generalized likelihood and conditional likelihood, which now take the form of primitives, to mixed systems, and show that the above dynamic programming procedure can be generalized accordingly. These primitives are implemented in SIG, and we demonstrate how simple graph manipulations can be used to convert the given system to a form which can be incrementally estimated. In fact, the graph transformations applied for both simulation and estimation turn out to be *identical*.

Finally, Section 4 contains some conclusions and perspectives.

It was not until recently, through discussions with A. P. Dempster, that we became aware that the work reported in this paper is in fact closely related to the Dempster-Shafer theory of belief functions [21, 22, 23] and belief networks [24, 25, 26, 27] in statistics and artificial intelligence. Although our work has independent origins, several aspects are common with belief network theories. First, like the Dempster-Shafer model of belief functions, the mixed systems we consider are not fully probabilized, and combine both random and unknown types of uncertainties. However, while the Dempster-Shafer approach relies on an axiomatic different from probability theory, we achieve comparable results by blending probabilistic methods with constraint analysis. The composition we employ for building complex systems from simpler ones takes a form analog to Dempster's "product-intersection" rule [22] for combining belief functions. Also, our incremental simulation scheme is similar in nature to the fusion/propagation mechanism of [25, 26]. However, there exists an important difference between the partly directed, partly undirected graphs that we use to compile the dependency relations existing between the variables of a compound system, and the standard viewpoint of artificial intelligence, where directed branches encode "subjective causality." Our graphs encode "objective causality" according to the terminology of [25], since they are used to "*direct and activate the data flow in the computations ...*" [25]. In addition, while artificial intelligence emphasizes Bayesian estimation, we show that similar ideas can be applied to the solution of ML estimation problems. Finally, the practical implementation of our model has the syntactic form of a data flow programming language, which differs from the network formalism of artificial intelligence.

2 CSS and the SIG mini-language

The CSS model relies on a formal definition of mixed stochastic/non-stochastic systems, which is used to express the composition rule $|$. To motivate our choice of combinator $|$, we first discuss several problems arising in the composition of mixed stochastic/non-stochastic models.

2.1 Issues in mixed systems composition

As noted in the introduction, the first requirement for the combinator $|$ is that, when applied to purely nonstochastic systems, it must behave as the conjunction of constraints (or, equivalently, as the intersection of legal behaviors). Let us examine this requirement in the context of automata. The Fig. 1 illustrates the problem one faces when attempting to extend the usual product of automata to the class of stochastic automata. Next, consider Fig. 2. It illustrates the difficulties arising in the description of mixed stochastic/non-stochastic systems. Suppose that instead of considering elementary events, such as state transitions, we seek to characterize more complex ones, such as complete behaviors. It becomes difficult to understand the status of such events. Should they be viewed as random, or just as nondeterministic? Also what are the consequences of the fact that certain variables are hidden? This is a natural consequence of modularity, since variables describing the inner interaction of subsystems are not visible when considering a large system.

Thus there is no doubt that a naive extension of nonstochastic composition will not work properly. Indeed, although the CCS composition rule has been extended to a probabilistic context by Jonsson et al. [28], it can only model specific systems but not general ones. Also, CCS composition does not perform the intersection of behaviours, and raises difficult questions of bisimulation.

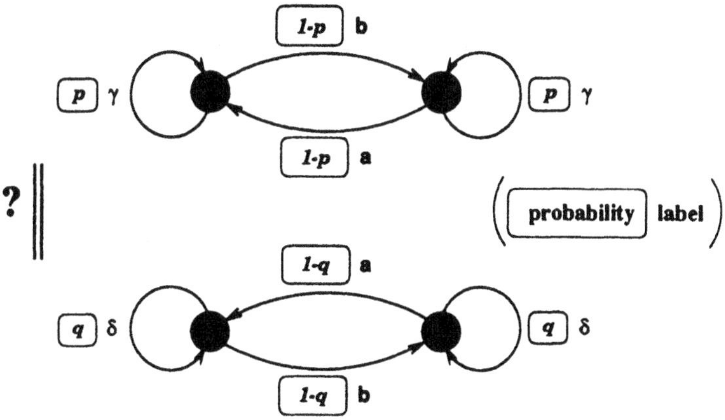

Fig. 1. *Combining stochastic automata.*Ignore temporarily the symbols within curved rectangles. Then the product of two automata is just the shuffle product of their associated infinite strings of symbols. This clearly satisfies our requirement concerning the intersection of behaviors. Next, take into account the transition probabilities appearing within the curved rectangles, so that we are now considering two interacting stochastic automata. QUESTION 1: how would you define the composition of these two automata?

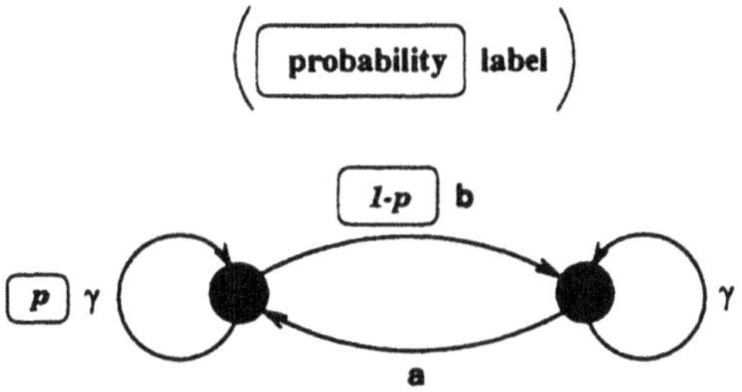

Fig. 2. *Mixed stochastic/non-stochastic systems.* The figure shows a system, where transitions from the state on the left are governed by probabilities, but transitions from the state on the right are just nondeterministic, i.e., unspecified. QUESTION 2: in the language $(a\gamma^* b\gamma^*)^*$, which events would you view as random, or as nondeterministic (unspecified)? Next, let us make the labels a and b more concrete by assuming they represent the actions of generating two random variables X and Y, respectively, with values in a finite alphabet. Then, set $Z = f(X, Y)$ for some function f with values in a finite alphabet, and assume we hide a, b, X, Y, while keeping only γ, Z visible. QUESTION 3: in the language $(Z\gamma^* Z\gamma^*)^*$, what events would you view as random, or as nondeterministic (unspecified)?

On the other hand, the difficulties we have highlighted, concerning the nature of complex events or behaviours in mixed stochastic/non-stochastic systems, together with the presence of hidden variables, were the motivation, in Dempster-Shafer and related theories, for the development of new axioms to define and manipulate events in large complex systems. Having demonstrated the need for a careful look at system composition, we now proceed with the introduction of the CSS model.

2.2 CSS

Model of mixed stochastic/non-stochastic systems The mixed systems we consider are described by a quadruple

$$\pi = \{\mathbf{X}, \Omega, \mathbf{W}, \mathrm{p}\} \tag{2.1}$$

where

- $\mathbf{X} = \{X_1, \ldots, X_p\}$ denotes a finite set of *variables* whose values are written as $x = (x_1, \ldots, x_p)$. The variables are the observable objects of our model. The domain of each variable X_i is denoted as V_{X_i}, so that the domain of the vector $\mathbf{X}$ can be expressed as $V_{\mathbf{X}} = \prod_i V_{X_i}$.
- $\mathbf{W} = \{W_1, \ldots, W_q\}$ denotes a finite set of *random variables* or simply *randoms* for short. Values of W_j are written as w_j, and we refer to the complete set of values $w = (w_1, \ldots, w_q)$ as a *random experiment*. The domain of W_i (resp. $\mathbf{W}$) is denoted by V_{W_i} (resp. $V_{\mathbf{W}}$); in this article we assume $\mathbf{W}$ is finite. Randoms are hidden, i.e., not visible from outside the system. The reason for this property will become clear below. $\mathbf{W}$ models the random part of the system, so that if $\mathbf{W}$ is empty, the system is completely non-stochastic.
- p constitutes an unnormalized probability distribution for $\mathbf{W}$[4]. Specifically, we only require $\mathrm{p} \geq 0$ and $0 < \sum_w \mathrm{p}(w) < \infty$.
- Ω denotes a relation on the pair $(\mathbf{X}, \mathbf{W})$. We shall sometimes write it more explicitly as $\Omega(X_1, \ldots, X_p \,;\, W_1, \ldots, W_q)$.

A system $\pi = \{\mathbf{X}, \Omega, \mathbf{W}, \mathrm{p}\}$ is observable only through its variables. Randoms cannot be seen, but transfer their behaviour to the system variables $\mathbf{X}$ through the relation Ω. In doing so, some predicates over the variables become random, namely those which are completely expressible in terms of $\mathbf{W}$. In the purely non-stochastic case, we may consider that $V_{\mathbf{W}}$ consists of a single point w_{triv}, with $\mathrm{p}(w_{\mathrm{triv}}) = 1$ and $\mathrm{p}(\emptyset) = 0$. Our notion of system is depicted in Fig. 3.

The unique system for which $\mathbf{X} = \emptyset$ is called NIL. Finally, given an arbitray system $\pi = \{\mathbf{X}, \Omega, \mathbf{W}, \mathrm{p}\}$, the system obtained by replacing its distribution p by a uniform one, say equal to one, is denoted by FLAT(π), so that we have FLAT$(\pi) \stackrel{\Delta}{=} \{\mathbf{X}, \Omega, \mathbf{W}, 1\}$.

[4] Handling unnormalized distributions may seem unusual, but has several advantages. It simplifies the definition of the composition | and the specification of conditional probabilities, and significantly decreases the computational cost of incremental simulation and estimation. Furthermore for many applications where the space $V_{\mathbf{W}}$ has a very large cardinality, such as for the study of Markov random fields in statistical mechanics [29, 30, 31], the computation of the normalizing constant (the partition function) which transforms p into a true probability is often unnecessary.

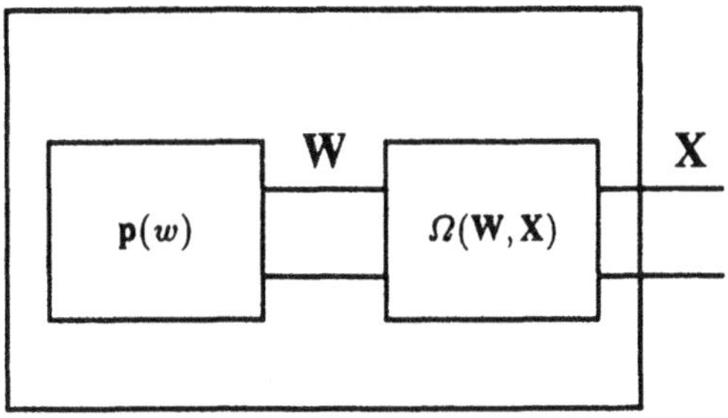

Fig. 3. *A system in CSS.* The outer box is the external boundary of the system, thus only $\mathbf{X}$ is visible from outside for further interaction.

EXAMPLES:

1. The above mixed systems contain as a subclass purely non-stochastic systems described by a set of relations, without any randoms. Another subclass corresponds to purely stochastic systems, for which we have $\mathbf{X} = \mathbf{W}$, and where the relation Ω is defined by $X_i = W_i$ for all i, so that all randoms are observed as variables.
2. A system $\pi = \{(X_1, X_2), W, \Omega, \mathrm{p}\}$, with $\Omega : f(X_1, X_2) = W$ for some function f, is a system with two variables. For instance, take $X_1 + X_2 = W$. In this case, each variable X_i cannot be viewed as random since its probability distribution is not defined, but the sum $X_1 + X_2$ is random. For a general function f, not all predicates on (X_1, X_2) are random, only those which involve $f(X_1, X_2)$.
3. For a system
$$\pi = \{X, (W_1, W_2), \Omega, \mathrm{p}\} \qquad \text{with} \qquad \Omega : X = f(W_1, W_2) \tag{2.2}$$
where f is a non injective function, the variable X is random. However, because f is not injective, there are "too many" randoms; for instance, if f depends only on W_1, we can remove W_2. This operation, called "compression," is described below in Secs. 3.1.
4. The class of linear Gaussian mixed stochastic/non-stochastic systems of the form $EY = AX + BW$ was studied in detail in [19]. For such systems, E, A, B are matrices of suitable dimensions, W is a Gaussian random vector with zero mean and unit covariance matrix, and the variables correspond to the vector pair (X, Y).

Our model of mixed systems is closely related to the one employed by Dempster and Shafer [21, 22, 23] to formulate their theory of belief functions. Like the systems examined here, Dempster's belief functions are specified by a quadruple consisting of a probability space $(V_{\mathbf{W}}, \mathrm{p})$, which is not directly visible, and a pair $(V_{\mathbf{X}}, \Gamma)$ formed by a set of system configurations, and a mapping Γ associating to each element $w \in V_{\mathbf{W}}$ a set $\Gamma(w) \subset V_{\mathbf{X}}$. For our model, the set-valued mapping Γ is specified implicitly by the relation Ω, which associates to each random w the set

$$\Gamma(w) = \{x : \Omega(x; w)\}, \tag{2.3}$$

of variables x which, together with w, satisfy the relation Ω.

Let us examine the modelling implications of the mixed system specification $\pi = \{\mathbf{X}, \Omega, \mathbf{W}, \mathrm{p}\}$. First, observe that by eliminating the randoms $\mathbf{W}$ from the relation Ω, we obtain a family of hard constraints for the variables $\{X_1, \ldots, X_p\}$. These constraints are often called "parity checks" in the failure detection literature [7]. The subset $V_{\mathbf{X}}^{\Omega}$ of $V_{\mathbf{X}}$ satisfying these constraints can be used to test the validity of the model π, by checking whether the visible variables belong to this set.

Next, we note that each set $B \subseteq V_{\mathbf{W}}$ of random experiments admits the prior probability

$$\mathrm{P}(B) = \frac{\sum_{w \in B} \mathrm{p}(w)}{\sum_{w} \mathrm{p}(w)} . \tag{2.4}$$

Eliminating the variables $\mathbf{X}$ from the relation Ω yields hard contraints that must be satisfied by the randoms $\{W_1, \ldots, W_q\}$. Let $V_{\mathbf{W}}^{\Omega}$ be the set of w's satisfying these constraints. The posterior probability on the randoms, given the set $V_{\mathbf{W}}^{\Omega}$ of allowable configurations, takes the form

$$\mathrm{P}^{\Omega}(B) = \frac{\sum_{w \in B \cap V_{\mathbf{W}}^{\Omega}} \mathrm{p}(w)}{\sum_{w \in V_{\mathbf{W}}^{\Omega}} \mathrm{p}(w)} = \frac{\mathrm{P}(B \cap V_{\mathbf{W}}^{\Omega})}{\mathrm{P}(V_{\mathbf{W}}^{\Omega})} = \mathrm{P}(B \mid V_{\mathbf{W}}^{\Omega}) . \tag{2.5}$$

The posterior probability P^{Ω} is the result of the interaction of the relation Ω with the prior distribution p in the system specification $\pi = \{\mathbf{X}, \Omega, \mathbf{W}, \mathrm{p}\}$. Unfortunately, this new probability cannot be transferred to the variables $\mathbf{X}$ because, since Ω is a relation, the sets $\Gamma(w)$ specified by (2.3) are not singletons, and may not be disjoints for different w's. This is just a manifestation of the fact that, because projection is a monotonic operator on sets, but is not additive, projecting a probability from one space onto another does not yield a probability, but a different object, called a *Choquet capacity* [32]. On $V_{\mathbf{X}}$, this capacity provides a partial probabilistic knowledge which was described by Dempster [21, 22] in terms of upper and lower probabilities for the subsets of $V_{\mathbf{X}}$. These upper and lower probabilities provide bounds describing the limits of our information concerning predicates of the $\mathbf{X}$ variables. In this paper, instead of adopting the Dempster-Shafer upper/lower probability framework, we shall remain within the realm of standard probability theory by considering exclusively probabilities over the set $V_{\mathbf{W}}$ of randoms.

The " | " system combinator The composition of mixed systems can be performed in the same manner as the combination of belief functions described in [21]. The main aspect of the combination operation is that different systems are allowed to share common variables, which describe their interaction, but not randoms. In other words, randoms are always private, and do not play a role in the combination of systems. For two systems $\pi_i = \{\mathbf{X}_i, \Omega_i, \mathbf{W}_i, \mathrm{p}_i\}$ with $i = 1, 2$, the combinator $\pi_1 \mid \pi_2 = \{\mathbf{X}, \Omega, \mathbf{W}, \mathrm{p}\}$ is defined as

$$\mathbf{X} = \mathbf{X}_1 \cup \mathbf{X}_2 \tag{2.6a}$$

$$\mathbf{W} = \mathbf{W}_1 \times \mathbf{W}_2 \tag{2.6b}$$

$$\mathbf{p}(w) = \mathbf{p}(w_1, w_2) = \mathbf{p}_1(w_1) \times \mathbf{p}_2(w_2) \tag{2.6c}$$

$$\Omega = \Omega_1 \wedge \Omega_2 , \tag{2.6d}$$

where $\Omega_1 \wedge \Omega_2$ denotes the conjunction of relations Ω_1 and Ω_2, which is the usual way of defining systems of equations in mathematics. The expressions (2.6a) and (2.6b) indicate respectively that variables may be shared, but not randoms.

The identities (2.6a–2.6d) show that systems interact only through their shared variables. The NIL system is a neutral element for the combinator " | ". Our notion of composition is illustrated in Fig. 4.

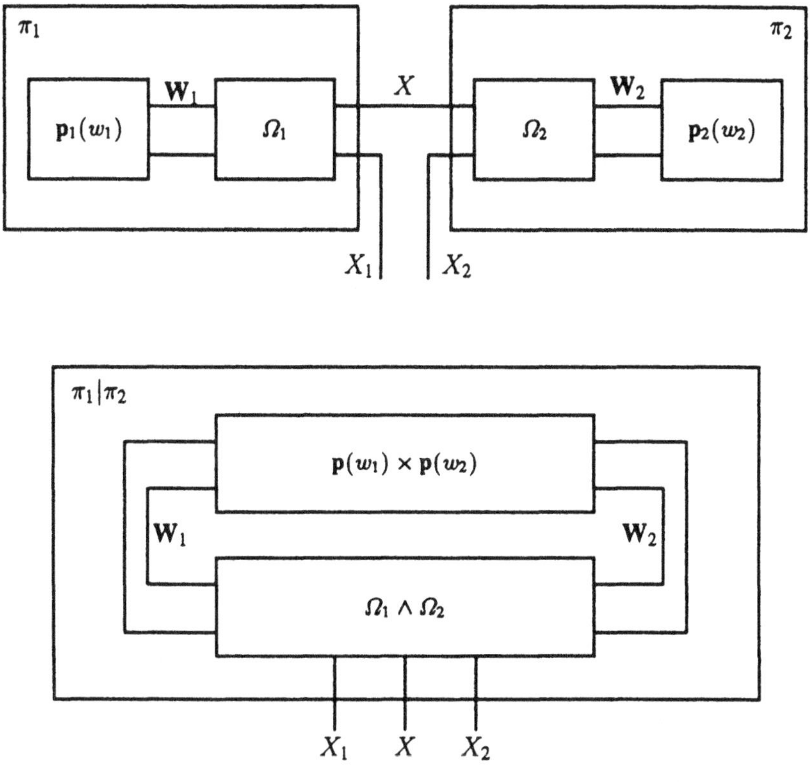

Fig. 4. *The | composition.* We consider two systems π_1, π_2 sharing the visible variable $\mathbf{X}$, but not $\mathbf{X}_1, \mathbf{X}_2$. The first picture illustrates how interaction occurs, and the second one shows its result.

EXAMPLES :

1. Consider two systems $\pi_i = \{\mathbf{X}_i, \Omega_i, \mathbf{W}_i, \mathbf{p}_i\}$ with $i = 1, 2$, where π_1 is purely stochastic, so that $\mathbf{X}_1$ and $\mathbf{W}_1$ have same cardinality and $\Omega_1 : X_1 = W_1, \dots X_p =$

W_p, and π_2 is purely non-stochastic, i.e., $\mathbf{W}_2 = \emptyset$, with the nontrivial relation Ω_2. Assume also that $\mathbf{X}_1 = \mathbf{X}_2$. Then, it is easy to check that $\pi_1 \mid \pi_2 = \{\mathbf{X}_1, (\Omega_1 \wedge \Omega_2), \mathbf{W}_1, \mathbf{p}_1\}$. The combined system $\pi_1 \mid \pi_2$ has still the feature that randoms are visible through the variables, since $\Omega_1 : X_1 = W_1, \ldots, X_p = W_p$. However, the variables $X_1, \ldots, X_p$ behave now according to the conditional distribution $\mathbf{p}_1^{\Omega_2}$ of $\mathbf{p}_1$ based on the constraint Ω_2. Thus the composition | provides a simple mechanism for specifying conditional probabilities, which will be used extensively in the SIG examples presented below.

2. Let $\pi_i = \{\mathbf{X}_i, \Omega_i, \mathbf{W}_i, \mathbf{p}_i\}$ with $i = 1, 2$ be two systems which do not interact, so that $\mathbf{X}_1 \cap \mathbf{X}_2 = \emptyset$. Then, in the combination $\pi_1 \mid \pi_2$, the randoms $\mathbf{W}_1$ and $\mathbf{W}_2$ are independent[5].

2.3 The SIG mini-language

We now proceed to describe a syntax, in the form of the langage SIG, which implements both the modelling format and composition rule of CSS.

The primitives of SIG The SIG language has the following primitives:

```
  (i) R(x1,...,xp)
 (ii) potential U(x1,...,xp)
(iii) P | Q
```

They admit the following informal interpretation.

(i) `R(x1,...,xp)` specifies a relation among the variables `x1,...,xp`. The corresponding system in the sense of (2.1) admits the `xi`'s as variables, has no randoms, and Ω is the relation R. Thus, `R(x1,...,xp)` is a purely non-stochastic system.

(ii) `potential U(x1,...,xp)`, where U is a function taking values over the line $(-\infty, +\infty]$, specifies random variables with unnormalized joint distribution

$$\exp -U(x_1, \ldots, x_p) \,. \tag{2.7}$$

The corresponding system in the sense of (2.1) has `x1,...,xp` as variables, its randoms $\mathbf{W} = (W_1, \ldots, W_p)$ have the distribution $\exp -U(w_1, \ldots, w_p)$, and Ω relates variables and randoms via the relations $\mathbf{x1} = W_1, \ldots, \mathbf{xp} = W_p$. Thus, `potential U(x1,...,xp)` is a purely stochastic system.

(iii) `P | Q` denotes the application of the "|" combinator to systems P and Q.

---

[5] According to the "maximum entropy principle" [30], among all joint distributions $\mathbf{p}(w_1, w_2)$ with prescribed marginals $\mathbf{p}_1(w_1)$ and $\mathbf{p}_2(w_2)$, the one which maximizes the entropy is given by $\mathbf{p}_1(w_1) \times \mathbf{p}_2(w_2)$, in which case the two components are independent.

Specifying systems with SIG A system in the sense of CSS and definition (2.1) can be declared as shown below, where we omit variable type declarations of the form "`integer`", etc:

```
system PI =
    { variable X,Y,Z }          % declaration of variables
    (| potential U(X,Y)         % distribution of (X,Y)
     | Z = f(X,Y)               % constraint on (X,Y,Z)
     |)
end
```

Several examples of SIG programs are now presented. The program

```
system LINEAR = (real E,A,B) % declaration of parameters
    { variable X,Y,U }
    (| potential (U**2)/2      % gaussian noise
     | E*Y = A*X + B*U         % constraint on (X,Y,U)
     |)
end
```

specifies a "linear observation" $EY = AX + BU$ of the form introduced and studied in detail in [19]. To generate a hidden Markov model (HMM) of the type discussed in [3, 19], we can employ the following program:

```
system  HMM_0 = (integer N)
    { variable X[i] i=0 to N, Y[i] i=1 to N }
    (| X[0] = 0
     | loop i=1 to N
         (| potential U(X[i-1],X[i])
          | potential V(X[i-1],X[i],Y[i])
          |)
       end
     |)
end
```

The first constraint fixes the initial condition, and the `loop` statement specifies the joint distribution of the internal states X_i and outputs Y_i. The resulting system `HMM_0` is a HMM with state X and output Y. It has 0 for initial state, and its state transitions and outputs are specified by the interactions U and V, so that

$$\mathrm{p}(x_0, \ldots, x_N \; ; \; y_1, \ldots, y_N) \propto \delta_0(x_0) \; \exp - \sum_{i=1}^{N} \left[U(x_{i-1}, x_i) + V(x_{i-1}, x_i, y_i) \right] \tag{2.8}$$

where $\propto$ denotes "proportional to", and $\delta_0(x) = 1$ if $x = 0$, $= 0$ otherwise. If we want to consider the same HMM *given that the final condition* `X[N] = X_MAX` *also holds*, one needs only to add the final constraint to the previous SIG program, thus yielding

```
system  HMM = (integer N)
    { variable X[i] i=0 to N, Y[i] i=1 to N }
    (| X[0] = 0
     | X[N] = X_MAX
     | loop i=1 to N
          (| potential U(X[i-1],X[i])
           | potential V(X[i-1],X[i],Y[i])
           |)
        end
     |)
end
```

To explain the interest of this simple trick, suppose X models the occupation level of a buffer, which behaves according to HMM_0. Assume X_MAX corresponds to a critical level, and we want to know the conditional distribution of the buffer evolution given that level X_MAX is reached at instant N. Then we only need to include the conditioning event X[N] = X_MAX as an additional constraint in our original program HMM_0 in order to obtain the desired behavior HMM. This mechanism can be employed whenever one seeks to concentrate on the set of experiments satisfying a condition of interest. Note for example that a common technique of risk analysis involves tracking cascades of events leading to a specific failure.

3 Simulation

We now turn to the simulation of hybrid systems. Simulation is operational in nature. In contrast, the system specification provided by CSS relies on relations, which are intrinsically nonoperational. This raises the issue of converting a system specification into an equivalent simulation. In this context, since we naturally wish to generate efficient simulations, we restrict our attention to *incremental simulations*. For instance, Markov chains or stochastic automata can be simulated incrementally by employing the Kolmogorov chain rule to generate the states one at a time. Such a feature is obviously mandatory for real-time applications.

Consider a pair (X, Y) of standard random variables with joint distribution $\mathrm{p}(x, y)$. These two random variables can be simulated incrementally by employing the following procedure.

1. Compute the marginal
$$\mathrm{p}(x) = \sum_y \mathrm{p}(x, y) \tag{3.1}$$
of p with respect to X.
2. Compute the conditional distribution $\mathrm{p}(y|x) = \mathrm{p}(x, y)/\mathrm{p}(x)$ of Y given X, so that we obtain the following factorization, also known as Bayes rule:
$$\mathrm{p}(x, y) = \mathrm{p}(y|x)\mathrm{p}(x)\,. \tag{3.2}$$
3. Draw X at random following the marginal $\mathrm{p}(x)$, and then, for a given X, draw Y at random according to the conditional distribution $\mathrm{p}(y|X)$.

We now generalize this technique to the case of hybrid systems.

3.1 Compressing the random part of a system

Since randoms are hidden, only their visible effect upon the system variables is of interest. But as we have already seen in example (2.2), the domain $V_{\mathbf{W}}$ of all randoms may include too many details. For example, consider a pair (W_1, W_2) and assume that W_1 is visible but not W_2. The corresponding CSS model has a single variable X_1 and constraint $X_1 = W_1$. Since W_2 is unneeded, it can be removed from the original system by computing the compressed distribution $\mathrm{p}_{co}(w_1) = \sum_{w_2} \mathrm{p}(w_1, w_2)$, which for this simple case reduces to the marginal distribution with respect to w_1. This is just an elementary case of the compression operation we now introduce.

To a system π, we can associate the following equivalence relation between randoms

$$w \sim_\pi w' \text{ iff } \forall x : \Omega(x; w) \Leftrightarrow \Omega(x, w') , \tag{3.3}$$

which just indicates that two randoms w and w' are equivalent if they cannot be distinguished by the variables. Accordingly, a set $B \subseteq V_{\mathbf{W}}$ is visible through the variables if and only if it satisfies the property

$$\left.\begin{array}{r} w \in B \\ w' \sim_\pi w \end{array}\right\} \Rightarrow w' \in B . \tag{3.4}$$

It is natural to restrict p to the sets of randoms satisfying this condition. Note in this respect that the family $\mathcal{W}$ of all sets B satisfying the condition (3.4) forms a σ-algebra, since it is closed under intersection and complementation, and contains the empty set. Hence, in order to characterize the random behavior of the system π, we only need to specify the conditional probability $\mathbf{P}(.|\mathcal{W})$ of $\mathbf{P}$ given $\mathcal{W}$. This can be accomplished by constructing what we shall call the *compression* π_{co} of π. The compression is obtained from π and the equivalence relation $\sim_\pi$ in the following manner.

1. First, we compress the set $V_{\mathbf{W}}$ of random experiments by retaining only the equivalence classes of the relation $\sim_\pi$. Thus, an experiment w belongs to an equivalence class w_{co}, and the set of all equivalence classes forms the compressed domain V_{co}.
2. Compress the relation Ω accordingly, by setting

$$\Omega_{co}(x ; w_{co}) \triangleq \Omega(x ; w) \text{ for } w \in w_{co} . \tag{3.5a}$$

3. Finally, to each equivalence class w_{co} of randoms, we assign the probability

$$\mathrm{p}_{co}(w_{co}) \triangleq \sum_{w \in w_{co}} \mathrm{p}(w) . \tag{3.5b}$$

Two systems π and π' admitting the same compressed form are said to be equivalent, which is denoted as

$$\pi \equiv \pi' . \tag{3.6}$$

Since the procedure employed to compress a system does not affect its external behavior as seen from the variables, we have the following result.

Theorem 3.1 *If* $\pi_i = \{\mathbf{X}_i, \Omega_i, \mathbf{W}_i, \mathrm{p}_i\}$ $i = 1,2$ *are equivalent in the sense of (3.6), they cannot be distinguished under simulation. In particular, they have*

1. *the same variables:* $\mathbf{X}_1 = \mathbf{X}_2$ *;*
2. *the same parity checks:* $V_{\mathbf{X}_1}^{\Omega_1} = V_{\mathbf{X}_2}^{\Omega_2}$ *;*
3. *the probability spaces* $\{V_{\mathbf{W}_i}^{\Omega_i}, \mathcal{W}_i, \mathbf{P}_i\}$ *are isomorphic, so that there exists a one-to-one map* ϕ *from the* σ*-algebra* $\mathcal{W}_1$ *onto* $\mathcal{W}_2$ *such that* $\forall B_1 \in \mathcal{W}_1$, $\mathbf{P}_2(\phi(B_1)) = \mathbf{P}_1(B_1)$.

The property

$$\pi \equiv \mathrm{FLAT}(\pi) \mid \pi , \tag{3.7}$$

which is proved in [1], is a straightforward consequence of the notion of system equivalence. This identity generalizes to hybrid systems the idempotence of composition property $\pi \mid \pi = \pi$ of purely non-stochastic systems.

Although the factorization (3.2) cannot be extended directly to hybrid stochastic/non-stochastic systems, by employing Theorem 3.1, we develop below a general procedure for decomposing an arbitrary hybrid system π into marginal and conditional components which extends the factorization (3.2) of standard probability distributions. This decomposition will provide the key element required for incremental system simulation.

3.2 Two primitives

Consider a system $\pi = \{\mathbf{X}, \Omega, \mathbf{W}, \mathbf{p}\}$ and a subset of variables $\mathbf{X}' \subset \mathbf{X}$. The concepts of marginal and conditional distributions can be extended to hybrid systems by constructing the *marginal* and *conditional* systems

$$\overline{\mathcal{S}}_{\mathbf{X}'}(\pi) \quad \text{and} \quad \mathcal{S}_{\mathbf{X}'}(\pi)$$

respectively, where $\mathcal{S}$ represents here a mnemonic for $\mathcal{S}$imulation.

The marginal. It consists of eliminating from π the variables not in $\mathbf{X}'$, which gives

$$\overline{\mathcal{S}}_{\mathbf{X}'}(\pi) = \{\mathbf{X}', \Omega', \mathbf{W}, \mathbf{p}\} , \tag{3.8}$$

where Ω' denotes the relation obtained by employing the existential qualifier $\exists$ to eliminate from Ω the variables not in $\mathbf{X}'$, so that

$$\Omega'(x'; w) \triangleq \exists x" : \Omega\left((x', x"); w\right) . \tag{3.9}$$

Note that neither the set of randoms $\mathbf{W}$ nor the density $\mathbf{p}$ are changed by this construction, which involves only tracking the effect of the projection of $\mathbf{X}$ onto $\mathbf{X}'$ in the relation Ω.

The conditional. The conditional system has the structure

$$\mathcal{S}_{\mathbf{X}'}(\pi) = \{\mathbf{X}, \Omega, \mathbf{W}, \mathbf{p}"\} , \tag{3.10a}$$

where the distribution $\mathbf{p}''$ is selected such that the factorization

$$\pi \equiv \overline{\mathcal{S}}_{\mathbf{X}'}(\pi) \mid \mathcal{S}_{\mathbf{X}'}(\pi) \tag{3.10b}$$

holds. Note that the relation $\equiv$ indicates that both sides have the same compressed form. The decomposition (3.10b) represents the extension to hybrid systems of the factorization (3.2) of a probability distribution into marginal and conditional components. A constructive proof of existence of the conditional is given in [1].

Notation. In the following, it will be convenient to extend the definition of $\overline{\mathcal{S}}_{\mathbf{Z}}(\pi)$ and $\mathcal{S}_{\mathbf{Z}}(\pi)$ to the case where $\mathbf{Z}$ is not necessarily a subset of the variables $\mathbf{X}$ of π, by denoting

$$\overline{\mathcal{S}}_{\mathbf{Z}}(\pi) \triangleq \overline{\mathcal{S}}_{\mathbf{Z}\cap\mathbf{X}}(\pi) \quad , \quad \mathcal{S}_{\mathbf{Z}}(\pi) \triangleq \mathcal{S}_{\mathbf{Z}\cap\mathbf{X}}(\pi) \; . \tag{3.11}$$

3.3 Properties of the primitives

The operations that we have just introduced admit a number of algebraic properties which are stated in [1]. We shall only discuss here the very nature of system interaction. In what follows, $f \circ g(x)$ denotes the composition of maps $f(g(x))$.

Systems with no shared variables have no interaction, and involve independent families of randoms. Hence if π_1 and π_2 have no shared variables, in order to simulate the composition $\pi_1 \mid \pi_2$, we only need to simulate π_1 and π_2 separately. This corresponds to the easiest, but trivial, case of incremental simulation. But our discussion at the beginning of this section indicates that incremental simulation can be performed under more general circumstances. Specifically, the reason why the factorization (3.2) allows the simulation of first $\mathbf{X}$ followed by $\mathbf{Y}$ is that combining $\mathrm{p}(y|x)$ with $\mathrm{p}(x)$ does not modify the behavior of $\mathbf{X}$. In other words, $\mathrm{p}(y|x)$ represents totally new information with no bearing on $\mathbf{X}$. This feature leads us to introduce the notion of *innovation* which extends to hybrid systems the familiar concept of innovations process in filtering and detection theory.

Definition 3.1 (innovation) *Let π_i and $\mathbf{X}_i$ $i = 1, 2$ be two systems and their variables. If $\mathbf{Y}$ denotes an arbitrary set of variables, the system π_2 is said to be a $\mathbf{Y}$–innovation of π_1, which we denote as*

$$\pi_2 \perp\!\!\!\perp_{\mathbf{Y}} \pi_1 \, ,$$

if

$$\overline{\mathcal{S}}_{\mathbf{X}_1 \cup \mathbf{Y}} \circ \mathcal{S}_{\mathbf{Y}}(\pi_2) \mid \mathcal{S}_{\mathbf{Y}}(\pi_1) \equiv \mathcal{S}_{\mathbf{Y}}(\pi_1) \; . \tag{3.12}$$

Thus, π_2 represents a $\mathbf{Y}$–innovation of π_1 if composing $\mathcal{S}_{\mathbf{Y}}(\pi_2)$ with $\mathcal{S}_{\mathbf{Y}}(\pi_1)$ does not modify $\mathcal{S}_{\mathbf{Y}}(\pi_1)$. More intuitively, this means that given $\mathbf{Y}$, the interaction between π_1 and π_2 is oriented from π_1 to π_2. For the special case when $\mathbf{Y}$ is empty, we just say that π_2 is an innovation *of π_1, which is written as $\pi_2 \perp\!\!\!\perp \pi_1$.*

From the above definition and comments it is clear that the relations $\perp\!\!\!\perp_{\mathbf{Y}}$ and $\perp\!\!\!\perp$ are not commutative. Also the selection of the conditioning set $\mathbf{Y}$ affects strongly whether a system constitutes an innovation of another. For example, if $\mathbf{X}_1 \cap \mathbf{X}_2 \subseteq \mathbf{Y}$, the two relations $\pi_1 \perp\!\!\!\perp_{\mathbf{Y}} \pi_2$ and $\pi_2 \perp\!\!\!\perp_{\mathbf{Y}} \pi_1$ hold trivially. The concept of innovation will form the basis for the derivation of compilation rules for decomposing a system into an *ordered* sequence of subsystems which can be simulated in accordance to this order. The compilation rules will rely on the following properties of innovations.

Lemma 3.1 *Given an arbitrary system π, and a subset $\mathbf{X}'$ of its variables, we have*

$$\mathcal{S}_{\mathbf{X}'}(\pi) \perp\!\!\!\perp \overline{\mathcal{S}}_{\mathbf{X}'}(\pi) \, , \tag{3.13}$$

i.e. the conditional innovates with respect to the marginal. Furthermore, if $\pi_2 \perp\!\!\!\perp_{\mathbf{Y}} \pi_1$, *i.e.* π_2 *is a* **Y**–*innovation of* π_1, *the following identities hold:*

$$\overline{\mathcal{S}}_{\mathbf{X}_1 \cup \mathbf{Y}}(\, \mathcal{S}_{\mathbf{Y}}(\pi_2) \;\mid\; \mathcal{S}_{\mathbf{Y}}(\pi_1)\,) = \mathcal{S}_{\mathbf{Y}}(\pi_1) \tag{3.14}$$

$$\mathcal{S}_{\mathbf{Y}}(\pi_1 \;\mid\; \pi_2) \equiv \mathcal{S}_{\mathbf{Y}}(\pi_1) \;\mid\; \mathcal{S}_{\mathbf{Y}}(\pi_2) \tag{3.15}$$

$$\overline{\mathcal{S}}_{\mathbf{Y}}(\pi_1 \;\mid\; \pi_2) \equiv \overline{\mathcal{S}}_{\mathbf{Y}}(\pi_1) \;\mid\; \overline{\mathcal{S}}_{\mathbf{Y}}(\pi_2)\;. \tag{3.16}$$

3.4 Incremental system simulation

Consider now a compound system of the form

$$\pi = \;\mid_{i \in I}\; \pi_i \tag{3.17}$$

where I denotes a finite index set. We seek to develop an incremental simulation procedure for such a system, so as to be able to evaluate progressively the probabilities of complex events.

Graphical representation. Let π_1 and π_2 be two systems admitting a nonempty set **X** of common variables. For these two systems, we employ the graphical notation

$$\pi_1 \;\text{——}\; \mathbf{X} \;\text{——}\; \pi_2 \quad \text{if} \quad \begin{cases} \pi_2 \text{ is not an innovation of } \pi_1\text{, and} \\ \pi_1 \text{ is not an innovation of } \pi_2\,. \end{cases} \tag{3.18a}$$

Similarly, we write

$$\pi_1 \longrightarrow \mathbf{X} \longrightarrow \pi_2 \quad \text{if} \quad \begin{cases} \pi_2 \text{ is an innovation of } \pi_1\text{, but} \\ \pi_1 \text{ is not an innovation of } \pi_2\,, \end{cases} \tag{3.18b}$$

and

$$\pi_1 \longleftrightarrow \mathbf{X} \longleftrightarrow \pi_2 \quad \text{if} \quad \begin{cases} \pi_2 \text{ is an innovation of } \pi_1 \text{ and} \\ \pi_1 \text{ is an innovation of } \pi_2\,. \end{cases} \tag{3.18c}$$

Obviously, it is rather uncommon that two systems should be mutual innovations, and still share common variables. However, this situation may occur in certain instances, such as when $\pi_1 = \pi_2 = \pi$ with π non-stochastic, since in this case the composition rule $\pi \mid \pi \equiv \pi$ implies π is its own innovation.

Next, consider each pair (π_i, π_j) of components of the compound system π given by (3.17). If π_i and π_j share common variables, we say they are *neighbors* and draw a branch between them. The choice of branch orientation or the lack thereof depends on which of the three cases (3.18a)–(3.18c) holds. In this manner, we generate a bipartite graph, where systems and variables alternate, which we call the *execution graph* of π, and denote by

$$\text{EXECGRAPH}\,(\pi)\;.$$

This graph has the effect of visualizing all the statistical dependency relations existing between the variables of subsystems π_i, $i \in I$.

It is worth noting that graphs of a similar nature have been introduced recently by a number of authors under the name of influence diagrams, or belief networks, to perform local computations on large networks of interconnected conditional probability distributions [27, 25, 26] or belief functions [24, 33]. Such networks, as well as the execution graphs described above, find their root in the standard graphical representation of Markov random fields in terms of cliques of neighbors [29]. However, while the graphs of Markov random fields are undirected, like the branches produced in (3.18a), the goal of belief networks is to perform local computations in a causal manner, which as will be shown below, requires a directed acyclic graph. At this stage, the execution graph associated to a compound system π of the form (3.17) is in general partly undirected, and partly directed. Our objective is now to develop compilation rules for transforming this graph into a directed one.

Graph compilation. The structure of the execution graph of a compound system provides all the information required to determine whether this system can be simulated incrementally, as shown by the following result.

Lemma 3.2 *Consider a partition $I = J \cup J^c$ with $J \cap J^c = \emptyset$, for which we write*

$$\pi_J = \mid_{j \in J} \pi_j \qquad \pi_J^c = \mid_{j \in J^c} \pi_j$$

$$\begin{aligned} \mathbf{X}_J &= \textit{set of private variables of } \pi_J \\ \mathbf{X}_J^c &= \textit{set of private variables of } \pi_J^c \\ \partial\mathbf{X} &= \textit{set of shared variables of } \pi_J \textit{ and } \pi_J^c . \end{aligned}$$

1. *We have*

$$\overline{\mathcal{S}}_{\mathbf{X}_J}(\pi) = \overline{\mathcal{S}}_{\mathbf{X}_J}\left(\pi_J \mid \overline{\mathcal{S}}_{\partial\mathbf{X}}(\pi_J^c) \right) . \tag{3.19}$$

2. *Under the stronger assumption*

$$\pi_J \longrightarrow \partial\mathbf{X} \longrightarrow \pi_J^c \quad \textit{or} \quad \pi_J \longleftrightarrow \partial\mathbf{X} \longleftrightarrow \pi_J^c , \tag{3.20}$$

the identity (3.19) reduces to

$$\overline{\mathcal{S}}_{\mathbf{X}_J}(\pi) = \overline{\mathcal{S}}_{\mathbf{X}_J}(\pi_J) , \tag{3.21}$$

which indicates that to simulate the variables $\mathbf{X}_J$ of the compound system π, we only need to simulate the subsystem π_J.

Based on the above lemma, we can readily determine from the execution graph of a compound system π whether this system can be simulated incrementally.

Theorem 3.2 *A compound system π of the form (3.17) admits an incremental simulation if and only if* EXECGRAPH (π) *is an acyclic directed graph. This means that this graph contains no undirected branch of the form (3.18a), and no directed cycle. When determining whether the graph contains cycles, all bidirectional branches of the form (3.18c) can be used as "wild cards" whose orientation can be selected so as to break potential cycles.*

As a side remark, note that fixed-point equations of the form (3.19) can be solved iteratively by employing stochastic relaxation methods such as the Metropolis algorithm or the Gibbs sampler [4]. However, such schemes fall outside the scope of the incremental simulation procedures described here.

Next, since most compound systems of the form (3.17) usually give rise to execution graphs which contain either undirected branches or cycles, it is of interest to develop transformation/compilation rules, which when applied to a given system π, will yield a new system which can be incrementally simulated. In doing so, we restrict our attention to transformations which preserve the local connectivity of EXECGRAPH (π). Otherwise, we could always aggregate all the subsystems π_i and their variables into the full π system which contains only one increment, and thus admits a trivial, but uninteresting, incremental simulation. Consequently, we require for the time being that the transformations applied to π should preserve the structure of the interaction graph obtained by removing all branch orientations from EXECGRAPH (π), as well as the variables $\mathbf{X}_i$ of the subsystems forming its vertices.

Theorem 3.3 *Given a compound system π whose interaction graph forms a tree, we can transform π into an equivalent system π' such that* EXECGRAPH (π') *is a directed tree, and is thus amenable to incremental simulation.*

PROOF: Since the interaction graph of π forms a tree, the index set I admits a natural distance, where for $i, j \in I$, $d(i,j) = k$ if the unique path linking π_i to π_j has k branches. Select now an arbitrary node $i_0 \in I$ as the root of the tree. A partial partial order can be defined over I by considering the distance of i to i_0. Thus we write $i \prec j$ if i is closer to i_0 than j. Consider the following rules:

RULE 1: Select $i \in I$, and let i_- be the unique neighbour of i such that $i_- \prec i$, i.e. i_- denotes the parent of i. If π_i and π_{i_-} share variables, and π_i is not already an innovation of π_{i_-}, then factor π_i as

$$\pi_i \equiv \overline{\mathcal{S}}_{\mathbf{X}_{i_-}}(\pi_i) \mid \mathcal{S}_{\mathbf{X}_{i_-}}(\pi_i) \,, \tag{3.22}$$

otherwise do nothing. Here, $\mathbf{X}_{i_-}$ denotes the set of variables of π_{i_-}.

RULE 2: If the factorization (3.22) has been performed, reorganize the compound system π by rewriting

$$\pi_{i_-} \mid \pi_i \equiv \pi'_{i_-} \mid \pi'_i \tag{3.23a}$$

with

$$\pi'_{i_-} \triangleq \pi_{i_-} \mid \overline{\mathcal{S}}_{\mathbf{X}_{i_-}}(\pi_i) \qquad \pi'_i \triangleq \mathcal{S}_{\mathbf{X}_{i_-}}(\pi_i) \,. \tag{3.23b}$$

This reorganization clearly preserves the structure of the interaction graph of π, as well as the variables of subsystems π_{i_-} and π_i.

The index set I can be ordered so that successive indices i are *nonincreasing* with respect to the partial order $\prec$. By successively applying RULE 1 and RULE 2 to this sequence, we find that once the transformation (3.23a)–(3.23b) has been applied to node i, the new system π' includes the branch

$$\pi'_{i_-} \longrightarrow \mathbf{X}_{i_-,i} \longrightarrow \pi'_i \tag{3.24}$$

in its execution graph, where $\mathbf{X}_{i_-,i}$ represents the set of shared variables of π'_{i_-} and π'_i. Then when RULE 1 and RULE 2 are subsequently applied to system π'_{i_-}, π'_{i_-} may change, but the orientation of the branch (3.24) remains the same. Thus, to transform the given tree into a fully oriented tree, we need to apply the rules only once at each node of the tree, by moving gradually from its extremities towards its root i_0, so that the complexity of the compilation procedure is proportional to the cardinality of I. Note that in the above procedure, the choice of root i_0 is completely arbitrary. □

3.5 The SIG simulation compiler

We now implement the marginal and conditional primitives in the SIG language, and use them to incrementally simulate compound systems. Let `SYSTEM` denote a system and `X, Y` be two of its variables. The two operators

```
extract X,Y in SYSTEM
given X,Y SYSTEM
```

denote respectively the marginal $\overline{\mathcal{S}}_{X,Y}$ (SYSTEM) and conditional $\mathcal{S}_{X,Y}$ (SYSTEM).

To illustrate the application of these operators, we consider the HMM example. As a first step, examine the system

```
system  HMM_inc = (integer N)
    { variable X[i] i=0 to N, Y[i] i=1 to N }
    (| X[0] = 0
     | loop i=1 to N
        (| given X[i-1] potential U(X[i-1],X[i])
         | given X[i-1],X[i] potential V(X[i-1],X[i],Y[i])
         |)
       end
     |)
end .
```

Since only " `given ... potential ...` " statements are used, the SIG program `HMM_inc` admits the execution graph

$$\begin{array}{ccccccccccc} \pi_0 \rightarrow x_0, x_1 \rightarrow \pi_1 \rightarrow x_1, x_2 \rightarrow \pi_2 & & & \pi_{N-1} \rightarrow x_{N-1}, x_N \rightarrow \pi_N \\ \quad\downarrow \qquad\qquad \downarrow & \cdots\cdots & & \downarrow \\ \quad\sigma_1 \qquad\qquad \sigma_2 & & & \sigma_N \end{array} \qquad (3.25)$$

where the subsystems appearing in the graph are defined by

```
   PI[i] ::= given X[i-1] potential U(X[i-1],X[i])
SIGMA[i] ::= given X[i-1],X[i] potential V(X[i-1],X[i],Y[i])
```

Since this execution graph is an oriented tree, according to Theorem 3.2, we can simulate `HMM_inc` "on-line" for increasing values of the index `i`. On the other hand, this is not the case if we consider the original HMM program, even if it contains only `given ...` statements, because of the presence of the two-point boundary-value condition

```
(| X[0] = 0
 | X[N] = X_MAX
 |) .
```

In fact, the execution graph of HMM takes the form

$$
\begin{array}{ccccccccccc}
\pi_0 & - & x_0, x_1 & - & \pi_1 - & x_1, x_2 & - & \pi_2 & & \pi_{N-1} & - & x_{N-1}, x_N & - & \pi_N \\
 & & | & & & | & & & \cdots\cdots & & & | & & \\
 & & \sigma_1 & & & \sigma_2 & & & & & & \sigma_N & &
\end{array}
\qquad (3.26)
$$

It is a nonoriented tree, which can be transformed into a directed one by employing the two compilation rules described in the proof of Theorem 3.3. The algorithm proceeds in two phases: we first apply the rules to the vertical branches of the tree, which model the HMM observations, and then perform a right to left sweep over the horizontal branches, which model the Markov chain dynamics.

1. Applying RULE 1, the `potential V(X[i-1],X[i],Y[i])` can be decomposed as follows, where `<=>` means $\equiv$:

```
    potential V(X[i-1],X[i],Y[i])
<=>
 (| extract X[i-1],X[i] in potential V(X[i-1],X[i],Y[i])
  | given X[i-1],X[i] potential V(X[i-1],X[i],Y[i])
  |) .
```

   For each index `i`, the subsystem

```
SIGMA[i] ::= given X[i-1],X[i] potential V(X[i-1],X[i],Y[i])
```

   is an innovation with respect to all other subsystems, and is executable as soon as `X[i-1]` and `X[i]` have been simulated.

2. Applying RULE 2, define

```
PI[i] ::=
(| extract X[i-1],X[i] in potential V(X[i-1],X[i],Y[i])
 | potential U(X[i-1],X[i])
 |) ,
```

   where the boundary constraints `X[0] = 0` and `X[N] = X_MAX` need also to be included for `i = 1` and `i = N`, respectively.

3. Recursively, for `i` decreasing from `N` to `1`,

   (a) apply RULE 1 and decompose

```
PI[i] <=> (| extract X[i-1] in PI[i]
           | given X[i-1] PI[i]
           |) ;
```

   (b) apply RULE 2 and redefine

```
PI[i]    ::= (| given X[i-1] PI[i]
              |)

PI[i-1] ::= (| extract X[i-1] in PI[i]
             | PI[i-1]
             |) .
```

The resulting system is equivalent to the original one, and has the execution graph (3.25), so that it is ready for simulation.

4 Discussion and Conclusions

We have introduced the CSS model and associated SIG minilanguage for describing stochastic/non-stochastic systems. CSS is a relational model where systems are defined by relations and unnormalized probability densities. This feature has several advantages. First, it makes the definition of the composition operation " | " relatively easy. Second, it provides us with a simple mechanism for specifying the conditional behavior of a system given that certain contraints are satisfied, which has the potential to be very useful when tracking cascades of events leading to system failures.

However, the system specification provided by CSS is generally not executable, i.e., it does not readily lead to a system implementation. To convert it to a form which can be simulated, we rely on a compilation, which examines the dependency relations, both non-stochastic and statistical, existing between the system variables. This compilation employs two operations. The marginal $\overline{\mathcal{S}}(.)$ and conditional $\mathcal{S}(.)$ extend to hybrid systems the standard marginal and conditional probability distributions of fully probabilized systems. With their help, we were able to introduce the notion of innovation, whereby π' is an innovation of π if, roughly speaking, π' does not influence π in the composition $\pi \mid \pi'$, but π may influence π', so that the interaction between π and π' is *oriented*, and $\pi \mid \pi'$ is amenable to incremental simulation. In general, systems interact in a non-oriented way. When the interaction graph of a system forms a tree, we have presented rules which can be used to convert the tree into a directed one while preserving equivalence of the compound system. In combination with the results of [27] for aggregating a triangulated graph into a tree, these rules can be used to compile arbitrary interaction graphs. A SIG implementation of the compilation rules was presented. Finally, it turns out that our simulation results can be adapted, with minor modifications, to the hidden state estimation of hybrid systems. We only need to replace the $\sum$ by the max in performing random compression.

Although CSS is obviously related to the theory of belief functions and belief networks developed in [21, 22, 23, 24], it differs from it in several respects. First, as mentioned earlier, unlike the Dempster-Shafer approach which relies on upper and lower probabilities in the space V_X of visible variables, we keep track of probability distributions on the random configurations. Second, through the introduction of the concept of innovation, which does not appear in the belief networks literature, CSS provides concrete solutions to basic problems such as hybrid system simulation and estimation. To our knowledge, no other approach offers this range of facilities.

The research presented here can be extended in several directions, we discuss only two of them, see [1] for more details.

- A first issue involves the introduction of two features currently missing from CSS, namely the specification of timing information, or the absence thereof, and the ability to define hybrid systems over infinite time intervals. These two features are already present in the previously introduced SIGNA*lea* language [18], which represents an an extension of the SIGNAL synchronous real-time language [34, 35, 36]. The SIGNA*lea* language generalizes stochastic Büchi automata, Petri nets, and our SIG minilanguage. But the mathematical foundations of SIGNA*lea* in [18] are somewhat shaky and estimation is not included. Thus, generalizing CSS to SIGNA*lea* is a high priority task, particularly since SIGNA*lea* is currently under implementation.

- Also, our results need to be tested on real applications. Two applications of SIGN*alea* are now under consideration. The first one involves the implementation for Electricté de France of the nonintrusive appliance load monitoring scheme proposed in [6], which presents strong similarities with speech recognition, and for which Viterbi-style estimation algorithms are expected to be successful. A second potential application in the area of power generation concerns the design of a monitoring and diagnostic system from its risk analysis description. In this context, we would like to determine whether our relational model, because of its ability to track cascades of events leading to specific failures, presents advantages for risk analysis.

References

1. A. Benveniste, B. Levy, E. Fabre, and P. L. Guernic, "A calculus of stochastic systems: specification, simulation, and hidden state estimation," Tech. Rep. 2465, INRIA, Rocquencourt, France, january 1995.
2. N. Viswanadham and Y. Narahari, *Performance Modeling of Automated Manufacturing Systems*. Englewood Cliffs, NJ: Prentice Hall, 1992.
3. L. R. Rabiner and B. H. Juang, "An introduction to hidden Markov models," *IEEE ASSP Magazine*, vol. 3, pp. 4–16, Jan. 1986.
4. S. Geman and D. Geman, "Stochastic relaxation, Gibbs distribution, and the Bayesian restoration of images," *IEEE Trans. on Pattern Analysis and Machine Intelligence*, vol. 6, pp. 721–741, Nov. 1984.
5. R. C. Dubes and A. K. Jain, "Random field models in image analysis," *J. Applied Stat.*, vol. 12, pp. 131–164, 1989.
6. G. W. Hart, "Nonintrusive applicance load monitoring," *Proc. IEEE*, vol. 80, pp. 1870–1891, Dec. 1992.
7. M. Basseville and I. V. Nikiforov, *Detection of Abrupt Changes : Theory and Applications*. Englewood Cliffs, NJ: Prentice Hall, 1993.
8. T. Soderstrom and P. Stoica, *System Identification*. Englewood Cliffs, NJ: Prentice Hall, 1989.
9. M. Molloy, "Performance analysis using stochastic Petri nets," *IEEE Trans. Computers*, vol. 31, pp. 913–917, Sep. 1982.
10. B. Plateau and K. Atif, "Stochastic automata network for modeling parallel systems," *IEEE Trans. on Software Engineering*, vol. 17, pp. 1093–1108, Oct. 1991.
11. B. Plateau and J. Fourneau, "A methodology for solving Markov models of parallel systems," *J. Parallel and Distributed Comput.*, vol. 12, pp. 370–387, 1991.
12. R. van Glabbeek, S. A. Smolka, B. Steffen, and C. Tofts, "Reactive, generative, and stratified models of probabilistic processes," in *Proc. 5th IEEE Int. Symp. on Logic in Computer Science*, (Philadelphia,PA), pp. 130–141, June 1990.
13. H. Hansson and B. Jonsson, "A calculus for communicating systems with time and probabilities," in *Proc. of the 11th IEEE Real-Time Systems Symposium*, (Los Alamitos), pp. 278–287, Dec. 1990.
14. B. Jonsson and K. Larsen, "Specification and refinement of probabilistic processes," in *Proc. 6th IEEE Int. Symp. on Logic in Computer Science*, (Amsterdam), pp. 266–277, July 1991.
15. A. Giacalone, C. Jou, and S. Smolka, "Algebraic reasoning for probabilistic concurrent systems," in *Proc. IFIP TC2 Working Conference on Programming Concepts and Methods*, (), p. , 1989.

16. S. Hart and M. Sharir, "Probabilistic propositional temporal logic," *Information and Control*, vol. 70, pp. 97–155, 1986.
17. R. Alur, C. Courcoubetis, and D. Dill, "Model checking for probabilistic real-time systems," in *Proc. 18th Int. Coll. on Automata Languages and Programming (ICALP)*, (), p. , 1991.
18. A. Benveniste, "Constructive probability and the SIGNAL*alea* language: building and handling random processes with programming," Tech. Rep. 1532, Institut National de Recherche en Informatique et Automatique, Rocquencourt, France, Oct. 1991.
19. B. C. Levy, A. Benveniste, and R. Nikoukhah, "High-level primitives for recursive maximum likelihood estimation," Tech. Rep. 767, IRISA, Rennes, France, Oct. 1993.
20. G. D. Forney, "The Viterbi algorithm," *Proc. IEEE*, vol. 61, pp. 268–278, March 1973.
21. A. P. Dempster, "Upper and lower probabilities induced by a multivalued mapping," *Annals Math. Statistics*, vol. 38, pp. 325–339, 1967.
22. A. P. Dempster, "A generalization of Bayesian inference (with discussion)," *Royal Stat. Soc., Series B*, vol. 30, pp. 205–247, 1968.
23. G. Shafer, *A Mathematical Theory of Evidence*. Princeton, NJ: Princeton Univ. Press, 1976.
24. P. P. Shenoi and G. Shafer, "Axioms for probability and belief function propagation," in *Uncertainty in Artificial Intelligence*, (R. D. Shachter, T. S. Levitt, L. N. Kanal, and J. F. Lemmer, eds.), pp. 169–198, Amsterdam: North-Holland, 1990.
25. J. Pearl, "Fusion, propagation, and structuring in belief networks," *Artificial Intelligence*, vol. 29, pp. 241–288, Sep. 1986.
26. M. A. Peot and R. D. Shachter, "Fusion and propagation with multiple observations in belief networks," *Artificial Intelligence*, vol. 48, pp. 299–318, 1991.
27. S. L. Lauritzen and D. J. Spiegelhalter, "Local computations with probabilities on graphical structures and their application to expert systems (with discussion)," *J. Royal Stat. Soc., Series B*, vol. 50, pp. 157–224, 1988.
28. B. Jonsson, C. Ho-Stuart, and Y. Wang, "Testing and refinement for nondeterministic and probabilistic processes," in *Lecture Notes in Computer Science*, pp. 418–430, Berlin: Springer Verlag, 1994.
29. R. Kindermann and J. L. Snell, *Markov Random Fields and their Applications*. Providence, RI: American Mathematical Society, 1980.
30. C. Robert, *Modèles Statistiques pour l'Intelligence Artificielle*. Paris: Masson, 1991.
31. B. Prum and J. Fort, *Stochastic Processes on a Lattice and Gibbs Measure*. Boston, MA: Kluwer Acad. Publ., 1991.
32. C. Dellacherie and P. Meyer, *Probabilités et Potentiels*. Paris: Hermann, 1976.
33. A. P. Dempster, "Construction and local computation aspects of network belief functions," in *Influence Diagrams, Belief Nets, and Decision analysis*, (R. M. Oliver and J. Q. Smith, eds.), ch. 6, pp. 121–141, Chichester, England: J. Wiley, 1990.
34. P. Le Guernic, T. Gauthier, M. Le Borgne, and C. Le Maire, "Programming real-time applications with SIGNAL," *Proc. IEEE*, vol. 79, pp. 1321–1336, Sep. 1991.
35. A. Benveniste and P. Le Guernic, "Hybrid dynamical systems theory and the SIGNAL language," *IEEE Trans. Automat. Contr.*, vol. 35, pp. 535–546, May 1990.
36. A. Benveniste, M. Le Borgne, and P. Le Guernic, "Hybrid systems: the SIGNAL approach," in *Lecture Notes in Computer Science*, pp. 230–254, Berlin: Springer Verlag, 1993.

Condensation of Information from Signals for Process Modeling and Control

J. D. Birdwell and B. C. Moore

Data Refining Technologies, Inc.
P. O. Box 893
Plaquemine, LA 70765–0893

Abstract. This paper deals with processing and storage methods for large volume industrial process data measurements which enable the utilization of historical data for process understanding. In the context of hybrid systems, the focus is upon the steps commonly lumped under the category of analog to digital conversion (ADC), which is the interface between the continuous time world of the industrial process and the discrete time world of digital processing. In both cases a hybrid systems viewpoint prevails where continuous variable, discrete variable, and discrete event activities occur.
A fundamentally different paradigm for signal processing is present based upon interval, rather than point set, mathematics. The advantages of this approach include: (1) a hierarchical signal processing structure with decreasing resolution at higher levels of the hierarchy, yet with a common mathematical structure at all levels of the hierarchy; (2) orders of magnitude reduction in the storage and processing requirements of signals as one moves up the hierarchy, and (3) efficient localization of and access to detailed information from higher level information. Because of the paradigm shift to interval mathematics and the structure of signal data (signal objects), we refer to these methods as Object Oriented Signal Processing.
Signal objects can be placed in categories which are automatically derived from the characteristics of the measurement data. This lends a symbolic character to this "signal material" and provides a linkage between the largely numeric processing steps of data collection and the primarily symbolic processing steps of data analysis using methods grouped under the general heading of "intelligent signal processing" or "intelligent control", such as automata, fuzzy sets, rule based expert systems, and linguistic processing. In effect, the relationships between condensed signals can be characterized by discrete event models, with access to underlying details. The same approach can be taken in the reverse direction, and provides linkage from intelligent control schemes to manipulated variables of the industrial process. The result is an abstract model and control strategy implemented within a discrete event, or hybrid systems, framework which utilizes condensed process information.

1 Introduction

Industrial process control applications suffer from a deluge of information. Implementations have evolved from small– to medium–scale automation, where tens to hundreds of signals are measured and controlled, into large–scale and plant–wide information processing systems involving thousands of signals measured as frequently as once per second or more. Where the structure of the supporting organizations was adequate at smaller scales, with information either periodically observed by operators or analyzed in detail by engineering staff when process upsets occurred, this structure is inadequate in the present environment. There is simply far too much information, continually supplemented by new measurements, for the operators and engineers to understand. Features in the information which could allow preemptive strikes to repair failing equipment before breakdowns cause either downtime or poor quality products are often not observed simply because they hide in the volume of data. When breakdowns do occur, the volume of information hinders fault isolation; again, the necessary information is there, but relevant portions are difficult to locate.

Our technology provides support to plant organizations by significantly reducing the volume of data which must be maintained and observed. In addition, the reduced volume enables automated processing methods which assist process support personnel who are responsible for using measured information to isolate and recover from degraded or failed equipment. Many of the methods we propose appear simple on the surface, and in fact, they are designed to provide a simple and intuitive interface to users, who are not specialists in data processing or control, but rather are experts in management of specific processes. It is neither realistic nor appropriate to provide, for example, tools such as Matlab to process support organizations. The people within these groups have neither the time to use these tools nor the interest in learning how to use them — they are simply not relevant to their problems, which center upon rapid isolation of problems with process equipment with minimal impact upon the time and resources of the organization.

Typical industrial processes generate very large quantities of measurement data, and a significant amount of information about a process can be derived from these data values, however, this information is rarely utilized because the volume of data precludes both long term storage and processing. It is imperative that data volume be reduced as a first stage in any information processing procedure based upon historical process data. For example, industrial processes which have 1,000 signals monitored and sampled once per second are common, which corresponds to 345 Mbytes of accumulated data per day of operation, 10.4 Gbytes/month, and 126 Gbytes/year. In our experience, systems that provide useful historical data collection and process understanding require at least a month's data, and more typically 3–6 month's. It is clear that even though data storage is possible at the one month to one year level, it is impractical and usually not possible to process these data sets in a timely manner.

As an example, this paper focuses on discovery and tracking of relationships between pairs of variables. To illustrate the difficulties, and why current methods are not adequate, we use data provided by Shell from a distillation column [3]. (Note that to differentiate the distillation column from other uses of the word "column" in this paper, we capitalize "Column" when it refers to the shell process.) Two variables, the tray temperature and the reflux, have an interesting nonlinear relationship. The data set provides 21 days of data, taken at one minute intervals (30,219 data points). The raw data are shown in Figs. 1 and 2.

We will revisit these data later; for now we note only two characteristics of the data: First, there are a lot of data; while 30,219 values is not excessive for a single variable, in a typical application there can be hundreds to thousands of such signals. Second, the collected data contain noise, and, as is typical with process data, the

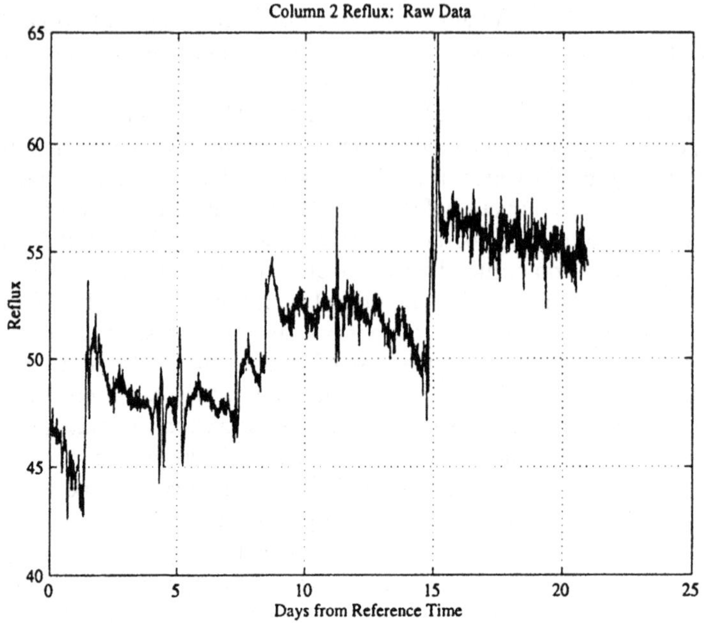

Fig. 1. Raw data for Shell Column 2: Reflux.

noise processes do not satisfy standard assumptions (such as Gaussian). When a person views this data, information is filtered from the noise: spikes tend to be ignored, and regions of relatively steady operation can be identified. These operations are not trivial; it has proven extremely difficult for the industry to provide methods which correctly separate these regions of steady, transient, and invalid data without *apriori* information. Even when information such as ranges of validity is known, current technology has difficulty; witness, for example, the various "solutions" proposed to deal with high false alarm rates in process measurement equipment in fields such as electric power transmission and distribution.

To illustrate the difference between human and machine data processing, Fig. 3 is a simple x-y plot of tray temperature versus reflux. There is an easily discernible inverse relationship, but there are also many data points which detract from this relationship. A child, given this plot, can trace an outline of the data features; however, with a computer this is difficult, primarily because it is difficult for a computer method to decide which data points to discard in order to sharpen the relationship. Therefore, a simple statement of our problem is: Without *apriori* information, find a region of validity for the relationship between two variables. This operation must be performed reliably and without human involvement, and it must be done efficiently using raw data sources which contain, for any pair of variables, 10's to 100's of thousands of data values.

It is instructive to note, as well, that the relationship depicted in Fig. 3 is not linear. It is also "fat": since many other factors contribute to the relationship between the variables, it is not acceptable to model it as a curve. Therefore, standard statis-

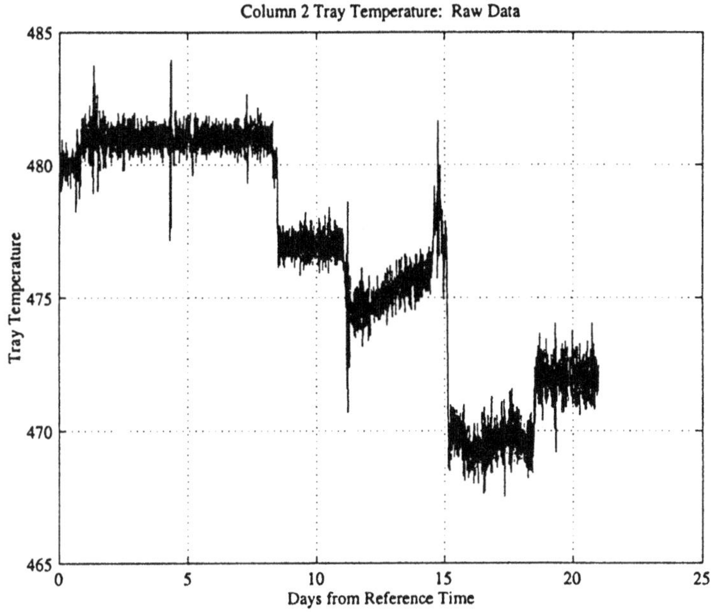

Fig. 2. Raw data for Shell Column 2: tray temperature.

tical methods, such as ARMA or nonlinear parametric models, are not appropriate (although they are often applied in this context!). Once a relationship is established and adequately described, if it is in an appropriate form, it can be used to sense if new measurements conform to behaviors predicted by historical data. Thus, our methods can be used to construct advanced alarm systems, which adapt their tolerance to measurement uncertainty, nonlinear effects, and unmeasured influences to historical patterns rather than to engineering analysis or guesswork. The result is dramatically lowered false alarm and failure to detect incidents in systems which utilize our technology.

1.1 Interval measurements and information condensation.

A fundamental shift in perspective is the basis for our work; signals are viewed as sets of interval measurements rather than as point values. This allows a uniform viewpoint at all levels of detail in hierarchical data measurement and storage systems with common methods utilized throughout. The shift in perspective also admits a mechanism for condensation of measurement data which supports hierarchical data organizations with rapid retrieval of detailed data relevant to specific needs. We differentiate here between "compression" and "condensation" because the condensed information is of the same form as the detailed signal data, allowing it to be used in place of detailed data in information processing applications. Typical levels of volume reduction on industrial processes of 500:1 have been achieved.

This method of data condensation works because typical industrial process signals contain significant periods of time when steady state or consistently varying

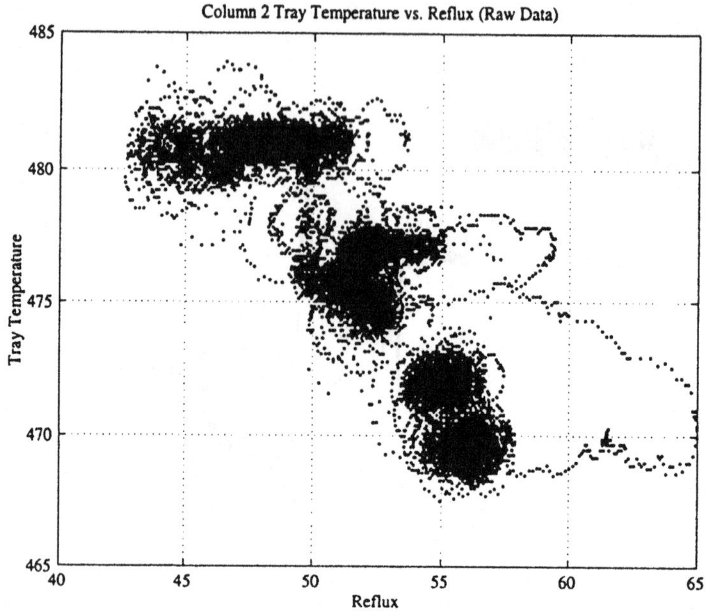

Fig. 3. Column 2 relationship between tray temperature and reflux based upon raw data.

behaviors are observed, with relatively rapid transients interspersed. Methods based upon, for example, moving averages or intermittent sampling can achieve data reduction, but lose time resolution of transient events. This does not occur with interval-based methods, and the causal relationships between variables are preserved.

Higher level signal objects can be placed in categories which are automatically derived from the characteristics of the measurement data. This lends a symbolic character to this "signal material" and provides a linkage between the largely numeric processing steps of data collection and the primarily symbolic processing steps of data analysis using methods grouped under the general heading of "intelligent signal processing" or "intelligent control", such as automata, fuzzy sets, rule based expert systems, and linguistic processing. In effect, the relationships between condensed signals can be characterized by discrete event models, with access to underlying details. The same approach can be taken in the reverse direction, and provides linkage from intelligent control schemes to manipulated variables of the industrial process. The result is an abstract model and control strategy implemented within a discrete event, or hybrid systems, framework which utilizes condensed process information.

2 A Mathematical Basis for Object Oriented Signal Processing.

2.1 Interval scales and triangular sets.

Fundamental to our approach to signal processing is the shift in perspective from point values to interval values: The value of a variable is defined to lie within an interval over an interval of time. This provides a uniform base for operations upon signals; at any level of detail, signals have the same form, and thus, the same software implementations can be utilized. Signal values are always defined with respect to an interval scale, which is a partition on the real line. Interval scales induce an algebraic structure upon the set of possible values a signal may assume; this structure can be exploited for efficient implementation of signal processing methods.

An object oriented signal is a set of objects, which can be viewed as rows in a sheet, each of which contains one or more intervals, where all objects within a sheet have the same structure. Columns of these sheets correspond to variables which assume interval values from a triangular set of intervals, to be defined in the sequel. Object structures may also define weights, which are variables which assume non-negative integral values. Table 1 is the first portion of a signal sheet which describes a condensed (interval) form of the Shell Column 2 reflux variable graphed in Fig. 1. Each signal object defines two intervals, one in the independent (time) variable, and the other in the data variable (reflux). This signal is defined to contain the "most significant[1]" portion of the data variable's values over each time interval. The elements of this table provide an approximation to the original signal's behavior in a significantly reduced volume of data. For example, the signal sheets for the reflux and tray temperature measurement signals shown in Figs. 1 and 2 contain 1,021 and 525 signal objects (rows), constituting reductions in the number of data elements by factors of 30 and 55, respectively.

It is evident that object oriented signals have significant potential for data volume reduction; however, it is essential that an interval representation for a signal be chosen which maximizes the retained information from the original signal while keeping its volume within reasonable limits. This is achieved through a "scrubbing" process which groups sequences of similar data within the same object and preserves resolution in the independent variable (time) of changes in signal behavior. The portion of the sheet illustrated in Table 1 was created using this scrubbing process. Key to the scrubber's operation is the mathematics of interval algebras and triangular sets, which are summarized here.

Definition 1. An *interval scale* on the reals is a set of $N+2$ intervals, with N inner half-open intervals of the form $[a,b)$, a lower interval of the form $(-\infty,a)$, and an upper interval of the form $[b,\infty)$, where every real value is contained in one, and only one interval in the set. An element in an interval scale is called a *basis interval*. Corresponding to these intervals is the integer representation set $\{0,1,\ldots,N+1\}$, where 0 and $N+1$ represent the two outer intervals, and $\{1,\ldots,N\}$ represent the

[1] We defer the discussion of how these intervals are formed to a future paper. An overly simplified explanation would be to say that the value interval contains at least x% of the data, where x is usually in the range of 95–100.

Table 1. A portion of the signal sheet for Shell Column 2 reflux data.

t_left	t_right	v_bot	v_top	weight
0.000	0.088	46.432	47.240	1
0.088	0.108	46.342	47.060	1
0.108	0.158	46.432	47.778	1
0.158	0.181	46.342	46.701	1
0.181	0.231	46.432	46.970	1
0.231	0.263	45.983	46.701	1
0.263	0.287	46.611	47.329	1
0.287	0.404	45.983	46.791	1
0.404	0.435	46.432	46.701	1
0.435	0.448	45.714	46.701	1
0.448	0.517	44.637	45.803	1
0.517	0.549	45.714	46.611	1
0.549	0.556	46.342	46.970	1
0.556	0.649	45.534	46.791	1

inner intervals, ordered on the reals. The basis intervals can be represented by either a single integer i from the representation set or by the interval notation $[i, i+1)$.

The interval scale not only provides a discretization of the reals into a finite set of intervals; it also provides a representation of elements of this set by integers in the range $\{0, 1, \ldots, N+1\}$. An interval scale is depicted in Fig. 4 with the representations of the basis intervals and a representative interval.

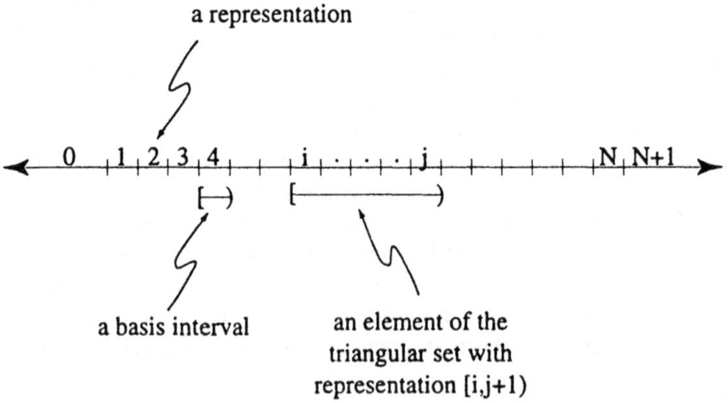

Fig. 4. An interval scale and interval representations.

The interval scale, with set union and complement operators, defines an algebra of subsets of the reals called the <u>*interval algebra*</u>[2]. Elements of the interval algebra

[2] We note that there are strong connections between these concepts and those related to interval mathematics and described in the literature, including interval orders [2] and lattices [1]. While a portion of the mathematical framework is similar, however, the goals within the interval mathematics research community are quite different from the objectives expressed herein.

are representations of subsets of the reals, such that for any subset of the reals, there exists exactly one element of the interval algebra which represents that subset.

Definition 2. There is one subset of the interval algebra which is of fundamental importance to object oriented signal processing. This is the interval scale *triangular set*, which is the set of all intervals which are members of the interval algebra. The triangular set is a lattice with a partial ordering relation defined by set inclusion and a maximal element corresponding to the entire set of reals.

Each interval in the triangular set can be represented by two integers $[i, j+1)$, $j \geq i$, in the set $\{0, 1, \ldots, N+1\}$ and can be formed from two elements of the interval scale using the rule described as follows: The interval in the triangular set represented by $[i, j+1)$ is the interval of least measure which contains both elements of the interval scale corresponding to representation integers i and j, as shown in Fig. 4. For our purposes, measure of an interval is defined as the positive difference between the two endpoints. Thus if i represents $[0, 5.0)$, and j represents $[7.0, 8.0)$, then $[i, j+1)$ represents the interval $[0, 8.0)$ of measure 8.0.

We call the set of all intervals in the interval algebra the triangular set because its elements can be displayed graphically in triangular form, as shown in Fig. 5. In the figure, each column of triangles represents all intervals of a specified width, and this width increases as one moves to the right. The vertical position of a triangle corresponds to an interval's center, which increases in value as one moves down a column.

Triangular sets play an important role in object oriented signal processing. Most computations, such as signal scrubbing, are formulated as optimizations over the elements of these sets, and the progress of an optimization method can be viewed as a set of paths through triangular sets. Design of efficient methods which operate upon object oriented signals is equivalent to design of methods which determine solutions to an optimization problem with minimum path length. The graphical representation of the members of triangular sets is a useful tool in design of implementations; with it, one can visualize the progress of computation.

The following observations can be made:

- Basis intervals are elements of the triangular set.
- Given an interval scale, every real value is mapped into a unique basis interval in its associated triangular set.
- For every subset of the reals, there is a unique representation in the interval algebra: This is the minimal element of the interval algebra which contains the subset, where "minimal" is with respect to the partial order defined by set inclusion.
- Every interval which is a subset of the reals has a unique representation in the triangular set.
- For an interval scale with N inner elements, the triangular set has $\frac{(N+2)(N+3)}{2}$ elements. In contrast, the interval algebra has 2^M elements, where $M = N + 2$, which is normally a much larger number.

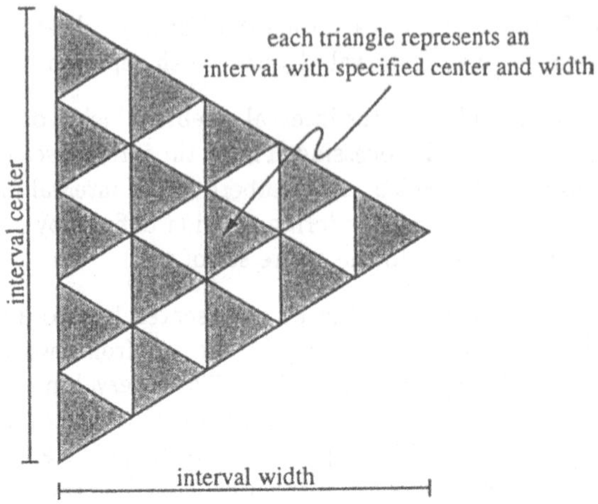

Fig. 5. A graphical depiction of a triangular set.

Thus, any subset of the reals has a unique representation in the interval algebra. This also holds for elements of a second interval algebra since they are subsets of the reals. Thus, there is a representation map from any interval algebra A to a second interval algebra B, and vice versa, although these maps are not in general inverses of one another[3].

2.2 Fundamental relations on interval scale triangular sets.

In this subsection we shall develop a set of fundamental relations between subsets of an interval scale triangular set which will provide the foundation of object oriented signal processing. For this discussion, we shall let R and T represent two such subsets.

Result 1. *There exists a set of 13 relations*[4] *between R and T such that for any pair (r,t) in $R \times T$, exactly one of these relations holds. This is a generalization of the set of ordering relations $\{<,=,>\}$ defined on the reals.*

Consider [bot, top) to be an element of R, and [low, high) to be an element of T. These thirteen relations (each with a label attached) are given in Table 2. Note that by definition top > bot and high > low.

[3] The class of interval algebras, together with the maps induced by the representations, is a category [5]. A unique representation map (arrow) exists from any interval algebra to any other interval algebra; this categorical structure is one possible view of the theoretical foundations of object oriented signal processing.

[4] Terminology on relations is taken from [4]. These 13 relations initiate an alternate and unique line of inquiry in the theory of interval orders, which is different than that which is found in literature such as [2], and whose concepts date to N. Wiener's paper [6].

Whereas previous work partitioned interval relations according to properties, such as reflexivity and symmetry, this work partitions sets which map (using the representation map) to elements of the interval algebra according to 13 specific relations. The emphasis is upon a partition of interval sets rather than a decomposition of relations.

Table 2. The 13 interval relations.

Relation	Mnemonic	Conditions on interval endpoints
R_0	LGTHGT	low>top ; high > top
R_1	LETHGT	low=top ; high > top
R_2	LICHGT	bot < low < top (low in center) ; high > top
R_3	LEBHGT	low = bot ; high > top
R_4	LLBHGT	low < bot ; high > top
R_5	LICHET	bot < low < top ; high = top
R_6	LEBHET	low = bot ; high = top
R_7	LLBHET	low < bot ; high = top
R_8	LICHIC	bot < low < top ; bot < high < top
R_9	LEBHIC	low = bot ; bot < high < top
R_{10}	LLBHIC	low < bot ; bot < high < top
R_{11}	LLBHEB	low < bot ; high = bot
R_{12}	LLBHLB	low < bot ; high < bot

The mnemonic code associated with each relation consists of two three–letter codes. In each of these codes, the first letter designates either the low or high endpoint of an element of T, and the next two letters indicate either a comparison with an endpoint of an element of R, or that the designated endpoint of T is "in center", meaning it is in the interior of the element of R. Fig. 6 depicts the relationships between the various endpoints for each interval relation.

2.3 Illumination relations formed from the fundamental relations.

The interval relations are used to construct relationships between sets of interval objects. For a specified variable, a reference interval and a subset of the interval relations induce a partition on a set of intervals over that variable's domain. The partition contains two subsets of the intervals: Those intervals which, together with the reference interval in R, satisfy one of the specified interval relations, and all other intervals. Multiple disjoint subsets of the interval relations can be specified, and induce a partition with one more element than the number of subsets. For a set of reference intervals R, this process defines a relation between elements of R and T. Through this relation, one may answer two questions: Given a reference interval, what is the set of all related target intervals (the "forward correspondence"), and given a target interval, what is the set of all related reference intervals (the "reverse correspondence")?

Operations upon object oriented signals become optimizations over subsets of a triangular set which satisfy specified (problem–dependent) relations. The weight, which is a non-negative integer–valued variable attached to each signal object, plays the role of an objective function, and for each interval is defined as a function of the subset of intervals which satisfies a specified relation for that interval (in either R or T). An overview of the theory is presented here.

Definition 3. An <u>*illumination selector*</u> IS is a subset of the set $\{0, \ldots, 12\}$. An

Relation	Mnemonic	Locations of Left and Right Endpoints of T
R_0	: LGTHGT	
R_1	: LETHGT	
R_2	: LICHGT	
R_3	: LEBHGT	
R_4	: LLBHGT	
R_5	: LICHET	
R_6	: LEBHET	
R_7	: LLBHET	
R_8	: LICHIC	
R_9	: LEBHIC	
R_{10}	: LLBHIC	
R_{11}	: LLBHEB	
R_{12}	: LLBHLB	
	Location of R :	

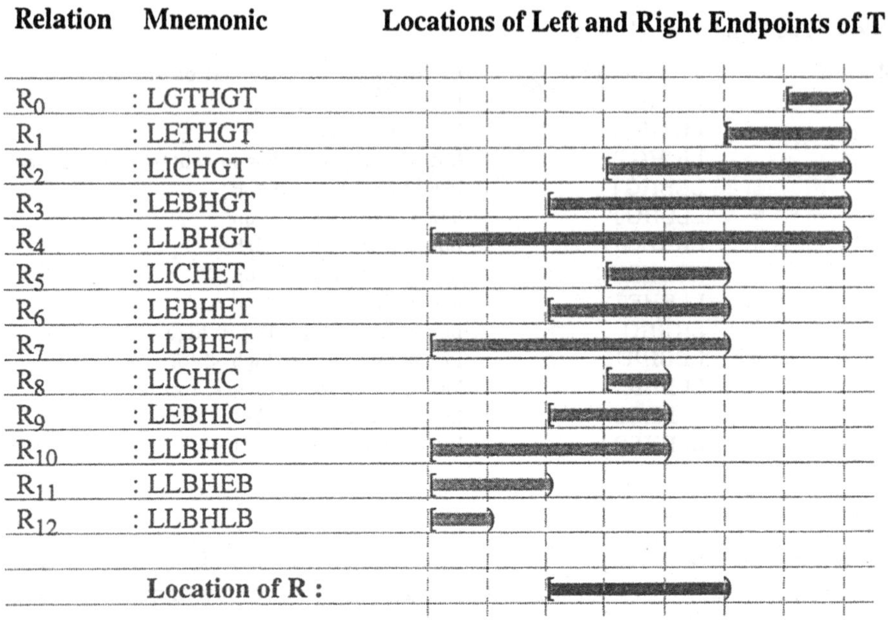

Fig. 6. Relationships between intervals defined by the 13 interval relations.

illumination selector selects a subset of the fundamental relations. This selected subset defines a corresponding _illumination relation_ **IR**: For any pair (r, t) in $R \times T$ the illumination relation **IR** is true if and only if one relation from its selected subset of relations is true.

Definition 4. Given a set of K disjoint illumination selectors $\mathbf{IS}_i$, $i \in \{1, \ldots, K\}$, the <u>dark selector</u> $\mathbf{IS}_{K+1}$ is the complement in $\{0, \ldots, 12\}$ of the union of the illumination selectors. The dark selector may be empty.

Observation. A set of K disjoint illumination selectors, together with the corresponding dark selector, defines a partition of the set of thirteen fundamental relations into $K + 1$ subsets of relations. Note that subset $K + 1$ may be empty since the dark selector may be empty.

Result 2. _Given any set of K disjoint illumination selectors, there exists a corresponding set of $K + 1$ illumination relations $\mathbf{IR}_i$ between R and T such that for any pair (r, t) in $R \times T$ exactly one of the illumination relations is true._

As observed, disjoint selectors partition the fundamental relations into $K + 1$ subsets. The relation $\mathbf{IR}_i(r, t)$ is true (by construction) if and only if one of the fundamental relations in the corresponding subset is true for (r, t). The dark selector may be empty, in which case the relation $\mathbf{IR}_{K+1}(r, t)$ is false for all pairs (r, t). Since exactly one of the fundamental relations holds for every pair (r, t), it follows that for any partition, exactly one subset contains a relation which holds. Thus exactly one illumination relation holds.

Terminology. If an illumination relation $\mathbf{IR}_i(r,t)$ holds true, then we say that the target element t is illuminated by the reference element r through the illumination relation $\mathbf{IR}_i$.

Definition 5. A *set of illumination relations* is a set of relations constructed as described above from a set of K disjoint illumination selectors.

2.4 Illumination maps induced by illumination relations.

Definition 6. The *illumination map* $\mathbf{IM}(r,t)$ from $R \times T$ to $\{1, \ldots, K+1\}$ induced by a set of illumination relations is constructed as follows: For each pair (r,t) the integer i is assigned for which the illumination relation $\mathbf{IR}_i$ holds true.

The illumination map induces two dual maps from the interval scale triangular set to the set $\{1, \ldots, K+1\}$.

Definition 7. The *forward illumination map* $\mathbf{FIM}_r(t) = \mathbf{IM}(r,t)|_{\text{r fixed}}$ is a map from the target set T to the set $\{1, \ldots, K+1\}$. This map partitions T into $K+1$ subsets corresponding to the illumination relations. The illumination map is said to be controlled by selection of the reference interval r.

Definition 8. The forward illumination map defines, for each element r of R, a set of $K+1$ *forward illumination correspondences* $\mathbf{FIC}_i(r)$ in $R \times T$. $\mathbf{FIC}_i(r)$ is the set of all pairs (r,t) for which $\mathbf{FIM}_r(t) = i$. Using the terminology of illumination, $\mathbf{FIC}_i(r)$ associates with r all of the target intervals illuminated through relation $\mathbf{IR}_i$.

Definition 9. The *reverse illumination map* $\mathbf{RIM}_t(r) = \mathbf{IM}(r,t)|_{\text{t fixed}}$ is a map from the reference set R to the set $\{1, \ldots, K+1\}$. This map partitions R into $K+1$ subsets, corresponding to the illumination relations.

Definition 10. The reverse illumination map defines, for each element t in T, a set of $K+1$ *reverse illumination correspondences* $\mathbf{RIC}_i(t)$ in $R \times T$. $\mathbf{RIC}_i(t)$ is the set of all pairs (r,t) for which $\mathbf{RIM}_t(r) = i$. Using the terminology of illumination, $\mathbf{RIC}_i(t)$ associates with a target interval t all of the reference intervals for which the target element is illuminated through relation $\mathbf{IR}_i$.

2.5 The role of triangular sets.

Triangular sets are used to select subsets of intervals with desirable properties. Given the illumination maps, we note that each set of intervals, R and T, has an associated interval scale. These interval scales in turn define triangular sets, Δ_R and Δ_T. For each element of the triangular sets, the associated illumination correspondence (forward or reverse) induces an ordered set of subsets of $R \times T$, and a function can be defined upon these subsets which evaluates a weight. Thus, over each triangular set, a set of distributions can be constructed, which measures the correspondence, with respect to the set of interval relations, between each element of the triangular set and either R or T.

If, in addition to the distribution, the correspondence is maintained between each element of the triangular set and the set of intervals, then elements of the triangular set can be used to select subsets of intervals with desirable properties. For example, properly constructed interval relations form the basis for signal scrubbing, which is our term for refining detailed signal information into lower resolution and condensed information. The condensed signals also provide accessibility to the detailed signal information. This provides our ability to construct very large hierarchically-organized historical signal data stores with efficient access and retrieval methods.

A second example is the ability to segregate signal information into categories, such as "transient", "steady", and "invalid" (TSI). A similar approach was used to construct the following example, which removes a small portion of signal information in order to form more precise relationships between measured variables.

3 Extraction of Relationships between Variables.

To illustrate our methods, the Shell Column 2 data shown in Fig. 1 and Fig. 2 are processed to obtain a representation of their relationship. The emphasis is upon automated processing; no one had to view and interpret the data in order to construct this relationship. The first step is to scrub the signal in order to reduce its volume and provide information which can be used to categorize its behavior. Specifically, we wish to isolate rapid transients and potentially invalid data points and exclude them from the relationship. Thus, while the relationship is not based specifically on quasi–steady state information, it is biased in that direction.

Figs. 7 and 8 portray the scrubbed data. For the reflux signal, the number of data objects is reduced by a factor of 30 from the original source; for the tray temperature signal, the reduction is by a factor of 55. This allows rapid inspection of long historical segments of data, should this be desired. Although it is not evident in a paper, the time required to graph signals such as those in Fig. 1 or Fig. 2 is excessive; on a Macintosh Quadra, presentation of these graphs consumed several minutes. In contrast, presentation of the scrubbed data is much faster, roughly in proportion to the reduction factors given above. Thus, while it is simply not practical for a plant operator or engineer to view long historical segments of more than a very few process variables using raw data, it is quite feasible to view the scrubbed representation of the same information, and as can be seen from the figures, the significant information characterizing the signal is retained.

To illustrate the relationship between scrubbed data and the detailed data layer upon which it is based, Figs. 9 and 10 graph a short (5 day) segment of both scrubbed and raw data overlaid. As is evident, the signal objects in the scrubbed data are rectangles defined by time intervals and value intervals. These intervals are chosen using distributions over triangular sets to contain all but a small fraction of the raw information within their borders. The optimization problem which is solved to obtain the scrubbed data objects is designed to preserve the time resolution of transient behavior, and in fact, of any significant change in behavior in the high resolution signal. Thus, raw data signal segments within each rectangle have similar characteristics relative to segments in adjacent boxes, and the rectangular boxes have differing widths (time intervals) to preserve the time resolution of transient information. We

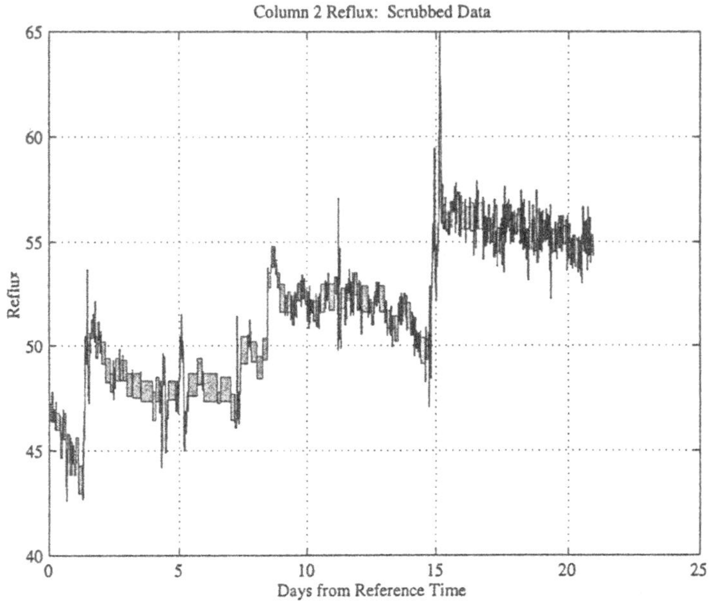

Fig. 7. Reflux (scrubbed data).

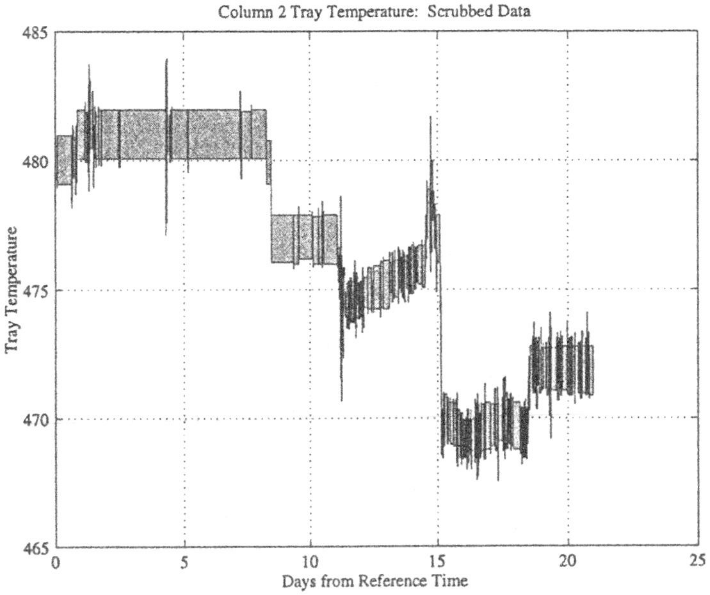

Fig. 8. Tray temperature (scrubbed data).

note as an aside that all scrubbed signal information presented in these figures is composed of sequences of objects represented by rectangles. Because of the resolution of the underlying interval scale for the time variable, some of these intervals are depicted as vertical lines of varying width. Fig. 10 is especially interesting because it demonstrates what occurs when a long sequence of similar high resolution signal information is encountered: The objects which contain this similar information can become arbitrarily wide. The same effect can be observed in the value axis in transients: Very high rectangles are generated.

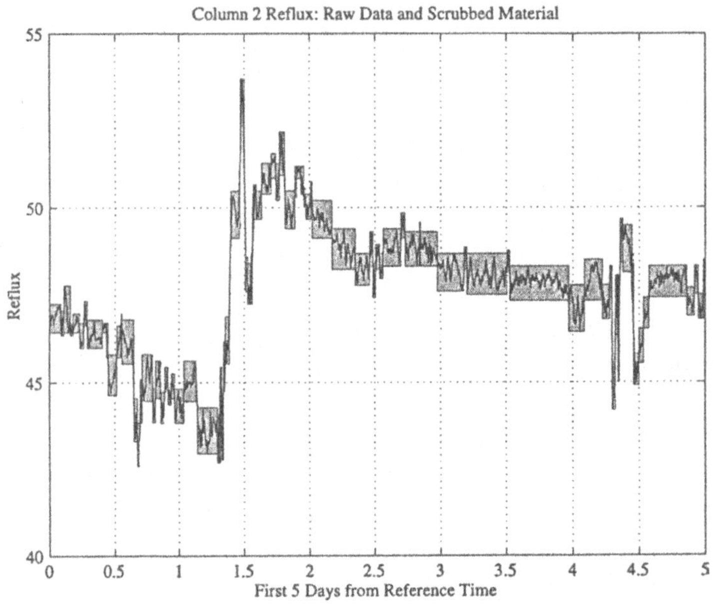

Fig. 9. Reflux: Comparison between raw data and scrubbed material.

The scrubbed signal data has another interpretation in a discrete event (hybrid systems) framework. Each data object belongs to a category, determined by its membership in correspondence sets within a triangular set. Thus, symbols can be attached to the objects, and an object oriented signal stream can be interpreted linguistically as a sequence of symbols. This opens the possibility of symbolic signal processing, using a variety of methods such as automata and formal languages and discrete event methodologies to filter, recognize, or otherwise process the information.

Although we have illustrated only one level of the hierarchy of object oriented signal streams, the data objects at all levels have the same structure. In fact, the first step our implementations take is conversion of raw (external) data to an object oriented (interval) form. Thus, scrubbed signal data can be further scrubbed to generate still higher level (lower resolution) information; the same methods apply. The scrubbing process has a single parameter which corresponds roughly to a time

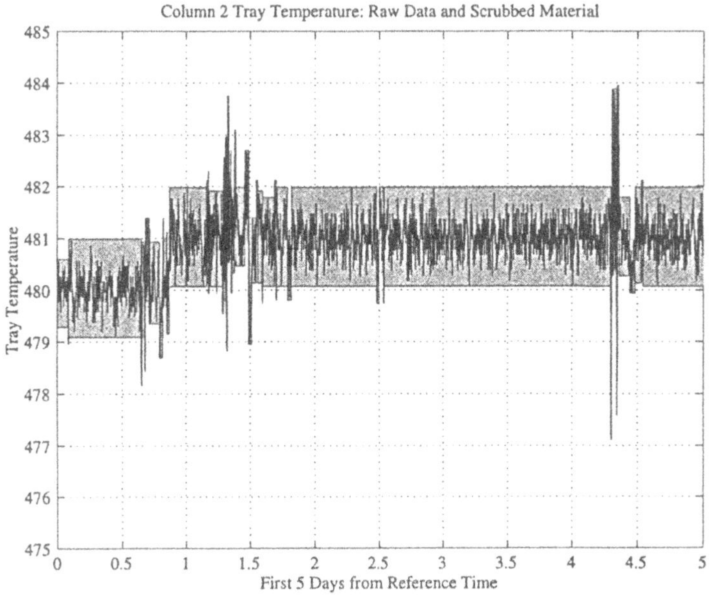

Fig. 10. Tray temperature: Comparison between raw data and scrubbed material.

constant; the larger this parameter, the more condensed and the less detailed the result. In addition, scrubbed objects carry sufficient detail to locate the more detailed information upon which they are based, so a very natural information retrieval strategy emerges.

In this example, our focus is upon the relationship between the two variables, and this relationship can be generated using any level of object oriented signal information, and any subset of the available information at that level. We point out, however, that this process is not entirely trivial, and in this paper many of the details are left out; note in particular that the data objects in two signal streams do not have a common time base, so objects do not naturally align. Thus, generation of relationship information is not performed by simple comparison of corresponding signal objects! The relationship which is produced by our method is also an object oriented signal, formed by objects containing at least two intervals, one for each underlying variable. A rough relationship can be formed from the scrubbed signal information in Figs. 7 and 8; this is illustrated in Fig. 11. In this figure the contribution of each signal object is shown (where it is not obscured by other objects) as a rectangular region. Note that as with the x-y scatter plot of the two raw data variables (temperature vs. reflux) shown in Fig. 3, the relationship is "fat"; it is normally not satisfactory to represent such information by a parametric model such as a line.

The relationship illustrated in Fig. 11 combines the effect of almost all of the raw data variables. For the purpose of process behavior monitoring, for example, in predictive maintenance applications, this is not satisfactory because the region

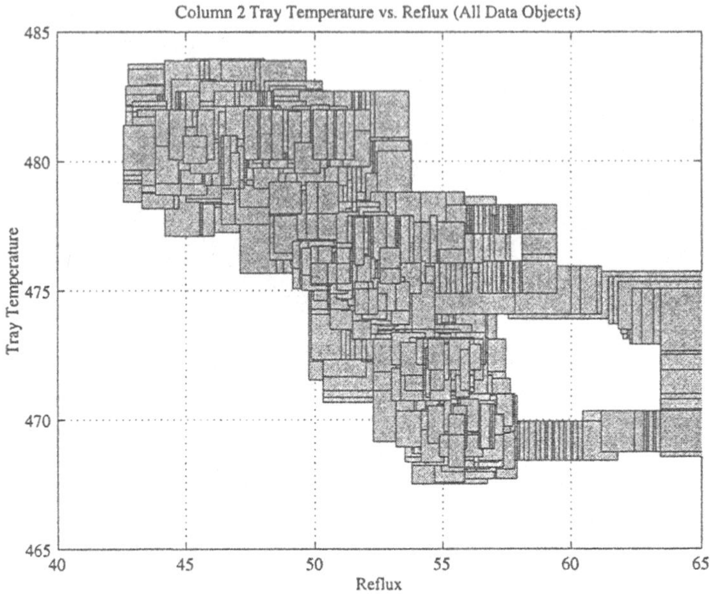

Fig. 11. Contribution of all data objects to the tray temperature vs. reflux relationship.

is broad enough to encompass all observed behaviors. It is rather more appropriate to remove a small percentage of the observed information which corresponds to seldom seen events, caused either by noise, equipment maintenance, or short–term component outages. This yields a relationship which may be used more along the lines of statistical process control (SPC): Process information can be monitored online, and alarms can be generated if the information migrates sufficiently far away from the region defined by historical operation. However, where SPC has difficulty with multivariate relationships, unless an accurate system model exists, this method handles the situation in a very natural manner. Fig. 12 illustrates the objects which remain after isolated data objects are removed; note that the relationship between tray temperature and reflux is sharper, although it still contains some information which may be due to transient behavior in the isolated regions on the right hand side. Fig. 13 shows the region which is defined using the filtered data. We note, again, that where Fig. 3 took tens of minutes to generate and display on a Macintosh Quadra, Fig. 13 can be displayed in seconds. This is a consequence of information condensation due to scrubbing, object filtering, and finally, merging into a single region.

Plots of regions are quite fine for people to look at, but one must ask the question whether the underlying information can be efficiently and effectively utilized for automated process monitoring. Again, the fact that the relationship is maintained as a set of data objects, with the same interval–based underlying structure as other object oriented signals, implies that the information can be rapidly accessed. In particular, it is quite easy to determine, for any interval value in either variable,

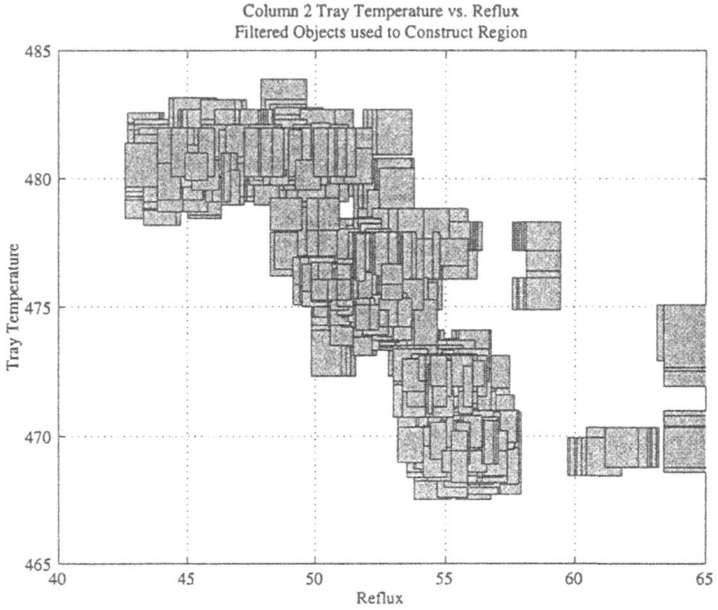

Fig. 12. Filtered data objects used to construct relationship.

the set of intervals in the other variable for which the relationship holds. Thus, it is feasible, and efficient, to utilize a constructed relationship in a manner similar to table look-up to determine if data generated and measured on-line are adequately explained by the historical relationship.

4 Summary.

An object oriented signal processing methodology has been introduced which facilitates operations involving very large on-line process measurement data stores. Relationship models between variables within the data store can be constructed and utilized in an on-line framework in a manner analogous to statistical process control. The methodology supports hierarchical data stores with rapid location and retrieval. An introduction to the underlying theory of interval-based signal processing was presented and placed in the context of an example using data provided by Shell Development Company from a distillation column. The procedures described in this paper are implemented in the Data Refining™ Software Library, a C++ software library used within Data Refining Technologies, Inc. as the basis for their products and services.

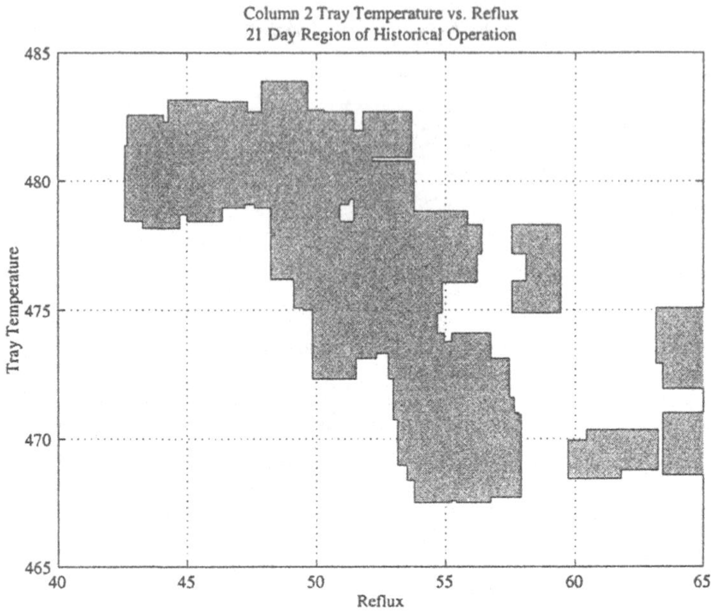

Fig. 13. Tray temperature vs. reflux: 21 day region of historical operation.

References

1. Beran, Ladislav. *Orthomodular Lattices*. D. Reidel, Boston, 1984.
2. Fishburn, Peter C. *Interval Orders and Interval Graphs: A Study of Partially Ordered Sets*. Wiley, New York, 1985.
3. Kozub, D. J. and Garcia, C. E., "Monitoring and diagnosis of automated controllers in the chemical process industries," AIChE meeting, St. Louis, MO, November, 1993.
4. Lax, R. F. *Modern Algebra and Discrete Structures*. Harper Collins, 1991.
5. McLarty, Colin. *Elementary Categories, Elementary Toposes*. Oxford Univ. Press, Oxford, 1992.
6. Wiener, N., "A contribution to the theory of relative position," Proc. Camb. Philos. Soc., **17**, 441–449, 1914.

On the Automatic Verification of Systems with Continuous Variables and Unbounded Discrete Data Structures*

Ahmed Bouajjani[1**] Rachid Echahed[2] Riadh Robbana[1**]

[1] VERIMAG, Miniparc-Zirst, Rue Lavoisier, 38330 Montbonnot St-Martin, France.
email: Ahmed.Bouajjani@imag.fr, Riadh.Robbana@imag.fr
[2] LGI-IMAG, CNRS, BP53, 38041 Grenoble cedex, France.
email: Rachid.Echahed@imag.fr

Abstract. We address the verification problem of invariance properties for hybrid systems. We consider as general models of hybrid systems finite automata supplied with (unbounded) discrete data structures and continuous variables. We focus on the case of systems manipulating discrete counters and one pushdown stack, and on the other hand, constant slope continuous variables. The use of unbounded discrete data sturcture allows to consider systems with a powerful control, and to reason about important notions as the number of occurrences of events in some computation. The use of constant slope continuous variables allows to reason for instance about the time separating events and the durations of phases within some computation interval. We present decidability results for several subclasses of such models of hybrid systems; this provides automatic verification procedures for these systems.

1 Introduction

Hybrid systems consist of interacting continuous and discrete components. Typically, a hybrid system is obtained as a composition of some *dynamic system*, whose parameters (variables) change continuously with time progress, and some *controller*, which is a digital device (computer) that manipulates discrete data structures. Many examples of hybrid systems appear in various domains as in control of chemical processes, robotics, multimedia communication protocols, etc. The applications where hybrid systems are involved are in general *safety critical*, which means that their correctness is crucial, even vital. Thus, a challenging problem is to provide conception methods and analysis techniques for such systems.

The first step is to chose suitable models for hybrid systems. Roughly speaking, the nature of these systems suggests that their models should be combinations of models of dynamic systems (continuous variables + differential equation systems) with models of discrete machines (automata + discrete data structures).

* This paper is based on results previously presented in [BER94a] and [BER94b].

** Partially supported by the ESPRIT-BRA project REACT.

Hence, adequate models for hybrid system are obtained by considering finite control location graphs supplied with discrete data structures (counters, stacks, etc), and real valued variables that change continuously at each control location. The transitions between control locations are conditioned by constraints on the values (or configurations) of the (discrete and/or continuous) variables and data structures of the system: the execution of these transitions updates or resets the dicrete data structures and the continuous variables of the system.

Following this scheme, automata-based computational models of hybrid systems have been recently proposed as in [MMP92, NSY92, MP93, ACHH93, NOSY93]. Globally, these works consider systems with only continuous variables, and particularly, *linear hybrid systems* where the variables are constant slope (with constant derivatives) at each control location. Interesting special cases of such systems are *timed systems* (or timed graphs) [ACD90] (resp. *integrator systems* [Cer92, ACHH93, KPSY93]) corresponding to systems such that all their variables are *clocks* (resp. *integrators*), i.e., their rates are always 1 (resp. 0 or 1). The use of clocks allows to reason about the time elapsed in some computation interval, whereas integrators are useful for the reasoning about *durations* of phases, i.e., the accumulated time the computation is in some particular phase.

In this paper we consider more general models by introducing discrete data stuctures. There are several reasons that motivate the investigation of such extended models. First of all, these models allow to strengthen the control part of hybrid systems by considering systems manipulating powerful data structures. Moreover, the consideration of discrete variables (counters) allows to reason about interesting notions as the *number of occurrences* of events in some computation interval. Finally, it is interesting to explore the (simulation) links between systems with discrete variables and systems with continuous variables. When such links can be established, the analysis of hybrid systems can be reduced to the analysis of discrete systems, and reasoning on delays and durations becomes reasoning about number of occurrences of events.

The systems we investigate are linear hybrid systems extended with counters and a pushdown stack. We call these systems *pushdown counter linear hybrid systems*. Then, our aim is to identify subclasses of such systems that can be analysed automatically. The verification problem we tackle consists in deciding whether some system satisfies some given *invariance property* describing the set of correct (safe) behaviours. In fact, as we have mentioned earlier, safety requirements are the most important part of the specification of a hybrid system. Invariance properties are expressed as constraints on the values of the variables of the system. We distinguish two kinds of variables: *control variables* which are tested by the system to determine its transitions between control locations, and *observation variables* that are permanently updated without any influence on the dynamic of the system: observation variables are used to record relevant informations about the computations of the system, and to express invariance properties on its behaviours.

It is well known that invariance properties are duals of the *reachability properties* asserting that some (dangerous) configuration is reachable. Hence, the verification problem of invariance properties reduces to reachability problem in the considered models of hybrid systems.

Clearly, the reachability problem is in general undecidable for the systems we consider since it subsumes the halting problem of counter machines. Actually, the reachability problem is undecidable even for linear hybrid systems (without counters) since discrete counters can be simulated by continuous variables with rates −1, 0 or 1. Moreover, it has been shown that the reachability problem is undecidable even for integrator graphs [Cer92, HKPV95]. On the other hand, the reachability problem is decidable for timed graphs [ACD90], as well as for some restricted forms of integrator graphs [ACH93, KPSY93].

In this paper, we present gradually decidability results concerning several special cases of pushdown counter linear hybrid systems. We show that the verification problem is decidable for pushdown timed systems, and that this problem remains decidable when these systems are extended with monotonic control counters and observation counters. Moreover, we show that under some conditions, the verification problem is also decidable when these systems are extended with either one observation constant slope continuous variable, or by one control integrator. The general approach we adopt to establish these results is in two steps: the first one consists of reducing the reachability problem in the considered dense-time model to the reachability problem in some discrete structure. This step uses either a digitization or a partition (region graph) technique. Then, the second step consists of solving the reachability problem in the obtained discrete structure; this stucture being infinite in general due to the fact that we consider unbounded data structures.

The remainder of the paper is organized as follows. In Section 2 we introduce the pushdown counter linear hybrid systems. In Section 3 we introduce the invariance and reachability formulas we consider. In Section 4 we present the verification problem and its reduction to a reachability problem, give the existing results in the literature concerning this problem, describe our approach to solve it for the systems we consider, and then, in several subsections, we expose our different decidability results. Concluding remarks are given in Section 5.

2 Linear Hybrid Systems with a Stack and Counters

We introduce in this section models of hybrid systems, we call *pushdown counter linear hybrid systems* (PCLHS's for short), that are extensions of *linear hybrid systems* introduced in [NSY92, MMP92, ACHH93, NOSY93]. Roughly speaking, they consist of automata with a finitely many control locations, supplied with one pushdown stack, a finite set of discrete (integer valued) counters, and a finite set of (real valued) constant slope continuous variables, i.e., variables that change continuously at each control location with some integer rate.

2.1 Simple linear constraints

Let us start by introducing *simple linear constraints* that are used as enabling guards for the systems presented in the next subsection.

Let V be a set of real valued variables. A *simple linear constraint* over V is a boolean combination of constraints of the form $x \sim c$ where $x \in V$, c is an integer constant ($c \in \mathbb{Z}$), and $\sim \in \{<, \leq\}$, or of the form $x \equiv_n c$ where $x \in V$, $n \in \mathbb{N} - \{0\}$, and c is a natural constant ($c \in \mathbb{N}$). The symbols $<$ and $\leq$ represent the usual (strict and nonstrict) ordering relations over reals, and for every $n \in \mathbb{N} - \{0\}$, $\equiv_n$ stands for the relation between reals and positive reals defined by $\forall r_1 \in \mathbb{R}. \forall r_2 \in \mathbb{R}_{\geq 0}. r_1 \equiv_n r_2$ iff $0 \leq r_2 < n$, and $\exists q \in \mathbb{Z}. r_1 = q \cdot n + r_2$. For every variable x, we write simply $int(x)$ instead of $x \equiv_1 0$. Let C_V be the set of simple linear constraints over V.

A *valuation* over V is a function in $[V \to \mathbb{R}]$. We say that a valuation ν is *integer* if $\forall x \in V. \nu(x) \in \mathbb{Z}$. A satisfaction relation is defined as usual between valuations and constraints. Given valuation ν and $f \in C_V$, we denote by $\nu \models f$ the fact that ν satisfies f.

Given a valuation ν and a set of variables $X \subseteq V$, we denote by $\nu[X \mapsto 0]$ the valuation which associates with each variable in X the value 0, and coincides with ν on all the other variables.

2.2 Definition

Let $\mathcal{P}$ be a finite set of atomic propositions and $\Sigma = 2^{\mathcal{P}}$. A pushdown counter linear hybrid system $\mathcal{H}$ over Σ consists of the following components:

- $\mathcal{L}$, a finite set of control locations,
- Π, a function in $[\mathcal{L} \to \Sigma]$, associating a set of atomic propositions with each location,
- $\mathcal{X}$, a finite set of continuous variables,
- ∂, a function in $[\mathcal{L} \times \mathcal{X} \to \mathbb{Z}]$, associating with each location ℓ and continuous variable x, a rate at which x changes continuously while the computation is at ℓ,
- $\mathcal{Y}$, a set of counters,
- Γ, a finite stack vocabulary,
- δ, a set of edges. Each edge is a tuple $e = (\ell, g, A, \gamma, \kappa, R, \ell')$ where $\ell, \ell' \in \mathcal{L}$ are the source and target locations, $g \in C_{\mathcal{X} \cup \mathcal{Y}}$ is an enabling guard, $A \in \Gamma$ is the top of the stack, $\gamma \in \Gamma^*$ is the sequence of symbols replacing A, κ is a function in $[\mathcal{Y} \to \mathbb{Z}]$ associating with each counter an integer value that should be added to it, and $R \subseteq \mathcal{X} \cup \mathcal{Y}$ is the set of reseted variables.

We define the semantics of PCLHS's as sets of computation sequences. This is done by means of the notion of timed configuration.

A *configuration* of $\mathcal{H}$ is a tuple $\langle \ell, \lambda, \nu, \mu \rangle$ where ℓ is a control location, $\lambda \in \Gamma^*$ represents the stack (the first element from the left of λ being the top of the stack), ν is a valuation over $\mathcal{X}$, and μ is an integer valuation over $\mathcal{Y}$. A *timed configuration* of $\mathcal{H}$ is a tuple $\langle \ell, \lambda, \nu, \mu, t \rangle$ where $\langle \ell, \lambda, \nu, \mu \rangle$ is a configuration and $t \in \mathbb{R}_{\geq 0}$ is a time stamp.

To define the sequencing between timed configurations, we need to introduce some notations. Given a valuation ν over $\mathcal{X}$, a control location ℓ, and $t \in \mathbb{R}_{\geq 0}$, we denote by $[\nu + t]_\ell$ the valuation ν' such that $\forall x \in \mathcal{X}.\ \nu'(x) = \nu(x) + \partial(\ell, x) \cdot t$, i.e., the valuation of the variables obtained from ν by staying at location ℓ for an amount of time equal to t. Given a valuation μ over $\mathcal{Y}$, and a function $\kappa \in [\mathcal{Y} \rightarrow \mathbb{Z}]$, we denote by $\mu + \kappa$ the valuations $\mu' \in [\mathcal{Y} \rightarrow \mathbb{Z}]$ such that $\forall y \in \mathcal{Y}.\ \mu'(y) = \mu(y) + \kappa(y)$.

A *computation sequence* of $\mathcal{H}$ starting from a given configuration $\langle \ell, \lambda, \nu, \mu \rangle$ is an infinite sequence of timed configurations $\{\langle \ell_i, \lambda_i, \nu_i, \mu_i, t_i \rangle\}_{i \in \omega}$ such that

- $\langle \ell, \lambda, \nu, \mu, 0 \rangle = \langle \ell_0, \lambda_0, \nu_0, \mu_0, t_0 \rangle$,
- $\forall i \in \omega.\ t_{i+1} \geq t_i$,
- $\lim_{i \rightarrow \infty} t_i = \infty$, and
- $\forall i \in \omega$,
  - either $\ell_i = \ell_{i+1}$, $\lambda_{i+1} = \lambda_i$, $\mu_{i+1} = \mu_i$, and $\nu_{i+1} = [\nu_i + (t_{i+1} - t_i)]_{\ell_i}$,
  - or $t_i = t_{i+1}$ and $\exists (\ell_i, g, A, \gamma, \kappa, R, \ell_{i+1}) \in \delta.\ (\nu_i, \mu_i) \models g$, $\lambda_i = A \cdot \lambda'$, $\lambda_{i+1} = \gamma \cdot \lambda'$, $\nu_{i+1} = \nu_i[(R \cap \mathcal{X}) \mapsto 0]$, and $\mu_{i+1} = (\mu_i + \kappa)[(R \cap \mathcal{Y}) \mapsto 0]$.

Given a configuration α, we denote by $\mathcal{C}(\alpha, \mathcal{H})$ the set of computation sequences starting from α.

Now, we define a *trail* as an infinite sequence $\{\langle \ell_i, t_i \rangle\}_{i \in \omega}$, such that $t_0 = 0$, $\forall i \in \omega.\ t_{i+1} \geq t_i$, and $\lim_{i \rightarrow \infty} t_i = \infty$. Every computation sequence generates a trail obtained by abstracting from the stack and the valuations of the variables, i.e., the trail generated by the computation sequence $\sigma = \{\langle \ell_i, \lambda_i, \nu_i, \mu_i, t_i \rangle\}_{i \in \omega}$ is the sequence $Trail(\sigma) = \{\langle \ell_i, t_i \rangle\}_{i \in \omega}$.

Finally, we call *integer computation sequence* (resp. *integer trail*) any computation sequence $\{\langle \ell_i, \lambda_i, \nu_i, \mu_i, t_i \rangle\}_{i \in \omega}$ (resp. $\{\langle \ell_i, t_i \rangle\}_{i \in \omega}$) where all the t_i's are natural numbers. An *integer configuration* is any configuration $\langle \ell, \lambda, \nu, \mu \rangle$ where ν is an integer valuation over $\mathcal{X}$. Then, given an integer configuration α, we denote by $\mathcal{C}_{\mathbb{Z}}(\alpha, \mathcal{H})$ the set of integer computation sequences of $\mathcal{H}$ starting from α.

2.3 Particular cases

Different subclasses of PCLHS's can be considered depending on the nature of their variables, their guards, and the moments at which the variables are reset. We introduce in the sequel subclasses of PCLHS's we consider for the verification problem issue in the next sections.

First of all, we distinguish between variables that are involved in the dynamic of the systems, i.e., that are tested in the guards of the system, from those that are never tested by the system. We call the formers *control variables* and the latters *observation variables*. The aim of observation variables is to record informations about the evolution of the system; this is useful in the expression of invariance properties on the behaviours of the system.

We call *pushdown linear hybrid system*, PLHS for short, (resp. *counter linear hybrid system*, CLHS for short) a PCLHS without counters (resp. without stack). A *linear hybrid system* (LHS) is PCLHS without stack and counters. In each of these cases, we omit in the description of the system and its semantics the nonrelevant components.

It can be verified that every (P)(C)LHS can be transformed into an equivalent one such that each of its guards is of the form

$$\bigwedge_{x \in \mathcal{X} \cup \mathcal{Y}} ((a_x \sim_1^x x \sim_2^x b_x) \wedge \bigwedge_{i=1}^{m_x} x \approx_i^x c_i^x) \qquad (1)$$

where the $\sim_j^x$'s are in $\{<, \leq\}$, the $\approx_i^x$'s in $\{\equiv_n, \not\equiv_n\}$, the a_x's in $\mathbb{Z} \cup \{-\infty\}$, the b_x's in $\mathbb{Z} \cup \{+\infty\}$, and the c_i^x's in $\mathbb{N}$. We say that a (P)(C)LHS is *weak* if in every guard appearing in one of its edges (given by some expression as (1)), all the $\sim_j^x$'s are $\leq$'s (i.e., nonstrict inequalities), and all the $\approx_i^x$'s are $\equiv_n$'s.

A discrete (resp. continuous) variable x is *monotonic* if for every edge $e = (\ell, g, A, \gamma, \kappa, R, \ell')$ (resp. location ℓ), we have $\kappa(x) \geq 0$ (resp. $\partial(\ell, x) \geq 0$). Now, we say that a variable x is a *clock* (resp. *integrator*) if for every control location ℓ, we have $\partial(\ell, x) = 1$ (resp. $\partial(\ell, x) \in \{0, 1\}$). Then, a (pushdown) (counter) timed system, (P)(C)TS for short, is a (P)(C)LHS such that all its continuous variables are clocks, whereas a (pushdown) (counter) integrator system, (P)(C)IS for short, is a (P)(C)LHS such that all its variables are integrators.

A n-(P)(C)IS is a (P)(C)IS with any number of clocks and exactly n continuous variables that are not clocks but integrators. Moreover, a *single-clock* n-(P)(C)IS is a n-(P)(C)IS which has exactly one clock. We say also that a n-(P)(C)IS is *simple* if for every edge $(\ell, g, A, \gamma, \kappa, R, \ell')$, if R contains some clock, then the guard g contains a constraint of the form $x \equiv_m c$ or $x = c$, for some clock x (not necessarily in R). Intuitively, this condition on the moments of the resets of the clocks implies that all the clocks have always the same fractional parts (we suppose that their initial values are natural numbers) and hence, reach integer values simultaneously. Notice that this condition does not concern the n variables that are not clocks.

3 Invariance/Reachability Properties

Let $\mathcal{H}$ be a PCLHS over $\Sigma = 2^{\mathcal{P}}$. We use letters $P, Q, \ldots$ to range over the set of atomic propositions $\mathcal{P}$, and letters $f, g, \ldots$ to range over simple linear constraints over $\mathcal{X} \cup \mathcal{Y}$. Consider the set of formulas defined by the following grammar:

$$\varphi ::= P \mid f \mid \neg\varphi \mid \varphi \vee \varphi \mid \forall\Box\varphi$$

The semantics of these formulas is defined using a *satisfaction relation* $\models$ by the configurations of $\mathcal{H}$. For every configuration $\alpha = \langle \ell, \lambda, \mu, \nu \rangle$, the relation $\models$ is inductively defined by:

$\alpha \models P$ iff $P \in \Pi(\ell)$
$\alpha \models f$ iff $(\nu, \mu) \models f$
$\alpha \models \neg\varphi$ iff $\alpha \not\models \varphi$
$\alpha \models \varphi_1 \vee \varphi_2$ iff $\alpha \models \varphi_1$ or $\alpha \models \varphi_2$
$\alpha \models \forall\Box\varphi$ iff $\forall \sigma = \{\langle \ell_i, \lambda_i, \nu_i, \mu_i, t_i \rangle\}_{i \in \omega} \in \mathcal{C}(\alpha, \mathcal{H})$. $\forall i \in \omega$. $\langle \ell_i, \lambda_i, \nu_i, \mu_i \rangle \models \varphi$

An *integer satisfaction relation* $\models_{\mathbb{Z}}$ can be defined similarly between the integer configurations of $\mathcal{H}$ and the set of formulas above by considering the set of integer computations $\mathcal{C}_{\mathbb{Z}}(\alpha, \mathcal{H})$ instead of the whole set of computations $\mathcal{C}(\alpha, \mathcal{H})$.

Let us introduce the abbreviation $\exists\Diamond\varphi = \neg\forall\Box\neg\varphi$. An *invariance property* (resp. *reachability property*) is described by a formula of the particular form $\forall\Box\phi$ (resp. $\exists\Diamond\phi$) where ϕ is a boolean combination of atomic propositions and simple linear constraints.

Notice that, for every integer configuration α and every invariance formula φ, $\alpha \models \varphi$ implies that $\alpha \models_{\mathbb{Z}} \varphi$. On the other hand, for every reachability formula φ, $\alpha \models_{\mathbb{Z}} \varphi$ implies that $\alpha \models \varphi$. However, in both cases, the converse does not hold in general.

Using standard laws of the boolean connectives, together with the fact that any formula $\exists\Diamond(\phi_1 \vee \phi_2)$ is equivalent to $(\exists\Diamond\phi_1) \vee (\exists\Diamond\phi_2)$, it can be easily seen that every invariance formula is equivalent to the negation of a disjunction of reachability formulas of the form

$$\exists\Diamond(\pi \wedge \bigwedge_{x \in \mathcal{X} \cup \mathcal{Y}} ((a_x \sim_1^x x \sim_2^x b_x) \wedge \bigwedge_{i=1}^{m_x} x \approx_i^x c_i^x)) \tag{2}$$

where π is a boolean combination of atomic propositions, the $\sim_j^x$'s are in $\{<, \leq\}$, the $\approx_i^x$'s in $\{\equiv_n, \not\equiv_n\}$, the a_x's in $\mathbb{Z} \cup \{-\infty\}$, the b_x's in $\mathbb{Z} \cup \{+\infty\}$, and the c_i^x's in $\mathbb{N}$. We say that a reachability formula (2) is *weak* if all the inequalities it contains are nonstrict (i.e., all the $\sim_j^x$'s are $\leq$'s), and all the $\approx_i^x$'s are $\equiv_n$'s.

Finally, a *propositional* invariance (resp. reachability) formula is a formula of the form $\forall\Box\pi$ (resp. $\exists\Diamond\pi$) where π is a boolean combination of atomic propositions.

4 Verification Problem

The verification problem we consider is whether some given configuration α of a system $\mathcal{H}$ satisfies some given invariance formula φ, i.e., whether $\alpha \models \varphi$. This verification problem reduces to the verification problem of propositional reachability formulas which is equivalent to the control location reachability problem in PCLHS's. Indeed, as we have seen in the previous section, each invariance formula φ is the negation of a union of reachability formulas $\psi_1, \cdots, \psi_n$. Then, the verification problem $\alpha \models \varphi$ reduces the checking that for every $i \in \{1, \cdots, n\}$, $\alpha \models \psi_i$ does not hold. Deciding whether $\alpha \models \psi_i$, with $\psi_i = \exists\Diamond(\pi \wedge \bigwedge_{x \in \mathcal{X} \cup \mathcal{Y}}((a_x \sim_1^x x \sim_2^x b_x) \wedge \bigwedge_{j=1}^{m_x} x \approx_j^x c_j^x))$ reduces to deciding whether $\alpha \models \exists\Diamond$ at-*target*, where at-*target* is true only at a special location *target* in the system obtained by extending $\mathcal{H}$ with *target* and edges that lead to it from all the locations ℓ such that $\Pi(\ell) = \pi_i$, and that are guarded by $\bigwedge_{x \in \mathcal{X} \cup \mathcal{Y}}((a_x \sim_1^x x \sim_2^x b_x) \wedge \bigwedge_{j=1}^{m_x} x \approx_j^x c_j^x)$.

It is well known that the (control location) reachability problem is decidable for timed systems [ACD90]. On the other hand, this problem can be shown to be undecidable in general for LHS's. Indeed, the halting problem of n-counter machines can be reduced to the reachability problem of single-clock n-LHS's

[KPSY93]: each counter can be simulated by a continuous variable with rates in $\{-1, 0, 1\}$, the additional clock is used to let a variable increase (resp. decrease) for exactly one time unit at each incrementation (resp. decrementation) of the corresponding counter. It can be seen that this reduction uses weak and simple single-clock n-LHS's. Thus, the reachability problem is undecidable for simple weak single-clock 2-LHS's. Concerning integrator systems, their reachability problem has been shown to be undecidable in [Cer92]. Furthermore, it has been shown that even for weak 1-IG's, the reachability problem is undecidable [HKPV95]. Decidable particular cases of linear hybrid systems has been identified as in [ACH93, KPSY93, PV94]. In particular, it is shown in [ACH93] (resp. [KPSY93]) that the verification problem of reachability formulas for timed systems (resp. weak timed systems) with one additional observation integrator (resp. observation variable) is decidable.

In this section, we present gradually decidability results concerning several special cases of linear hybrid systems with a stack and counters. We extend the results in [ACD90] to the case of a system with a stack, and any number of observation counters and monotonic control counters. Then, we show that the reachability problem is also decidable for weak such systems augmented by one continuous observation variable (extending the result given in [KPSY93]), as well as for for weak simple systems augmented by one additional control integrator.

The general principle we adopt for proving these results is in two steps. The first step consists of reducing the reachability problem in the original dense system to a reachability problem in some discrete structure. Then, the second step consists of solving the reachability problem in the discrete structure, which is infinite in general due to the consideration of nonbounded discrete data stuctures (stack and counters).

The first step is carried out using either a *digitization* or a *partition* technique. Digitization consists of proving that the reachability problem in a dense-time system can be solved by considering only the integer computations of the system. This approach has been used for instance in [HMP92] to decide the verification problem for a class of timed linear-time properties and weak timed systems. On the other hand, the partition technique consists of finding a countable partition of the dense space of valuations of the continuous variables, such that configurations with the same control location and valuations in the same class (region) satisfy the same reachability properties. This approach has been introduced in [ACD90] to decide the verification problem for the logic TCTL and timed systems.

In all the cases we consider, the reachability problem can be reduced, using either digitization or partition, to a reachability problem in a pushdown automaton, and in some case to a pushdown automaton with observation counters. This reachability problem can be solved as a nonemptiness problem of context-free languages, or as an integer linear programming problem when observation counters are considered.

4.1 Pushdown Timed Systems

Let $\mathcal{H}$ be a pushdown timed system. We show in the sequel that the verification problem of reachability formulas for these systems is decidable. As we have said in the beginning of Section 4, the verification problem of reachability formulas reduces to the verification problem of propositional reachability formulas, which is equivalent to solve a control location reachability problem. More precisely, deciding whether some reachability formula is satisfied starting from some configuration α of $\mathcal{H}$ is equivalent to decide whether some special location *target* is reachable from α in an appropriately extended system. To solve this reachability problem, we extend the partition (region graph) method introduced in [ACD90]. Indeed, we consider a finite index equivalence relation on the set of valuations of clocks which preserves the propositional reachability properties. This allows to reduce the control location reachability problem on a dense system to a control location reachability problem on a discrete structure. This discrete structure is however infinite since we consider a system with a stack. Then, we show that our reachability problem can be reduced to a control location reachability problem in a pushdown automaton, which is equivalent to the nonemptiness problem of context-free languages.

Let us start by considering a partition of the set of valuations of $\mathcal{X}$; let $\mathcal{E}$ denotes this set, i.e., $\mathcal{E} = [\mathcal{X} \rightarrow \mathbb{R}]$. In [ACD90], an equivalence relation $\simeq$ on $\mathcal{E}$ is introduced so that from any pair of configurations of a (finite) timed system having the same control location and $\simeq$-equivalent valuations, the set of reachable control locations are the same. The basic idea behind the definition of this equivalence is that the exact value of every clock x is only relevant while it is less than the greater constant which is compared with it in the system, say c_x. Indeed, beyond c_x, all the values the clock x can take satisfy the same guards appearing in the system.

So, let us recall the definition of the relation $\simeq$. For every $\nu, \nu' \in \mathcal{E}$, we have $\nu \simeq \nu'$ iff the following conditions hold:

- $\forall x \in \mathcal{X}$, if $\nu(x) \leq c_x$, then $\lfloor \nu(x) \rfloor = \lfloor \nu'(x) \rfloor$ and $fract(\nu(x)) = 0$ iff $fract(\nu'(x)) = 0$.
- $\forall x, y \in \mathcal{X}$, if $\nu(x) \leq c_x$ and $\nu(y) \leq c_y$, then $fract(\nu(x)) \leq fract(\nu(y))$ iff $fract(\nu'(x)) \leq fract(\nu'(y))$.

Actually, to take into account the fact that $\mathcal{H}$ may contain contraints of the form $x \equiv_n c$, we consider the equivalence $\cong$ which is defined as the refinement of the equivalence $\simeq$ w.r.t. all the (finite-index) $\equiv_n$'s appearing in $\mathcal{H}$. Then, it is easy to verify that the result of [ACD90] mentioned above holds also for pushdown timed systems.

Lemma 4.1 *Let ν, $\nu' \in \mathcal{E}$. Then, $\nu \cong \nu'$ implies that, for every propositional reachability formula φ, every $\ell \in \mathcal{L}$, and every $\lambda \in \Gamma^*$, $\langle \ell, \lambda, \nu \rangle \models \varphi$ iff $\langle \ell, \lambda, \nu' \rangle \models \varphi$.*

Lemma 4.1 allows to reduce the (control location) reachability problem in a dense configuration graph to a reachability problem in a discrete (countable) structure. This structure corresponds to what is called *region graph* in [ACD90] in the case of timed systems. This region graph is obtained by considering the quotient of the configuration graph of $\mathcal{H}$ w.r.t. the equivalence $\cong$. Let us see how the region graph is defined. Let $[\mathcal{E}]$ be the quotient set of $\mathcal{E}$ under the relation $\cong$. The set $[\mathcal{E}]$ can be supplied by a successor function *succ* between equivalence classes which captures time progress. The function *succ* is defined in the following manner: For every $\nu_1, \nu_2 \in \mathcal{E}$, $succ([\nu_1]) = [\nu_2]$ iff $\nu_1 \not\cong \nu_2$, and $\exists t \in I\!R^+$ such that $\nu' \cong \nu + t$, and $\forall t' \in I\!R^+$, $0 \leq t' < t$, either $\nu + t' \cong \nu$ or $\nu + t' \cong \nu'$. Now, we define a *region* as a triplet $\langle \ell, \lambda, [\nu] \rangle$ where $\ell \in \mathcal{L}$, $\lambda \in \Gamma^*$, and $[\nu] \in [\mathcal{E}]$. Let Reg be the set of such regions. Then, the region graph of $\mathcal{H}$, denoted $\mathcal{R}(\mathcal{H})$, is defined as the graph $\langle Reg, Edg \rangle$ where the vertex set is Reg defined above, and the set of edges Edg is the smallest set such that:

- Each vertex (region) $\langle \ell, \lambda, [\nu_1] \rangle$ has an edge to $\langle \ell, \lambda, succ([\nu_2]) \rangle$.
- Each vertex $\langle \ell, A \cdot \lambda, [\nu_1] \rangle$ has an edge to $\langle \ell', \gamma \cdot \lambda, [\nu_2[R \longmapsto 0]] \rangle$ for each edge $(\ell, g, A, \lambda, R, \ell') \in \delta$ such that $\nu_1 \models g$.

Clearly, the region graph $\mathcal{R}(\mathcal{H})$ is infinite due to the use of the stack. However, recall that the reachability problem we consider on this graph is whether some configuration with a particular control location *target* is reachable starting from some configuration $\alpha = \langle \ell, \lambda, \nu \rangle$. In other words, the value of the stack is not relevant in the target configurations; only the fact that they correspond to the control location *target* is relevant. Then, we can reduce this reachability problem to a control location reachability problem in the input-free pushdown automaton $\mathcal{A}$ having $\mathcal{R}(\mathcal{H})$ as configuration graph. The automaton $\mathcal{A}$ is such that:

- the stack vocabulary is Γ,
- the set of control locations is $\mathcal{L}' = \mathcal{L} \times \mathcal{E}$,
- the set of final control locations is the set of locations $\langle target, \varepsilon \rangle \in \mathcal{L}'$, for any $\varepsilon \in [\mathcal{E}]$,
- the transition relation is given by the set of edges $(\langle \ell_1, [\nu_1] \rangle, A, \gamma, \langle \ell_2, [\nu_2] \rangle)$ such that
  - either $\ell_1 = \ell_2$, $\gamma = A$, and $succ([\nu_1]) = [\nu_2]$,
  - or $\exists (\ell_1, g, A, \gamma, R, \ell_2) \in \delta$ such that $\nu_1 \models g$, and $[\nu_2] = [\nu_1[R \longmapsto 0]]$.

Then, it can be easily seen that the location *target* in $\mathcal{H}$ is reachable from α iff the automaton $\mathcal{A}$ accepts some word starting from the location $\langle \ell, [\nu] \rangle$ with a stack equal to λ. This problem is equivalent to the nonemptiness problem of context-free languages. Hence, we obtain the following decidability result.

Theorem 4.1 *The verification problem of reachability formulas for PTS's is decidable.*

4.2 Pushdown Timed Systems + Counters

In this subsection, we investigate the verification problem of reachability properties for extensions of pushdown timed systems with some kind of counters. We start by the simple case of PTS's with monotonic control counters, and then, we consider the case of PTS's extended with observation counters. We show that for these systems, the verification of reachability formulas is decidable.

PTS's + Monotonic Control Counters Let $\mathcal{H}$ be a PTS's augmented by monotonic control counters. Actually, in this case, we can handle the counters in the same manner as clocks. Indeed, since a monotonic counter never decreases, its exact value is only relevant while it is less than the greater constant which is compared with it in the system. Then, in the same manner as in the subsection 4.1, we can define a finite-index equivalence relation $\cong$ on valuations of counters and clocks so that configurations with the same location, the same stack, and $\cong$-equivalent valuations, satisfy the same propositional reachability formulas.

Let us give the definition of the relation $\cong$. For each variable y of the system $\mathcal{H}$, let c_y denotes the greater constant which is compared with y in the guards of $\mathcal{H}$. Then, consider the equivalence $\simeq$ such that $(\nu, \mu) \simeq (\nu', \mu')$ iff the following conditions hold:

- $\forall x \in \mathcal{X}$, if $\nu(x) \leq c_x$, then $\lfloor \nu(x) \rfloor = \lfloor \nu'(x) \rfloor$ and $fract(\nu(x)) = 0$ iff $fract(\nu'(x)) = 0$.
- $\forall x, y \in \mathcal{X}$, if $\nu(x) \leq c_x$ and $\nu(y) \leq c_y$, then $fract(\nu(x)) \leq fract(\nu(y))$ iff $fract(\nu'(x)) \leq fract(\nu'(y))$.
- $\forall z \in \mathcal{Y}$, if $\mu(z) \leq c_z$, then $\mu(z) = \mu'(z)$.

As in the subsection 4.1, we define the relation $\cong$ as the refinement of the equivalence $\simeq$ defined above w.r.t. all the $\equiv_n$'s that appear in the guard of $\mathcal{H}$. Therefore, we can reduce the reachability problem in $\mathcal{H}$ to a reachability problem in a pushdown automaton, and decide it as a nonemptiness problem of a context-free language. Then, we obtain the following result.

Theorem 4.2 *The verification problem of reachability formulas for PTS's with monotonic counters is decidable.*

PTS's + Observation Counters We consider now the verification problem of reachability formulas for PTS's with observation counters. Let $\mathcal{H}$ be such a system, $\alpha = \langle \ell, \lambda, \nu, \mu \rangle$ a configuration of $\mathcal{H}$, and φ a reachability formula given by:

$$\exists \Diamond (\pi \wedge f \wedge \bigwedge_{y \in \mathcal{Y}} ((a_y \sim_1^y y \sim_2^y b_y) \wedge \bigwedge_{i=1}^{m_y} y \approx_i^y c_i^y)) \tag{3}$$

where f involves only clocks, and suppose that we want to decide whether $\alpha \models \varphi$.

We have shown in the subsection 4.1 that if we consider only the reachability formula $\exists \Diamond (\pi \wedge f)$, then the verification problem reduces to check the nonemptiness of the language accepted by an input-free pushdown automaton $\mathcal{A}$ constructed from the system $\mathcal{H}$. In the present case, the verification problem reduces to a *constrained nonemptiness problem* which consists of checking whether the automaton $\mathcal{A}$ accepts some sequence which satisfies the constraints on the observation counters involved in φ. In other words, we have to decide whether there exists some sequence accepted by $\mathcal{A}$ such that, for each observation counter y, if we accumulate its initial value (i.e., $\mu(y)$) with all the effects of the taken transitions in the sequence, then the obtained value, say v_y, should satisfy $(a_y \sim_1^y v_y \sim_2^y b_y) \wedge \bigwedge_{i=1}^{m_y} v_y \approx_i^y c_i^y$. The effect of each transition of the pushdown automaton is either null if the transition corresponds to time progress, or given by the function κ of the system $\mathcal{H}$ if it corresponds to the execution of a guarded transition of $\mathcal{H}$.

To solve this constrained nonemptiness problem, we actually transform the automaton $\mathcal{A}$ to an equivalent context-free grammar $\mathcal{G} = (\Sigma, \mathcal{N}, Prod, S)$ where $\Sigma = \emptyset$ is the vocabulary, $\mathcal{N}$ is the set of nonterminals, $Prod$ is the set of productions, and S is the starting symbol. Notice that all the productions are of the form $A \rightarrow \gamma$ where $\gamma \in \mathcal{N}^*$. Hence, if $L(\mathcal{G}) \neq \emptyset$, then $L(\mathcal{G}) = \{\epsilon\}$ where ϵ is the empty sequence. The transformation we use is the standard one given in [HU79]. Each production (may be several) in $\mathcal{G}$ is obtained in some way from a transition in the automaton $\mathcal{A}$. Then, we can define a function κ' which describes the effect of the application of the productions in $\mathcal{G}$ on the observation variables, i.e., κ' associates with each production p of $\mathcal{G}$ and each observation counter y an integer which should be added to y at each application of p.

In the sequel we denote by $(\Longrightarrow)$ the derivation relation in the grammar $\mathcal{G}$, and by $(\overset{*}{\Longrightarrow})$ its reflexive-transitive closure. We use subscripts to precise the set of productions or the sequences of productions used in the derivation.

Then, our constrained nonemptiness problem consists of deciding whether there exists some derivation σ in $\mathcal{G}$ given by

$$S \Longrightarrow_{p_1} \lambda_1 \cdots \Longrightarrow_{p_n} \lambda_n \Longrightarrow_{p_{n+1}} \epsilon \tag{4}$$

where the pi_i's are in $Prod$ and the λ_i's are nonempty sequences of nonterminals (in $\mathcal{N}^+$), and such that $\bigwedge_{y \in \mathcal{Y}}((a_y \sim_1^y v_y \sim_2^y b_y) \wedge \bigwedge_{i=1}^{m_y} v_y \approx_i^y c_i^y)$ where $v_y = \mu(y) + \sum_{j=1}^{n+1} \kappa'(p_j, y)$.

We show in the sequel that this decision problem can be solved as an integer linear programming problem, i.e., we construct a linear formula Ω which is satisfiable if and only if a derivation like σ above exists.

Let us define the sets of variables that are involved in the formula Ω. We associate with every observation counter y a variable v_y which stands for its value at the end of σ. Moreover, with every production $p \in Prod$ we associate a variable w_p which stands for the number of applications of p in σ.

First of all, we have to express that obviousely, the v_y's and the w_p's are positive by the formula POS:

$$(\bigwedge_{y \in \mathcal{Y}} v_y \geq 0) \wedge (\bigwedge_{p \in Prod} w_p \geq 0)$$

Then, for every observation counter y, the value of v_y is linked with the values of the w_p's by the formula ACC_y:

$$v_y = \mu(y) + \sum_{p \in Prod} \kappa'(p, y) \cdot w_p$$

Then, we have to consider the constraints on the observation counters involved in the formula φ (3). Hence, we define for every counter y the linear constraint COND_y:

$$(a_y \sim_1^y v_y \sim_2^y b_y) \wedge \bigwedge_{i=1}^{m_y} v_y \approx_i^y c_i^y$$

Now, let us define the constraints on the w_p's. First of all, These constraints must express the fact that any occurrence of a nonterminal appearing along σ must be reduced so that we get the empty sequence ϵ at the end of σ. Thus, in the derivation σ, for any nonterminal A, the number of the reductions of A, i.e., applications of productions p having A as left-hand side (A-productions), must be equal to the number of the A-introductions, i.e., the number of the occurrences of A in the right-hand sides of the applied productions, plus one if A is equal to S.

For every production $p : A \rightarrow \gamma$, let us denote by $lhs(p)$ and $rhs(p)$ respectively the left-hand side and the right-hand side of p, i.e., $lhs(p) = A$ and $rhs(p) = \gamma$. Given $\gamma \in N^*$, and $A \in N$, we denote by $|\gamma|_A$ the number of occurrences of A in γ.

Then, for every nonterminal A we define the formula REDUCT_A:

$$\sum_{p \in Prod} |lhs(p)|_A \cdot w_p = |S|_A + \sum_{p \in Prod} |rhs(p)|_A \cdot w_p$$

Another fact we have to express is that a nonterminal is reduced if and only if it is reachable by applying productions participating to the derivation σ. Indeed, consider some production $p = B \rightarrow B$ and suppose that B never appears in σ. Then, we can asssign any positive value to w_p and still satisfy the constraints REDUCT_A for every A. This is not acceptable since the values v_y's calculated using the constraints ACC_y's will not necessarily correspond to values that can be obtained from existing derivation in $\mathcal{G}$. Thus, we must express the fact that for any nonterminal A, there exists some A-production p with $w_p > 0$ if and only if A is S or it appears in the λ_i's along σ.

Let us introduce first some notation. We say that a sequence of productions $\theta \in Prod^*$ is *elementary* if all its productions apply to different nonterminals. Given a nonterminal A we define Θ_A to be the set of elementary sequences θ on $Prod$ such that $\exists \gamma, \gamma' \in N^*,\ S \overset{*}{\Longrightarrow}_\theta \gamma \cdot A \cdot \gamma'$. Notice that the set Θ_A is finite.

Then, we define for every nonterminal A the formula REACH_A:

$$\sum_{p \in Prod} |lhs(p)|_A \cdot w_p > 0 \Leftrightarrow \bigvee_{\theta \in \Theta_A} \bigwedge_{p \in \theta} w_p > 0.$$

Finally, the formula Ω is defined by:

$$\mathsf{POS} \wedge (\bigwedge_{A \in N} \mathsf{REDUCT}_A \wedge \mathsf{REACH}_A) \wedge (\bigwedge_{y \in \mathcal{Y}} \mathsf{ACC}_y \wedge \mathsf{COND}_y)$$

Since the satisfiability problem of linear formulas on integers is deciable, we obtain the following result.

Theorem 4.3 *The verification problem of reachability formulas for PTS's with observation counters is decidable.*

Actually, it can be shown that the verification problem of reachability formulas is also decidable for PTS's with monotonic control counters and observation counters. For that, it suffices to contruct the grammar $\mathcal{G}$ from the pushdown automaton used for the justification of Theorem 4.2 (see Section 4.2).

4.3 Pushdown Timed Systems + Continuous Variables

This subsection concerns extensions of pushdown timed systems with continuous variables. We consider either weak PTS's with one observation control variable or weak simple PTS's with one control integrator. We show that for these systems, the verification of weak reachability formulas is decidable. For that, we use digitization techniques to reduce their verification problems to the reachability problem in pushdown automata with either one observation counter or one monotonic control counter, respectively.

Digitization We introduce a notion of digitization which extends the notion defined in [HMP92]. We call *digitization quantum* any value $\epsilon \in [0,1)$. We call *digitization sequence* any infinite sequence of digitization quanta $\vec{\epsilon} = \{\epsilon_i\}_{i\in\omega}$. In particular, we say that the digitization sequence is *uniform* if all the ϵ_i's are equal. Given some real time amount $t \in \mathbb{R}_{\geq 0}$, we define, for every digitization quantum $\epsilon \in [0,1)$, the integer time amount $[t]_\epsilon =$ if $t \leq \lfloor t \rfloor + \epsilon$ then $\lfloor t \rfloor$ else $\lceil t \rceil$.

Now, consider a trail $\rho = \{\langle \ell_i, t_i \rangle\}_{i\in\omega}$, and a digitization sequence $\vec{\epsilon} = \{\epsilon_i\}_{i\in\omega}$. Then, the *digitization* of ρ w.r.t. $\vec{\epsilon}$ is the integer trail $[\rho]_{\vec{\epsilon}} = \{\langle \ell_i, [t_i]_{\vec{\epsilon}(t_i)} \rangle\}_{i\in\omega}$ where $\forall t \in \mathbb{R}_{\geq 0}.\ \vec{\epsilon}(t) = \epsilon_{\lfloor t \rfloor}$. A digitization of some trail is *uniform* if it uses a uniform digitization sequence.

The notion of digitization defined in [HMP92] corresponds actually to the notion of uniform digitization. This notion is useful for the reasoning about weak timed systems, since for each such a system, the set of trails generated by its computation sequences is closed under uniform digitization. This notion of uniform digitization is also useful for the case of weak timed systems with one observation continuous variable as it has been shown in [KPSY93]; we extend this fact to pushdown timed systems. However, this notion does not allow to reason about integrator systems. Consider for instance the weak simple 1-IG $\mathcal{H}$, with a clock x and an integrator y, which is represented by the following picture:

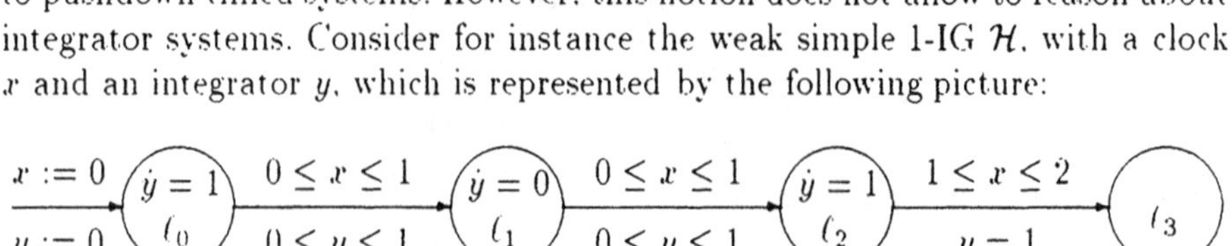

and let σ be the computation sequence $\langle \ell_0, (x=0, y=0), 0\rangle \langle \ell_0, (x=0.5, y=0.5), 0.5\rangle \langle \ell_1, (x=0.5, y=0.5), 0.5\rangle \langle \ell_1, (x=1, y=0.5), 1\rangle \langle \ell_2, (x=1, y=0.5), 1\rangle \langle \ell_2, (x=1.5, y=1), 1.5\rangle \langle \ell_3, (x=1.5, y=1), 1.5\rangle$.

Then, it can be seen that the system $\mathcal{H}$ above has no integer computation that generates some uniform digitization of $Trail(\sigma)$. Indeed, if we take any uniform digitization sequence such that all its elements are equal to some $\epsilon \in [0, 0.5)$, then we obtain the integer trail

$$\rho_1 = \langle \ell_0, 0\rangle\langle \ell_0, 1\rangle\langle \ell_1, 1\rangle\langle \ell_1, 1\rangle\langle \ell_2, 1\rangle\langle \ell_2, 2\rangle\langle \ell_3, 2\rangle$$

whereas if we consider that $\epsilon \in [0.5, 1)$, we obtain the integer trail

$$\rho_2 = \langle \ell_0, 0\rangle\langle \ell_0, 0\rangle\langle \ell_1, 0\rangle\langle \ell_1, 1\rangle\langle \ell_2, 1\rangle\langle \ell_2, 1\rangle\langle \ell_3, 1\rangle$$

Then, it can be verified that these trails cannot be generated by computation sequences, since from the time stamps of ρ_1 (resp. ρ_2) we can deduce that the transition from ℓ_2 to ℓ_3 is executed when the value of y is 2 (resp. 0), which falsifies the guard $y = 1$.

However, the system $\mathcal{H}$ has two integer computations generating trails that are *nonuniform* digitizations of $Trail(\sigma)$. These digitizations are

$$\rho_3 = \langle \ell_0, 0\rangle\langle \ell_0, 0\rangle\langle \ell_1, 0\rangle\langle \ell_1, 1\rangle\langle \ell_2, 1\rangle\langle \ell_2, 2\rangle\langle \ell_3, 2\rangle$$

and

$$\rho_4 = \langle \ell_0, 0\rangle\langle \ell_0, 1\rangle\langle \ell_1, 1\rangle\langle \ell_1, 1\rangle\langle \ell_2, 1\rangle\langle \ell_2, 1\rangle\langle \ell_3, 1\rangle$$

the trail ρ_3 (resp. ρ_4) is obtained when the time values in the interval $[0, 1]$ are digitized w.r.t. some $\epsilon_1 \in [0.5, 1)$ (resp. $\epsilon_1 \in [0, 0.5)$), and the time values in the interval $[1, 2]$ w.r.t. some $\epsilon_2 \in [0, 0.5)$ (resp. $\epsilon_2 \in [0.5, 1)$).

PTS's + One Continuous Observation Variable Let $\mathcal{H}$ be a weak PTS extended with one continuous observation variable; let y be this variable. We show in the sequel that the verification of weak reachability formulas on such a system is decidable (when the starting configuration is integer). For that, we use the digitization result proved in [KPSY93] which says that if a constraint $y \leq c$ is satisfied along some trail ρ at some position k, and assuming that its initial value is 0, then $y \leq c$ is also satisfied along some uniform digitization of ρ at the same position k.

Lemma 4.2 *Let $\rho = \{\langle \ell_i, t_i\rangle\}_{i\in\omega}$ be a trail. Then, if there is some $k \in \omega$ such that $\sum_{i=0}^{k-1} \partial(\ell_i, y) \cdot (t_{i+1} - t_i) \leq c$, then there exists some digitization quantum $\epsilon \in [0, 1)$ such that $\sum_{i=0}^{k-1} \partial(\ell_i, y) \cdot ([t_{i+1}]_\epsilon - [t_i]_\epsilon) \leq c$.*

It has been shown in [HMP92] that for weak timed systems, for every computation sequence σ, and every uniform digitization sequence $\vec{\epsilon}$, there exists a computation sequence σ' such that $[Trail(\sigma)]_{\vec{\epsilon}} = Trail(\sigma')$. Hence, by Lemma 4.2, it can be deduced that for timed systems with one observation variable y, for every computation sequence σ which starts from some integer configuration, and where the constraint $y \leq c$ is satisfied at some position k, there exist an integer computation σ' and a digitization sequence $\vec{\epsilon}$ such that $[Trail(\sigma)]_{\vec{\epsilon}} = Trail(\sigma')$, and $y \leq c$ is satisfied in σ' at position k. This fact can be extended straightforwardly to systems with a stack and counters, since digitization leads to computations that perform the same sequence of operations on these discrete data stuctures than the original one. Then, we obtain the following fact.

Proposition 4.1 *Let $\mathcal{H}$ be a weak PTS with one additional continuous observation variable, α an integer configuration of $\mathcal{H}$, and $\sigma \in \mathcal{C}(\alpha, \mathcal{H})$. Then, there exists a digitization sequence $\vec{\epsilon}$, and $\sigma' \in \mathcal{C}_{\mathbb{Z}}(\alpha, \mathcal{H})$, such that $[Trail(\sigma)]_{\vec{\epsilon}} = Trail(\sigma')$.*

By Proposition 4.1, we can deduce that the verification of reachability formulas for weak PTS's with one continuous observation variable can be solved by considering only integer valuations of the continuous variables (clocks + the observation variable y). Then, all the clocks become monotonic control counters whereas the variable y becomes an observation counter. Hence, using a similar reasoning as in Section 4.2, the verification of a reachability formula on $\mathcal{H}$ reduces to a reachability problem in a pushdown automaton with constraints on observation counters, which is solvable as an integer linear programming problem.

Theorem 4.4 *Let $\mathcal{H}$ be a weak PTS with one additional continuous observation variable, α an integer configuration of $\mathcal{H}$, and φ a weak reachability formula. Then, the verification problem $\alpha \models \varphi$ is decidable.*

PTS's + One Control Integrator Let $\mathcal{H}$ be a weak simple 1-PIG. We prove in the sequel that given any integer configuration α of $\mathcal{H}$, for every computation sequence $\sigma \in \mathcal{C}(\alpha, \mathcal{H})$, there exists an integer computation sequence $\sigma' \in \mathcal{C}_{\mathbb{Z}}(\alpha, \mathcal{H})$ and a digitization sequence $\vec{\epsilon}$ such that $[Trail(\sigma)]_{\vec{\epsilon}} = Trail(\sigma')$. Let us give an informal description of the proof.

Assume without loss of generality that the computation $\sigma = \langle \ell_i, \lambda_i, \nu_i, \mu_i, t_i \rangle$ is such that for every $k \in \omega$, there exists some index i_k, such that $t_{i_k} = k$. Indeed, each computation sequence can be *completed* into a computation satisfying this condition, and which visits the same locations. Let $\rho = Trail(\sigma)$.

Then, remember that since $\mathcal{H}$ is a simple system, all its clocks have the same fractional parts along the sequence σ, and notice that this fractional part is equal at each timed configuration of σ to the fractional part of the time stamp t_i. Hence, since any digitization associates with each noninteger time stamp t either $\lfloor t \rfloor$ or $\lceil t \rceil$, and keeps unchanged integer time stamps, and since $\mathcal{H}$ is a weak system, then it suffices to consider the effect of the digitization on the integrator y. The basic idea is to prove that there exists a digitization sequence $\vec{\epsilon}$ (i.e., a digitization quantum ϵ_u for each time interval $[u, u+1]$) such that, at every position k where a guarded transition is taken by σ, the distance between the (original) value of y, and its integer value according to $[\rho]_{\vec{\epsilon}}$, is strictly less than 1. Let us call this distance the *error* on y at position k. Indeed, if such a digitization sequence exists, then, since $\mathcal{H}$ is a weak system, the integer trail $[\rho]_{\vec{\epsilon}}$ respects all the guards of the transitions fired by σ, and hence it corresponds to a computation sequence of $\mathcal{H}$. We prove the existence of $\vec{\epsilon}$ by showing that for every time interval $\mathcal{I}_u = [u, u+1]$, if it is possible to digitize ρ in all the time intervals before $\mathcal{I}_u$ in such a manner that the error on y at the beginning of $\mathcal{I}_u$ is < 1, then it is always possible to choose a digitization quantum ϵ_u such that the error on y remains < 1 at each position in $\mathcal{I}_u$. Then, we obtain the following digitization result (see the detailed proof in the appendix).

Proposition 4.2 *Let $\mathcal{H}$ be a weak simple 1-PIS, α an integer configuration of $\mathcal{H}$, and $\sigma \in \mathcal{C}(\alpha, \mathcal{H})$. Then, there exists a digitization sequence $\vec{\epsilon}$, and an integer computation $\sigma' \in \mathcal{C}_{\mathbb{Z}}(\alpha, \mathcal{H})$, such that $[Trail(\sigma)]_{\vec{\epsilon}} = Trail(\sigma')$.*

By Proposition 4.2, the verification problem of weak reachability formulas for weak simple 1-PIS's can be solved by considering only integer valuations of the continuous variables. Hence, all these continuous variables can be seen as monotonic control counters, and then, the verification problem of reachability formulas reduces to the reachability problem in pushdown automata with monotonic control counters, which is decidable (see Section 4.2).

Theorem 4.5 *Let $\mathcal{H}$ be a weak simple 1-PIS, α an integer configuration of $\mathcal{H}$, and φ a weak reachability formula. Then, the verification problem $\alpha \models \varphi$ is decidable.*

5 Conclusion

We have investigated the verification of invariance/reachability properties for hybrid systems modeled by finite control location graphs supplied with unbounded discrete data structures and constant slope continuous variables. We have shown that this problem is decidable for various special cases of such systems. The decision results we present are based on two facts. The first one is that, in all the cases we consider, we can show that the verification problem is reducible to a verification problem on a discrete structure. This is done either using some adequate partitioning of the dense configuration space preserving the satisfaction of the reachability properties, or by using digitization principles in order to show that for every computation sequence of the system, there exists some *integer* computation that visits the same control locations and fires the same transitions. Then, the second fact, is that the verification problem on the discrete structure we obtain reduces to a control location (constrained) reachability problem in a pushdown automaton, which can be solved as a nonemptiness problem of a context-free language, or as an integer linear programming problem in the constrained case.

References

[ACD90] R. Alur, C. Courcoubetis, and D. Dill. Model-Checking for Real-Time Systems. In *5th Symp. on Logic in Computer Science (LICS'90)*. IEEE, 1990.

[ACH93] R. Alur, C. Courcoubetis, and T. A. Henzinger. Computing Accumulated Delays in Real-time Systems. In *Proc. Intern. Conf. on Computer Aided Verification (CAV'93)*. LNCS 697, 1993.

[ACHH93] R. Alur, C. Courcoubetis, T. Henzinger, and P-H. Ho. Hybrid Automata: An Algorithmic Approach to the Specification and Verification of Hybrid Systems. In *Hybrid Systems*. LNCS 736, 1993.

[BER94a] A. Bouajjani, R. Echahed, and R. Robbana. Verification of Context-Free Timed Systems using Linear Hybrid Observers. In *Proc. Intern. Conf. on Computer Aided Verification (CAV'94)*. LNCS 818, 1994.

[BER94b] A. Bouajjani, R. Echahed, and R. Robbana. Verifying Invariance Properties of Timed Systems with Duration Variables. In *Intern. Symp. on Formal Techniques in Real Time and Fault Tolerant Systems (FTRTFT'94)*. LNCS 863, 1994.

[Cer92] K. Cerans. Decidability of Bisimulation Equivalence for Parallel Timer Processes. In *Proc. Intern. Conf. on Computer Aided Verification (CAV'92)*. LNCS 663, 1992.

[HKPV95] T. Henzinger, P. Kopke, A. Puri, and P. Varaiya. What's Decidable about Hybrid Automata. In *STOC'95*, 1995.

[HMP92] T. Henzinger, Z. Manna, and A. Pnueli. What Good are Digital Clocks? In *19th. Intern. Coll. on Automata, Languages and Programming (ICALP'92)*. LNCS 623, 1992.

[HU79] J. E. Hopcroft and J. D. Ullman. *Introduction to Automata Theory, Languages and Computation*. Addison-Wesley Pub. Comp., 1979.

[KPSY93] Y. Kesten, A. Pnueli, J. Sifakis, and S. Yovine. Integration Graphs: A Class of Decidable Hybrid Systems. In *Hybrid Systems*. LNCS 736, 1993.

[MMP92] O. Maler, Z. Manna, and A. Pnueli. From Timed to Hybrid Systems. In *REX workshop on Real-Time: Theory and Practice*. LNCS 600, 1992.

[MP93] Z. Manna and A. Pnueli. Verifying Hybrid Systems. In *Hybrid Systems*. Springer-Verlag, 1993. LNCS 736.

[NOSY93] X. Nicollin, A. Olivero, J. Sifakis, and S. Yovine. An Approach to the Description and Analysis of Hybrid Systems. In *Hybrid Systems*. LNCS 736, 1993.

[NSY92] X. Nicollin, J. Sifakis, and S. Yovine. From ATP to Timed Graphs and Hybrid Systems. In *REX workshop on Real-Time: Theory and Practice*. LNCS 600, 1992.

[PV94] A. Puri and P. Varaiya. Decidability of Hybrid Systems with Rectangular Differential Inclusions. In *Proc. Intern. Conf. on Computer Aided Verification (CAV'94)*. LNCS 818, 1994.

Appendix

Proof of Proposition 4.2

Let α be an integer configuration of $\mathcal{H}$, $\sigma = \{\langle \ell_i, \lambda_i, \nu_i, \tau_i \rangle\}_{i \in \omega}$ a computation sequence of $\mathcal{H}$ starting from α, and $\rho = \{\langle \ell_i, \tau_i \rangle\}_{i \in \omega}$ the trail generated by σ. We assume without loss of generality that for every $k \in I\!N$, there exists some index i_k, such that $t_{i_k} = k$. To simplify the exposition of the proof, we assume that the variable y is never reset along σ, and that its initial value is 0. Taking into account resets of y, and the consideration of other integer initial values for y, do not present any difficulty.

Let $\vec{\epsilon} = \{\epsilon_i\}_{i \in \omega}$ be a digitization sequence. For every $k \in \omega$, we define

$$\Delta^{\rho}(k) = \tau_{k+1} - \tau_k$$

and

$$\Delta^{\rho}_{\vec{\epsilon}}(k) = [\tau_{k+1}]_{\epsilon_{\lfloor \tau_{k+1} \rfloor}} - [\tau_k]_{\epsilon_{\lfloor \tau_k \rfloor}}.$$

Intuitively, $\Delta^{\rho}(k)$ is the time taken by the kth transition in the real trail ρ (i.e., the time spent at location ℓ_k) whereas $\Delta^{\rho}_{\vec{\epsilon}}(k)$ is the time taken by the same transition in the integer trail $[\rho]_{\vec{\epsilon}}$.

Given $k \in \omega$, we define $E^k_{\vec{\epsilon}}$ to be the difference between the values of the integrator y in $[\rho]_{\vec{\epsilon}}$ and ρ at position k, i.e., just after reaching the kth configuration (if it exists). Let us give the formal definition. For every $k \in \omega$, it is clear that

$$\nu_k(y) = \sum_{i=0}^{k-1} \partial(\ell_i, y) \cdot \Delta^{\rho}(i).$$

Then, let us define

$$\nu^k_{\vec{\epsilon}}(y) = \sum_{i=0}^{k-1} \partial(\ell_i, y) \cdot \Delta^{\rho}_{\vec{\epsilon}}(i)$$

Intuitively, $\nu^k_{\vec{\epsilon}}(y)$ is the value of of y at position k in the integer trail $[\rho]_{\vec{\epsilon}}$. So, we have

$$E^k_{\vec{\epsilon}} = \nu^k_{\vec{\epsilon}}(y) - \nu_k(y)$$

First, it is easy to check that the following fact holds.

Lemma 5.1 *Let $k \in \omega$. Then, $|E^k_{\vec{\epsilon}}| < 1$ implies that for every $n \in I\!N$,*

- *$\nu_k(y) = n$ implies that $\nu^k_{\vec{\epsilon}}(y) = n$.*
- *$n < \nu_k(y) < n + 1$ implies that $\nu^k_{\vec{\epsilon}}(y) \in \{n, n+1\}$.*

Since $\mathcal{H}$ is a weak system, we can deduce that when $|E^k_{\vec{\epsilon}}| < 1$, then $\nu_k(y)$ and $\nu^k_{\vec{\epsilon}}(y)$ are equivalent w.r.t. the guards of $\mathcal{H}$. Hence, if $|E^k_{\vec{\epsilon}}| < 1$ for every position k where some guarded transition is taken in σ, then the trail $[\rho]_{\vec{\epsilon}}$ corresponds to a computation sequence of $\mathcal{H}$. We prove hereafter that it is always possible to find $\vec{\epsilon}$ such that $|E^k_{\vec{\epsilon}}| < 1$ for every $k \in \omega$.

Let $u \in I\!N$, $\mathcal{I}_u = [u, u+1]$, and $J = \{k, k+1, \ldots, k+m\}$ be the maximal set of indices such that $\forall j \in J,\ \tau_j \in \mathcal{I}_u$. Notice that, by the assumptions on the computation σ mentioned above, we have necessarily $\tau_k = u$ and $\tau_{k+m} = u+1$.

Then, we show that if there exists some digitization sequence $\vec{\epsilon}$ such that $\epsilon_0, \cdots, \epsilon_{u-1}$ allows to digitize ρ until position k in such a way that the distance between the real value of y and its digitization (i.e., the error on y) is always less than 1, then it is possible to find a digitization quantum ϵ allowing to prolong this digitization to the interval $\mathcal{I}_u$ and keep the error on y less than 1 along the interval $\mathcal{I}_u$ too.

Lemma 5.2 *For every $\vec{\epsilon}$, if $|E^k_{\vec{\epsilon}}| < 1$, then there exists $\vec{\epsilon}'$ such that*

- *$\vec{\epsilon}'_i = \vec{\epsilon}_i$ for every $i < u$, and*
- *$\forall j \in J,\ |E^j_{\vec{\epsilon}'}| < 1$.*

Proof. Let us denote by J^+ the subset of $J - \{k+m\}$ such that for every $j \in J^+$, we have $\Delta^\rho(j) = \tau_{j+1} - \tau_j \neq 0$. Intuitively, J^+ is the set of indices of transitions corresponding to time progress in the interval $\mathcal{I}_u$, in other words, to transitions that (may) modify the values of the variables. Let i_1 and i_2 denote respectively the minimum and the maximum of J^+. Notice that $\tau_{i_1} = \tau_k = u$ and $\tau_{i_2+1} = \tau_{k+m} = u+1$. Notice also that $E^k_{\vec{\epsilon}} = E^k_{\vec{\epsilon}'}$.

Now, assume that $|E^k_{\vec{\epsilon}}| < 1$. We show that we can chose a value in $\epsilon \in [0, 1[$ such that, for any sequence $\vec{\epsilon}' = \vec{\epsilon}(0) \cdots \vec{\epsilon}(u-1) \cdot \epsilon \cdots$, we have for every $j \in J$, $|E^j_{\vec{\epsilon}'}| < 1$. Actually, it suffices to consider the values $E^{j+1}_{\vec{\epsilon}'}$'s with $j \in J^+$, i.e., just after some transition corresponding to time progress, since the variables are not modified by the other transitions. We recall that, by definition, we have

$$E^{j+1}_{\vec{\epsilon}'} = E^k_{\vec{\epsilon}'} + \sum_{i \in J^+, i \leq j} \partial(\ell_i, y) \cdot (\Delta^\rho_{\vec{\epsilon}'}(i) - \Delta^\rho(i)) \tag{5}$$

The proof is carried out by considering two cases according to the fact that $E^k_{\vec{\epsilon}}$ is positive or negative.

Let us start with the case $0 \leq E^k_{\vec{\epsilon}} < 1$. Again, there are tree subcases to consider, according to the fact that all of the rates of y in the locations ℓ_j with $j \in J^+$ are 1, or all of them are 0, or there are some of them that are positive and others that are null.

1.1. Suppose that for every $j \in J^+$, we have $\partial(\ell_j, y) = 0$. Then, we can take any $\vec{\epsilon}'$ such that as $\vec{\epsilon}'(u)$ any value in $[0, 1[$. Indeed, in this case, from (5), we have for every $j \in J^+$,

$$E^{j+1}_{\vec{\epsilon}'} = E^k_{\vec{\epsilon}'}$$

1.2. Suppose that for every $j \in J^+$, we have $\partial(\ell_j, y) = 1$. Then, we show that we can take any $\vec{\epsilon}'$ such that $\vec{\epsilon}'(u) = \epsilon \in [fract(\tau_{i_2}), 1[$.
Now, since the rates of y are 1 in all the locations ℓ_j's, by definition of the digitization, we have for every $i \in J^+$, $[\tau_i]_\epsilon = u$. Thus, for every $i \in J^+$ with $i \neq i_2$, we have $\Delta^{\rho}_{\vec{\epsilon}'}(i) = 0$, and $\Delta^{\rho}_{\vec{\epsilon}'}(i_2) = 1$.
Then, from (5), we have for every $j \in J^+$ such that $j \neq i_2$,

$$E^{j+1}_{\vec{\epsilon}'} = E^k_{\vec{\epsilon}'} - \sum_{i \in J^+, i \leq j} \Delta^{\rho}(i)$$

and since

$$0 < \sum_{i \in J^+, i \leq j} \Delta^{\rho}(i) = \tau_{j+1} - \tau_{i_1} < 1$$

and since $0 \leq E^k_{\vec{\epsilon}} < 1$ by hypothesis, we have necessarily

$$-1 < E^{j+1}_{\vec{\epsilon}'} < E^k_{\vec{\epsilon}} < 1 \tag{6}$$

On the other hand, we obtain also

$$E^{i_2+1}_{\vec{\epsilon}'} = E^k_{\vec{\epsilon}'} + 1 - \sum_{i \in J^+} \Delta^{\rho}(i)$$

then, since

$$\sum_{i \in J^+} \Delta^{\rho}(i) = \tau_{i_2+1} - \tau_{i_1} = 1$$

we have

$$E^{i_2+1}_{\vec{\epsilon}'} = E^k_{\vec{\epsilon}'} \tag{7}$$

Thus, from (6) and (7), we obtain that for every $j \in J^+$,

$$-1 < E^{j+1}_{\vec{\epsilon}'} \leq E^k_{\vec{\epsilon}} < 1$$

1.3. Suppose that y has some rates equal to 1 and others equal to 0. Then, let J_1^+ (resp. J_0^+) be the set of indices $j \in J^+$ such that $\partial(\ell_j, y) = 1$ (resp. $\partial(\ell_j, y) = 0$).
Let p be any index in J_0^+. We show that we can take $\vec{\epsilon}'$ such that $\vec{\epsilon}'(u)$ in $[fract(\tau_p), fract(\tau_{p+1})[$ if $\tau_{p+1} \neq u + 1$, otherwise in $[fract(\tau_p), 1[$.
Indeed, from (5), we have for every $j \in J^+$,

$$E^{j+1}_{\vec{\epsilon}'} = E^k_{\vec{\epsilon}'} - \sum_{i \in J_1^+, i \leq j} \Delta^{\rho}(i)$$

Then, it is clear that we have

$$0 \leq \sum_{i \in J_1^+, i \leq j} \Delta^{\rho}(i) < 1$$

and thus, since $0 \leq E^k_{\vec{\epsilon}'} < 1$, we deduce that for every $j \in J^+$,

$$-1 < E^{j+1}_{\vec{\epsilon}'} \leq E^k_{\vec{\epsilon}'} < 1$$

This terminates the proof in the case $0 \leq E^k_{\vec{\epsilon}} < 1$. So, let us consider now the case $-1 < E^k_{\vec{\epsilon}'} \leq 0$.

Then, we have to consider the same tree subcases as above according to the rates of y.

The first case is exactly the same as previously (see case 1.1). Then, let us consider the other cases.

2.1. Suppose that for every $j \in J^+$, we have $\partial(\ell_j, y) = 1$. We show that we can take $\vec{\epsilon}'(u) = \epsilon \in [0, fract(\tau_{i_1+1})[$ if $\tau_{i_1+1} \neq u+1$, otherwise we take $\vec{\epsilon}'(u) = \epsilon \in [0, 1[$.
Indeed, we have in this case $[\tau_{i_1}]_\epsilon = u$, and for every $i \in J^+$ such that $i \neq i_1$, we have $[\tau_i]_\epsilon = u + 1$. Then, from (5), for every $j \in J^+$, we have

$$E^{j+1}_{\vec{\epsilon}'} = E^k_{\vec{\epsilon}'} + 1 - \sum_{i \in J^+, i \leq j} \Delta^\rho(i)$$

thus, since

$$0 < \sum_{i \in J^+, i \leq j} \Delta^\rho(i) \leq 1$$

and since $-1 < E^k_{\vec{\epsilon}'} \leq 0$ by hypothesis, we have necessarily for every $j \in J^+$,

$$-1 < E^k_{\vec{\epsilon}'} \leq E^{j+1}_{\vec{\epsilon}'} < 1$$

2.2. Suppose now that y has some rates equal to 1 and others equal to 0. Then, let J^+_1 and J^+_0 be the partition of J^+ defined exactly as in the case (1.3) above. Let p be the minimum of the set J^+_1. Then, we have to take $\vec{\epsilon}'$ such that $\vec{\epsilon}'(u)$ is in the interval $[fract(\tau_p), fract(\tau_{p+1})[$ if $\tau_{p+1} \neq u+1$, otherwise in $[fract(\tau_p), 1[$.
Indeed, in this case, for every $j \in J^+$ such that $j < p$ (hence, j is necessarily in J^+_0 if it exists), we have

$$E^{j+1}_{\vec{\epsilon}'} = E^k_{\vec{\epsilon}'} \tag{8}$$

On the other hand, for every $j \in J^+$ such that $j \geq p$, we have

$$E^{j+1}_{\vec{\epsilon}'} = E^k_{\vec{\epsilon}'} + 1 - \sum_{i \in J^+_1, i \leq j} \Delta^\rho(i)$$

thus, since it can be seen that

$$0 < \sum_{i \in J^+_1, i \leq j} \Delta^\rho(i) < 1$$

and since $-1 < E^k_{\vec{\epsilon}'} \leq 0$ by hypothesis, we have necessarily for every $j \in J^+$ with $j \geq p$,

$$-1 < E^k_{\vec{\epsilon}'} < E^{j+1}_{\vec{\epsilon}'} < 1 \tag{9}$$

Hence, from (8) and (9), we obtain that for every $j \in J^+$,

$$-1 < E^k_{\vec{\epsilon}'} \leq E^{j+1}_{\vec{\epsilon}'} < 1$$

This terminates the proof in the case $-1 < E^k_{\vec{\epsilon}} \leq 0$ too.

On Dynamically Consistent Hybrid Systems*

Peter E. Caines** and Yuan-Jun Wei

Department of Electrical Engineering
and
Centre for Intelligent Machines
McGill University
3480 University St, Montreal, Quebec, Canada H3A 2A7

Abstract. The notion of dynamical consistency [3] is extended to hybrid systems so as to define the set of dynamically consistent hybrid partition machines associated with a continuous system $\mathcal{S}$. After the presentation of some basic results on the algebraic structure of discrete partition machines, dynamical consistency is defined for hybrid systems, and the lattice $HIBC(\mathcal{S})$ of hybrid in-block controllable partition machines is investigated. A sufficient condition for the existence of a non-trivial $HIBC(\mathcal{S})$ is presented.

1 Introduction

An important class of hybrid systems is that characterised by a continuous system in a feedback configuration with a discrete controller (see e.g. [1], [2], [9], [10], [11]). These works present decompositions of the state spaces of continuous systems with the ultimate objective of the effective design of discrete controllers. If successful, the advantage of such an approach would evidently be the enhanced realizability of controllers by computer systems.

It is a familiar perception that a controller should have a certain minimum amount of information concerning the internal structure of a system under control. One formalization of this is the *internal model principle* of Wonham [14]. In the case of hybrid systems with a discrete controller, such a property might be translated into the requirement that a controller shall be designed based upon a discrete version of the dynamics of the controlled system.

A common approach to hybrid control is to decompose the state space into a countable set of cells and represent each cell by a symbol that plays the role of a state of the discrete system as viewed by the controller. In this paper, we shall call the application of a single control action on each such block a *block invariant* control law.

* Work supported by NSERC grant A1329 and project B5 of NSERC-NCE-IRIS program.
** Also affiliated with the Canadian Institute for Advanced Research.

A. Nerode and W. Kohn introduced the notion of a *small topology* [9], which is the information space of the declarative controller. The quantization process takes place in the product space of (1) the state space; (2) the goal space; (3) the sensor state space; (4) the controller state space; (5) the control action space; and (6) the message space (communication space). The statement that a certain state trajectory lies in some set of the small topology is all the information available to a discrete controller [9] in terms of the location of the state. Further, in their recent work ([6], [7]), an optimization problem (with ϵ-tolerance) is converted into a linear programming problem based upon the discreet topology. In this framework, the notion of the state of a discrete event system (DES) is defined as a segment of a trajectory generated by a fixed control function within a pre-decided time interval, and a discrete control action is interpreted as the switch from one control function to the next one.

Motivated by Liapunov stability theory, M. Glaum [4] related algebraic properties of the vector fields associated with a control system to topological obstructions to achievable closed-loop dynamics produced as solutions to the stabilization problem. Specifically, algebraic properties of the vector fields induce a natural partition of the state manifold. The transversality of the fields to the edges of this partition influences the global stabilizability of the system, and this is expressed in an associated graph. Successful design of stabilizers, nevertheless, relies on refinements of the partition. Note that the approach naturally abandons block invariant control since the stabilizers are state dependent.

In this paper, we approach the quantization problem from a hierarchical control point of view. This work views a discrete controller as a discrete system supervising its continuous subsystems via their feedback relations with the discrete controller, while each of the continuous subsystems is itself (internally) subject to feedback control.

In order to formulate the notion of a dynamically consistency (DC) for a hybrid system, we define an abstract transition from one block to another whenever every state in the first block can be driven directly into the second block. This condition does not link each such abstract transition of the discrete controller to a fixed, block-wise constant, control function. A typical situation in hierarchical control is that several control options are available when some high level decision is taken based upon the aggregated information available to the high level agent. Naturally, in order to complete a given control task, this aggregated information should be consistent with all detailed information concerning the system state.

An input-state-output dynamical system is such that a high level agent can only process an abstracted version of the dynamics of the system. Hence, while a high level controller makes supervisory decisions, low level controllers will actually generate the required state trajectory. In terms of these considerations, the notion of dynamical consistency plays a significant role. The main contribution of this paper is to show that the notion of dynamical consistency introduced originally for discrete event systems in [3] can be extended to continuous systems in a completely natural way.

2 Dynamically Consistent Partition Machines: Discrete Case

This section contains a brief review of [3], where the notion of dynamical consistency was introduced. We will not give a detailed description of that work here, and the reader is referred to [3], [13] and [12] for details.

In this paper, a finite system is modelled by a finite state machine (FSM) $\mathcal{M} =< X, U, \Phi >$, where X is a finite set of states, U a finite set of inputs and $\Phi : X \times U \to X$ a partially defined transition function.

A partition π of X is a collection of pairwise disjoint subsets (called blocks) of X such that the union of this collection equals X. A partition π_1 is said to be *stronger than* a partition π_2, denoted $\pi_1 \preceq \pi_2$, if every π_1-block is a subset of some π_2-block. $\pi_1 \prec \pi_2$ shall mean $\pi_1 \preceq \pi_2$ and $\pi_1 \neq \pi_2$. The greatest lower bound and the least upper bound of two partitions are defined respectively via *intersection* $\cap$ and *chain union* $\cup^c$.

Definition 1. Dynamically Consistent Partition Machines.
Consider $\pi = \{X_1, \cdots, X_k\} \in \Pi(X)$ and $\langle X_i, X_j \rangle \in \pi \times \pi$. The pair $\langle X_i, X_j \rangle$ is *dynamically consistent* (DC) if for each element x_i of X_i, there exists at least one direct trajectory from X_i to X_j which has initial state x_i, i.e.

$$\forall x \in X_i, \exists \sigma \in U^*, (\forall \sigma' < \sigma,\ \Phi(x, \sigma') \in X_i) \wedge (\Phi(x, \sigma) \in X_j,) \text{ for } i \neq j,$$

and

$$\forall x \in X_i, \exists \sigma \in U^*, \forall \sigma' \leq \sigma,\ (\Phi(x, \sigma') \in X_i), \text{ for } i = j;$$

where $\sigma' \leq \sigma$ means that σ' is an initial segment of σ. Let $\mathcal{U} \underline{\Delta} \{\overline{U}_i^j;\ 1 \leq i, j \leq k\}$ be the set of input sequences associated above with the block X_i, X_j, the transition function $\Phi^\pi : \pi \times \mathcal{U} \to \pi$ of the *(dynamically consistent) partition machine* $\mathcal{M}^\pi =< \pi, \mathcal{U}, \Phi^\pi >$ is defined by the equality relation $\Phi^\pi(\overline{X}_i, \overline{U}_i^j) = \overline{X}_j$ if and only if $\langle X_i, X_j \rangle$ is DC, $1 \leq i, j \leq |\pi|$. $\Pi(\mathcal{M})$ shall denote the set of all partition machines of $\mathcal{M}$.

Example 1. An 8 state machine and some of its partition machines are displayed in Figure 1.

A partial order of the set of partition machines $\Pi(\mathcal{M})$ is defined by the partial order of the partitions. We say $\mathcal{M}^{\pi_2}$ is *weaker* than $\mathcal{M}^{\pi_1}$, written as $\mathcal{M}^{\pi_1} \preceq \mathcal{M}^{\pi_2}$, if $\pi_1 \preceq \pi_2$. Under this definition, the partition machine $\mathcal{M}^{\pi_{tr}}$, defined on the trivial partition $\pi_{tr} = \{\overline{X}\}$, and the identity partition machine $\mathcal{M}^{\pi_{id}}$ defined on

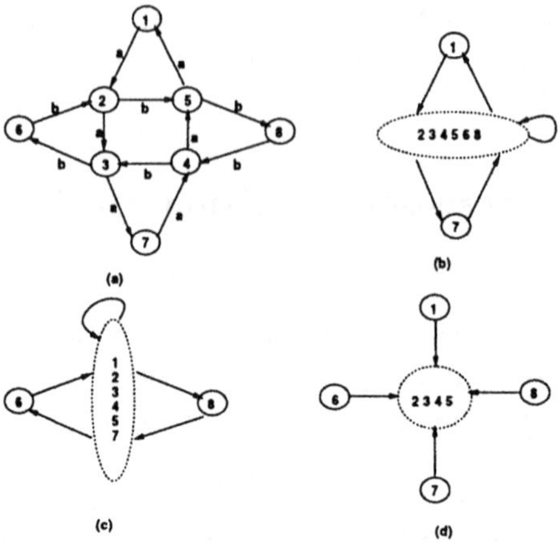

Fig. 1. (a) The base machine $\mathcal{M}_8$. (b) $\mathcal{M}^{\pi_6}$, (c) $\mathcal{M}^{\pi_5}$, (d) $\mathcal{M}^{\pi_5 \cap \pi_6}$.

the partition $\pi_{id} = \{\{x^i\} : x^i \in X\}$ are, respectively, the weakest and strongest partition machines of $\mathcal{M}$. The lattice of partitions induces the associated *lattice of dynamically consistent machines.*

2.1 In-block Controllable Partition Machines and their Associated Lattices

Definition 2. Finite Machine Controllability. A machine $\mathcal{M}$ is called *controllable* if $|X| = 1$ or if for any $(x, y) \in X \times X$, there exists a control sequence $\sigma \in U^*$ which gives rise to a trajectory from x to y, i.e.

$$\forall x \in X, \forall y \in X, \exists s \in U^*, (\Phi(x, s) = y).$$

□

Definition 3. In-block and Between-block Controllability. $\mathcal{M}^\pi$ is called *in-block controllable* if the submachine associated to each block $X_i, 1 \leq i \leq |\pi|$, $\mathcal{M}_i = < X_i, U_i, \Phi_{|X_i \times U} >$ of the base machine is controllable, i.e.

$$\forall X_i \in \pi, \forall x, y \in X_i, \exists s \in U_i^*, \forall s' \leq s, (\Phi^i(x, s') \in X_i \wedge \Phi^i(x, s) = y).$$

$\mathcal{M}^\pi$ is called *between-block controllable* if

$$\forall \overline{X}_i, \overline{X}_i \in \pi, \exists S \in \mathcal{U}^*, (\Phi^\pi(\overline{X}_i, S) = \overline{X}_j).$$

$IBC(\mathcal{M})$ (respectively, $BBC(\mathcal{M})$) shall denote the set of all in-block controllable (respectively, between-block controllable) partition machines.

□

Theorem 4. *[3]*
$IBC(\mathcal{M})$ is closed under chain union.

□

From the above theorem, $\mathcal{M}^{\pi_1 \cup^c \pi_2}$ is the least upper bound of $\mathcal{M}^{\pi_1}$ and $\mathcal{M}^{\pi_2}$ in $IBC(\mathcal{M})$. Intersection does not preserve in-block controllability, hence $IBC(\mathcal{M})$ cannot be a lattice with respect $\cap$ and $\cup^c$. However we obtain a lattice as follows:

Definition 5. Greatest Lower Bound in $IBC(\mathcal{M})$.
Given a finite machine $\mathcal{M}$ and two in-block controllable DC partitions π_1 and π_2, let

$$\pi_1 \sqcap \pi_2 \underline{\Delta} \bigcup^c \{\pi^{'};\ \pi^{'} \preceq \pi_1, \pi^{'} \preceq \pi_2\},$$

and

$$\mathcal{M}^{\pi_1} \sqcap \mathcal{M}^{\pi_2} \underline{\Delta} \mathcal{M}^{\pi_1 \sqcap \pi_2}.$$

□

Theorem 6. *[3]*
If $\mathcal{M}$ is controllable, then $IBC(\mathcal{M}) \underline{\Delta} \langle HIBC(\mathcal{M}), \sqcap, \cup^c, \preceq \rangle$ forms a lattice bounded above and below by $\mathcal{M}^{\pi_{tr}}$ and $\mathcal{M}^{\pi_{id}}$ respectively. □

□

$IBC(\mathcal{M})$ will be referred to as the *in-block controllable lattice of partition machines.* It may be verified that this lattice is in general non-distributive.

Example 2. All partitions in $IBC(\mathcal{M}_8)$, except π_5, π_6, can be generated by chain union operations on π_i, $1 \leq i \leq 4$:

$$\begin{aligned}
\pi_1 &= \{\overline{\{2,3,6\}}, \overline{\{1\}}, \overline{\{4\}}, \overline{\{5\}}, \overline{\{7\}}, \overline{\{8\}}\},\\
\pi_2 &= \{\overline{\{1,2,5\}}, \overline{\{3\}}, \overline{\{4\}}, \overline{\{6\}}, \overline{\{7\}}, \overline{\{8\}}\},\\
\pi_3 &= \{\overline{\{4,5,8\}}, \overline{\{1\}}, \overline{\{2\}}, \overline{\{3\}}, \overline{\{6\}}, \overline{\{7\}}\},\\
\pi_4 &= \{\overline{\{3,4,7\}}, \overline{\{1\}}, \overline{\{2\}}, \overline{\{5\}}, \overline{\{6\}}, \overline{\{8\}}\},\\
\pi_5 &= \{\overline{\{2,3,4,5,6,8\}}, \overline{\{1\}}, \overline{\{7\}}\},\\
\pi_6 &= \{\overline{\{1,2,3,4,5,7\}}, \overline{\{6\}}, \overline{\{8\}}\}\overline{\{6\}}, \overline{\{8\}}\},
\end{aligned}$$

□

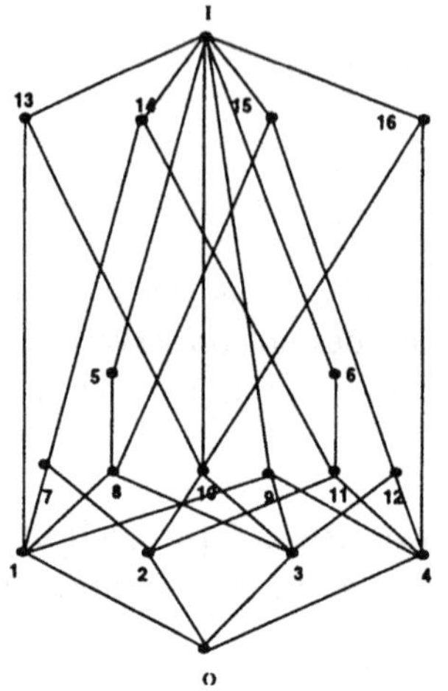

Fig. 2. $IBC(\mathcal{M}_8)$

3 Hybrid DC Partition Machines

In this section we extend the notion of dynamical consistency to a hybrid version in which the base system is continuous in time and state. Consider the differential system

$$\mathcal{S}: \quad \dot{x} = f(x, u), \qquad x \in \Re^n, u \in \Re^m, u(\cdot) \in \mathcal{U}, f : D \subset \Re^{n+m} \to \Re^n, \quad (1)$$

where we assume that $\mathcal{U}$ is the set of all piecewise $C^q(\Re^1)$ ($q \geq 1$) functions of time (i.e. q times continuously differentiable functions except possibly at a finite number of points in any compact interval) which are non-zero on bounded intervals. We further assume throughout this paper that $f \in C^1(\Re^{n+m})$ and that it satisfies a global Lipschitz condition on D. D is assumed to be a closed set and to have a non-empty, path-connected, interior. We shall refer to these conditions on f and $\mathcal{U}$ as the *standard ODE conditions.* Subject to these conditions, a unique solution exists for any initial condition $x^0 \in in(D)$ and control function $u \in \mathcal{U}$.

For $A \subset \Re^n$, we shall use, respectively, the notation $cl(A), in(A)$ and $\partial(A)$ to denote the *closure, interior*, and *boundary* of A. From the standard definitions, $cl(A) = in(A) \cup \partial(A)$. A is open if and only if $A = in(A)$. For any $A, B \subset \Re^n$, $d(A, B) = \inf\{||x - y||; x \in A, y \in B\}$ shall indicate the distance between A and B.

Definition 7. A *finite* (C^r) *partition* of the state space $D \subset \Re^n$ of (1) is a pairwise disjoint collection of subsets $\pi = \{X_1, X_2, \cdots, X_p\}$ such that each X_i is open, path-connected, and the following holds:

$$D = \bigcup_{i=1}^{p} cl(X_i) = \bigcup_{i=1}^{p} X_i \cup \partial X_i,$$

and, further, is such that the boundary of every block X_i is a finite union of $n-1$ dimensional C^r manifolds. The non-empty intersections $\partial X_i \cap \partial X_j, 1 \leq i, j \leq |\pi|$, are called *edges* and are denoted $E_{i,j}$. $\Pi(D)$ shall denote the set of all (C^r) partitions over D. A partition $\pi \in \Pi(D)$ is said to be *non-trivial* if $|\pi| > 1$.

□

We observe that there is only one trivial finite partition $\{in(D)\}$ in $\Pi(D)$.

A straightforward generalization of Definition 7 is to permit the partitions to be of infinite cardinality but to require them to be locally finite, i.e., for all $x \in D$, there exists a neighbourhood $N(x, \epsilon)$ such that $N(x, \epsilon)$ meets only a finite number of members of π.

Due to the potential complexity of system behaviour along the boundaries, the DC condition for a continuous system is defined in two steps. The first states that each interior state of the first block can be locally driven to an appropriate boundary; the second states that each boundary state can be driven into the second block.

Definition 8. For $\pi \in \Pi(D)$, $\langle X_i, X_j \rangle \in \pi \times \pi$ is said to be a *dynamically consistent* (DC) pair (with respect to $\mathcal{S}$) if and only if:
For all x in X_i, there exists $u_x(\cdot) \in \mathcal{U}$, defined upon $[0, T_x], 0 < T_x < \infty$, such that:

(i) $\forall t \in [0, T_x), \phi(t, x, u_x) \in X_i$, and $\lim_{t \to T_x^-} \phi(t, x, u_x) = y \in \partial X_i \cap \partial X_j$;

and for the state y in (i) there exists $u_y \in \mathcal{U}$ defined on $[0, T_y), 0 < T_y < \infty$, such that

(ii) $\forall t \in [0, T_y), \phi(t, y, u_y) \in X_j$;

where $\phi(\cdot, \cdot, \cdot)$ in (i) and (ii) are the integral curves of the vector field $f(\cdot, \cdot)$ with respect to the control functions $u_x, u_y \in \mathcal{U}$ and the initial conditions x, y, respectively.

□

The notion of a DC pair $\langle X_i, X_j \rangle$ has the following interpretation. For any $x^0 \in X_i$, there exists a control function $u_{x^0}(\cdot)$ such that:

1. the initial value problem

$$\dot{x}(t) = f(x(t), u_{x^0}(t)), \quad x(0) = x^0 \in X_i,$$

has a solution ϕ in $X_i \cup (\partial X_i \cap \partial X_j)$; and
2. there is an interval $[0, T_x]$ such that the trajectory will reach X_j at $t = T_x < \infty$. Immediately after this, the system has a solution trajectory for some control which enters X_j.

Although, in general, trajectories may lie in the boundaries of partitions, for the sake of simplicity, this is forbidden in Definition 8. However, condition (ii) can be easily modified to deal with boundary trajectories and most of the discussion in this paper still remains valid.

An obvious necessary condition for the pair $\langle X_i, X_j \rangle$ to be a DC pair is $\partial X_i \cap \partial X_j \neq \emptyset$.

The definition of dynamical consistency obviously depends upon the class of control functions. For a given function f in (1), partition π of D, and class of control functions, the existential statement in (i) of Definition 8 may or may not hold.

Definition 9. Given $\pi \in \Pi(D)$, the *hybrid DC partition machine*

$$\mathcal{M}^{\pi} = \langle X^{\pi} = \{\overline{X}_1, \cdots, \overline{X}_p\}, U = \{\overline{U}_i^j; 1 \leq i, j \leq p\}, \Phi^{\pi} \rangle,$$

based upon the system $\mathcal{S}$, is the finite state machine defined via: $\Phi^{\pi}(\overline{X}_i, \overline{U}_i^j) = \overline{X}_j$ if and only if $\langle X_i, X_j \rangle$ is DC for all i, j, $1 \leq i, j \leq |\pi|$. $\Pi(\mathcal{S})$ shall denote all hybrid partition machines of $\mathcal{S}$. $\mathcal{M}^{\pi}$ is called *non-trivial* if π is non-trivial.

□

By the semi-group property of the flow of $\mathcal{S}$, it is straightforward to prove that $\mathcal{M}^{\pi}$ above satisfies the semi-group property. We shall call $\mathcal{M}^{\pi}$ the DC partition machine of $\mathcal{S}$.

Remark. The hybrid partition machine $\mathcal{M}^{\pi}$ may be regarded as the system model employed by a high level controller. Furthermore, Definition 9 has the interpretation that if it is the case that $\Phi^{\pi}(\overline{X}_i, \overline{U}_i^j) = \overline{X}_j$, then, whenever the high level controller is required to guide the system from block X_i to X_j, it can always to do so.

Definition 10. A Partial Ordering of Partition Machines
Let $\mathcal{M}^{\pi_1}, \mathcal{M}^{\pi_2} \in \Pi(\mathcal{S})$, we say that $\mathcal{M}^{\pi_2}$ is *weaker* than $\mathcal{M}^{\pi_1}$, denoted $\mathcal{M}^{\pi_1} \preceq \mathcal{M}^{\pi_2}$, if, for every $X_i \in \pi_1 = \{X_1, \cdots, X_n\}$, there exists $Y_j \in \pi_2 = \{Y_1, \cdots, Y_m\}$ such that $X_i \subset Y_j$.

□

Lemma 11. *Let $\pi_1 \preceq \pi_2$, then for any $Y_j \in \pi_2$,*

$$cl(Y_j) = \bigcup_{i \in J} cl(X_i),$$

where $J \underline{\Delta} \{i; i \in \{1, \cdots, |\pi_1|\} \wedge X_i \in \pi_1 \wedge X_i \subset Y_j\}$.

Proof
First, $A \subset B$ implies $cl(A) \subset cl(B)$ and so

$$\bigcup_{i \epsilon J} cl(X_i) \subset cl(Y_j).$$

For the reverse inclusion, consider $y \epsilon Y_j$. Either (i) $y \epsilon X_i \cup \partial X_i$, in which case $y \epsilon X_i$ for some $i \epsilon J$, or (ii) $y \epsilon \partial Y_j$, and in this case, there exists a sequence $\{y_n; n \geq 1\} \subset Y_j$ such that $y_n \to y$ as $n \to \infty$. Since J is finite, there exists at least one index, say $q \epsilon J$, such that $X_q \cup \partial X_q$ contains a subsequence $\{y_{n_k} : k \geq 1\}$ of $\{y_n : n \geq 1\}$. It is obvious that $y_{n_k} \to y$ as $k \to \infty$ and hence $y \epsilon cl(X_q)$. This leads to the reverse inclusion. □

Definition 12. Chain Union of Hybrid Partition Machines.
The *chain union machine* $\mathcal{M}^{\pi_1} \cup^c \mathcal{M}^{\pi_2}$ of $\mathcal{M}^{\pi_1}$ and $\mathcal{M}^{\pi_2} \epsilon \Pi(\mathcal{S})$ is the partition machine defined on the partition $\pi_1 \cup^c \pi_2$, where each block Z_k of $\pi_1 \cup^c \pi_2$ is defined as follows: Suppose $\pi_1 = \{X_1, \cdots, X_p\}, \pi_2 = \{Y_1, \cdots, Y_q\}$. Set $Z_{k,0} = X_{i_1} \epsilon \pi_1$, where i_1 is arbitrary. Suppose $Z_{k,n}$ is defined, then

$$Z_{k,n+1} = \begin{cases} Z_{k,n} \cup Y_j & \text{if } Z_{k,n} \cap Y_j \neq \emptyset, n = even, \\ Z_{k,n} \cup X_j & \text{if } Z_{k,n} \cap X_j \neq \emptyset, n = odd, \end{cases}$$

and let $Z_k = \cup_{n=0}^{p+q} Z_{k,n}$.

□

We observe that since $X_i, 1 \leq i \leq p$, $Y_j, 1 \leq j \leq q$, are open sets, so are $Z_k, 1 \leq k \leq min\{|\pi_1|, |\pi_2|\}$, and that it is possible for a set of Z_k to contain the elements of ∂X_i, ∂Y_j for some i, j.

4 Controllabilities of Hybrid Partition Machines

In this section, we define two types of controllability for a partition machine. The first one is block-wise controllability, referring to global controllability, and the second is in-block controllability, referring to local controllability.

Definition 13. Hybrid Between-Block and In-block Controllability
A hybrid partition machine $\mathcal{M}^\pi$ is called *hybrid between-block controllable* (HBBC) if any block state $\overline{X}_i$ can be reached from any other block state $\overline{X}_j$ by applying a finite number of block transitions. $\mathcal{M}^\pi$ is called *hybrid in-block controllable* (HIBC) if for every $X_i \epsilon \pi$, and for all $x, y \epsilon X_i$, the following holds:

$$\exists u(\cdot) \epsilon \mathcal{U}, \exists T \geq 0, \; (\forall t < T, \phi(t, x, u) \epsilon X_i) \wedge \phi(T, x, u) = y. \qquad (2)$$

In other words, each block $X_i \epsilon \pi$ is *controllable for the system* $\mathcal{S}$. We shall use $HBBC(\mathcal{S})$ (respectively, $HIBC(\mathcal{S})$) to denote the set of all HBBC (respectively, HIBC) machines of $\mathcal{S}$.

□

Notice that, in contrast to the set of all partitions of the state set of a finite machine (each with its set of associated sub-machines), there is, in general, an infinite set of partitions (and hence partition machines) for each particular cardinality $|\pi| < \infty$, and, further, $|\pi|$ is unbounded. In particular, it can be seen that an uncountable number of partitions can be generated by displacements of the boundaries of any given partition.

It is worth pointing out that the existence of the control function in the HIBC definition is only indirectly dependent on the index of the block through the dynamics of sub-systems; this information may be viewed as global information which need not be directly available to a local agent. On the other hand, HBBC is entirely dependent upon global information.

The following theorems have counterparts in the discrete case (see [3], [12]). Hence only sketches of the proofs are given here.

Theorem 14. *For $\mathcal{M}^\pi \in HIBC(\mathcal{S})$, $\mathcal{M}^\pi$ is HBBC if and only if $\mathcal{S}$ is controllable in $in(D)$.*

Proof

$\Longrightarrow$. For any $x, y \in in(D)$, there are blocks X_x, X_y such that $x \in X_x$ and $y \in X_y$. HBBC implies that a block trajectory exists from X_x to X_y, say,

$$X_x = X_{i_0} \to X_{i_1} \to \cdots \to X_{i_{n-1}} \to X_{i_n} = X_y.$$

Clearly, the DC condition which holds between each consecutive pair of blocks ensures that x can eventually be driven into X_y along this block trajectory. Since $\mathcal{M}^\pi \in HIBC(\mathcal{S})$, X_y forms a controllable subsystem, and we conclude that x can be driven to y.

$\Longleftarrow$. For two arbitrary blocks X_1 and X_2, take any $x \in X_1$ and $y \in X_2$. Then the controllability of $\mathcal{S}$ implies that there is a trajectory from x to y. The HIBC hypothesis on $\mathcal{M}^\pi$ assures that the DC condition holds between successive blocks along the trajectory. Hence a block trajectory can be found along this state trajectory.

□

The DC definition for hybrid systems does not give an effective way to construct a hybrid partition machine for a given partition. However, for any given an HIBC partition machine $\mathcal{M}^\pi$, the DC test for any weaker (and hence necessarily finite) partition machine can be carried out with respect to $\mathcal{M}^\pi$, instead of $\mathcal{S}$. Hence the DC testing procedure can be made into an effective procedure when based upon the notion defined below.

Definition 15. Let $\pi_1 = \{X_1, \cdots, X_p\}$ and $\pi_2 = \{Y_1, \cdots, Y_q\}$ be two partitions of D with $\pi_1 \preceq \pi_2$. We say the pair $\langle Y_i, Y_j \rangle$ is *DC with respect to* $\mathcal{M}^{\pi_1}$, which is denoted $\langle Y_i, Y_j \rangle_{\pi_1}$, if and only if every π_1-block $X_{i_1} \in Y_1$ can be driven directly to some π_1-block in Y_j. In this case, we also call the pair $\langle Y_i, Y_j \rangle$ *relatively DC* (RDC) *with respect to* $\mathcal{M}^{\pi_1}$.

□

Lemma 16. *Let $\mathcal{M}^{\pi_1} \preceq \mathcal{M}^{\pi_2}$ be two members of $HIBC(\mathcal{S})$. For a $Y_j \in \pi_2$, let $J = \{i; X_i \subset Y_j\}$. Then $\{X_i; i \in J\}$ forms a controllable submachine of $\mathcal{M}^{\pi_1}$. In other words, $\mathcal{M}^{\pi_2} \in IBC(\mathcal{M}^{\pi_1})$.*

Proof
Since $\mathcal{M}^{\pi_2} \in HIBC(\mathcal{S})$, the open set Y_j is a controllable set for the system. Consider $X_{i_1}, X_{i_2}, i_1, i_2 \in J$, and take any $x_{i_1} \in X_{i_1}, x_{i_2} \in X_{i_2}$. Then there exists a control u_1^2 taking x_{i_1} to x_{i_2} along a trajectory in Y_j. Using the HIBC property of each block $X_{i_s} \in \pi_1$ which meets this trajectory, we obtain the DC relation between a sequence of π_1-blocks begining with X_{i_1}, and ending with X_{i_2}. Since each of these π_1-blocks lie in Y_j, there exists a π_1-block trajectory from X_{i_1} to X_{i_2} within Y_j, and the result follows. □

Theorem 17. *Let $\mathcal{M}^{\pi_1} \preceq \mathcal{M}^{\pi_2}$ be two members of $HIBC(\mathcal{S})$. Then any block pair $\langle Y_1, Y_2 \rangle \in \pi_2 \times \pi_2$ is DC if and only if it is RDC with respect to $\mathcal{M}^{\pi_1}$.*

Proof
Let J_1 and J_2 be defined as in Lemma 16.
$\Longrightarrow$.
Suppose $\langle Y_1, Y_2 \rangle$ is DC. Then for any $y_1 \in Y_1$, there is $y_{1,2} \in \partial Y_1 \cap \partial Y_2$, and $y_2 \in Y_2$ such that control functions in $\mathcal{U}$ perform the successive transformations

$$y_1 \longrightarrow y_{1,2} \longrightarrow y_2.$$

Suppose the initial segment $tr(y_1, y_{1,2})$ of the above trajectory passes through the set of π_1-blocks $\{X_l; l \in J_1\}$ with X_{i_k} the last block in Y_1 along the trajectory. Let $X_{j_1}, j_1 \in J_2$, be the first block along the trajectory in Y_2. Then it is obvious that $\langle X_{i_k}, X_{j_1} \rangle$ is DC by the assumption that $\mathcal{M}^{\pi_1}$ is HIBC. Hence, a π_1 transition is defined from X_{i_k} to X_{j_1}. But by Lemma 16, all other π_1-blocks can be driven to X_{i_k} by π_1 block transitions. Hence $\langle Y_1, Y_2 \rangle$ is RDC with respect to $\mathcal{M}^{\pi_1}$.

$\Longleftarrow$.
Suppose $\langle Y_1, Y_2 \rangle \in \pi_2 \times \pi_2$ is RDC with respect to $\mathcal{M}^{\pi_1}$. For any $y \in Y_1$, there are two cases to consider:
Case 1. $y \in X_{i_l} \subset Y_1$ for some $i_l \in J_1$. By the RDC property of $\mathcal{M}^{\pi_2}$, there exists a π_1 block $X_{j_1} \subset Y_2$ and a π_1 block trajectory in Y_1 from X_{i_l} to X_{j_1}. Then the DC conditions among the consecutive blocks along this π_1-trajectory ensure that y can be driven to some state y' in Y_2.

Case 2. $y \in \partial X_{i_k} \subset Y_1$ for some $i_k \in J_1$, and hence, necessarily, $X_{i_k} \subset Y_1$. The HIBC hypothesis on $\mathcal{M}^{\pi_2}$ implies that Y_1 is a controllable set for the system $\mathcal{S}$ and so insures that y can be driven into some state in X_{i_k}. Then this case is converted into Case 1.

Since y in (i) and (ii) above is an arbitrary state in Y_1, it follows that $\langle Y_1, Y_2 \rangle$ is DC. This concludes the proof.

□

5 The Lattices of Hybrid In-block Controllable Machines

The partially ordered set $HIBC(\mathcal{S})$ is the main object of interest in this paper and the following theorem is the starting point of our investigation; its discrete version is presented in Theorem 4 above.

Theorem 18. *Hybrid in-block controllability is closed under chain union, i.e.* $\mathcal{M}^{\pi_1}, \mathcal{M}^{\pi_2} \in HIBC(\mathcal{S})$ *implies* $\mathcal{M}^{\pi_1 \cup^c \pi_1} \in HIBC(\mathcal{S})$.

Proof

The argument is similar to that when the base system is a finite state machine ([3], [12]). Consequently, we only give an outline of the proof here. Consider the basic step in the definition of chain union: say $Z_k = X_i \cup Y_j$ with $X_i \cap Y_j \neq \emptyset$. For any $x, y \in Z_k$, if both x and y are in X_i (respectively, Y_j), then the hypothesis of $\mathcal{M}^{\pi_1}$ (respectively, $\mathcal{M}^{\pi_2}$) will directly lead to the conclusion. If the state $x \in X_i$ and $y \in Y_j$, then x can be reached from $y \in Y_j$ through some state in $X_i \cap Y_j$. The combinatoric set of cases where at least one of x or y lies in a boundary of X_i and Y_j reduces the proof to the previous case.

□

Lemma 19. Upper Semi-Lattice of HIBC Machines *Suppose $in(D)$ is controllable for $\mathcal{S}$, then the tuple $\langle HIBC(\mathcal{S}), \cup^c, \preceq \rangle$ forms a non-empty upper semi-lattice, called the hybrid in-block controllable semi-lattice of $\mathcal{S}$.*

Proof

Since $\mathcal{S}$ itself is a controllable system on $in(D)$, $\mathcal{M}^{\pi_{tr}} \in HIBC(\mathcal{S})$, where $\pi_{tr} = \{in(D)\}$. Hence $HIBC(\mathcal{S}) \neq \emptyset$.

□

In [12], we showed that every non-trivial controllable FSM has a non-trivial in-block controllable lattice This is not the case for hybrid systems, since in general we may have infinitely many elements below two elements in $HIBC(\mathcal{S})$ or, at the other end of the spectrum, there may be none. As in the discrete case, set intersection for two HIBC partitions may destroy the HIBC property too. Furthermore, it may be the case that there is no bottom (i.e. unique minimum) element to $HIBC(\mathcal{S})$. Hence the *meet* $\sqcap$ operation, which is dual to the *join* operation $\cup^c$, may not exist. This implies that some controllable systems may not have an in-block controllable lattice, in contrast to the discrete case.

However, if for any pair of elements in $HIBC(\mathcal{S})$ there exists at least one HIBC partition below this pair, we can define a lattice structure for the set of all HIBC machines.

For this purpose, we give the following properties concerning the partial ordered set $HIBC(\mathcal{S})$.

Lemma 20. *Let $\mathcal{M}^{\pi_1}, \mathcal{M}^{\pi_2} \in HIBC(\mathcal{S})$,*
(i) $\mathcal{M}^{\pi_1} \preceq \mathcal{M}^{\pi_2}$ and $|\pi_1| = |\pi_2|$ implies $\mathcal{M}^{\pi_1} = \mathcal{M}^{\pi_2}$.
(ii) Any chain (linearly ordered subset) $\mathcal{C} = \{\mathcal{M}^{\pi_\alpha}; \alpha \in \mathcal{J}\} \subset HIBC(\mathcal{S})$ contains only countably many distinguished elements.
(iii) Any chain of $HIBC(\mathcal{S})$ has a top element (i.e. unique maximal element) in $HIBC(\mathcal{S})$.
(iv) (i)-(iii) hold for any upper-bounded subset of $HIBC(\mathcal{S})$.

Proof
(i) The proof consists of a sequence of steps:

(a) It is obvious from the hypotheses that each π_2-block contains one and only one π_1-block, and so, without loss of generality, we assume that $X_i \subset Y_j$, $1 \leq i \leq n$.
(b) From (a), for any $i \in \{1, \cdots, n\}$, if $Y_i \setminus X_i \neq \emptyset$, then $Y_i \setminus X_i$ contains no interior point of any π_1-block. Otherwise, suppose $y \in (Y_i \setminus X_i) \cap X_j$ for some j, $j \neq i$. Then $Y_i \cap X_j \neq \emptyset$, hence $Y_i \cap Y_j \neq \emptyset$, which is impossible. This implies that $Y_i \setminus X_i \subset \partial X_i$
(c) For any $j, j \neq i$, $(Y_i \setminus X_i) \cap (\partial X_i \cap \partial X_j) = \emptyset$. Otherwise, for some y in the intersection, there exists an $\epsilon > 0$ and an open ball $B(y, \epsilon)$ such that

$$B(y,\epsilon) \subset Y_i \quad \text{and} \quad B(y,\epsilon) \cap X_j \neq \emptyset.$$

Hence Y_i contains some interior point of X_j, which gives the same contradiction as in (b).
(d) $\partial X_i \cap \partial X_j = \partial Y_i \cap \partial Y_j$, for $i \neq j, 1 \leq i, j \leq n$. This is the case since, by (a)-(c), the only set difference between Y_i and X_i is the set of internal boundary states of X_i, i.e. states in ∂X_i which are contained in neighbourhoods not meeting any other block $X_j, j \neq i$.
(e) We may now establish that $\langle X_i, X_j \rangle$ is DC if and only if $\langle Y_i, Y_j \rangle$ is DC. Suppose $\langle X_i, X_j \rangle$ is DC. Then any state x in $X_i \setminus Y_i$ can be driven into some state in $X_j \subset Y_j$ through the same non-interior boundary. Further, for $y \in Y_i \setminus X_i$, the hypothesis that $\mathcal{M}^{\pi_2}$ is HIBC allows y to be driven into some state in X_i, and then to some other state in X_j. Hence $\langle Y_i, Y_j \rangle$ is DC. The reverse implication is obvious by (d).

(ii) Let $\mathcal{C} \subset HIBC(\mathcal{S})$ be a chain and define

$$N(\mathcal{C}) = \{n_i; \exists \mathcal{M}^{\pi} \in \mathcal{C}, s.t. |\pi| = n_i\}.$$

Define $\xi(\pi) = |\pi|, \mathcal{M}^{\pi} \in \mathcal{C}$. Then ξ is onto $N(\mathcal{C})$. Further, by (i), $\pi_1 \prec \pi_2$ implies $|\pi_1| < |\pi_2|$. Hence $\xi(\pi) \neq \xi(\pi')$ if and only if $\mathcal{M}^{\pi} \neq \mathcal{M}^{\pi'}$. So ξ is a one-to-one mapping from the set of all distinct elements of $\mathcal{C}$ into $N(\mathcal{C})$. The latter is at most a countable set. This gives (ii).

(iii) By (ii), the set $N(\mathcal{C})$ forms a monotonic decreasing integer set bounded below by 1 (the cardinality of π_{tr}). Hence there is a minimum element in $N(\mathcal{C})$, say n_0, and hence $\xi^{-1}(n_0)$ is a maximal element of $\mathcal{C}$ with respect to $\preceq$.

(iv) Repeat the proofs of (i)-(iii) with $\mathcal{C}$ properly restricted. □

We now present the main result of this section.

Theorem 21. Existence of a glb for HIBC Machines.
If for every pair $\mathcal{M}^{\pi_1}, \mathcal{M}^{\pi_2} \in HIBC(\mathcal{S})$, the set

$$LOW(\mathcal{M}^{\pi_1}, \mathcal{M}^{\pi_2}) \underline{\Delta} \{\mathcal{M}^{\pi}; \mathcal{M}^{\pi} \in HIBC(\mathcal{S}), \mathcal{M}^{\pi} \preceq \mathcal{M}^{\pi_1}, \mathcal{M}^{\pi} \preceq \mathcal{M}^{\pi_2}\},$$

is not empty, then the greatest lower bound partition machine $\mathcal{M}^{\pi_1 \sqcap \pi_2}$ exists and is given by the machine defined on the following partition:

$$\pi_1 \sqcap \pi_2 = \bigcup^{c} \{\pi; \pi \in LOW(\mathcal{M}^{\pi_1}, \mathcal{M}^{\pi_2})\}.$$

□

The proof needs the following well-known lemma which is equivalent to the Axiom of Choice.

Lemma 22. Maximality Principle *[8]*
For every partially ordered set $\mathcal{Q}$, the collection $\mathcal{C}$ of all chains (linearly ordered subsets of $\mathcal{Q}$) has a maximal member with respect to the partial ordering of $\mathcal{C}$ by the inclusion $\subset$.

□

Proof (of Theorem 21).
Since $LOW(\mathcal{M}^{\pi_1}, \mathcal{M}^{\pi_2}) \neq \emptyset$, we only need to prove that $LOW(\mathcal{M}^{\pi_1}, \mathcal{M}^{\pi_2})$ has a unique maximal element. First by the Maximality Principle, we can find a maximal chain, say $\mathcal{C}^*$. Then by (i) and (ii) of Lemma 20, $\mathcal{C}^*$ has at most a countable number of distinct elements. Let $\mathcal{C}_1$ denote this subset of $\mathcal{C}^*$. Obviously, if the top element of $\mathcal{C}^*$ exists, then it must be in $\mathcal{C}_1$. By (iii), $\mathcal{C}_1$, and hence $\mathcal{C}^*$, contain the top element. But $\mathcal{C}^*$ is an arbitrary maximal chain, so every maximal chain of $LOW(\mathcal{M}^{\pi_1}, \mathcal{M}^{\pi_2})$ has a top element in $LOW(\mathcal{M}^{\pi_1}, \mathcal{M}^{\pi_2})$. Now we prove that all maximal chains share the same top element. If not, let $\mathcal{C}^*$ and $\mathcal{C}^{**}$ be two maximal chains containing, respectively, $\mathcal{M}^{\pi^*}$ and $\mathcal{M}^{\pi^{**}}$ as distinct top elements. Then by Theorem 18, $\mathcal{M}^{\pi^* \cup^c \pi^{**}} \in LOW(\mathcal{M}^{\pi_1}, \mathcal{M}^{\pi_2})$. If $|\pi^* \cup^c \pi^{**}| = |\pi^*|$, then, by Lemma 20, $\mathcal{M}^{\pi^*} = \mathcal{M}^{\pi^* \cup^c \pi^{**}}$. If it is also true that $|\pi^* \cup^c \pi^{**}| = |\pi^{**}|$, then we will have $\mathcal{M}^{\pi^{**}} = \mathcal{M}^{\pi^* \cup^c \pi^{**}}$, which means that $\mathcal{M}^{\pi^{**}} = \mathcal{M}^{\pi^*}$, a contradiction. But if $|\pi^* \cup^c \pi^{**}| \neq |\pi^{**}|$, then $\mathcal{M}^{\pi^* \cup^c \pi^{**}} \neq$

$\mathcal{M}^{\pi^{**}}$. So $\{C^{**}, \mathcal{M}^{\pi^* \cup^c \pi^{**}}\}$ is also a chain in $LOW(\mathcal{M}^{\pi_1}, \mathcal{M}^{\pi_2})$, which implies C^{**} is not a maximal chain. The argument is identical in the case involving $|\pi^* \cup^c \pi^{**}|$ and $|\pi^{**}|$. These contradictions imply that there is a unique maximal element $\mathcal{M}^{\pi_0} \in LOW(\mathcal{M}^{\pi_1}, \mathcal{M}^{\pi_2})$ which, by definition, equals the glb of $\mathcal{M}^{\pi_1}$ and $\mathcal{M}^{\pi_2}$.

□

Theorem 18 guarantees the closure of $HIBC(\mathcal{S})$ under $\cup^c$ (i.e. existence of lub) and Theorem 21 guarantees the closure under $\sqcap$, so we have:

Corollary 23. *Under the conditions of Lemma 19 and Theorem 21, the tuple* $\langle HIBC(\mathcal{S}), \cup^c, \sqcap, \preceq \rangle$ *forms a lattice, called the hybrid in-block controllable (HIBC) lattice of* $\mathcal{S}$, *with top element* $\mathcal{M}^{\pi_{tr}}$.

□

With a slight abuse of notation, we shall use the $HIBC(\mathcal{S})$ to denote the lattice itself.

In the following section, we will show (Theorem 26) that the condition in Theorem 21 holds under by the local controllability condition.

6 Existence of Non-trivial HIBC Lattices

In the previous section we showed that it is possible to construct a hierarchical lattice (HIBC) for a system. The question remains as to what kind of system may possess such a hierarchical structure which is non-trivial. Here we present some initial results concerning this question. For this purpose, we need the following concept.

Definition 24. ([15], [5]) A dynamic system $\mathcal{S}$ is said to be *locally controllable* (LC) at $\overline{x}$ if for any $\epsilon > 0$, there is $\delta \in (0, \epsilon)$ such that for all $x, y \in B(\overline{x}, \delta)$, there exists $T < \infty$ and a control function $u \in \mathcal{U}$ such that

$$\forall t < T, (\phi(t, x, u) \in B(\overline{x}, \epsilon)) \wedge (\phi(T, x, u) = y). \tag{3}$$

We shall call $B(\overline{x}, \delta)$ a *locally controllable region with respect to* ϵ.

□

Notice that for a given neighbourhood of a locally controllable point $\overline{x}$, HIBC actually requires complete locality in the neighbourhood, by which we mean that as long as the initial state and target state are located within the neighbourhood, the reachability problem can be solved completely within that neighbourhood. Local controllability only maintains the reachability property between all points in a yet smaller neighbourhood. Hence HIBC is stronger than LC. However, one can always find a controllable set (not necessarily open) around a locally controllable point $\overline{x}$, and this forms an HIBC block if it is open.

Suppose $\mathcal{S}$ is locally controllable at $\overline{x}$, then for each pair x, y in the locally controllable region, there exists a non-empty subset $\mathcal{U}_{x,y}$ of $\mathcal{U}$ which generates a set of trajectories in the open ball $B(\overline{x}, \epsilon)$. For each such control function u, we define

$$tra(x, y, u) = \{z; \exists t \le T < \infty, \phi(t, x, u) = z \in B(\overline{x}, \epsilon)\},$$

and define

$$W(\overline{x}, \epsilon, \delta) = \bigcup_{x,y \in B(\overline{x},\delta)} \bigcup_{u \in \mathcal{U}_{x,y}} tra(x, y, u)$$

Lemma 25. *The facts below are straightforward to prove:*

1. *$W(\overline{x}, \epsilon, \delta)$ is arcwise connected.*
2. *Every point in $W(\overline{x}, \epsilon, \delta)$ can be driven to $\overline{x}$ and vise versa, and $B(\overline{x}, \delta) \subset W(\overline{x}, \epsilon, \delta)$.*
3. *For any $x \in W(\overline{x}, \epsilon, \delta)$, there exist $x_0, y_0 \in B(\overline{x}, \delta)$ such that $x \in tra(x_0, y_0, u)$.*

□

We will call the above set $W(\overline{x}, \epsilon, \delta)$ the *locally controllable bundle associated with $\overline{x}$.*

Theorem 26. *Assume that (1) is locally controllable everywhere in $in(D)$. Then any open and path-connected subset $X \subset D$ forms an HIBC block.*

Proof

For any given $\overline{x}, \overline{y} \in X$, there is a continuous curve l connecting $\overline{x}$ and $\overline{y}$ and $l \subset X$. The set of all points on l forms a compact subset of X. For any $y \in l \subset X$, there is an open ball $B(y, \epsilon_y) \subset X$. Form the balls $B(y, \epsilon_y/2), y \in l$. By local controllability, at each y, there is a strictly smaller ball $B(y, \delta_y)$ which is a locally controllable region with respect to $\epsilon_y/2$. Then the following is an open covering of l:

$$\{B(y, \delta_y); y \in l\}.$$

Hence there is finite covering of l, say

$$\{B(y_j, \delta_{y_j}); 1 \le j \le N\}.$$

It is obvious that each ball in the above finite covering set must intersect at least one other ball. Suppose $\overline{x} \in B(y_1, \delta_{y_1})$ and $\overline{y} \in B(y_N, \delta_{y_N})$. Then $\overline{x}$ can be driven to a state in the intersection of $B(y_1, \delta_{y_1})$ and another ball, say $B(y_2, \delta_{y_2})$, and then to y_2. This trajectory obviously lies in $B(y_1, \epsilon_{y_1}/2) \cup B(y_2, \epsilon_{y_2}/2)$. Repeating the same argument, y_2 can be driven to y_3 within $B(y_2, \epsilon_{y_2}/2) \cup B(y_3, \epsilon_{y_3}/2)$, and so on. Finally, y_N can be driven to $\overline{y}$ along a trajectory within $B(y_N, \epsilon_{y_N}/2)$. Hence $\overline{x}$ can be driven to $\overline{y}$ within X (see Figure 3), as follows:

$$\overline{x} \overset{u_{x,y_1}, \Delta_1}{\longrightarrow} y_1 \overset{u_{y_1,y_2}, \Delta_2}{\longrightarrow} y_2 \longrightarrow \cdots \longrightarrow \overset{u_{y_{N_1},y_N}, \Delta_N}{\longrightarrow} y_N \overset{u_{y_N,y_{N+1}}, \Delta_{N+1}}{\longrightarrow} \overline{y}.$$

□

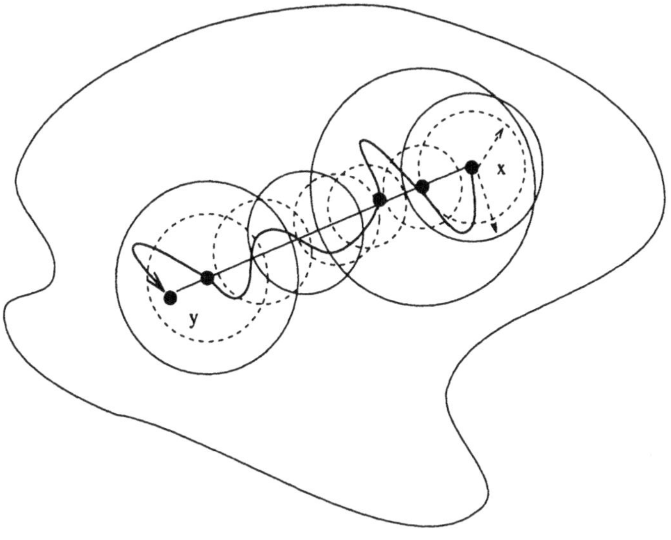

Fig. 3.

Corollary 27. *Under the conditions of Theorem 26, $HIBC(\mathcal{S})$ is closed under set intersection and the tuple $\langle \Pi(D), \cup^c, \cap, \preceq \rangle$ forms the HIBC lattice of $\mathcal{S}$.*

Proof
For any $\pi = \{X_1, \cdots, X_p\} \in \Pi(D)$, repeat the proof of Theorem 26 for each block. Further notice that the intersection of two HIBC partition machine is also HIBC. So $\sqcap$ coincides with $\cap$.

□

Remark. The above theorem displays the link between the existence of non-trivial HIBC lattice and local controllability. But the LC condition in the theorem is too strong to hold even for the entire set of controllable linear time-invariant systems (see Example 3). Hence one seeks weaker conditions for existence of non-trivial HIBC lattices. Under such conditions, the HIBC lattices would not, in general, be the set of all partitions.

Theorem 26 also indicates that if a system is locally controllable everywhere, then the DC test can be restricted so that it is only carried out along the boundaries. This may simplify the DC test. It is perhaps worth remarking that the following well-known sufficient condition (see [15]) for local controllability holds: If for $\overline{x} \in D$, there exists $\overline{u} \in U$ such that $f(\overline{x}, \overline{u}) = 0$, and the linearized system at $\overline{x}, \overline{u}$ is controllable, then (1) is locally controllable at $\overline{x}$ with respect to $\mathcal{U}$.

References

1. R. Brockett. Language driven hybrid systems. In *Proceedings of the 33rd IEEE Conference on Decision and Control*, pages 4210–4214, Lake Buena Vista, FL, 1994.

2. P. E. Caines and R. Ortega. The semi-lattice of piecewise constant controls for non-linear systems: A possible foundation for fuzzy control. To be presented at NOLCOS 95, Tahoe, CA, June 1995.
3. P.E. Caines and Y-J. Wei. The hierarchical lattices of a finite machine. To appear in *System and Control Letters*, 1995.
4. M. Glaum. Smooth partitions, transversality and graphs. Master's thesis, Department of Mathematics, The University of British Columbia, Vancouver, Canada, 1993.
5. R. Hermann and A. J. Krener. Nonlinear controllability and observability. *IEEE Trans. on Automatic Control*, AC-22(5):728–740, October 1977.
6. W. Kohn, J. James, and A. Nerode. A hybrid systems approach to integration of medical models. In *Proceedings of the 33nd IEEE Conference on Decision and Control*, pages 4247–4252, Lake Buena Vista, FL, 1994.
7. W. Kohn, A. Nerode, J. Remmel, and X. Ge. Multiple agent hybrid control: Carrier manifolds and chattering approximations to optimal control. In *Proceedings of the 33nd IEEE Conference on Decision and Control*, pages 4221–4227, Lake Buena Vista, FL, 1994.
8. E. J. McShane and T. A. Botts. *Real Analysis*. D. Van Nostrand Campany, Inc, Princeton, New Jersey.
9. A. Nerode and W. Kohn. Models for hybrid systems: Automata, topologies, stability. Technical report, Cornell University, Ithaca, NY, 1992.
10. J. A. Stiver and P.J. Antsaklis. On the controllability of hybrid control systems. In *Proceedings of the 32nd IEEE Conference on Decision and Control*, pages 294–299, San Antonio, Texas, 1993.
11. J. A. Stiver, P.J. Antsaklis, and M.D. Lemmon. Digital control from a hybrid perspective. In *Proceedings of the 33rd IEEE Conference on Decision and Control*, pages 4241–4246, Lake Buena Vista, FL, 1994.
12. Y-J Wei. *Logic Control: Markovian Fragments, Hierarchy and Hybrid Systems*. PhD thesis, Department of Electrical Engineering, McGill University, Montreal, Canada, 1995. in preparation.
13. Y-J. Wei and P.E. Caines. Lattice structure and hierarchical cocolog for finite machines. In *Proceedings of the 33rd IEEE Conference in Decision and Control*, pages 3125–3130, Lake Buena Vista,FL, December 1994.
14. W.M. Wonham. Towards an abstract internal model principle. *IEEE Trans. on Systems, Man, and Cyberneics*, SMC-6(11):735–739, November 1976.
15. J. Zabczyk. *Mathematical Control Theory: An Introduction*. Birkhäuser, Boston, USA, 1992.

7 Appendix: Examples

In the examples below, we take D to be the standard campactification of $\Re^2$.

Example 3.

$$\begin{bmatrix} \dot{x}_1 \\ \dot{x}_2 \end{bmatrix} = \begin{bmatrix} 0 & 0 \\ 1 & 0 \end{bmatrix} \begin{bmatrix} x_1 \\ x_2 \end{bmatrix} + \begin{bmatrix} 1 \\ 0 \end{bmatrix} u \tag{4}$$

Consider partition $\pi_1 = \{X_i; 0 \leq i \leq 4\}$, where

$$X_0 = \{(x_1, x_2); x_1^2 + x_2^2 < 1\},$$
$$X_1 = \{(x_1, x_2); x_1 > 0, x_2 > 0\} \setminus X_0,$$
$$X_2 = \{(x_1, x_2); x_1 < 0, x_2 > 0\} \setminus X_0,$$
$$X_3 = \{(x_1, x_2); x_1 < 0, x_2 < 0\} \setminus X_0,$$
$$X_4 = \{(x_1, x_2); x_1 > 0, x_2 < 0\} \setminus X_0.$$

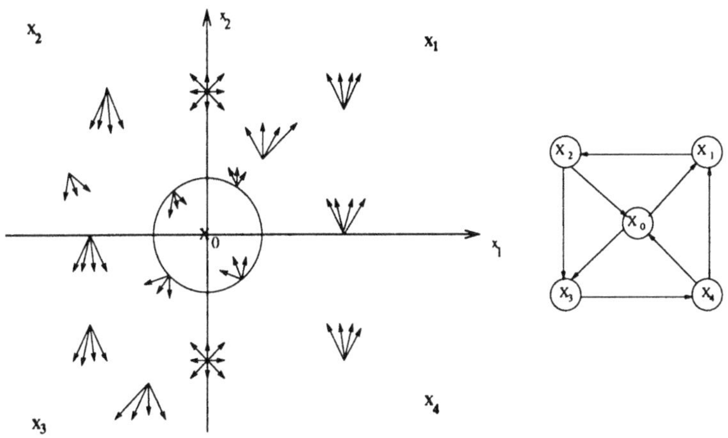

Fig. 4. The vector fields of the system and partition π_1. $\mathcal{M}^{\pi_1}$ (not HIBC)

Example 4.

$$\begin{bmatrix} \dot{x}_1 \\ \dot{x}_2 \end{bmatrix} = \begin{bmatrix} 0 & 0 \\ 1 & 0 \end{bmatrix} \begin{bmatrix} x_1 \\ x_2 \end{bmatrix} + \begin{bmatrix} 1 & 0 \\ 0 & 1 \end{bmatrix} \begin{bmatrix} u_1 \\ u_2 \end{bmatrix} \tag{5}$$

π_2:

$$X_0 = \{(x_1, x_2); x_1^2 + x_2^2 < 1\},$$
$$X_1 = \{(x_1, x_2); x_1 > x_2 > 0\} \setminus X_0,$$
$$X_2 = \{(x_1, x_2); x_2 > x_1 > 0\} \setminus X_0,$$
$$X_3 = \{(x_1, x_2); x_1 < 0, x_2 > 0\} \setminus X_0,$$
$$X_4 = \{(x_1, x_2); x_1 < 0, x_2 < 0\} \setminus X_0,$$
$$X_5 = \{(x_1, x_2); x_1 > 0, x_2 < 0\} \setminus X_0.$$

π_3:

$$X_0 = \{(x_1, x_2); x_1^2 + x_2^2 < 1\},$$
$$X_1 = \{(x_1, x_2); x_1 > 0, x_2 > 0\} \setminus X_0,$$
$$X_2 = \{(x_1, x_2); x_1 < 0, x_2 > 0, x_2 > x_1\} \setminus X_0,$$
$$X_3 = \{(x_1, x_2); x_1 < 0, x_2 > 0, x_1 > x_2\} \setminus X_0,$$
$$X_4 = \{(x_1, x_2); x_1 < 0, x_2 < 0, x_1 > x_2\} \setminus X_0,$$
$$X_5 = \{(x_1, x_2); x_1 > 0, x_2 < 0, x_1 < x_2\} \setminus X_0,$$
$$X_6 = \{(x_1, x_2); x_1 > 0, x_2 < 0\} \setminus X_0.$$

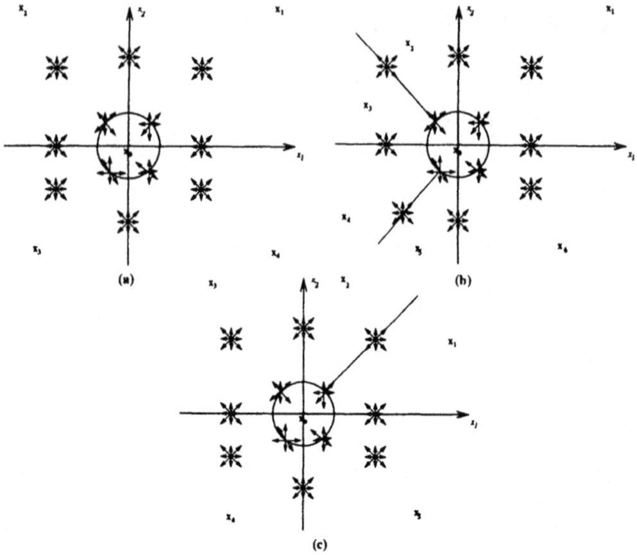

Fig. 5. The vector fields and (a) partition π_1, (b) partition π_2, (c) partition π_3.

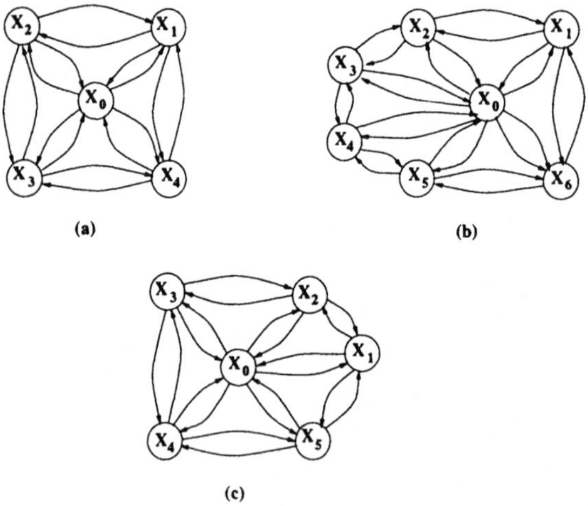

Fig. 6. The DC partition machines: (a) $\mathcal{M}^{\pi_1}$. (b) $\mathcal{M}^{\pi_2}$. (c) $\mathcal{M}^{\pi^3}$

The partitions in Figure 4 above yield HIBC machines in Figure 5. Notice that, according to Theorem 17, the partition machine $\mathcal{M}^{\pi_1}$ can be constructed directly from $\mathcal{M}^{\pi_2}, \mathcal{M}^{\pi_3}$.

A Self-Learning Neuro-Fuzzy System

Nicholas DeClaris[1] and Mu-Chun Su[2]

[1] School of Medicine and College of Engineering, University of Maryland in Baltimore and College Park, U.S.A.
[2] Department of Electrical Engineering, Tamkang University, Tamsui, Taiwan, R.O.C.

Abstract. Often a major difficulty in the design of rule-based expert systems is the process of acquiring the requisite knowledge in the form of production rules. In most of expert systems, crisp or fuzzy if-then rules are generally derived from human experts using linguistic information. However, the initial linguistic rules are invariably rather crude and, although qualitatively correct, need to be refined to achieve better performance. In this paper, we present an innovative approach to the rule extraction directly from experimental numerical data for pattern recognition and system identification. This paper presents a neural network-based system which is trained in such a way that it provides an appealing solution to the problem of knowledge acquisition based on experimental numerical data. The values of the network's parameters, after sufficient training , are then utilized to generate both crisp and fuzzy if-then rules. The concept and method presented in this paper are illustrated through one traditional academic pattern recognition problem and one highly nonlinear system identification problem.

1 Introduction

Pattern recognition is a key element to many engineering applications–medical diagnosis , man-machine communications (speech recognition and image understanding), and control . There are two different approaches to pattern recognition. Conventional approach attempts to estimate an exact boundary that contains patterns of the same class. In such cases features characterizing input vectors are precise and complete (i. e. numerical) in nature. In some cases, it is not appropriate to give an exact numerical representation to uncertain feature data. Rather, it is reasonable to represent uncertain feature information by fuzzy sets. Fuzzy sets introduced by Zadeh [1] are a generalization of conventional set theory as means of representing and manipulating data that are not precise, but rather fuzzy. Therefore, fuzzy classification provides another solution to pattern classification. No matter which approach (fuzzy or crisp) we take, a pattern recognition machine can be implemented by a expert system based on a set of fuzzy or crisp classification rules.

Traditionally, the design of (crisp or fuzzy) rule-based expert systems involves a process of interactions between a domain expert and a knowledge engineer who formalizes the expert's knowledge as inference rules and encodes it in a computer. However, there are several difficulties in obtaining an adequate set of rules from human experts. First, human experts may not know, or may be unable to articulate, what knowledge they actually used in solving their problems. In addition, it takes long time to implement an expert system. One way to alleviate these problems is to use neural networks to automate the process of knowledge acquisition [2].

Recently, neural networks and fuzzy systems have been widely applied to many fields, yet in practice each has its own advantages and disadvantages. One of the most appealing aspect of neural networks is that they can inductively learn concepts from given experimental data. Furthermore, a two-layer backpropagation network with sufficient hidden nodes has been proven to be a universal approximator. Nevertheless, backpropagation networks suffer from lengthy training time. Besides, it is very difficult to interpret, in physically meaningful way, trained networks. That is , the knowledge is encoded in the parameters. This kind of knowledge representation is in a sense not acceptable to human users. On the other hand, the knowledge acquisition is the bottleneck of implementing fuzzy systems. Conventional approaches to fuzzy system designs either presume that fuzzy rules be given by human experts or involve dividing the input and the output space into fuzzy regions. However, in many applications, the initial linguistic rules are too crude for engineering purposes. On the other hand, the major restriction of the second approach is that the number of division of each input and output variables must be defined in advance. Besides, it is very difficult to apply such fuzzy systems to problems in which the number of input variables is large. Therefore, to overcome the bottleneck of the knowledge acquisition, several methods have been proposed for extracting fuzzy rules directly from numerical data [3] [4] [5].

One approach uses back-propagation networks to extract fuzzy rules [3]. The antecedent and consequent of the fuzzy if-then rules are represented by two separate back-propagation networks. The knowledge is encoded in the parameters of the trained networks. Thus such approach does not give a meaningful expression of the qualitative aspects of human reasoning. Another approach uses Gaussian potential functions for the construction of the input membership functions. Then the least mean squared (LMS) algorithm or back-propagation algorithm is used to train such fuzzy neural networks [4], [5]. The major disadvantage of this approach is that the initialization of weights. The initial values of the parameters are set either by the k-means algorithm or in such a way that the membership functions along each axis can cover the operating range totally with sufficient overlapping with each other. Input data clustered by the k-means algorithm may exhibit very different output responses. The initial membership functions set by the second method are decided too heuristically. Therefore, such initialization may not really speed up the learning process.

This paper is organized as follows. The architecture of the class of neural networks with hyperrectangular neural-type junctions is discussed in section 2. Section 3 introduces the novel class of fuzzy degraded hyperellipsoidal composite neural networks (FDHECNN's). The hybrid training algorithm is given in section 4. The simulation results are demonstrated in section 5. Finally some concluding remarks are given in section 6.

2 Neural Networks with Hyperrectangular Neural-Type Junctions

The construction of an rule-based expert system involves the process of acquiring production rules. Production rules are often represented as "IF condition THEN act". The backpropagation networks may not arrive at an inference structure close to this kind of high level knowledge representation. Here a novel class of hyperrectangular composite neural networks is presented. The class of hyperrectangular composite neural networks provides a new tool for machine learning. The classification knowledge is easily extracted from the weights in a hyperrectangular composite neural network. The symbolic representation of a neural node is with hyperrectangular neural-type junctions shown in Fig. 1(a) and it is described by the following equations :

$$net_j(\underline{x}) = \sum_{i=1}^{n} f((M_{ji} - x_i)(x_i - m_{ji})) - n \tag{1}$$

$$out_j(\underline{x}) = f(net_j(\underline{x})) \tag{2}$$

where

$$f(x) = \begin{cases} 1 & if\ x \geq 0 \\ 0 & otherwise \end{cases} \tag{3}$$

M_{ji} and $m_{ji} \in R$ are adjustable weights of the jth neural node, n is the dimensionality of input variables, and $Out(\underline{x}) : R^n \rightarrow R$ is an output function of a neural node with hyperrectangular neural-type junctions.

2.1 Extraction of Crisp rules

According to Eq. 1, 2, and 3, we know the neural node with hyperrectangular neural-type junctions outputs one only under the following condition:

$$\begin{cases} m_{j1} \leq x_1 \leq M_{j1} \\ m_{j2} \leq x_2 \leq M_{j2} \\ \quad \cdots \\ m_{jn} \leq x_n \leq M_{jn} \end{cases} \tag{4}$$

Therefore, the classification knowledge can be described in the form of a production rule. The IF...THEN...rule has an antecedent (premise) consisting of the one condition as well as a single consequent.

$$\begin{aligned} &IF\ (\ (m_{j1} \leq x_1 \leq M_{j1}) \cap \ \cdots \ \cap (m_{jn} \leq x_n \leq M_{jn})) \\ &\qquad THEN\ \ OUT = 1. \end{aligned} \tag{5}$$

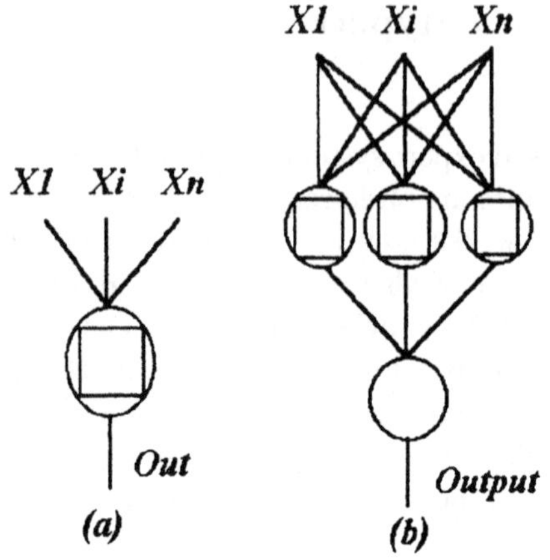

Fig. 1. (a) A neural node with "hyperrectangular neural-type" junctions and (b) A hyperrectangular composite neural network

The domain defined by the antecedent of Eq.(5) is a hyperrectangle in n-dimensional space. Owing to characteristics of data, such as dispersion characteristic, data may present an existence of many distinct clusters in input space. It is efficient to cope with this situation by using a set of hyperrectangles with different sizes and locations to fit the data. Each hyperrectangle corresponds to an "intermediate rule". The final classification rule is the aggregation of the intermediate rules. This kind of expression is then mapped to a tow-layer hyperrectangular composite neural network. In the composite neural network configuration shown in Fig. 1(b), input variables are assigned to input nodes, intermediate rules correspond to hidden nodes, and the final classification rule is assigned to an output node. Each hidden node is connected, with weight valued 1.0, to an output node. The output node implements the "OR" function. Therefore, if there are K hidden nodes in the composite neural network, the classification rule is expressed in the following form:

$$\begin{aligned}
&IF\ (\ (m_{11} \leq x_1 \leq M_{11}) \cap \cdots \cap (m_{1n} \leq x_n \leq M_{1n})) \\
&\qquad \text{THEN OUTPUT} = 1; \\
&\cdots \\
&ELSEIF((m_{k1} \leq x_1 \leq M_{k1}) \cap \cdots \cap (m_{kn} \leq x_n \leq M_{kn})) \\
&\qquad \text{THEN OUTPUT} = 1; \\
&ELSE\ OUTPUT = 0;
\end{aligned} \tag{6}$$

2.2 Extraction of Fuzzy rules

After having found a set of crispy if-then rules, we may fuzzyify these crispy rules by using a reasonable fuzzy membership function. The membership function $m_j(\underline{x})$ for the jth hyperrectangle, $0 \leq m_j(\underline{x}) \leq 1$, must measure the degree to which the input pattern x falls outside of the hyperrectangle. As $m_j(\underline{x})$ approaches 1, the patterns should be more contained by the hyperrectangle, with the value 1 representing complete hyperrectangle containment. In addition, the measurement should reflect the degree of importance of the input feature on a dimension by dimension basis. The membership function that meets all these criteria is defined as

$$m_j(\underline{x}) = \frac{V_j}{V_j + s(Vol(\underline{x}) - V_j)} (\text{or} \quad e^{-S_j^2(Vol_j(\underline{x}) - V_j)^2}) \tag{7}$$

where

$$lV_j = \prod_{i=1}^{n}(M_{ji} - m_{ji}) \tag{8}$$
$$= \text{volume of the jth hyperrectangle,}$$

$$Vol_j(\underline{x}) = \prod_{i=1}^{n}(M_{ji} - m_{ji}, x_i - m_{ji}, M_{ji} - x_i) \tag{9}$$

and s_j = the sensitivity parameter that regulates how fast the membership values decreases as the distance between x and the jth hyperrectangle .

Fig. 2 shows the relationship between $m_j(\underline{x})$ and $Vol(\underline{x})$. An example of this membership function for two dimensional case is shown in Fig.3. The connection between the jth hidden node and the output node represents the fraction of positive examples that fall within the jth hyperrectangle . The "maximum-membership defuzzification scheme " is utilized at the last stage of fuzzy systems [6]. This scheme was motivated by the popular probabilistic methods of maximum-likelihood and maximum-aposteriori parameter estimation. The other approach to find wj is to use the LMS or the backpropagation algorithm.

2.3 The Supervised Decision-Directed Learning (SDDL) Algorithm

The supervised decision-directed learning (SDDL) algorithm generates a two-layer feedforward network in a sequential manner by adding hidden nodes as needed. As long as there are no identical data over different classes, We can obtain 100% recognition rate for training data. First of all, training patterns are divided into two classes : (1) a "positive class" from which we want to extract the "concept" and (2) a "negative class" which provides the counterexamples with respect to the "concept". A "seed" pattern is used as the base of the "initial concept" (a hyperrectangle with arbitrarily small size). The seed pattern can be arbitrarily chosen from the positive class. Then we try to generalize (expand)

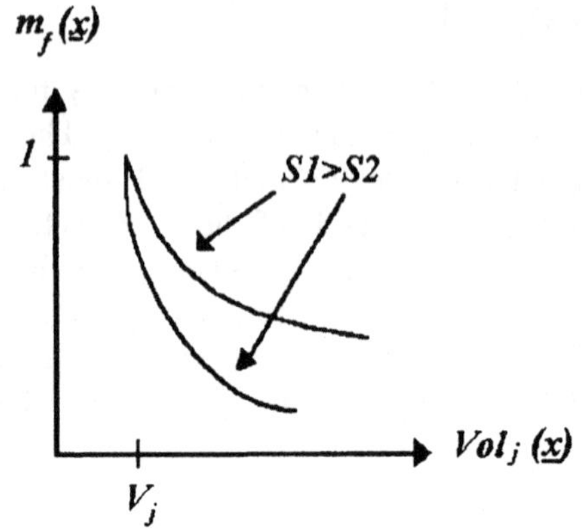

Fig. 2. The relation between mj(x) and Volj(x), s represents the sensitivity parameter

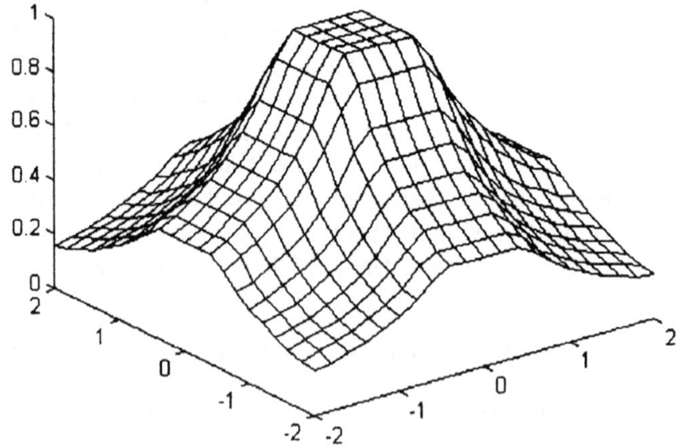

Fig. 3. An example of mj(x) when sj is small for two-dimensional case

the initial concept (hyperrectangle) to explain next positive pattern . After this , we have to check whether there is any negative pattern falling inside the present hyperrectangle in order to prevent the occurrence of "overgeneralization". Fig. 4 illustrates the training procedure. The flow chart of the SDDL algorithm is given in Fig. 5. A more detailed description of the training procedure is given in [2].

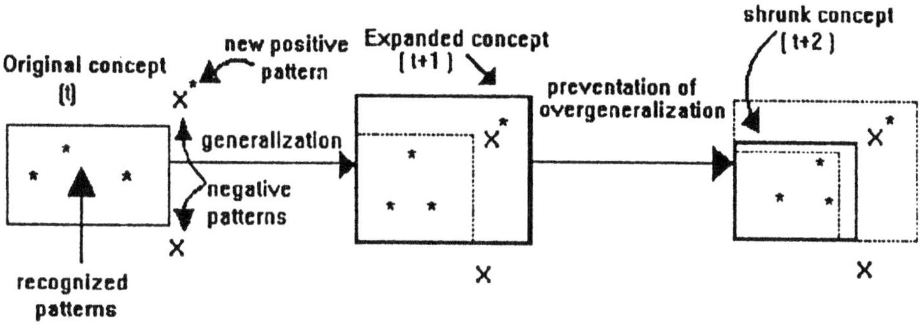

Fig. 4. The training procedure

Given all description of the training algorithm, we now give pseudo-code description of two important procedures:

- Procedure of generalization

```
begin (x is a positive pattern)
      for i from 1 to dimensions-of-input
          begin
                if x_i >= M_i(t)
                then M_i(t+1) = x_i + ε;
                else if x_i <= m_i(t)
                then m_i(t+1) = x_i - ε;
          end;  end; end;
```

- Procedure of prevention-of-overgeneralization

```
begin ( x is a counterexample)
      for i from 1 to dimensions-of-input
```

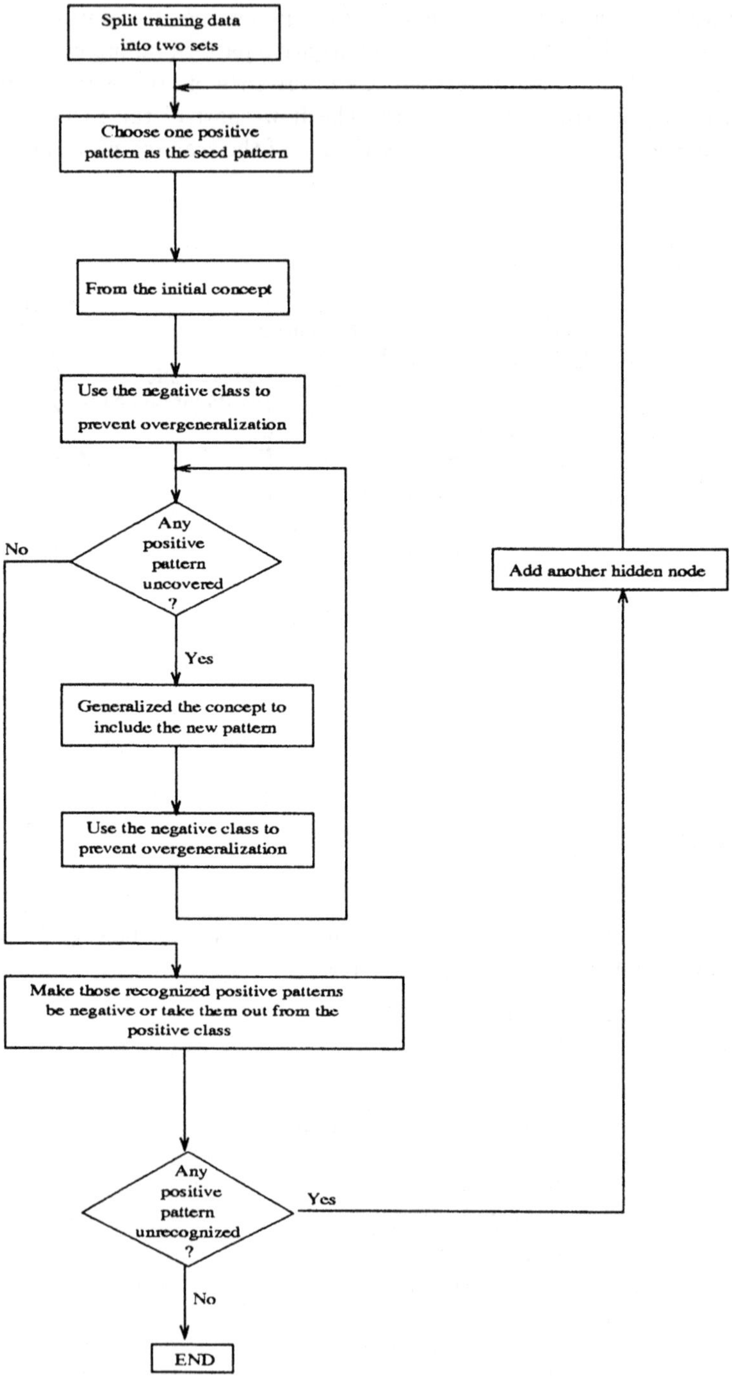

Fig. 5. The flow chart of the SDDL algorithm

```
        begin
            if x_i < M_i(t)
            then M_i(t+1) = x_i - δ(δ should be chosen to ensure x_i + δ ≤ M_i(t));
            else if x_i > m_i(t)
            then m_i(t+1) = x_i + δ(δ should be chosen to ensure x_i + δ ≤ m_i(t));
        end;
    end;
end;
```

There are similarities between the fuzzy min-max networks classifiers [7] and the hyperrectangular composite neural networks. Both approaches utilize hyperrectangles as rule representation elements. However, there are still many major differences between these two classes of neural networks [2]. First, the hyperrectangular composite neural networks was originally developed to generate crispy if-then rules. After crispy rules have been generated, a reasonable membership function is utilized to fuzzyify these crispy rules. The fuzzy min-max neural networks focus on finding fuzzy rules. Second, the fuzzy min-max neural network classifiers find a set of hyperrectangles under a special expansion criterion so that the sizes of hyperrectangles are limited. On the contrary, hyperrectangular composite neural networks find hyperrectangles whose sizes are as large as possible. Third, the contraction procedures are different in these two classes of neural networks. This means that he training algorithms are very different in these two approaches.

3 Fuzzy Degraded Hyperellipsoidal Composite Neural Networks (FDHECNN's)

Ordinarily, fuzzy system designers often assume input variables are independent, therefore, the membership function is assigned variable by variable.

Fig. 6 shows an example of conventional fuzzy rule partitions. Most of the membership functions are assumed to be triangular, trapezoidal or bell-shaped. In fact, there is no straightforward method for choosing membership functions. Either trial-and-error tuning or backpropagation algorithm is used to find the right number of division of inputs and appropriate membership functions. Since it is likely that there exist correlations among input variables, in this case, the fuzzy rule partitions should not be divided variable by variable. The fuzzy regions should be arbitrarily shaped, as shown in Fig. 7.

A neural network approach to fuzzy rule partitions was proposed in [3]. However, the extracted knowledge is still too implicit. Therefore, we propose to use the aggregations of hyperellipsoids to approximate arbitrary fuzzy rule partitions as shown in Fig. 8. This new scheme is implemented as an two-layer FHRCNN. The novel class of FDHECNN's is a modified version of the class of neural networks with quadratic junctions. The inherent architectural efficiency of this kind of neural networks was demonstrated in [2], [8], [9].

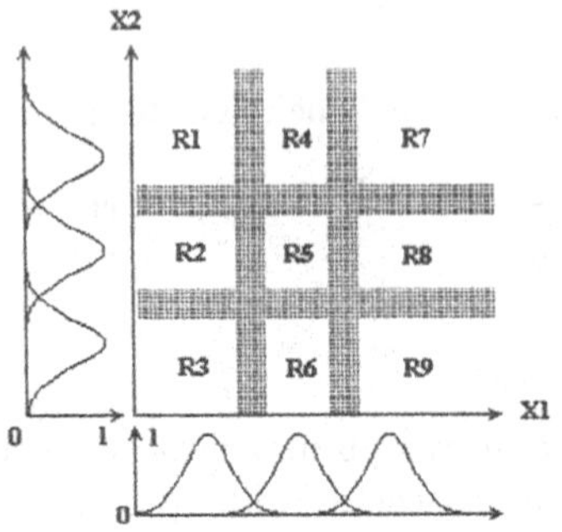

Fig. 6. An example of traditional fuzzy rule partitions

A two-layer feedforward fuzzy degraded hyperellipsoidal composite neural network (FDHECNN) shown in Fig. 9 is described by

$$net_j(\underline{x}) = d_j^2 - \sum_{i=1}^{n}(w_{ji}x_i + b_{ji})^2, \tag{10}$$

$$m_j(\underline{x}) = exp\{-s_j^2[max(d_j^2, d_j^2 - net_j(\underline{x})) - d_j^2]\}, \tag{11}$$

and

$$Out = \sum_{j=1}^{J} w_j m_j(\underline{x}) + \theta, \tag{12}$$

where b_{ji}, w_{ji} and d_j are adaptive weights, s_j is the slope value which regulates how fast $m_j(\underline{x})$ goes down, w_j is the connection weight from hidden node j to the output, and θ is a small real number.

If we assume d_j be zero then the FDHECNN's are isomorphic to the radial basis function (RBF) networks. The RBF networks using Gaussian potential functions have been proven to be universal approximators [10]. The essential difference between the Gaussian function and the $m_j(\underline{x})$ can be illustrated in Fig.10 and 11. Obviously, the $m_j(\underline{x})$ offers more flexibility. The functionality of the $m_j(\underline{x})$ can change from the Gaussian function depicted in Fig. 10 to the steplike depicted in Fig. 11.

Therefore the FDHECNN's offer significantly more flexibility in approximating functions. Most fuzzy systems employ "center average defuzzifier"

$$Out = \sum_{j=1}^{J}(out_j mem_j(\underline{x}) / \sum_{j=1}^{J}(mem_j(\underline{x}) \tag{13}$$

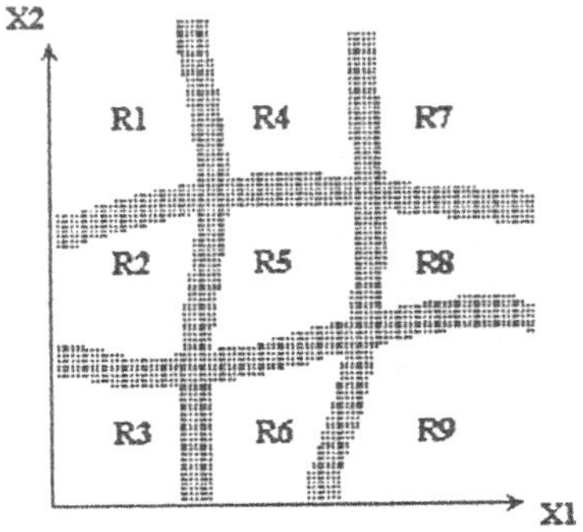

Fig. 7. An example of arbitrarily shaped fuzzy rule partitions

However each fuzzy rule plays a different role in contributing to the computation of the crispy final output. Therefore, we propose the "weighted center average defuzzifier"

$$out = \sum_{j=1}^{J} cf_j out_j m_j(\underline{x}) + \theta = \sum_{j=1}^{J} (w_j m_j(\underline{x}) + \theta \tag{14}$$

where cf_j is the certainty factor of the jth fuzzy rule. The weighted center average defuzzifier in a sense takes the linear combination of the firing strengths of J fuzzy rules.

4 Hybrid Training Algorithm for FDHECNN's

Basically, the FDHECNN's provide more flexible local functions (the output functions of hidden nodes) than the RBF networks do. Therefore, it can be shown to be a universal approximator. The proof is given elsewhere. The performance of a FDHECNN depends on the chosen fuzzy hyperellipsoids defined by $d_j^2 \geq \sum_{i=1}^{n}(w_{ji}x_i + b_{ji})^2$ and the connection weights. The fuzzy hyperellipsoids should suitably sample the input domain and reflect the data distribution. Traditional approaches used the k-means algorithm to cluster input data directly on the input space. They do not take account for the output response of each input pattern. Besides, in the process of system identification, input data are often chosen to scatter evenly over the input space. The k-means algorithm does not reflect any important data distribution in these cases. Based on above discussions, a hybrid learning algorithm consists of the following steps:

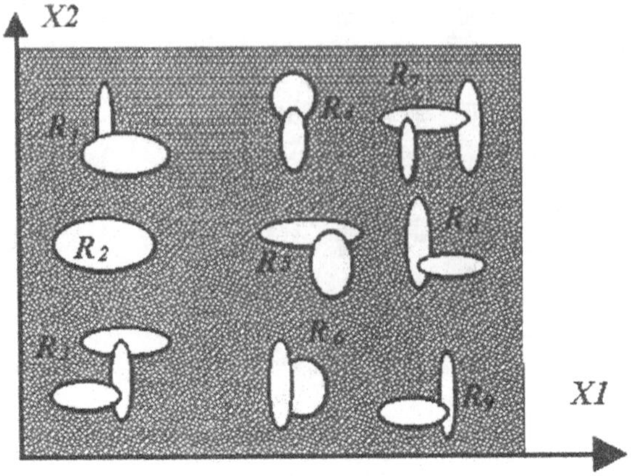

Fig. 8. Aggregations of hyperellipsoids for approximating arbitrary rule partitions

Step 1. Partition the Output Space into Fuzzy Regions

Divide the output into regions (the length of these divided intervals can be equal or unequal), denoted by SN (Small N), ..., S1 (Small 1), CE (Center), L1 (Large 1),..., LN (Large N). At this time, we do not need to assign a fuzzy membership function to each region since the defuzzifier employs the "weighted center average defuzzifier".

Step 2. Transform a Function Approximation Problem into a Pattern Recognition Problem

The original input-output pairs are then transformed into quantized input-output pairs, that is, a function approximation problem is converted to a (2N+1)-class pattern recognition problem. By the SDDL algorithm, (2N+1) two-layer HRCNN's (hyperrectangular composite neural networks) are created to recognize these patterns. After sufficient training, (2N+1) sets of hyperrectangles are created. Assume there be The m_i^c hidden nodes in the ith trained HRCNN, i=1,.., 2N+1. The m_i^c hidden nodes are then ranked in the order of respective representativeness.

Step 3. Initialize and Update Weights

For output class i, the first n_i out of The m_i^c hidden nodes are selected to initialize the weights $(b_{ji}, d_j, and w_{ji})$ of a FDHECNN with $(n_1+n_2+...+n_{2N+1})$ hidden nodes. Fig.12 illustrates such initialization scheme.

The connection weights, $w_j^{'}s$, are initialized to be the mean values of the (2N+1) segmented intervals, respectively. There are two approaches to adjust the weights. The first one is to fix the weights, b_{ji}, d_j, and w_{ji}, and to update only the connection weights, w_j, in real-time using the recursive least square error algorithm. However, it is advantageous to update all weights, b_{ji}, d_{ji}, w_{ji} and w_j, simultaneously because this will significantly improve both the modeling

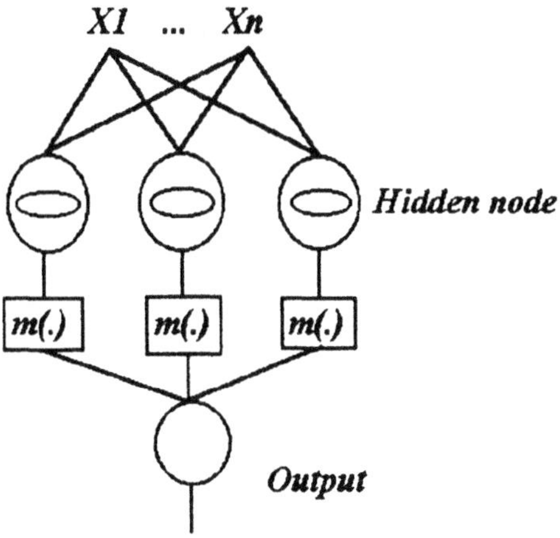

Fig. 9. The symbolic representations of a two-layer FDHECNN

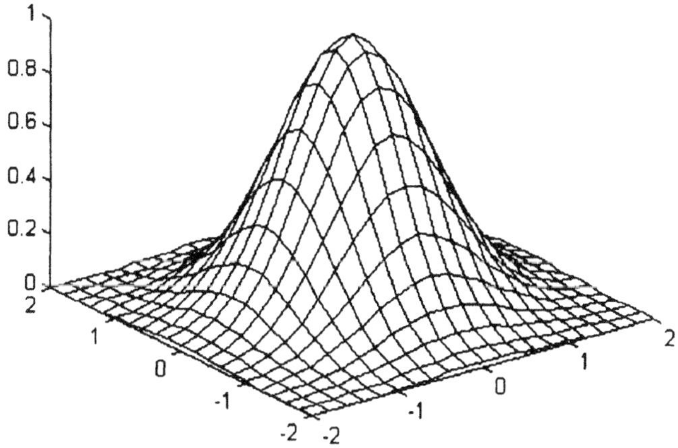

_ **ig. 10.** An example of mj(x) when dj=0

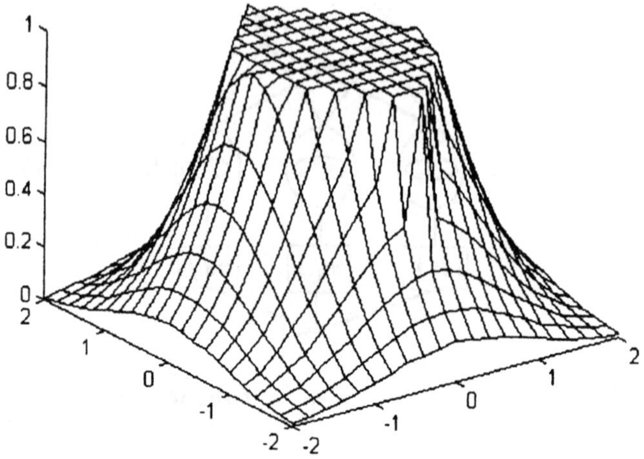

Fig. 11. An example of mj(x) when dj is a small number

$$\frac{b_2}{w_2} = -\frac{1}{2}(M_2+m_2)$$

$$w_2 = 1\Big/ \tfrac{1}{2}(M_2-m_2)$$

(m_1, M_2) $\qquad$ (M_1, M_2)

$$\frac{b_1}{w_1} = -\frac{1}{2}(M_1+m_1)$$

$$w_1 = 1\Big/ \tfrac{1}{2}(M_1-m_1)$$

(m_1, m_2) $\qquad$ (M_1, m_2)

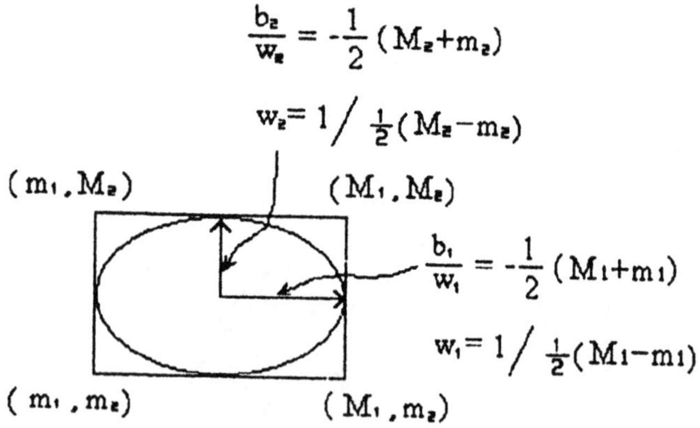

Fig. 12. Parameters initialization procedure

capability. The reason why the modeling capability can be improved is because the boundary of hyperellipsoid can be adjusted to efficiently sample the data distribution. Assuming the given training data set has p entries, we can define an error function as

$$E = \sum_p E_p = \sum_p \frac{1}{2}(Out_p - t_p)^2 \qquad (15)$$

where t_p is the pth desired output and Out_p is the actual output of the FHRCNN with respect to the pth input. By use of the recursive LMS algorithm or back-propagation algorithm, the weights can be adjusted to minimize the total error E. All weights are adapted according to the following equations:

$$\begin{aligned}\frac{\partial E}{\partial \theta} &= \sum_p \frac{\partial E_p}{\partial Out_p}\frac{\partial Out_p}{\partial \theta} \\ &= \sum_p (Out_p - t_p)\end{aligned} \qquad (16)$$

$$\begin{aligned}\frac{\partial E}{\partial w_j} &= \sum_p \frac{\partial E_p}{\partial Out_p}\frac{\partial Out_p}{\partial w_j} \\ &= \sum_p (Out_p - t_p) m_j(\underline{x})\end{aligned} \qquad (17)$$

$$\frac{\partial E}{\partial b_{ji}} = \frac{\partial E}{\partial Out_p}\frac{\partial Out_p}{\partial m_j}\frac{\partial m_j}{\partial b_{ji}} \qquad (18)$$

$$= \begin{cases} 0 & if z_j > 0 \\ \sum_p (Out_p - t_p) w_j (-2) s_j^2 (w_{ji} x_i + b_{ji}) e^{s_j^2 z_j} & otherwise \end{cases}$$

$$\frac{\partial E}{\partial d_j} = \frac{\partial E}{\partial Out_p}\frac{\partial Out_p}{\partial m_j}\frac{\partial m_j}{\partial d_j} \qquad (19)$$

$$= \begin{cases} 0 & if z_j > 0 \\ \sum_p (Out_p - t_p) w_j s_j^2 2 d_j e^{s_j^2 z_j} & otherwise \end{cases}$$

$$\frac{\partial E}{\partial w_{ji}} = \frac{\partial E}{\partial Out_p}\frac{\partial Out_p}{\partial m_j}\frac{\partial m_j}{\partial w_{ji}} \qquad (20)$$

$$= \begin{cases} 0 & if z_j > 0 \\ \sum_p (Out_p - t_p) w_j s_j^2 (-2x_i)(w_{ji} x_i + b_{ji}) e^{s_j^2 z_j} & otherwise \end{cases}$$

Step 4. Interpret the trained FDHECNN's

Let HE_k be the hyperellipsoid, $\sum_{i=1}^{n}(w_{ki}x_i + b_{ki})^2 \leq d_k^2$. From the trained FHRCNN, a set of fuzzy rules can be extracted and represented as

$$\begin{aligned} Rule1 &: IF\ (\underline{x} \text{ is near } HE_1) \\ &\quad THEN \text{ the system output is } LN \\ &\quad \ldots \\ Rule2N+1 &: ELSEIF(\underline{x} \text{ is near } HE_{2N+1}) \\ &\quad THEN \text{ the system output is } LP \end{aligned} \tag{21}$$

5 Simulation Results

Example 1 : Pattern Recognition

The approach can be illustrated easily on two-dimensional data because the data as well as the decision region are easily portrayed graphically. The two sets of spiral data shown in Fig. 13 consists of 192 patterns, with 96 patterns for each spiral. Spiral 1 and 2 are represented by "+" and "-" points, respectively.

This pattern recognition problem is highly not linearly separable. One spiral is generated as a reflection of another, namely (x1, y1)=(-x2, -y2), and the formulas used to generate the spirals are given below [11]:

$$\rho = \alpha\theta \quad \alpha = 10.0 \quad \theta \leq 6\pi \tag{22}$$

$$spiral1 : \begin{cases} x_1 = \rho cos(\theta) \\ y_1 = \rho sin(\theta) \end{cases} \tag{23}$$

$$spiral2 : \begin{cases} x_1 = -\rho cos(\theta) \\ y_1 = -\rho sin(\theta) \end{cases} \tag{24}$$

To recognize these spiral patterns is a very difficult task for backpropagation networks. Lang and Withbrock built a four-layer neural network (with three hidden layers and one output layer) with total of 16 nodes by means of the trial-and-error method [12]. Cios and Liu used CID3 algorithm to self-generate a six-layer neural network with total of 31 nodes. Both these two networks accomplished the learning task. The decision regions formed by these two networks were shown in [11]. Fig. 14 and 15 illustrate the decision regions formed by expert systems using crispy and fuzzy classification rules, respectively.

Example 2 : System Identification

The plant to be identified is governed by the differential equation

$$y(t+1) = (1 + exp(-4u^2(t)))sin(\pi y(t)) \tag{25}$$

where u and y are the input and output of the plant, respectively. The training patterns are in two-dimensional input space (y(k) and u(k)) and one-dimensional output space, as shown in Fig. 16. The input parts of the training data are

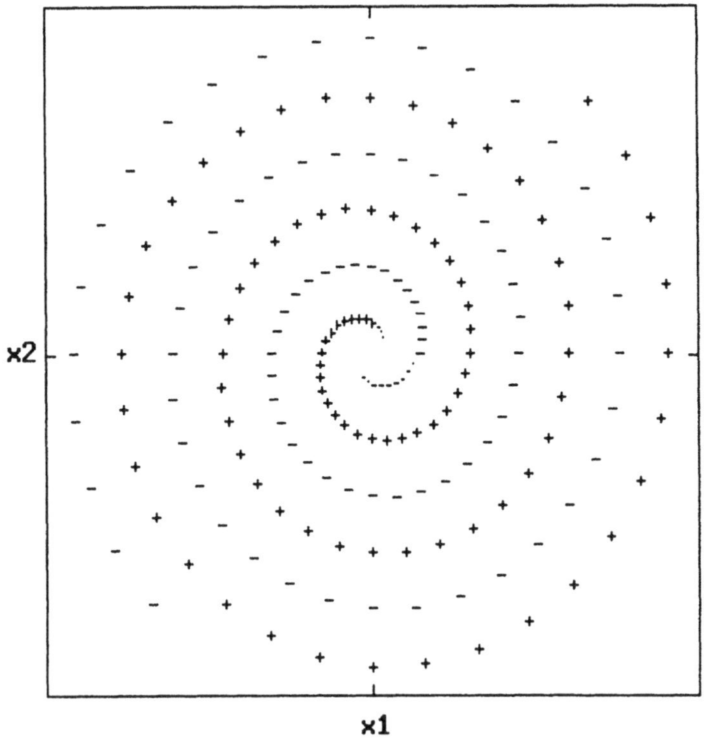

Fig. 13. The two sets of spiral data : "+" represents the positive patterns "-" represents the negative patterns

chosen to scatter evenly over a square region [-1, 1]x[-1, 1]. The total number of training data is 441. First, we partition the output variables into five regions, [-2.0, -1.2](large negative), [-1.2, -0.4](small negative), [-0.4, 0.4] (near zero), [0.4, 1.2](small positive) and [1.2, 2.0](large positive). In the following step, five HRCNN's were trained to recognize the 5-class pattern recognition problem. After training, five HRCNN's with 3, 12, 9, 12 and 4 hidden nodes, corresponding to 5 different classes, were created. A FDHECNN with (2+2+0+4+2) hidden nodes was trained to identify the plant. Fig. 17 depicts the three dimensional representatives learned by the FHRCNN's using the back-propagation algorithm.

The error curve is shown in Fig. 18. The ten extracted fuzzy rules are represented as

IF (y(k),u(k)) is near the ellipse : $[(y(k)+0.5)^2/0.2^2]+[(u(k)+0.02)^2/0.54^2]=1$
THEN y(k+1) is large negative;
IF (y(k),u(k)) is near the ellipse : $[(y(k)+0.51)^2/0.24^2]+[(u(k)+0.0)^2/0.45^2]=1$
THEN y(k+1) is large negative.
IF (y(k),u(k)) is near the ellipse : $[(y(k)+0.5)^2/0.4^2]+[(u(k)+0.85)^2/0.25^2]=1$
THEN y(k+1) is small negative.

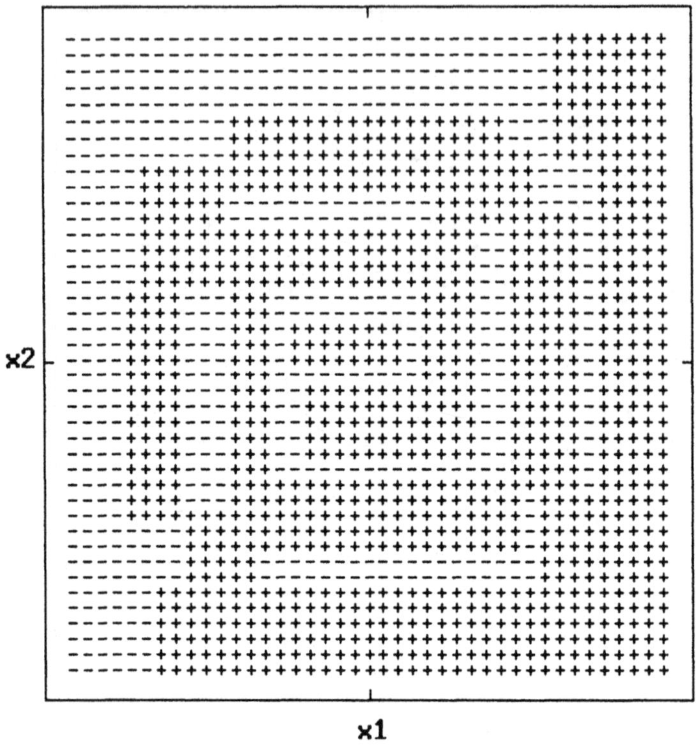

Fig. 14. The decision region formed by the expert network using extracted crispy rule

IF (y(k),u(k)) is near the ellipse : $[(y(k)+0.83)^2/0.12^2]+[(u(k)-0.02)^2/0.57^2]=1$
THEN y(k+1) is small negative.
IF (y(k),u(k)) is near the ellipse : $[(y(k)-0.5)^2/0.3^2]+[(u(k)+0.85)^2/0.25^2]=1$
THEN y(k+1) is small positive.
IF (y(k),u(k)) is near the ellipse : $[(y(k)-0.2)^2/0.1^2]+[(u(k)-0.05)^2/0.55^2]=1$
THEN y(k+1) is small positive;
IF (y(k),u(k)) is near the ellipse : $[(y(k)-0.8)^2/0.1^2]+[(u(k)-0.03)^2/0.53^2]=1$
THEN y(k+1) is small positive.
IF (y(k),u(k)) is near the ellipse : $[(y(k)-0.3)^2/0.1^2]+[(u(k)-0.74)^2/0.3^2]=1$
THEN y(k+1) is small positive.
IF (y(k),u(k)) is near the ellipse : $[(y(k)-0.5)^2/0.1^2]+[(u(k)+0.03)^2/0.47^2]=1$
THEN y(k+1) is large positive.
IF (y(k),u(k)) is near the ellipse : $[(y(k)-0.5)^2/0.2^2]+[(u(k)-0.02)^2/0.42^2]=1$
THEN y(k+1) is large positive.

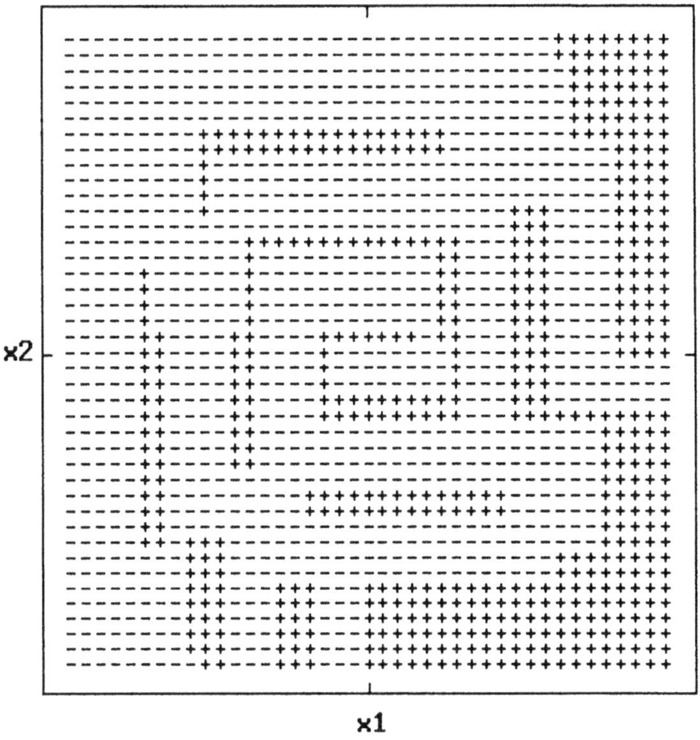

Fig. 15. The decision region formed by the expert network using fuzzy rules

6 Concluding Remarks

In this paper, a novel class of hyperrectangular composite neural networks has been proposed and presented. After training, the parameters of the trained networks could be utilized to generate both crispy and fuzzy rules. The SDDL algorithm generates a two-layer feedforward HRCNN in a sequential manner by adding hidden nodes as needed. As long as there are no identical data over different classes, 100% recognition rate for training data can be achieved. Besides, we also have proposed a novel approach to the rule extraction directly from experimental numerical data for function approximation without interviewing domain experts. This approach employs the fuzzy degraded hyperellipsoidal composite neural networks as building blocks and the LMS or the backpropagation algorithm as a training procedure. A special preprocessing scheme, transforming a function approximation problem into a pattern recognition problem, is proposed to find a set of good initializations in order to speed up the learning steps. Most of all, the correlations among input features have been considered. The fuzzy rule partitions are no longer parallel to input features. The aggregation of a set of hyperellipsoids are utilized to represent a fuzzy rule.

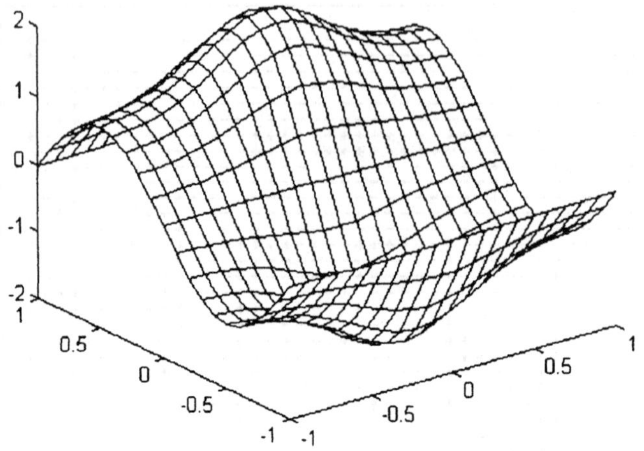

Fig. 16. The training patterns presented to the FDHECNN for learning

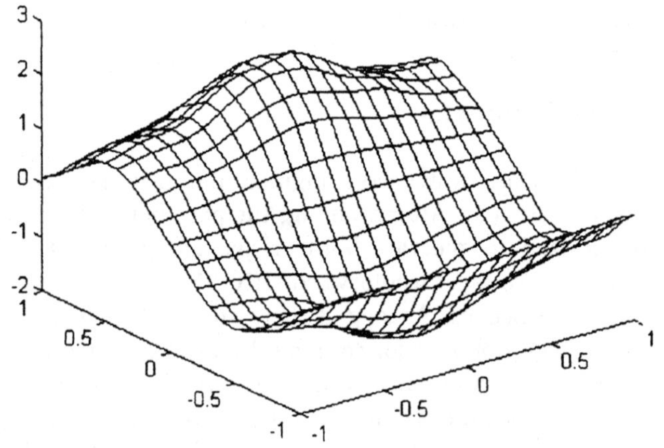

Fig. 17. Three dimensional representatives learned by the FDHECNN

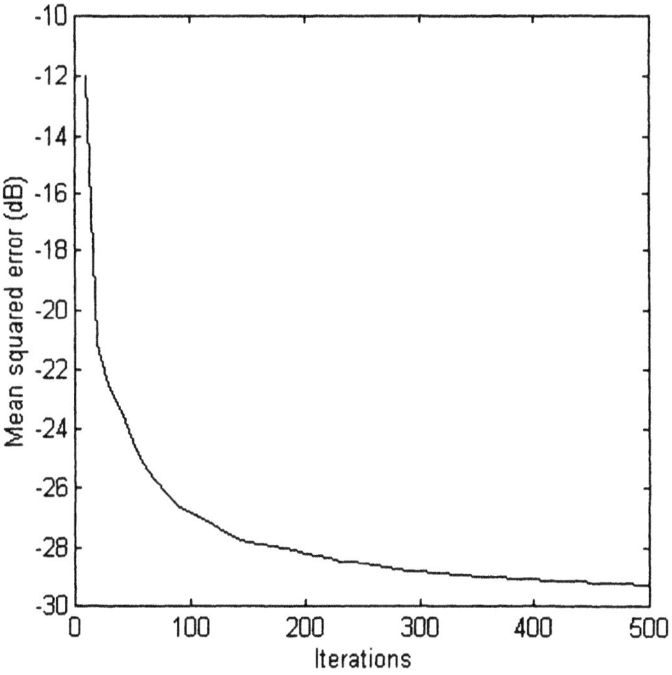

Fig. 18. Evolution of the mean squared error

Acknowledgments

This research has been supported in part by NIH Grant 1-S10RR06460-01, N. DeClaris, Principal Investigator.

References

1. L.A. Zadeh ," Fuzzy Sets ," Information and Control, vol. 8 , pp.338-353 , 1965.
2. M. C. Su, A Neural Network Approach to Knowledge Acquisition, Ph.D. thesis, University of Maryland, August, 1993.
3. H. Takagi and I. Hayashi, "NN-driven fuzzy reasoning", Int. J. Approximate Reasoning, vol.5, No3, PP. 191-212, May, 1991.
4. L. X. Wang, Adaptive Fuzzy Systems and Control: Design and Stability Analysis, Prentice Hall, Inc. New Jersey, 1994.
5. C. T. Lin C. S. G. Lee, Neural network-based fuzzy logic control and decision system, "IEEE Trans. Computers, vol.40, No.12, PP.1320-1336, 1991.
6. B. Kosko, Neural Networks and Fuzzy Systems : A Dynamical Systems Approach to Machine Intelligence, Prentice-Hall, NJ, 1992.
7. P. K. Simpson, "Fuzzy min-max neural networks-part 1 : Classification," IEEE Trans. on Neural Networks, vol. 3, pp. 776-786, Sep. 1992.

8. N. DeClaris and M. C. Su, "A novel class of neural networks with quadratic junctions, "IEEE Int. Conf. on systems, Man, and Cybernetics, PP. 1557-1562, 1991 (Received the 1992 Franklin V. Taylor Award).
9. N. DeClaris and M. C. Su, "Introduction to the theory and application of neural networks with quadratic junctions, "IEEE Int. Conf. on systems, Man, and Cybernetics, PP. 1320-1325, 1992.
10. T. Poggio and F. Girosi, "Networks for approximation and learning ," Proc. of the IEEE, vol.78, No.9, PP.1481-1497, 1990.
11. K. J. Cios and N. Liu, "A machine learning method for generation of a neural architecture : a continuous ID3 algorithm," IEEE Trans. on Neural Networks, vol. 3, pp. 280-291, March 1992.
12. K. Lang and M. J. Witbrock, "Learning to tell two spirals apart," Proc. Connectionist Models summer school, pp. 52-59, 1988.

Viable Control of Hybrid Systems[1]

Akash Deshpande and Pravin Varaiya
akash@eclair.eecs.berkeley.edu and varaiya@eclair.eecs.berkeley.edu
Department of Electrical Engineering and Computer Sciences
University of California at Berkeley
Berkeley, California 94720

Abstract. Hybrid systems are continuous variable, continuous time systems with a phased operation. Within each phase the system evolves continuously according to the dynamical law of that phase; when an event occurs, the system makes a transition from one phase to the next. In this paper, we study hybrid systems with nondeterministic discrete and continuous behaviors. We use nondeterministic finite automata to model the discrete behavior and state dependent differential inclusions to model the continuous behavior. By viability we mean the system's ability to take an infinite number of discrete transitions. Viability can be used to express safety and fairness properties over the system's state trajectories. To ensure viability, the system's evolution must be restricted so that discrete transitions occur within specific subsets of their enabling conditions. Following Aubin [3], we call these subsets the system's viability kernel. We give results pertaining to certain continuity properties of the viability kernel, we give conditions under which the viability kernel can be computed in a finite number of steps, and we synthesize a hybrid controller that yields all viable trajectories.

1 Introduction

Hybrid systems are continuous variable, continuous time systems with a phased operation. Within each phase the system evolves continuously according to the dynamical law of that phase; when an event occurs, the system makes a transition from one phase to the next. The state of the system is a pair—the discrete state and the continuous state. The discrete state identifies a flow, and the continuous state identifies a position in it. For a transition to take place, the discrete state and the continuous state together must satisfy a condition that enables the transition. Once a transition occurs, the discrete state and the continuous state are changed abruptly. Then, until the next transition, the continuous state evolves according to the flow identified by the new discrete state. Thus, the discrete dynamics and the continuous dynamics influence each other. In this paper, we study hybrid systems with nondeterministic discrete and continuous behaviors.

By viability we mean the system's ability to take an infinite number of transitions. Viability places a restriction on the system's evolution since at each time some discrete transition must remain unblocked as the system evolves. To meet this restriction, one must compute for each transition a subset of its enabling condition

[1] Research supported by NSF under grants ECS 9111907 and IRI 9120074, and by the PATH program, University of California, Berkeley.

where alone the transition is permitted. We call these restricted enabling conditions the viability kernel of the system.

Further, to ensure viability one must devise a control strategy that prohibits, based on the system's state, certain transitions from occuring, and that guides the continuous state into the viability kernel of some transition and then forces that transition to occur. We describe such a strategy as the 0-lag nonblocking control strategy. We synthesize a two-level hybrid controller that realizes this strategy: at the "higher" or supervisory level the controller makes discrete or planning decisions and sets a reachability problem to be solved by the "lower-level" controller, and at the "lower" or regulatory level the controller solves this reachability problem accounting for the nonlinear system dynamics. Our results are also described in Deshpande [4].

In section 2 we present the hybrid automaton model as a generalization of recent models proposed in literature. In section 3 we define the graph nonblocking operator and the viability kernel, and in Theorem 4 we give continuity conditions under which the viability kernel is a fixed point of the nonblocking operator. In section 4, we give conditions under which the viability kernel can be computed in a finite number of steps. The main computability result, given in Theorem 5, can be paraphrased as follows: if in each simple loop of the hybrid automaton there exists at least one discrete state in which the continuous variable dynamics are controllable (in a sense made precise in the result), then the viability kernel can be computed in a finite number of steps. In section 5 we define the 0-lag nonblocking control strategy. In section 6 we give the hybrid controller synthesis and the conditions under which it yields this strategy.

Our approach to viability in hybrid systems is a generalization of Aubin's [3] approach to viability in systems described by differential inclusions. In the latter setting, a system is said to be viable if its state remains in a specified subset, called the viability set, of the state space—i.e., the viability set must remain invariant under the system dynamics. For this to obtain, it is necessary to maintain the state within a subset of the viability set, called the viability kernel, so that an "inward" velocity is available at each point on the boundary of this kernel. Kohn, Nerode, *et. al.* [6, 9] propose a graph-theoretic approach for constructing the viability kernel in Aubin's setting.

Our approach is a generalization of Aubin's in the following sense. Aubin considers a single differential inclusion; we consider a family of inclusions and the system can switch between them. Viability, as described by Aubin, is a particular infinitary property which, in automata theory, can be termed a safety property of the system. Infinitary properties of automata comprise both safety and fairness; these are typically given over infinite traces of edge transitions, as we have done. Thus, in our setting, it is natural to describe viability in terms of the enabling conditions of the edges, and to seek the viability kernel as a subset of the edge enabling conditions.

In section 7 we give a process control application based on our theory. In this application, we consider a plant whose dynamics are given by nonlinear differential equations. The process requires cycling the plant repeatedly through three operating points. The process is robust so that it is sufficient to visit some specified neighborhoods of these operating points. A cost function is imposed on control trajectories and optimal state feedback control laws are synthesized using standard optimal control approaches. It should be noted that these optimal laws do not exploit the knowledge of process robustness. Further, in case of disturbances, these laws place a prohibitive computational load on the control computers. We synthesize viable control laws that incorporate the optimality considerations as well as process robustness information. Viable laws typically require additional computation only during a short portion of each cycle and always in a small region of the state space. In fact, for the displayed application, there is no additional computational load. Thus, the viable control theory described by us is an approach to synthesize controllers that yield close to optimal behavior in a robust manner.

2 The Hybrid Automaton Model

The hybrid automaton model given below is a generalization of the models recently introduced in literature (see Alur *et. al.* [1, 2], and Puri *et. al.* [10]).

A hybrid transition system is a tuple

$$H = (Q, \mathbf{R}^n, \Sigma, E, \Phi).$$

Q is the finite set of discrete states and $\mathbf{R}^n$ is the set of continuous states. Σ is the finite set of discrete events.

$$E \subset Q \times \mathcal{P}(\mathbf{R}^n) \times \Sigma \times \{\mathbf{R}^n \to \mathbf{R}^n\} \times Q$$

is the finite set of edges.[2] The edges model the discrete event dynamics of the system. An edge $e \in E$ is denoted as

$$(q_e, X_e, V_e, r_e, q'_e),$$

and is enabled when the discrete state is q_e and the continuous state is in X_e. When a transition through e is taken, the event $V_e \in \Sigma$ is accepted by the system. The continuous state is then reset according to the map r_e, and the system enters the discrete state q'_e.

$$\Phi = \{F_q : \mathbf{R}^n \to \mathcal{P}(\mathbf{R}^n)\backslash\emptyset \mid q \in Q\}$$

is a set of differential inclusions that model the continuous dynamics of the system. When the discrete state is q, the continuous state evolves according to the differential inclusion[3]

$$\dot{x}_c(t) \in F_q(x_c(t)).$$

When the continuous state evolution is given as a controlled vector field $f_q : \mathbf{R}^n \times \mathbf{R}^m \to \mathbf{R}^n$ with

$$\dot{x}_c(t) = f_q(x_c(t), u(t)),$$

we can write

$$F_q(\cdot) = \{f_q(\cdot, w) \mid w \in \mathbf{R}^m\}.$$

Sometimes the system description is extended by giving set-valued reset maps on the edges and state invariance conditions in the discrete states. See sections 4.2 and 4.3 respectively for these extensions.

[2] Throughout, $\mathcal{P}(\cdot)$ denotes the power set (or, the set of all subsets) of $(\cdot)$.

[3] Guckenheimer and Johnson [5] analyze deterministic switched systems; their concerns are related to the study of singularity properties of planar hybrid systems.

The hybrid transition system executes in a sequence of phases. In each phase it first evolves continuously by allowing time to pass, and then it evolves discretely by making an instantaneous transition. Let

$$\tau = [\tau_0', \tau_1], [\tau_1', \tau_2], [\tau_2', \tau_3], \ldots, \; \tau_0' = 0 \text{ and } \forall i \; (\tau_i = \tau_i' \text{ and } \tau_{i+1} \geq \tau_i')$$

be a sequence of intervals of $\mathbf{R}_+$.[4] The sequence τ may be finite or infinite, and it may cover $\mathbf{R}_+$ or only an initial segment of it. Let T be the set of all such interval sequences. For notational convenience, we refer to $\cup_i[\tau_i', \tau_{i+1}]$ also as τ.

For each i, H evolves continuously in the interval (τ_i', τ_{i+1}), and makes an instantaneous transition at τ_{i+1}. If $\tau_i' = \tau_{i+1}$ then it makes a transition at τ_{i+1} without any continuous evolution in that interval; in this case if $i > 0$, there are two transitions at $\tau_i = \tau_{i+1}$, and the order in which they occur is given by the index.

Formally, the semantics of a hybrid transition system are given over traces $s = (x_d, x_c, v)$ where x_d is the discrete state trace, x_c is the continuous state trace, and v, is the edge transition trace. Figure 1 depicts a system trace. Let the set of all such

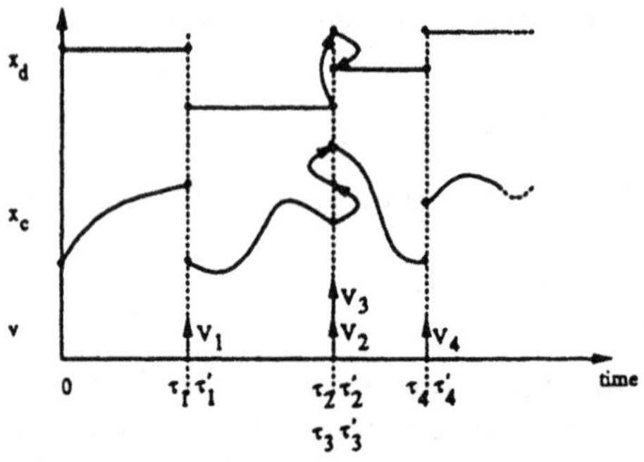

Fig. 1. A system trace $s = (x_d, x_c, v)$.

traces be

$$S = \{(\tau \to Q, \tau \to \mathbf{R}^n, \tau \to E \cup \{null\}) \mid \tau \in T\},$$

where *null* denotes the absence of a transition. $s \in S$ given over τ is a *run* of H if the following two conditions hold.

1. Continuous evolution.
   For each i
   (a) $\forall t \in (\tau_i', \tau_{i+1}) \; (v(t) = null)$,
   (b) $x_d(\cdot)$ is constant over the interval $[\tau_i', \tau_{i+1}]$, and
   (c) $\forall t \in (\tau_i', \tau_{i+1}) \; (\dot{x}_c(t) \in F_{x_d(t)}(x_c(t)))$.
2. Discrete evolution.
   At each boundary point $\tau_i = \tau_i'$
   (a) $v(\tau_i)$ is an edge, say e, in E,

[4] Throughout, we use parentheses (or brackets) to disambiguate the scope of quantified variables; thus, $\forall x \; (P)$ or $\exists x \; (P)$, where x is a variable and P is a predicate.

(b) $x_d(\tau_i) = q_e$ and $x_c(\tau_i) \in X_e$, i.e., the edge e is enabled at τ_i, and
(c) $x_d(\tau_i') = q_e'$ and $x_c(\tau_i') = r_e(x_c(\tau_i))$, i.e., the state is reset instantaneously.
Note that x_d and x_c are multivalued at the boundary points $\tau_i = \tau_i'$. v may also be multivalued if $\tau_i' = \tau_{i+1}$ for some $i > 0$.

The set of ω-runs of H (i.e., runs that have infinitely many transitions) is

$$S_H = \{s \text{ run of } H \mid s \text{ is defined over an infinite sequence } \tau \in T\}.$$

The hybrid transition system can be augmented to form a hybrid automaton when initial states, final states, and an accepting condition are specified.

3 The Viability Kernel

For $F_q \in \Phi$, let the state evolution map be $\phi_q : \mathbf{R}_+ \times \mathbf{R}^n \to \mathcal{P}(\mathbf{R}^n)$, i.e., $\phi_q(t, x_0)$ is the set of continuous states that can be reached from x_0 at time t under the inclusion

$$\dot{x}_c(t) \in F_q(x_c(t)), \; x_c(0) = x_0.$$

Define the forward projection of x_0 under ϕ_q as

$$\phi_q(x_0) = \cup_{t \geq 0} \phi_q(t, x_0).$$

$\phi_q(x_0)$ is the set of continuous states that can be reached from x_0 under the dynamical law of q. Define the backward projection of $x \in \mathbf{R}^n$ under ϕ_q as

$$\beta_q(x) = \{x_0 \mid x \in \phi_q(x_0)\},$$

and for $X \subset \mathbf{R}^n$ define

$$\beta_q(X) = \cup_{x \in X} \beta_q(x).$$

$\beta_q(X)$ is the set of continuous states that can reach X under the dynamical law of q.

For $e \in E$ define the nonblocking switch-and-coast operator $n_e : \mathcal{P}(\mathbf{R}^n) \to \mathcal{P}(\mathbf{R}^n)$ by

$$n_e(Y) = X_e \cap r_e^{-1}(\beta_{q_e'}(Y)).$$

$n_e(Y)$ is the set of continuous states from which a transition through e can be taken immediately (given the discrete state is q_e) and then Y can be reached after the passage of some time in the discrete state q_e'. Evidently, n_e is monotone, i.e.,

$$Y_1 \subset Y_2 \Rightarrow n_e(Y_1) \subset n_e(Y_2).$$

Lemma 1. *n_e distributes over $\cup$, i.e., $n_e(Y_1 \cup Y_2) = n_e(Y_1) \cup n_e(Y_2)$.*

Proof.

$$\begin{aligned}
\beta_{q'_e}(Y_1 \cup Y_2) &= \cup_{y \in Y_1 \cup Y_2}\{x \mid \exists t \geq 0\ (y \in \phi_{q'_e}(t,x))\} \\
&= [\cup_{y \in Y_1}\{x \mid \exists t \geq 0\ (y \in \phi_{q'_e}(t,x))\}] \cup [\cup_{y \in Y_2}\{x \mid \exists t \geq 0\ (y \in \phi_{q'_e}(t,x))\}] \\
&= \beta_{q'_e}(Y_1) \cup \beta_{q'_e}(Y_2). \\
r_e^{-1}(Y_1 \cup Y_2) &= \{x \mid r_e(x) \in Y_1 \cup Y_2\} \\
&= \{x \mid r_e(x) \in Y_1\} \cup \{x \mid r_e(x) \in Y_2\} \\
&= r_e^{-1}(Y_1) \cup r_e^{-1}(Y_2). \\
n_e(Y_1 \cup Y_2) &= X_e \cap r_e^{-1}(\beta_{q'_e}(Y_1 \cup Y_2)) \\
&= X_e \cap r_e^{-1}(\beta_{q'_e}(Y_1) \cup \beta_{q'_e}(Y_2)) \\
&= X_e \cap [r_e^{-1}(\beta_{q'_e}(Y_1)) \cup r_e^{-1}(\beta_{q'_e}(Y_2))] \\
&= X_e \cap r_e^{-1}(\beta_{q'_e}(Y_1)) \cup X_e \cap r_e^{-1}(\beta_{q'_e}(Y_2)) \\
&= n_e(Y_1) \cup n_e(Y_2). \square
\end{aligned}$$

For $q \in Q$ define $out(q)$ as the edges (including self-loops) that exit q,

$$out(q) = \{e \in E \mid q_e = q\}.$$

Let E be ordered as $\{e_1, \ldots, e_{|E|}\}$ and let

$$X_H = (X_{e_1}, \ldots, X_{e_{|E|}})^T \in [\mathcal{P}(\mathbf{R}^n)]^{|E|}$$

be the vector of edge enabling conditions.

Define $N : [\mathcal{P}(\mathbf{R}^n)]^{|E|} \to [\mathcal{P}(\mathbf{R}^n)]^{|E|}$ as

$$[N(Z)]_i = n_{e_i}(\cup_{j | e_j \in out(q'_{e_i})} Z_j).$$

$[N(Z)]_i$ is the set of continuous states from which a transition through e_i can be taken immediately (given the discrete state is q_{e_i}) and then some Z_j can be reached after the passage of some time in q'_{e_i}, where e_j is an edge that exits q'_{e_i}. From Lemma 1 we obtain

$$[N(Z)]_i = \cup_{j | e_j \in out(q'_{e_i})} n_{e_i}(Z_j).$$

Definition 2. An edge sequence $e_1, \ldots, e_l$ is said to be a *path* if $\forall i = 1, \ldots, l-1\ (q'_{e_i} = q_{e_{i+1}})$. It is said to be a *loop* if it is a path and $q_{e_1} = q'_{e_l}$. It is a *simple loop* if it is a loop and $\forall i, j = 1, \ldots, l\ (i \neq j \Rightarrow q_{e_i} \neq q_{e_j})$.

The following example illustrates repeated application of N to X_H.

Example 1. Consider the hybrid system shown in Figure 2. The paths of length 4 beginning with edge e_2 are $e_2e_3e_4e_1$ and $e_2e_3e_4e_2$. For one of these, say $e_2e_3e_4e_1$, to be taken with possibility of extending to a fifth transition, say e_4, the continuous state must be in $n_2n_3n_4n_1(X_4)$ when e_2 is first taken. We call $n_2n_3n_4n_1(\cdot)$ as the *path nonblocking operator* for the path $e_2e_3e_4e_1$. It can be verified that

$$[N^4(X_H)]_2 = n_2n_3n_4n_1(X_4) \cup n_2n_3n_4n_2(X_3).$$

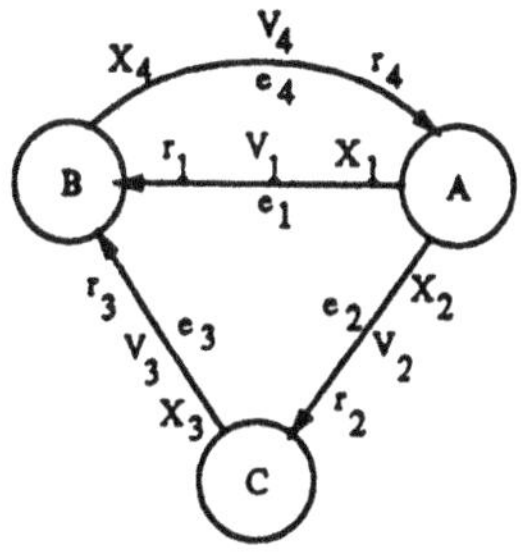

Fig. 2. A sample hybrid system.

In general, for k transitions to be taken beginning with edge e_i with possibility of extending to some $(k+1)$st transition, the continuous state must be in $[N^k(X_H)]_i$ when e_i is first taken. From Lemma 1, $[N^k(X_H)]_i$ can be written as the union of the path nonblocking operators over all paths of length k beginning with e_i and extending to some (k+1)st edge. □

Note that $N^{k+1}(X_H) \subset N^k(X_H)$ for all k.[5]

Definition 3. Define the viability kernel of H as $X^* = \cap_k N^k(X_H)$.

3.1 Continuity Results

For an edge e, X_e^* can be thought of as follows. For any $x \in X_e^*$ and any k, there is a run $s = (x_d, x_c, v)$ of H with at least k transitions, and with $v(0) = e$ and $x_c(0) = x$.

However, there may not exist any *infinite* run starting at $v(0) = e$ and $x_c(0) = x$. If $N(X^*) = X^*$ then an infinite run starting at $v(0) = e$ and $x_c(0) = x$ does exist for any $x \in X_e^*$ and any e. The following theorem gives conditions under which $N(X^*) = X^*$. For $q \in Q$, analogous to $out(q)$, define $in(q) = \{e \mid q'_e = q\}$. Further, define

$$X_q^{out} = \cup_{e \in out(q)} X_e \quad \text{and} \quad X_q^{in} = \cup_{e \in in(q)} r_e(X_e).$$

Theorem 4. *Assume that*

1. *for each $e \in E$, X_e is compact and r_e is continuous, and*
2. *for each $q \in Q$ the set $\{(x, x') \mid x \in X_q^{in}, x' \in X_q^{out} \cap \phi_q(x)\}$ is closed.*

Then $N(X^*) = X^*$.

Proof. Condition (2) along with the compactness of X_q^{out} and X_q^{in} implies that for a closed set $Y \subset X_q^{out}$,

$$\{x \in X_q^{in} \mid Y \cap \phi_q(x) \neq \emptyset\} = X_q^{in} \cap \beta_q(Y) \text{ is closed.}$$

[5] For notational convenience, for $Z, Z' \in [\mathcal{P}(\mathbb{R}^n)]^{|E|}$, let $Z \subset Z'$ denote $\forall i\ (Z_i \subset Z'_i)$ and similarly for $\cup$, $\cap$, etc. Further, for an edge e_i, we use the notations Z_i and Z_{e_i} interchangeably.

Since r_e is continuous for each $e \in in(q)$, $r_e^{-1}(X_q^{in} \cap \beta_q(Y))$ is also closed.

Consider any edge e. Denote $[N^k(X_H)]_e$ by X_e^k. By definition,

$$X_e^{k+1} = \cup_{e' \in out(q'_e)} X_e \cap r_e^{-1}(\beta_{q'_e}(X_{e'}^k))$$
$$= \cup_{e' \in out(q'_e)} X_e \cap r_e^{-1}(X_{q'_e}^{in} \cap \beta_{q'_e}(X_{e'}^k)) \text{ since } X_e \subset r_e^{-1}(X_{q'_e}^{in}).$$

Since each X_e is compact, each X_e^k is also compact and hence X_e^* is compact.

$x \in X_e^* \Rightarrow \forall k\ [\phi_{q'_e}(r_e(x)) \cap (\cup_{e' \in out(q'_e)} X_{e'}^k) \neq \emptyset]$

$\Rightarrow \phi_{q'_e}(r_e(x)) \cap (\cup_{e' \in out(q'_e)} X_{e'}^*) \neq \emptyset$

since $X_{q'_e}^{out} \cap \phi_{q'_e}(r_e(x))$ is closed, $X_{q'_e}^{out} \supset \cup_{e' \in out(q'_e)} X_{e'}^*$, and each $X_{e'}^k$ is compact

$\Rightarrow n_e(\cup_{e' \in out(q'_e)} X_{e'}^*) \supset X_e^*$.

Combining, we obtain $N(X^*) \supset X^*$ and hence $N(X^*) = X^*$. □

Now we give conditions on F_q so that condition (2) of Theorem 4 is obtained. We consider the case where for each q $F_q(x) = \{f_q(x, w) \mid w \in \mathbf{R}^m\}$ for some $f_q : \mathbf{R}^n \times \mathbf{R}^m \to \mathbf{R}^n$. Let $\phi_q^T(x) = \cup_{t \leq T} \phi_q(t, x)$. Suppose

1. $F_q(x)$ is convex and closed for all x and F_q is upper semicontinuous,
2. $\exists K_1, K_2\ \forall x\ \forall u\ (|f(x,u)| \leq K_1 + K_2|x|)$,
3. $\exists T\ \forall x \in X_q^{in}\ \forall t > T\ (X_q^{out} \cap \phi_q(t, x) = \emptyset)$, and

Then from Theorem 2.1 of Varaiya [11] we deduce that $\phi_q^T(x)$ is closed.

Further, consider a sequence $\{(x_k, y_k)\}$, $x_k \in X_q^{in}$, $y_k \in X_q^{out} \cap \phi_q(x_k)$, with $x_k \to x^*(\in X_q^{in})$, $y_k \to y^*$. From condition (3), each $y_k \in X_q^{out} \cap \phi_q^T(x_k)$. Then, again by Theorem 2.1 of Varaiya [11], there is a subsequence of trajectories that converges to a trajectory from x^* to y^*. Hence $y^* \in X_q^{out} \cap \phi_q^T(x^*) \subset X_q^{out} \cap \phi_q(x^*)$. Thus the graph of $X_q^{out} \cap \phi_q(\cdot)$ restricted to X_q^{in} is closed as required.

4 Computability Results

In this section we give conditions on H so that

$$\exists K\ (N^K(X_H) = X^*).$$

Suppose that there exists K such that $N^{K+1}(X_H) = N^K(X_H)$. Then

$$N^{K+2}(X_H) = N(N^{K+1}(X_H)) = N^K(X_H),$$

and thus $N^K(X_H) = X^*$.

Example 2 illustrates a hybrid system for which $N^k(X_H) \neq X^*$ for any k.

Example 2. Consider a hybrid system consisting of a single discrete state q and a single self-loop edge e. This is shown in Figure 3. The continuous state space is $\mathbf{R}$. The dynamics in q are given as $\dot{x} = 0$. $X_e = [-1, 1]$ and $r_e(x) = 2x$. Then, $N^k(X_H) = [-\frac{1}{2^k}, \frac{1}{2^k}]$ and $X^* = \{0\}$. Thus, $\forall k\ (N^k(X_H) \neq X^*)$. □

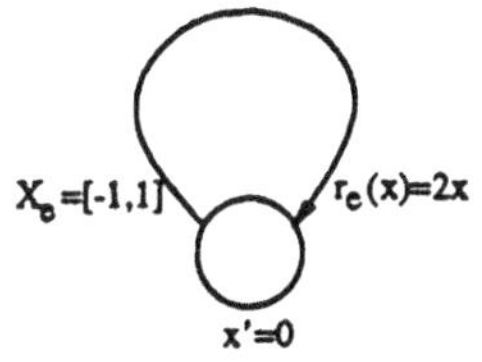

Fig. 3. $N^k(X_H) \neq X^*$ for any k.

Now we proceed to give conditions on H so that there is a K such that $N^K(X_H) = X^*$. This is done in Theorem 5 which is proved using Lemmas 7 and 8. Then Theorem 9 gives an extension of Theorem 5, and finally we give extensions to set-valued reset maps and state invariance conditions.

Theorem 5. *Suppose each simple loop $e_1, \ldots, e_l$ of H has an edge e_1 such that*

1. *$r_{e_1}(X_{e_1})$ is bounded, and*
2. *$\exists \rho_{e_1} > 0\ \forall x \in X_{e_2}\ \forall y \in \beta_{q'_{e_1}}(x)\ [\beta_{q'_{e_1}}(x) \cap \phi_{q'_{e_1}}(y)$ is a union of ρ_{e_1}-balls].*

Then $\exists K\ (N^K(X_H) = X^*)$.

Condition (1) states that the image of the enabling set for e_1 under its reset map is bounded. Condition (2) states that for any continuous state x that enables e_2, and any y that can reach x (under the dynamical law in q'_{e_1}), the set of continuous states that can both reach x and be reached from y is a union of ρ_{e_1}-balls. This is a controllability requirement on H.

For example, condition (2) is satisfied if each simple loop has a discrete state in which the continuous variable dynamics are given by a completely controllable linear system.

Definition 6. For a sequence of edges $p_k = e_1, \ldots, e_k$, define $n_{p_k} = n_{e_1} \circ \cdots \circ n_{e_k}$.

Lemma 7. *Let $p = e_1, e_2, e_3, \ldots$ be an infinite path and let $p_k = e_1, \ldots, e_k$ be its prefix of length k. If*

$$\forall p\ \exists M\ \forall k \geq M\ [n_{p_k}(X_{e_{k+1}}) = n_{p_{k+1}}(X_{e_{k+2}})] \tag{1}$$

then $\exists K\ (N^K(X_H) = X^*)$.

Proof. Let $ext(p_k)$ denote all infinite paths that have the prefix p_k. Consider the following prefix property of p_k:

$$\forall j \geq k\ [n_{p_{k-1}}(X_{e_k}) = \cup_{p' \in ext(p_k)} n_{p'_{j-1}}(X_{e'_j})]. \tag{2}$$

Then from (1) we can deduce that for every path p with M as given in (1), p_{M+1} satisfies (2). This is because for $k = 1, 2, \ldots$

$$n_{e_{M+1}} n_{e_{M+2}} \cdots n_{e_{M+k}}(X_{e_{M+k+1}}) \subset \cup_{p' \in ext(e_{M+1})} n_{p'_k}(X_{e'_{k+1}}) \subset X_{e_{M+1}},$$

and applying n_{p_M} to each term, we obtain

$$\forall j \geq M+1\ [\cup_{p' \in ext(p_{M+1})} n_{p'_{j-1}}(X_{e'_j}) = n_{p_M}(X_{e_{M+1}})].$$

Let $pre(k)$, defined inductively with $pre(1) \subset E$ and $pre(k+1) \subset pre(k)E$, be the set of all prefixes of length k for which (2) does not hold. Then

$$\forall k\ [ext(pre(k+1)) \subset ext(pre(k))].$$

From (1) we can deduce that

$$\cap_k ext(pre(k)) = \emptyset \tag{3}$$

for if $p \in \cap_k ext(pre(k))$ then (2) does not hold for every prefix of p, and hence (1) does not hold for it either.

Then we claim that $\cap_{k=0}^{K} ext(pre(k)) = \emptyset$ since a tree with bounded branching and with no infinite path in it must be finite.[6] □

Consider the following example which shows a system for which $\exists K\ (N^K(X_H) = X^*)$ and there exists an infinite path for which $\forall k\ [n_{p_k}(X_{e_{k+1}}) \neq n_{p_{k+1}}(X_{e_{k+2}})]$. Thus, the condition of Lemma 7 is sufficient but not necessary.

Example 3. Consider the hybrid system H shown in Figure 4. Then

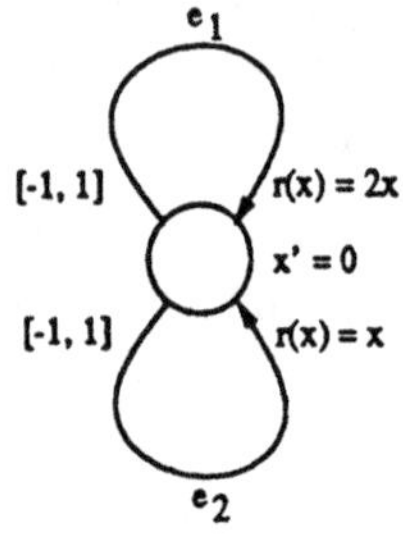

Fig. 4. A sample hybrid system.

$$X_H = ([-1,1],[-1,1])^T,$$
$$N(X_H) = ([-\frac{1}{2},\frac{1}{2}],[-1,1])^T, \text{ and}$$
$$X^* = N(X_H).$$

[6] To see this, endow the space $E \times E \times \cdots$ with the topology of pointwise convergence with the discrete topology on E. By Tychonoff's Theorem [7], this space is compact since E is finite. In this topology, the set of all infinite paths with a common finite prefix is a closed set, and finite unions of such sets are also closed. Evidently, then, each $ext(pre(k))$ is compact.

Suppose that $\forall K\ [\cap_{k=0}^{K} ext(pre(k)) \neq \emptyset]$. Then, by the finite intersection property applied to compact sets, $\cap_{k=0}^{\infty} ext(pre(k)) \neq \emptyset$ [8], yielding a contradiction.

Now consider the infinite path $p = e_1, e_1, e_1, \ldots$ Then

$$n_{p_k}(X_{e_1}) = [-\frac{1}{2^k}, \frac{1}{2^k}].$$

□

Lemma 8. *If for each simple loop $e_1, \ldots, e_l$ there is an edge, say e_1, such that for all sequences*

$$X_{e_2} = Y_0 \supset Y_1 \supset Y_2 \ldots$$

there exists M such that $\forall k \geq M$ $(n_{e_1}(Y_k) = n_{e_1}(Y_{k+1}))$ then $\exists K$ $(N^K(X_H) = X^*)$.

Proof. Let $p = e_{I_1}, e_{I_2}, e_{I_3}, \ldots$ be an infinite path. Then some simple loop $e_1, \ldots, e_l$ must occur infinitely often in it. Consider the sequence of prefixes of p

$$p_{l_0} = e_{I_1}, e_{I_2}, \ldots, e_1$$
$$p_{l_1} = e_{I_1}, e_{I_2}, \ldots, e_1, \ldots, e_l \ldots\ldots e_1$$
$$p_{l_2} = e_{I_1}, e_{I_2}, \ldots, e_1, \ldots, e_l \ldots\ldots e_1, \ldots, e_l, \ldots\ldots, e_1$$

i.e., p_{l_k} stops after k appearances of e_1 participating in the loop $e_1, \ldots, e_l$. Let p_{l_0}/p_{l_k} denote the path obtained by removing the prefix p_{l_0} from p_{l_k}. Let

$$Y_k = n_{p_{l_0}/p_{l_k}}(X_{e_2}).$$

Then, using the given property, every p satisfies the condition of Lemma 7, and hence the result follows. □

Proof. (of Theorem 5) Let $X_{e_2} = Y_0 \supset Y_1 \supset Y_2 \supset \ldots$ be a sequence of sets nesting inside X_{e_2}, and consider Y_j and Y_{j+1}. If $\beta_{q'_{e_1}}(Y_{j+1}) = \beta_{q'_{e_1}}(Y_j)$ then $n_{e_1}(Y_j) = n_{e_1}(Y_{j+1})$. Now suppose

$$\beta_{q'_{e_1}}(Y_{j+1}) \neq \beta_{q'_{e_1}}(Y_j).$$

Let

$$x \in Y_j \backslash \beta_{q'_{e_1}}(Y_{j+1}),\ y \in \beta_{q'_{e_1}}(x) \backslash \beta_{q'_{e_1}}(Y_{j+1}).$$

Note that both exist. Let

$$Z_j(x, y) = \beta_{q'_{e_1}}(x) \cap \phi_{q'_{e_1}}(y).$$

Then $Z_j(x, y) \subset \beta_{q'_{e_1}}(Y_j) \backslash \beta_{q'_{e_1}}(Y_{j+1})$ and $Z_j(x, y)$ is a union of ρ_{e_1}-balls. This is shown in Figure 5. Let

$$Z_j = \cup_x \cup_y Z_j(x, y),$$

with x and y as given above. Note that each Z_j is also a union of ρ_{e_1}-balls, and that $Z_0, Z_1, \ldots$ is a disjoint family. Since $r_{e_1}(X_{e_1})$ is bounded, only a finite number of sets from $Z_0, Z_1, \ldots$ can have a nonempty intersection with $r_{e_1}(X_{e_1})$. Hence, for the sequence $Y_0, Y_1, \ldots$ there exists M such that $\forall k \geq M$ $[n_{e_1}(Y_k) = n_{e_1}(Y_{k+1})]$. The theorem follows from Lemma 8. □

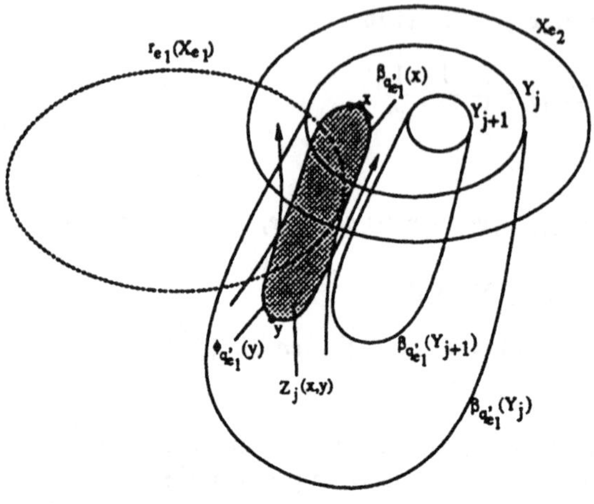

Fig. 5. $Z_j(x, y)$ is a union of ρ_{e_1}-balls.

4.1 Discretizing Reset Maps

Theorem 5 dealt with the controllability of the continuous dynamics of the hybrid system. We can extend the result to edges with "discretizing" reset maps, as formalized below. (For example, the map $r : \mathbf{R} \to \mathbf{R}$ which assigns the integer k to all points in the interval $[k - \frac{1}{2}, k + \frac{1}{2})$ is a discretizing map.)

Theorem 9. *Suppose each simple loop $e_1, \ldots, e_l$ of H has an edge e_1 such that either e_1 satisfies the conditions of Theorem 5 or*

1. *X_{e_1} is bounded, and*
2. *$\exists \rho_{e_1} > 0\ \forall y \in \beta_{q'_{e_1}}(X_{e_2})\ (r_{e_1}^{-1}(y) = \emptyset$ or $r_{e_1}^{-1}(y)$ is a union of ρ_{e_1}-balls).*

Then $\exists K\ (N^K(X_H) = X^*)$.

Condition (1) is a bound on the enabling set for e_1. Condition (2) states that for any continuous state y that can reach a continuous state that enables e_2 (under the dynamical law of q'_{e_1}), either no continuous state is reset to y after a transition through e_1 or the set of continuous states so reset is a union of ρ_{e_1}-balls.

Proof. (of Theorem 9) As before, for a simple loop $e_1, \ldots, e_l$ with e_1 as given consider the sequence $X_{e_2} = Y_0 \supset Y_1 \supset Y_2 \ldots$ If e_1 satisfies the conditions of Theorem 5, we follow the corresponding proof. Otherwise, consider Y_j and Y_{j+1} and suppose $\beta_{q'_{e_1}}(Y_{j+1}) \neq \beta_{q'_{e_1}}(Y_j)$. Let

$$Z_j = r_{e_1}^{-1}[\beta_{q'_{e_1}}(Y_j) \backslash \beta_{q'_{e_1}}(Y_{j+1})].$$

Then either $Z_j = \emptyset$ or Z_j is a union of ρ_{e_1}-balls.

Then we use the same argument as for the proof of Theorem 5 to claim the result.
□

4.2 Set-valued Reset Maps

Sometimes the reset map r_e for $e \in E$ may need to be set-valued, denoted R_e, i.e., the continuous state is reset nondeterministically when a transition through e is taken. If the controller can choose the value to which the continuous state is reset after a transition through e, then the reset is said to be controlled, and we define

$$R_e^{-1}(Y) = \{x \mid R_e(x) \cap Y \neq \emptyset\}.$$

If, on the other hand, the choice of the reset value is not available to the controller we define

$$R_e^{-1}(Y) = \{x \mid R_e(x) \subset Y\}.$$

Define

$$n_e(\cdot) = X_e \cap R_e^{-1}(\beta_{q'_e}(\cdot)).$$

Note that for uncontrolled set-valued reset maps R_e^{-1} does not distribute over $\cup$. Then we have the following extension.

Theorem 10. *If H satisfies the conditions of Theorem 9 and for all edges e with an uncontrolled set-valued reset map $out(q'_e)$ is a singleton, then $\exists\ (N^K(X_H) = X^*)$.*

4.3 State Invariance Conditions

Sometimes it is useful to associate to $q \in Q$ a set $G_q \subset \mathbf{R}^n$ and require that when the system is in phase q, the continuous state must be in G_q, i.e.,

$$\forall s \text{ run of } H\ \forall t\ (x_d(t) = q \Rightarrow x_c(t) \in G_q).$$

Then for each edge $e \in E$ we replace X_e by $X_e \cap G_{q_e}$ as the enabling set, and define for $q \in Q$

$$\beta_q(Y) = \{x \mid \exists t \geq 0\ [\phi_q(t,x) \in Y \text{ and } \forall t' \leq t\ (\phi_q(t',x) \in G_q)]\}.$$

Noting that β_q given above distributes over $\cup$, Theorem 9 continues to hold.

5 Control

In this section we describe the information structure for hybrid control and define the 0-lag nonblocking control strategy. Henceforth we assume that the differential inclusions arise from the control variables, i.e., for each discrete state q,

$$F_q(\cdot) = \{f_q(\cdot, w) \mid w \in \mathbf{R}^m\}$$

for some $f_q : \mathbf{R}^n \times \mathbf{R}^m \to \mathbf{R}^n$.

5.1 The Information Structure

Now we specify what the controller can observe and what it can control. E is the union of E_p, E_f and E_u where $E_p \cap E_u = E_f \cap E_u = \emptyset$. Edges from E_p are prohibitable and edges from E_f are forcible, while edges from E_u are uncontrollable. Let

$$Y = Q \times \mathbf{R}^n$$

be the set of possible observations, and let $y : \tau \to Y$ be an observation trace. Let S_Y be the set of all observation traces. Let

$$U = \mathcal{P}(E_p) \times \mathcal{P}(E_f) \times \mathbf{R}^m$$

be the set of possible controls, and let $u = (u_p, u_f, u_c) : \tau \to U$, with u_p and u_f piecewise constant and u_c measurable, be a control trace. Let S_U be the set of all control traces.

Let a special forcible edge called *block* which is always enabled be available to the controller. Let

$$U^B = \mathcal{P}(E_p) \times [\mathcal{P}(E_f) \cup \{block\}] \times \mathbf{R}^m$$

be the set of extended controls and let S_{U^B} be the set of all extended control traces such that

$$\forall u \in S_{U^B}\ \forall t\ [u_f(t) = \{block\} \Rightarrow \forall t' > t\ (u_f(t') = \{block\})].$$

Let $S^B \supset S$ be the set of all extended traces (x_d, x_c, v) where v is allowed to have *block* transitions in addition to E and *null* transitions.

Now we give conditions so that the observation trace is consistent with the system trace, and the system trace is consistent with the control trace. Given an observation trace $y = (x_d, x_c) \in S_Y$ the set of extended system traces that can generate y is given by the inverse projection

$$G_y = \{(x_d, x_c, v) \in S^B\}$$

Given a control trace $u = (u_p, u_f, u_c) \in S_{U^B}$ the set of extended system traces that can be produced in response to u is given by

$$\begin{aligned} G_u = \{(x_d, x_c, v) \in S^B \mid \forall t\ [&v(t) \notin u_p(t), \\ &\exists e \in u_f(t)\ (x_d(t) = q_e \text{ and } x_c(t) \in X_e) \Rightarrow \exists e \in u_f(t)\ (v(t) = e) \text{ and} \\ &v(t) \neq block \Rightarrow\ (\dot{x}_c(t) = f_{x_d(t)}(x_c(t), u_c(t))\ )]\}. \end{aligned}$$

This means that the system cannot take prohibited transitions, it must take some forced transition if any forced transition is enabled, and if it is not blocked, it must follow the continuous state dynamical law.

For a trace s let s_t denote the prefix of s up to t and for a prefix s_t let $ext(s_t) = \{s' \mid s'_t = s_t\}$ denote the set of traces that extend s_t.

5.2 Control Strategy

The control strategy formulation given below is patterned after Varaiya and Lin [12]. A control strategy is a point-to-set map

$$\kappa : S_Y \to \mathcal{P}(S_{U^B})\backslash\emptyset.$$

The extended closed loop behavior of κ is

$$G_\kappa = \{G_u \cap G_y \mid u \in \kappa(y)\},$$

and the closed-loop behavior of κ is

$$S_\kappa = G_\kappa \cap S.$$

Definition 11. Fix $\alpha > 0$ and a hybrid system H. A point-to-set map $\kappa_H^\alpha : S_Y \to \mathcal{P}(S_{U^B})\backslash\emptyset$ is an *α-lag possibly blocking control strategy* for H iff

1. $S_{\kappa_H^\alpha} \subset S_H$,
2. $\forall y, y' \in S_Y\ \forall t\ (y_{(t+\alpha)} = y'_{(t+\alpha)} \Rightarrow \{u_t \mid u \in \kappa_H^\alpha(y)\} = \{u_t \mid u \in \kappa_H^\alpha(y')\})$, and
3. $\forall y\ \forall u\ \forall t\ [\forall y' \in ext(y_{t+\alpha})\ \forall u' \in ext(u_t)$

$$(G_{u'} \cap G_{y'} \cap S_H = \emptyset) \Rightarrow \forall u'' \in ext(u_t) \cap \kappa_H^\alpha(y)\ \forall r > t\ (u''_f(r) = \{block\})].$$

Condition (1) means that the traces generated by κ_H^α must be permitted by H. Condition (2) means that the choice of control at time t can depend on the observations of the plant over $[0, t+\alpha]$, i.e., the controller lags the plant by time α. Condition (3) means that if no future evolution after t is viable in H, the strategy blocks at t.

If for an α-lag possibly blocking control strategy κ, $G_\kappa \subset S$, then we call κ an *α-lag nonblocking control strategy* for H, denoted as $\tilde{\kappa}_H^\alpha$. For some index set I, if $\tilde{\kappa}_i$, $i \in I$ is a family of α-lag nonblocking strategies for H then $\cup_{i \in I}\tilde{\kappa}_i$ is also an α-lag nonblocking strategy. Hence there is a largest (in the sense of set inclusion) α-lag nonblocking strategy, $\tilde{U}_H^\alpha$. However, a nonblocking strategy may not exist for H, or if it exists, it may be only a partial map. Define the 0-lag nonblocking control strategy as

$$\tilde{U}_H = \cap_{\alpha > 0}\tilde{U}_H^\alpha.$$

$\tilde{U}_H$ assumes observations of the plant behavior in an infinitesimal future.[7]

6 Hybrid Controller Synthesis

Now we give a hybrid controller for H that achieves $\tilde{U}_H$. First we introduce some notation.

For $Y \subset \mathbf{R}^n$, let $K_Y : \mathbf{R}^n \to \mathcal{P}(\mathbf{R}^n)$ be the contingent cone (see Aubin [3]) of Y, i.e.,

$$K_Y(y) = \{h \in \mathbf{R}^n \mid \lim_{\delta \to 0+} y + \delta h \in Y\}.$$

For a controlled vector field $f_q : \mathbf{R}^n \times \mathbf{R}^m \to \mathbf{R}^n$ and a measurable control law $u : \mathbf{R}_+ \to \mathbf{R}^m$, let the state evolution map be denoted as $\phi_q^u : \mathbf{R}_+ \times \mathbf{R}^n \to \mathbf{R}^n$

[7] 0-lead nonblocking strategies can be defined analogously (see [4]).

where $x(t) = \phi_q^u(t, x_0)$ is the state reached at time t starting from $x(0) = x_0$ under the control trace u.

The hybrid controller is given as a (partial) map

$$\gamma_H = (\gamma_p, \gamma_f, \gamma_c) : Q \times \mathbf{R}^n \to \mathcal{P}(E_p) \times \mathcal{P}(E_f) \times [\mathcal{P}(\{\mathbf{R}_+ \to \mathbf{R}^m\})\backslash\emptyset]$$

where

$$\gamma_p(q,x) = \{e \in E_p \mid e \in out(q) \text{ and } x \notin X_e^\bullet\},$$
$$\gamma_f(q,x) = \{e \in E_f \mid e \in out(q),\ x \in X_e^\bullet, \text{ and } \forall e' \in out(q)\ [F_q(x) \cap K_{\beta_q(X_{e'}^\bullet)}(x) = \emptyset]\},$$
$$\gamma_c(q,x) = \{u \mid u \text{ measurable and } \exists e \in out(q)\ \exists t \geq 0\ (\phi_q^u(t,x) \in X_e^\bullet)\}.$$

γ_p means that a prohibitable outgoing edge e is prohibited unless the continuous state is in $X_e^\bullet$. γ_f means that a forcible outgoing edge e is forced if the continuous state is in $X_e^\bullet$ and it is imminently exiting $\beta_q(X_{e'}^\bullet)$ for all outgoing edges e'. γ_c means that a continuous control trajectory is permitted if it is measurable and under it $X_e^\bullet$ is reached for some outgoing edge e.

The hybrid controller can be thought of as having two levels: the "higher" or supervisory level and the "lower" or regulatory level. The supervisory controller takes the discrete or planning actions through γ_p and γ_f and sets a reachability problem to be solved by the regulatory controller which deals with the continuous dynamics. The pragmatic understanding of this interaction is that on entering a discrete state, the supervisor sets the above reachability problem, the regulatory controller solves the problem as γ_c, and then the supervisory controller chooses a selection from γ_c to be applied in that discrete state.

Now we give conditions under which this hybrid controller achieves the 0-lag nonblocking control strategy. We call H viable under the following conditions: (1) the viability kernel is closed and nonempty and it is a fixed point of the nonblocking operator N; further, if an edge is not both forcible and prohibitable, then the enabling set for that edge is also its viability kernel; and (2) at the state (q,x), at time t, if edge e is enabled but imminently it can nevermore be taken and every other outgoing edge is either prohibited by the control $u_p(t)$ or can nevermore be taken, then H takes e at t. These are formalized below.

Definition 12. *H is said to be viable iff*

1. $\exists e \in E\ (X_e^\bullet \neq \emptyset)$, $\forall e \in E\ (X_e^\bullet$ is closed $)$, $N(X^\bullet) = X^\bullet$, and $\forall e \in E\ (e \notin E_p \cap E_f \Rightarrow X_e^\bullet = X_e)$, and
2. in any state (q,x), at any time t, given $u_p(t)$ if there exists an outgoing edge $e \notin u_p(t)$ such that

   $$x \in X_e,\ F_q(x) \cap K_{\beta_q(X_\bullet)}(x) = \emptyset \text{ and } \forall e' \neq e\ (e' \in out(q) \Rightarrow e' \in u_p(t) \text{ or } x \notin \beta_q(X_{e'}))$$

   then e is taken at t.

Theorem 13. *If H is viable then*

$$\{(x_d, x_c, v) \in G_{\gamma_H} \mid x_c(0) \in \beta_{x_d(0)}(\cup_{e \in out(x_d(0))} X_e^\bullet)\} = S_{\bar{U}_H}$$

The theorem states that, over the given initial conditions, controller γ_H realises $\tilde{U}_H$, the largest nonblocking control strategy for H.

Proof. Consider $s \in LHS$. $\forall t\ (v(t) \in E \Rightarrow x_c(t) \in X^*_{v(t)})$ since for $e \in E_p$, $\gamma_p(t)$ prohibits otherwise and for $e \notin E_p$ $(X^*_e = X_e)$. Thus, since $N(X^*) = X^*$, at each t where $t = 0$ or t is a transition time, there exists some edge e' such that $X^*_{e'}$ is reachable. Let some control in $\gamma_c(t)$ be followed so that either some other edge is taken or $X^*_{e'}$ is reached. In the latter case, either because it is forced or because of condition (2) of definition 12, e' is taken. Thus we have shown that with initial conditions as given γ_H does not block at any t. It is clear that $S_{\gamma_H} \subset S_H$, and that γ_H is 0-lag. Thus, $LHS \subset RHS$.

Consider $s \notin LHS$. Then $\forall t\ (v(t) \in E \Rightarrow x_c(t) \notin X^*_{v(t)})$. Thus, s can have only finitely many transitions, and after them the controller must block. Hence $s \notin RHS$. Thus, $LHS \supset RHS$, and the theorem follows. □

7 Process Control Application

Consider a plant given by the differential equations

$$\dot{x}_1 = \frac{-x_1}{1 + \frac{1}{2}\sin(10u_1)} + u_1,$$
$$\dot{x}_2 = -x_2 + (1 + x_1^2)u_2.$$

The state of the plant is given by $(x_1, x_2) \in \mathbf{R}^2$, and the control inputs are given by $(u_1, u_2) \in \mathbf{R}^2$. The state variable x_1 is interpreted as the temperature of the plant and the state variable x_2 as its pressure. Their units of measurement are normalized so that $(0,0)$ represents ambient conditions. The control input u_1 is interpreted as the rate at which heat is supplied to (or removed from) the plant, and the control input u_2 as the rate at which a pneumatic piston is displaced. Note that all zero input state trajectories are attracted by the origin, the input u_1 affects the thermal inertia of the plant, and the plant's temperature affects its sensitivity to the pneumatic input.

The plant is to be used in a process in which the plant state must be cycled repeatedly through three operating points at which three steps of the process occur. These steps are denoted as e_1, e_2, and e_3, and the corresponding operating points are $P_1 = (0,0)$, $P_2 = (2.5, 2)$, and $P_3 = (1,3)$. The process is known to be robust, and it is sufficient for the state to be in specified neighborhoods of these operating points for the corresponding reaction steps to occur. The step e_1 can occur if the L_2 norm of the state deviation from P_1 is bounded by 0.2. The step e_2 can occur if plant temperature is above that in P_2, pressure below that in P_2, and the L_1 norm of the state deviation from P_2 is bounded by 0.2. The step e_3 can occur if the plant pressure is above that in P_3, and the L_2 norm of the state deviation from P_3 is bounded by 0.3. These neighborhoods for the steps e_1, e_2, and e_3 are denoted by X_1, X_2, and X_3, respectively. For the process to execute as desired, the plant state must visit these neighborhoods so as to describe the sequence

$$X_1\ X_2\ X_3\ X_1\ X_2\ X_3\ X_1\ X_2\ X_3\ \ldots$$

Three phases of operation of the plant can be identified, q_1, q_2, and q_3. In phase q_1, the plant state is controlled to reach the set X_1 that enables the step e_1, in phase q_2 to reach X_2 that enables e_2, and in phase q_3 to reach X_3 that enables e_3. This hybrid model of the plant is shown in Figure 6. Note that the reset maps on the edges are the identity maps.

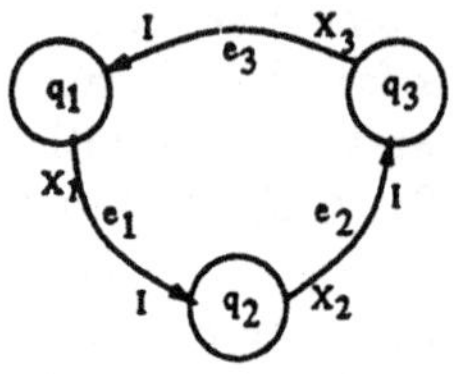

Fig. 6. Hybrid Application.

Consider the plant's operation in phase q_2. The control law of this phase must move the plant state from any point in X_1 to some point in X_2. Synthesizing a state feedback law for such a behavior requires the controller to solve on-line for u_1 that satisfies

$$\frac{-x_1}{1+\frac{1}{2}\sin(10u_1)} + u_1 = v_1(x)$$

where v_1 is some state dependent parameter known to the controller. Similar is true in the other phases of operation. This on-line computational load can be quite formidable.

Off-line computational techniques can be used to synthesize state feedback control laws that move the plant state from one operating point to the next. (These laws can reflect optimality considerations if they exist.) It can be verified that the following laws are adequate for this purpose.

$$\text{Phase } q_1 \ (1,3) \text{ to } (0,0) \quad \begin{cases} u_1(x) = x_1 - 1 \\ u_2(x) = \frac{1}{1+x_1^2}(x_2 - 3) \end{cases}$$

$$\text{Phase } q_2 \ (0,0) \text{ to } (2.5,2) \quad \begin{cases} u_1(x) = x_1 + 3 \\ u_2(x) = \frac{1}{1+x_1^2}(x_2 + 2) \end{cases}$$

$$\text{Phase } q_3 \ (2.5,2) \text{ to } (1,3) \quad \begin{cases} u_1(x) = x_1 - 1.5 \\ u_2(x) = \frac{1}{1+x_1^2}(x_2 + 1) \end{cases}$$

The state trajectories under these laws starting at the corresponding initial conditions are shown in Figure 7. However, these laws do not use the knowledge of process robustness, and the system behavior under them is quite sensitive to variations in initial conditions or other disturbances.

We propose that the control laws for phases q_1 and q_3 are fixed as the ones given above. In phase q_2 the control law given above is considered as the reference law, but other laws are also permitted. We then proceed to calculate X_1^*, X_2^*, and X_3^* in an off-line manner. These are shown in Figure 8 as the shaded regions of the

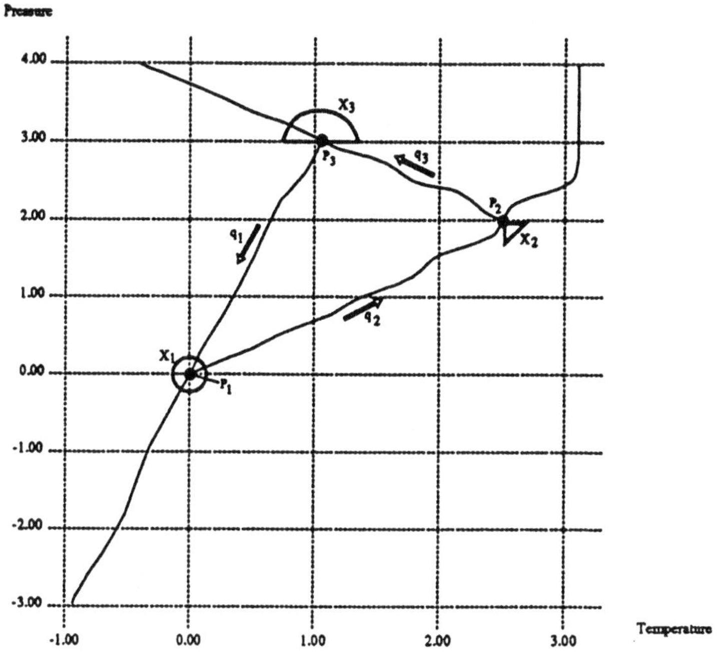

Fig. 7. Reference trajectories between operating points.

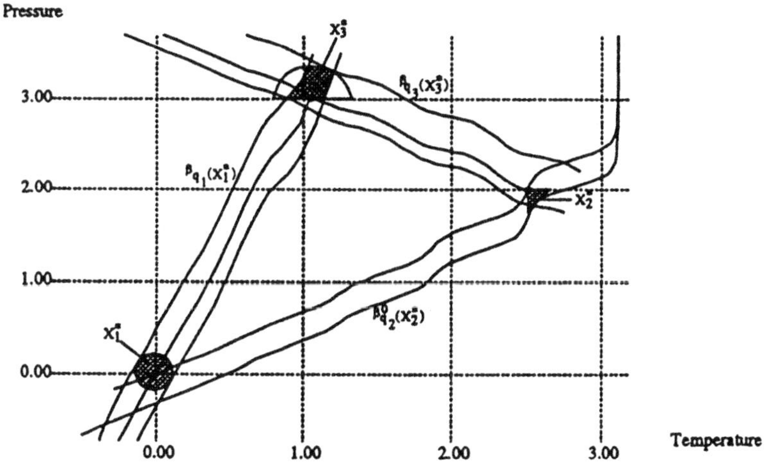

Fig. 8. Backward projection of X^* under reference control laws.

state space. The backward projections of X_1^* and X_3^* are shown as $\beta_{q_1}(X_1^*)$ and $\beta_{q_3}(X_3^*)$, respectively, and the backward projection of X_2^* under the reference law in q_2 is shown as $\beta_{q_2}^0(X_2^*)$. Whenever step e_1 occurs in $X_1^* \backslash \beta_{q_2}^0(X_2^*)$, the controller must compute a law under which $\beta_{q_2}^0(X_2^*)$ is reached; afterwards the reference law in q_2 is followed. Thus, the on-line computational load on the controller is reduced while guaranteeing that the system performance is desired and close to the reference (or optimal) behavior. In fact, in the given application, no on-line computation is necessary since the law of phase q_1 can be followed to reach $\beta_{q_2}^0(X_2^*)$.

8 Conclusion

We considered hybrid systems with nondeterministic discrete continuous behaviors. We used nondeterministic finite automata to model the discrete behavior and state dependent differential inclusions to model the continuous behavior. We defined viability as the system's ability to take an infinite number of discrete transitions. We defined the concepts of the viability kernel of a hybrid system and of the 0-lag nonblocking control strategy. We gave analysis results pertaining to continuity and computability of the viability kernel. We gave the synthesis of a hybrid controller and specified conditions under which it realizes the 0-lag nonblocking control strategy. Finally we illustrated our approach using a process control application with nonlinear system dynamics.

References

1. R. Alur, C. Courcoubetis, and D. Dill. Model-checking for real time systems. In *Proc. 5th IEEE Symp. on Logic in Computer Science*, IEEE Computer Society Press, 1990.
2. R. Alur, C. Courcoubetis, T. Henzinger, and P. Ho. Hybrid Automata: An Algorithmic Approach to the Specification and Verification of Hybrid Systems. In *Hybrid Systems*, Lecture Notes in Computer Science 736, Springer-Verlag, 1993, pp. 209-229.
3. J.-P. Aubin. *Viability Theory*, Birkhauser, 1991.
4. A. Deshpande. Control of Hybrid Systems. *Ph.D. Thesis*. University of California at Berkeley. 1994.
5. J. Guckenheimer and S. Johnson. Planar Hybrid Systems. *Hybrid Systems II*, Lecture Notes in Computer Science, Springer-Verlag. 1995.
6. W. Kohn, A. Nerode, J. Remmel, and A. Yakhnis. Viability for Hybrid Systems. *MSI Technical Report 94-37*, Mathematical Sciences Institute, Cornell University, 1994. To appear in *Hybrid Systems*, Journal of Theoretical Computer Science. 1995.
7. J. Munkres. *Topology*. Prentice-Hall, 1975. pp 232.
8. J. Munkres. *Topology*. Prentice-Hall, 1975. pp 170.
9. A. Nerode and A. Yakhnis. Control Automata and Fixed Points of Set-Valued Operators for Discrete Sensing Hybrid Systems. *MSI Technical Report 93-105*, Mathematical Sciences Institute, Cornell University. 1993.
10. A. Puri and P. Varaiya. Decidability of hybrid systems with rectangular differential inclusions. *Proc. 6th Workshop Computer-Aided Verification*, 1994.
11. P. Varaiya. On the trajectories of a differential system. In *Mathematical Theory of Control*, Academic Press, 1967.
12. P. Varaiya and J. Lin. Existence of saddle points in differential games. *SIAM J. Control*. Vol. 7, No. 1, 1969.

Modeling and Stability Issues in Hybrid Systems

Murat Doğruel and Ümit Özgüner
Department of Electrical Engineering
The Ohio State University
2015 Neil Avenue
Columbus, OH 43210

Abstract

Classical modeling of physical systems relies on differential-difference type equations. However many systems today contain logical quantities and discrete decision makers. For these type of systems the classical modeling approach is either not sufficient or only represents some portion of the system. We develop a hybrid system formulation to study systems with both discrete and continuous quantities.

Here we describe discrete state and hybrid systems, and attempt to further understand their construction and behavior. We specifically consider stability issues for hybrid systems. Furthermore we discuss how to control hybrid systems by possibly using hybrid controllers so that we will be able to build more and more sophisticated intelligent machines.

1. Introduction

The mathematical models of physical systems can be obtained in many ways such that available analysis and design methods are applied. Often two kinds of variables exist in a model. Some variables take on values which are real numbers, while others assume values from a finite set. In an automobile, for example the speed may be represented as a real number, and the position of the transmission as a member of the set of all possible (finite) positions of the transmission. Another example is a robot. The position of the robot in a 3-dimensional space can be represented by a combination of real numbers, while the task may be represented as an element of a finite set, such as "going left" or "holding object." We shall call systems where the state variables can assume only a finite, discrete set of values Discrete State Systems (DSS) and systems with only continuous states Continuous State Systems (CSS). In this work we shall develop models that lead to the analysis of the combination mentioned above, which we call Hybrid Systems.

We use the term "Discrete State Systems" (DSS) as a model for a sub class of Discrete Event Systems (DES) which have been studied widely in the literature [6]. The main advantage of using the DSS model is to analyze DES in a simpler and more familiar way. DSS can be seen as a pure automaton so that reachability, controllability and stabilizability properties can be studied [2]. The DSS model which we propose to develop, is similar to state equations. We can also consider a discrete state system as a nonlinear discrete time system. Thus discrete state system construction provide us a bridge to work on both logical and differentiation-difference dynamics. It plays a dual role connecting the theories of some DES and Continuous State Systems. Furthermore, as in continuous state systems, cascade and feedback connections of DSS can be analyzed easily and the resulting systems can be modeled as DSS [1]

Hybrid system model consists of discrete and continuous state systems. The combination of discrete and continuous state variables defines the hybrid state space. We consider a model of hybrid systems in which the discrete and the continuous state systems are directly connected. Since we have to deal with logical and differential-difference dynamics at the same time, some complexity arises in the analysis and the design. We can not use the standard analysis methods in continuous or discrete state system analysis for hybrid systems since in the combination many different and interesting behavior patterns occur which one does not encounter in classical system models.

In section 2, discrete state systems are defined. Since an important part of hybrid systems is the discrete state part, it is essential to understand the construction well. In section 3, we consider hybrid systems. First, we describe the models we use. A discrete time model is obtained by a discrete time discrete state system and either a discrete or continuous time continuous state system. In this case the discrete state can change only at certain points on time axis. In continuous time model, however, the discrete states can change any time according to discrete inputs or continuous states. In section 3.2 we define reachability and stabilizability properties of hybrid systems. In section 4, we study the hybrid system state space and consider stability. We try to understand the construction and behavior of hybrid systems. We define classical stability concepts for hybrid systems. Then using Lyapunov theory we provide some new results on stability. In section 5, two examples of hybrid systems are provided. First we model a robot with a gripper as a hybrid system and study some hybrid system properties. We use a controller for the robot to obtain a desirable behavior. Second, we consider a continuous time hybrid system numerical example and study stability concepts.

2. Discrete State Systems

In many cases, it is convenient to consider automata in "vector" form, which we shall call a Discrete State System (DSS). A *Discrete State System* is an automaton of the form:

$$\begin{aligned} X(k+1) &= F(X(k), U(k)), \\ Y(k) &= G(X(k), U(k)), \quad k = 0, 1, 2, \cdots \end{aligned} \tag{2.1}$$

where

$$X(k) = \begin{bmatrix} X_1(k) \\ X_2(k) \\ \vdots \\ X_N(k) \end{bmatrix}, \ U(k) = \begin{bmatrix} U_1(k) \\ U_2(k) \\ \vdots \\ U_R(k) \end{bmatrix}, \ Y(k) = \begin{bmatrix} Y_1(k) \\ Y_2(k) \\ \vdots \\ Y_M(k) \end{bmatrix} \tag{2.2}$$

where

$$X_i(\cdot) \in \mathcal{X}_i \triangleq \{0, 1, \cdots, l_{X_i} - 1\}, \tag{2.3}$$

$$U_i(\cdot) \in \mathcal{U}_i \triangleq \{0, 1, \cdots, l_{U_i} - 1\}, \tag{2.4}$$

$$Y_i(\cdot) \in \mathcal{Y}_i \triangleq \{0, 1, \cdots, l_{Y_i} - 1\}. \tag{2.5}$$

As we see the state (X_i) , the input (U_i) , and, the output (Y_i) variables take on values from finite sets with cardinal numbers l_{X_i}, l_{U_i}, and, l_{Y_i} correspondingly.

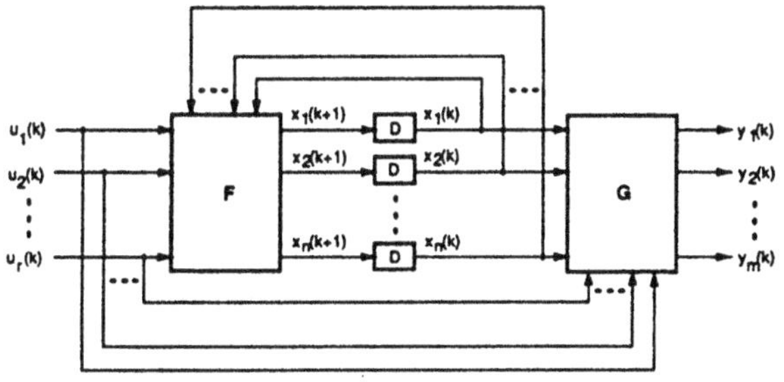

Figure 2.1: Discrete State System block diagram.

Note that, there are N state , R input, and, M output variables. The description of DSS above is similar to the description of discrete time systems in the literature but here each variable takes values from a finite set. The state space of (2.1) is $\mathcal{X} = \mathcal{X}_1 \times \mathcal{X}_2 \times \ldots \times \mathcal{X}_N$. The system can be represented as a block diagram as shown in Figure 2.1.

Since the ith state variable can be written as

$$X_i(k+1) = F_i(X_1(k), X_2(k), \cdots, X_i(k), \cdots, X_N(k), U(k)), \tag{2.6}$$

we can consider it as a state of the individual automaton with other variables as the inputs to that automaton. So, DSS can be considered as a network of automata as in Figure 2.2.

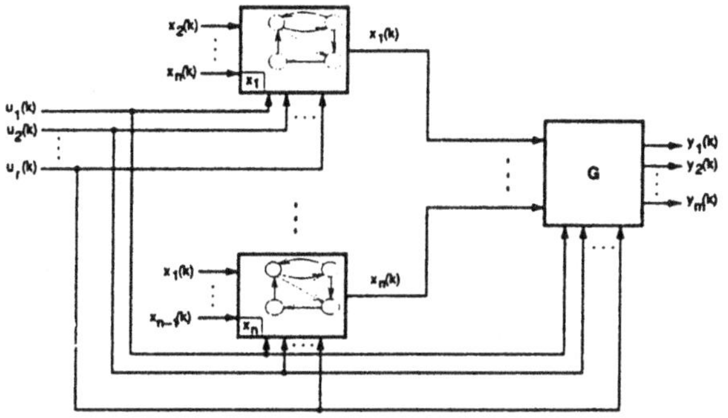

Figure 2.2: DSS, a network of automata.

DSS (2.1) can also be considered as a single automaton. This can be done by defining a single state variable $\hat{X}(k)$ of the single automaton to be obtained, which represents the vector $X(k)$. So we need a mapping from a vector set to a single variable set. This mapping can be defined in many ways, one of them is to choose

$$\hat{X}(k) = p(X(k)) \triangleq X_1(k) + l_{X_1} X_2(k) + \cdots + \left(\prod_{i=1}^{N-1} l_{X_i} \right) X_N(k). \tag{2.7}$$

Note that $\hat{X}(k)$ is a single variable whereas $X(k)$ is a vector of variables. The mapping (2.7) is one-to-one and onto, that is p^{-1} exists, since we can recover $X_i(k)$ using

$$X_i(k) = p_i^{-1}(\hat{X}(k)) = \left(\hat{X}(k) \; div \prod_{j=1}^{i-1} l_{X_j} \right) mod \; l_{X_i} \tag{2.8}$$

where div is integer division and mod is modulo operator, and,

$$X(k) = p^{-1}(\hat{X}(k)) = [\; p_1^{-1}(\hat{X}(k)) \;\; p_2^{-1}(\hat{X}(k)) \;\; \ldots \;\; p_N^{-1}(\hat{X}(k)) \;]'. \tag{2.9}$$

Similar mappings for the input $\hat{U}$ and the output $\hat{Y}$ of the single automaton can also be defined.

3. Hybrid Systems

3.1 Models of Hybrid Systems

We construct hybrid systems as a combination of Discrete State Systems [1, 3] and Continuous State Systems. Let $X(t)$ and $x(t)$ denote the discrete and the continuous state variables in a hybrid system. Similarly let $U(t)$ and $u(t)$ be the discrete and the continuous inputs, and, $Y(t)$ and $y(t)$ be the discrete and the continuous outputs at the time instant t.

3.1.1 Discrete Time Model

In this model discrete state system part is modeled as a discrete time system whereas the continuous state system can be either discrete or continuous time system. If we use a continuous time configuration for continuous state system we have the equations as

$$\begin{aligned} X(k+1) &= F(X(k), s(kT), U(k)), && (3.1)\\ \frac{d}{dt}x(t) &= f(x(t), S(k), u(t)), && (3.2)\\ Y(k) &= G(X(k), s(kT), U(k)), && (3.3)\\ y(t) &= g(x(t), S(k), u(t)), && (3.4) \end{aligned}$$

where

$$\begin{aligned} s(t) &= q(x(t), S(k), u(t)), && (3.5)\\ S(k) &= Q(X(k), s(kT), U(k)), && (3.6) \end{aligned}$$

$$t \geq 0, \quad k = \lfloor t/T \rfloor, \quad T \in R^+ \text{ is known.}$$

$$X = \begin{bmatrix} X_1 \\ \vdots \\ X_N \end{bmatrix} \in \mathcal{X}, \quad U = \begin{bmatrix} U_1 \\ \vdots \\ U_R \end{bmatrix} \in \mathcal{U}, \quad Y = \begin{bmatrix} Y_1 \\ \vdots \\ Y_M \end{bmatrix} \in \mathcal{Y}, \quad S = \begin{bmatrix} S_1 \\ \vdots \\ S_H \end{bmatrix} \in \mathcal{S},$$

$$X_I \in \{0, 1, \cdots, l_{X_I}\}, \quad U_I \in \{0, 1, \cdots, l_{U_I}\}, \quad Y_I \in \{0, 1, \cdots, l_{Y_I}\}, \quad S_I \in \{0, 1, \cdots, l_{S_I}\},$$

$$x = \begin{bmatrix} x_1 \\ \vdots \\ x_n \end{bmatrix} \in R^n, \quad u = \begin{bmatrix} u_1 \\ \vdots \\ u_r \end{bmatrix} \in R^r, \quad y = \begin{bmatrix} y_1 \\ \vdots \\ y_m \end{bmatrix} \in R^m, \quad s = \begin{bmatrix} s_1 \\ \vdots \\ s_h \end{bmatrix} \in R^h,$$

$$\begin{aligned} F &: \mathcal{X} \times \mathcal{U} \times R^h \to \mathcal{X}, & G &: \mathcal{X} \times \mathcal{U} \times R^h \to \mathcal{Y},\\ f &: R^n \times R^r \times \mathcal{S} \to R^n, & g &: R^n \times R^r \times \mathcal{S} \to R^m,\\ q &: R^n \times R^r \times \mathcal{S} \to R^h, & Q &: \mathcal{X} \times \mathcal{U} \times R^h \to \mathcal{S}. \end{aligned}$$

$X(0)$ and $x(0)$ are initial conditions. The block diagram of a hybrid system is shown in Figure 3.1.

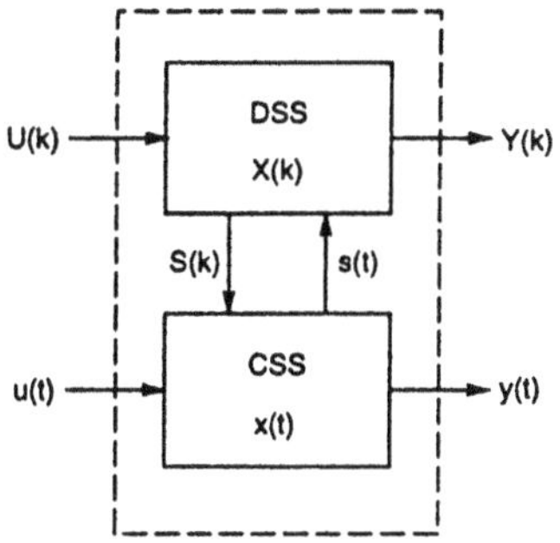

Figure 3.1: Hybrid system block diagram.

As we see from the model, the information inputs (outputs) S and s are used between continuous and discrete state systems. Continuous inputs to discrete state system are sampled with the sampling time T, and, discrete inputs to continuous state system are held for T seconds.

First, we observe that $s(t)$ does not need to have information about $S(k)$ because $S(k)$ is obtained from $X(k)$ and $U(k)$ which are known in the discrete state block. Thus q can only depend on $x(t)$ and $u(t)$. The same thing applies for $S(k)$, in this case Q can only depend on $X(k)$ and $U(k)$. So we have

$$s(t) = q(x(t), u(t)), \tag{3.7}$$
$$S(k) = Q(X(k), U(k)). \tag{3.8}$$

Furthermore if the information between discrete and continuous blocks is not been affected by the hybrid system inputs directly then S and s only depend on the discrete and continuous states respectively. For this case we have a hybrid system model as

$$X(k+1) = F(X(k), x(kT), U(k)), \tag{3.9}$$
$$\frac{d}{dt}x(t) = f(X(k), x(t), u(t)), \tag{3.10}$$
$$Y(k) = G(X(k), x(kT), U(k)), \tag{3.11}$$
$$y(t) = g(X(k), x(t), u(t)). \tag{3.12}$$

The configuration of a hybrid system for this case is shown in Figure 3.2.

We can also model continuous state system as a discrete time system. In this case the system equations are

$$X(k+1) = F(X(k), s(k), U(k)), \tag{3.13}$$
$$x(k+1) = f(x(k), S(k), u(k)), \tag{3.14}$$
$$Y(k) = G(X(k), s(k), U(k)), \tag{3.15}$$
$$y(k) = g(x(k), S(k), u(k)), \tag{3.16}$$
$$s(k) = q(x(k), u(k)), \tag{3.17}$$
$$S(k) = Q(X(k), U(k)), \quad k = 0, 1, 2, \cdots. \tag{3.18}$$

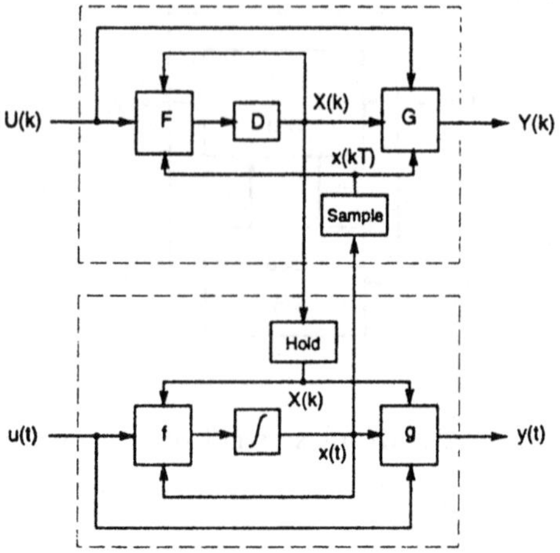

Figure 3.2: Hybrid system configuration.

3.1.2 Continuous Time Model

The second model is obtained by letting $T \to 0$ and simplifying the equations as shown below,

$$\begin{aligned} X(t) &= F(X(t^-), x(t^-), U(t^-)), && (3.19)\\ \frac{d}{dt}x(t) &= f(X(t), x(t), u(t)), && (3.20)\\ Y(t) &= G(X(t^-), x(t^-), U(t^-)), && (3.21)\\ y(t) &= g(X(t), x(t), u(t)). && (3.22) \end{aligned}$$

Then the whole state space of a hybrid system is

$$\mathcal{X}_H = \mathcal{X}_1 \times \mathcal{X}_2 \times \cdots \times \mathcal{X}_N \times R^n = \mathcal{X} \times R^n. \qquad (3.23)$$

We assume that the continuous state dynamic function, f, is piecewise continuous, and $x(t)$ is absolutely continuous. Here we use Flippov's construction to find the equivalent dynamics for the discontinuity surfaces in continuous state space [7, 8].

3.2 Some Properties of Hybrid Systems

In this section we define reachability and stabilizability for hybrid systems. These definitions are similar to the ones in classical system theory and automata theory.

Definition 3.2.1 The set $\begin{bmatrix} \mathcal{X}^b \\ \Omega^b \end{bmatrix} \subseteq \begin{bmatrix} \mathcal{X} \\ R^n \end{bmatrix}$ is *reachable from* the set $\begin{bmatrix} \mathcal{X}^a \\ \Omega^a \end{bmatrix} \subseteq \begin{bmatrix} \mathcal{X} \\ R^n \end{bmatrix}$ if $\forall X(0) \in \mathcal{X}^a, x(0) \in \Omega^a, \exists U(k), k = 0, 1, \cdots, K; u(t), t \in [0, KT]$; such that $X(K) \in \mathcal{X}^b$ and $x(KT) \in \Omega^b$.

Definition 3.2.2 The set $\begin{bmatrix} \mathcal{X}^a \\ \Omega^a \end{bmatrix} \subseteq \begin{bmatrix} \mathcal{X} \\ R^n \end{bmatrix}$ is *1-step returnable* if when $X(0) \in \mathcal{X}^a$ and $x(0) \in \Omega^a, \exists U(0) \in \mathcal{U}$ and $u(t), t \in [0, T]$; such that $X(1) \in \mathcal{X}^a$ and $x(t) \in \Omega^a, t \in [0, T]$.

Definition 3.2.3 The set $\begin{bmatrix} \mathcal{X}^a \\ \Omega^a \end{bmatrix} \subseteq \begin{bmatrix} \mathcal{X} \\ R^n \end{bmatrix}$ is *stabilizable* if it is reachable from $\begin{bmatrix} \mathcal{X}^0 \\ \Omega^0 \end{bmatrix}$ and 1-step returnable. The sets $\mathcal{X}^0$ and Ω^0 consist of the possible initial conditions.

4. Hybrid System State Space and Stability

In this section we consider stability concepts in hybrid systems, hence we restrict ourselves to hybrid systems with no input. Here we consider continuous time model. Note that since the states in the discrete state system are finite we can label them and provide a model with only one state variable. Here we choose $\mathcal{X} = \{1, 2, \cdots, N\}$. Then we consider the system

$$\begin{aligned} X(t) &= F(X(t^-), x(t^-)), \\ \tfrac{d}{dt}x(t) &= f(X(t), x(t)). \end{aligned} \tag{4.1}$$

Let us first define

$$\Omega_{ij} \stackrel{\Delta}{=} \{x \in R^n \mid j = F(i, x)\}, \quad i, j \in \mathcal{X}. \tag{4.2}$$

Figure 4.1 illustrates a discrete state transition occurring due to continuous states being in a certain region of the state space. As shown in Figure 4.1, when the continuous state is in the region Ω_{ij} the discrete state jumps from i to j. Note that the union of all regions Ω_{ij} for any state i fills the continuous state space,

$$\bigcup_{j=1}^{N} \Omega_{ij} = R^n \text{ for all } i \in \mathcal{X}. \tag{4.3}$$

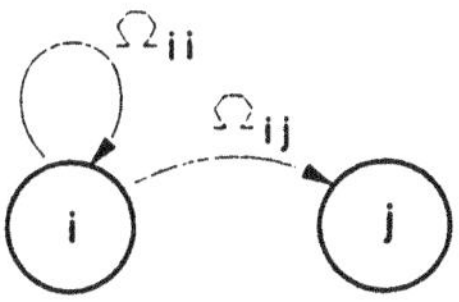

Figure 4.1: Hybrid System discrete state transition.

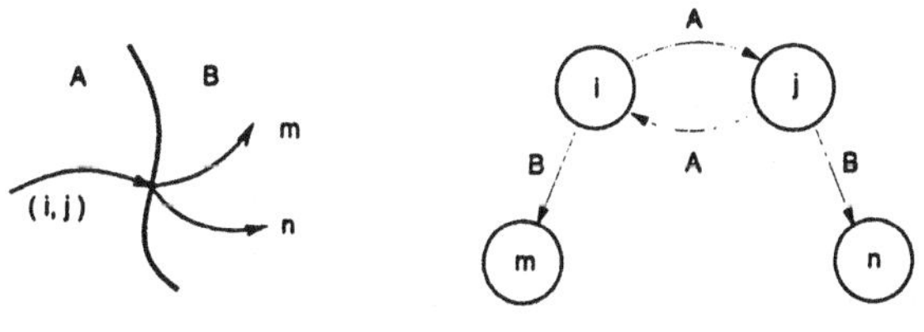

Figure 4.2: The uncertainty in hybrid systems.

We also define

$$\Omega_j \stackrel{\Delta}{=} \bigcup_{i=1}^{N} \Omega_{ij}, \quad j \in \mathcal{X}, \tag{4.4}$$

which denotes the only region in which the system function $f(j, x)$ may participate in the continuous dynamics of hybrid system.

In the hybrid system representation (4.1) we may have some situations where the next states may not be determined. As shown in Figure 4.2 when the continuous state is in region A the discrete system will switch between the discrete states i and j. Since the derivative of the continuous state variable will not be continuous in this situation in general, the continuous dynamics is not identified in this region. However we assume that in a regional switching situation as mentioned above, the hybrid system spends equal times in the switching discrete states.

Thus, the continuous state dynamics in a regional switching situation of k states will be assumed to be

$$\frac{d}{dt}x(t) = \frac{(f(i,x(t)) + \cdots + f(l,x(t)))}{k} . \tag{4.5}$$

However after occurrence of a switching, the system trajectory may not be determined, as illustrated in Figure 4.2. In these cases we need to supply some additional information to eliminate the uncertain behavior.

Given an N dimensional row vector $C \in \mathcal{X}^N$ let us define the region

$$\Omega_C \triangleq \Omega_{1C(1)} \cap \Omega_{2C(2)} \cap \cdots \cap \Omega_{NC(N)} . \tag{4.6}$$

Ω_C denotes the region in which the discrete state i will immediately go to $C(i)$ for $i \in \mathcal{X}$. Thus we partition the continuous state space into N^N regions. Some of the regions may be empty. Note that we have

$$\bigcup_{C \in \mathcal{X}^N} \Omega_C \;=\; R^n. \tag{4.7}$$

We provide an example of a partition in Figure 4.3. As we see in Figure 4.3(c) the representative functions in each region are found. In $\Omega_{[11]}$ regardless of the discrete state we use the function $f(1,x)$ since the discrete state 2 becomes 1 immediately and stays as 1 in this region. In $\Omega_{[21]}$ the discrete state switches between 1 and 2 thus we use the arithmetical average of the two functions $f(1,x)$ and $f(2,x)$. In $\Omega_{[12]}$, on the other hand, the behavior of the continuous state system depends on the discrete state of the system. Thus hybrid systems differ from a general kind of nonlinear systems since they have discrete memory. We need to track the discrete state variable also to find out about the system behavior in general.

In general in each region Ω_C the representative functions can be found as follows. Let $C = [C(1)C(2)\cdots C(N)]$, $C(i) \in \mathcal{X}$, $i \in \mathcal{X}$. For the discrete state $i \in \mathcal{X}$, start from $C(i)$ and then find $C(C(i))$ and take the resulting state as the initial state and continue this process. After at most N steps the resulting sequence of states will repeat themselves since there are only N states in $\mathcal{X}$. Take these repeating sequence as a set of states $\mathcal{X}_C(i)$ which denotes the actual following states of i in region Ω_C. Thus the representative function for the region Ω_C and the initial state i is

$$f_C(i,x) \triangleq \frac{1}{|\mathcal{X}_C(i)|} \sum_{j \in \mathcal{X}_C(i)} f(j,x) . \tag{4.8}$$

Thus we obtain all the functions to represent the continuous state space. If there is no regional switching, $\mathcal{X}_C(i)$ will have just one element and the original system functions will be used. A condition for no regional switchings in a hybrid system can be given as

$$\bigcup_{i,j,k,\cdots,m \in \mathcal{X}} \Omega_{ij} \cap \Omega_{jk} \cap \cdots \cap \Omega_{mi} \;=\; \emptyset. \tag{4.9}$$

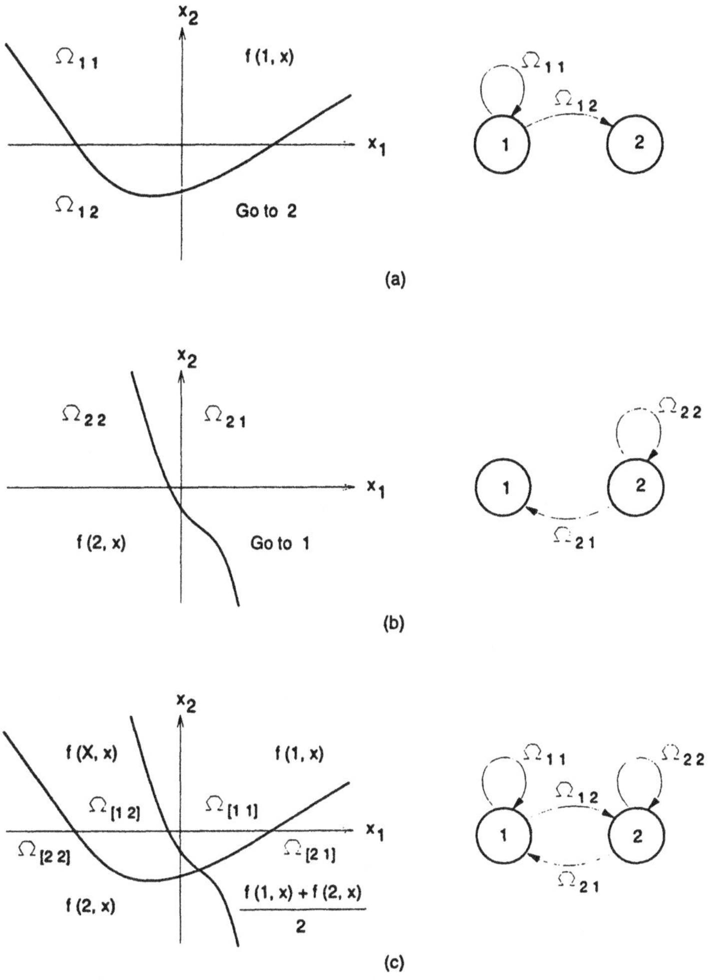

Figure 4.3: Obtaining the partition of a hybrid system.

4.1 Stability definitions

We consider the classical stability concept of continuous state systems for hybrid systems [7].

Definition 4.1.1 $\tilde{x} \in R^n$ is an *equilibrium state* of a hybrid system if $f_C(i, \tilde{x}) = 0$ for all $i \in \mathcal{X}$.

Definition 4.1.2 The equilibrium state $x = 0$ is *stable* if for any $\varepsilon > 0$ there exists a $\delta > 0$ such that for any $\|x(0)\| < \delta$ and for any $X(0) \in \mathcal{X}$ we have $\|x(t)\| < \varepsilon$ for all $t \geq 0$. Otherwise the equilibrium state is unstable.

Definition 4.1.3 The equilibrium state $x = 0$ is *globally asymptotically stable* if it is stable and for any initial conditions $x(0) \in R^n$ and $X(0) \in \mathcal{X}$ we have $x(t) \to 0$ as $t \to \infty$.

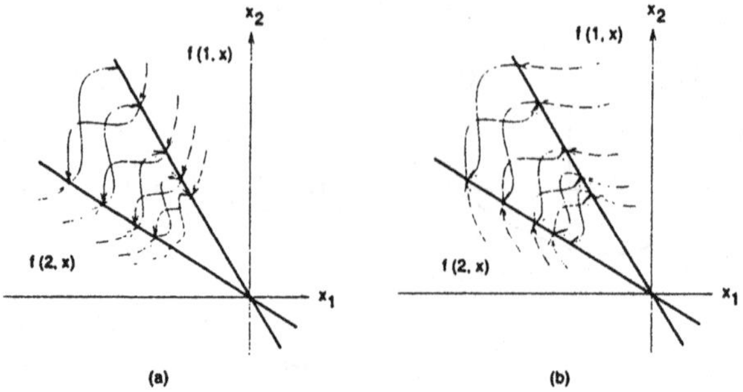

Figure 4.4: Unstable and stable hybrid systems.

From Figure 4.3(c), let us obtain two nonlinear systems by choosing $f(1, X)$ and $f(2, X)$ for the region $\Omega_{[12]}$. One may wonder if it is sufficient or necessary for the global asymptotic stability of a hybrid system to have both nonlinear systems to be globally asymptotically stable. However as shown in Figure 4.4(a) although both nonlinear systems are stable, the hybrid system is not. Also as in Figure 4.4(b) although both nonlinear systems are unstable, the hybrid system is stable. Hence we can not analyze hybrid systems by just splitting a hybrid system into a collection of nonlinear systems.

4.2 Lyapunov Approach

Here we give sufficient conditions for the stability of a hybrid system following classical Lyapunov stability theory.

Theorem 4.2.1 If there exists a continuous scalar function $V(x) : R^n \to R$ of the continuous state x, with continuous first order partial derivatives, such that

(i) $V(x)$ is positive definite,

(ii) $\dot{V}(x, i) = \nabla V \cdot f_C(i, x) \leq -w(x)$, where $w(x)$ is a positive definite function, for all $i \in \mathcal{X}$, in all the regions $x \in \Omega_C, C \in \mathcal{X}^N$,

(iii) $V(x) \to \infty$ as $\|x\| \to \infty$,

then the equilibrium point at the origin of the continuous state space of the hybrid system is globally asymptotically stable.

The above theorem uses the fact that the Lyapunov function will decrease along any hybrid system trajectory. The proof follows from stability theory of time varying systems with discontinuous dynamics[7]. Note that this is a sufficient condition. Furthermore we can not always find a Lyapunov function even if the hybrid system is globally asymptotically stable (consider Figure 4.4(b)). Note also that we do not need to consider discrete state dynamics since the necessary information for hybrid system stability is contained in $f_C(X, x)$.

If there is no regional switchings in the hybrid system, using the above theorem, we can give simpler conditions for the hybrid system Lyapunov function as stated in the theorem below.

Theorem 4.2.2 If there exists a continuous scalar function $V(x) : R^n \to R$ of the continuous state x, with continuous first order partial derivatives, such that

(i) $V(x)$ is positive definite,

(ii) $\dot{V}(x, i) = \nabla V \cdot f(i, x) \leq -w(x)$, where $w(x)$ is a positive definite function, in all the regions $x \in \Omega_{ii}$, for $i \in \mathcal{X}$,

(iii) $V(x) \to \infty$ as $\|x\| \to \infty$,

then the equilibrium point at the origin of the continuous state space of the hybrid system with no regional switchings is globally asymptotically stable.

Let us define the region

$$\tilde{\Omega}_j \triangleq \{x \in R^n \mid x \in \Omega_C,\ C(p) = j,\ j \in \mathcal{X}_C(j),\ p \in \mathcal{X},\ C \in \mathcal{X}^N\}, \quad j \in \mathcal{X}, \tag{4.10}$$

which denotes the only region in which the system function $f(j,x)$ actually participates in the continuous dynamics of hybrid system. Note that $\tilde{\Omega}_j \subset \Omega_j$. Following Theorem 4.2.1 and Theorem 4.2.2, we can give the corollary below.

Corollary 4.2.1 If there exists a continuous scalar function $V(x) : R^n \to R$ of the continuous state x, with continuous first order partial derivatives, such that

(i) $V(x)$ is positive definite,

(ii) $\dot{V}(x,i) = \nabla V \cdot f(i,x) \leq -w(x)$, where $w(x)$ is a positive definite function, in all the regions $x \in \tilde{\Omega}_i$, for $i \in \mathcal{X}$,

(iii) $V(x) \to \infty \quad \text{as} \quad \|x\| \to \infty$,

then the equilibrium point at the origin of the continuous state space of the hybrid system is globally asymptotically stable.

The difference between the above corollary and Theorem 4.2.2 is that Corollary 4.2.1 considers the switching regions also. Here we use the fact that if the Lyapunov function decreases for all the dynamic functions participating in the switching, then it must decrease along the switching trajectory also.

As we mentioned, using a single Lyapunov function may not be enough to prove stability due to the discrete states of a hybrid system. To overcome this difficulty we may want to use multiple Lyapunov functions as given in the theorem below.

Theorem 4.2.3 If there exist continuous scalar functions $V_1(x), V_2(x), \cdots V_k(x) : R^n \to R$ of the continuous state x, with continuous first order derivatives in each region $\Omega_C, C \in \mathcal{X}^N$, and $V_1(x) = V_2(x) = \cdots = V_k(x)$ for all $x \in \partial\Omega_C, C \in \mathcal{X}^N$ such that

(i) $V_j(x)$ is positive definite for all $j \in [1,k]$,

(ii) For all regions $\Omega_C, C \in \mathcal{X}^N$ and for all discrete states $i \in \mathcal{X}$, there is a $j \in [1,k]$ such that $V_j(x)$ is negative definite along the system function $f_C(i,x)$ in Ω_C,

(iii) $V_j(x) \to \infty \quad \text{as} \quad \|x\| \to \infty \quad \text{for all } j \in [1,k]$,

then the equilibrium point at the origin of the continuous state space of the hybrid system is globally asymptotically stable.

In the above theorem we require that the Lyapunov functions have the same value on the boundary of each region Ω_C, but possibly different values inside. Thus if the conditions of the above theorem are satisfied we guarantee that the value of the Lyapunov function in effect decreases and the system is stabilized. Since we have N different discrete state values the result given below is obvious.

Corollary 4.2.2 Using at most N Lyapunov functions is enough to prove the global asymptotic stability of a globally asymptotically stable hybrid system.

5. Applications and Examples

5.1 A Robot with a Gripper

Consider the robot with the gripper shown in Figure 5.1. The robot can go to positions 0, 1, 2, 3. It has two input variables. One of the input variables commands the robot to go to the left or right. The other one commands the gripper to hold or to release the object. The duty of the robot is to take the objects from position 0 to position 3 and to continue to do this job.

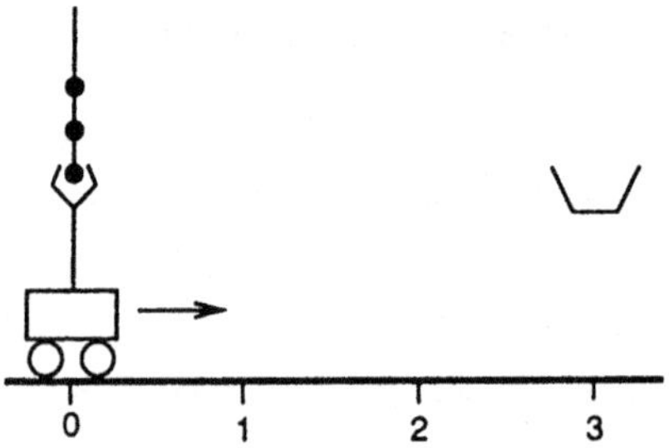

Figure 5.1: A robot with a gripper.

Let us construct a discrete time hybrid system model for this robot. The hybrid system equations are

$$
\begin{aligned}
X_1(k+1) &= U_1(k), \\
X_2(k+1) &= F_2(X_2(k), U_2(k), s(kT)), \\
\frac{d}{dt}x(t) &= -ax(t) + aS(k), \\
Y_1(k) &= X_1(k), \\
Y_2(k) &= X_2(k), \\
y(t) &= x(t), \\
s(t) &= x(t), \\
S(k) &= Q(X_2(k), U_2(k)),
\end{aligned}
$$

where

$$X_1(k) = \begin{cases} 0 & \text{means the gripper holds,} \\ 1 & \text{the gripper does not hold,} \end{cases} \qquad U_1(k) = \begin{cases} 0 & \text{means to hold,} \\ 1 & \text{means to release,} \end{cases}$$

$X_2(k)$ is the sampled position,

$$U_2(k) = \begin{cases} 0 & \text{means to go to the right,} \\ 1 & \text{means to go to the left,} \end{cases}$$

$x(t)$ is the real position, $S(k)$ is the desired position.

Q is given in Table 1, a is a positive real constant, and,

$$F_2(X_2, U_2, s) = \begin{cases} X_2 & \text{If } |Q(X_2, U_2) - s| > 0.05 \\ Q(X_2, U_2) & \text{otherwise} \end{cases}$$

$Q(X_2(k), U_2(k))$ is the desired position of the robot while $s(kT)$ represents the real position. When $|Q(X_2(k), U_2(k)) - s(kT)| > 0.05$ the robot is not called to be at the desired position, hence, X_2 does not change. As soon as $|Q(X_2(k), U_2(k)) - s(kT)| \leq 0.05$, we let $X_2(k+1) = Q(X_2(k), U_2(k))$.

Table 1.

U_2	X_2	$Q(X_2, U_2)$
0	0	1
0	1	2
0	2	3
0	3	3
1	0	0
1	1	0
1	2	1
1	3	2

Table 2.

U_3	X_3	$F_3(X_3, U_3)$
0	0	0
0	1	0
1	0	0
1	1	1
2	0	0
2	1	1
3	0	1
3	1	1

Now, let us check the reachability and stabilizability of some sets for this hybrid system. It is trivial to see that the set

$$\left[\frac{\mathcal{X}^b}{\Omega^b}\right]_j \triangleq \left\{\left[\begin{array}{c} i \\ j \\ \hline l \end{array}\right] ; i \in \{0,1\}, l \in [j-0.05, j+0.05]\right\} \tag{5.1}$$

for any $j \in \{0,1,2,3\}$ is reachable from the set

$$\left[\frac{\mathcal{X}^a}{\Omega^a}\right] \triangleq \left\{\left[\begin{array}{c} i \\ j \\ \hline l \end{array}\right] ; i \in \{0,1\}, j \in \{0,1,2,3\}, l \in [0,3]\right\}. \tag{5.2}$$

And note that $\left[\frac{\mathcal{X}^b}{\Omega^b}\right]_3$ and $\left[\frac{\mathcal{X}^b}{\Omega^b}\right]_0$ are stabilizable but $\left[\frac{\mathcal{X}^b}{\Omega^b}\right]_1$ and $\left[\frac{\mathcal{X}^b}{\Omega^b}\right]_2$ are not.

The set

$$\left[\frac{\mathcal{X}^c}{\Omega^c}\right] \triangleq \left\{\left[\begin{array}{c} i \\ 1 \\ \hline l \end{array}\right] ; i \in \{0,1\}, l \in [2.95, 3.05]\right\} \tag{5.3}$$

is not reachable from $\left[\frac{\mathcal{X}^a}{\Omega^a}\right]$ because if you choose $X(0) = 0$ and $x(0) = 0$, $x(t)$ follows $X_2(k)$ and always we get $|x(t) - X_2(k)| \leq 1$. So for $x(t) \in [2.95, 3.05]$ we get $2 \leq X_2(k) \leq 3$ but not $X_2(k) = 1$.

For these sets, it is not difficult to check the reachability and stabilizability conditions. This is not always the case. For example, checking if the set

$$\left[\frac{\mathcal{X}^d}{\Omega^d}\right] \triangleq \left\{\left[\begin{array}{c} 1 \\ 2 \\ \hline l \end{array}\right] ; l \in [2.37, 2.38]\right\} \tag{5.4}$$

is reachable from $\left[\frac{\mathcal{X}^a}{\Omega^a}\right]$ is not trivial and may require too many numerical calculations since the inputs to give to the system to reach the set can not be simply obtained.

As a result, we can say that sometimes it may be too difficult to check the reachability for hybrid systems.

The duty of the robot is to take the objects from position 0 to position 3. One possible controller having two states is

$$
\begin{aligned}
X_3(k+1) &= F_3(X_3(k), U_3(k)), & (5.5)\\
Y_3(k) &= X_3(k), & (5.6)\\
Y_4(k) &= F_3(X_3(k), U_3(k)), & (5.7)
\end{aligned}
$$

where

$$X_3(k) = \begin{cases} 0 & \text{robot goes to the right,} \\ 1 & \text{robot goes to the left,} \end{cases} \qquad U_3(k) \quad \text{is the position of the robot,}$$

and F_3 is given in Table 2.

The connection equations are

$$U_1(k) = Y_4(k), \quad U_2(k) = Y_3(k), \quad U_3(k) = Y_2(k). \tag{5.8}$$

The whole system is shown in Figure 5.2 and the hybrid system equations are given below.

$$
\begin{aligned}
X_1(k+1) &= F_3(X_3(k), X_2(k)), & (5.9)\\
X_2(k+1) &= F_2(X_2(k), x_3(k), s(kT)), & (5.10)\\
X_3(k+1) &= F_3(X_3(k), X_2(k)), & (5.11)\\
& & (5.12)
\end{aligned}
$$

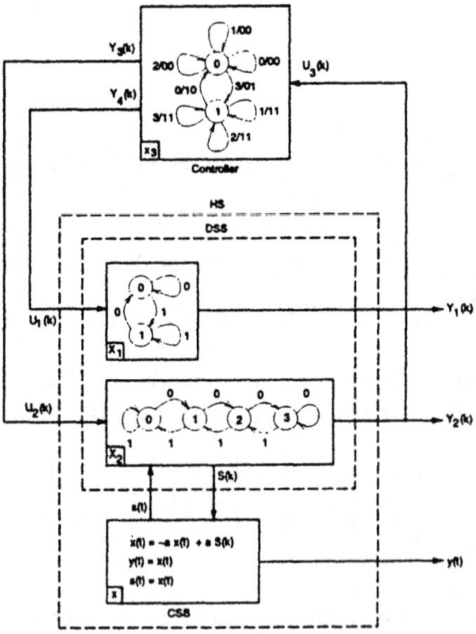

Figure 5.2: Hybrid system model of the controlled robot with the gripper.

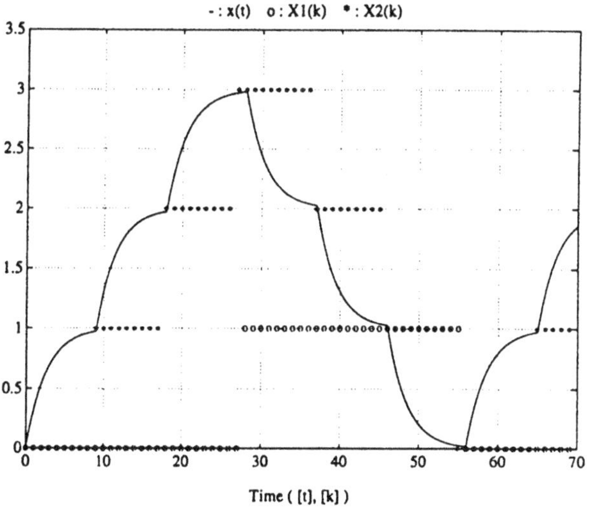

Figure 5.3: The outputs of the hybrid system when $a = 0.4$ with zero initial conditions.

$$\frac{d}{dt}x(t) = -ax(t) + aS(k), \quad (5.13)$$
$$Y_1(k) = X_1(k), \quad (5.14)$$
$$Y_2(k) = X_2(k), \quad (5.15)$$
$$y(t) = x(t), \quad (5.16)$$
$$s(t) = x(t), \quad (5.17)$$
$$S(k) = F_{21}(X_2(k), X_3(k)). \quad (5.18)$$

Let us choose $a = 0.4$ and $T = 1$ with zero initial conditions. The output of the system is shown in Figure 5.3. As we see the robot performs as desired.

5.2 Hybrid System Numerical Example

Let us consider the hybrid system given by equations

$$\begin{aligned} X(t) &= F(X(t^-), x(t^-)), \\ \tfrac{d}{dt}x(t) &= f(X(t), x(t)), \end{aligned} \quad (5.19)$$

where $x \in R^2$, $X \in \{1, 2\}$, and,

$$F(1,x) = \begin{cases} 1, & x_2 \geq 0 \\ 2, & x_2 < 0 \end{cases} \qquad F(2,x) = \begin{cases} 1, & x_1 < 0 \\ 2, & x_1 \geq 0 \end{cases} \quad (5.20)$$

$$f(X,x) = \begin{cases} A_1 x, & X = 1 \\ A_2 x, & X = 2 \end{cases} \quad (5.21)$$

$$A_1 = \begin{bmatrix} -0.1 & 1 \\ -0.5 & -1 \end{bmatrix}, \quad A_2 = \begin{bmatrix} 0.15 & -1.25 \\ 1.5 & -0.2 \end{bmatrix}. \quad (5.22)$$

The eigenvalues of A_1 and A_2 are given as

$$\sigma_{A_1} = \{-0.55 \pm j0.545\}, \quad \sigma_{A_2} = \{-0.025 \pm j1.358\}.$$

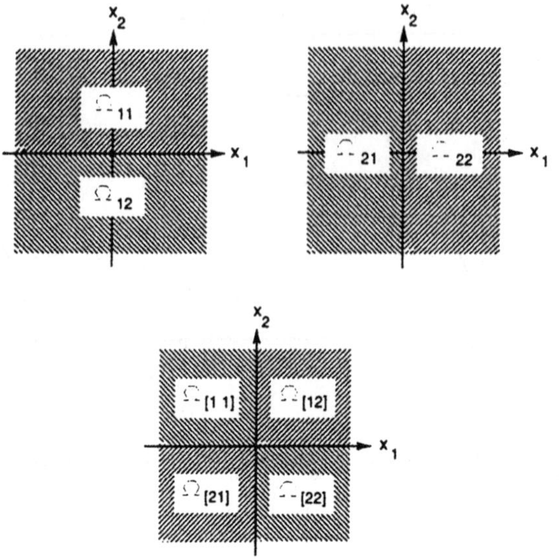

Figure 5.4: Hybrid system continuous state regions.

The regions defined in section 4 for this example are shown in Figure 5.4. We simulate this hybrid system by a computer choosing the sampling time $T = 0.01$. The continuous state trajectories for different initial conditions are shown in Figure 5.5.

As we see, in the third quadrant of the continuous state space ($\Omega_{[21]}$) there are discrete state switchings. Actually, as claimed before, the dynamics in this region is obtained as the arithmetical average of the participating dynamics. Thus we have

$$\tilde{A} = (A_1 + A_2)/2 = \begin{bmatrix} 0.025 & -0.125 \\ 0.5 & -0.6 \end{bmatrix}. \tag{5.23}$$

The eigenvalues of the matrix $\tilde{A}$ are $\{-0.1, -0.475\}$, so the behavior in the region $\Omega_{[21]}$ is as expected.

In the first quadrant of the continuous state space ($\Omega_{[12]}$) the dynamics is chosen according to the discrete states. Because of the loops of the trajectory in this region, we can not find any Lyapunov function to prove the stability of hybrid system. However by the help of multiple Lyapunov functions (Theorem 4.2.3), we could prove that the hybrid system is actually stable. First we choose the Lyapunov functions as

$$V_1(x) = 0.5{x_1}^2 + 0.425{x_2}^2, \tag{5.24}$$

$$V_2(x) = 0.5(x_1 - 0.12x_2)^2 + 0.4178{x_2}^2, \tag{5.25}$$

which are both positive definite. Note also that for $x_1 = 0$ or $x_2 = 0$ (for the boundaries of the regions) we have $V_1(x) = V_2(x)$.

The derivative of V_1 along the first dynamics and the derivative of V_2 along the second dynamics are

$$\dot{V}_1(x) = -0.1(x_1 - 2.875x_2)^2 - 0.0234{x_2}^2, \tag{5.26}$$

$$\dot{V}_2(x) = -0.03(x_1 - 0.5167x_2)^2 - 0.012{x_2}^2, \tag{5.27}$$

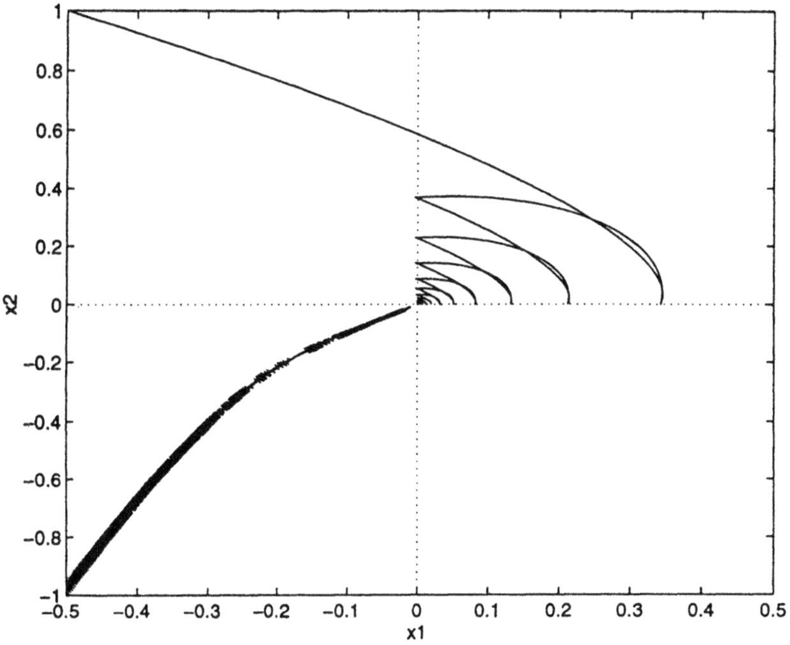

Figure 5.5: Hybrid system continuous state trajectories.

which are both negative definite. Thus we conclude that the hybrid system given in this example is stable.

6. Conclusions

In this work we introduced two models of hybrid systems, a discrete time and a continuous time model. In continuous time hybrid system model there may be infinite frequency discrete state switchings as discussed. In these cases the state trajectory of the hybrid system can not be found without an additional information. We use Flippov's construction [7] and apply and modify it to hybrid systems to find out the state trajectory. We investigate the uncertain behavior occurring after the infinite frequency discrete state switchings and their effects on the stability. In continuous time model we defined representative functions, that is, we showed how to obtain another hybrid system so that the effect of the discrete state dynamics is minimized when considering stability.

In discrete time hybrid system model, the continuous dynamics also changes according to discrete states. Thus here also, we need to use Flippov's construction. However, since the discrete state can only change with a given rate, infinite frequency discrete state switchings will not occur in this model.

In this study we consider the stability of the continuous state system in hybrid system configuration. This approach may be useful especially if our goal is to reach to an equilibrium point in continuous state space and stabilize the system at that point regardless the discrete states. Following Lyapunov theory we provide some theorems to check global asymptotic stability of hybrid systems. As discussed, using a single Lyapunov function may not be sufficient to prove the stability of a hybrid system. We investigate using multiple Lyapunov functions to prove the stability.

We consider reachability and stabilizability properties for hybrid systems and show that it may be too difficult to prove these properties in general. We can design hybrid controllers for hybrid systems. Here we follow the results of hybrid system stability theory. We consider a Lyapunov function candidate and then the hybrid feedback system is to be designed such that the candidate satisfies the conditions for the desired type of stability. A hybrid system feedback may be necessary to control a hybrid system in general.

Finally we provide two examples of hybrid systems. First a robot with a gripper is modeled as a hybrid system and some hybrid system properties are investigated. Furthermore a controller is designed for the robot to obtain a desirable behavior. Second, a continuous time hybrid system numerical example is provided and stability concepts are studied.

REFERENCES

[1] M. Doğruel, "Discrete State Control Systems," M.S. Thesis, Dept. of Electrical Eng., The Ohio State University, June 1992.

[2] M. Doğruel and Ü. Özgüner, "Controllability, Reachability, Stabilizability and State Reduction in Automata," *Proceedings of 1992 IEEE International Symposium on Intelligent Control*, (Scotland, UK), Aug. 1992.

[3] M. Doğruel, S. Drakunov and Ü. Özgüner, "Sliding Mode Control of Discrete State Systems," *Proceedings of the 32nd Conference on Decision and Control*, (San Antonio, Texas), Dec. 1993.

[4] S. Drakunov, M. Doğruel and Ü.Özgüner, "Sliding Mode Control in Hybrid Systems," *Proceedings of 1993 IEEE International Symposium on Intelligent Control*, (Chicago, Illinois), Aug. 1993.

[5] M. Doğruel and Ü. Özgüner, "Discrete State and Hybrid Systems," *The Third Turkish Symposium on Artificial Intelligence and Neural Networks*, (Ankara, Turkey), June 1994.

[6] X. R. Cao and Y. C. Ho, "Models of Discrete Event Dynamic Systems," *IEEE Control Systems Magazine*, June 1990.

[7] A. F. Filippov, *Diferential Equations with Discontinuous Righthand Sides*, Kluwer Acedemic Publishers, 1988.

[8] D. Shevitz and B. Paden, "Lyapunov Stability Theory of Nonsmooth Systems," *IEEE Trans. Aut. Control*, v. 39, n. 9, Sep. 1994.

[9] K. M. Passino and Ü. Özgüner, "Modeling and Analysis of Hybrid Systems: Examples," *Proceedings of 1991 IEEE International Symposium on Intelligent Control*, (Arlington, VA), Aug. 1991.

Hierarchical Hybrid Control: a Case Study*

Datta N. Godbole, John Lygeros and Shankar Sastry
Intelligent Machines and Robotics Laboratory
Department of Electrical Engineering and Computer Sciences
University of California, Berkeley
Berkeley, CA 94720
godbole, lygeros, sastry@robotics.eecs.berkeley.edu

Abstract. A case study of the difficulties encountered in the design of hierarchical, hybrid control systems is presented. As our example we use the Intelligent Vehicle Highway System (IVHS) architecture proposed for vehicle platooning, a system that involves both continuous state and discrete event controllers. We point out that even though conventional analysis tools suggest that the proposed design should fulfill certain performance requirements, simulation results show that it does not. We consider this as an indication that the conventional tools currently in use for the design and verification of control systems may be inadequate for the design of hierarchical controllers for hybrid systems. The analysis also indicates certain shortcomings of the current IVHS design. We propose solutions to fix these problems.

Keywords: Intelligent Vehicle Highway Systems, Hybrid Control Systems, Safety, Verification.

1 Introduction

The term hybrid system is used to describe a large and varied class of systems. A large class of hybrid systems can be described by the architecture of Figure 1. A typical hybrid system is arranged in two (or more) layers [10, 24]. Different levels of abstractions of the plant model are used at each layer of the hierarchy. In the bottom layer the plant model is usually described by means of differential and/or difference equations. This level contains the actual plant and any conventional controllers working at the same level of abstraction. In the top layer the plant description is more abstract. Typical choices of description language at this level are finite state machines, fuzzy logic, Petri nets etc. Typically the controllers designed at this level are discrete event supervisory controllers (see e.g., [30]). The two levels communicate by means of an interface that plays the role of a translator between signals and symbols. As the techniques for control design and verification are well developed for the continuous and discrete systems, the design of the interface is of the utmost importance because it determines the way in which the combined system behaves.

Of course in a general hierarchical structure more than two levels may exist. If this is the case, the structure of Figure 1 can be viewed as the interaction between

* Research supported by the PATH program, Institute of Transportation Studies, University of California, Berkeley, under MOU-100

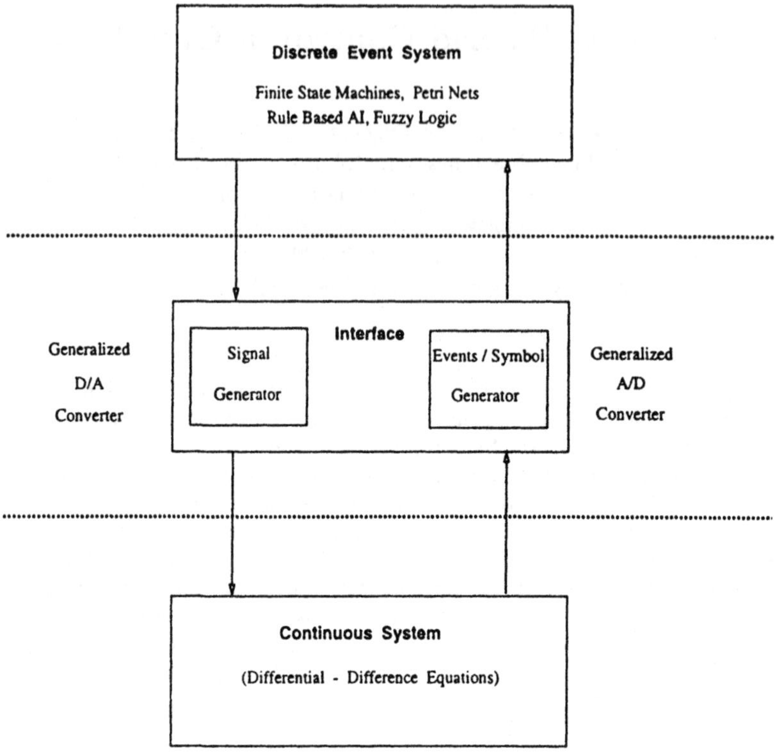

Fig. 1. Hybrid System Architecture

any two of the layers. Alternatively the Discrete Event layer can be assumed to be a lumped version of all the layers that are not part of the continuous domain. Typically in a multilayered hierarchy, as we move up the hierarchy system description gets more abstract (e.g., closer to linguistic), information is condensed (i.e., a signal in the higher levels encodes many facts about the lower levels) and commands become more descriptive (i.e., a single command at a high level induces many actions at the lower levels).

The control architecture described above appears in wide variety of applications and forms the heart of most hybrid system formalisms. Switching controllers [22, 23], Expert Control [27], Intelligent Control [19, 26], Motion Control [5], among others, make use of the structure of Figure 1. For most of these systems the design approach has been "divide and conquer", that is the continuous and discrete controllers are designed independently and then combined by an interface which is designed for the specific problem. This is not a rule however, as the literature also contains examples of systems that show active (on-line) design of the hybrid controllers (e.g., [4] and [25]) as well as attempts of formulating a consistent interface that is not case specific (e.g., [6]). In addition to theoretical formalisms a lot of work has also been done on techniques for simulating hierarchical, hybrid systems (e.g., [3] and [33]).

In this paper we present a case study of the design of hybrid control systems. We illustrate the problems associated with their design and verification by means of an example, the Intelligent Vehicle Highway System (IVHS) designed in the framework of the PATH project. Our goal is to illustrate that the "divide and conquer" approach to hybrid system design is not always effective. The results we present indicate that for multilayered control structures there is a need for an independent proof of performance for the combined system, in addition to the usual proofs in the discrete and continuous domains. In the next section we outline the overall structure of the IVHS architecture that is used in this study. More details on the parts of the design that are relevant to this study will be given in Section 3.

2 Intelligent Vehicle Highway System Architecture

One example where the hybrid system structure of Figure 1 can be found is the Intelligent Vehicle Highway System described in [34, 35]. The goal is to design a system that can significantly increase safety and highway capacity without having to build new roads, by adding intelligence to both the vehicles and the roadside. In order to achieve this, the concept of "platooning" is introduced in [35]. It is assumed that traffic on the highway is organized in groups of tightly spaced vehicles, called platoons. Intuition suggests that this should lead to an increase in the capacity and throughput of the highway; indeed theoretical studies indicate that the capacity increase if such a scheme is implemented successfully will be substantial (as high as four times the current capacity). What may be more surprising is that this will be done without a negative impact on passenger safety. By having the vehicles within a platoon follow each other with a small intra-platoon separation of about 1 meter, we guarantee that, if there is a failure and an impact is unavoidable, the relative speed of the vehicles involved in the collision will be small, hence the damage will be minimized. The inter-platoon separation, on the other hand, is large (of the order of 30 meters) to physically isolate the platoons from each other. This ensures that, if needed, the platoons will have enough time to come to a stop before they collide. In addition a large separation guarantees that transient decelerations will be attenuated as they propagate up the freeway.

2.1 Control hierarchy

Clearly, implementation of such a scheme would require the vehicles to be automatically controlled, as human drivers are not fast and reliable enough to be able to form platoons. The design of such a large scale control system poses a formidable problem. In the architecture outlined in [35] the system is organized in five layers. The block diagram of Figure 2 shows different layers of the control hierarchy.

The top layer, called the **network layer**, is responsible for the flow of traffic on the entire highway system[2]. Its task is to prevent congestion and maximize throughput by dynamic routing of traffic.

[2] The highway system might consist of the interconnection of several highways around an urban metropolis.

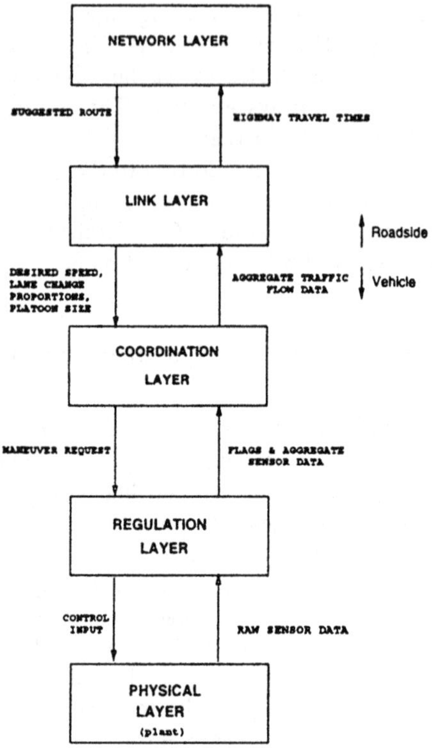

Fig. 2. IVHS Architecture

The second layer, called the **link layer**, coordinates the operation of whole sections (links) of the highway. Its primary task is to maximize throughput while maintaining safe conditions of operation. With these criteria in mind, it calculates an optimum platoon size and an optimum velocity for each highway section. It also decides which lanes the vehicles should follow to get to their destination as fast as possible. Finally, it monitors incidents on the highway and diverts traffic in order to minimize the impact of the incident on traffic flow and safety. Because the link layer bases its control actions on large numbers of vehicles, it inevitably has to use some form of aggregate information. Therefore it treats the vehicles in a section statistically, rather than considering the state of individual vehicles or platoons. Likewise, the commands it issues are not addressed to individual vehicles but rather to all the vehicles in the section; a typical command would be "30% of the vehicles that wish to get off the highway at the next exit should change lanes now" or "all platoons in this section should try to be 10 vehicles long".

The next level below the link layer is the **coordination layer**. Its task is to coordinate the operation of platoons with their neighbors. It receives the link layer commands and translates them to specific maneuvers that the platoons need to carry out. For example, it will ask two platoons to merge to a single platoon whose size is closer to the optimum or, given a command like "30% of the vehicles going to the next exit change lanes now", it will decide which vehicles will comprise this 30% and split the platoons accordingly to let them out. The current design [16] uses protocols, in the form of finite state machines, to organize the maneuvers in a systematic way. They receive the commands of the link layer and aggregated sensor information from the individual vehicles (of the form "there is a vehicle in the adjacent lane"). They then use this information to decide on a control policy and issue commands to the regulation layer. The commands are typically of the form "accelerate to merge to the preceding platoon" or "decelerate so that another vehicle may move into your lane ahead of you".

Below the coordination layer lies the **regulation layer**. Its task is to receive the coordination layer commands and translate them to throttle, steering and braking input for the actuators on the vehicle. For this purpose it utilizes a number of continuous time feedback control laws ([9, 12, 28, 32]) that use the readings provided by the sensors to calculate the actuator inputs required for a particular maneuver. The regulation layer occasionally needs to communicate with the coordination layer to inform it of the outcome of the maneuver.

The bottom layer is not part of the controller. It is called the **physical layer** and it contains the actual plant (in this case the vehicles with their sensors, actuators and communication equipment and the highway topology). For the purposes of simulation it can be assumed that the physical layer contains models of the actual physical quantities. From a hybrid systems point of view, the physical layer can be merged with the regulation layer.

2.2 Hybrid control problem

The current paper focuses on the coordination and regulation layers and their interaction. Our aim is to demonstrate that the behavior of the overall hybrid control system may display characteristics, possibly undesirable, that are not predicted by analyzing the individual layers. In order to do this we will first describe briefly a possible IVHS design for which a dedicated simulation tool has been developed. In Section 3.1, the results in [9] that describe a possible design for the regulation layer are presented, while in Section 3.2 we give an outline of the results in [16] where the design of the coordination layer is described. Finally in Section 3.3 we discuss the results in [20] where an interface between these two layers is presented. References to alternative designs are also provided.

The references provide proofs of performance bounds for the individual layers. Still however no proof of performance for the overall design exists. The only means available for testing it at the moment is the SmartPath simulator [7]. In Section 4 of this report we present the results of extensive simulations performed using this tool. It turns out that the performance of the combined system is not quite the expected. In particular, situations arise indicating limitations of the design that were not predicted by the individual layer analysis. Even though ways of designing around these shortcoming exist in most cases, it becomes apparent that if any faith is to be placed in the overall design some way of systematically determining its performance and limitations must be found. The failure of the standard tools (both in the finite state and the continuous domain) to predict the limitations discovered by means of simulation indicates that new, more powerful tools are needed to achieve this goal. Our current work focuses on the development of such tools.

3 Design of Individual Layers

In this section we will give a brief description of the multilayered control design that was implemented in the SmartPath simulator. We focus our attention on the regulation (continuous) and coordination (discrete) layers and their interface.

It should be noted that extensive work has also been done in the link layer design (see [31]) while work is already in progress for the network layer. The link layer control scheme has also been added to the SmartPath simulator. However, the simulation results presented here were obtained using a simplified version of SmartPath where the detailed link layer design is substituted by a small number of simple rules; for example such a rule might be "all vehicles coming in the freeway should move to the fast lane and stay there until they are one mile from their exit". The justification for this simplification is that we are interested in the interaction of the two bottom layers of the control architecture and the problems it pauses as a hybrid system and do not want the link layer intelligence to "pollute" this behavior. A few more thoughts on the merits of a simplistic abstraction for the link layer are given in Section 4.5.

It should also be noted that the IVHS design presented here is by no means unique or optimal, but is sufficient to illustrate our point. References to possible alternative designs will be given throughout the paper.

3.1 Continuous layer

Conventional differential equation models are used to design continuous control laws at this level, which contains both the regulation and the physical layer of Figure 2. The advantage of this framework is that it supports well established verification techniques in the form of mathematical proofs. The procedure used for the design of most of the control laws that appear in the IVHS architecture in question is to approximate the dynamics of a vehicle by a linear or feedback linearizable system. The controllers designed for this system are robust enough to take care of any unmodeled dynamics. This process simplifies the task of analyzing the stability and performance of the algorithms. The regulation layer contains six such control laws:

- Leader Control law:
  We assume that the platoon leaders will implement an Autonomous Intelligent Cruise Control (AICC) law like the one presented in [9][3]. The goal of this controller is to maintain safe inter-platoon spacing (which is taken as a constant time headway of 1 second in [9]) and, if possible, track the optimal velocity determined by the link layer (typically 60-65 miles per hour). Because the objectives of this controller are very similar to the objectives of human drivers, this control law is used as the default for a leader, i.e. it is the law implemented unless there is a specific command to do otherwise.

[3] See [17] for a different design of AICC control law.

- Follower control law:
  This is the default control law for the followers in a platoon (see [12, 32]). Its task is simply to track the velocity of the vehicle in front while keeping a tight spacing (1 meter in this case).
- Merge control law:
  Merge is the action taken by two platoons that want to become one. The trailing platoon, that requests the merge, accelerates to catch up with the leading platoon and joins it. This acceleration is done according to a reference trajectory which is calculated based on the assumption that the preceding platoon will be moving at constant velocity throughout the entire maneuver. State feedback is then used to stabilize the closed loop system about this trajectory and hence eliminate the effect of any acceleration or deceleration of the leading platoon (see [9]).
- Split control law:
  Split is exactly the opposite of merge. A follower becomes the leader of all the cars that follow it (in the same platoon) and decelerates until it reaches safe inter-platoon distance from the mother platoon. Like the merge control law, a reference open loop trajectory is calculated and then state feedback is used to guarantee asymptotic tracking.
- Change lane control law:
  Apart from the obvious lateral (steering) action, change lane also requires longitudinal action in terms of aligning the vehicle that wishes to change lane with a proper gap in the target lane. This action is in principle similar to a split as discussed in [9]. Both the longitudinal and lateral components of this maneuver are carried out by controllers that are designed in the same way as the merge or split controllers.
- Lane keeping lateral control law:
  The objective of this law is to keep the vehicle in the center of the lane. This is achieved by means of a frequency shaped linear quadratic (FSLQ) regulator [28].

It can be shown that each of the control laws described above leads to a closed loop system that is stable and capable of performing the desired task adequately. The proofs and simulation results of this claim can be found in the appropriate references ([9, 12, 28]). In addition, some of these controllers have been tested experimentally on actual vehicles with satisfactory results.

The continuous layer in the sense of Figure 1 also contains the physical layer of Figure 2. The work presented here was carried out under the assumption that the operation of the sensors and actuators of the physical layer of all vehicles is perfect, i.e. there are no faults, no time delays, no steady state tracking error, no false measurements, etc. The only restrictive assumption we impose is limits on the ranges of the sensors and actuators. These limits are based on experimental data and assume ideal environmental conditions (perfect road surface and perfect weather). Therefore they are, if anything, optimistic given the current technology. They are summarized in Table 3.1.

As will become apparent in Section 4, these limits play a crucial role in the behavior of the combined system. In particular, they imply the existence of a region

Max. Acceleration	3 m/s^2
Max. Deceleration	-5 m/s^2
Distance Sensor Front	60 m
Distance Sensor Rear	30 m
Distance sensor Adjacent Lane, Front	30 m
Distance sensor Adjacent Lane, Rear	30 m

Table 1. Constraints on Actuators and Sensors

in the state space of the lead vehicle, corresponding to certain severe disturbances, from which the leader control law can not recover (see [9] for details).

3.2 Discrete layer

The discrete layer works at a more abstract level than the continuous layer. Many different formalisms exist for describing this layer; standard choices include finite state machines, Petri or Neural Nets and Fuzzy Logic among others. The framework of most of these techniques supports methods of carrying out proofs of correctness. For example, a very effective proof method is the one used for the verification of finite state machines. It is based on the fact that, because of the finiteness of the state space, it is possible to enumerate all possible traces that the machine accepts and hence verify that all possible sequences of events possess certain desirable properties. Tools exist in the form of computer programs that perform this verification task automatically. Of course the actual tools are a lot more sophisticated than the simplistic description given above and are designed to be computationally efficient, a very useful property as the machines in question often have hundreds of thousands or even millions of states.

In our example we assume that the discrete layer contains only the coordination layer. As mentioned in the introduction a simple abstraction will be used for the link layer. The task of this discrete layer is to provide a consistent way of coordinating the maneuvers of adjacent platoons so that they are effective (vehicles get to their destination) and efficient (capacity is maximized) without compromising safety. To facilitate the analysis and keep the problem tractable, the design in [16] distinguishes only three maneuvers: *Merge* to form a single platoon from two platoons, *Split* to do the opposite and *Change* to move a vehicle from one lane to the other. To further simplify the situation it is assumed that each platoon can only be involved in one maneuver at a time and that only free agents, that is one car platoons, can change lane. Clearly these assumptions lead to a rather restricted set of possible behaviors. It is possible to obtain alternative designs by relaxing some of them (see for example [14]). However, the restrictions allow us to keep track of the problem and prove that the design possesses certain desirable properties.

The coordination layer controller is supposed to carry out these three maneuvers efficiently and safely. To accomplish this, the coordination layer of each vehicle exchanges messages with neighboring vehicles according to certain protocols (which in

[16] are modeled by finite state machines). When mutual agreement (coordination) is reached, the coordination layer controller commands the regulation layer to carry out the appropriate maneuver. The protocols were verified to posses certain desirable properties using an automated software verification tool, COSPAN [11]. For the purpose of this verification the behavior of the continuous layer was abstracted by a set of finite state machines, that are supposed to model the behavior of the sensors and the conventional controllers from the protocol point of view. So the verification proved that the protocol logic is "correct" when coupled with the abstraction of the continuous layer. How closely the abstractions match the behavior of the actual system is still an open question.

3.3 Interface

Interface design is the most challenging part of a hybrid system since it straddles both the continuous and the discrete world. A good interface design is very important, as it determines to a large extend what one can prove about the combined system. It may also help us extend advantages of the discrete event system theory (e.g., ease of computation) to the continuous domain and conversely (e.g., derive mathematical proofs in discrete space from equivalent proofs on continuous state spaces).

In applications such as ours, the interface is usually designed last. The disadvantage of this approach is that it leads to interfaces that are case specific and whose performance is limited by the limitations of the rest of the design. One of the goals of this research is to develop a technique for systematically designing interfaces with desirable properties for general systems.

In our example, the interface is a finite state machine whose transitions depend upon the commands from the coordination layer, the readings of the sensors (physical layer responses) and the state of the continuous controllers. It plays a dual role. On the one side it acts as a symbol to signal translator and therefore directly influences the evolution of the continuous system. It receives the coordination layer commands (symbols) and uses them to switch between different continuous layer controllers (signals). In addition it keeps track of which of these controllers need to be initialized (symbol) and carries out this initialization by directly changing the controller state (signal). In the other direction the interface acts as a signal to symbol translator. It processes the sensory information (signal) and presents it to the coordination layer in an aggregate form compatible with the finite state machine formalism (symbol). It also monitors the evolution of the continuous system (signal) and decides if the maneuver in progress is safe or not. If at any stage the maneuver becomes hazardous it aborts the maneuver, notifies the coordination layer of its decision (symbol) and switches to a different continuous control law that will get the system back to a safe configuration.

Unfortunately there is no systematic way of verifying the complete interface. [20] describes the verification of the discrete part of the interface, where the continuous state considerations are either ignored or abstracted. However, the addition of the actual continuous effects to the framework in a consistent way is not supported by the current theory.

The only way of testing the behavior of the combined system at the moment is by simulation. Even though successful simulation results are not nearly as good as a mathematical proof of the properties of the system, a failure in the simulation can be considered as a proof of a shortcoming of the design. This approach is used in the next section to point out problems in the IVHS design outlined above.

4 Properties of Combined System

The design described above was put to test by implementing it in simulation. The result was the dedicated simulator SmartPath, described in [7]. An effort was made to make the simulator as flexible and modular as possible so that changes in the design (e.g., alternative regulation layer controllers) can be implemented easily. Using this tool, long simulations for a variety of initial conditions and inputs and a few choices of simplified link layer design were carried out. Contrary to our expectation we observed quite a few scenarios where the performance of the system was inferior to the one predicted by the individual level analysis, for example, unpredicted car crashes occurred. We now document these different scenarios.

4.1 Changing from a slow to a fast lane

According to the coordination layer design, only free agents (single vehicle platoons) are allowed to change lane. Before a vehicle initiates a lane change it looks (through its sensors) to the adjacent lane to make sure that there is room for it there. If no vehicle is visible in the sensor range the move is initiated immediately. If a vehicle is found and its distance is less than the safe inter-platoon spacing, communication is established to coordinate the maneuver. This goes on until a gap twice as large as the safe inter-platoon spacing is found. Then the lane change takes place in the middle of this gap.

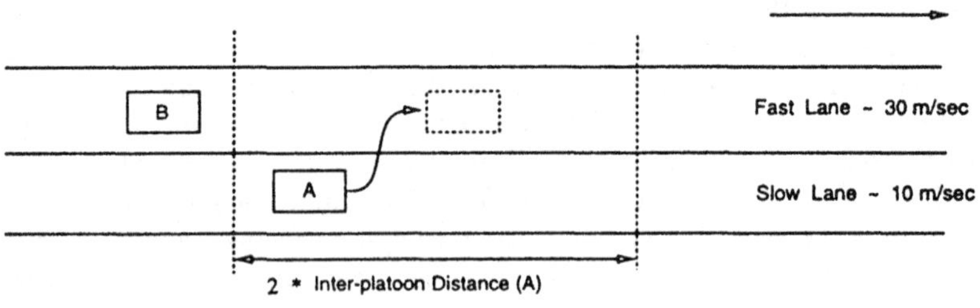

Fig. 3. Change lane from a slow to a fast lane

In most situations this arrangement should cause no problems. Indeed both the protocol that coordinates the maneuver and the regulation layer controllers that align the free agent with a gap in the next lane have been proven to perform well. Consider however the scenario shown in Figure 3. Free agent A switches from a slow lane to a fast lane. During the change a gap big enough for A to move into is present in the fast lane (for example no vehicle is visible in the sensor range). It is conceivable however that a vehicle (denoted by B) is present in the fast lane behind A, which, after the lane change is complete, finds itself just outside A's lateral sensor range (say $35m$) and moving a lot faster than A (say $30m/s$ as opposed to $10m/s$). It turns out that the AICC lead controller is incapable of recovering from such drastic initial conditions, so a crash is inevitable.[4]

We were able to create this kind of crash in a relatively light traffic situation by asking many vehicles to change lanes at the same time in order to leave at an intersection. As the vehicles need to be free agents before they change lane, many splits were carried out which inevitably lead to a large deceleration in the lane of origin. This deceleration was not present in the target lane however, so, after a few seconds, the vehicles in the target lane found themselves moving a lot faster than the vehicles in the origin lane. Crashes of the kind described above were then observed. An example of this scenario is given in Figure 7 (where each point represents the position of a vehicle at the given time instant).

4.2 Changing from a fast to a slow lane

This is a mirror image of the situation of previous case. In this case (Figure 4),

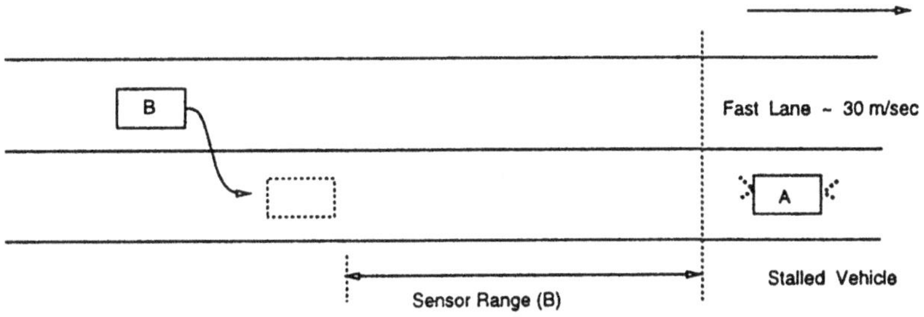

Fig. 4. Change lane from a fast to a slow lane

vehicle B changes lane and hits a slow moving (or stalled) vehicle A from behind. During the lane change vehicle A was just outside the range of the forward adjacent lane sensor of B. After the change was complete the front sensor of B reveals the presence of A but it is already too late to stop. The scenarios that might lead to such a crash are the same as the ones of Section 4.1, with the origin and target lanes interchanged.

It turns out that it is very difficult to create a crash like this using the current version of SmartPath. Figure 13 shows an extreme case where a stalled vehicle is seen at the edge of the front sensor range (60 m). The optimum velocity commanded by the link layer is too high for the second vehicle to stop in time and therefore a crash is unavoidable.

[4] Note that the safe inter-platoon spacing for a vehicle is affine in its velocity. In Figure 3, the safe inter-platoon spacing is 20 m for vehicle A but 40 m for vehicle B.

4.3 Merging to a decelerating platoon

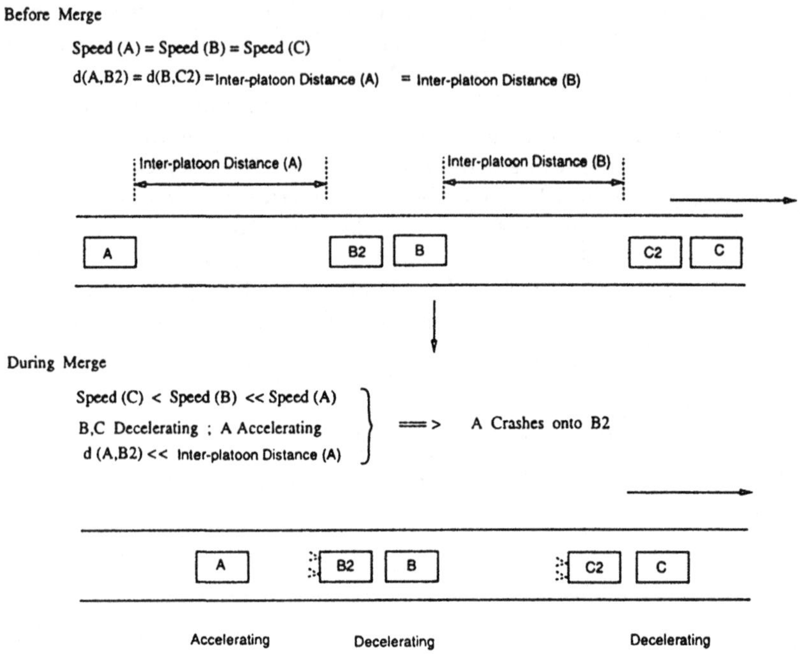

Fig. 5. Crash during Merge maneuver

With the merge control law described in Section 3.1, changes in the velocity of the front platoon should cause no problem as the state feedback takes care of any deviations from the desired trajectory. However, the actual trajectory *will* deviate from the desired one if the limits of the actuators (throttle and brake) are reached. To avoid this possibility, the interface aborts the maneuver when it detects the danger of actuator saturation (see [20]). After aborting the maneuver, the system should find itself in a position from which it can continue safely under the AICC lead control law. The simulation indicates that under extreme conditions this may not be true and merging may cause a major hazard. We were able to recreate two such scenarios.

In the first case we created a large deceleration in a lane by making the simplified link layer ask many of the vehicles in the adjacent lanes to move to the lane in question. The vehicles already in the lane were then forced to decelerate to create space for the incoming vehicles. In the second scenario we caused a large deceleration by asking many of the vehicles in a densely occupied lane to exit. As already discussed this caused a number of splits in the platoons as the vehicles tried to become free agents in order to change lanes. In both situations the deceleration built up enough to cause saturation of the actuators. Therefore merging vehicles upstream of the disruption were forced to abort their maneuvers. This led to a situation where vehicles (like vehicle A in Figure 5) found themselves close to the preceding platoon (B-B2) which is already decelerating at the maximum rate, while moving faster than it. Hence, even though A decelerates at saturation level as well, a crash is unavoidable. A situation where such a crash is observed is shown in Figures 9 and 10.

Note: Vehicle crashes associated with merge maneuver were also predicted by A. Hitchcock (see [15]) using *fault tree analysis* method.

4.4 Multiple lane changes from a single platoon

The situation described above, where many vehicles wish to change out of a lane thus creating multiple splits, caused yet another kind of crash. The scenario is outlined in Figure 6.

A platoon breaks up so that vehicles B, C and D can leave the lane. The current interface declares a split maneuver complete the moment the deceleration begins. This allows the coordination layer to initiate a lane change maneuver while the split deceleration is taking place. This is done to improve the rate at which vehicles leave the lane. However, it is conceivable that three splits will occur before a lane change is initiated. So B, C and D will all become free agents one after the other. This will cause no problem as long as the vehicles stay in the lane. Suppose though that all three of them want to move to the next lane. As discussed above they will all look at the target lane through their sensors and, assuming they find it empty, will all start moving over at roughly the same time. If D reaches the target lane first it will start accelerating to get back to the optimum speed. In the meantime C, which has been decelerating because of the split maneuver, reaches the target lane. As a result D finds itself very close to C and moving slightly faster. It is very likely that this situation might lead to a crash. Indeed Figure 11 shows the results of a simulation where such a crash was observed. It should be noted that crashes like these are typically not severe, as the relative velocity at impact is likely to be very small.

4.5 Crash Analysis

The common feature of all the crashes described above is that they are caused by continuous layer performance not accounted for by the discrete layer. In all cases this leads to the discrete layer making requests that are incompatible with the current situation of the continuous layer, such as a potentially dangerous maneuver. The situations in Sections 4.1 and 4.2 arise because the limits on the sensor range and the continuous time controller performance are not represented adequately by the corresponding coordination layer abstractions. The same is true for the situation in 4.3, only instead of the sensor limits the problem here arises because of the actuator limits. Finally, the problem in Section 4.4 is caused by the fact that the coordination layer implicitly assumes that, during a lane change, vehicles in the origin lane will already be at a safe distance, an assumption related to the continuous layer. As a result it does not establish communication with them and therefore is oblivious of their intention to change lanes.

Clearly most of these crashes could have been avoided by changes in the controller design (see Section 4.6). This fact however is besides the point. For one, ad-hoc

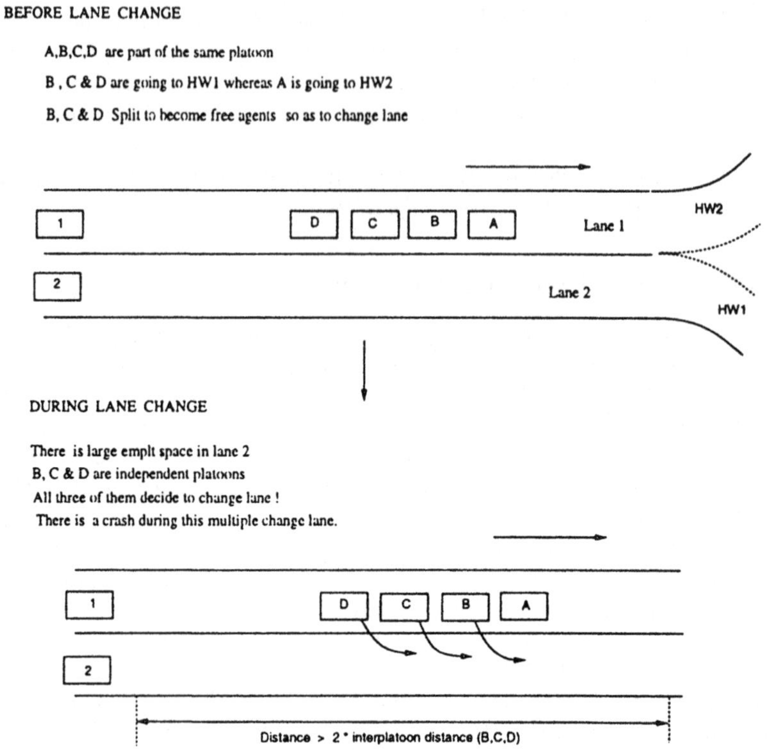
BEFORE LANE CHANGE
A,B,C,D are part of the same platoon
B , C & D are going to HW1 whereas A is going to HW2
B, C & D Split to become free agents so as to change lane
1
D
C
B
A
Lane 1
HW2
2
Lane 2
HW1
DURING LANE CHANGE
There is large emplt space in lane 2
B, C & D are independent platoons
All three of them decide to change lane !
There is a crash during this multiple change lane.
1
D
C
B
A
2
Distance > 2 * interplatoon distance (B,C,D)

Fig. 6. Multiple Change Lane

changes like that might have unpredictable effect on other aspects of the system performance (e.g., capacity). More importantly, the simulation results demonstrated that, even though the individual layers have been verified independently to exclude certain undesirable kinds of behavior, the combined system is capable of producing them. It should also be noted here that the list of faults presented above is by no means exhaustive. The fact that we were able to identify only the above scenarios as dangerous does not mean that there do not exist other situations that may lead to unpredictable catastrophes. These observations naturally lead to the question to what extent can one trust the conventional discrete and continuous verification techniques when it comes to hybrid systems. Clearly if any faith is to be placed in the IVHS design presented here, a proof of its performance claims is needed. The above discussion indicates that such a proof is not possible using conventional tools.

As already mentioned the link layer designs used in the simulation were extremely simple. One might feel that the crashes could have been avoided with a more realistic and intelligent link layer design. To make the design less sensitive to faults such as roadside to vehicle communication breakdowns, we would like to avoid assigning such a safety critical role to the link layer. In any case, the performance claims made by the verification of the individual layers about the coordination-regulation layer interaction are independent of the link layer design.

It should be noted here that the above discussion should not give the impression that car crashes are a common occurrence in the proposed IVHS design. In fact they are rather rare and occur only under special, extreme conditions, which we mostly set up specifically in order to observe the crashes. Statistically speaking the crash described in Section 4.3 is the most common. Crashes of the kind described in Section 4.1 are rather rare and only happen under extreme conditions. The same is true for the crashes described in Section 4.4 which are even more rare. Finally crashes like the ones in Section 4.2 are very rare indeed and are only observed under scenarios that were artificially created.

4.6 Redesign of IVHS controller for safety

Some of the crashes have simple solutions. For example to exclude the scenario in Section 4.4, we could ask the interface to declare the split maneuver complete (thus allowing the initiation of a new maneuver) *after* the splitting platoon has decelerated all the way to safe spacing.[5] This will increase the amount of time needed for a lane change and the amount of disturbance to upstream vehicles. The link layer design has to take this into account in deciding how far away it has to command lane changes so that vehicles do not miss their exits.

In the case of merge, the merging vehicle goes from one safe state (large inter-platoon distance) to another (small intra-platoon distance) via an unsafe region (close to the vehicle ahead and moving faster). If a large disturbance is created downstream when the merging vehicle is in the unsafe region, then aborting the merge maneuver may result in a crash. There are two ways of avoiding this problem. The total number of maneuvers in a section can be restricted so that a large disturbance is never created for vehicles upstream. This is not a good solution as it involves controlling traffic downstream of the vehicle, it assigns a safety critical role to the link layer and its effectiveness is difficult to prove formally. The second solution, presented in [8], involves redesigning the merge maneuver trajectory so that regardless of the disturbance downstream, the merge maneuver is never aborted. The merge trajectory of [8] is designed to avoid any high[6] relative velocity collisions. Obviously, this trajectory takes longer to complete than the minimum time trajectory of [9]. The redesigned merge maneuver guarantees safe operation but it has an adverse effect on capacity.

Lane change introduces the most complicated disturbance. With the bound on maximum deceleration given by $-5m/s^2$, a simple calculation shows that it takes $90m$ for a vehicle to stop from $30m/sec$. One way to avoid crashes due to lane changes across lanes with speed differential is to increase the range of lateral communication devices and lateral distance and velocity sensors to $90m$ in all directions (lateral front, lateral rear, left and right). Implementation of this solution could be possible

[5] Another alternative is to extend the lane change protocol to include communication between vehicles in your own lane. This will have less impact on capacity but it increases the complexity of the communication protocol.

[6] A mapping between relative velocity of collision and the severity of an accident is obtained in [13]. An accident is assumed to be safe if the resulting passenger injury does not require hospitalization. Based on real accident data, it is concluded that all accidents with a relative crash velocity less than 3.3 m/sec are safe in the above sense.

using cameras for lateral distance and velocity sensing [18]. The interface should also be changed to include a safety check that looks at relative velocity along with relative distance from vehicles in the next lane before deciding to change lane. This also means that lane changes should be prohibited at places on the highway which does not allow wide range of lateral sensing, e.g. near sharp curves. This scheme will reduce the number of successful lane changes and consequently affect the capacity of the overall system.

5 Concluding Remarks

Summarizing the above results, the regulation and coordination (continuous and discrete) layers of the IVHS architecture shown in Figure 2 were independently developed and then combined by an interface. At each level (discrete, continuous and interface) verification of desirable properties was carried out using the appropriate tools. However, contrary to expectations, problems were detected in the combined system performance by simulation. The problems were due to the fact that the discrete layer can only comprehend abstractions of the continuous layer. As a result it does not take sufficiently into account certain continuous layer parameters, which in our case were the sensor and actuator ranges and the controller performance bounds. These observations led to the conclusion that in order to fully trust the design we need verification tools that test the performance of the combined *hybrid* system.

Computer aided verification of timed systems [1] and hybrid systems [2, 29] is currently an active area of research. For verification, the actual behaviors of the hybrid systems are usually abstracted into a finite set of behaviors. Thus one trajectory of the abstraction will include many possible system trajectories which are "close" to it in some sense. This is a promising development as, for example, it could be used in the context of IVHS to prove that a particular design will not produce any crashes in normal mode. It should be noted however that the application of such a technique to the current design may be inconclusive as the failure of the verification may be attributed to either a faulty design or over-abstraction.

In terms of progress in the IVHS problem, the simulation results may also suggest considerations to be taken into account by the link layer. For example, while dealing with an accident downstream, it might seem natural to declare all vehicles in a platoon to be free agents in order to try and get as many of them out of this lane as possible. But this will create successive splits and excessive decelerations for vehicles further upstream. Under these conditions, crashes like the ones described in Sections 4.1 and 4.3 will be possible. Similarly, downstream from an accident, vehicles will want to move from the congested (free) lane to the nearly empty lane (where the accident was). This may again lead to the same type of crashes. Finally, the risks associated with the merge maneuver may be unavoidable without extensive redesign, as the cause of deceleration (e.g. splitting of platoons downstream etc.) are uncontrollable and can occur any time during the merge [7]. These results also indicate that the performance of the current design, even with the link layer, may not be

[7] There exists a different scheme of organizing traffic without merge maneuver due to A. Hitchcock [14]. This architecture is currently under investigation. It will probably still be vulnerable to crashes during lane changes.

robust enough to deal with certain situations. We collectively refer to such extreme conditions as "degraded modes of operation". More work is currently in progress to provide alternatives to the current design in order to deal with these scenarios [21].

Acknowledgment: The authors would like to thank Farokh Eskafi, Delnaz Khorramabadi, Dr. Bobby Rao and Prof. Pravin Varaiya for helpful discussions providing insight into the problem.

References

1. R. Alur, C. Courcoubetis, and D. Dill. Model checking for real-time systems. *Logic in Computer Science*, pages 414–425, 1990.
2. R. Alur, C. Courcoubetis, T. A. Henzinger, and P. H. Ho. Hybrid automaton: An algorithmic approach to the specification and verification of hybrid systems. In Robert L. Grossman, Anil Nerode, Anders P. Ravn, and Hans Rischel, editors, *Hybrid System*, pages 209–229. Springer Verlag, New York, 1993.
3. Allen Back, John Guckenheimer, and Mark Myers. A dynamical simulation facility for hybrid systems. Technical Report 92-6, Cornell University, April 1992.
4. V. Borkar, M. S. Branicky, and S. K. Mitter. A unified framework for hybrid control. (preprint).
5. R. W. Brockett. Hybrid models for motion control systems. In H.L.Trentelman and J.C.Willems, editors, *Essays on Control: Perspectives in the Theory and its Applications*, pages 29–54. Birkhauser, Boston, 1993.
6. R. W. Brockett. Pulse driven dynamical systems. In A. Isidori and T.J. Tarn, editors, *Systems, Models and Feedback: Theory and Applications*, pages 73–80. Birkhauser, Boston, 1993.
7. Farokh Eskafi, Delnaz Khorramabadi, and Pravin Varaiya. SmartPath: An automated highway system simulator. PATH Technical Report UCB-ITS-94-4. Institute of Transportation Studies, University of California, Berkeley, 1994.
8. J. Frankel, L. Alvarez, R. Horowitz, and P. Li. Robust platoon maneuvers for AVHS. (preprint), November 1994.
9. Datta N. Godbole and John Lygeros. Longitudinal control of the lead car of a platoon. *IEEE Transactions on Vehicular Technology*, 43(4):1125–1135, 1994.
10. Robert L. Grossman, Anil Nerode, Anders P. Ravn, and Hans Rischel. *Hybrid Systems*. Springer Verlag, 1993.
11. Z. Har'El and R.P. Kurshan. *Cospan User's Guide*. AT&T Bell Laboratories, 1987.
12. J. K. Hedrick, D.McMahon, V. Narendran, and D. Swaroop. Longitudinal vehicle controller design for IVHS system. In *American Control Conference*, pages 3107–3112, 1991.
13. A. Hitchcock. Intelligent vehicle/highway system safety: Multiple collisions in AHS systems. PATH Technical Report, Institute of Transportation Studies, University of California, Berkeley, CA, 1994.
14. A. Hitchcock. A specification of an automated freeway with vehicle-borne intelligence. PATH Technical Report UCB-ITS-PRR-92-18, Institute of Transportation Studies, University of California, Berkeley, 1994.
15. Anthony Hitchcock. Casualties in accidents occuring during split and merge maneuvers. PATH Technical Memorandum 93-9, Institute of Transportation Studies, University of California, Berkeley, 1993.

16. Ann Hsu, Farokh Eskafi, Sonia Sachs, and Pravin Varaiya. Protocol design for an automated highway system. *Discrete Event Dynamic Systems*, 2(1):183–206, 1994.
17. P. Ioannou, Z. Xu, S. Eckert, D. Clemons, and T. Sieja. Intelligent cruise control: Theory and experiments. In *IEEE Control and Decision Conference*, pages 1885–1890, 1993.
18. D. Koller, T. Luong, and J. Malik. Binocular stereopsis and lane marker flow for vehicle navigation: Lateral and longitudinal control. Technical Report UCB/CSD-94-804, University of California, Berkeley, 1994.
19. M. Lemmon, J. A. Stiver, and P. J. Antsaklis. Event identification and intelligent hybrid control. In Robert L. Grossman, Anil Nerode, Anders P. Ravn, and Hans Rischel, editors, *Hybrid System*, pages 268–296. Springer Verlag, New York, 1993.
20. John Lygeros and Datta N. Godbole. An interface between continuous and discrete-event controllers for vehicle automation. In *American Control Conference*, pages 801–805, 1994.
21. John Lygeros, Datta N. Godbole, and Mireille E. Broucke. Design of an extended architecture for degraded modes of operation of IVHS. In *American Control Conference*, 1995. To Appear.
22. A. S. Morse. Supervisory control of family of linear set point controllers. In *IEEE Control and Decision Conference*, pages 1055–1060, 1993.
23. Kumpati S. Narendra and Jayendran Balakrishnan. Improving transient response of adaptive control systems using multiple models and switching. In *IEEE Control and Decision Conference*, pages 1067–1072, 1993.
24. A. Nerode and W. Kohn. Models for hybrid systems: Automata, topologies, controllability, observability. In Robert L. Grossman, Anil Nerode, Anders P. Ravn, and Hans Rischel, editors, *Hybrid System*, pages 317–356. Springer Verlag, New York, 1993.
25. A. Nerode and W. Kohn. Multiple agent hybrid control architecture. In Robert L. Grossman, Anil Nerode, Anders P. Ravn, and Hans Rischel, editors, *Hybrid System*, pages 297–316. Springer Verlag, New York, 1993.
26. K. M. Passino and P. J. Antsaklis. Modeling and analysis of artificially intelligent planning systems. In Panos J. Antsaklis and Kevin M. Passino, editors, *An Introduction to Intelligent and Autonomous Control*, pages 191–214. Kluwer Academic Publishing, Boston, 1993.
27. Kevin M. Passino and Alfonsus D. Lunardhi. Stability analysis of expert control systems. In *IEEE Control and Decision Conference*, pages 765–770, 1993.
28. H. Peng and M. Tomizuka. Vehicle lateral control for highway automation. In *American Control Conference*, pages 788–794, 1990.
29. Anuj Puri and Pravin Varaiya. Decidebility of hybrid systems with rectangular differential inclusions. In *Computer Aided Verification*, pages 95–104, 1994.
30. P. J. G. Ramadge and W. M. Wonham. The control of discrete event dynamical systems. *Proceedings of IEEE*, Vol.77(1):81–98, 1989.
31. B. S. Y. Rao and Pravin Varaiya. Roadside intelligence for flow control in an IVHS. *Transportation Research - C*, 2(1):49–72, 1994.
32. Shahab Sheikholeslam and Charles A. Desoer. Longitudinal control of a platoon of vehicles. In *American Control Conference*, pages 291–297, 1990.
33. Lucio Tavernini. Differential automata and their simulators. *Nonlinear Analysis, Theory, Methods and Applications*, Vol.11(6):665–683, 1987.
34. P. Varaiya and Steven E. Shladover. Sketch of an IVHS systems architecture. Technical Report UCB-ITS-PRR-91-3, Institute of Transportation Studies, University of California, Berkeley, 1991.
35. Pravin Varaiya. Smart cars on smart roads: problems of control. *IEEE Transactions on Automatic Control*, AC-38(2):195–207, 1993.

A Simulation Results

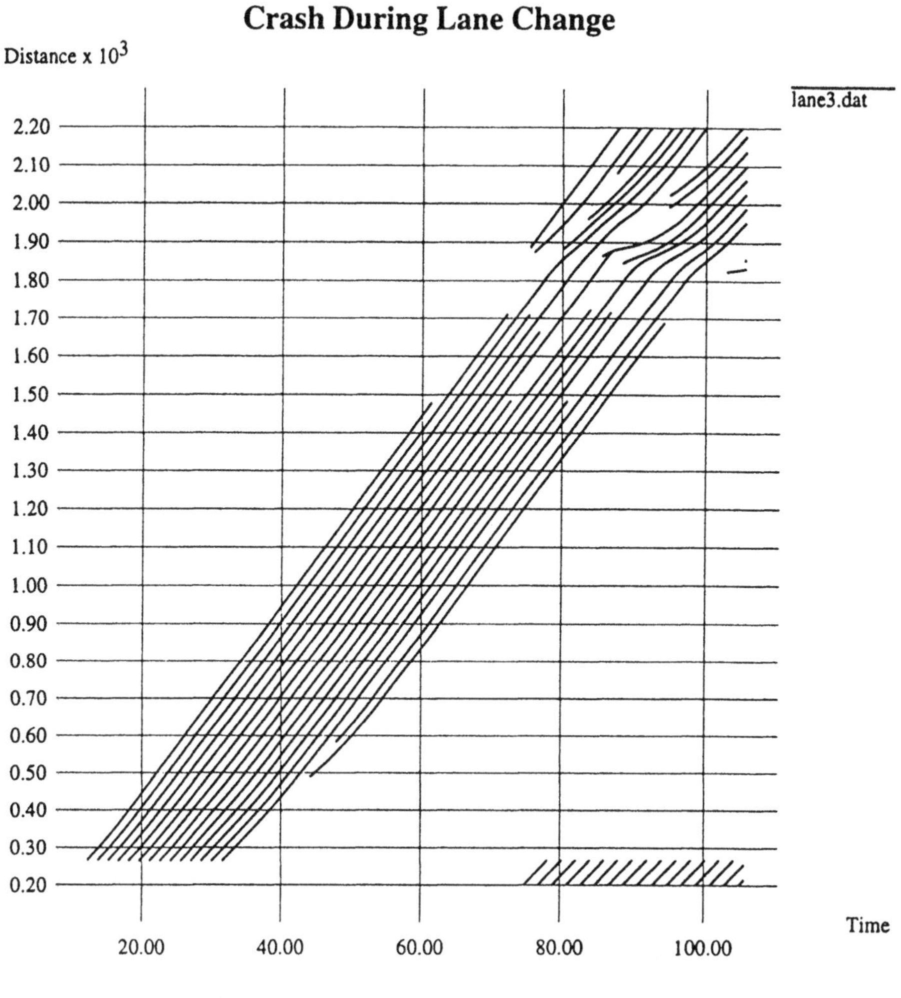

Fig. 7. Lane change from slow to fast lane: the crash occurs at 86.91 seconds at a distance of 1870 meters. The next figure shows a close up view of the crash

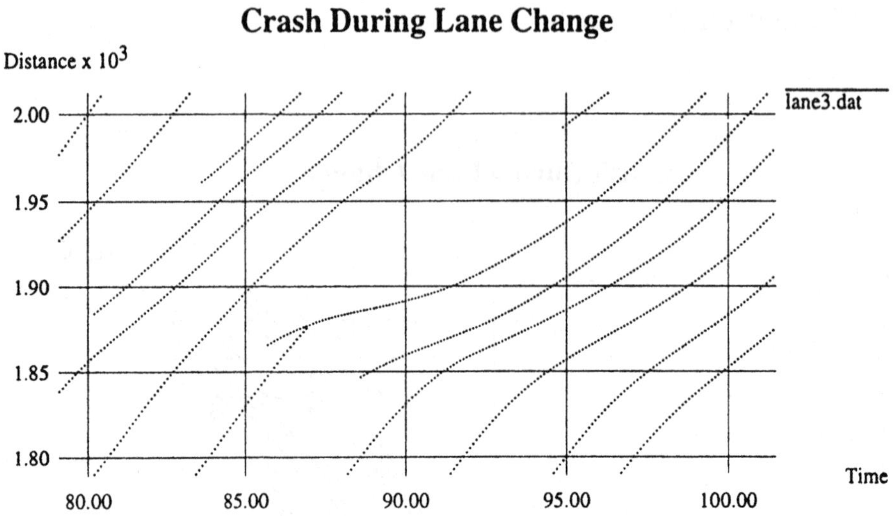

Fig. 8. Lane change from slow to fast lane: close up view of the crash

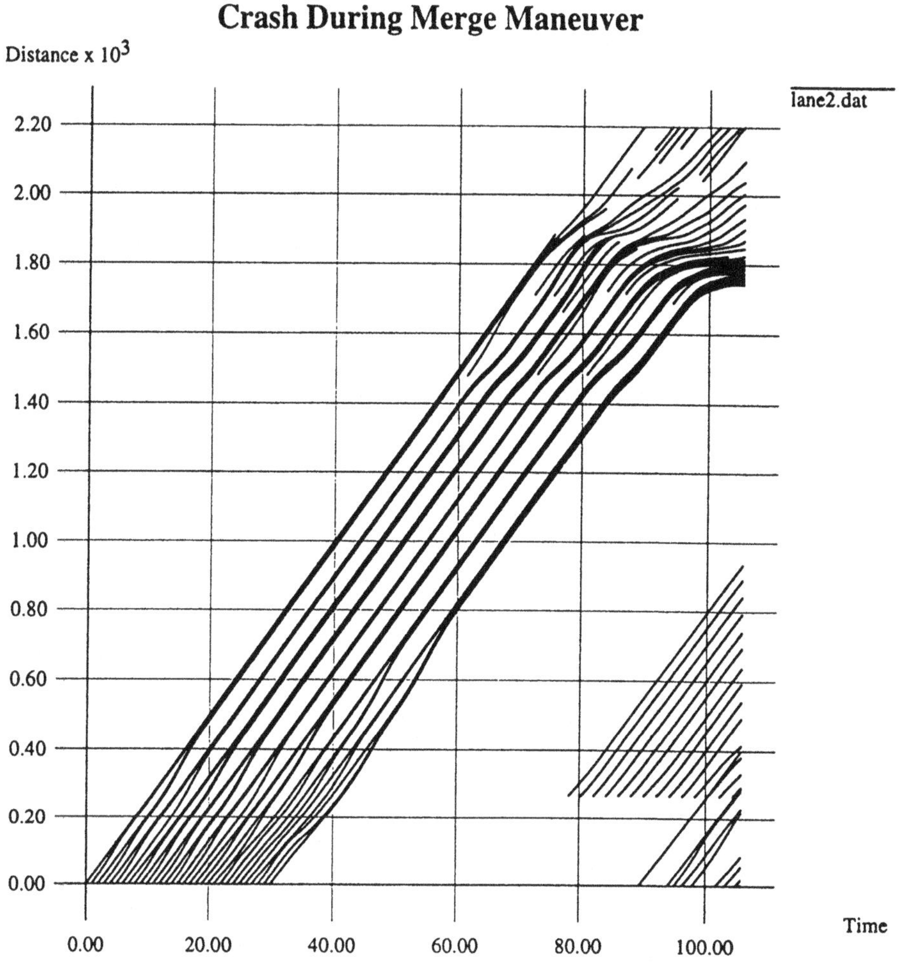

Fig. 9. Distance-Time plot of lane 2 showing crashes related to merge maneuvers: there is a four car pile-up at 82 seconds and 1880 meters whose close up is presented in the next plot. There is another crash at 94.8 seconds and 1780 meters

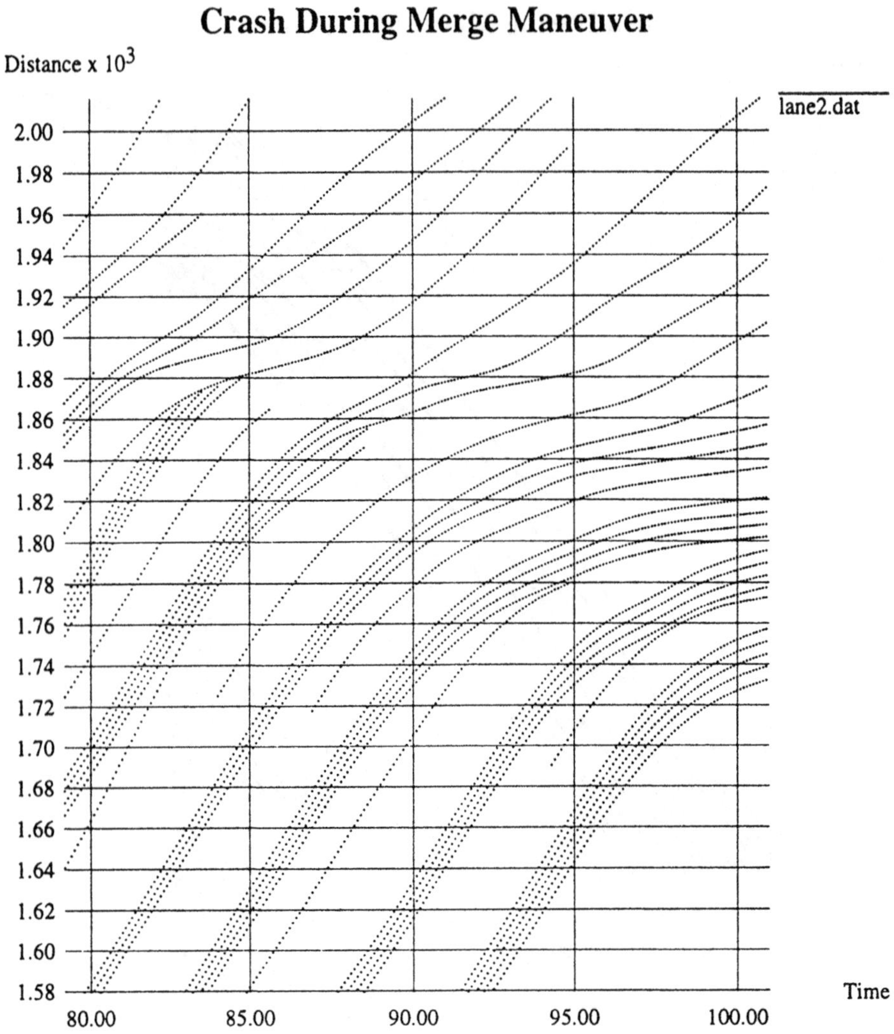

Fig. 10. Both the crashes mentioned above can be seen in this plot clearly. The crashes during the merge maneuver occur because of a large deceleration created downstream by the highway junction

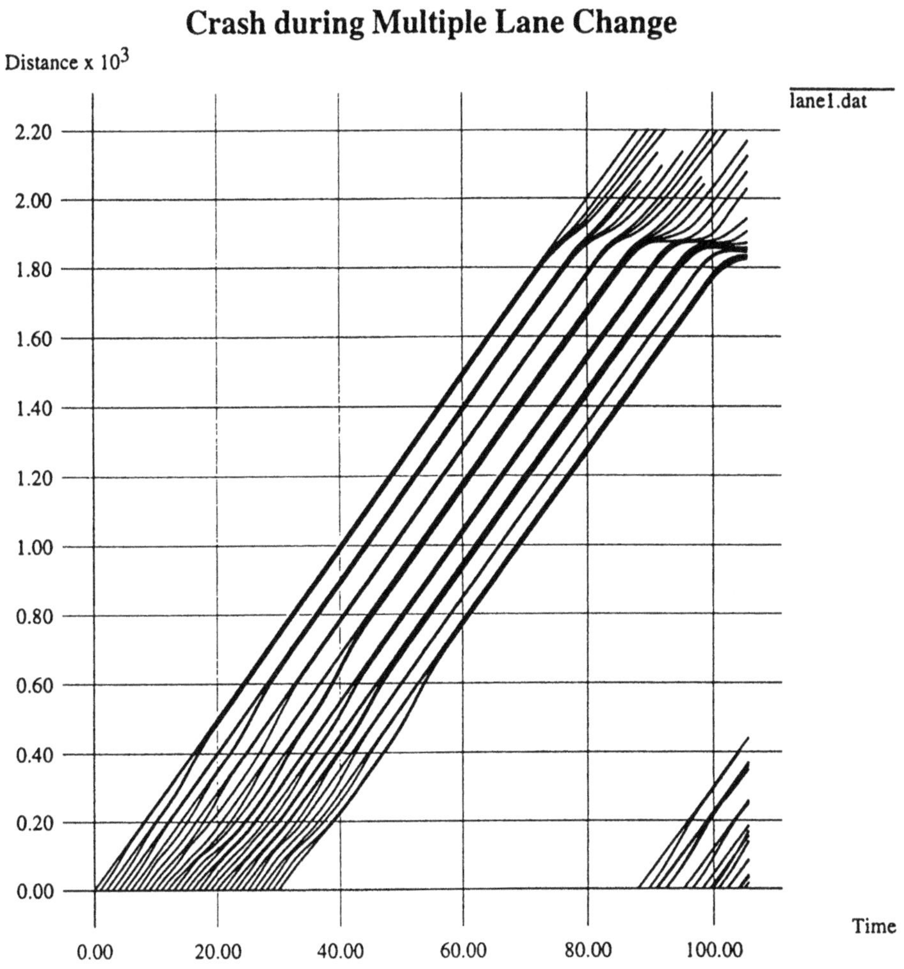

Fig. 11. Distance-Time plot of lane 1 showing problems associated with multiple lane change

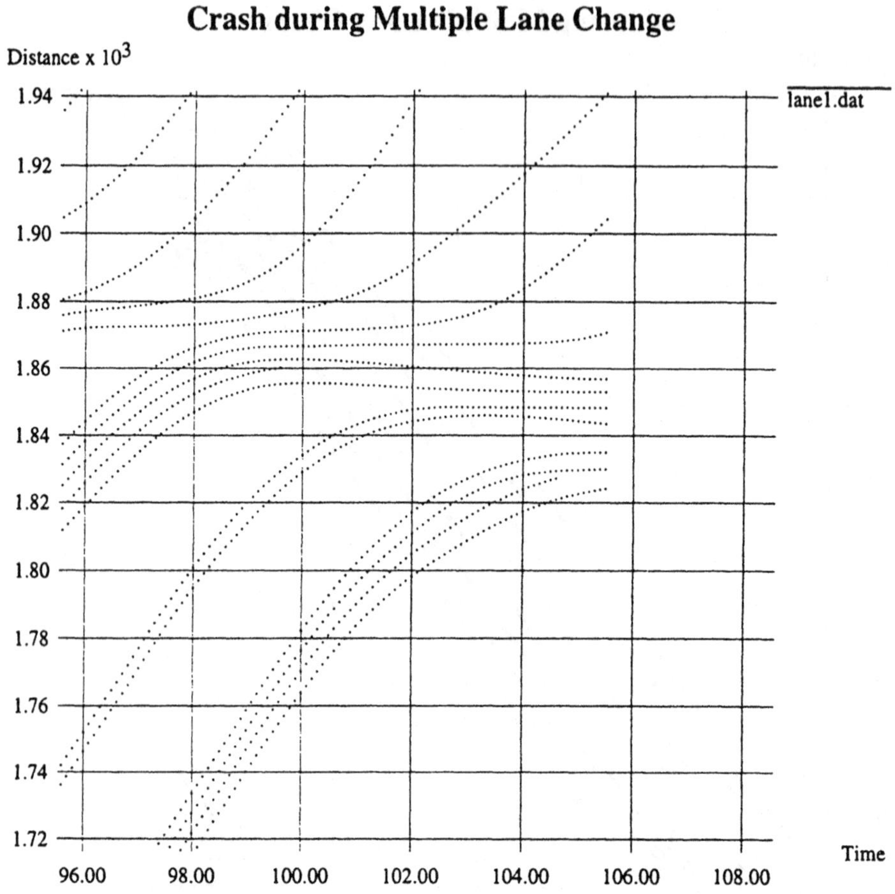

Fig. 12. Multiple lane change close up: All three followers of last platoon want to change lane around 101 seconds. All of them split from each other successively and decide to change lane in the same space on adjacent lane. As they are independent platoons, they are not communicating with each other. The third car in this platoon can be seen to crash into the second car at 104.61 seconds and 1825 meters

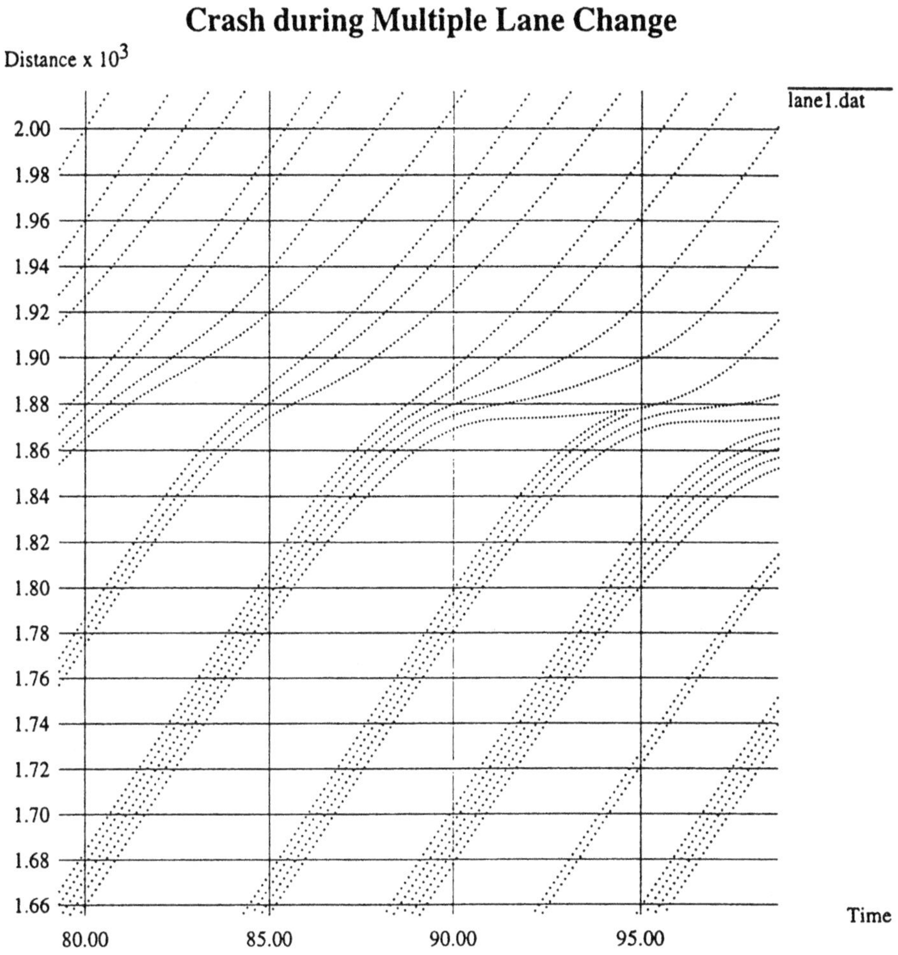

Fig. 13. Distance-Time plot of lane 1: crash at 94.31 seconds and 1880 meters is similar in nature to the case of "changing from a fast to a slow lane"

Hybrid Systems and Quantum Automata: Preliminary Announcement

R. L. Grossman* and M. Sweedler**

University of Illinois at Chicago and Cornell University

Abstract. Let H denote an algebra of input symbols or events. If X is the state space for a system, then one can form the space R of observations of X. Under suitable conditions, both X and R are H-modules. Loosely speaking, the formal systems studied in this paper consists of a bialgebra H describing the input symbols and two H-modules describing the states and observations of the system. Finite automata and input-output systems are concerned with commutative R, while quantum systems, such as quantum automata, are concerned with non-commutative R arising from Hermitian operators on the state space. Of special interest are those systems consisting of interacting networks of classical systems and automata. These types of systems have become known as hybrid systems and are examples of formal systems with commutative R. In this paper, we present a number of examples of hybrid systems and quantum automata and point out some relationships between them. This is a preliminary announcement: a detailed exposition, including proofs, will appear elsewhere.

1 Introduction

Loosely speaking, by a formal system we mean a system which accepts discrete input symbols, updates its state by flowing in a state space, and which is observed through outputs. A hybrid system as viewed in [7] is a formal system in which the observations have the structure of an algebra. A quantum automaton as viewed in [9] is a formal system in which the observations are Hermitian operators on the state space.

Automata, like other systems, have external and internal descriptions. The external description is given by the map from inputs to outputs, while the internal description views the system as a flow on suitable state space. In the latter description, the inputs determine the flow, while the outputs are a function of the flow. A realization of a system takes an external input-output description and produces an internal description involving a state space.

* This research was supported in part by NASA grant NAG2-513, DOE grant DE-FG02-92ER25133, and NSF grants IRI 9224605 and CDA 9303133. Part of the work was done while visiting the Design Research Institute and Computer Science Department at Cornell University.

** Sponsored in part by Army Research Office contract DAAL03-91-C-002.

The purpose of this paper is to give some interesting examples of formal systems and their realizations and to point out some relationships between quantum automata and hybrid systems.

This work grew out of [6] and [7] whose goal was to use algebraic methods to study physical and engineering systems. Both systems modeling physical processes, such as a mechanical device, and engineering processes, such as the evolution of a computer program, can be understood using the basic concepts of inputs, outputs and states. Since we are concerned with the algebraic structure of such systems rather than the analytic structure, we speak of *formal systems*. For example, with the latter, attention would be paid to the convergence of series; with the former, it is sufficient whether the series are defined.

Formal Systems. Let k denote a field of characteristic zero. Let Ω denote the space of input symbols or input events, which we call the input alphabet. Let Ω^* denote the monoid of words formed from the alphabet Ω. The product in Ω^* is concatenation. Let $H = k\Omega^*$ be the algebra over k formed by taking finite formal sums of words. The product in H arises from the product in Ω^*. H is a bialgebra, which gives the dual an algebra structure.

If X is the state space for a system, then one can form the space R of observations of X. Under suitable conditions, both X and R are H-modules. Loosely speaking, a formal system consists of a bialgebra H describing the input symbols and two H-modules describing the states and observations of the system. Classical systems, such as those studied in [6] and [7], are concerned with commutative R, while quantum systems, such as quantum automata, are concerned with non-commutative R arising from Hermitian operators on the state space.

Of special interest are those systems consisting of interacting networks of classical systems and automata. These types of systems have become known as hybrid systems [5]. It turns out that hybrid systems can also be viewed as formal systems [7] and that the observation space of a hybrid system has a natural noncommutative structure, so that hybrid systems are naturally associated with quantum automata [8]. This suggests that a better understanding of quantum automaton would lead to a better understanding of hybrid systems, which is one of the motivations for this present work.

Related work. In this paper, we view a hybrid system as an interacting collection of nonlinear input-output systems, each corresponding to a different mode of the hybrid system: the role of the automaton is to switch modes. There are several other view points and many questions one can pose about hybrid systems. See [5] for a collection of papers discussing some of these.

The basic idea of quantum computing was introduced by Feynmam [4]. Important foundations were provided by the papers of Benioff [1] and Deutsch [3]. Recently, quantum computation has invaded complexity theory [2] and algorithm design [11].

Organization of this paper. In Section 2, we introduce our approach to quantum automata and hybrid systems with two simple examples. In Section 3, we provide some background material. Formal systems are defined in Section 4. Quantum automata recognizing arbitrary languages are described in Section 5 and those recognizing context free languages are defined in Section 6.

This is a preliminary announcement: a detailed version with proofs will appear later.

2 Two Examples

In this section, we give two important motivating examples. First we define define a quantum automaton which can recognize expressions which contain balanced parentheses. In the second example, we will define a simple hybrid system with two modes, each specifying a different nonlinear system in the plane. In both cases, we will define a space of input symbols H, a state space and an observation space.

It turns out that the following differences between these two examples are fundamental:

> For hybrid systems, the space of observations is an algebra R as well as an H-module, and these two structures are compatible.

> For quantum automata, the space of observations is still an H-module, but not necessarily an algebra.

For both examples, let k denote a field of characteristic zero.

Example 1. *A quantum automata recognizing parentheses expressions.* In this example, which is adapted from [9], the input alphabet Ω consists of two symbols l and r, denoting left and right parentheses. Let Ω^* denote the monoid consisting of words formed from the alphabet $\{l, r\}$. Let $H = k\Omega^*$ denote the k-algebra whose basis consists of words $w \in \Omega^*$ and whose multiplication is induced from the multiplication in Ω^*. In this way, we have defined the algebra of H of input symbols or input events.

Let X denote the vector space whose basis consist of the elements x_0, x_1, x_2, ..., and let x_i^* denote the dual basis of X^*. Assume that $H = k\Omega^*$ acts on basis elements of X as follows:

$$
\begin{aligned}
r \cdot x_i &= x_{i+1}, \qquad & i \geq 0\\
l \cdot x_i &= x_{i-1}, \qquad & i \geq 1\\
l \cdot x_0 &= 0
\end{aligned}
$$

and is extended to act linearly on X. This defines an action of H on X which codes the dynamics of the quantum automaton.

Define a quantum observation $S : X \longrightarrow X^*$ via

$$
\begin{aligned}
Sx_i &= x_0^*, \qquad i \neq 0\\
Sx_0 &= \sum_{i \geq 0} x_i^*
\end{aligned}
$$

and extend S to X so that $[Sx](y) = \overline{[Sy](x)}$, for all $x, y \in X$ so that S is Hermitian, as defined below.

It is also easy to check that the map

$$w \in \Omega^* \mapsto [S(w \cdot x_0)](w \cdot x_0)$$

is one precisely when the word w is a balanced expression in left and right parentheses and zero otherwise.

Let $L \subset \Omega^*$ denote the language consisting of well balanced parentheses expressions. Let $p \in H^*$ denote the characteristic series of the language L so that $p(w)$ is one precisely when $w \in L$. We can write the equation above

$$p(w) = [S(w \cdot x_0)](w \cdot x_0).$$

This can be interpreted as a realization. The left hand side consists of data involving the inputs and outputs, while the right hand side consists of an action of input symbols w on states coding the dynamics and a quantum observation S.

To summarize, we have constructed a quantum automaton which recognizes well balanced parentheses expressions. Recall that a finite automaton is not capable of recognizing such expressions.

Example 2. *A two mode hybrid system, switching between planar control systems.* In this example, which is adapted from [7], we are given two nonlinear control systems on k^2

$$\dot{x}^{<\beta>}(t) = (u_1(t)E_1^{<\beta>} + u_2(t)E_2^{<\beta>})x(t), \qquad \beta = 1, 2$$

and an automaton which switches between them. Here $E_k^{<\beta>}$ for $\beta = 1, 2$ and $k = 1, 2$ are vector fields in the plane; that is derivations of some ring R_j of functions on the plane k^2, and $t \mapsto u_k(t)$ are controls. For example, one can take the ring of polynomials $R_j = k[x_1, x_2]$ or formal power series $R_j = k[[x_1, x_2]]$, for $j = 1, 2$.

The goal is to describe the structure of the entire ensemble consisting of the two nonlinear systems and the automaton switching between them. We begin by describing the space of observations of the systems. Since the space of observations of the nonlinear system for mode j is simply R_j, it is natural to take the space R of observations for the entire hybrid system to be $R = R_1 \oplus R_2$, [6].

The space of input events Ω for the automaton consists of two symbols β_1, which we interpret as "move to mode 1" and β_2, which we interpret as "move to mode 2."

In this example, the algebra of input events for each of the nonlinear input-output systems is the free associative algebra $k{<}\xi_1, \xi_2, \beta_1, \beta_2{>}$ in the indeterminates ξ_1, ξ_2, β_1, and β_2.

We specify the action of ξ_1 and ξ_2 on R by specifying its actions on R_i, for $i = 1, 2$: on R_1

$$\xi_1 \text{ acts as } E_1^{<1>}$$

$$\xi_2 \text{ acts as } E_2^{<1>}$$

on R_2

$$\xi_1 \text{ acts as } E_1^{<2>}$$
$$\xi_2 \text{ acts as } E_2^{<2>}$$

We next specify the action of β_1 and β_2 on R:

$$\beta_1(f \oplus g) = f \oplus f$$
$$\beta_2(f \oplus g) = g \oplus g.$$

Intuitively, β_1 maps all states into State 1, and β_2 maps all states into State 2. The action of β_1 on R is the transpose of this map. For the element

$$(u_1\xi_1 + u_2\xi_2)\beta_2(v_1\xi_1 + v_2\xi_2) \in H,$$

and state i, this is to be interpreted as flowing along $v_1 E_1^{<i>} + v_2 E_2^{<i>}$, making a transition to State 2, and then flowing along $u_1 E_1^{<2>} + u_2 E_2^{<2>}$.

To summarize, the space of input events for this two mode hybrid system is the algebra

$$H = k{<}\xi_1, \xi_2, \beta_1, \beta_2{>},$$

while the algebra of observations is

$$R = R_1 \oplus R_2.$$

We have showed how R has an H-module structure. It turns out that the algebra structure of R is compatible with the H-module structure so that R has the structure of what is called an H-module algebra. Note that R is commutative. The space of observation functions R captures the state space of the hybrid system, while the action of H on R captures the dynamics. Any element $f \in R$ can be thought of as an observation of the hybrid system. For more details, including a description of the realization of this system, see [7].

3 Preliminary Material

To describe more complicated examples requires some preparation, which is the subject of this section. This section is adapted from [9], which provides additional background material. Let k denote a field with conjugation. Let X denote a vector space over k and X^* its k-linear dual.

Hermitian forms and operators. A linear operator

$$S : X \longrightarrow X^*$$

naturally defines an inner product on X

$$<x, y>_S = [S(x)](y), , \qquad x, y \in X.$$

Conversely, for any inner product $<<x, y>>$, we can define a linear operator $T : X \longrightarrow X^*$ via the formula

$$[T(x)](y) = <<x, y>>.$$

The operator S is called Hermitian in case

$$[S(x)](y) = \overline{[S(y)](x)}, \qquad x, y \in X.$$

This is equivalent to $<\cdot, \cdot>_S$ being a Hermitian inner product.

Bialgebras. In this section, which is adapted from [7], we briefly cover some facts about bialgebras and H-modules, which we require in order to a give a careful definition of formal systems. This information is repeated here to make this paper self-contained. For more details, see [12].

An *algebra* A over the field k is a k-vector space A equipped with a multiplication $A \otimes A \longrightarrow A$ mapping $a \otimes b \mapsto ab$ and a unit $k \longrightarrow A$ mapping $1 \in k \mapsto 1 \in A$. The algebra is called *augmented* if there is an algebra homomorphism $A \longrightarrow k$.

A *coalgebra* C over the field k is a k-vector space C equipped with a comultiplication $C \longrightarrow C \otimes C$ and a counit $C \longrightarrow k$. A *bialgebra* H over the field k is a k-vector space H which has both an algebra and a coalgebra structure such that the comultiplication and the counit maps are algebra homomorphisms, or equivalently, such that the multiplication and unit maps are coalgebra morphisms.

Coproduct Notation. If H is a coalgebra, it is convenient to write the comultiplication $H \longrightarrow H \otimes H$ as the map $h \mapsto \sum_{(h)} h_{(1)} \otimes h_{(2)}$. This notation will be used throughout the remainder of the paper.

Here are two examples of bialgebras which occur in hybrid systems. Let G be a semigroup with unit. Then the semi-group algebra kG consisting of all formal finite linear combinations of elements of G with comultiplication defined by $g \mapsto g \otimes g$ and counit defined by $g \mapsto 1$ for $g \in G$, is a bialgebra. In this paper, this case arises when we have an input alphabet Ω and form the semigroup Ω^* of all words by concatenating input symbols from Ω. As another example, let L be a Lie algebra. Then the universal enveloping algebra $U(L)$ with comultiplication defined by $x \mapsto 1 \otimes x + x \otimes 1$ and counit defined by $x \mapsto 0$ for $x \in L$, is a bialgebra. In this paper, this example arises by viewing the free associative algebra $k<\xi_1, \xi_2, \ldots, \xi_n>$ as the universal enveloping algebra on the free Lie algebra generated by the symbols ξ_j and the Lie bracket defined by the commutator $[\xi_i, \xi_j] = \xi_i\xi_j - \xi_j\xi_i$.

Let H be a bialgebra. A vector space X is called an *H-module* in case there is an action of H on X denoted $h \cdot x$, for $h \in H$ and $\in X$, which is k-linear and which satisfies $(h_2 h_1) \cdot x = h_2 \cdot (h_1 \cdot x)$, where $h_2, h_1 \in H$ and $x \in X$. In this paper, the action of a H-module codes the dynamics of a formal system.

An algebra R with augmentation $\epsilon : R \longrightarrow k$ is called a left *H-module algebra* in case R is a left H-module and

$$h \cdot (ab) = \sum_{(h)} (h_{(1)} \cdot a)(h_{(2)} \cdot b),$$

and

$$h \cdot 1 = \epsilon(h) \cdot 1$$

for all $a, b \in R$, $h \in H$.

4 Formal Systems

After this preparatory material, we can now give definitions of formal systems, hybrid systems, and quantum automata.

A *formal system* over a field k with involution $\alpha \mapsto \bar{\alpha}$ consists of

1. a bialgebra H containing elements which we interpret as input symbols or events and the words formed from them;

2. an H-module $\mathcal{O}$, which we interpret as the space of observables.

A *hybrid system* is the special case in which the observables also form an algebra, while a *quantum automaton* is the special case in which the observables can be identified with Hermetian operators on an inner product space.

Here are two very basic examples. Let X denote a state space and assume there is an action of H on X, coding the dynamics. A hybrid system arises by taking classical state space observations

$$\mathcal{O} = \{f : X \longrightarrow k\}.$$

A quantum automaton arises by taking Hermetian observations

$$\mathcal{O} = \{\text{Hermetian } S : X \longrightarrow X^*\}.$$

In both cases, the action of H on X induces an action of H on $\mathcal{O}$.

5 Quantum automata and languages

Algebra of input events. Consider a language defined over an input alphabet Ω. As usual, let Ω^* denote the monoid of words formed by concatenating input symbols and let $H = k\Omega^*$ denote the corresponding k-algebra formed from the vector space whose basis is the set of words $w \in \Omega^*$. In this way, we define a bialgebra of input events H.

State space. Let $X = k\hat{\Omega}^*$ denote infinite formal linear combinations over k of elements of Ω^*. This is the algebra of formal noncommutative power series in the alphabet Ω over the field k and turns out to be the state space of the quantum automaton. There is a conjugation, denoted $x \mapsto \bar{x}$, defined on X which for each power series, conjugates the coefficients and replaces words by its palindrome.

Suffix action. We now define an action of H on X, called the *suffix* action, which codes the dynamics For each word $v \in \Omega^*$, the suffix action, denoted $v \rightharpoonup$ is a linear map from H to H. For w a word in Ω^*, thought of a lying in X,

$$v \rightharpoonup w = \begin{cases} u, & uv = w \\ 0, & \text{otherwise} \end{cases}$$

Here u is a word in Ω^*. The action $v \rightharpoonup$ extends to all of X by linearity. Also, by linearity the action $h \rightharpoonup$ extends to all $h \in H$.

One can check that the suffix action makes X an H-module in that $(uv) \rightharpoonup w = u \rightharpoonup (v \rightharpoonup w)$, for .

Hermetian observable. We now define a Hermitian form on the state space X: Let $\lambda \in \Omega^*$ denote the empty word. Define

$$<x, y> = \chi_\lambda(x\bar{y}), \qquad x, y \in X$$

where $\chi_\lambda : k\Omega^* \longrightarrow k$ is defined by

$$\chi_\lambda(w) = \begin{cases} 1, & w = \lambda \\ 0, & \text{otherwise.} \end{cases}$$

Here $w \in \Omega^*$ and the map is extended to all of X by linearity.

Of course, we can view this as a Hermitian operator

$$S : X \longrightarrow X^*$$

via $[S(x)](y) = <x, y>$.

Realizations of an arbitrary language. Given a language $L \subset \Omega^*$, let $p \in H^*$ denote the characteristic series of the language, where

$$p(w) = \begin{cases} 1 & w \in L \\ 0 & \text{otherwise} \end{cases}$$

We also define the element

$$x_L = \sum_{l \in L} l \in X.$$

Note that this makes sense even if the language is infinite since X contains infinite sums.

For consistency with the notation above, think of $x_0 = x_L$, which we view as an initial condition. One verifies directly [9] that

$$p(h) = \sum_h <h_{(1)} \rightharpoonup x_0, h_{(2)} \rightharpoonup x_0>.$$

Note that this equation uses the coalgebra structure of H. In the case that $h \in k\Omega^*$ is a word in Ω^*, this simplifies to

$$p(h) = <h \cdot x_0, h \cdot x_0> = \begin{cases} 1 & h \in L \\ 0 & \text{otherwise.} \end{cases}$$

Summary. To summarize, given a language with input alphabet Ω, we defined a bialgebra of input symbols $H = k\Omega^*$, a state space $X = k\hat{\Omega}^*$, and an action (the suffix action) of H on X. We also defined a Hermitian observable $S : X \longrightarrow X^*$ and wrote down a quantum recognizer for the language. In other words, given an arbitrary language, we constructed a quantum automaton which recognizes it.

6 Quantum automata and context free languages

In this section, we consider quantum automata which recognize context free languages. Classically, these are recognized by push down automaton. This is related to the fact that a realization of a context free language satisfies a finiteness condition, which we describe in this section. This is analogous to the fact that a natural equivalence relation on words in Ω^* has finite index for regular automata [10]. This is also analogous to the notion of finite Lie rank defined in [6] for formal series associated to nonlinear input-output systems. See [7].

We proceed as above: given an input alphabet Ω, we define the bialgebra $H = k\Omega^*$ and the state space $X = k\hat{\Omega}^*$.

Finiteness Condition. For an integer $n > 0$, and define the ring of noncommuting polynomials in the indeterminates $Z_1, \ldots, Z_n$ with coefficients from H

$$R_n = H\{Z_1, \ldots, Z_n\}.$$

It is important to note that the Z_j do not commute with the coefficients in H. We can also write this as

$$R_n = k(\Omega^* \cup \{Z_1, \ldots, Z_n\})^*.$$

Let A denote a ring containing a homomorphic image of H. We need to define what it means for an element $a \in A$ to be algebraic over H.

Equations. A *PDA equation* in R_n is an expression of the form

$$Z_j - \alpha_j(Z_1, Z_2, \ldots, Z_n) = 0,$$

where $\alpha_j \in R_n$ and α_j does not contain the empty word λ nor the single term Z_i, for any i. Of course α_j can contain a term of the form wZ_i, where $w \in \Omega^*$ and $w \neq \lambda$.

Algebraic elements. An element $a \in A$ is *algebraic* over H in the sense of formal languages in case there exists $n > 0$ and elements $a_1, \ldots, a_n$ and equations $Z_1 - \alpha_1, \ldots, Z_n - \alpha_n \in R_n$ such that

1. $a_i = \alpha_i(a_1, \ldots, a_n)$
2. a lies in the subalgebra of A generated by $a_1, \ldots, a_n$ and (the homomorphic image of) H.

Algebraic elements of X. Recall that X is the algebra of formal power series in the noncommuting variables in the alphabet Ω. Note that there is a subalgebra, which is a copy of $H \in X$ consisting of those power series which are only finite sums, and hence noncommuting polynomials. Having of H in X gives an action of H on X by left multiplication, but it is important to note that this does not give the suffix action. However, having H as a subalgebra of X, we can speak of elements of X as being *algebraic* over H.

In this approach to recognizing languages using quantum recognizers, there is a finiteness condition characterizing context free languages which is based upon the notion of algebraic elements. We have

Theorem 1 *Fix an alphabet Ω, the monoid of words Ω^* and the corresponding bialgebra of input events $H = k\Omega^*$. Define the state space $X = k\hat{\Omega}^*$. With the notation above, a language $L \subset \Omega^*$ with characteristic series $p \in H^*$ is context free iff there is a $x_0 \in X$ giving the quantum automata recognizer*

$$p(h) = \sum_h < h_{(1)} \rightharpoonup x_0, h_{(2)} \rightharpoonup x_0 >,$$

where the element $x_0 \in X$ is algebraic over H.

See [9] for a proof.

7 Conclusion

In this paper, we have defined formal systems and considered two cases: hybrid systems and quantum automata. We have also given several examples of formal systems and their realizations.

We view a hybrid system as an interacting collection of nonlinear input-output systems, each corresponding to a different mode, and an automaton which switches between them. As the example of the hybrid system in Section 2 illustrates, a hybrid system may be given by defining an algebra of input events H containing variables corresponding to both the continuous and discrete components of the hybrid system. An important advantage of our approach is that by viewing the space of observations $\mathcal{O}$ as fundamental, the variables corresponding to the discrete and continuous components of the hybrid system may be treated symmetrically. Each induces an action on the observation space $\mathcal{O}$ coding the dynamics. Variables corresponding to continuous components act as derivations; those corresponding to discrete components act as homomorphisms. The reason for viewing H as bialgebra and $\mathcal{O}$ as an H-module is to to treat these two actions symmetrically.

Finite automata recognize regular languages and correspond to formal systems in which the H-module $\mathcal{O}$ of observations is a commutative algebra. Push down automata recognize context free languages and correspond to formal systems in which the H-module $\mathcal{O}$ of observations consists of Hermetian operators. We identified the finiteness condition associated with quantum automata which recognize context free languages.

This illustrates the power of the fundamental idea of understanding dynamics by focusing on the action induced on observations rather than on the underlying state space itself.

References

1. P. Benioff, "Quantum mechanical Hamiltonian models of Turing machines," *Journal of Statistical Physics*, Volume 29, pp. 515 546, 1982.
2. A. Berthiaume and G. Brassard, "The quantum challenge to structural complexity theory," *Proceedings of the Seventh IEEE Conference on Structure in Complexity Theory*, IEEE, pp. 132 137, 1992.
3. D. Deutsch, "Quantum theory, the Church-Turing principle and the universal quantum computer," *Proceedings of the Royal Society*, Volume A425, pp. 73 90, 1985.
4. R. P. Feynman, "Simulating physics with computers," *International Journal of Theoretical Physics*, Voluem 21, pp. 467 488, 1982.
5. R. L. Grossman, A. Nerode, A. Ravn, and H. Rischel, editors, "Hybrid Systems", Springer Lecture Notes in Computer Science, Springer-Verlag, Bonn, 1993.
6. R. Grossman and R. G. Larson, "The realization of input-output maps using bialgebras," *Forum Mathematicum*, Volume 4, pp. 109 121, 1992.
7. R. L. Grossman and R. G. Larson, "An algebraic approach to hybrid systems," *Journal of Theoretical Computer Science*, Volume 138, pp. 101 112, 1995.
8. R. L. Grossman, A. Nerode, and M. Sweedler, "A hilbert space approach to hybrid systems using formal dynamical systems," submitted for publication.
9. R. L. Grossman and M. Sweedler, "Quantum automata and their realizations," submitted for publication.
10. J. E. Hopcroft and J. D. Ullman, *Formal Languages and their Relation to Automata*, Addison-Wesley, Reading, 1969.
11. P. W. Shor, "Algorithms for the quantum computation: Discrete log and factoring," submitted for publication.
12. M. E. Sweedler, *Hopf algebras*, W. A. Benjamin, New York, 1969.

Planar Hybrid Systems

John Guckenheimer
Mathematics Department,Cornell University, Ithaca, NY 14853
gucken@cam.cornell.edu

Stewart Johnson
Department of Mathematics,Williams College, Williamstown, MA 01267
sjohnson@williams.edu

Introduction

The design of nonlinear dynamical systems to realize the desired behavior of devices and processes is a fundamental problem that is often approached from the perspective of control theory. Control provides a large set of tools for tackling these design problems, but some classes of problems have remained resistant to these methods. This paper discusses a theory of **hybrid systems** that extends the concept of a smooth dynamical system to a context in which there is a mixture of continuous time and discrete time elements. This is particularly appropriate for the description of "switching systems" in which digital control is applied to change continuous control laws on a time scale much faster than the dynamical evolution of the device being controlled. Digital control makes it (relatively) easy to build devices in which the applied control is any computable function of system measurements. This enormously widens the scope for the system designs that can be implemented, but the additional design freedom brings with it many more design parameters. Consequently, we seek better theory to provide guiding principles that can help us cope with this new found freedom. This paper is a step in this direction. We consider hybrid systems with two dimensional phase spaces subject only to mild genericity requirements, and describe the types of long term dynamical phenomena displayed by these systems.

Our focus on two dimensional systems is a severe restriction, but we shall see that there is very rich dynamical behavior that is already present in this context. We are able to build upon two of the great successes of "qualitative theory" for dynamical systems, the theory of two dimensional flows in the plane [AP] and the theory of iterations of one dimensional mappings [MS]. From these we derive a classification of the generic behaviors displayed by two dimensional hybrid systems. The limiting behavior displayed by a trajectory with a finite number of switches is determined by the two dimensional flow in the domain that contains that part of the trajectory after its last switch. For trajectories with an infinite number of switches, we analyze their asymptotic dynamics in terms of an **induced transformation** defined on switching curves that advance the trajectory from one switch to the next. For generic systems, this produces a piecewise smooth map defined on a finite collection of one dimensional curves. Over the past twenty years, an extensive theory of one dimensional iterations has produced detailed understanding of the dynamics of these systems. We summarize aspects of the theory that are relevant to the classification of asymptotic behavior for two dimensional hybrid systems.

From a design perspective, one seeks to build systems that maintain their desired behavior when subjected to noise or other perturbations. If the desired behavior is an "attractor" for the underlying dynamical system, then the size of its domain of attraction is closely related to its persistence following perturbations. The ultimate achievement in this direction is the design of systems that have a globally attracting state. "Unimodal" one dimensional maps that satisfy mild additional restrictions [G1] have a single attracting state. Koditschek [K] has emphasized the use of these one dimensional mappings in constructing controllers for periodic orbits that have a large basin of attraction. In systems with chaotic attractors, we show how any periodic orbit in the attractor can be made stable by a (C^0 small) continuous perturbation of a switching curve that intersects the periodic orbit. By merely altering the time at which a switch from one continuous controller to another is applied, we achieve a "control of chaos."

We end this introduction with a simple example, a thermostat, that illustrates how hybrid systems can be used for solving control problems. This example also points to a significant difference between mode switching and the use of "variable structure" as a control strategy. A thermostatically controlled furnace has two modes, ON and OFF. In the ON mode the furnace will heat up a room in a predictable manner until the temperature reaches a set point for switching from ON to OFF, say 70°. The furnace/thermostat system then switches to the OFF mode. Assuming a constant environment, the room will then lose heat in a predicable manner until the temperature reaches a set point for switching from OFF to ON, say 66°. The system then switches to ON and the cycle repeats. The thermostat controls the temperature so that it remains in the range between 66° and 70°. Ideally, one would like to control the temperature to be constant. One can approach this ideal by adjusting the mode switches closer to one another, but there is a penalty: switching induces wear on the system. Moreover, keeping the temperature of the thermostat constant may not meet the desired objective of keeping the temperature of the room at the desired uniform temperature. As the set points for switching are allowed to approach each other, the thermostat switches modes more frequently. There are physical limits that place an upper bound on the switching frequency, so the switching strategy of switching both off and on at the same temperature is problematic.

This difficulty is related to the variable structure approach to control in which controls are sought that constrain a system to lie on a submanifold of its state space. Without a furnace that is able to vary its heat output continuously, this is impossible to achieve. By relaxing our requirements on system performance, we are able to achieve acceptable system behavior (i.e., temperature in a small range) from a low-cost hybrid strategy that is very reliable.

Our goal is to develop a general theory that embodies the features of the thermostat example. When faced with nonlinear systems whose dynamical behavior we desire to control, we would like to make logical decisions that switch control modes at times that are triggered by observations of the system state. The effect of these decisions should be to control the system behavior so that it evolves in approximately the desired manner. Rather than specifying completely the path the system should take, we would like to specify a target region or trajectory we would like the system to approach and then develop a switching strategy that will lead it there. There are many similarities between this approach and the strategies that are used by animals (including humans) to solve control problems. Hybrid control strategies have the advantage that they can rely on simpler physical mechanisms for implementing control than those that require the ability to steer a system along specified trajectories throughout the state space.

The next section gives a definition of hybrid systems that is intended to be as general as possible, but precise enough to support a clear definition of trajectory. The following sections restrict attention to a generic class of planar systems. After defining this class, we show how the complexity of a planar hybrid system can be reduced to a one-dimensional transformation by inducing the system onto the set of switching curves. We then investigate the properties of this induced system, including the structure of the domain, the number and types of singularities, and an explicit calculation of the derivative.

Definition of hybrid systems

In this section we give a basic definition of a planar hybrid system.

A hybrid system is a collection of continuous flows and discrete transformations that switch between these flows events are triggered by "threshold" functions. We make these concepts more precise.

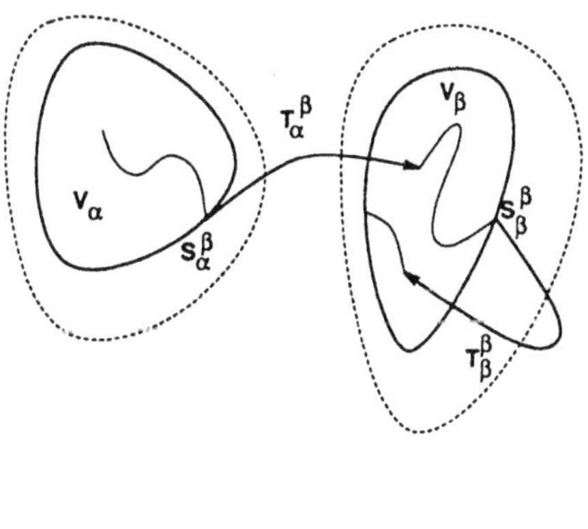

Figure 1: Flows and switches

Let $\mathtt{I}$ be a discrete indexing set and $\mathtt{D}_\alpha$, $\alpha \in \mathtt{I}$ a collection of euclidean domains (usually $\mathtt{R}^n$). Let $\mathcal{F}_\alpha$ be a (possibly time-dependent) flow on an open subset V_α of $\mathtt{D}_\alpha$, thus $\mathcal{F}_\alpha : V_\alpha \times \mathtt{R} \to \mathtt{D}_\alpha$. The flow domain V_α is called **a chart.** The collection of charts is called an **atlas.** The flows constitute the continuous time portion of a hybrid system. Flows are frequently defined indirectly by systems of nonlinear ordinary differential equations, so that numerical integration is required to recover the flow from the differential equations.

Switching occurs at **events** that are described by **threshold functions.** For each $\alpha \in \mathtt{I}$ we have a collection of real-valued C^2 functions $h_\alpha^\beta : V_\alpha \to \mathtt{R}$ indexed by β in some set J_α. As points flow across a chart, the threshold functions h_α^β monitor whether a transition to a (possibly) new chart should occur at that instant. When the function h_α^β drops below zero the trajectory switches to a chart determined by the index β. To simplify notation, we shall consider all the threshold functions at which switches from patch α to patch β occur as a single piecewise smooth threshold function h_α^β. Thus the index set J_α is regarded as I and h_α^β is the domain for switches from chart α to chart β. Switching between charts occurs via mappings $\mathcal{T}_\alpha^\beta$ with domains in V_α and ranges in V_β. These maps are also assumed to be piecewise smooth. The hybrid system is defined by the collections V_α, $\mathcal{F}_\alpha$, h_α^β and $\mathcal{T}_\alpha^\beta$. Points within the system are specified by a location $x \in V_\alpha$ in chart $\alpha \in I$ at a time $t \in \mathtt{R}$, as (x, α, t). The level curves $h_\alpha^\beta = 0$ are denoted by S_α^β

Trajectories of a hybrid system are defined as follows. Let (x, α, t) be an initial point. If $h_\alpha^\beta(x) > 0$ for all $\beta \in \mathtt{I}$ then the point flows in V_α as $\mathcal{F}_\alpha(x, t)$ until the minimal time t_1 at which $h_\alpha^\beta(x(t)) = 0$ for some β. An event occurs at time t_1. Taking $x_1 = x(t_1)$, the transformation $\mathcal{T}_\alpha^\beta$ transports the trajectory from $(x_1, \alpha, t_1) \in V_\alpha$ to the point $(T_\alpha^\beta(x_1), \beta, t_1) \in V_\beta$. The point $(T_\alpha^\beta(x_1), \beta, t_1)$ is regarded as a new initial point and the process continues. This naturally yields maximal solutions of the initial value problem. If one of the threshold functions is nonpositive at the initial point then a switch occurs immediately. Note that this does not give unique futures in the event that several threshold functions are non-positive.

In parameterizing trajectories, we want to keep track of pre-event and post-event states associated with each switch. Multiple switches can occur instantaneously; note that the above trajectory had two locations at the same time t_1. To accommodate this, we use a construct of time that has both continuous and discrete components. We let a **time sequence** be a sequence of abutting closed intervals $e_0 = [t_0, t_1]$, $e_1 = [t_1, t_2], \ldots$ which is either infinite, or finite with the last interval of the form $e_n = [t_n, \infty)$. Points in the time sequence are given by (t, n) with $n \in \mathtt{Z}^+$ and $t \in e_n$. These points are simply ordered, $(t_a, n_a) < (t_b, n_b)$ iff $t_a < t_b$ or $t_a = t_b$ and $n_a < n_b$. The time intervals e_n are called **epochs.**

Trajectories are thus mappings from a time sequence $\{e_0, e_1, \ldots\}$ to an atlas $\{V_\alpha\}$ defined by flowing $\mathcal{F}_\alpha$ and switching $\mathcal{T}_\alpha^\beta$ at certain events $h_\alpha^\beta = 0$. We parameterize the trajectory as $(x(t, n), \alpha(n))$, noting that the chart location α depends only on the epoch n.

This completes our definition of hybrid systems and trajectories. We have four comments regarding the definition, followed by a more careful recursive construction of trajectories and time sequences.

1. This definition of hybrid systems and trajectories is sufficiently general that the initial value problem is not well posed: the same initial conditions may produce multiple trajectories, and trajectories may bifurcate. There does not appear to be a satisfactory way of defining a restricted class of systems which will have unique trajectories, good analytic properties and be general enough to include basic examples.

2. Our definitions allow for multiple switches to occur instantaneously, creating epochs that consist of a single point. In most applications this is to be avoided. Let $U_\alpha \subset V_\alpha$ be the set for which $h_\alpha^\beta > 0$ for all $\beta \in \mathrm{I}$, called a **patch.** If each patch has compact closure and the range of each $\mathcal{T}_\alpha^\beta$ is contained in the patch U_β for all $\beta \in \mathrm{I}$ then trajectories with initial points in a patch can not have multiple switches occurring at the same instant. All epochs will have an interior and the switching transformation $\mathcal{T}_\alpha^\beta$ need only be defined on a subset of ∂U_α.

3. We shall consider long time properties of a system, where "long time" can be interpreted as unbounded continuous time or as an infinite number of switches (or both). Formally, a time sequence $[t_n, t_{n+1}]$ is **c-bounded** if the t_{n+1} are bounded and **d-finite** if it contains a finite number of intervals.

4. We require flows to be time-independent, but there are several possible choices for how to incorporate the dependence of a system upon time. For the purpose of defining systems, non-autonomous behavior can be incorporated by including a phase space variable that represents time. However, one of our principal concerns is recurrence of a system and time does not recur. Many real systems are built with clocks and events that are triggered at regular intervals of time. Such clocks and periodic time dependence can be incorporated with a variable that flows at a uniform rate and is reset to its initial value at regular intervals.

Solving the initial value problem for a hybrid system can be done iteratively, as indicated above. There are a number of technicalities that arise in the process.

Assume that a hybrid system is given by collections V_α, $\mathcal{F}_\alpha$, h_α^β, $\mathcal{T}_\alpha^\beta$, and that (x_0, α_0, t_0) is an initial condition in V_{α_0}. The solution begins with the maximal time interval $[t_0, t_1)$ for which the trajectory $\mathcal{F}_{\alpha_0}(x_0, t)$ stays inside U_{α_0}. If the initial point is not interior to the patch U_{α_0}, then $t_0 = t_1$ and the first epoch consists of a single point $[t_0, t_1] = \{t_0\}$. If the initial point is interior to the patch and the forward trajectory remains in U_{α_0} forever, we take $[t_0, \infty)$ as the first and only epoch in the time sequence. Otherwise there is some finite maximal time $t_1 > t_0$ for which the trajectory $x(t, 0) = \mathcal{F}_{\alpha_0}(x_0, t)$ stays in U_{α_0} for $t_0 < t < t_1$, and the first epoch is taken as $e_0 = [t_0, t_1]$. In any case, we take $\alpha(0) = \alpha_0$, and this defines the trajectory $(x(t, n), \alpha(n))$ for the first epoch $n = 0$.

To continue the trajectory beyond t_1, two properties need to hold. First, $x_1 = x(t_1, 0)$ must be well defined. If $t_1 > t_0$ then the trajectory must have a limit x_1 at time t_1. Second, there must exist an $\alpha_1 \in \mathrm{I}$ with $h_{\alpha_0}^{\alpha_1}(x_1) \leq 0$ and x_1 in the domain of $\mathcal{T}_{\alpha_0}^{\alpha_1}$. If these assumptions are satisfied, then we start the next epoch at time t_1 with the point $(\mathcal{T}_{\alpha_0}^{\alpha_1}(x_1), \alpha_1, t_1)$. We use the notation $x_1^+ = \mathcal{T}_{\alpha_0}^{\alpha_1}(x_1)$

Inductively, assume there is a time sequence $e_0, \ldots, e_m$ with $e_n = [t_n, t_{n+1}]$ and a trajectory $(x(t, n), \alpha(n))$ defined for $t \in e_n$ and $0 \leq n \leq m - 1$. We assume $x_m = x(t_m, m-1) \in V_{\alpha_{m-1}} - U_{\alpha_{m-1}}$. If there is an α_m with $h_{\alpha_{m-1}}^{\alpha_m}(x_m) \leq 0$, then set $\alpha(m) = \alpha_m$ (this may not be uniquely defined), and set $x(t_m, m) = x_m^+ = \mathcal{T}_{\alpha_{m-1}}^{\alpha_m}(x_m)$ at the start of epoch m.

Take $t_{m+1} = \min\{t \geq t_m : \mathcal{F}_{\alpha_m}(x_m^+, t) \in V_{\alpha_m} - U_{\alpha_m}\}$. If $t_{m+1} = \infty$ we take $e_m = [t_m, \infty)$ and the induction stops, otherwise we take $e_m = [t_m, t_{m+1}]$ and continue (note that $t_m = t_{m+1}$ is allowed). This recursively defines a time sequence for a trajectory with the given initial condition.

$$
\begin{array}{ccccccccc}
& \overbrace{\quad} V_{\alpha_0} \overbrace{\quad} & & & \overbrace{\quad} V_{\alpha_1} \overbrace{\quad} & & & \overbrace{\quad} V_{\alpha_2} \overbrace{\quad} & \cdots \\
x_0 & \xrightarrow{\mathcal{F}_{\alpha_0}} x_1 & \xrightarrow{T_{\alpha_0}^{\alpha_1}} & x_1^+ & \xrightarrow{\mathcal{F}_{\alpha_1}} x_2 & \xrightarrow{T_{\alpha_1}^{\alpha_2}} & x_2^+ & \xrightarrow{\mathcal{F}_{\alpha_3}} x_3 & \cdots
\end{array}
$$

$$
\begin{array}{ccccccc}
[t_0 & t_1] & [t_1 & t_2] & [t_2 & t_3] & \cdots \\
\underbrace{\quad\; e_0 \;\quad} & & \underbrace{\quad\; e_1 \;\quad} & & \underbrace{\quad\; e_2 \;\quad} & \cdots \\
| & & | & & | & \\
h_{\alpha_0}^{\alpha_1} & & h_{\alpha_1}^{\alpha_2} & & h_{\alpha_2}^{\alpha_3} & \cdots \\
=0 & & =0 & & =0 & \cdots
\end{array}
$$

$$
\begin{array}{cccccc}
\cdots & \overbrace{\quad} V_{\alpha_{m-1}} \overbrace{\quad} & & & \overbrace{\quad} V_{\alpha_m} \overbrace{\quad} & \cdots \\
\cdots & x_{m-1}^+ \xrightarrow{\mathcal{F}_{\alpha_{m-1}}} x_m & \xrightarrow{T_{\alpha_{m-1}}^{\alpha_m}} & x_m^+ & \xrightarrow{\mathcal{F}_{\alpha_m}} x_{m+1} & \cdots
\end{array}
$$

$$
\begin{array}{ccccc}
\cdots & [t_{m-1} & t_m] & [t_m & t_{m+1}] & \cdots \\
\cdots & \underbrace{\quad\; e_{m-1} \;\quad} & & \underbrace{\quad\; e_m \;\quad} & \cdots \\
& | & & | & \\
& h_{\alpha_{m-1}}^{\alpha_m} & & h_{\alpha_m}^{\alpha_{m+1}} & \cdots \\
& =0 & & =0 & \cdots
\end{array}
$$

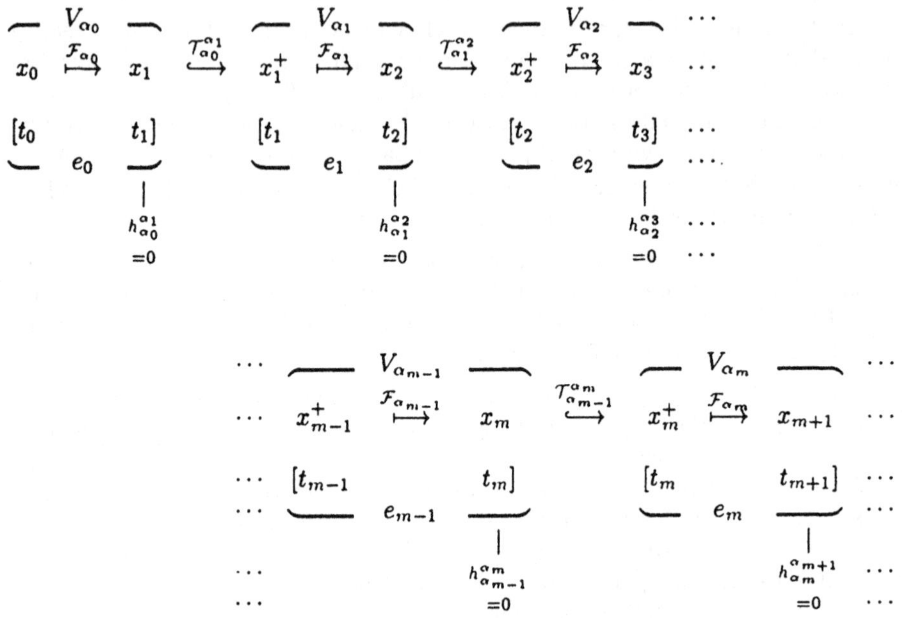

The following theorem gives a setting in which this construction can always be continued.

Theorem Let X be a hybrid system defined by V_α, $\mathcal{F}_\alpha$, h_α^β, and T_α^β. Assume the following:

1. The closure of each patch $\bar{U}_\alpha$ is a compact subset of the corresponding chart V_α.

2. If $x(t)$ is the maximal trajectory of $\mathcal{F}_\alpha$ in U_α with initial condition $(x_0, t_0) \in U_\alpha \times \mathbb{R}$, either there is $(x_1, t_1) \in (\bar{U}_\alpha - U_\alpha) \times \mathbb{R}$ with $x(t) \to x_1$ as $t \to t_1$ or $x(t)$ is defined and remains in U_α for all $t > t_0$.

3. At each point x of $\cup_{\alpha \in I} U_\alpha$ in the boundary $\bar{U}_\alpha - U_\alpha$, there is a β with $h_\alpha^\beta(x) = 0$ and $T_\alpha^\beta(x) \in V_\beta$ defined.

Then every point lies in a trajectory that is c-unbounded or d-infinite.

Generic Autonomous Systems with Interior Transformations

In this section we delineate a class of hybrid systems for which global structure theorems hold

We now restrict our attention to planar systems, $D_\alpha = \mathbb{R}^2$. We also assume that the systems have switches whose images are interior to patches and that the systems satisfy a set of "genericity" hypotheses described below. The interior switching assumption guarantees that all epochs have an interior and prohibits multiple instantaneous switches. It also avoids geometric problems of "deadlock" in which there is a cycle of discrete instantaneous switches that blocks future continuous time evolution of a trajectory.

The term generic is used here in the context of transversality and singularity theory. We make a list of requirements that singular behavior within the system should satisfy. Each of these requirements is to be preserved by C^2 perturbation, and every system should be C^2 close to one that satisfies the assumptions. The first genericity assumption is that each patch U_α has a boundary that is a finite union of smooth curves in the level sets of the threshold functions h_α^β. The domains of the switching transformations T_α^β can be restricted to the boundary of U_α. Level curves of distinct threshold functions are assumed to intersect only at their endpoints, so there are only a finite number of points in the boundary of the patch U_α which lie in the allowed domains of more than one switching transformation. The intersection points will act as boundary points in the domain components of the induced map. The second assumption is that all tangencies between the flow and the switching curves or images of switching curves are quadratic; this is satisfied if the curvature of the flow and the curve are not equal. Next, we assume that a point of tangency does not flow to another point of tangency. Finally, we assume that the images of switching transformations intersect certain periodic orbits transversely. These periodic orbits are characterized by the following definition: a periodic orbit contained in U_α is **patch-maximal** if it is not interior to another cycle contained in U_α.

For emphasis, we call a two dimensional hybrid system V_α, $\mathcal{F}_\alpha$, h_α^β, T_α^β satisfying the following properties a **GAIT** system (generic, autonomous, interior transforms):

Flow properties:

The flows $\mathcal{F}_\alpha$ are autonomous, C^2, with isolated fixed points.

Switching properties:

The switching transformation T_α^β is piecewise C^2 with domain a piecewise C^2 curve $S_\alpha^\beta \cap \partial U_\alpha$ and range $T_\alpha^\beta(S_\alpha^\beta \cap \partial U_\alpha)$ interior to patch U_β.

Generic properties:

1. Equilibrium points of the flows are interior to the patches and hyperbolic. Saddle points are not resonant.

2. Periodic orbits contained inside a patch are hyperbolic.

3. The union of the sets S_α^β is a finite collection of C^2 curves and all intersections of two distinct curves are endpoints.

4. Points of tangency of ∂U_α and $\mathcal{F}_\alpha$ are isolated and locally quadratic.

5. Points of tangency of $T_\alpha^\beta(S_\alpha^\beta \cap \partial U_\alpha)$ and $\mathcal{F}_\beta$ are isolated and locally quadratic.

6. The forward trajectory under $\mathcal{F}_\beta$ of a point of tangency of $\mathcal{F}_\beta$ and $T_\alpha^\beta(S_\alpha^\beta \cap \partial U_\alpha)$ is never tangent to ∂U_β

7. The curve $T_\alpha^\beta(S_\alpha^\beta \cap \partial U_\alpha)$ is transverse to patch maximal periodic orbits.

Induced Transformations and Switches

In this section we demonstrate how a hybrid system can be reduced to a one dimensional transformation

The long term behavior of dynamical systems is frequently studied by the introduction of cross-sections and transition maps for the flow of trajectories between cross-sections. The cross-sections of planar vector fields are one dimensional, and the transition maps between them are monotonic. Hybrid systems have natural submanifolds of codimension one in their switching surfaces, but these need not be transversal to a flow. Moreover, the images of switching transformations do not need to be transverse to the underlying flow of the image domain, even if the switching transformations are diffeomorphisms. The dynamics of d-infinite trajectories can be analyzed in terms of the successive switching transformations and transition maps that move trajectories from patch to patch and then flow to the next switching surface. We use the term **epoch analysis** to describe the dynamics of these discrete systems.

An **induced map** of a hybrid system is a discrete dynamical system constructed by collapsing the continuous time segments of trajectories. That is the trajectory is projected onto its sequence of pre-event points.

More specifically, the patch boundaries ∂U_α are called **switching curves** and the union of all switching curves $M = \cup_{\alpha \in I} \partial U_\alpha$ as the **switching manifold.** Each ∂U_α can be written as a finite union of smooth arcs, each a subset of level curve S_α^β for some $\beta \in I$ and each arc intersecting another only at an end point. For each $x \in \partial U_\alpha$ in the interior of an arc, we construct a trajectory $(x(t,n), \alpha(n))$ and time sequence $[t_0, t_1], [t_1, t_2], \ldots$ as above with initial point $x_0 = x$, $\alpha_1 = \alpha$, and $t_0 = 0$. The induced map is then defined by $\mathcal{M}(x) = x(t_1, 1)$. That is, if x is in $\partial U_\alpha \cap S_\alpha^\beta$ then we first apply a switch $T_\alpha^\beta(x)$ and then flow under $\mathcal{F}_\beta$ until hitting the boundary $\partial U_\beta \subset M$ at a point defined to be $\mathcal{M}(x)$.

This map is undefined if $T_\alpha^\beta(x)$ remains in U_β forever. The map is not uniquely defined on the intersections of curves in $S_\alpha^\beta \cap \partial U_\alpha$.

Local Singularities of Planar Hybrid Systems

In this section, we restrict our discussion to GAIT systems and describe the types of behavior we expect to find in induced maps.

There are three types of singularities that result in singular behavior of the induced maps for generic systems. These are (1) tangency of the image of a switching transformation with the vector field, (2) tangency of the flow with a patch boundary and (3) saddle points in the interior of a patch. The tangencies of GAIT systems are quadratic: i.e., the flow trajectories have different curvature than the image of the switching transformation or the boundary curve. The saddle points of GAIT systems are **non-resonant** with eigenvalues of distinct magnitude.

Local models of the singularities can be described readily by introducing the appropriate coordinate systems. To analyze tangencies, we use flow box coordinates in which the flow trajectories are curves parallel to the first coordinate axis. If a switching transformation T_α^β has an image that is tangent to the flow at $T_\alpha^\beta(x)$, the induced map at x will have a quadratic turning point. On the other hand, if the flow has a point of tangency with a switching curve at x, then the flow leaves the patch on one side of x and enters on the other side of x. Thus x will be an endpoint for the switching transformation. The point x will be a singular value for the induced map with a square root singularity.

Flows are smoothly linearizable near non-resonant equilibria [S]. This implies that the transition maps near a non-resonant saddle have power law singularities. Let $\lambda_1 < 0 < \lambda_2$ be the eigenvalues of the saddle, and let $\alpha = -\lambda_1/\lambda_2$. If x is a coordinate on a cross-section to the stable manifold of the saddle, its transition map will be a smooth function of x^α where α is the ratio of the magnitudes of the eigenvalues at the saddle point. If $x \in \partial U_\alpha$ is in the stable manifold of a saddle s in U_α and an unstable separatrix of s exits U_α, then the induced map has a discontinuity with a power law singularity at x. Note that this singularity can have any exponent $\alpha \neq 1$.

Using this information about singularities, we obtain the following theorem:

> **Theorem** Let X be a generic planar hybrid system defined by V_α, $\mathcal{F}_\alpha$, h_α^β, and $\mathcal{T}_\alpha^\beta$. The induced map $\mathcal{M}$ of X is well defined, smooth and piecewise monotone on a disjoint union of open intervals. The limit of $\mathcal{M}$ at each endpoint exists and is given by a smooth function or a power law singularity. Interior turning points of $\mathcal{M}$ are non-degenerate critical points.

Global structure

The previous section delineated the types of singularities that can occur and the structure of $\mathcal{M}$ on an interval of definition. This section presents methods for counting the singularities and assessing the structure of the collection of intervals on which $\mathcal{M}$ is defined.

The reason for inducing a hybrid system onto the switching manifold is to reduce the complexity of the system. One goal of the theory is to be able to determine as much information as possible about the induced transformation without actually constructing it. Two important features of the induced map are (1) its discontinuities and maximal intervals of continuity, and (2) the number and types of singularities.

To examine discontinuities, we take a point $x \in \partial U_\alpha$ mapped to $y = \mathcal{T}_\alpha^\beta(x)$ and flowing to $\mathcal{M}(x) \in \partial U_\beta$. If x is interior to an arc in $\partial U_\alpha \cap S_\alpha^\beta$, then y is well defined and any singularity will stem from the flow. We say y is an **extremal point** of the **image curve** $\mathcal{T}_\alpha^\beta(\partial U_\alpha \cap S_\alpha^\beta)$ if every planar neighborhood of y contains trajectories that do not intersect the image curve. A discontinuity can occur at x only if x is an endpoint of $\partial U_\alpha \cap S_\alpha^\beta$, y is an extremal point, or y flows to a fixed point. To see that these are the only possible discontinuities, we take the flow time function $\tau(x)$ for x on an image curve, and examine the continuity of $\mathcal{F}(x, \tau(x))$. If τ is defined only on a half-neighborhood then x is an extremal point. If the two one-sided limits of τ at x exist and are unequal then $\mathcal{F}(x, \tau(x))$ is an extremal point on the curve determined by the lesser limit (here we allow ∞ as a limit). If $\tau \to \infty$ near x and the patch has compact closure then trajectories near x are getting close to an equilibrium point, so x must be on the stable manifold of a saddle. That the limit of τ exists follows by compactness and smoothness. Thus, in the generic case the induced transformation is defined and continuous on intervals with interior.

Singularities arise from switching curve tangencies (infinite one-sided derivatives), image curve tangencies (zero derivatives), and saddle separatrices (power law singularities). A switching curve tangency creates a single critical value at the point of tangency. While several points may map to this value, the order of the singularity of a GAIT system is quadratic. An image curve tangency creates a single critical point and value. Disallowing non-generic tangencies of switching curves and stable/unstable manifolds, a saddle point connection can create two critical values, one for each unstable separatrix, and four critical points, one on each side of each stable separatrix.

A **domain interval** for the induced transformation is a maximal curve segment $C \subset S_\alpha^\beta \cap \partial U_\alpha$ such that $\mathcal{M}$ is continuous on C. The induced map of a GAIT system is differentiable on the interior of domain intervals with one-sided derivatives existing for each endpoint. The turning points of this map occur at points whose images are tangent to the flow at the post-event point.

Domain intervals can accumulate only at points whose image under the switching transformation lies on a cycle. If x is a point where domain intervals accumulate, then every neighborhood of $y = \mathcal{T}_\alpha^\beta(x)$ has flow discontinuities associated with fixed or extremal points. If the patch has compact closure, there are only finitely many such points. Therefore some trajectory has a reverse-time monotone sequence approaching y and so y must lie on a (one-sided repelling) cycle.

Such an accumulation can occur only if switching curves do not cut across the cycle associated with y. A connected component of a patch cannot therefore contain two concentric cycles that create accumulation. A cycle is **patch-maximal** if it is not interior to another cycle contained in the patch. The number of accumulation points of domain intervals is then bounded by the number of points on the switching manifold that map into patch-maximal cycles. In particular, if no entrance curve intersects a cycle, then the induced transformation has a finite number of branches. Also, disallowing non-generic tangencies of entrance curves and cycles, the accumulation must be one sided.

Theorem The domain of $\mathcal{M}$ consists of a countable number of domain intervals with interior, and $\mathcal{M}$ is differentiable in each domain interval with one-sided derivatives existing at each endpoint (allowing ∞).

1. Every discontinuity in $\mathcal{M}$ occurs at a flow preimage of an extremal point or the intersection of M and the stable manifold of a saddle

2. Every critical value of $\mathcal{M}$ occurs at the flow image of an entrance tangency or the intersection of M and the stable manifold of a saddle.

3. The domain intervals of $\mathcal{M}$ can accumulate only at points where the image curve intersects a patch-maximal cycle. The accumulating domains eventually have identical images.

For current purposes, a topological attractor is a forward invariant set containing the omega limit set of an open (non-empty) set of points. Topological attractors fall into four types: (1) periodic orbits, (2) cantor sets that are generalized adding machines, (3) quasi-periodic intervals/circles, and (4) chaotic attractors – intervals with sensitive dependence to initial conditions. For C^2 maps with non-degenerate critical points, attractors of type 2, 3, and 4 must attract a critical point. This result extends to the class of maps that arise as induced maps from GAIT systems, and we have the following.

Theorem Each attractor of an induced GAIT system is one of:

1. Periodic orbit.

2. Cantor sets that are generalized adding machines.

3. Quasi-periodic intervals/circles.

4. Chaotic attractors: intervals with sensitive dependence to initial conditions.

Each GAIT system has a finite number of attractors of type 2, 3, and 4.

Under mild conditions, a C^2 map must also have a bounded number of periodic orbits. We believe this result will extend to the induced maps from GAIT systems, and offer the following

Conjecture For each GAIT system, there is a bound on the number of attracting periodic orbits that pass through switches.

Induced derivative

This section contains a precise calculation of the derivative of the induced map.

The derivative of the induced map can be separated into four components: switching expansion, entrance refraction, divergence, and exit refraction. We consider a point $x \in \partial U_\alpha \cap S_\alpha^\beta$ with image $\mathcal{M}(x) \in \partial U_\beta$. Taking s_1 as arclength on $S_\alpha^\beta \cap U_\alpha$ near x and s_2 as arclength on U_β near $\mathcal{M}(x)$, then the derivative is $\mathcal{M}'(x) = \frac{ds_2}{ds_1}$.

The first component of the derivative is from switching. If s is arclength along the image curve $\mathcal{T}_\alpha^\beta(\partial U_\alpha \cap S_\alpha^\beta)$ near $x_1^+ = \mathcal{T}_\alpha^\beta(x)$ then with some abuse of notation, $\frac{ds}{ds_1} = \mathcal{T}_\alpha^{\beta\prime}(x)$.

The next component is from entrance refraction. If a trajectory crosses a curve at x at an angle of θ from the normal to the curve, then a nearby trajectory, a distance ϵ away, will cross the curve at an arclength of about $|\epsilon\cos(\theta)|$ away. This is the refractive effect. Taking τ_1 as distance transverse to the flow $\mathcal{F}_\beta$ at $x_1 = \mathcal{T}_\alpha^\beta(x)$ and Υ_1 as the unit normal to the image curve at x_1^+, then the refraction is given by:

$$\frac{d\tau_1}{ds} = \frac{|\mathcal{F}_\beta(x_1^+)|}{|\mathcal{F}_\beta(x_1^+)\cdot\Upsilon_1|}.$$

The trajectory then flows under $\mathcal{F}_\beta$ until reaching $x_2 = \mathcal{M}(x)$, and nearby trajectories may converge or diverge. Taking τ_2 as transversal distance at x_2, we have

$$\frac{d\tau_2}{d\tau_1} = \frac{|\mathcal{F}_\beta(x_1^+)|}{|\mathcal{F}_\beta(x_2)|}\exp\int_{x_1^+}^{x_2}\nabla\cdot\mathcal{F}_\beta'\,dt. \qquad \text{TD-FLOW}$$

Finally we have the exit refraction

$$\frac{ds_2}{d\tau_2} = \frac{|\mathcal{F}_2(x_2) \cdot \Upsilon_2|}{|\mathcal{F}_2'(x_2)|}$$

where Υ is the unit normal to ∂U_β at x_2. The net derivative is then

$$\begin{aligned} \mathcal{M}'(x) &= \frac{ds_2}{ds_1} \\ &= \frac{ds}{ds_1} \frac{d\tau_1}{ds} \frac{d\tau_2}{d\tau_1} \frac{ds_2}{d\tau_2} \\ &= |\mathcal{T}_\alpha^\beta(x_1)| \frac{|\mathcal{F}_2(x_2) \cdot \Upsilon_2|}{|\mathcal{F}_2(x_1^+) \cdot \Upsilon_1|} \exp \int_{x_1^+}^{x_2} \nabla \cdot \mathcal{F}' \, dt \end{aligned} \qquad \text{M DERIV}$$

Theorem If each of the four component derivatives exist and are finite then the derivative of $\mathcal{M}$ exists and is finite. For GAIT systems, if the derivative of $\mathcal{M}$ exists and is finite then each of the four component derivatives exist and are finite.

Example: Flows on a torus

Our definition of hybrid system is general enough to represent smooth vector fields on compact manifolds as hybrid systems. This example provides a simple illustration. The two dimensional torus can be constructed from the unit square by identifying opposite sides. To form hybrid system representations of vector fields on the torus, we define the square as a single patch and four switching transformations that are translations mapping the opposite sides of the square to themselves.

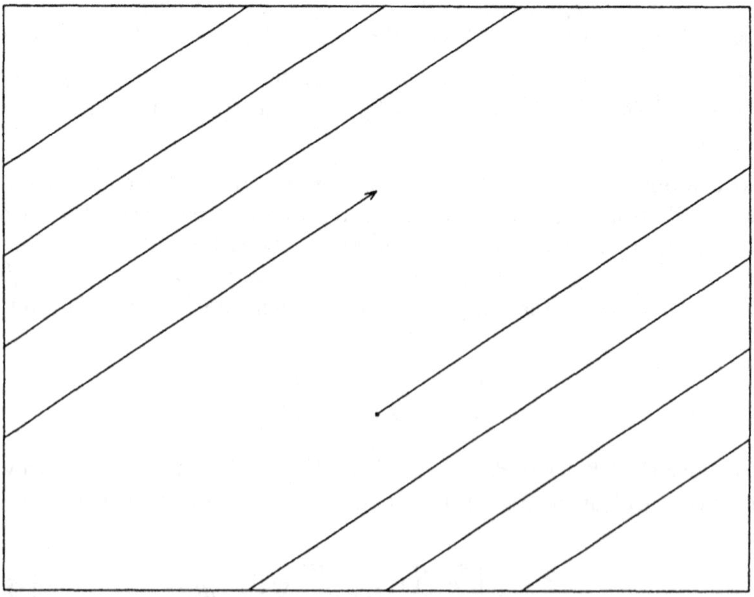

Figure 2: Typical toroidal flow

Constant vector fields on the two dimensional torus T^2 are periodic if their slope is rational and quasiperiodic if the slope is irrational. Lebesgue measure is invariant for affine flows. The hybrid representation of these torus flows preserve these properties. A constant vector field with irrational slope produces an induced transformation that is piecewise linear. It is most easily related to the theory of rotation numbers by considering the return map for one side of the square. This return map is an **interval exchange transformation.** The side of the square is partitioned into two intervals and these intervals are mapped back onto the side of the square by translations whose images are disjoint.

There is one aspect of this example that is problematic in terms of our definition of GAIT hybrid system: the images of the switching transformations lie on the boundary of a patch rather than in its interior. At the expense of adding complexity to the example and the interpretation of its dynamics, one can define a hybrid system representing toral flows that does satisfy the genericity hypotheses we have imposed. Here is one way to do so. Take a square of side length three. Define twelve switching transformations on the sides of the square that are integer translations mapping points onto the opposite side of the middle unit square. Regarding the torus as quotient of the plane by the group action of the integer lattice of translations, the patch of our hybrid system is the union of nine fundamental domains. As we leave the patch the switching transformation sends us to the boundary of the middle fundamental domain. To interpret the hybrid system as a toral flow, we identify points whose coordinates differ by an integer vector. The return map to one side of the boundary of the middle unit square is still an interval exchange transformation.

Example: Graphical analysis

The simplest method of embedding a discrete dynamical system into a hybrid system is to use the "suspension" construction familiar from dynamical systems theory. Let $f : \mathbb{R}^n \to \mathbb{R}^n$ be a discrete system and consider the flow on $\mathbb{R}^{n+1}$ with uniform velocity 1 parallel to the x_{n+1} axis. We define a threshold function $x_{n+1} - 1$ and take the switching transformation to be $T(x_1, \ldots, x_n, 1) = (f(x_1, \ldots, x_n), 0)$. Clearly, the induced transformation on the plane $x_{n+1} = 1$ is given by the map f.

We define in this example a somewhat more complex procedure for embedding a discrete dynamical system in a hybrid system. "Graphical analysis" is a common technique for tracing the orbits of a one dimensional mapping f. It consists of alternately tracing horizontal curves from the graph of f to the diagonal $y = x$ and vertical curves from the diagonal to the graph of f.

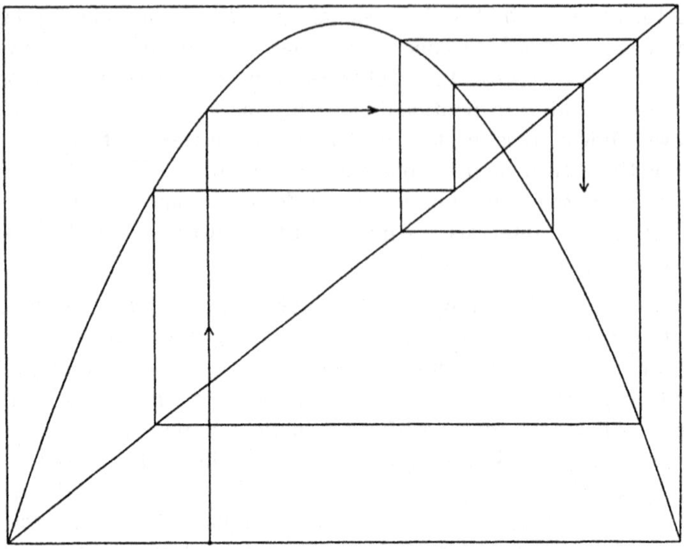

Figure 3A: Typical graphical analysis

This method can be represented as a hybrid system with horizontal and vertical flows. We consider four vector fields on the planes V_1, V_2, V_3 and V_4, as in figure 3B:

$$\mathcal{F}_1 = \begin{pmatrix} 0 \\ 1 \end{pmatrix} \qquad \mathcal{F}_2 = \begin{pmatrix} 0 \\ -1 \end{pmatrix}$$

$$\mathcal{F}_3 = \begin{pmatrix} 1 \\ 0 \end{pmatrix} \qquad \mathcal{F}_4 = \begin{pmatrix} -1 \\ 0 \end{pmatrix}$$

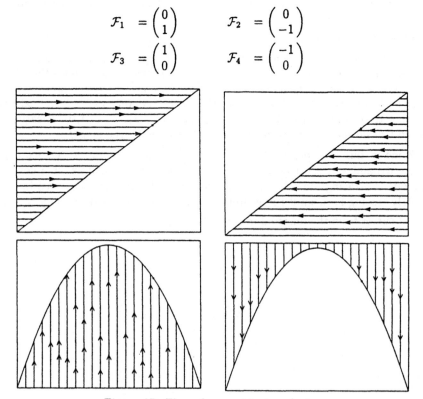

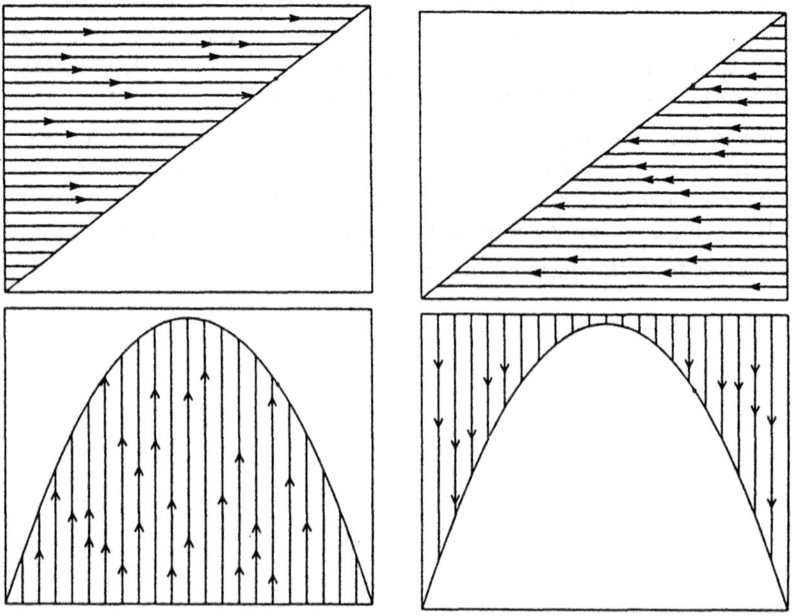

Figure 3B: Flows for graphical analysis

The following switching logic:

$$
\begin{array}{llllll}
\text{switch} & 1 \to 3 & \text{if} & y > f(x) & \text{and} & x < f(x) \\
\text{switch} & 1 \to 4 & \text{if} & y > f(x) & \text{and} & x > f(x) \\
\\
\text{switch} & 2 \to 3 & \text{if} & y < f(x) & \text{and} & x < f(x) \\
\text{switch} & 2 \to 4 & \text{if} & y < f(x) & \text{and} & x > f(x) \\
\\
\text{switch} & 3 \to 1 & \text{if} & y < x & \text{and} & x < f(x) \\
\text{switch} & 3 \to 2 & \text{if} & y < x & \text{and} & x > f(x) \\
\\
\text{switch} & 4 \to 1 & \text{if} & y > x & \text{and} & x < f(x) \\
\text{switch} & 4 \to 2 & \text{if} & y > x & \text{and} & x > f(x)
\end{array}
$$

is employed with the threshold functions:

$$
\begin{aligned}
h_1^3 &= \max\{f(x) - y, x - f(x)\} \\
h_1^4 &= \max\{f(x) - y, f(x) - x\}
\end{aligned}
$$

$$
\begin{aligned}
h_2^3 &= \max\{y - f(x), x - f(x)\} \\
h_2^4 &= \max\{y - f(x), f(x) - x\}
\end{aligned}
$$

$$
\begin{aligned}
h_3^1 &= \max\{y - x, x - f(x)\} \\
h_3^1 &= \max\{y - x, f(x) - x\}
\end{aligned}
$$

$$
\begin{aligned}
h_3^1 &= \max\{x - y, x - f(x)\} \\
h_3^1 &= \max\{x - y, f(x) - x\}
\end{aligned}
$$

All of the switching transformations are the identity on the plane.

We describe this more completely by following the evolution of a trajectory starting in V_1. The initial point is below the graph of f and flows up until it hits the graph at a point p. If p is above the diagonal $y = x$, we switch to patch V_3 and flow right to the diagonal, arriving there at a point q. If p is below the diagonal $y = x$, we switch to patch V_4 and flow left to the diagonal, again arriving at a point we call q. If q is below the graph of f, we switch to patch V_1 and flow upwards once more to the diagonal. If q is above the graph of f, we switch to patch V_2 and flow down until we hit the diagonal. The alternation of vertical motion to the graph of f and horizontal motion to the diagonal produces a piecewise smooth curve whose successive intersections with the graph of f are successive points in an orbit of f regarded as a discrete time dynamical system.

We make one further remark about the example. There is an ambiguity at fixed points of f, where we are at a boundary point between domains of switching transformations that go to different patches. Furthermore, a linearly stable fixed point will attract trajectories whose segments decrease in length geometrically. These trajectories make an infinite number of switches in finite time. Thus they are d-infinite but c-bounded.

We use this example to demonstrate "control of chaos" by making a periodic orbit globally attracting. For $f(x) = 4x(1-x)$, it is well known that almost every trajectory in $I = [0,1]$ is dense in I. The points $x_\pm = (5 \pm \sqrt{5})/8$ form a repelling orbit of period two. To calculate the transversal derivative of the cycle in the hybrid system through $x_\pm$, we note that all flows have zero divergence and switches on the diagonal have no refractive expansion. At switches on the graph of f there are contributions $f'(x_\pm)$. Thus the net transversal derivative for one loop around the cycle is $f'(x_+)f'(x_-) = -4$.

This orbit in the hybrid system is stabilized as follows. We modify the diagonal in an ϵ neighborhood of the point $r = (x_+, x_+)$ so as to change its slope at r. For example, define a deformation g of the diagonal that is non-decreasing and has $g'(r) > 4$. Using g as a threshold function retains the periodic orbit through the point r; however, the transversal derivative of the orbit has a new contribution $1/g'(r)$. This stabilizes the periodic orbit, giving it a transversal derivative between -1 and 0.

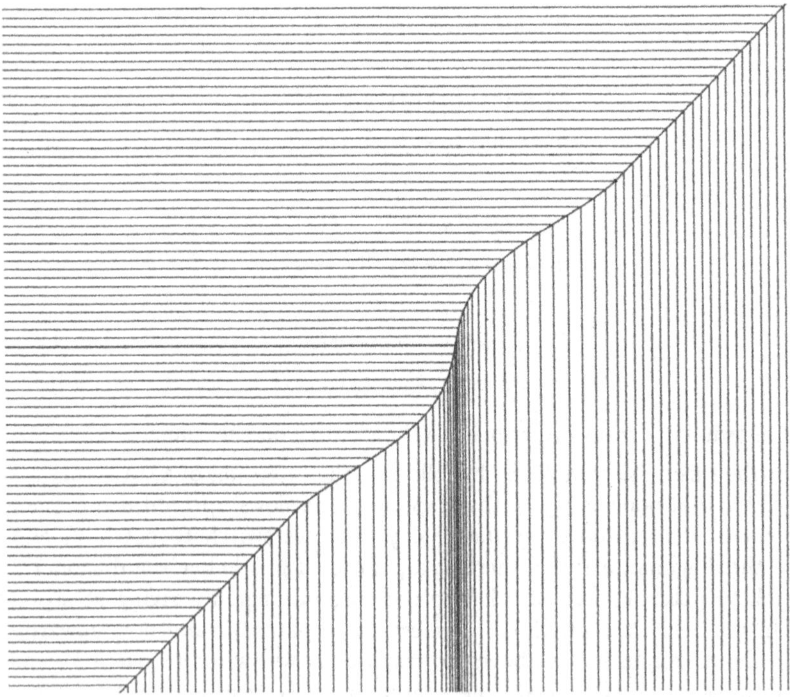

Figure 4: Blow up of modified curve.

The symbolic dynamics of the one dimensional quadratic map are well understood, and are essentially the same for the hybrid system, both before and after its modifications. The periodic orbit is used to divide the itineraries into three sets: those that stay a bounded distance away from the orbit, preimages of the orbit, and everything else. In the original map, the first corresponds to a union of Cantor sets and has zero measure, the second corresponds to a countable dense set of points, and the third to a set of full measure.

If the perturbation g is taken to be smooth and monotone then the set of itineraries remains unchanged in the perturbed system. The first set still corresponds to a Cantor set of measure zero, with trajectories that stay ϵ away from the orbit remaining unchanged since the perturbation is restricted to an ϵ-interval of r. The second set now corresponds to a countable number of intervals and has full measure, and the third set is a cantor set of zero measure.

Example: Balanced pendulum

Balancing an inverted pendulum subject to horizontal acceleration is a classical example of control theory, and applying a linear control to stabilize the top is a necessary exercise. Such stability, however, is local and succumbs to non-linearity at about 23°. Hybrid system theory offers a simple way of making the upright fixed point globally attracting.

We consider a bob of mass m_2 attached by a rigid rod of length 1 to a frictionless pivot of mass m_1. The clockwise angle of the rod from the upright position is taken as x. We consider the pivot to have motion restricted to a horizontal line. The control of the system is to apply a constant lateral force, either to the right or left, to the pivot.

The gravitational force on the bob has magnitude gm_2, and decomposes into a tangential component of size $B_{tang} = gm_2 \sin(x)$ and a radial component of size $B_{rad} = gm_2 \cos(x)$. If the location of the pivot is fixed, the resulting system is:

$$\begin{aligned} \dot{x} &= y \\ \dot{y} &= gm_2 \sin(x) \end{aligned}$$

We now allow the pivot to move on a horizontal line, and subject the pivot to a constant acceleration in the amount a. By considering the accelerated reference frame, this is synonymous to introducing a lateral force of size $-am_2$ to the bob, yielding the system:

$$\begin{aligned} \dot{x} &= y \\ \dot{y} &= gm_2 \sin(x) - am_2 \cos(x) \end{aligned}$$

A constant lateral control force F will not generally produce constant acceleration, since this force must interact with the radial component of the gravitational force on the bob, $B_{rad} = gm_2 \cos(x)$. As the pivot is restricted to the horizontal line, the lateral component on this radial force is $gm_2 \cos(x) \sin(x)$, and the net acceleration on the pivot is

$$a = \frac{F}{m_1} - \frac{gm_2}{m_1} \cos(x) \sin(x) = \frac{F}{m_1} - \frac{gm_2}{2m_1} \sin(2x)$$

Thus the resulting system is

$$\dot{x} = y$$

$$\dot{y} = gm_2 \sin(x) - \left(\frac{F}{m_1} + \frac{gm_2}{m_1}\cos(x)\sin(x)\right) m_2 \cos(x)$$
$$= gm_2 \sin(x)\left(1 - \frac{m_2}{m_1}\cos^2(x)\right) + F\frac{m_2}{m_1}\cos(x)$$

If m_1 is large compared to m_2 and F/m_1 is small, this is approximated by

$$\dot{x} = y$$
$$\dot{y} = \sin(y + \epsilon)$$

which corresponds to rotation of the gravity vector.

This system consists of three phase portraits, indexed as L (left) for $F = -\delta$, R (right) for $F = +\delta$, and S (steady) for the uncontrolled system $F = 0$.

The goal of navigating the pendulum to the upright position is reformulated in terms of energy levels. Energy measured in the steady plane is

$$E = \cos(x) + \frac{1}{2}y^2.$$

To reach the upright position, it suffices to attain an energy level of $E = 1$.

Now the strategy is stated simply in local terms: if the energy level is too high and either field R or L is heading down the potential then switch to it. Continue in the R or L plane until energy reaches 1, or until the potential is no longer decreasing, and switch back to S. Conversely if the energy is too low. While this sounds good in principle, questions of global dynamics and proof of feasibility are in the domain of hybrid systems.

This strategy is implemented with identity switching transformations and the following threshold functions:

$$h_S^L = \max\{\max\{(E-1),\ \nabla E \cdot \mathcal{F}_L\},\ \max\{(1-E),\ -\nabla E \cdot \mathcal{F}_L\}\}$$

$$h_S^R = \max\{\max\{(E-1),\ \nabla E \cdot \mathcal{F}_R\},\ \max\{(1-E),\ -\nabla E \cdot \mathcal{F}_R\}\}$$

$$h_R^S = \max\{\min\{(1-E),\ \nabla E \cdot \mathcal{F}_R\},\ \min\{(E-1),\ -\nabla E \cdot \mathcal{F}_R\}\}$$

$$h_L^S = \max\{\min\{(1-E),\ \nabla E \cdot \mathcal{F}_L\},\ \min\{(E-1),\ -\nabla E \cdot \mathcal{F}_L\}\}$$

The corresponding patches for these thresholds are shown in figure 5. Figure 5A depicts the S (steady) plane: the light grey regions are where $h_S^L < 0$ and a switch to the L plane will decrease energy, the darker grey regions are where $h_S^R < 0$, the very dark regions are where both thresholds are negative, and the white regions are where neither threshold is negative. Figure 5B depicts the R plane, with the greyed regions indicating where $H_R^L < 0$.

By examination of these patches, it is possible to argue that the upright position attracts almost every trajectory. It is possible to modify the patches so that every trajectory approaches the upright position.

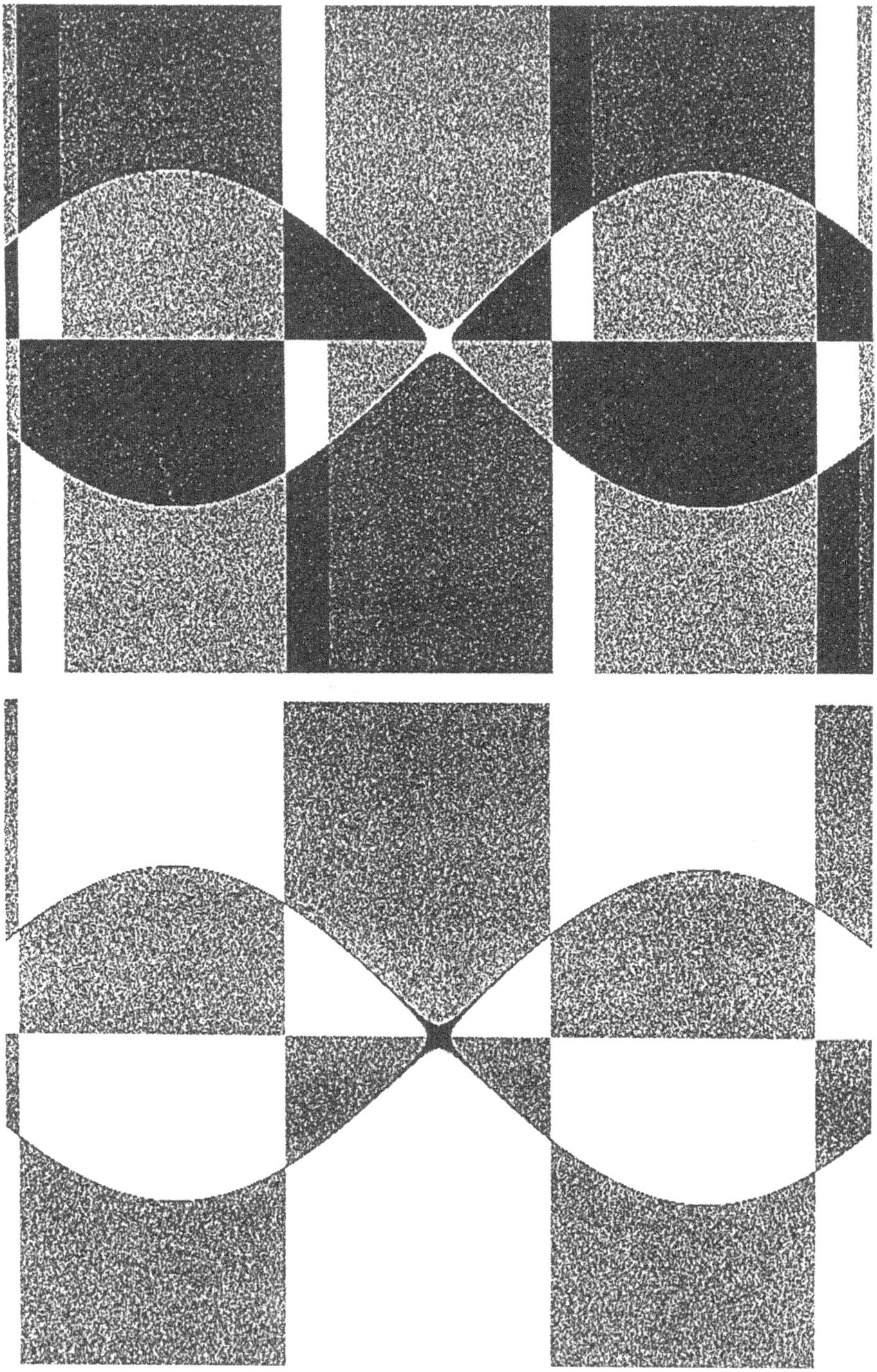

Figure 5A & 5B: Patches for pendulum

The gradient technique in this example is readily generalized. Suppose we have any system S_0 with unstable fixed point p and controls that yield perturbed systems $S_1, S_2 \ldots S_n$. Suppose there is a function H such that $H = 0$ corresponds to a stable manifold in S_0 of p. The gradient strategy is as above: if H is positive, employ whatever control may lessen H, and conversely if H is negative. This will tend to bring trajectories to the stable manifold.

Example: Lorenz system

The Lorenz attractor is a robust example of a chaotic attractor for a three dimensional vector field. Williams introduced a semiflow on a two dimensional branched manifold as a model that simplifies the analysis of the (geometric) Lorenz attractor. This branched manifold corresponds to the evolution of a family of one dimensional strong stable manifolds on the Lorenz attractor. In this example, we give a simple realization of the two dimensional Lorenz semiflow as a hybrid system. There is one patch for the system and two threshold functions.

Consider the linear vector field defined by

$$\dot{x} = x$$
$$\dot{y} = -ay$$

on the square $S = [-1,1] \times [-1,1]$. Trajectories enter the unit square from top and bottom and leave from the left and right sides. We define two threshold functions $h(x,y) = x \pm 1$ that vanish on the two sides of the square. The switching transformations will be "translations" that map segments of the left and right sides of the square onto horizontal segments at the top of the square, preserving length: $T_{\pm}(\pm 1, y) = (\pm(y-b), 1)$. A typical trajectory in the resulting system is depicted in figure 6.

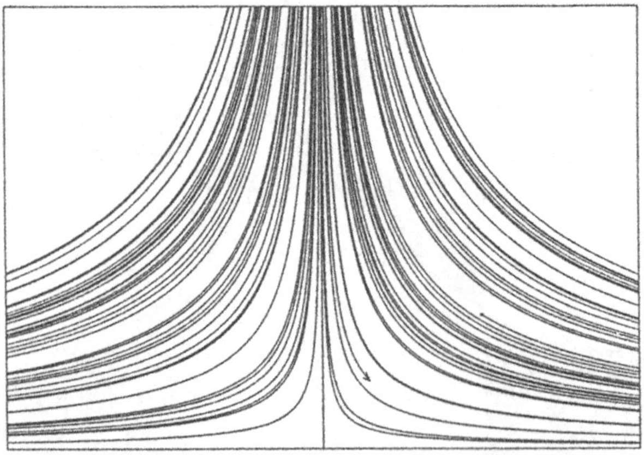

Figure 6: Lorenz system

The induced transformation for the top of the square has a discontinuity at 0 and is monotone increasing to the right and left of 0. By selecting the parameters a and b in the appropriate ranges, the induced transformation is expanding and maps an interval into itself. Specifically, we claim that there is an open set of parameters a, b such that the interval $[-b, b]$ on the top of the square returns into itself, and the return map is expanding.

First, we note that the passage map from $(x, 1)$ to $(\pm 1, y)$ along the flow is given by $y = sgn(x)|x|^a$ and has derivative $sgn(x)a|x|^{a-1}$. We want this derivative to be larger than 1 in the interval $[0, b]$, so we require $a < 1$ and $ab^{a-1} > 1$. The second requirement is that $b^a - b < b$, so that the intervals $[0, b^a]$ on the left and right sides of the square will be mapped by the switching transformations into $[-b, b]$ on the top of the square. For each $1/2 < a < 1$, the inequalities $2a > ab^{a-1} > 1$ are satisfied by an interval of values for b since $a - 1 < 0$.

There is a chaotic attractor for this hybrid system that is a "Lorenz semiflow." The boundary of the attractor is formed by the interval $[-b, b]$ at the top of the square together with the unstable separatrices of the saddle point at the origin followed once through the switching transformations and extended until they make their second intersections with the sides of S. The theory of maps of the interval implies that the induced map of this attractor has an absolutely continuous invariant measure supported on the segment $[-b, b]$ at the top of the square. It is also readily seen that both periodic orbits are dense in the attractor and that there are dense orbits in the attractor as well. These properties persist under C^1 perturbations of the system

Example: Swingsets and an example of Feng & Loparo

Parents save a significant amount of repetitive labor by teaching their kids how to push themselves on a swing. The actual technique is a complicated interaction of biodynamics and angular momenta, and is surprisingly efficient. A simplified version consists of starting a small cycle in the swing with one's feet on the ground, then at the posterior apex kick the feet forward and throw the torso back, and at the anterior apex kick the feet back and throw the torso forward. This quick action at the apexes has the marvelous effect of pumping energy into the system.

The quick pumping action is taken to be instantaneous, so we simplify the system and model its dynamics with a hybrid system of two patches. We assume a spherical frictionless child, and since children rarely if ever go all the way over the top, we use linear oscillators $y' = v$, $v' = -ay$ for both phase portraits. Threshold functions $h_1^2 = -v$, $h_2^1 = v$ utilize the upper half-plane of the first portrait and the lower half-plane of the second. Switching transformations $\mathcal{T}_1^2 : (y, v) \mapsto (y + \rho, v)$, and $\mathcal{T}_2^1 : (y, v) \mapsto (y - \rho, v)$ serve to bump up the energy level.

This is equivalently modeled using identity switching transformations and offsetting the equilibria of the two oscillators, $\mathcal{F}_1 : (y, v)' = (v, -a(y - \rho))$, $\mathcal{F}_2 : (y, v)' = (v, -a(y + \rho))$, and using the same thresholds.

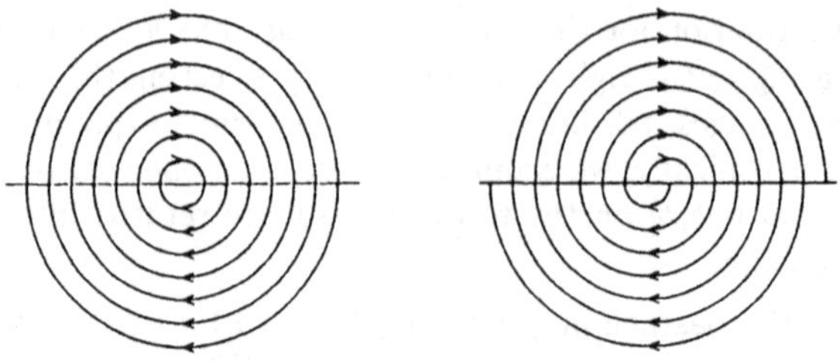

Figure 7: Outward spirals by offset

Either case produces outward spirals. Allowing friction in the system will contain the outward spiral and produce a globally stable cycle.

Feng & Loparo [FL] study an example with two patches and flows defined by linear equations having unstable focal equilibrium points. $\mathcal{F}_1 : (y, v)' = (v, -a(y - \rho) + \delta v)$, $\mathcal{F}_2 : (y, v)' = (v, -a(y - \rho) + \delta v)$. Switching transformations were employed to keep trajectories of the system bounded. This is modeled by using identity transforms and thresholds $h_1^2 = v$ and $h_2^1 = -v$. Trajectories either get stuck on the horizontal $[-\rho, \rho]$ or spiral outwards to ∞. This is the reverse-time picture of the swingset with friction.

The trajectories that get stuck are not c-infinite. This can be avoided with threshold functions of the form $h_1^2 = max\{y, x - \rho\}$. The only avenue for switching from 1 to 2 is on the horizontal $x < -\rho$.

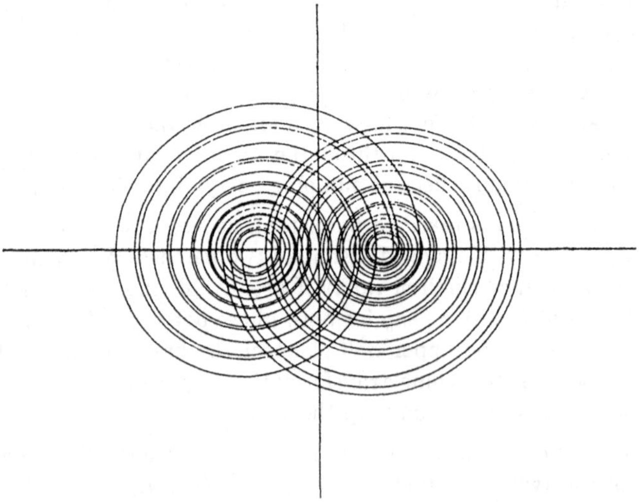

Figure 8: Typical Feng & Loparo trajectory

Trajectories beginning in the interval $v = 0$, $|y| < \rho(1 + \frac{2}{e^{\kappa}-1})$ with $\kappa = \frac{\pi\delta}{\sqrt{4a-\delta^2}}$ stay bounded. Since both flows have a positive divergence and switches are parallel, the induced map is expanding. Induced maps have a finite number of image intervals, and so a result of Lasota and Yorke [LY] implies that the induced map supports an absolutely continuous invariant measure. The intervals $[-\rho(1+2e^{\kappa}), -\rho]$,and $[\rho, \rho(1+2e^{\kappa})]$ are ergodic components of this map, each dense with periodic orbits.

Pathology Example: Double tangencies

A trajectory that encounters two tangencies can fail to have a transverse derivative, even when the system and curves are smooth. Two basic situations to consider are (a) when the switching transformation maps a tangency to another tangency and (b) when a point of tangency flows to another point of tangency.

For example, consider two unit flows to the right, the first switching at the curve $y = f_1(x)$, $f_1(0) = f_1'(0) = 0$, to the second flow at the curve $y = f_2(x)$, $f_2(0) = f_2'(0) = 0$, where the switching transformation $\mathcal{T}$ is defined by $\mathcal{T}(0,0) = (0,0)$ and $s_1 = s_2$. The resulting transversal flow map is $\tau_2 = f_2(f_1^{-1}(\tau_1))$

Let $f_1(x) = h(x)\,g(x)$ with $h(0) = h'(0) = 0$ and $g'(x)$ failing to converge as $x \to 0$ and such that $f(x)$ is monotone. For example, $h(x) = x^2$ and $g(x) = \sin(\ln(x)) + 2$. Let $f_2(x) = \int h\,dx$. Then both f_1 and f_2 are legitimate tangent switches, mapping a transverse half-neighborhood at $x_1 = 0$ onto a transverse half-neighborhood at $x_2 = 0$. A direct calculation yields $\frac{d\tau_2}{d\tau_1} = 1/g(f_1^{-1}(\tau_1))$, and so the derivative fails to converge as τ_1 approaches 0.

Similar examples can be constructed for situation (b), and are left to the reader.

Acknowledgments

This research of John Guckenheimer was partially supported by the National Science Foundation, the Department of Energy and the Air Force Office of Scientific Research.

List of symbols:

R reals, Z integers, D some euclidean domain.

I a discrete indexing set, $\alpha, \beta, \rho \in$ I

Sets: $\underbrace{U_\alpha}_{\text{patch}} \subset \underbrace{V_\alpha}_{\text{chart}} \subseteq \underbrace{D_\alpha}_{\text{domain}}$, $V = \cup_{\alpha \in I} V_\alpha$

Flows: $\mathcal{F}_\alpha : V_\alpha \times R \to D_\alpha$

Transformations: $\mathcal{T}_\alpha^\beta : V_\alpha \mapsto V_\beta$

Threshold functions and level sets: $h_\alpha^\beta : V_\alpha \mapsto R \quad S_\alpha^\beta = \{x : \; h_\alpha^\beta(x) = 0\}$

Hybrid system: $X = \{V_\alpha, \mathcal{F}_\alpha, h_\alpha^\beta, \mathcal{T}_\alpha^\beta\}$

Epochs (closed time intervals): $e_m = [t_m, t_{m+1}]$

Induced transformation on the switching manifold: $\mathcal{M} : \; M \mapsto M$

References:

[AP] A. A. Andronov, & L. Pontryagin, Systèms Grossiers, *Dokl. Akad. Nauk. SSSR,* **14,** 247-251, (1937).

[BGM] L. Back, J. Guckenheimer, & M. Meyers, "A Dynamical Simulation Facility for Hybrid Systems," *Springer Lecture notes in Comp. Sci.,* **736,** 255-267, (1993).

[MS] de Melo & van Strien, "One Dimensional Dynamics," *Springer.*

[FL] X. Feng, & K. Loparo, "Chaotic Motion and its Probabilistic Description in a Family of Two Dimensional Nonlinear Systems with Hysteresis," *J. of Nonlinear Science,* **12** #8, (1991).

[G1] J. Guckenheimer, "Sensitive Dependence on Initial Conditions for One Dimensional Maps," *Comm. Math. Phys.,* **70,** 133-160, (1980).

[G2] J. Guckenheimer, "A Robust Stabilization Strategy for Equilibria," *IEEE Trans. Automatic Control,* in press, (1995).

[J] S. Johnson, "Simple Hybrid Systems," *Int. J. of Bifurcation & Chaos,* to appear.

[K] D.E. Koditscheck & M. Buhler, "An Analysis of a Simplified Hopping Robot," *Int. J. Robotics Research,* to appear.

[LY] A. Lasota & J. Yorke, "On the Existence of invariant Measures for Piecewise Monotonic Transformations," *Trans. Amer. Math. Soc.,* **186** #12, (1973).

[S] S. Sternberg, "On the Structure of Local Homeomorphisms of Euclidean n-Space, II," *Amer. J. Math.,* 80, 623-631, (1958).

Programming in hybrid constraint languages

Vineet Gupta * Radha Jagadeesan ** Vijay Saraswat* Daniel G Bobrow*

Abstract. We present a language, Hybrid cc, for modeling hybrid systems compositionally. This language is declarative, with programs being understood as logical formulas that place constraints upon the temporal evolution of a system. We show the expressiveness of our language by presenting several examples, including a model for the paperpath of a photocopier. We describe an interpreter for our language, and provide traces for some of the example programs.

1 Introduction and Motivation

The constant marketplace demand of ever greater functionality at ever lower price is forcing the artifacts our industrial society designs to become ever more complex. Before the advent of silicon, this complexity would have been unmanageable. Now, the economics and power of digital computation make it the medium of choice for gluing together and controlling complex systems composed of electro-mechanical and computationally realized elements.

As a result, the construction of the software to implement, monitor, control and diagnose such systems has become a gargantuan, nearly impossible, task. For instance, traditional product development methods for reprographics systems (photo-copiers, printers) involve hundreds of systems engineers, mechanical designers, electrical, hardware and software engineers, working over tens of months through several hard-prototyping cycles. Software for controlling such systems is produced by hand – separate analyses are performed by the software, hardware and systems engineers, with substantial time being spent in group meetings and discussions coordinating different design decisions and changes, and assessing the impact these decisions will have on the several teams involved. Few, if any, automated tools are available to help in the production of documentation (principles of operation, module/system operating descriptions), or in the analysis and design of the product itself. A different team — often starting from little else than the product in a box, with little "documentation" — produces paper and computational systems to help field-service representatives diagnose and fix such products in the field.

Such work practices are unable to deal with the increasing demand for faster time to market, and for flexibility in product lines. Instead of producing one product, it is now necessary to produce a family of "plug-and-play" generic components — finishers (stackers, staplers, mailboxes), scanners, FAX modules, imaging systems, paper-trays, high-capacity feeders — that may come together for the very first time at customer-site.

* Xerox Palo Alto Research Center, 3333 Coyote Hill Road, Palo Alto Ca 94304; {vgupta, saraswat, bobrow}@parc.xerox.com
** Dept. of Mathematical Sciences, Loyola University-Lake Shore Campus, Chicago, Il 60626; radha@math.luc.edu

The software for controlling such systems must be produced in such a way that it can work even if the configuration to be controlled is known only at run-time.

To address some of these problems, we are investigating the application of *model-based computing* techniques. The central idea is to develop compositional, declarative models of the various components of a photo-copier product family, at different levels of granularity, customized for different tasks. For instance, a marker is viewed as a transducer that takes in (timed) streams of sheets S and video images V, to produce (timed) streams of prints P. A model for the marker, from the viewpoint of the scheduler, is a set of constraints that capture precisely the triples $\langle S, V, P\rangle$ which are in fact physically realizable on this marker. The marker model is itself constructed from models for nips, rollers, motors, belts, paper baffles, solenoids, control gates etc.; the models are hooked together in exactly the same way as the corresponding physical components are linked together. Together with this, software architectures are developed for the tasks that need to be accomplished (such as scheduling, simulation, machine control, diagnostic tree generation). Finally, linking the two are special-purpose reasoners (operating at "configuration-time") that produce information of the right kind for the given task architecture, given the component models and system configuration.

We expect this approach to be useful for a variety of tasks:

- **Scheduling.** A model of the system specifies what outputs will be produced given (perhaps continuous) input and control commands. In principle, the same model can be used to search the space of inputs and control commands, given a description of system output. In practice, the design of such an inference engine is made complex by the inherent combinatorial complexity of the search process. With a formal model in hand, it becomes possible to use automated techniques to better understand the search-space and design special-purpose reasoners. This approach is currently being deployed, in collaboration with several other teams, in the development of schedulers for a new generation of products from Xerox Corporation.
- **Code generation.** A physical model of a system can be used for *envisionment* [dKB85] — studying the possible paths of evolution of the system. Since some of the parameters of a system can be controlled, it becomes possible to develop automatically controllers which would specify these parameters leading the system to a desirable state, and away from paths which lead to unsafe states.
- **Diagnostic-tree generation.** Given a model of components, their interconnection, and their correct (and possibly faulty) behavior — perhaps with other information such as prior probabilities for failure — it is possible to construct off-line repair-action procedures[FBB+94]. These can be used by service technicians as guides in making probes to determine root cause for the manifest symptom.
- **Explanation.** Given a simulation-based model, it is possible to annotate behavioral rules with text in such a way that the text can be systematically composed to provide natural-language explanations for why an observable parameter does or does not have a particular value.
- **Productivity Analysis.** Models may be analyzed to determine how corresponding product designs will perform on different job mixes (e.g., a sequence of all single-sided black-&-white jobs, followed by all double-sided, color, stapled jobs).

The model-based computing approach places a set of requirements on the nature of the modeling language. We motivate these demands by considering a concrete example — the paper path of a simple photocopier, see Figure 1.

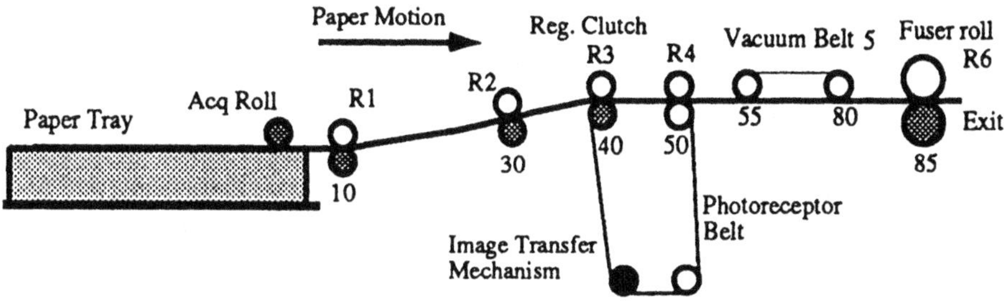

Fig. 1. A simple photocopier

Example: Paper Transportation in a Photocopier. In this photocopier, paper is loaded in a paper tray at the left of the machine. When a signal is received, the acquisition roll is lowered onto the paper and pulls the top sheet of paper towards the first set of rollers (R1). After the paper is grasped by the first set of rollers, the acquisition roll is lifted, and the rollers pull the paper forward, till it reaches the registration clutch (R3). This starts at a precise moment for perfect alignment with the toner image on the belt. The toner is transferred electrostatically onto the paper by the image transfer mechanism. The vacuum belt (5) transports it to the fuser roll (R6) which fuses the toner into the paper, and from where it exits.

Hybrid modeling. The photocopier above has a collection of components with continuous behavior, for example, the rollers and belts. The control program of the photocopier is a discrete event driven system. Therefore, the modeling language should be able to describe the interaction of the control program and the continuous components — thus, the modeling language has to fall in the framework of hybrid systems [BG90,NSY91,MMP92,GNRR93].

Executability. Given a model of the system, it should be possible to predict the behavior of the system when inputs are supplied. Thus the model should be executable, *i.e.* it should also be possible to view models as *programs*. This would allow hybrid control programs to be written using the same notation as component models. If sensors and actuators coupling the language implementation with the physical environment are provided then, in fact, it should be possible to use programs in this notation to drive physical mechanisms.

Compositional modeling. Out of considerations of reuse, it seems clear that models of composite systems should be built up from models of the components, and that models of components should reflect their physics, without reflecting any pre-compiled knowledge of the structure (configurations) in which they will be used (the "no function in structure principle", [dKB85]). Concretely, this implies that the modeling language must be expressive enough to modularly describe and support extant control architectures and techniques for compositional design of hybrid systems, see for example [HMP93, Hoo93,NK93]. From a programming language standpoint, these modularity concerns are addressed by the analysis underlying synchronous programming languages [BB91, Hal93,BG92,HCP91,GBGM91,Har87,CLM91,SJG95], (adapted to dense discrete domains in [BBG93]); this analysis leads to the following demands on the modeling language.

The modeling should support the interconnection of different components, and allow for the hiding of these interconnections. For example, the model of the copier is the composition of the the control program and the model of the paper path. The model of the paper path in turn, is the composition of the models of the rollers, clutches etc.

Furthermore, in the photocopier example, if paper jams, the control program should cause the rollers to stop. However, the model of the rollers, in isolation, does not need to have any knowledge of potential error conditions. Thus, the modeling language should support *orthogonal preemption* [Ber93] — any signal can cause preemption.

Finally, the language should allow the expression of multiple notions of *logical time* — for example, in the photocopier the notion of time relevant to the paper tray is (occurrences of) the event of removing paper from the tray; the notion of time relevant to the acquisition roll is determined by the rotation rate of the roller; the notion of time of the image transfer mechanism is the duration of the action of transferring images etc.

Thus, we demand that the modeling language be an *algebra of processes, that includes concurrency, hiding, preemption and multiform time.*

Declarative view. It must be possible for systems engineers — people quite different in training and background from software engineers — to use such a formalism. Typically, systems engineers understand the physics of the system being designed or analyzed, and are used to mathematical or constraint-based formalisms (equational and algebraic models, transfer functions, differential equations) for expressing that knowledge. This suggests that it must be possible to view a model (fragment) expressed in the language as a declaration of facts in a (real-time) (temporal) logic — see for example [MP91,AH92].

Reasoning. Given a model of the paper-path, and the control procedures governing it, it should be possible to perform a *tolerance analysis* — establish the windows on early/late arrival of sheets of papers at various sensors on the paper-path, given that a particular physical component may exhibit any behavior within a specified tolerance. This, and other such engineering tasks, suggest that the modeling language must be *amenable* to (adapting) the methodology developed in the extensive research on reasoning about hybrid and real-time systems — for example, specification and verification of properties of hybrid systems [GNRR93,dBHdRR92], qualitative reasoning about physical systems [Wd89], and envisionment of qualitative states [Kui94].

1.1 This paper

This paper describes programming in the modeling language, Hybrid cc — hybrid concurrent constraint programming. Intuitively, Hybrid cc is obtained by "freely extending" an untimed non-monotonic language Default cc over continuous time. Hybrid cc has the following key features:

- The notion of a *continuous* constraint system describes the continuous evolution of system trajectories.
- Hybrid constraint languages — developed generically over continuous constraint systems — are obtained by adding a single temporal construct, called **hence**. Intuitively, a formula **hence** A is read as asserting that A holds continuously beyond the current instant.
- Continuous variants of preemption-based control constructs and multiform timing constructs are definable in Hybrid cc.

The formal foundations of Hybrid cc are discussed in [GJSB95].

The rest of this paper is organised as follows. First, we describe the computational intuitions underlying Hybrid cc. We follow with a brief description of an interpreter for a language in the Hybrid cc framework. We then describe a series of examples, and show traces of the execution of the interpreter. These examples illustrate the programming idioms and expressiveness of Hybrid cc.

2 Hybrid cc: Computational intuitions

2.1 Background

Concurrent Constraint Programming As mentioned in the previous section, some of the crucial characteristics in a hybrid programming language are easy specification and composition of model fragments. This led us to consider Concurrent Constraint Programming languages [Sar93,SRP91] as a starting point, since these languages are built on top of constraint systems, which can be made as expressive as desired, and they have very fine-grained concurrency, so compositionality is achieved without effort. Recently, several concrete general-purpose programming languages have been implemented in this paradigm [Deb93,JH91,SHW94].

Concurrent Constraint Programming (cc) languages are declarative concurrent languages, where each construct is a logical formula. cc replaces the traditional notion of a store as a valuation of variables with the notion of a store as a constraint on the possible values of variables. Thus the store consists of pieces of information which restrict the possible values of the variables. A cc program consists of a set of agents[3] running concurrently and interacting with the shared store. Agents are of two basic kinds — *tell* agents which add information to the store (written a), and ask agents, or conditionals

[3] We use the words *agent* and *program* interchangeably in this paper, usually referring to fragments of a program as agents.

(written **if** a **then** A), which query the store about the validity of some information, and reduce to other agents if it is valid[4]. This yields the grammar:

$$A ::= a \mid \textbf{if } a \textbf{ then } A \mid \textbf{new } X \textbf{ in } A \mid A, A$$

Computation is monotonic — information can only be added to the store. Ask actions are used for synchronization — if a query is answered positively, then the agent can proceed, otherwise it waits (possibly forever) till there is enough information in the store to entail the information in the query. When no more computation is being performed (a state of *quiescence* is reached), the store is output.

The information that is added to the store consists of constraints which are drawn from a *constraint system*. Formally, a constraint system C is a system of partial information, consisting of a set of primitive constraints or tokens[5] D with minimal first order structure — variables and existential quantification. Associated with a constraint system is an entailment relation (denoted $\vdash_C$) which specifies when a token a can be deduced from some others $b_1, \ldots, b_n$, denoted by $b_1, \ldots, b_n \vdash_C a$. Examples of such systems are the system Herbrand, underlying logic programming — here tokens are equalities over terms which are finite trees with variables ranging over trees — and FD [HSD92] or finite domains, its tokens are equalities of variables and expressions saying that the range of a variable is some finite set.

A salient aspect of the cc computation model is that agents may be thought of as imposing constraints on the evolution of the system. The agent a imposes the constraint a. (A, B) imposes the constraints of both A and B — logically, this is the conjunction of A and B. **new** X **in** A imposes the constraints of A, but hides the variable X from the other agents — logically, this can be thought of as a form of existential quantification. The agent **if** a **then** A imposes the constraints of A provided that the rest of the system imposes the constraints a — logically, this can be thought of as intuitionist implication.

This declarative way of looking at programs is complemented by an operational view. The basic idea in the operational view is that of a network of agents interacting with a shared store of primitive constraints. The agent a is viewed as adding a to the store instantaneously. (A, B) behaves like the simultaneous execution of both A and B. **new** X **in** A starts A but creates a new local variable X, so no information can be communicated on it outside. The agent **if** a **then** A behaves like A if the current store entails a.

The main difficulty in using cc languages in modeling is that the cc programs can detect only the presence of information, not its absence. However, in reactive systems, it is important to handle information saying that a certain event did not occur — examples are timeouts in UNIX or the presence of jams in a photocopier (usually inferred by the paper not reaching a point by some specified time). The problem in assuming the absence of information (called negative information) during a computation is that it may be invalidated later when someone else adds that information, leading to invalidation of the subsequent computation.

---

[4] There are also hiding agents **new** X **in** A which hide any information about X from everyone else, these are needed for modularity.

[5] Tokens are denoted by $a, b, \ldots$ in this paper.

Our first attempt at fixing this problem was the addition of a sequence of phases of execution to the cc paradigm [SJG94b,SJG94a]. At each phase we executed a cc program, and this gave the output for the phase, and also produced the agents to be executed in subsequent phases. At the end of each phase we detected the absence of information, and used it in the *next* phase, giving us a reactive programming language, tcc. Each phase was denoted by a time tick, so we had a discrete model of time.

Default cc. The addition of time to cc however still left us with a problem—how can we detect negative information *instantaneously*? This is necessary because in some applications we found that it was not acceptable to wait until the next phase to utilize negative information [SJG95].

Since the negative information is to be detected in the same time step, it must be done at the level of the untimed language, cc. However, the cc paradigm is inherently opposed to instantaneous negative information detection. Thus in order to allow negative information to be detected within the same computation cycle, we have to modify cc — we have to use Default cc. The basic idea behind Default cc is — if the final output of a computation is known, then any negative information can be obtained from this output, and this negative information can never be invalidated, as the final output has all the information ever added. So a Default cc program executes like a cc program, except that before the beginning of the execution, it guesses the output store. All negative information requests are resolved with respect to this guess, other than that the Default cc program adds information to the store and queries the store for positive information just like a cc program. At the end of the computation, if the store is equal to the guess, then the guess was correct, and this is a valid answer. Otherwise, this branch of execution is terminated (or we can do backtracking). In an actual implementation, this backtracking is done at compile time, so at runtime we have to do a simple table lookup.

The Default cc paradigm augments the cc paradigm with the ability to detect negative information, so the only new syntax we need to add to the cc syntax is the negative ask combinator **if** a **else** A, this reduces to A if the guessed output e does not imply a. Thus the Default cc syntax is given by the grammar

$$A ::= a \mid \textbf{if } a \textbf{ then } A \mid A, A \mid \textbf{new } X \textbf{ in } A \mid \textbf{if } a \textbf{ else } A$$

The agent **if** a **else** A imposes the constraints of A unless the rest of the system imposes the constraint a — logically, this can be thought of as a *default* [Rei80].[6] (In the timed contexts discussed below, the subtlety of the non-monotonic behavior of programs — intimately related to the causality issues in synchronous programming languages — arises from this combinator.) To see this, note that A may itself cause further information to be added to the store at the current time instant; and indeed, several other agents may simultaneously be active and adding more information to the store. Therefore requiring

[6] Note that **if** a **else** A is quite distinct from the agent **if** $\neg a$ **then** A (assuming that the constraint system is closed under negation). The former will reduce to A iff the final store does not entail a; the latter will reduce to A iff the final store entails $\neg a$. The difference arises because, by their very nature, stores can contain partial information; they may not be strong enough to entail either a or $\neg a$.

that information a be absent amounts to making a demand on "stability" of negative information.

Thus we now have a language that permits instantaneous negative information detection. This enables us to write strong timeouts — if a signal is not produced at a certain time, the execution of the program can be terminated at that very time, not at some later time as we did for timed cc. Now in order to get a language for modeling discrete reactive systems, we again introduce phases in each of which a Default cc program executes. To extend Default cc across time, we need just one more construct, **hence** A, which starts a copy of A in each phase *after* the current one. We extend Default cc across real time to get a language for hybrid reactive systems in the next subsection.

2.2 Continuous evolution over time.

We follow a similar intuition in developing Hybrid cc: the continuous timed language is obtained by uniformly extending Default cc across real (continuous) time. This is accomplished by two technical developments. For the formal development, see [GJSB95].

Continuous Constraint Systems. First, we enrich the underlying notion of constraint system to make it possible to describe the continuous evolution of state. Intuitively, we allow constraints expressing initial value (integration) problems, e.g. constraints of the form `X = 0, hence dot(X) = 1`; from these we can infer at time `t` that `X = t`. The technical innovation here is the presentation of a *generic* notion of continuous constraint system (ccs), which builds into the very general notion of constraint sytems just the extra structure needed to enable the definition of continuous control constructs (without committing to a *particular* choice of vocabulary for constraints involving continuous time). As a result subsequent development is *parametric* on the underlying constraint language: for each choice of a ccs we get a hybrid programming language. Later we will give an example of a simple constraint system and the language built over it.

Program Combinators. Second we add to the untimed Default cc the same temporal control construct: **hence** A, but now interpreting it over real time. Declaratively, **hence** A imposes the constraints of A at every *real* time instant after the current one. Operationally, if **hence** A is invoked at time t, a new copy of A is invoked at each instant in (t, ∞).

Agents	Propositions
a	a holds now
if a **then** A	if a holds now, then A holds now
if a **else** A	if a will not hold now, then A holds now
new X **in** A	exists an instance $A[t/X]$ that holds now
A, B	both A and B hold now
hence A	A holds at every instant after now

Intuitively, **hence** might appear to be a very specialized construct, since it requires repetition of the *same* program at every subsequent time instant. However, **hence** can combine in very powerful ways with positive and negative ask operations to yield rich patterns of temporal evolution. The key idea is that negative asks allow the instantaneous preemption of a program — hence, a program **hence if** P **else** A will in fact not execute A at all those time instants at which P is true.

Let us consider some concrete examples. Suppose that we require that an agent A be executed at every time point beyond the current one until the time at which a is true. This can be expressed as **new** X **in** (**hence** (**if** X **else** A, **if** a **then always** X)). Intuitively, at every time point beyond the current one, the condition X is checked. Unless it holds, A is executed. X is local — the only way it can be generated is by the other agent (**if** a **then always** X), which, in fact generates X continuously if it generates it at all. Thus, a copy of A is executed at each time point beyond the current one upto (and excluding) the time at which a is detected.

Similarly, to execute A precisely at the first time instant (assuming there is one) at which a holds, execute: **new** X **in hence** (**if** X **else if** a **then** A, **if** a **then hence** X).

In particular, the continuous version of the general preemption control construct **clock** (introduced in [SJG94b]) is definable within Hybrid cc. The patterns of temporal behavior described above are obtainable as specializations of **clock**. Further, programs containing the **clock** combinator may be equationally rewritten into programs not containing the combinator.

While conceptually simple to understand, **hence** A requires the execution of A at every subsequent real time instant. Such a powerful combinator may seem impossible to implement computationally. For example, it may be possible to express programs of the form **new** T **in** ($T = 0$, **hence** $dot(T) = 1$, **hence if** $rational(T)$ **then** A) which require the execution of A at every rational $q > 0$. Such programs are not implementable. To make Hybrid cc computationally realizable, the basic intuition we exploit is that, in general, physical systems change slowly, with points of discontinuous change, followed by periods of continuous evolution. This is captured by a *stability* condition on continuous constraint systems that guarantees that for every constraint a and b there is a neighborhood around 0 in which a either entails or disentails b at every point. This rules out constraints such as `rational(T)` as inadmissible.

With this restriction, computation in Hybrid cc may be thought of as progressing in alternating phases of computation at a time point, and in an open interval. Computation at the time point establishes the constraint in effect at that instant, and sets up the program to execute subsequently. Computation in the succeeding open interval determines the length of the interval r and the constraint whose continuous evolution over $(0, r)$ describes the state of the system over $(0, r)$.

3 The implementation

We will now present a simple implementation for Hybrid cc built on top of a simple continuous constraint system. The interpreter is built on top of Prolog, and consideration has been given to simplicity rather than efficiency. We have already identified several ways in which performance can be significantly enhanced using standard logic programming techniques.

3.1 The continuous constraint system

Basic tokens are formulas d, **hence** d or $prev(d)$, where d is either an atomic proposition p or an equation of the form $dot(x, m) = r$, for x a variable, m a non-negative integer and r a real number. Tokens consist of basic tokens closed under conjunction and existential quantification. Some rudimentary arithmetic can be done using the underlying prolog primitives. The inference relations are defined in the obvious way under the interpretation that p states that p is true and $dot(x, m) = r$ states that the mth derivative of x is r. $prev(d)$ asserts that d was true in the limit from the left, and **hence** d states that for all time $t > 0$ d holds. The inference relations are trivially decidable: the functions of time expressible are exactly the polynomials. The only non-trivial computation involved is that of finding the smallest non-negative root of univariate polynomials (this can be done using one of several numerical methods).

3.2 The basic interpreter

The interpreter simply implements the formal operational semantics discussed in detail in [GJSB95]. It operates alternately in point and interval phases, passing information from one to the other as one phase ends. All discrete change takes place in the point state, which executes a simple Default cc program. In a continuous phase computation progresses *only* through the evolution of time. The interval state is exited as soon as the status of one of the conditionals changes — one which always fired does not fire anymore, or one starts firing.

The interpreter takes a Hybrid cc program and starts in the point phase with time $t = 0$.

The point phase. The input to the point phase is a set of Hybrid cc agents and a *prev* store, which carries information from the previous interval phase — the atomic propositions and the values of the variables at the right endpoint of the interval (initially at time 0 this is empty). The outputs are the store at the current instant, which is printed out and passed on to the next interval phase as the initial store, and the agent or the *continuation*, to be executed in the interval phase.

The interpreter takes each agent one by one and processes it. It maintains the store as a pair of lists — the first list contains the atomic propositions that are known to be true at the point, and the second contains a list of variables and the information that is known about the values of their derivatives. It also maintains lists of suspended conditionals (**if** ...**then** and **if**...**else** statements whose constraints are not yet known to be true or false) and a list of agents to be executed at the next interval phase.

When a tell action (a constraint a) is seen, it is added to the store and a check for consistency is made. If the constraint is of the form $dot(x, m) = r$, then it means that all derivatives of x of order $< m$ are continuous, and thus their values are copied from the *prev* store. Any conditional agents that have been suspended on a are reactivated and placed in the pool of active agents.

When any conditional agent appears, its constraint is immediately checked for validity. If this can be determined, the appropriate action is taken on the consequence of the agent — it is discarded or added to the pool of active agents. Otherwise it is added to an indexed list of suspended conditionals for future processing. A **hence** action is placed in the list of agents to be executed in the next interval.

When no more active agents are left, the interpreter looks at the suspended agents of the form **if** a **else** A, the *defaults*. On successful termination, one of two conditions must hold for each such agent: either the final store entails a, or else the final store incorporates the effect of executing A. Thus all that the interpreter has to do is to non-deterministically choose one alternative for each such default. In the current implementation, this non-deterministic choice is implemented by backtracking, with execution in the current phase terminating with the first successful branch.

The interval phase. This is quite similar to the point phase, except that it also has to return the duration of the phase. The input is a set of agents and an initial store which determines the initial values for the differential equations true in the store. The output is a set of agents to be executed in the next point phase, a store giving the values of the variables at the *end* of the interval phase, and the length of the phase. Initially we assume that the length will be infinite, this is trimmed down during the execution of the phase.

A *tell* is processed as before — it is added to a similar store, and reactivates any suspended conditionals. Conditional agents are added to the suspended list if their conditions are not known to be valid or invalid yet, otherwise the consequence of the agent is added to the active pool or discarded. In addition, if the condition can be decided, then we compute the duration for which the status will hold. For example if we have $dot(x, 1) = 3$ in the store, and $x = 4$ in the initial store, then we know that $x = 3t + 4$. Now if we ask **if** $(x = 10)$ **then** A, then we immediately know that this is not true in the current interval, so A can be discarded for now. However, we also know, by solving the equation $3t + 4 = 10$ that this status will hold for only 2 seconds, so this interval phase cannot be longer than two seconds. (This is where we need to solve the polynomial equations.)

At the end of the processing, the defaults are processed as before, and are used to trim the length of the interval phase further. Also, the remaining suspended conditionals **if** a **then** ... are checked to when a could first become valid, to reduce the duration of the phase if necessary. Finally a computation is done to find the store at the end of the interval, and this information is passed to the next point phase. Since all agents are of the form **hence** A in an interval phase, the agent to be executed at the next point phase must be A, **hence** A.

If any interval phase has no agents to execute, the program terminates.

3.3 Combinators

This basic syntax of Hybrid cc has been quite enough for us to write all the programs that we have. However, we have noticed several recurrent patterns in our programs, and some of these can be written up as definable combinators, and make the job of programming considerably easier. Note that these are defined combinators, *i.e.* they are definable in terms of the basic combinators. Thus, these defined combinators are merely syntactic sugar, and can be compiled away into basic combinators. The formal principles underlying the compilation process are discussed in a companion paper [GJSB95].

Always. **always** $A = A$, **hence** A is read logically as $\Box A$, the necessity modality. Parameterless recursion can be mimicked — replace a recursive process $P :: body$ by **always if** P **then** $body$. Parameters can be handled by textual substitution.

Waiting for a condition. **whenever** a **do** A reduces to A at the first time instant that a becomes true — if there is a well-defined notion of first occurrence of a. Note that there is no "first" occurrence of a in the situation when the a occurs in the interval $(0, t)$, as happens in the agent **hence** a. In such cases, A will not be invoked. **whenever** can be defined in terms of the basic combinators as mentioned in the previous section.

In the following discussion on definable combinators, and in the rest of this paper, we shall not repeat the caveat about the the subtlety of the well-definedness of the "first" occurrences of events.

Watchdogs. **do** A **watching** a, read logically as "A until a", is the strong abortion interrupt of ESTEREL [Ber93]. The familiar `cntrl-C` is a construct in this vein. **do** A **watching** a behaves like A until the first time instant when a is entailed; when a is entailed A is killed instantaneously. It can be defined using the basic constructs as follows.

$$\begin{aligned}
\textbf{do } b \textbf{ watching } a &= \textbf{if } a \textbf{ else } b \\
\textbf{do if } b \textbf{ then } A \textbf{ watching } a &= \textbf{if } b \textbf{ then do } A \textbf{ watching } b \\
\textbf{do if } b \textbf{ else } A \textbf{ watching } a &= \textbf{if } b \textbf{ else do } A \textbf{ watching } a \\
\textbf{do } A, B \textbf{ watching } a &= \textbf{do } A \textbf{ watching } a, \textbf{do } B \textbf{ watching } a \\
\textbf{do new } X \textbf{ in } A \textbf{ watching } a &= \textbf{new } X \textbf{ in do } A \textbf{ watching } a \quad \text{X not free in } a \\
\textbf{do hence } A \textbf{ watching } a &= \textbf{new } X \textbf{ in (hence [if } X \textbf{ else do } A \textbf{ watching } a, \\
&\qquad \textbf{if } a \textbf{ then hence } X])
\end{aligned}$$

Multiform time. **time** A **on** a denotes a process whose notion of time is the occurrence of a — A evolves only the store entails a. It can be defined by structural induction as above.

Implementation of the defined constructs. The laws given above suffice to implement all these constructs in the interpreter. However for the sake of efficiency, we have implemented the constructs directly, at the same level as basic constructs. The **always** construct is of course trivial, and has been implemented using the given rule. **whenever** is implemented similarly to an **if** a **then** A construct, except that if a is not entailed, then it is carried across to the next phase. It has no effect in an interval phase for the reason mentioned above.

do A **watching** a is very similar to an **if** b **else** B, and is implemented analogously. The only thing to be noted here is that whenever something from the A is to be passed on to the next phase, it must be passed inside a **do** ... **watching** a, so that the proper termination of all spawned agents can be achieved when a becomes true. **time** A **on** a is similar — it is like an **if** a **then** A, remembering the same rule about passing spawned agents through phases.

4 Programming examples

We illustrate the programming/model description style of Hybrid cc via a few examples. In each of these examples, we write each of the conceptually distinct components separately. We then exploit the expressive power of Hybrid cc to combine the subprograms (submodels) via appropriate combinators, to get the complete program/model. We also present the trace of the interpreter on some of these examples.

4.1 Sawtooth function

We start off with a simple example, the sawtooth function defined as $f(y) = y - \lfloor y \rfloor$, where $\lfloor y \rfloor$ denotes the greatest integer in y.

As mentioned above, prev(x=1.0) is read as asserting that the left limit of x equals 1.

```
sawtooth ::
    x = 0,
    hence (if prev(x = 1.0) then x = 0),
    hence (if prev(x = 1.0) else dot(x) = 1)).
```

The trace of this program as executed by our interpreter is seen below. The execution in the point and interval phases is displayed separately. For both phases of execution, the contents of the store are displayed. In addition, for the interval phase, the interval is also displayed, and the continuous variables are displayed as polynomials in time. In the interval, note that time is always measured from the beginning of the interval, not from 0. Also, this program has infinite output, so we aborted it after some time.

```
Time = 0.
Store contains [dot(x,0)=0].

Time Interval is (0, 1.0)
Store contains [x=1*t+0].

Time = 1.0.
Store contains [dot(x,0)=0].

Time Interval is (1.0, 2.0)
Store contains [x=1*t+0].

Time = 2.0.
Store contains [dot(x,0)=0].

Time Interval is (2.0, 3.0)
Store contains [x=1*t+0].
```

```
Time = 3.0.
Store contains [dot(x,0)=0].
Prolog interruption (h for help)? a
{Execution aborted}
| ?-
```

4.2 Temperature Controller.

We model a simple room heating system which consists of a furnace which supplies heat, and a controller which turns it on and off. The temperature of the furnace is denoted `temp`. The furnace is either on (modeled by the signal `furnace_on`) or off (modeled by the signal `furnace_off`). The actual switching is modeled by the signals `switch_on` and `switch_off`. When the furnace is on, the temperature rises at a given rate, `HeatR`. When the furnace is off, the temperature falls at a given rate, `CoolR`. The controller detects the temperature of the furnace, and switches the furnace on and off as the temperature reaches certain pre-specified thresholds: `Cut_out` is the maximum temperature and `Cut_in` is the minimum temperature.

The heating of the furnace is modeled by the following program. The multiform time construct ensures that heating occurs only when the signal `furnace_on` is present.

```
furnace_heat(HeatR) ::
   time (always dot(temp) = HeatR)  on furnace_on.
```

The cooling of the furnace is modeled by the following program. The multiform time construct ensures that heating occurs only when the signal `furnace_off` is present.

```
furnace_cool(CoolR) ::
   time (always dot(temp) = - CoolR) on furnace_off.
```

The furnace itself is the parallel composition of the heating and cooling programs.

```
furnace(HeatR, CoolR) ::
   furnace_heat(HeatR), furnace_cool(CoolR).
```

The controller is modeled by the following program — at any instant, the program watches for the thresholds to be exceeded, and turns the appropriate switch on or off.

```
controller(Cut_out, Cut_in) ::
    always (if switch_on then do (always furnace_on) watching switch_off,
            if switch_off then do (always furnace_off) watching switch_on,
            if prev(temp = Cut_out) then switch_off,
            if prev(temp = Cut_in) then switch_on).
```

The entire assembly is defined by the parallel composition of the furnace and the controller.

```
controlled_furnace(HeatR, CoolR, Cut_out, Cut_in) ::
     (furnace(HeatR, CoolR),
      controller(Cut_out, Cut_in)).
```

The trace of this program with parameters `HeatR = 2, CoolR = -0.5, Cut_Out = 30, Cut_in = 26` and initial conditions `temp = 26` and the signal `switch_on`, as executed by our interpreter is seen below.

```
Time = 0.
Store contains [furnace_on,switch_on,dot(temp,1)=2.0,
dot(temp,0)=26.0].

Time Interval is (0, 2.0)
Store contains [furnace_on,temp=2.0*t+26.0].

Time = 2.0.
Store contains [furnace_off,switch_off,dot(temp,1)= -0.5,
dot(temp,0)=30.0].

Time Interval is (2.0, 10.0)
Store contains [furnace_off,temp= -0.5*t+30.0].

Time = 10.0.
Store contains [furnace_on,switch_on,dot(temp,1)=2.0,
dot(temp,0)=26.0].

Time Interval is (10.0, 12.0)
Store contains [furnace_on,temp=2.0*t+26.0].
Prolog interruption (h for help)? a
{Execution aborted}
```

4.3 Cat and Mouse.

Next, we consider the example of a cat chasing a mouse. A mouse starts the origin, running at speed 10 metres/second for a hole 100 metres away. After it has traveled 50 metres, a cat is released, that runs at speed 20 metres/second after the mouse. The positions of the cat and the mouse are modeled by `c` and `m` respectively. The cat wins (modeled by the signal `wincat`) if it catches the mouse before the hole, and loses otherwise (modeled by the signal `winmouse`).

The cat and the mouse are modeled quite simply. The positions of the cat and the mouse change according to the respective velocities. Note that there is no reference to the cat or the position of the hole in the mouse program; similarly for the cat program. This is essential for *reuse* of these models.

```
mouse :: M = 0, always dot(M) = 10.
cat :: C = 0, always dot(C) = 20.
```

The entire system is given by the assembly of the components. The mouse wins if it reaches the hole `m=100`; the cat wins if `c=100` and the mouse has not reached the hole. The combination

do (...) **watching** `prev(m = 100)`, **whenever** `m = 100` **do** ...

is used as an exception handler; it handles the case of the mouse reaching the hole. Similarly, the combination

do (...) **watching** `prev(c = 100))`, **whenever** `prev(c = 100)` **do** ...

is used as an exception handler; it handles the case of the cat reaching hole. The higher priority of **do** ... **watching** `prev(c=100)` relative to **do** `(...)` **watching** `prev(m=100)` is captured by lexical nesting — in this program the cat catches the mouse if both reach the hole simultaneously.

```
system_configuration ::
   do (do (mouse,
           whenever prev(m = 50) do cat)
        watching prev(m = 100),
       whenever prev(m = 100) do winmouse)
    watching prev(c = 100),
   whenever prev(c = 100) do wincat.
```

A trace of this program as executed by the interpreter is given below:

```
Time = 0.
Store contains [dot(m,1)=10,dot(m,0)=0].

Time Interval is (0, 5.0)
Store contains [m=10*t+0].

Time = 5.0.
Store contains [dot(m,1)=10,dot(m,0)=50.0,dot(c,1)=20,
dot(c,0)=0].
```

```
Time Interval is (5.0, 10.0)
Store contains [c=20*t+0,m=10*t+50.0].

Time = 10.0.
Store contains [wincat].

Termination after Time 10.0.
 Initial Conditions: [wincat].
 Store has [].

yes
```

The other alternative, *i.e.* the mouse wins if if the cat and the mouse reach the hole simultenously. is captured by reversing the lexical nesting of **do** `(...)` **watching** `prev(m=100)` and **do** `(...)` **watching** `prev(c=100)`. Concretely, the program is:

```
system_configuration ::
   do (do (mouse,
            whenever prev(m = 50) do cat)
         watching prev(c = 100),
        whenever prev(c = 100) do wincat)
    watching prev(m = 100),
   whenever prev(m = 100) do winmouse.
```

A trace of this program as executed by the interpreter is given below:

```
Time = 0.
Store contains [dot(m,1)=10,dot(m,0)=0].

Time Interval is (0, 5.0)
Store contains [m=10*t+0].

Time = 5.0.
Store contains [dot(m,1)=10,dot(m,0)=50.0,dot(c,1)=20,
dot(c,0)=0].

Time Interval is (5.0, 10.0)
Store contains [c=20*t+0,m=10*t+50.0].

Time = 10.0.
Store contains [winmouse].

Termination after Time 10.0.
 Initial Conditions: [winmouse].
 Store has [].
 yes
```

4.4 A simple game of Billiards

We model a billiards (pool) table with several balls. The balls roll in a straight line till a collision with another ball or an edge occurs. When a collision occurs the velocity of the balls involved changes discretely. When a ball falls into a pocket, it disappears from the game. For simplicity, we assume that all balls have equal mass and radius (called R). We model only two ball collisions, and assume that there is no friction.

Impulses, denoted I, are assumed to be vectors. Velocities, positions are assumed to be pairs with an x-component and a y-component.

The structure of the program is that each ball, each kind of collision, and the check for pocketing for each ball are modeled by programs. A ball is basically a record with fields for name, position and velocity.

The agent `ball` maintains a given ball (`Ball`) with initial position (`InitPos`) and velocity (`InitVel`).[7] As the game evolves, position changes according to velocity (`dot(Ball.pos) = Ball.InitVel`) and velocity changes according to the effect of collisions. A propositional constraint `Change(Ball)`, shared between the collision and ball agents, communicates occurrences of changes in the velocity of the ball named `Ball`. The `ball` process is terminated when the ball is removed, *eg.* it falls in a pocket — as before, the higher priority of **do** ... **watching** `pocketed(Ball)` relative to **do** (...) **watching** `Change(Ball)` is captured by lexical nesting.

```
ball(Ball, InitPos, InitVel) ::
   do (Ball.pos = InitPos,
        do hence Ball.vel = InitVel watching Change(Ball),
        whenever Change(Ball) do ball(Ball, Ball.pos, Ball.newvel)
        hence dot(Ball.pos) = Ball.vel)
     watching pocketed(Ball).
```

The table is assumed to start at $(0,0)$, with length *xMax* and breadth *yMax*. If a ball hits the edge, one velocity component is reversed in sign and the other component is unchanged.

```
edge_collision(B) ::
   always (if (B.pos.x = B.r) or (B.pos.x = xMax - B.r)
             then (Change(B), B.newvel = ( - B.vel.x, B.vel.y)),
           if (B.pos.y = B.r) or (B.pos.y = yMax - B.r)
             then (Change(B), B.newvel = (B.vel.x, - B.vel.y))).
```

[7] The program is given here in a syntax that allows the declaration of procedures. The syntax $p(X_1,\ldots,X_n)$ :: A is read as asserting that for all $X_1,\ldots,X_n$, $p(X_1,\ldots,X_n)$ is equivalent to A.

Ball-ball collisions involve solutions to the quadratic conservation of energy equation. "**if** $|I| = 0$ **else** $|I| > 0$" chooses the correct solution, $I \neq 0$. The solution $I = 0$ makes the balls go through each other! We use distance(P1,P2) as short hand for the computation of the distance between the P1 and P2.

```
two_ball_collision(B1, B2) ::
   always if (B1.pos = P1, B2.pos = P2, B1.vel = V1, B2.vel = V2
              distance(P1, P2) = 2 * R)
       then (I_{B1, B2}.y * (P2.x - P1.x) = (P2.y - P1.y) * I_{B1, B2}.x,
             M * | V1 |^2 + M * | V2 |^2 = M * | B1.newvel |^2 + M * | B2.newvel |^2,
             M * V1 + I_{B1, B2} = M * B1.newvel, M * V2 - I_{B1, B2} = M * B2.newvel,
             if | I_{B1, B2} | = 0 else | I_{B1, B2} | > 0,
             change(B1), change(B2)).
```

A ball is pocketed if its center is within distance p from some pocket.

```
pocket(Ball) ::
   always if in_pocket(xMax, yMax, Ball.pos) then pocketed(Ball).
in_pocket(xMax, yMax, P) ::
   (distance(P, (0, 0)) < p or distance(P, (0, yMax)) < p or
    distance(P, (xMax, 0)) < p or distance(P, (xMax, yMax)) < p or
    distance(P, (xMax/2, 0)) < p or distance(P, (xMax/2, yMax)) < p).
```

4.5 The Copier Paper Path

We now present a larger example, which utilizes some of the programming ideas illustrated by the previous "toy" examples. We model the copier paper path that we mentioned in Section 1. As we discussed there, we will model each of the components of the paper path in Hybrid cc, and put them to to produce a model for the entire paper path. This model and the underlying intuitions are discussed in detail in [GSS95]—here we present an overview.

The constraint system we use for this model is richer than the constraint system described before — besides simple propositions and first-order formulas on them, we can assert arithmetic inequalities between variables, their derivatives and real constants. We will also allow the expression of attribute-value lists — given a list L we can use $L.a$ to refer to the value of a in the list L, if it exists. $L[a = r, b = s, \ldots]$ adds the attributes $a, b, \ldots$ to L with value $r, s, \ldots$.

The Intuitions. Our basic idea is to construct models for each of the individual components of the paperpath — the rollers and belts etc. We will also build a model for the sheets of paper going through the paperpath. In addition, we need to model the *local interaction* between a sheet of paper and a transportation element, i.e. a belt or a roller.

This interaction results in the transportation elements applying forces to the sheet, and the sheet then transmits these forces and moves under their influence. The local interactions will naturally induce a partition of the sheet into segments, which we view as individual entities, analyzing their freebody diagrams.

We make a number of modeling assumptions to simplify our model. For example, we do not model the acceleration of the sheet of paper, instead we view its velocity as changing discretely when a change takes place in the influences on it. We also assume that the sheet is homogeneous, the paperpath is straight, and some other things, these are discussed in detail in [GSS95].

The Model. The paperpath is modeled as a segment of the real line, and each component occupies a segment of the real line. All forces and velocities are oriented with respect to the paperpath. The first model fragment describes a basic transportation element—this can be further specialized into models for different types of belts and rollers. The properties and initial state of the element are specified using an attribute-value list `Init`. Each element has a nominal velocity V_{nom}, which is the velocity along the paperpath supplied by the motor. It also has an actual velocity V_{act}, which may differ from the nominal velocity, as a sheet of paper may be pulling or pushing the element. The force exerted on the transportation element is denoted F_{act}, and the force exerted by the motor is denoted F_{resist}. The freebody equation relates these two. F_{resist} is bounded by F_{fast} and F_{slow}. The remaining constraints show the relation between the forces and the velocities — they assert that if the element is going faster than the nominal velocity V_{nom}, then the force F_{act} must be at its maximum value and similarly for the minimum value.

```
transportationElement(T, Init) ::
   always (transport(T),
         T.Loc = Init.Loc, T.surfacetype = Init.surfacetype,
         T.F_fast = Init.F_fast, T.F_slow = Init.F_slow,
         T.F_normal = Init.F_normal,
         T.F_act + T.F_resist = 0,
         − T.F_fast ≤ T.F_resist ≤ T.F_slow,
         if T.V_act > T.V_nom then T.F_act = T.F_fast,
         if T.V_act < T.V_nom then T.F_act = − T.F_slow).
```

This model is simply an enumeration of the constraints on the transportation element, these are **always** true. The value of V_{nom} is not set here, this is done during the specialization of this generic transportation element description into various particular kinds of elements.

The next "component" of the paperpath is the sheet of paper that travels along it. This again has a number of properties, and also an initial location. Furthermore, the sheet is partitioned into various segments — the `contactSegments` are those portions of the sheet in contact with some transportation element. Between two contact segments is an `internalSegment`, whereas beyond the extreme contact segments are the `TailSegments`. In order to create these segments, we use the new operator **forall**

$X : C[X].A[X]$. This is simply a parallel composition of *finitely* many $A[X]$'s, one for each X such that $C[X]$ is true. The declarative style of programming is particularly useful here — this way of asserting the existence of segments is much simpler than the dynamic creation and destruction of segments that would have to be done in an imperative program.

```
sheet(S, Init) ::
   S.Loc = Init.Loc,
   do always
      (sheet(S),
       S.width = Init.width, S.length = Init.length,
       S.thickness = Init.thickness, S.elasticity = Init.elasticity,
       S.surfacetype = Init.surfacetype, S.strength = Init.strength,
       if ∃I.(S.I.condition = tearing) then S.torn,
       forall T : transport(T). if (T.engaged, | T.Loc ∩ S.Loc | > 0)
                                   then contact(S, S.Loc ∩ T.Loc),
       forall I : contact(S, I). contactSegment(S, I),
       forall I : contact(S, I). forall J : contact(S, J).
         if I < J then if ∃K.(contact(S, K), I < K < J)
                        else internalSegment(S, (ub(I), lb(J))),
       forall I : contact(S, I). if ∃K.(contact(S, K), K < I)
                                   else leftTailSegment(S, (lb(S.Loc), lb(I))),
       forall I : contact(S, I). if ∃K.(contact(S, K), I < K)
                                   else rightTailSegment(S, (ub(I), ub(S.Loc))),
       dot(lb(S.Loc)) = S.vel.lb(S.Loc),
       dot(ub(S.Loc)) = S.vel.ub(S.Loc))
   watching S.torn.
```

For brevity, we omit the code for the segments of the sheet. The agent `interact` creates an interaction process `sheetTransportation` for each transportation element overlapping the sheet. The sheet transportation process models the interaction between the sheet and the transportation element. It asserts that the force exerted by the element on the sheet is equal and opposite to the force exerted by the sheet on the element. It gives the freebody equation for the contact segment, and then bounds the amount of force exerted by the transportation element by the amount of friction. Finally, the friction and the velocities are related.

```
interact(S) ::
   always forall T : transport(T).
     if (T.engaged, | T.Loc ∩ S.Loc | > 0)
        then sheetTransportation(S, T, T.Loc ∩ S.Loc).
```

```
sheetTransportation(S, T, I) ::
   new F_friction in
     (I = S.Loc ∩ T.Loc,
      F_friction = mu(S.surfacetype, T.surfacetype) × T.F_normal,
      S.I.F_te + T.F_act = 0,
      S.(lb(I)).F_left + S.(ub(I)).F_right + S.I.F_te = 0,
      | S.I.F_te | ≤ F_friction,
      if S.vel.(ub(I)) > T.V_act then S.I.F_te = − F_friction,
      if S.vel.(ub(I)) < T.V_act then S.I.F_te = F_friction).
```

Finally we provide the model for the paperpath shown in Figure 1. The various kinds of transportation elements are modeled first, followed by a typical sheet, then these are composed together to give the entire paperpath. There is one new combinator used here — **wait** T **do** A, which waits for T seconds before starting A. This can be expressed in terms of the basic combinators. A new sheet is placed at $(0, 21)$ every time a signal Sync is received from the scheduler.

```
cvElement(R, Init) ::
   transportationElement(R, Init), always (R.V_nom = Init.V_nom, R.engaged).
roller(R, Init) ::
   cvElement(R, Init[surfaceType = rubber, F_fast = 1000, F_slow = 1000,
     F_normal = 300]),
   if (R.F_act = R.F_slow ∨ R.F_act = − F_fast) then R.broken.
clutchedRoller(R, Init) ::
   cvElement(R, Init[surfaceType = rubber, F_fast = 3, F_slow = 1000,
     F_normal = 200]),
   if R.F_act = R.F_slow then R.broken.
belt(R, Init) ::
   cvElement(R, Init[surfaceType = steel, F_fast = 10000, F_slow = 10000,
     F_normal = 120]).
fuserRoll(R, Init) ::
   cvElement(R, Init[surfaceType = rubber, F_fast = 1000, F_slow = 1000,
     F_normal = 1000]).
regClutch(R, Init) ::
   transportationElement(R, Init), always R.engaged,
   always (if R.on then do always R.V_nom = Init.V_nom watching R.off,
           if R.off then do always R.V_nom = 0 watching R.on).

a4GlossySheet(S) ::
   sheet(S, [length = 21, width = 29.7, thickness = 0.015, elasticity = 15,
             surfaceType = glossy, strength = 500, Loc = (0, 21)]).

copierModel(Sync) :: new (R1, R2, R3, R4, R5, R6) in
   (roller(R1, [V_nom = 30, Loc = (10, 10.5)]),
```

```
roller(R2, [V_nom = 30, Loc = (30, 30.2)]),
regClutch(R3, [V_nom = 30, Loc = (39.9, 40.1), surfaceType = rubber,
                   F_fast = 1000, F_slow = 1000, F_normal = 300]),
clutchedRoller(R4, [V_nom = 0, Loc = (50, 50.1)]),
belt(R5, [V_nom = 30, Loc = (55, 80)]),
fuserRoll(R6, [V_nom = 35, Loc = (85, 86)]),

always (f(rubber, glossy) = 0.8, f(steel, glossy) = 0.1),
always if Sync then (new S in (a4GlossySheet(S), interact(S)),
                        R3.off, wait .8 do R3.on)).
```

The model can now be executed to simulate the photocopier. However, we would like to put to the various other uses mentioned in the introduction, by exploiting the fact that Hybrid tcc has a well-defined mathematical semantics.

One simple use for our model would be to verify whether some properties are true of it. For example we might want to assert that sheets will never tear — **always forall** `S:sheet(S)` . **if** `S.torn` **then** ERROR. Similarly, several other properties can be asserted, these can be now proved using various theorem-proving and model-checking techniques. We could also want to prove that successive sheets never overlap — this proof would give us a condition on how far apart the `Sync` signals must be. Similar techniques may be used for control code generation. The other applications mentioned in the introduction also make use of such formal models and their semantics.

4.6 Qualitative simulation.

Abstract interpretation of Hybrid tcc programs might provide an alternate conceptual framework for the qualitative modeling approaches of [Kui94].

Hybrid tcc can be used to model exactly systems with continuous and discrete change. For example, here is a model of a bathtub, with the tap on, and the drain unplugged [Kui94, p112].

```
bathtub(Inflow, F, Capacity) :: new (netflow, outflow, amount) in
   constant(F), monotone(F), F(0) = 0,
   constant(Inflow),
   constant(Capacity),
   do always (netflow = outflow + Inflow,
              outflow = F(amount),
              dot(amount) = netflow)
     watching (amount = Capacity ∧ dot(amount) > 0).
```

Here the first three lines assert that the three parameters do not vary with time, and that F is a monotone function in its argument, assuming the value 0 on input 0.

Essentially with this information, QSIM produces a graph that shows that from the initial state there are only three possible (qualitative) states that the program can transit to (for any possible input values of Inflow, F, and Capacity): namely those in which the tub overflows, the tub equilibrates at a level below Capacity, and the tub equilibrates at capacity. To do this reasoning, QSIM employs knowledge about the behavior of continuously-varying functions.

Since Hybrid tcc provides an exact description of the model, it should be possible to do an abstract interpretation of the program to get the qualitative reasoning of QSIM. This is useful in cases when a precise value for the inputs is not known, only a range is given. Abstract interpretation could also be useful in verification. For example, in the cat-and-mouse example, we may not know the exact velocities of the cat and the mouse, only a range of velocities. However it may still be possible to prove that the mouse always wins. The ability to run this program abstractly for all the possible values of the velocities provides an alternative way of verification of this property.

5 Future work

This paper presents a simple programming language for hybrid computing, extending the untimed computation model of Default cc uniformly over the reals. Conceptually, computation is performed at each real time point; however, the language is such that it can be compiled into finite automata of the style of [NOSY93]. We have an interpreter for the language by directly extending the interpreter for Default cc, with an integrator for computing evolution of a set of variables subject to constraints involving differentiation.

We demonstrate the expressiveness of the language through several programming examples, and show that the preemption-based control constructs of the (integer-time based) synchronous programming languages (such as "**do** ... **watching** ...) extend smoothly to the hybrid setting.

Much remains to be done. What are semantic foundations for higher-order hybrid programming, in which programs themselves are representable as data-objects? How should such a hybrid modeling methodology be combined with the program structuring ideas of object-oriented programming? What is the appropriate notion of types in such a setting? How do the frameworks for static analysis of programs (data- and control-flow analysis, abstract interpretation techniques) generalize to this setting? What are appropriate techniques for partial evaluation and program transformation of such programs? Besides these language issues, there are a number of modeling and implementation issues — What are appropriate continuous constraint systems which have fast algorithms for deciding the entailment relation? Can boundary value problems be solved in this framework? How does stepsize in numerical integration techniques affect execution and visualization of results? Some of these will be the subject of future papers.

Acknowledgements. Work on this paper has been supported in part by ONR through grants to Vijay Saraswat and to Radha Jagadeesan, and by NASA. Radha Jagadeesan has also been supported by a grant from NSF.

References

[AH92] R. Alur and T. Henzinger. Logics and models of real time: a survey. In J. W. de Bakker, C. Huizing, W. P. de Roever, and G. Rozenberg, editors, *REX workshop "Real time: Theory in Practice"*, volume 600 of *LNCS*, 1992.

[BB91] A. Benveniste and G. Berry. The synchronous approach to reactive and real-time systems. In *Special issue on Another Look at Real-time Systems*, Proceedings of the IEEE, September 1991.

[BBG93] A. Benveniste, M. Le Borgne, and Paul Le Guernic. *Hybrid Systems: The SIGNAL approach*. Number 736 in LNCS. Springer Verlag, 1993.

[Ber93] G. Berry. Preemption in concurrent systems. In *Proc. of FSTTCS*. Springer-Verlag, 1993. LNCS 781.

[BG90] A. Benveniste and P. Le Guernic. Hybrid dynamical systems and the signal language. *IEEE Transactions on Automatic control*, 35(5):535–546, 1990.

[BG92] G. Berry and G. Gonthier. The ESTEREL programming language: Design, semantics and implementation. *Science of Computer Programming*, 19(2):87 – 152, November 1992.

[CLM91] E. M. Clarke, D. E. Long, and K. L. McMillan. A language for compositional specification and verification of finite state hardware controllers. *Proceedings of the IEEE*, 79(9), September 1991.

[dBHdRR92] J. W. de Bakker, C. Huizing, W. P. de Roever, and G. Rozenberg, editors. *REX workshop "Real time: Theory in Practice"*, volume 600 of *LNCS*, 1992.

[Deb93] Saumya K. Debray. QD-Janus: A Sequential Implementation of Janus in Prolog. *Software—Practice and Experience*, 23(12):1337–1360, December 1993.

[dKB85] Johan de Kleer and John Seely Brown. *Qualitative Reasoning about Physical Systems*, chapter Qualitative Physics Based on Confluences. MIT Press, 1985. Also published in AIJ, 1984.

[FBB+94] M. Fromherz, D. Bell, D. Bobrow, et al. RAPPER: The Copier Modeling Project. In *Working Papers of the Eighth International Workshop on Qualitative Reasoning About Physical Systems*, pages 1–12, June 1994.

[GBGM91] P. Le Guernic, M. Le Borgne, T. Gauthier, and C. Le Maire. Programming real time applications with SIGNAL. In *Special issue on Another Look at Real-time Systems*, Proceedings of the IEEE, September 1991.

[GJSB95] Vineet Gupta, Radha Jagadeesan, Vijay Saraswat, and Daniel Bobrow. Computing with continuous change. Technical report, Xerox Palo Alto Research Center, May 1995. Submitted.

[GNRR93] Robert Grossman, Anil Nerode, Anders Ravn, and Hans Rischel, editors. *Hybrid Systems*. Springer Verlag, 1993. LNCS 736.

[GSS95] Vineet Gupta, , Vijay Saraswat, and Peter Struss. A model of a photocopier paper path. In *Proceedings of the 2nd IJCAI Workshop on Engineering Problems for Qualitative Reasoning*, August 1995.

[Hal93] N. Halbwachs. *Synchronous programming of reactive systems*. The Kluwer international series in Engineering and Computer Science. Kluwer Academic publishers, 1993.

[Har87] D. Harel. Statecharts: A visual approach to complex systems. *Science of Computer Programming*, 8:231 – 274, 1987.

[HCP91] N. Halbwachs, P. Caspi, and D. Pilaud. The synchronous programming language LUSTRE. In *Special issue on Another Look at Real-time Systems*, Proceedings of the IEEE, Special issue on Another Look at Real-time Systems, September 1991.

[HMP93] T. A. Henzinger, Z. Manna, and A. Pnueli. Towards refining temporal specifications into hybrid systems. In R. L. Grossman, A. Nerode, A. P. Ravn, and H. Rischel, editors, *Hybrid Systems*, volume 736 of *LNCS*, pages 209–229, 1993.

[Hoo93] J. Hooman. A compositional approach to the design of hybrid systems. In R. L. Grossman, A. Nerode, A. P. Ravn, and H. Rischel, editors, *Hybrid Systems*, volume 736 of *LNCS*, pages 209–229, 1993.

[HSD92] Pascal Van Hentenryck, Vijay A. Saraswat, and Yves Deville. Constraint processing in cc(fd). Technical report, Computer Science Department, Brown University, 1992.

[JH91] Sverker Janson and Seif Haridi. Programming Paradigms of the Andorra Kernel Language. In *Logic Programming: Proceedings of the 1991 International Symposium*. MIT Press, 1991.

[Kui94] Ben Kuipers, editor. *Qualitative Simulation*. MIT Press, 1994.

[MMP92] O. Maler, Z. Manna, and A. Pnueli. From timed to hybrid systems. In J. W. de Bakker, C. Huizing, W. P. de Roever, and G. Rozenberg, editors, *REX workshop "Real time: Theory and Practice"*, volume 600 of *Lecture notes in computer science*, pages 447–484, 1992.

[MP91] Z. Manna and A. Pnueli. *The Temporal Logic of Reactive and Concurrent Systems*. Springer-Verlag, 1991. 427 pp.

[NK93] A. Nerode and W. Kohn. *Multiple Agent Hybrid Control Architecture*. Number 736 in LNCS. Springer Verlag, 1993.

[NOSY93] X. Nicollin, A. Olivero, J. Sifakis, and S. Yovine. *An Approach to the Description and Analysis of Hybrid Systems*. Number 736 in LNCS. Springer Verlag, 1993.

[NSY91] X. Nicollin, J. Sifakis, and S. Yovine. From atp to timed graphs and hybrid systems. In J. W. de Bakker, C. Huizing, W. P. de Roever, and G. Rozenberg, editors, *REX workshop "Real time: Theory and Practice"*, volume 600 of *LNCS*, 1991.

[Rei80] Ray Reiter. A logic for default reasoning. *Artificial Intelligence*, 13:81 – 132, 1980.

[Sar93] Vijay A. Saraswat. *Concurrent Constraint Programming*. Logic Programming and Doctoral Dissertation Award Series. MIT Press, March 1993.

[SHW94] Gert Smolka, Henz, and J. Werz. *Constraint Programming: The Newport Papers*, chapter Object-oriented programming in Oz. MIT Press, 1994.

[SJG94a] V. A. Saraswat, R. Jagadeesan, and V. Gupta. *Constraint Programming*, chapter Programming in Timed Concurrent Constraint Languages. NATO Advanced Science Institute Series, Series F: Computer and System Sciences. Springer-Verlag, 1994.

[SJG94b] V. A. Saraswat, R. Jagadeesan, and V. Gupta. Foundations of Timed Concurrent Constraint Programming. In Samson Abramsky, editor, *Proceedings of the Ninth Annual IEEE Symposium on Logic in Computer Science*. IEEE Computer Press, July 1994.

[SJG95] V. A. Saraswat, R. Jagadeesan, and V. Gupta. Default Timed Concurrent Constraint Programming. In *Proceedings of Twenty Second ACM Symposium on Principles of Programming Languages, San Francisco*, January 1995.

[SRP91] V. A. Saraswat, M. Rinard, and P. Panangaden. Semantic foundations of concurrent constraint programming. In *Proceedings of Eighteenth ACM Symposium on Principles of Programming Languages, Orlando*, January 1991.

[Wd89] D.S. Weld and J. de Kleer. *Readings in Qualitative Reasoning about Physical Systems*. Morgan Kaufmann, 1989.

A Note on Abstract Interpretation Strategies for Hybrid Automata*

Thomas A. Henzinger** and Pei-Hsin Ho

Computer Science Department, Cornell University, Ithaca, NY 14853
(tah|ho)@cs.cornell.edu

Abstract. We report on several abstract interpretation strategies that are designed to improve the performance of HYTECH, a symbolic model checker for linear hybrid systems. We (1) simultaneously compute the target region from different directions, (2) conservatively approximate the target region by dropping constraints, and (3) iteratively refine the approximation until sufficient precision is obtained. We consider the standard abstract convex-hull operator and a novel abstract extrapolation operator.

1 Introduction

Hybrid automata extend finite automata with continuous activities [ACHH93]. Consider a communication protocol A with k propositional variables (values 0 or 1), and a robot B with k propositional control variables and m real-valued status variables (e.g., the position of the robot). Let $d = k + m$. A state of A is a point in k-dimensional *boolean* space; a state of B, a point in d-dimensional *euclidean* space. Each transition of A is a k-dimensional boolean transformation; if B is *linear*, then each discrete control transition of B is a d-dimensional linear transformation and each continuous activity of B is a d-dimensional piecewise-linear motion. This geometric intuition inspires the abstract interpretation strategies we develop.

For the finite-state system A, model-checking algorithms compute the state set (*target region*) that satisfy a given temporal requirement specification, by iterative approximation of the target region as a fixpoint. This is done either enumeratively [CES86] or symbolically, by representing state sets (*regions*) as boolean expressions [BCM+92]. Since the state space of the linear hybrid automaton B is infinite, enumerative model checking is no longer possible, and regions must be represented symbolically. A symbolic model-checking algorithm for linear hybrid automata has been developed [HNSY94, ACH+95], and implemented (HYTECH) by representing regions as real-valued linear expressions [AHH93]. The linearity condition is the key so that if the target region

* This research was supported in part by the National Science Foundation under grant CCR-9200794, by the Air Force Office of Scientific Research under contract F49620-93-1-0056, by the Office of Naval Research under YIP grant N00014-95-1-0520, and by the Defense Advanced Research Projects Agency under grant NAG2-892.

** Phone: (607) 255-3009. FAX: (607) 255-4428.

is a finitely approximable fixpoint, then it can be computed in the theory of the reals with addition, at "tolerable" cost. The most serious limitation of finite-state model checking, the state-explosion problem, reoccurs for the infinite state spaces of linear hybrid automata in three guises. The main theoretical limitation of the method is that termination (i.e., finite approximability) is not guaranteed (problem A); the main practical limitation of the implementation is its cost, which is largely due to two factors. First, continuous activities correspond to quantifier-elimination steps on real-valued linear expressions (problem B); second, if expressions are transformed into disjunctive normal form for further manipulation, then the number of disjuncts grows rapidly (problem C).

Having identified the three problems, we improve the performance of HYTECH by using algorithms that manipulate geometric objects rather than expressions, and by adding various abstract interpretation strategies, both exact and conservative, to guide the fixpoint computation of the target region. To address problem A, we approach the target region simultaneously from several directions, so that success is guaranteed even if only one direction terminates (Section 3). To address problem B, we represent regions as unions of convex polyhedra, so that quantifier elimination is replaced by linear operations on polyhedra. To address problem C, we conservatively approximate the union of several convex polyhedra by a single convex polyhedron. We discuss two abstract operators—convex-hull approximation and extrapolation (Section 4). Each approximation can be iteratively refined to achieve the desired precision (Section 5).

We wish to point out that the abstract operators and the iterative approximation process are well-known *abstract interpretation* techniques [CC77, Cou81] applied to a new domain—that of linear hybrid automata. The convex-hull operator [CH78] and the widening operator [CC77, CC92], which is similar to our extrapolation operator, are dynamic abstract operators used for convergence acceleration [CC92]. The idea of two-way iteratively refining approximations is also due to Patrick and Radhia Cousot [CC92], and it has been used for the analysis of real-time systems [WTD94]. Finally, we acknowledge the work of Nicolas Halbwachs, who has applied the convex-hull and widening operations to the analysis of synchronous programs [Hal93] and linear hybrid automata [HRP94]. The current implementation of HYTECH makes use of Halbwachs's polyhedron-manipulation library. We suggest a new extrapolation operator instead of Halbwachs's widening operator, which is sometimes too coarse for our purposes (on the other hand, widening, unlike extrapolation, guarantees the convergence of iterative application). We feel that the convex-hull and extrapolation operators are particularly suited for the analysis of the *continuous* and *linear* state spaces, because a polyhedron more naturally represents all of its interior points rather than just the grid of interior integer points, and because the linear regions are closed under both approximation operations.

2 Linear Hybrid Automata

Linear hybrid automata are formally defined in [HH95]; here we give only an informal review of the definition. A *linear hybrid automaton* A consists of a

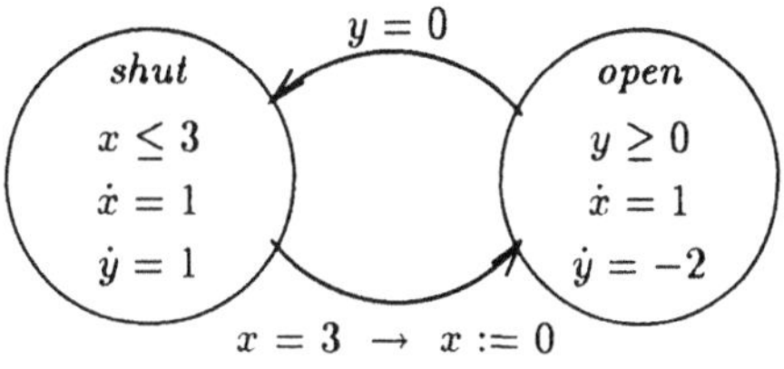

Fig. 1. The water-tank automaton

finite set X of m real-valued variables and a labeled graph (V, E). The edges in E represent discrete transitions and are labeled with guarded commands on X; the vertices in V represent continuous activities and are labeled with constraints on the variables in X and their first derivatives. A *state* (v, s) of the hybrid automaton A consists of a control location $v \in V$ (a vertex) and a data state $s \in \mathbb{R}^m$ (a valuation for the variables in X). A *data region* is a finite union of convex polyhedra in $\mathbb{R}^m$. A *region* $S = \bigcup_{v \in V}(v, S_v)$ is a collection of data regions $S_v \subseteq \mathbb{R}^m$, one for each control location $v \in V$.

The state (v, s) of the hybrid automaton A can change in two ways: (1) [time step] by a continuous activity that changes the data state s along a smooth function in accordance with the label of the control location v; or (2) [transition step] by an instantaneous discrete transition that changes both the control location v and the data state s according to the label of an edge with the source location v. The state σ can reach the state σ' in a *single step*, denoted $\sigma \Rightarrow \sigma'$, if σ can evolve into σ' by a time step followed by a transition step. The hybrid automaton A, then, determines a transition system $(\Sigma, \Rightarrow)$ with the infinite state space $\Sigma = V \times \mathbb{R}^m$ and the transition relation $\Rightarrow \subseteq \Sigma^2$. A *trajectory* of A is a finite path in $(\Sigma, \Rightarrow)$.

The *weakest precondition* of the region S, denoted $pre(S)$, is the set of all states σ such that for some state σ' in S, $\sigma \Rightarrow \sigma'$. The *strongest postcondition* of S, denoted $post(S)$, is the set of all states σ such that for some state σ' in S, $\sigma' \Rightarrow \sigma$. Both $pre(S)$ and $post(S)$ are again regions [AHH93]. In the following, it suffices to assume we are given, in place of the linear hybrid automaton A, the location space V, the dimension m, and algorithms for computing the functions *pre* and *post* on regions.

Example: water tank

The linear hybrid automaton of Figure 1 models a water-level controller that opens and shuts the outflow of a water tank. The automaton has two variables, x and y, and two locations, *shut* and *open*. The variable x represents a clock of the water-level controller and the variable y represents the water level in the tank. Since the clock x measures time, the first derivative of x is always 1 (i.e., $\dot{x} = 1$). In location *shut*, the outflow of the water tank is shut and the water level increases 1 inch per second ($\dot{y} = 1$); in location *open*, the outflow of the water tank is open and the water level decreases 2 inches per second ($\dot{y} = -2$).

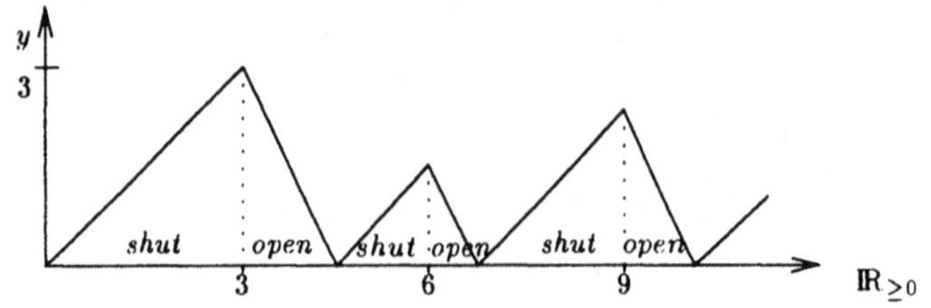

Fig. 2. A trace of the water level y

The transition from *open* to *shut* (shut the outflow) is taken as soon as the water tank becomes empty: the guard $y = 0$ on the transition ensures that the transition may be taken only when the water level is 0; the constraint $y \geq 0$ on *open* ensures that the transition to *shut* must be taken before the water level becomes negative. The transition from *shut* to *open* (open the outflow) is taken every 3 seconds: the transition is taken whenever $x = 3$, and the transition restarts the clock x at 0. If the automaton is started from the state $(shut, x = y = 0)$, then Figure 2 shows how the water level y changes as a piecewise-linear function of time.

The symbolic model checker HyTech

In this paper, we study the performance of HyTech, a fully automatic verifier for linear hybrid automata [HH95], on reachability problems. While the original implementation of HyTech represented each data region by boolean combinations of linear inequalities, the current version of HyTech represents each data region by a set of convex polyhedra. In the current version of HyTech, the user interface and the main program are implemented in Mathematica, which calls primitive region-manipulation procedures written in C++. These region-manipulation procedures call on a polyhedron-manipulation library [Hal93].

3 Forward versus Backward Reachability Analysis

The *reachability problem* for a hybrid automaton A asks, given an *initial region* S and a *final region* T, if there is a trajectory of A that leads from a state in S to a state in T. If T represents the set of "unsafe" states specified by a safety requirement, then the safety requirement can be verified by reachability analysis.

There are two possible approaches for solving the reachability problem by symbolic model checking [ACHH93]: we may compute the target region $post^*(S)$ of states that can be reached from the initial region S and check if $post^*(S) \cap T = \emptyset$ (forward reachability analysis), or we may compute the target region $pre^*(T)$ of states from which the final region T can be reached and check if $pre^*(T) \cap S = \emptyset$ (backward reachability analysis). For any given reachability problem, one approach may perform better than the other approach; indeed, it may be that one approach terminates and the other does not.

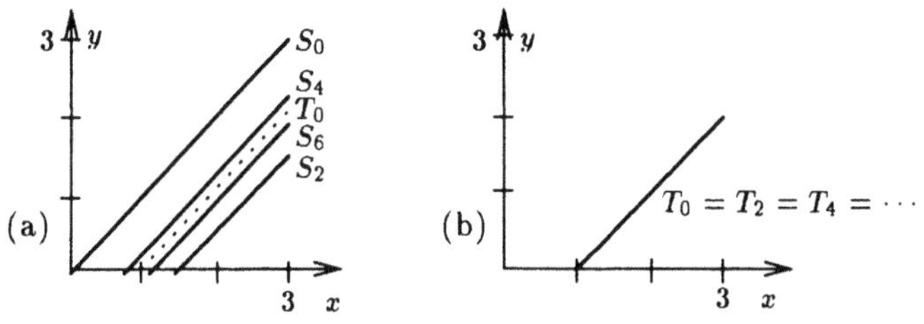

Fig. 3. Exact forward and backward analysis

Recall, for example, the water-tank automaton of Figure 1. We wish to check if the final region $T = (shut, 1 \leq x \leq 3 \wedge x = y + 1)$ can be reached from the initial region $S = (shut, x = y \wedge x \leq 3)$. The symbolic model-checking procedure computes the two regions $post^*(S)$ and $pre^*(T)$ iteratively as the limits of two infinite sequences of regions. Let $S_{i+1} = post(S_i)$ be the region of states that can be reached from S by $i+1$ single steps, and let $T_{i+1} = pre(T_i)$ be the region of states that can reach T by $i+1$ single steps. Then $post^*(S)$ is the limit $\bigcup_{i \geq 0} S_i$ of the infinite region sequence $\bigcup_{0 \leq i \leq n} S_n$, for $n \geq 0$, and $pre^*(T)$ is the limit $\bigcup_{i \geq 0} T_i$ of the infinite region sequence $\bigcup_{0 \leq i \leq n} T_n$, $n \geq 0$. It is clear that if $post^{n+1}(S) \subseteq \bigcup_{0 \leq i \leq n} post^i(S)$, then $post^*(S) = \bigcup_{0 \leq i \leq n} post^i(S)$, and an analogous termination condition holds for the computation of $pre^*(T)$. HYTECH checks if T is reachable from S on the fly; that is, if $post^n(S) \cap T \neq \emptyset$ or $pre^n(T) \cap S \neq \emptyset$, then T is reachable from S and the limit computation is aborted. Since for each $i \geq 0$,

$$S_{2i} = (shut, x - y = 1 + \frac{(-1)^{i+1}}{2^i} \wedge x \leq 3)$$

(Figure 3(a)) and

$$S_{2i+1} = (open, 2x + y = 2 + \frac{(-1)^i}{2^i} \wedge y \geq 0),$$

the forward computation of $post^*(S)$ does not terminate within any finite number of iterations. The backward computation, however, converges in a single iteration with the result

$$T_0 = T_2 = (shut, 1 \leq x \leq 3 \wedge x = y + 1)$$

(Figure 3(b)) and

$$T_1 = T_3 = (open, x + 2y = 2).$$

It follows that

$$pre^*(T) = (shut, 1 \leq x \leq 3 \wedge x = y + 1) \cup (open, x + 2y = 2).$$

Since $S \cap pre^*(T) = \emptyset$, we conclude that the final region T is not reachable from the initial region S. For optimal performance, we have implemented a strategy that dovetails both approaches by computing the alternating sequence $S_0, T_0, S_1, T_1, S_2, \ldots$ of regions.

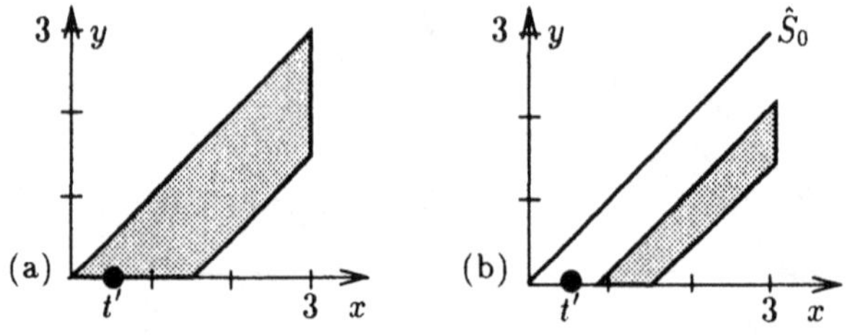

Fig. 4. Application of the naive and the refined convex-hull operators

4 Convergence Acceleration through Abstract Operators

Sometimes the exact computation of a target region is prohibitively expensive or does not terminate, independent of the verification strategy (forward vs. backward). In such cases, we approximate the target region and iteratively refine the approximation until sufficient precision is obtained.

We discuss two convergence acceleration operators, convex hull and extrapolation, which can be turned on or off by the user of HyTech. Both operators overapproximate a union of convex polyhedra by a single convex polyhedron and thus reduce the time and space requirements of the verifier; indeed, either operator, or the combination of both operators, may cause the termination of an otherwise infinite computation. If we overapproximate, by $\hat{S}$, the target region $post^*(S)$ of states that are reachable from the initial region S (i.e., $post^*(S) \subseteq \hat{S}$), and $\hat{S}$ contains no final state from T, then we can conclude that T is not reachable from S; if, on the other hand, $\hat{S}$ does contain a final state, then we cannot reach a valid conclusion and must refine the approximation.

An *abstract operator* $+$ is a binary operator on data regions such that for two data regions S and S', the result $S+S'$ is a convex data region with $S \cup S' \subseteq S+S'$. For the abstract operator $+$, the *approximated postcondition* of the region S is the region $post_+(S) = \cup_v(v, S_v + post(S)_v)$. Then $\cup_{0 \leq i \leq n} post^i(S) = post^n_\cup(S) \subseteq post^n_+(S)$, and we can overapproximate the target region $post^*(S)$ by the limit $\hat{S} = \cup_{i \geq 0} post^i_+(S)$. While we concentrate on the approximate forward analysis in this chapter, analogous observations hold for the approximate backward analysis using *pre*.

Convex hull

An obviously abstract operator is the *convex-hull operator* $\sqcup$ [Hal93, HRP94], which maps two data regions S and S' to the convex hull of the union $S \cup S'$. Recall once again the water-tank automaton of Figure 1. Now we wish to check if the final state $T' = (shut, x = \frac{1}{2} \wedge y = 0)$ can be reached from the initial region $S = (shut, x = y \wedge x \leq 3)$. Again, the forward computation of the target region $post^*(S)$ does not terminate. This time, however, we can make the forward approach work with the help of the convex-hull operator. Let $\tilde{S}_i = post^i_\sqcup(S)$. Then $\tilde{S}_0 = S$,

$$\tilde{S}_1 = (shut, x - y = 0 \wedge x \leq 3) \cup (open, y + 2x = 3 \wedge y \geq 0),$$

$$\begin{aligned}
\tilde{S}_2 &= (\mathit{shut}, (x - y = 0 \wedge x \leq 3) \sqcup (x - y = \tfrac{3}{2} \wedge x \leq 3)) \\
&\quad \cup (\mathit{open}, y + 2x = 3 \wedge y \geq 0) \\
&= (\mathit{shut}, 0 \leq x - y \leq \tfrac{3}{2} \wedge x \leq 3) \;\; (\mathit{Figure}\ 4(a)) \\
&\quad \cup (\mathit{open}, y + 2x = 3 \wedge y \geq 0), \\
\tilde{S}_3 &= (\mathit{shut}, 0 \leq x - y \leq \tfrac{3}{2} \wedge x \leq 3) \cup (\mathit{open}, \tfrac{3}{2} \leq y + 2x \leq 3 \wedge y \geq 0), \\
\tilde{S}_4 &= \tilde{S}_3.
\end{aligned}$$

The forward computation terminates with the overapproximation $\tilde{S}_3$ of the target region. Since $\tilde{S}_3$ contains T', however, we cannot conclude that T' is or is not reachable from S. Thus we refine our approximation strategy, and apply the convex-hull operator only if the resulting region does not contain any final states. This refinement is sensible for all abstract operators. Formally, we overapproximate the target region $post^*(S)$ by the limit $\hat{S} = \cup_{i \geq 0} \hat{S}_i$, where

$$\hat{S}_{i+1} = \begin{cases} post_+(\hat{S}_i) \text{ if } post_+(\hat{S}_i) \cap T = \emptyset, \\ post(\hat{S}_i) \quad \text{otherwise.} \end{cases}$$

For the water-tank example, we obtain $\hat{S}_0 = S$,

$$\begin{aligned}
\hat{S}_1 &= (\mathit{shut}, x - y = 0 \wedge x \leq 3) \cup (\mathit{open}, 2x + y = 3 \wedge y \geq 0), \\
\hat{S}_2 &= (\mathit{shut}, x - y = \tfrac{3}{2} \wedge x \leq 3) \cup (\mathit{open}, 2x + y = 3 \wedge y \geq 0), \\
\hat{S}_3 &= (\mathit{shut}, x - y = \tfrac{3}{2} \wedge x \leq 3) \\
&\quad \cup (\mathit{open}, (2x + y = 3 \wedge y \geq 0) \sqcup (2x + y = \tfrac{3}{2} \wedge y \geq 0)) \\
&= (\mathit{shut}, x - y = \tfrac{3}{2} \wedge x \leq 3) \cup (\mathit{open}, \tfrac{3}{2} \leq 2x + y \leq 3 \wedge y \geq 0), \\
\hat{S}_4 &= (\mathit{shut}, (x - y = \tfrac{3}{2} \wedge x \leq 3) \sqcup (\tfrac{3}{4} \leq x - y \leq \tfrac{3}{2} \wedge x \leq 3)) \\
&\quad \cup (\mathit{open}, \tfrac{3}{2} \leq 2x + y \leq 3 \wedge y \geq 0) \\
&= (\mathit{shut}, \tfrac{3}{4} \leq x - y \leq \tfrac{3}{2} \wedge x \leq 3) \;\; (\mathit{Figure}\ 4(b)) \\
&\quad \cup (\mathit{open}, \tfrac{3}{2} \leq 2x + y \leq 3 \wedge y \geq 0), \\
\hat{S}_5 &= \hat{S}_4.
\end{aligned}$$

The forward computaion terminates with the overapproximation $\cup_{0 \leq i \leq 4} \hat{S}_i$ of the target region. Since $\cup_{0 \leq i \leq 4} \hat{S}_i$ does not contain T', we conclude that the final state T' is not reachable from the initial region S.

Extrapolation

Convex-hull approximation is typically useful for systems with variables whose values are bounded, such as the water-tank automaton. By contrast, if the regions in the sequence $\cup_{0 \leq i \leq n} S_i$, $n \geq 0$, grow without bound, we may want an extrapolation operator that "guesses" an overapproximation of the limit $\cup_{i \geq 0} S_i$. Consider, for example, the hybrid automaton of Figure 5, which models the movement of a robot in the (x, y)-plane. The robot can be in one of two modes—heading roughly northeast (location A) or heading roughly southeast (location B):

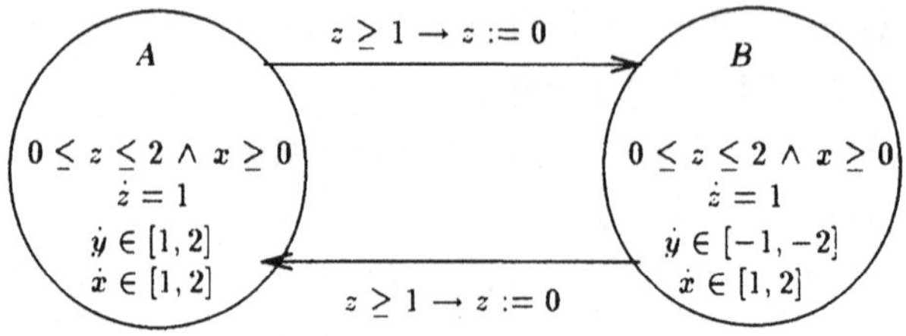

Fig. 5. The robot automaton

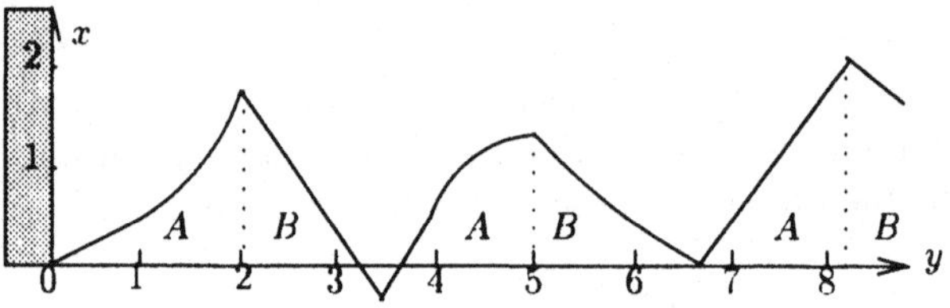

Fig. 6. A trajectory of the robot automaton

in mode A, the derivatives of the coordinates x and y vary within the closed interval $[1, 2]$; in mode B, the derivative of x remains between 1 and 2, while the derivative of y changes its sign and varies between -1 and -2. The robot changes its mode every 1 to 2 minutes, according to its clock z. Moreover, there is a wall at the $x = 0$ line, causing the system invariant $x \geq 0$. Figure 6 shows how the position of the robot in the (x, y)-plane may change with time, assuming the robot is started in the initial state $S = (shut, x = y = z = 0)$.

We wish to check if the robot can reach, from S, the final position $T = (x = 9 \wedge y = 12)$. At this point, we invite the ambitious reader to prove that T cannot be reached from S. Using HyTech, the backward computation of the region $pre^*(T)$ terminates within 74 seconds of CPU time[3] after seven iterations of pre, and $S \cap pre^*(T) = \emptyset$. The forward computation of the region $post^*(S)$, by contrast, does not terminate, because the robot keeps zig-zagging to the east and the limit $\cup_{i \geq 0} post^i(S)$ cannot be computed exactly.[4] As convex-hull approximation does not help with this example, we define an extrapolation operator that approximates the unbounded target region $post^*(S)$; using HyTech, the forward computation with extrapolation, then, terminates within 8 seconds of CPU time.

We first give an intuitive motivation for the extrapolation operator. Suppose that the iterative computation of the target region leads, for a given control state, to the data region (polyhedron) S and, in the subsequent iteration, to

---

[3] All performance figures are given for a SPARC 670MP station.

[4] True, if the target region T is reachable, then it can be reached within a finite number of discrete transitions, and if not, then the invariant $x > 9$ becomes true after a finite number of transitions; extrapolation, however, avoids the ad-hoc "guessing" of suitable invariants.

the polyhedron S'. Suppose, furthermore, that there is a function f such that f maps S to S', mapping extreme points to extreme points. A reasonable guess for the target region, then, would be $f^{\infty}(S)$.

To formally define an operator $\propto$ that extrapolates S and $f(S)$ to $f^{\infty}(S)$, we need to review some facts about convex polyhedra [NW88]. By Minkowski's theorem, every nonempty convex polyhedron R in $\mathbb{R}^n$ has a unique (within scalar multiplication) representation by its extreme points $\{x_i \mid i \in I\}$ and extreme rays $\{r_j \mid j \in J\}$ such that

$$S = \{x \in \mathbb{R}^n \mid x = \sum_{i \in I} \lambda_i x_i + \sum_{j \in J} \mu_j r_j, \text{ where } \lambda_i, \mu_j \in \mathbb{R}_{\geq 0} \text{ and } \sum_{i \in I} \lambda_i = 1\}.$$

Notice that if the point x and the ray r are in S, then so are all points of the form $x + \lambda r$, for $\lambda \in \mathbb{R}_{\geq 0}$. The nonempty convex polyhedron S has *dimension* k, denoted $dim(S) = k$, if there are $k+1$ points $x_1, x_2, \ldots, x_{k+1}$ in S such that the k differences $x_2 - x_1, x_3 - x_1, \ldots, x_{k+1} - x_1$ are linearly independent. Notice that if $dim(S) = 0$, then S is a point, and if $dim(S) = 1$, then S is a straight line, a ray, or a line segment. The set $F = \{x \in S \mid ax = b\}$, for two vectors a and b, is a *face* of S if every point x in S satisfies the inequality $ax \leq b$. Notice that every face of S is also a convex polyhedron. The face F is a *facet* of S if $dim(F) = dim(S) - 1$. The set F is a 1-dimensional face of S iff F is the intersection of $dim(S) - 1$ facets of S (Criterion (†)).

Now consider two polyhedra S and S'. We first compute the convex hull $S \sqcup S'$ of $S \cup S'$. All extreme points and extreme rays of the convex hull $S \sqcup S'$ are extreme points and extreme rays of S or S'. Let $\{x_k \mid k \in I\}$ be the extreme points of $S \sqcup S'$ and S, let $\{x_l \mid l \in I'\}$ be the remaining extreme points of $S \sqcup S'$, and let $\{r_j \mid j \in J\}$ be the extreme rays of $S \sqcup S'$. Suppose that $k \in I$ and $l \in I'$, and x_k and x_l are the endpoints of a 1-dimensional face of $R \sqcup S'$, which can be checked by Criterion (†). Then it is possible that $f(x_k) = x_l$ for the imaginary function f from S to S'. Moreover, if f is applied infinitely many times, then all points of the form $x_k + \lambda(x_l - x_k)$, for $\lambda \in \mathbb{R}_{\geq 0}$, are included in $f^{\infty}(R)$. We therefore add the ray $x_l - x_k$ into the generators of $S \propto S'$: the *extrapolation operator* $\propto$ maps the two polyhedra S and S' to $S \propto S'$ =

$$\{x \in \mathbb{R}_n \mid x = \sum_{i \in I \cup I'} \lambda_i x_i + \sum_{j \in J} \mu_j r_j + \sum_{(k,l) \in K} \mu_{k,l}(x_l - x_k),$$

$$\text{where } \lambda_i, \mu_j, \mu_{k,l} \in \mathbb{R}_{\geq 0} \text{ and } \sum_{i \in I \cup I'} \lambda_i = 1\},$$

where $(k, l) \in K$ iff $k \in I$ and $l \in I'$ and both x_k and x_l are in the intersection of $dim(R \sqcup S') - 1$ facets of $S \sqcup S'$. Notice that $\propto$ is an abstract operator, that $\propto$ is not symmetric (i.e., $S \propto S'$ and $S' \propto S$ may be different), and that $S \sqcup S' \subseteq S \propto S'$.

Figure 7 shows the application of the extrapolation operator to two convex 2-dimensional polyhedra S and S'. The result of the extrapolation $S \propto S'$ is shown as the left figure in Figure 8. As a comparison, the right figure in Figure 8 shows the result of applying the widening operator ∇ [Hal93, HRP94] to the same regions.

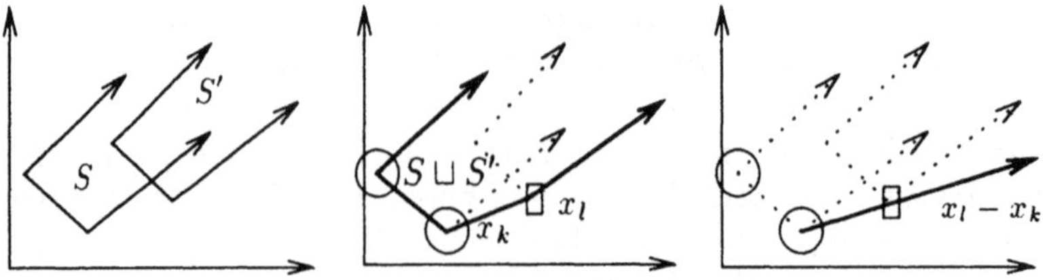

Fig. 7. Extrapolation operator

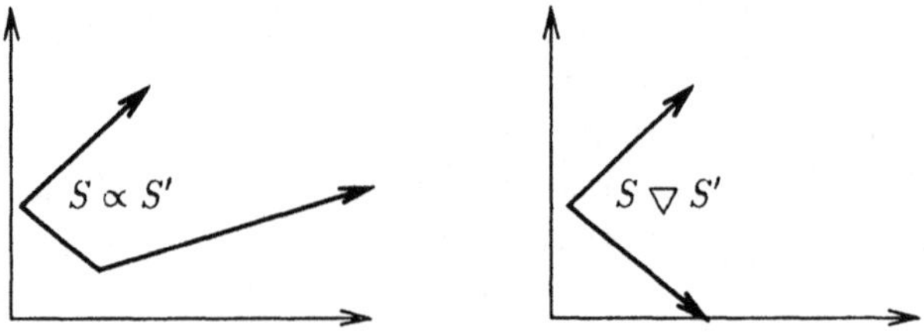

Fig. 8. Results of the extrapolation operator and the widening operator

5 Two-way Iterative Approximation

Suppose that an approximate forward reachability analysis is inconclusive; that is, the overapproximation $\hat{S}$ of the target region $post^*(S)$ contains some final states from T. If the intersection $\hat{S} \cap T$ is a proper subset of T, then we have nonetheless obtained new information—namely, that the states in $T - \hat{S}$ are not reachable from S—and we may proceed computing backward the new target region $pre^*(T \cap \hat{S})$, or an overapproximation $\hat{T}$ of $pre^*(T \cap \hat{S})$. If $\hat{T} \cap S = \emptyset$, then T is not reachable from S; if, on the other hand, $\hat{T}$ does contain some initial states from S, then we may continue the two-way iterative approximation process, this time computing forward from $\hat{T} \cap S$.

The two-way iterative approximation strategy is illustrated in Figure 9 and Figure 10, assuming a 2-dimensional state space and convex-hull approximation. The two shaded boxes in the lower left corner of the state space represent the initial region S; the two shaded boxes in the upper right corner represent the final region T. The derivatives of both variables x and y are 1. While an exact forward or backward analysis would show that T cannot be reached from S, both approximate forward analysis and approximate backward analysis are inconclusive (Figure 9). An approximate forward analysis followed by an approximate backward analysis is successful (Figure 10).

We now apply the two-way iterative approximation strategy to analyze the bouncing-ball automaton of Figure 11. The variable x represents the horizontal distance of the ball from a reference point, the variable y represents the distance of the ball from the floor, and the variable z represents the "energy" of the ball. Suppose that the ball is dropped at the position $x = 14 \wedge y = 4$ in the direction of the reference point; that is, $S = (down, x = 14 \wedge y = 4 \wedge z = 0)$. While the ball is falling (location *down*), its energy is increasing ($\dot{z} = 1$); when the ball hits

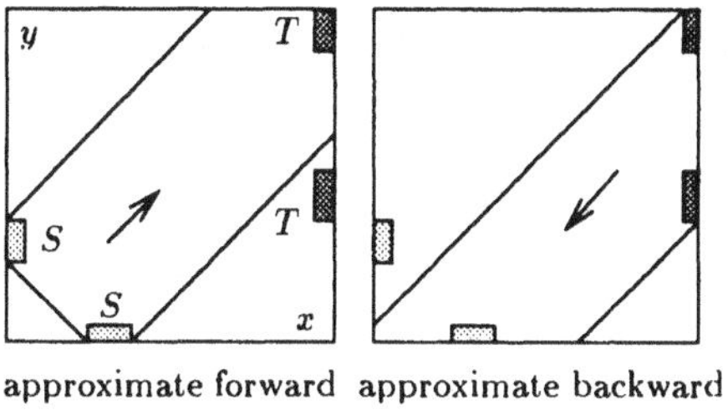

Fig. 9. One-way approximative analysis

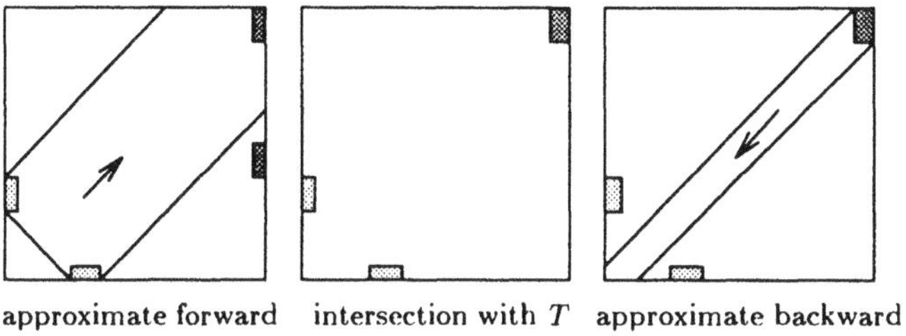

Fig. 10. Two-way iterative approximative analysis

the floor ($y = 0$), it loses half of its energy and bounces back up (location *up*); on the way up, the energy decreases ($\dot{z} = -1$) until it becomes 0 and the ball starts to fall again. The trajectory of the ball is shown in Figure 12.

We wish to prove that the ball never reaches the region $T = (x \leq 2 \land y = 0)$ of the floor. First notice that the exact forward computation of the target region $post^*(S)$ does not terminate, and convex-hull approximation is of no help. If we use extrapolation, then we obtain the overapproximation $\hat{S}$ of the target region $post^*(S)$, and $\hat{S}$ contains the final states $\hat{S} \cap T = (x = 2 \land y = 0 \land z = 0)$. Since $pre(\hat{S} \cap T) = \hat{S} \cap T$, a second, backward, pass terminates immediately without reaching the initial state S. HYTECH requires 12 seconds of CPU time for both passes. (Notice that if we were to begin with a backward pass, attempting to compute the target region $pre^*(T)$, neither the exact computation, nor the approximate computation with convex-hull approximation, nor the approximate computation with extrapolation, would terminate.)

References

[ACH+95] R. Alur, C. Coucoubetis, N. Halbwachs, T.A. Henzinger, P.-H. Ho, X. Nicollin, A. Olivero, J. Sifakis, and S. Yovine. The algorithmic analysis of hybrid systems. *Theoretical Computer Science*, 138:3–34, 1995.

[ACHH93] R. Alur, C. Courcoubetis, T.A. Henzinger, and P.-H. Ho. Hybrid automata: an algorithmic approach to the specification and verification of hybrid systems. In R.L. Grossman, A. Nerode, A.P. Ravn, and H. Rischel, editors, *Hybrid Systems*, Lecture Notes in Computer Science 736, pages 209–229. Springer-Verlag, 1993.

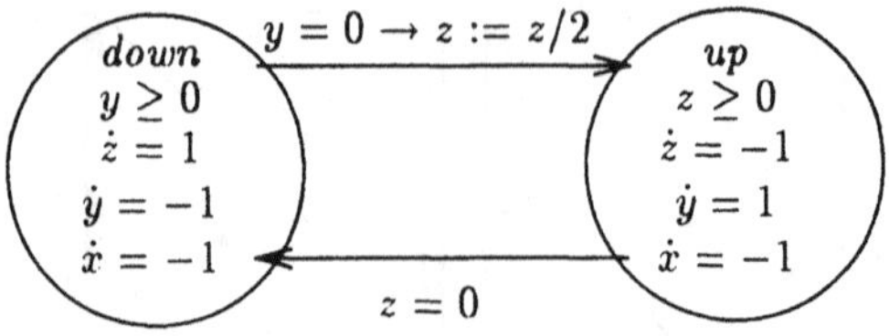

Fig. 11. The bouncing-ball automaton

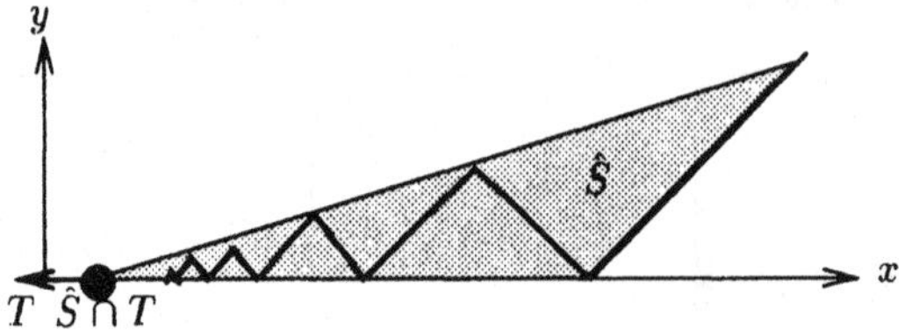

Fig. 12. A trajectory of the bouncing-ball automaton

[AHH93] R. Alur, T.A. Henzinger, and P.-H. Ho. Automatic symbolic verification of embedded systems. In *Proceedings of the 14th Annual Real-time Systems Symposium*, pages 2–11. IEEE Computer Society Press, 1993.

[BCM+92] J.R. Burch, E.M. Clarke, K.L. McMillan, D.L. Dill, and L.J. Hwang. Symbolic model checking: 10^{20} states and beyond. *Information and Computation*, 98(2):142–170, 1992.

[CC77] P. Cousot and R. Cousot. Abstract interpretation: a unified lattice model for the static analysis of programs by construction or approximation of fixpoints. In *Proceedings of the Fourth Annual Symposium on Principles of Programming Languages*, pages 238–252, Los Angeles, California, 1977. ACM Press.

[CC92] P. Cousot and R. Cousot. Abstract interpretation and application to logic programs. *Journal of Logic Programming*, 13(2/3):103–179, 1992.

[CES86] E.M. Clarke, E.A. Emerson, and A.P. Sistla. Automatic verification of finite-state concurrent systems using temporal-logic specifications. *ACM Transactions on Programming Languages and Systems*, 8(2):244–263, 1986.

[CH78] P. Cousot and N. Halbwachs. Automatic discovery of linear restraints among variables of a program. In *Proceedings of the Fifth Annual Symposium on Principles of Programming Languages*. ACM Press, 1978.

[Cou81] P. Cousot. Semantics fundations of program analysis. In S.S. Muchnick and N.D. Jones, editors, *Program Flow Analysis: Theory and Applications*, pages 303–342. Prentice-Hall, 1981.

[Hal93] N. Halbwachs. Delay analysis in synchronous programs. In C. Courcoubetis, editor, *CAV 93: Computer-aided Verification*, Lecture Notes in Computer Science 697, pages 333–346. Springer-Verlag, 1993.

[HH95] T.A. Henzinger and P.-H. Ho. HyTech: The Cornell Hybrid Technology Tool. This volume, 1995.

[HNSY94] T.A. Henzinger, X. Nicollin, J. Sifakis, and S. Yovine. Symbolic model checking for real-time systems. *Information and Computation*, 111(2):193–244, 1994.

[HRP94] N. Halbwachs, P. Raymond, and Y.-E. Proy. Verification of linear hybrid systems by means of convex approximation. In B. LeCharlier, editor, *International Symposium on Static Analysis, SAS'94*, Lecture Notes in Computer Science 864, Namur (belgium), September 1994. Springer-Verlag.

[NW88] G.L. Nemhauser and L.A. Wolsey. *Integer and Combinatorial Optimization*. Wiley, 1988.

[WTD94] Howard Wong-Toi and David L. Dill. Approximations for verifying timing properties. In Teo Rus and Charles Rattray, editors, *Theories and Experiences for Real-Time System Development (Proceedings First AMAST Workshop on Real-Time System Development)*, chapter 7. World Scientific Publishing, 1994.

HyTech : The Cornell HYbrid TECHnology Tool*

Thomas A. Henzinger and Pei-Hsin Ho

Computer Science Department, Cornell University, Ithaca, NY 14853
(tah|ho)@cs.cornell.edu

Abstract. This paper is addressed to potential users of HyTech, the Cornell Hybrid Technology Tool, an automatic tool for analyzing hybrid systems. We review the formal technologies that have been incorporated into HyTech, and we illustrate the use of HyTech with three nontrivial case studies.

1 Introduction

Hybrid systems are digital real-time systems that interact with the physical world through sensors and actuators. Due to the rapid development of digital processor technology, hybrid systems directly control much of what we depend on in our daily lives. Many hybrid systems, ranging from automobiles to aircraft, operate in safety-critical situations, and therefore call for rigorous analysis techniques.

HyTech[2] is a symbolic model checker for linear hybrid systems. The underlying system model is *hybrid automata*, an extension of finite automata with continuous variables that are governed by differential equations [ACHH93]. The requirement specification language is the *integrator computation tree logic* ICTL, a branching-time logic with clocks and stop-watches for specifying timing constraints. Safety, liveness, real-time, and duration requirements of hybrid systems can be specified in ICTL [AHH93]. Given a hybrid automaton describing a system and an ICTL formula describing a requirement, HyTech computes the state predicate that characterizes the set of system states that satisfy the requirement.

In this report we review the formal technologies that have been incorporated into HyTech. In Section 2, we define the syntax and semantics of linear hybrid automata, which were introduced in [ACHH93, NOSY93]. In Section 3, we give an introduction to ICTL model checking and the reachability analysis of linear hybrid automata, which was presented in [AHH93, ACH+95]. We concentrate on the analysis of systems with unknown delay parameters, and use HyTech to derive sufficient and necessary conditions on the parameters such that the system satisfies a given ICTL requirement. We also demonstrate the use of abstract-interpretation operators, which are discussed in greater detail in [HH95b]. In

* This research was supported in part by the National Science Foundation under grant CCR-9200794, by the Air Force Office of Scientific Research under contract F49620-93-1-0056, by the Office of Naval Research under YIP grant N00014-95-1-0520, and by the Defense Advanced Research Projects Agency under grant NAG2-892.

[2] HyTech is available by anonymous ftp from ftp.cs.cornell.edu, cd ~pub/tah/HyTech. See also http://www.cs.cornell.edu/Info/People/tah/hytech.html.

Section 4, we indicate how nonlinear hybrid systems can be translated into linear hybrid automata, so that linear analysis techniques apply [HH95a]. Throughout, we use a temperature controller for a toy nuclear reactor as a running example to illustrate the use of HYTECH. For the practitioners, we present the actual input language for describing linear hybrid automata and verification commands.

In Section 5, we apply HYTECH to three nontrivial benchmark problems. All three examples are taken from the literature, rather than devised by us. The first case study is a distributed control system introduced by Corbett [Cor94]. The system consists of a controller and two sensors, and is required to issue control commands to a robot within certain time limits. The two sensor processes are executed on a single processor, as scheduled by a priority scheduler. This scenario is modeled by linear hybrid automata with clocks and stop-watches. HYTECH automatically computes the maximum time difference between two consecutive control commands generated by the controller. It follows, for example, that a scheduler that gives higher priority to one sensor may meet the specification requirement, while a scheduler that gives priority to the other sensor may fail the requirement.

The second case study is a two-robot manufacturing system introduced by Puri and Varaiya [PV95]. The system consists of a conveyor belt with two boxes, a service station, and two robots. The boxes will not fall to the floor iff initially the boxes are not positioned closely together on the conveyor belt. HYTECH automatically computes the minimum allowable initial distance between the two boxes.

The third case study is the Philips audio control protocol presented by Bosscher, Polak, and Vaandrager [BPV94]. The protocol consists of a sender that converts a bit string into an analog signal using the so-called Manchester encoding, and a receiver that converts the analog signal back into a bit string. The sender and the receiver use clocks that may be drifting apart. In [BPV94], it was shown, by a human proof, that the receiver decodes the signal correctly if and only if the clock drift is bounded by a certain constant. HYTECH automatically computes that constant for input strings up to 8 bits. With some extra care in modeling, HYTECH can also be used to analyze the general case of input strings with arbitrary length [HW95].

2 Specification of Linear Hybrid Automata in HYTECH

The system modeling language of HYTECH is linear hybrid automata [AHH93]. Intuitively, a linear hybrid automaton is a labeled multigraph (V, E) with a finite set X of real-valued variables. The edges in E represent discrete system actions and are labeled with guarded assignments to X. The vertices in V represent continuous environment activities and are labeled with constraints on the variables in X and their first derivatives. The state of a hybrid automaton changes either through instantaneous system actions or, while time elapses, through continuous environment activities.

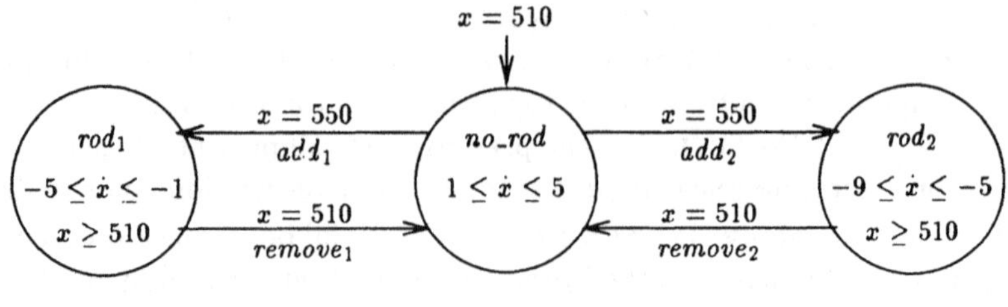

Fig. 1. The reactor core automaton

Example: reactor temperature control We use a variant of the reactor temperature control system from [NOSY93] as a running example. The system consists of a reactor core and two control rods that control the temperature of the reactor core. The reactor core is modeled by the linear hybrid automaton in Figure 1. The temperature of the reactor core is represented by the variable x. Initially the core temperature is 510 degrees and both control rods are not in the reactor core. In this case, the core temperature rises at a rate that varies between 1 and 5 degrees per second. We use $\dot{x}$ to denote the first derivative of the variable x. If the core temperature reaches 550 degrees, one of two control rods can be put into the reactor core to dampen the reaction. If control rod 1 is put in, the core temperature falls at a rate that may vary between -5 and -1 degrees per second. Control rod 2 has a stronger effect; if it is put in, the core temperature falls at a rate that varies between -9 and -5 degrees per second. Either control rod is removed once the core temperature falls back to 510 degrees.

2.1 Syntax

A *linear term* over a set X of real-valued variables is a linear combination of variables with integer coefficients. A *linear inequality* over X is a nonstrict inequality between linear terms over X.[3] A *linear hybrid automaton* A consists of the following components.

Data variables A finite ordered set $X = \{x_1, x_2, \ldots, x_n\}$ of real-valued *data variables*. For example, the reactor core automaton from Figure 1 has the single data variable x.

A *data state* is a point $(a_1, a_2, \ldots, a_n)$ in the n-dimensional real space $\mathbb{R}^n$ or, equivalently, a function that maps each variable x_i to a real value a_i. A *convex data region* is a convex polyhedron in $\mathbb{R}^n$, and a *data region* is a finite union of convex data regions. A *convex data predicate* is a conjunction of linear inequalities, and a *data predicate* is a disjunction of convex data predicates. Every (convex) data predicate ϕ defines a (convex) data region $[\![\phi]\!]$ of data states that satisfy ϕ.

[3] The restriction to *nonstrict* inequalities is not essential, but forced upon us by the polyhedron-manipulation library we use in the current implementation of HyTech.

Control locations A finite set V of vertices called *control locations*. For example, the reactor core automaton has the three control locations no_rod, rod_1, and rod_2.

A *state* (v, s) of the hybrid automaton A consists of a control location $v \in V$ and a data state $s \in \mathbb{R}^n$. A *region* $\bigcup_{v \in V}\{(v, S_v)\}$ is a collection of data regions $S_v \subseteq \mathbb{R}^n$, one for each control location $v \in V$. A *state predicate* is a collection $\bigcup_{v \in V}\{(v, \phi_v)\}$ of data predicates ϕ_v, one for each control location $v \in V$. When writing state predicates, we use the *location counter* l, which ranges over the set V of control locations. The location constraint $l = v$ denotes the state predicate $\{(v, true)\} \cup \bigcup_{v' \neq v}\{(v', false)\}$. Each state predicate $\bigcup_{v \in V}\{(v, \phi_v)\}$ defines the region $\bigcup_{v \in V}\{(v, [\![\phi_v]\!])\}$.

Location invariants A labeling function inv that assigns to each control location $v \in V$ a convex data predicate $inv(v)$, the *invariant* of v. The automaton control may reside in location v only as long as the invariant $inv(v)$ is true; so the invariants enforce progress in a hybrid automaton. The state (v, s) is *admissible* if the data state s satisfies the invariant $inv(v)$. We write Σ_A for the region $\bigcup_{v \in V}\{(v, [\![inv(v)]\!])\}$ of all admissible states of A.

In the graphical representation of a hybrid automaton, we suppress invariants of the form *true*. In the reactor core automaton, we have $inv(no\_rod) = true$, $inv(rod_1) = (x \geq 510)$, and $inv(rod_2) = (510 \leq x)$. In HyTech, we specify these invariants as follows:

```
inv[l[core] == norod] = True
inv[l[core] == rodone] = 510<=x
inv[l[core] == rodtwo] = 510<=x
```

Continuous activities A labeling function dif assigns to each control location $v \in V$ and each data variable $x_i \in X$ a *rate interval* $dif(v, x_i) = [a_i, b_i]$, where a_i and b_i are integer constants. The rate interval $dif(v, x_i) = [a_i, b_i]$ specifies that the first derivative of the data variable x_i may vary within the interval $[a_i, b_i] \subset \mathbb{R}$ while the automaton control resides in location v. If $a_i = b_i$, then $dif(v, x_i)$ is called the *slope* of x_i in location v. A data variable is a *discrete variable* if it has the slope 0 in all locations; a *clock*, if it has the slope 1 in all locations; and a *stop-watch*, if in each location it has either the slope 1 or the slope 0.

In the graphical representation of a hybrid automaton, we write $\dot{x} = a$ short for $\dot{x} \in [a, a]$, and we suppress rate intervals of the form $\dot{x} = 0$. In the reactor core automaton, $dif(no\_rod, x) = [1, 5]$, $dif(rod_1, x) = [-5, -1]$, and $dif(rod_2, x) = [-9, -5]$. In HyTech, we specify these rate intervals as follows:

```
dif[core,norod,x] = {1,5}
dif[core,rodone,x] = {-5,-1}
dif[core,rodtwo,x] = {-9,-5}
```

Transitions A finite multiset E of edges called *transitions*. Each transition (v, v') identifies a source location $v \in V$ and a target location $v' \in V$. The reactor core automaton has four transitions.

Synchronization letters A finite set L of letters called *synchronization alphabet*, and a labeling function *syn* that assigns to each transition $e \in E$ a letter from L. The synchronization letters are used to define the parallel composition of hybrid automata. In the graphical representation of a hybrid automaton, we suppress synchronization letters that do not occur in the alphabet of any other automata. The reactor core automaton has the four synchronization letters add_1, add_2, $remove_1$, and $remove_2$.

Discrete actions A labeling function *act* that assigns to each transition $e \in E$ a guarded command $act(e) = (\phi \rightarrow \alpha)$. The *guard* ϕ is a convex data predicate. The *command* α is a set of assignments $x_i := t_i$, at most one for each data variable $x_i \in X$, such that each t_i is a linear term over X. We write $dom(\alpha)$ for the set of variables that make up the left-hand sides of the assignments in α, and $t_i(s)$ for the value of the linear term t_i if interpreted in the data state s. The command α defines a function on data states that leaves the variables outside $dom(\alpha)$ unchanged: for all data states s, $\alpha_i(s) = t_i(s)$ if $x_i \in dom(\alpha)$, and $\alpha_i(s) = s_i$ if $x_i \notin dom(\alpha)$, where $\alpha_i(s)$ denotes the i-th component of the data state $\alpha(s)$. A *parameter* is a discrete variable that does not occur in the domain $dom(\alpha)$ of any command α.

In the graphical representation of a hybrid automaton, we write α for the guarded command $true \rightarrow \alpha$, and ϕ for the guarded command $\phi \rightarrow \emptyset$, and we suppress the guarded command $true \rightarrow \emptyset$. In the reactor core automaton, $act(no\_rod, rod_1) = (x = 550 \rightarrow \emptyset)$, etc. In HyTech, we specify the transitions, synchronization letters, and guarded commands of the reactor core automaton as follows:

```
act[core,1]={l[core]==norod && 550==x, add1, {l[core]->rodone}}
act[core,2]={l[core]==norod && 550==x, add2, {l[core]->rodtwo}}
act[core,3]={l[core]==rodone && 510==x, remove1, {l[core]->norod}}
act[core,4]={l[core]==rodtwo && 510==x, remove2, {l[core]->norod}}
```

Notice that we encode the source and target locations of a transition within a guarded command.

2.2 Semantics

At any time instant, the state of a hybrid automaton specifies a control location and the values of all data variables. The state can change in two ways: (1) by an instantaneous discrete transition that changes both the control location and the values of data variables, or (2) by a time delay that changes only the values of data variables in a continuous manner according to the rate intervals of the corresponding control location. Accordingly, we define the following two binary relations on the admissible states of the given automaton A.

Transition step For all admissible states (v, s) and (v', s') of A, and all synchronization letters σ, let $(v, s) \xrightarrow{\sigma} (v', s')$ iff there exists a transition e from v to v' such that (1) $syn(e) = \sigma$ and (2) $act(e) = (\phi \rightarrow \alpha)$ with $s \in [\![\phi]\!]$ and $s' = \alpha(s)$.

Time step For all admissible states (v, s) and (v, s') of A, and all nonnegative reals $\delta \geq 0$, let $(v, s)\overset{\delta}{\rightarrow}(v, s')$ iff there is a differentiable function $\rho\colon [0, \delta] \rightarrow \mathbb{R}^n$ such that (1) $f(0) = s$, (2) $f(\delta) = s'$, (3) for all reals $t \in [0, \delta]$, $\rho(t) \in [\![inv(v)]\!]$, and (4) for all reals $t \in (0, \delta)$ and each data variable x_i, $d\rho_i(t)/dt \in dif(v, x_i)$, where $\rho_i(t)$ denotes the i-th component of the data state $\rho(t)$.

The linear hybrid automaton A defines the labeled transition system $[\![A]\!] = \langle \Sigma_A, \mathcal{L}, \rightarrow_A \rangle$ that consists of the infinite state space Σ_A, the infinite label set $\mathcal{L} = L \cup \mathbb{R}_{\geq 0}$, and the binary transition relation $\rightarrow_A = \bigcup\{\overset{\sigma}{\rightarrow} \mid \sigma \in L\} \cup \bigcup\{\overset{\delta}{\rightarrow} \mid \delta \geq 0\}$ on Σ_A.

For a region S, we define $pre(S)$ to be the set of all states σ such that $\sigma \rightarrow_A \sigma'$ for some state $\sigma' \in S$. Similarly, we define $post(S)$ to be the set of all states σ such that $\sigma' \rightarrow_A \sigma$ for some state $\sigma' \in S$. Both $pre(S)$ and $post(S)$ are again regions [AHH93]. We write $pre^*(S)$ for the infinite union $\bigcup_{i \geq 0} pre^i(S)$, and $post^*(S)$ for the infinite union $\bigcup_{i \geq 0} post^i(S)$. In other words, $pre^*(S)$ is the set of all states that can reach a state in S by a finite sequence of transitions of the labeled transition system $[\![A]\!]$; and $post^*(S)$ is the set of all states that can be reached from a state in S by a finite sequence of transitions of $[\![A]\!]$.

2.3 Parallel Composition

A hybrid system typically consists of several components that operate concurrently and communicate with each other. We describe each component as a linear hybrid automaton. The component automata may coordinate either through shared variables or via synchronization letters. The linear hybrid automaton that models the entire system is then constructed from the component automata using a product operation.

Let $A_1 = (X_1, V_1, inv_1, dif_1, E_1, L_1, syn_1, act_1)$ and $A_2 = (X_2, V_2, inv_2, dif_2, E_2, L_2, syn_2, act_2)$ be two linear hybrid automata. The product automaton $A_1 \times A_2$ generally interleaves the transitions of the component automata A_1 and A_2. If, however, a transition e_1 of A_1 is labeled with a synchronization letter σ that is contained also in the alphabet of A_2, then e_1 can be executed only simultaneously with a σ-labeled transition of A_2. Formally, the *product* $A_1 \times A_2$ is the linear hybrid automaton $A = (X_1 \cup X_2, V_1 \times V_2, inv, dif, E, L_1 \cup L_2, syn, act)$:

- Each location (v, v') in $V_1 \times V_2$ has the invariant $inv(v, v') = (inv_1(v) \wedge inv_2(v'))$. For each variable $x \in X_1 \backslash X_2$, $dif((v, v'), x) = dif_1(v, x)$; for each variable $x \in X_2 \backslash X_1$, $dif((v, v'), x) = dif_2(v', x)$; and for each shared variable $x \in X_1 \cap X_2$, $dif((v, v'), x) = dif_1(v, x) \cap dif_2(v', x)$.
- E contains the transition $e = ((v_1, v_2), (v_1', v_2'))$ iff
  (1) $e_1 = (v_1, v_1') \in E_1$, $v_2 = v_2'$, and $syn_1(e_1) \notin L_2$; or
  (2) $e_2 = (v_2, v_2') \in E_2$, $v_1 = v_1'$, and $syn_2(e_2) \notin L_1$; or
  (3) $e_1 = (v_1, v_1') \in E_1$, $e_2 = (v_2, v_2') \in E_2$, and $syn_1(e_1) = syn_2(e_2)$.

  Suppose that $act_1(e_1) = (\phi_1 \rightarrow \alpha_1)$, and $act_2(e_2) = (\phi_2 \rightarrow \alpha_2)$. In case (1), $syn(e) = syn_1(e_1)$ and $act(e) = act_1(e_1)$. In case (2), $syn(e) = syn_2(e_2)$ and $act(e) = act_2(e_2)$. In case (3), $syn(e) = syn_1(e_1) = syn_2(e_2)$; moreover, $act(e) = (\phi_1 \wedge \phi_2 \rightarrow \alpha_1 \cup \alpha_2)$ if $dom(\alpha_1) \cap dom(\alpha_2) = \emptyset$, and $act(e) = (false \rightarrow \emptyset)$ if $dom(\alpha_1) \cap dom(\alpha_2) \neq \emptyset$,

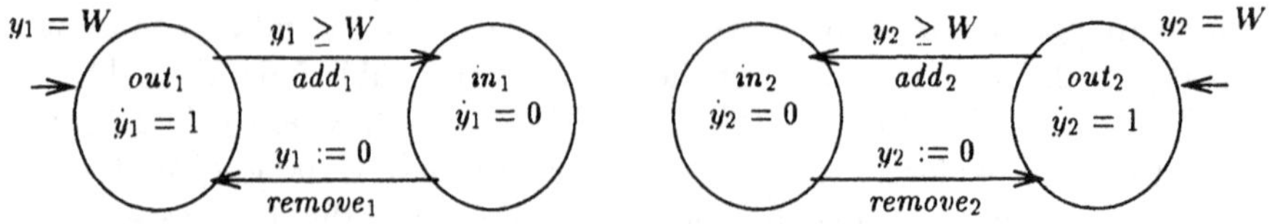

Fig. 2. The control rod automata

HYTECH automatically constructs the product automaton from a set of input automata.

For the reactor example, we use the two linear hybrid automata of Figure 2 to model the two control rods. Due to the mechanics of moving control rods, after a control rod is removed from the reactor core, it cannot be put back into the core for W seconds, where W is an unknown parameter. This requirement is enforced by the stop-watch y_1 that measures the time that has elapsed since control rod 1 was removed from the reactor core, and the stop-watch y_2 that measures the time that has elapsed since control rod 2 was removed. The rod automata synchronize with the core automaton through synchronization letters such as $remove_1$, which indicates the removal of control rod 1. The entire reactor system, then, is obtained by constructing the product of the core automaton of Figures 1 and the two rod automata of Figure 2.

We now show how the complete reactor temperature control system is specified in HYTECH. First we declare the data variables:

```
AnaVariables = {x, y1, y2}
DisVariables = {w}
```

The data variables x, $y1$, and $y2$ are analog variables, and the data variable W is a discrete variable. We have already defined the reactor core automaton. Now we define the two control rod automata:

```
inv[l[rod1] == out] = 0<=y1
inv[l[rod1] == in] = 0<=y1
inv[l[rod2] == out] = 0<=y2
inv[l[rod2] == in] = 0<=y2

dif[rod1,in,y1] = {0,0}
dif[rod1,out,y1] = {1,1}
dif[rod2,in,y2] = {0,0}
dif[rod2,out,y2] = {1,1}

act[rod1,1] = { l[rod1]==out && w<=y1, add1, {l[rod1] -> in}}
act[rod1,2] = { l[rod1]==in, remove1, {l[rod1] -> out, y1 -> 0}}
act[rod2,1] = { l[rod2]==out && w<=y2, add2, {l[rod2] -> in}}
act[rod2,2] = { l[rod2]==in, remove2, {l[rod2] -> out, y2 -> 0}}
```

The synchronization alphabet of each automaton is defined by declaring a scope for each synchronization letter. The *scope* of the letter σ is the set of automata that contain σ in their synchronization alphabet. For the reactor temperature control system, we specify

```
syn[remove1] = {rod1,core}
syn[remove2] = {rod2,core}
syn[add1] = {rod1,core}
syn[add2] = {rod2,core}
```

For example, the letter $remove_1$ is used by the reactor core automaton and by the first control rod automaton. This means that the core automaton and the rod 1 automaton must synchronize on transitions labeled with $remove_1$.

While we have given symbolic names like *core* and *no_rod* to automata and locations, the analysis procedures of HyTech require that all automaton names and location names are integers starting from 1. To replace the symbolic names with integers, HyTech calls a macro language preprocessor `m4` when it reads an input file. Therefore, we need to define the integer values of the symbolic names at the beginning of the input file. The symbolic names that we use for the reactor temperature control system may be defined as follows:

```
define(rod1,1)
define(rod2,2)
define(core,3)
define(rodone,1)
define(rodtwo,2)
define(norod,3)
define(out,1)
define(in,2)
```

We also must declare the number of input automata, and the number of locations and transitions of each automaton:

```
AutomataNo = 3
locationo = {2,2,3}
transitiono = {2,2,4}
```

The expression `locationo = {2,2,3}` means that the first (control rod 1), second (control rod 2), and third (reactor core) automaton has 2, 2, and 3 locations, respectively. The expression `transitiono = {2,2,4}` specifies the number of transitions in each input automaton.

Global invariants for modeling urgent transitions Although the product automaton is constructed automatically by HyTech, it is sometimes useful to specify global conjuncts of all invariants of the product automaton. Such global invariants permit, in particular, the modeling of *urgent transitions*, which are transitions that must be taken as soon as possible. In the graphical representation of hybrid automata, we use boldface synchronization letters to mark urgent transitions. HyTech allows the user to specify location invariants for locations of the product automaton using the command `GlobalInvar`. We will show how urgent transitions can be modeled with global invariants as we analyze the examples of Section 4. The reactor temperature control system does not have any urgent transitions, so we write:

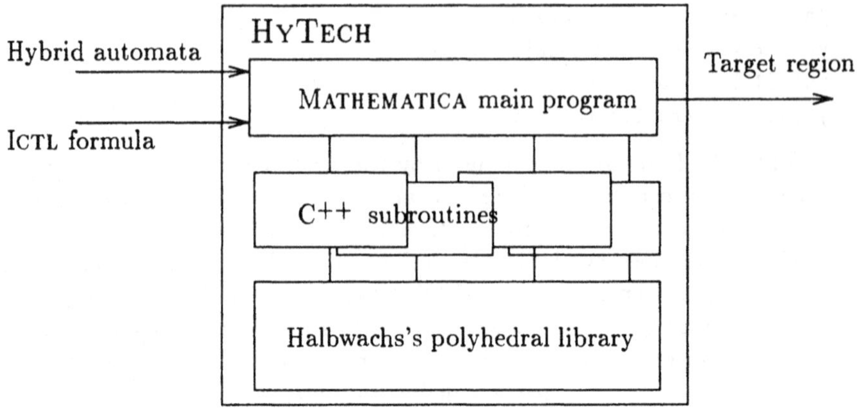

Fig. 3. The architecture of HyTech

```
GlobalInvar = {}
```

This completes the specification of the reactor temperature control system. Except for initial `define` statements, all HyTech input commands can be written in any order.

3 Symbolic Analysis of Linear Hybrid Automata in HyTech

The core of HyTech is a symbolic model-checking procedure, whose primitives are *pre*, *post*, and boolean operations on regions. The original implementation of HyTech represented regions as state predicates and manipulated regions by syntactic operations on formulas. We have improved the performance of HyTech by representing and manipulating regions geometrically: each data region is represented as a union of convex polyhedra. The current implementation of HyTech consists of a Mathematica main program and a collection of C++ subroutines that make use of a polyhedron-manipulation library by Halbwachs [Hal93, HRP94]. The architecture of HyTech is shown in Figure 3.

3.1 Reachability Analysis

The *reachability problem* $(A, \varphi_I, \varphi_F)$ for a linear hybrid automaton A, an initial state predicate φ_I, and a final state predicate φ_F, asks if the region $post^*([\![\varphi_I]\!]) \cap [\![\varphi_F]\!]$ is empty or, equivalently, if the region $[\![\varphi_I]\!] \cap pre^*([\![\varphi_F]\!])$ is empty. In other words, the reachability problem $(A, \varphi_I, \varphi_F)$ asks if there is no finite path in the underlying transition system $[\![A]\!]$ from some state in $[\![\varphi_I]\!]$ to some state in $[\![\varphi_F]\!]$. If $[\![\varphi_I]\!]$ represents the set of "initial" states of the automaton A, and $[\![\varphi_F]\!]$ represents the set of "unsafe" states specified by a safety requirement, then the safety requirement can be verified by reachability analysis: the automaton satisfies the safety requirement iff the reachability problem has the answer *yes* (i.e., $post^*(\llbracket\varphi_I\rrbracket) \cap \llbracket\varphi_F\rrbracket = \emptyset$).

Unfortunately, the computation of $post^*(\llbracket\varphi_I\rrbracket)$ or $pre^*(\llbracket\varphi_F\rrbracket)$ may not terminate within a finite number of *post* or *pre* operations, because the reachability problem for linear hybrid automata is undecidable [ACHH93]. HYTECH, in other words, offers a semidecision procedure for the reachability analysis. It is our experience, however, that for practical examples, including the examples in this paper, the computation does terminate and HYTECH solves the corresponding reachability problems. Indeed, as for the practitioner there is little difference between a nonterminating computation and one that runs out of time or space resources, we submit that decidability questions are mostly of theoretical interest.

Suppose that in the reactor temperature control system, the reactor needs to be shut down if the core temperature exceeds 550 degrees. We wish to check the safety requirement that *the reactor never needs to be shut down*; more precisely, whenever the core temperature reaches 550 degrees, then either y_1 or y_2 shows at least W seconds, thus allowing the corresponding control rod to be put into the reactor core. Let A denote the product of the reactor core automaton and the two control rod automata. We define the reachability problem $(A, \varphi_I, \varphi_F)$ as follows. The initial states are characterized by the state predicate

$$\varphi_I = (l[rod_1] = out \wedge l[rod_2] = out \wedge l[core] = no\_rod \wedge x = 510 \wedge y_1 = y_2 = W);$$

that is, initially no rod is in the reactor core, the initial temperature is 510 degrees, and $y_1 = y_2 = W$ (we write $l[c]$ for the component of the location counter l that is associated with the component automaton c; so $l[core]$ ranges over the locations of the reactor core automaton, etc.). The unsafe states are characterized by the state predicate

$$\varphi_F = (l[core] = no\_rod \wedge x = 550 \wedge y_1 \leq W \wedge y_2 \leq W);$$

that is, the unsafe situation is that the core temperature reaches 550 degrees and neither y_1 nor y_2 shows more than W seconds[4] (and, thus, none of the control rods is available). The answer to the reachability problem $(A, \varphi_I, \varphi_F)$ is *yes* iff the reactor temperature control system satisfies the safety requirement.

In HYTECH, the reachability problem is specified as follows:

```
InitialState = l[rod1]==out && l[rod2]==out && l[core]==norod &&
    510==x && w==y1 && w==y2
Bad = l[core]==norod && 550==x && y1<=w && y2<=w
```

Forward versus backward analysis HYTECH can attack a reachability problem by forward analysis or by backward analysis. Given the reachability problem $(A, \varphi_I, \varphi_F)$, the *forward analysis* computes the state predicate that defines the region $post^*(\llbracket\varphi_I\rrbracket)$, and then takes the conjunction with the final state predicate φ_F; the *backward analysis* computes the state predicate that defines the

[4] Remember that we are limited to *nonstrict* inequalities.

region $pre^*([\![\varphi_F]\!])$, and then takes the conjunction with the initial state predicate φ_I. For a given reachability problem, one direction may perform better than the other direction. In fact, it may be that one direction terminates and the other does not. For example, only the backward analysis terminates for the reactor temperature control system.

We ask HYTECH to perform a forward or backward analysis, respectively, by writing

```
Go := PrintTime[ Forward ]
```

or

```
Go := PrintTime[ Backward ]
```

These commands also print the CPU time consumed by the reachability analysis.

Parametric analysis The automatic derivation of parameters was introduced for real-time systems in [AHV93] and applied to hybrid systems in [AHH93]. We can use HYTECH to synthesize necessary and sufficient conditions on system parameters such that a hybrid automaton satisfies a requirement.

Recall that the reactor temperature control system contains the parameter w, which specifies the necessary rest time for a control rod. Clearly, the safety requirement will not be satisfied for large values of w. Indeed, the *target region* $[\![\varphi_I]\!] \cap pre^*([\![\varphi_F]\!])$ gives a sufficient and necessary condition on w such that the safety requirement is not satisfied. Typically the state predicate that defines the target region is too complex to see the conditions on the parameters clearly, but these can be isolated in HYTECH using *projection operators*. By writing

```
EliminateLocList = {rod1,rod2,core}
EliminateVarList = {x,y1,y2}
```

we eliminate all location information from the state predicate that defines the target region, and we project out all information about the data variables x, y_1, and y_2. Then the resulting projection of the target region, as computed by HYTECH using backward analysis, is

```
9w >= 184
```

In other words, the target region is empty if and only if $9W < 184$. It follows that $9W < 184$ is a necessary and sufficient condition on the parameter W that prevents the reactor from shutdown. The verification requires 17.27 seconds of CPU time.[5]

3.2 Abstract Interpretation

To expedite the reachability analysis and to force the termination of the analysis, HYTECH provides several abstract-interpretation operators [CC77, HH95b], including the convex-hull operator and the extrapolation operator. Our extrapolation operator is similar to the widening operator of [CH78, Hal93].

[5] All performance figures are given for a SPARC 670MP station.

An abstract-interpretation operator approximates a set of convex data regions with a single convex data region. The *convex-hull operator* overapproximates a union of convex data regions by its convex hull. The *extrapolation operator* overapproximates a directed chain $S \subset f(S) \subset f^2(S) \subset \cdots$ of convex data regions by a "guess" of the limit region $\bigcup_{i \geq 0} f^i(S)$. Either operator, or the combination of both operators, may cause the termination of a forward or backward reachability analysis that does not terminate otherwise. However, since the use of either operator results in an overapproximation of the target region, the abstract analysis is sound but not complete: if HyTech returns the answer *yes* to a reachability problem, then the approximate target region is empty, and therefore also the exact target region must be empty; but if the answer is *no*, then the exact target region may still be empty, and the correct answer to the reachability problem may be *yes*. In the latter case, we have to refine our approximation, by applying fewer abstract-interpretation operators, or by using two-way iterative approximation (see below).

In HyTech, we write

```
TakeConvex = True
```

or

```
TakeConvex = False
```

to turn the convex-hull operator on or off, respectively. The extrapolation operator can be turned on or off selectively for individual control locations. For example, if we want to apply the extrapolation operator only to data regions that correspond to the two locations $l[rod_1] = out \wedge l[rod_2] = in \wedge l[core] = no\_rod$ and $l[rod_1] = in \wedge l[rod_2] = out \wedge l[core] = no\_rod$ of the reactor temperature control system, then we write:

```
ExtraSet[lc_]=((lc === l[rod1]==out && l[rod2]==in && l[core]==norod)
|| (lc === l[rod1]==in && l[rod2]==out && l[core]==norod))
```

The commands

```
ExtraSet[lc_] = True
```

and

```
ExtraSet[lc_] = False
```

ask HyTech to apply the extrapolation operator to all or none of the control locations, respectively. (In our analysis of the reactor temperature control system, it was not necessary to use any abstract-interpretation operators.)

Two-way iterative approximation If inconclusive, the approximate reachability analysis can be refined by alternating approximate forward and backward analysis [CC92, DW95]. If any abstract-interpretation operators are used, HyTech automatically performs a two-way iterative analysis, beginning with the specified forward or backward pass. The readers should refer to [HH95b] for the details about the two-way iterative analysis of hybrid systems.

3.3 ICTL Model Checking

To check hybrid automata against more general requirements than reachability, we use the requirement specification language ICTL [AHH93]. ICTL is a branching-time logic in the tradition of CTL [CES86], with additional clock and stop-watch variables for specifying timing constraints. For a formal definition of ICTL, and a discussion of the model-checking algorithm, we refer the reader to [AHH93]; here we present only a couple of typical ICTL requirements for the reactor temperature control system.

First, recall the safety requirement that *the reactor never needs to be shut down*, which was characterized by a reachability problem $(A, \varphi_I, \varphi_F)$ in Section 3.1. In ICTL, the safety requirement is specified by the formula

$$\varphi_I \rightarrow \forall\Box\neg\varphi_F,$$

which asserts that in the labeled transition system $[\![A]\!]$, along all paths of infinite duration that start from an initial state, no unsafe state is visited.

In addition to safety requirements, in ICTL we can specify also liveness, real-time, and duration requirements of hybrid automata. Consider, for example, the duration requirement that *it is possible to keep the reactor running without using either control rod more than one third of the time.* To specify duration requirements, we use stop-watches. The *type* of a stop-watch z is a set U of control locations: the stop-watch z has the slope 1 whenever the automaton control is in a location of U, and otherwise z has the slope 0. Then z measures the accumulated amount of time that the automaton control spends in locations of U. We specify the type of a stop-watch by a state predicate that constrains the location counter. For example, we write $(z_1 : l[rod_1] = in)$ to declare that the variable z_1 is a stop-watch whose type is the set of all control locations where control rod 1 is in the reactor core. The given duration requirement uses a clock and two stop-watches. The clock z (a stop-watch of type *true*) measures the total elapsed time, the stop-watch z_1 of type $l[rod_1] = in$ measures the accumulated amount of time that control rod 1 spends in the reactor core, and the stop-watch z_2 of type $l[rod_2] = in$ measures the accumulated amount of time that control rod 2 spends in the reactor core. The ICTL formula

$$\varphi_I \rightarrow (z : true)(z_1 : l[rod_1] = in)(z_2 : l[rod_2] = in)\exists\Box(x \leq 550 \wedge 3z_1 \leq z \wedge 3z_2 \leq z)$$

asserts that in the labeled transition system $[\![A]\!]$, there is a path of infinite duration that starts from an initial state along which the core temperature does not exceed 550 degrees, and the accumulated time that either control rod spends in the reactor core is always at most a third of the total elapsed time.

While the original HYTECH prototype accepts ICTL input, we have not yet completed the implementation of ICTL model checking for the current version of HYTECH. However, like safety requirements, also many real-time and duration requirements can be reduced to reachability analysis, by moving clocks and stop-watches from the requirement specification to the system model. Consider, for example, the duration requirement of the reactor temperature control system that *independent of the control strategy that is used for deciding which control*

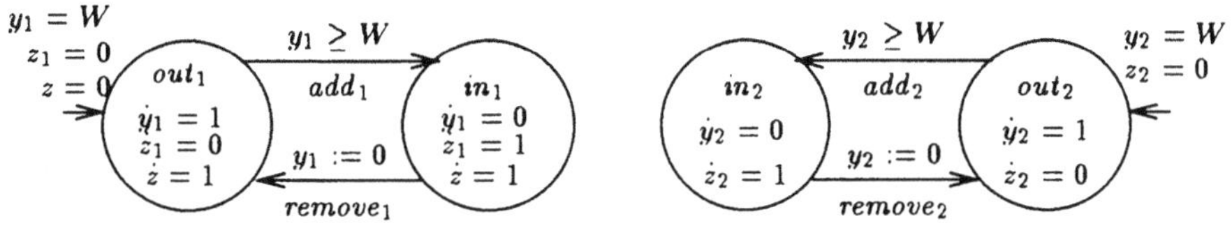

Fig. 4. The augmented control rod automata

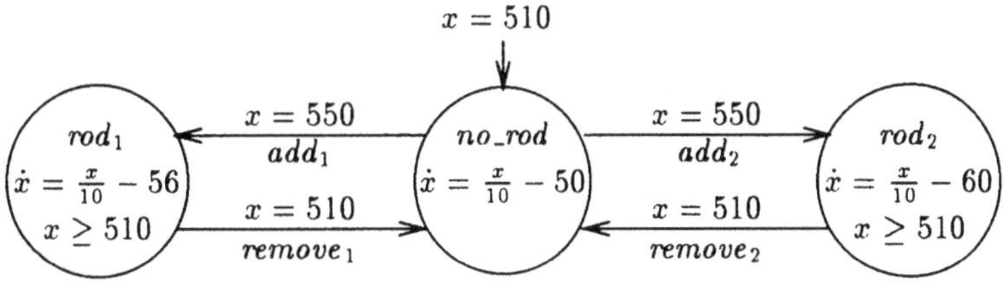

Fig. 5. The nonlinear reactor core automaton

rod to put into the reactor core, each control rod is used at most one third of the time:

$$\varphi_I \rightarrow (z\colon true)(z_1\colon l[rod_1] = in)(z_2\colon l[rod_2] = in)\ \forall\Box(3z_1 \leq z \wedge 3z_2 \leq z).$$

To verify this requirement, we move the clock z and the stop-watches z_1 and z_2 to the control rod automata as shown in Figure 4. Then we use HyTech to check if any state in the unsafe region $[\![3z_1 \geq z \vee 3z_2 \geq z]\!]$ is reachable from an initial state.

4 Analysis of Nonlinear Hybrid Systems

Many physical quantities, such as temperature, exhibit nonconstant derivatives. When using linear hybrid automata for modeling, these quantities need to be either translated into piecewise-linear quantities or approximated by rate intervals. Both options can be formalized as algorithmic translations—the *clock translation* and the *rate translation*—from nonlinear hybrid automata to linear hybrid automata. Both translations are currently being implemented in HyTech, and formal definitions of the translations can be found in [HH95a].

4.1 Clock Translation

Suppose that the reactor core of the reactor temperature control example is modeled by the hybrid automaton shown in Figure 5. The core temperature x increases according to the differential equation $\dot{x} = x/10 - 50$ if no control rod is in the reactor core; x decreases according to the differential equation $\dot{x} = x/10 - 56$ if control rod 1 is in the reactor core; and x decreases according to the differential equation $\dot{x} = x/10 - 60$ if control rod 2 is in the reactor

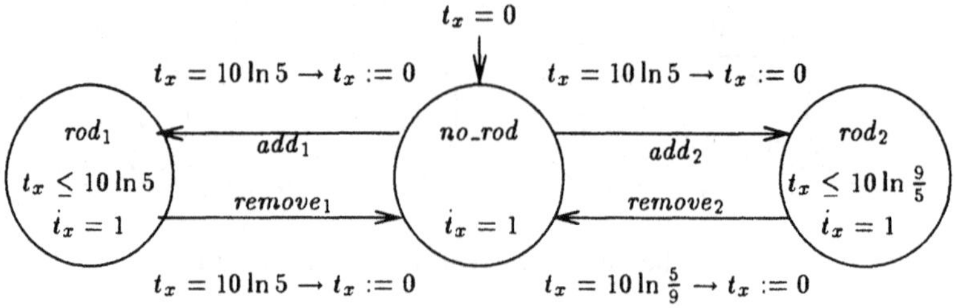

Fig. 6. The clock-translated reactor core automaton

core. Since the derivative of x is not governed by a rate interval, we say that x is a *nonlinear variable*, and the automaton containing x is a *nonlinear hybrid automaton.*

The clock translation replaces nonlinear variables by clocks. The nonlinear variable x can be replaced by a clock if its value is uniquely determined by the last assignment to x and the time that has expired since that assignment. For example, the core temperature x from Figure 5 can be replaced by a clock t_x. The resulting automaton, which is shown in Figure 6, is not a linear hybrid automaton, because real numbers occur in invariants and guards. In a second step, we overapproximate the automaton of Figure 6 by the linear hybrid automaton of Figure 7. For instance, since $16 \leq 10 \ln 5 \leq 16.1$, we overapproximate the the invariant $t_x \leq 10 \ln 5$ of the location *no_rod* by the invariant $10t_x \leq 161$, and we overapproximate the guard $t_x = 10 \ln 5$ of the transition from *no_rod* to rod_1 by the guard $160 \leq 10t_x \leq 161$.

Now suppose we wish to check if a final state in $[\![\varphi_F]\!]$ is reachable from an initial state in $[\![\varphi_I]\!]$ according to the nonlinear hybrid automaton of Figure 5. For simplicity, assume that neither φ_I nor φ_F contain the nonlinear variable x. If no final state in $[\![\varphi_F]\!]$ is reachable from an initial state in $[\![\varphi_I]\!]$ according to the linear hybrid automaton of Figure 6, which can be checked using HyTech, then this is also the case for the original nonlinear hybrid automaton of Figure 5. The converse, however, is not necessarily true, because of our overapproximation of invariants and guards. So if the reachability problem for the overapproximated automaton has a positive answer, we cannot conclude anything about the original reachability problem and must refine our approximation.

In summary, the clock translation of nonlinear variables is exact (i.e., it preserves all ICTL requirements) up to the representation of real numbers. The applicability of the clock translation depends on the solvability of differential equations, and we have begun to characterize sufficient conditions for the applicability of the clock translation [HH95a].

4.2 Rate Translation

The rate translation overapproximates nonlinear variables using linear variables that are governed by rate intervals. For the nonlinear variable x, and for each control location v, we compute the minimal and maximal derivative of x in v that

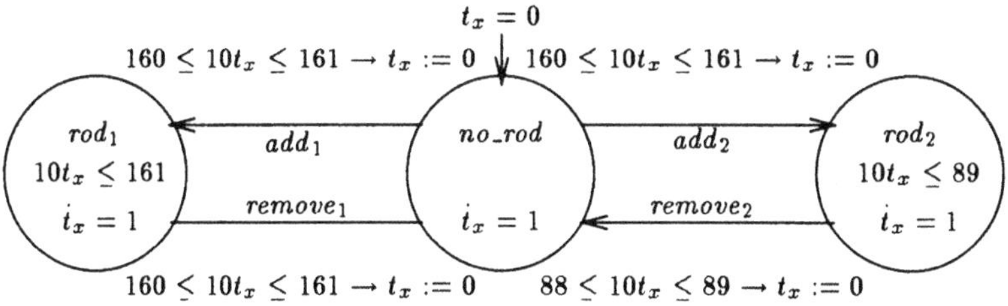

Fig. 7. The overapproximated clock-translated core automaton

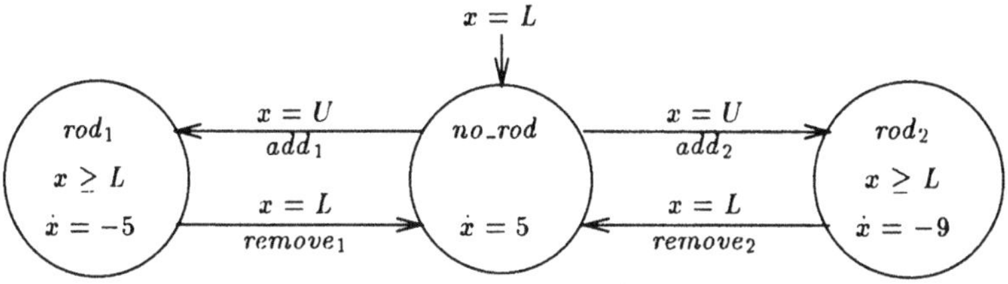

Fig. 8. The reactor core automaton with linear worst-case assumptions

observes the invariant of v. Consider, for example, the location rod_1 of the reactor core automaton from Figure 5, with the differential equation $\dot{x} = x/10-56$. Since we can strengthen the invariant to $510 \leq x \leq 550$, the derivative of x is bounded below by -5 and above by -1. Thus we can overapproximate the behavior of the nonlinear variable x in the location rod_1 by replacing the differential equation with the rate interval $dif(no\_rod, x) = [-5, -1]$. By treating the other locations similarly, we obtain the linear hybrid automaton of Figure 1.

Recall that HYTECH guarantees that the reactor core automaton of Figure 1 meets its safety requirement iff the parameter W satisfies the condition $9W < 184$. For the clock-translated reactor core automaton of Figure 7, HYTECH computes the weaker condition $5W < 189$. This condition is weaker, because the clock translation of Figure 7 gives a better approximation of the nonlinear system of Figure 5 than does the rate translation of Figure 1. Thus better approximations allow the design engineers to use slower mechanisms for moving the control rods. If desired, the approximation by rate translation can be refined by splitting control locations [HH95a].

Finally, suppose we replace the differential equations for the nonlinear variable x of the reactor core automaton from Figure 5 by worst-case constant-slope assumptions. By experimenting, we find that under these simplifying assumptions, additional parameters can be synthesized by HYTECH. In particular, we replace the lower bound of 510 and the upper bound of 550 for the core temperature by the parameters L and U, respectively. The resulting linear hybrid automaton is shown in Figure 8. HYTECH automatically synthesizes a necessary and sufficient condition on the three parameters W, L, and U such that the

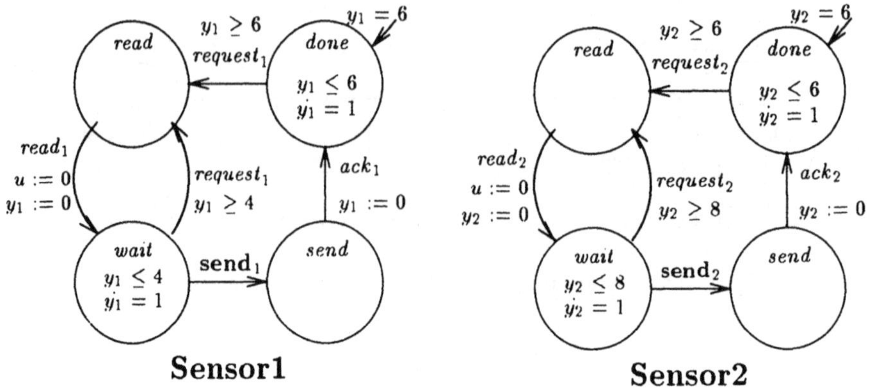

Fig. 9. The two sensors

reactor meets the safety requirement of never being shut down:

$$45W < 23(U - L)$$

(the computation uses 17.38 seconds of CPU time). Notice that if we replace the parameters L and U by the constants 510 and 550, respectively, we obtain exactly the condition $9W < 184$ that results from analyzing the rate-translated reactor core automaton of Figure 1.

5 Three Case Studies

We report on the application of HyTech to three nontrivial benchmark problems.

5.1 A Distributed Control System with Time-outs

The distributed control system of [Cor94] consists of two sensors and a controller that generates control commands to a robot according to the sensor readings. The programs for the two sensors and the controller are written in Ada. The two sensors share a single processor, and the priority of sensor 2 for using the processor is higher than the priority of sensor 1. In other words, if both sensor 1 and sensor 2 want to use the processor to construct a reading, only sensor 2 obtains the processor, and sensor 1 has to wait. The two sensors are modeled by the two linear hybrid automata in Figure 9 and the priorities for using the shared processor are modeled by the scheduler automaton in Figure 10.

Each sensor can be constructing a reading (location *read*), waiting for sending the reading (location *wait*), sending the reading (location *send*), or sleeping (location *done*). The processor can be scheduled idle (location *idle*), serving sensor 1 (location $sensor_1$), serving sensor 2 while sensor 1 is waiting (location $sensor_2\&wait_1$), or serving sensor 2 while sensor 1 is not waiting (location $sensor_2$).

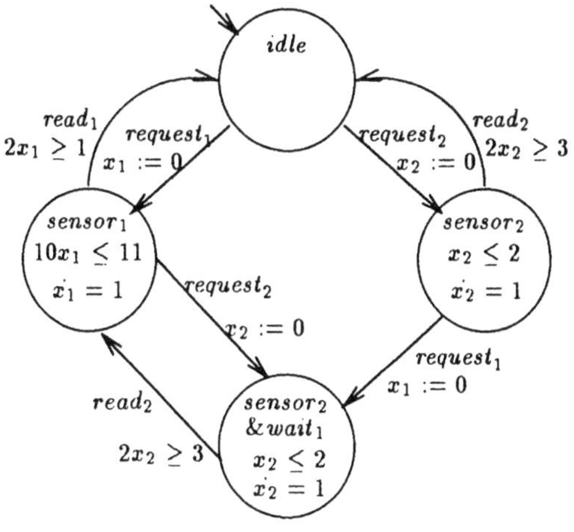

Fig. 10. The scheduler

Each sensor constructs a reading and sends the reading to the controller. The shared processor for constructing sensor readings is requested via *request* transitions, the completion of a reading is signaled via *read* transitions, and the reading is delivered to the controller via **send** transitions. Sensor 1 takes 0.5 to 1.1 milliseconds and sensor 2 takes 1.5 to 2 milliseconds of CPU time to construct a reading. These times are measured by the stop-watches x_1 and x_2 of the scheduler automaton. Notice that at most one of the two stop-watches x_1 and x_2 runs in a location of the scheduler automaton, which reflects the fact that only one sensor can use the shared processor at a time. If sensor 1 loses the processor because of preemption by sensor 2, it can continue the construction of its reading after the processor is released by sensor 2.

Once constructed, the reading of sensor 1 expires if it is not delivered within 4 milliseconds, and the reading of sensor 2 expires if it is not delivered within 8 milliseconds. These times are measured by the clocks y_1 and y_2 of the sensor automata. If a reading expires, then a new reading must be constructed. After successfully delivering a reading, a sensor sleeps for 6 milliseconds (measured again by the clocks y_1 and y_2), and then constructs the next reading.

The controller is modeled by the automaton in Figure 11. The controller is executed on a dedicated processor, so it does not compete with the sensors for CPU time. We use the clock z to measure the delays and time-outs of the controller. The controller accepts and acknowledges a reading from each sensor, in either order, and then computes and sends a command to the robot. The sensor readings are acknowledged via *ack* transitions, and the robot command is delivered via a *signal* transition. It takes 0.9 to 1 milliseconds to receive and acknowledge a sensor reading. The two sensor readings that are used to construct a robot command must be received within 10 milliseconds. If the controller receives a reading from one sensor but does not receive the reading from the other sensor within 10 milliseconds, then the first sensor reading expires (via an

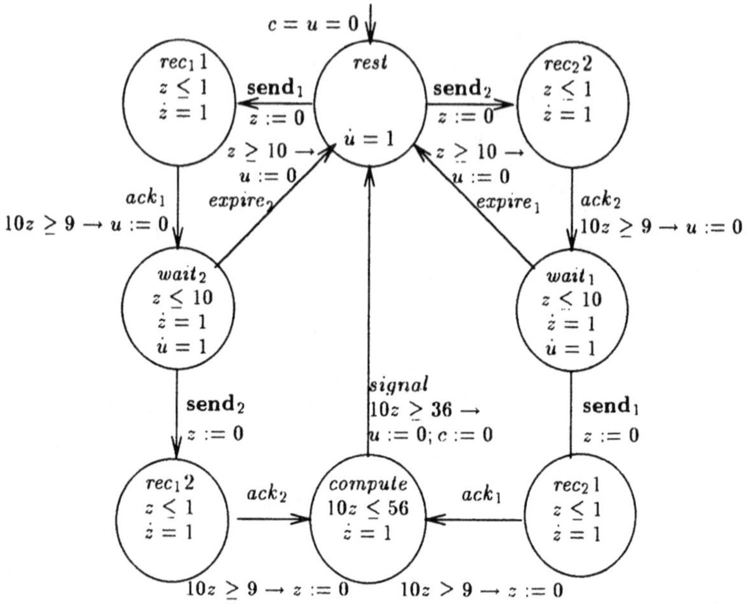

Fig. 11. The controller

expire transition). Once both reading are received, the controller takes 3.6 to 5.6 milliseconds to synthesize a robot command.

We want to know how often a robot command can be generated by the controller. For this purpose, we add a clock c to the controller automaton such that c measures the elapsed time since the last robot command was sent. The slope of the clock c is 1 in all locations of the controller automaton (this is omitted from Figure 11), and c is reset to 0 whenever a robot command is sent. We want to compute the maximum value of the clock c in all states that are reachable in the product of all four automata.

However, the product of the four automata does not model the system exactly according to Corbett's specification. This is because the **send** transitions should be urgent, that is, they should be taken as soon as they are enabled. We model the urgency of the **send** transitions by adding an additional clock, u, and global invariants. The clock u is reset whenever a sensor is ready to send a reading to the controller, and whenever the controller is ready to receive a sensor reading. Then we use the global invariant that $u = 0$ if both a sensor and the controller are ready for a transmission; that is,

$$
\begin{aligned}
&(l[sensor_1] = wait \wedge l[controller] = rest \;\rightarrow\; u = 0) \;\wedge\\
&(l[sensor_2] = wait \wedge l[controller] = rest \;\rightarrow\; u = 0) \;\wedge\\
&(l[sensor_1] = wait \wedge l[controller] = wait_1 \;\rightarrow\; u = 0) \;\wedge\\
&(l[sensor_2] = wait \wedge l[controller] = wait_2 \;\rightarrow\; u = 0).
\end{aligned}
$$

This invariant enforces whenever a transmission of a sensor reading is enabled, the transmission happens immediately.

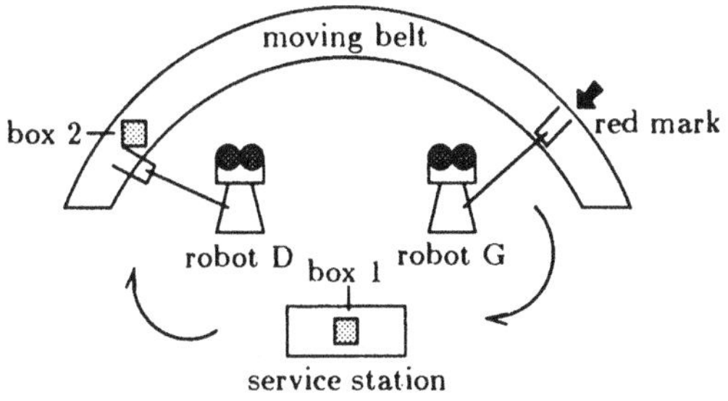

Fig. 12. The two-robot manufacturing system

In HYTECH, the global invariant is defined as follows:

```
GlobalInvar = {{l[sensor1]==wait && l[controller]==rest, 0==u},
    {l[sensor2]==wait && l[controller]==rest, 0==u},
    {l[sensor1]==wait && l[controller]==wait1, 0==u},
    {l[sensor2]==wait && l[controller]==wait2, 0==u)}}
```

To compute the range of possible values for the clock c in the reachable states, we write:

```
InitialState = l[sensor1]==done && l[sensor2]==done && l[scheduler]==
    idle && l[controller]==rest && 0==c && 6==y1 && 6==y2 && 0==u
Bad = True

EliminateLocList = {sensor1,sensor2,sched,gen}
EliminateVarList = {x1,x2,y1,y2,z,u}
```

Notice that, using the two projection operators, we ask HYTECH to print only information about the clock c. Using forward analysis without approximation, HYTECH returns, in 89.53 seconds of CPU time, the following answer:

```
0 <= c && -12 <= -5*c || -7 <= -2*c && 9 <= 10*c ||
-3 <= -c && 7 <= 10*c || -9 <= -2*c && 3 <= 2*c ||
12 <= 5*c && -28 <= -5*c || 5 <= 2*c && -18 <= -2*c ||
33 <= 10*c && -105 <= -10*c || 42 <= 5*c && -56 <= -5*c
```

From this result (the last disjunct is $42 \leq 5c \wedge -56 \leq -5c$), it follows that the maximum value of the clock c is 11.2; that is, a robot command is generated by the controller at least once every 11.2 milliseconds. We can also apply HYTECH to analyze the same system except that the priority of sensor 1 for using the shared processor is higher than the priority of sensor 2. In that case, a robot command is generated at least once every 11.0 milliseconds.

5.2 A Two-robot Manufacturing System

Puri and Varaiya [PV95] designed a manufacturing system that consists of a conveyor belt with two boxes, a service station, and two robots. The system is illustrated in Figure 12. This system has been also modeled and analyzed in [DY95].

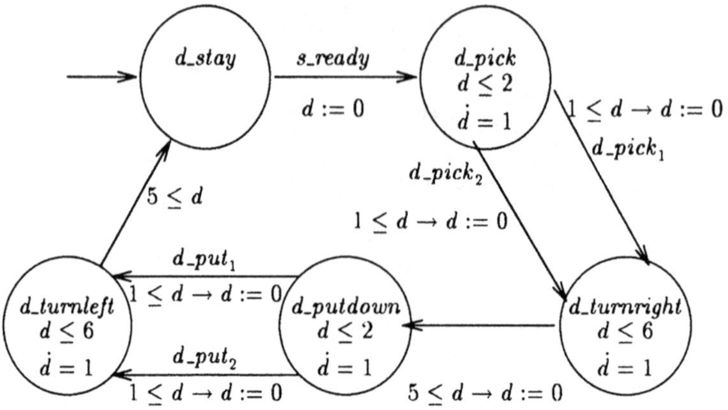

Fig. 13. Robot D

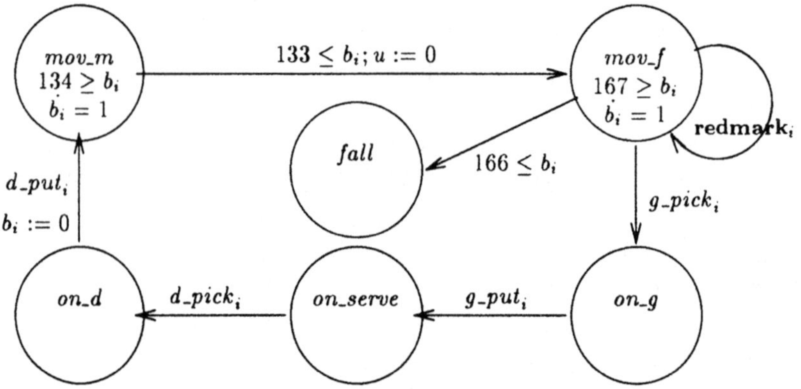

Fig. 14. Box i

Robot D, one of the two robots, is modeled by the linear hybrid automaton of Figure 13. The clock d is used to measure the time needed for the actions performed by robot D. Initially robot D is looking at the service station (location *d_stay*). When it sees an unprocessed box in the service station, it picks up that box from the service station in 1 to 2 seconds (location *d_pick*), makes a right turn in 5 to 6 seconds (location *d_turnright*), puts the box at one end of the conveyor belt in 1 to 2 seconds (location *d_putdown*), makes a left turn back to the service station in 5 to 6 seconds (location *d_turnleft*), and stays at there waiting for the next unprocessed box (location *d_stay*).

The two boxes, box 1 and box 2, are modeled by the indexed linear hybrid automaton in Figure 14, where the index i is either 1 (for box 1) or 2 (for box 2). A box may be in the service station (location *on_serve*), held by robot D (location *on_d*), moving on the conveyor belt before a red mark (location *mov_m*), moving on the conveyor belt beyond the red mark (location *mov_f*), held by robot G (location *on_g*), or falling off the end of the conveyor belt (location *fall*). A box

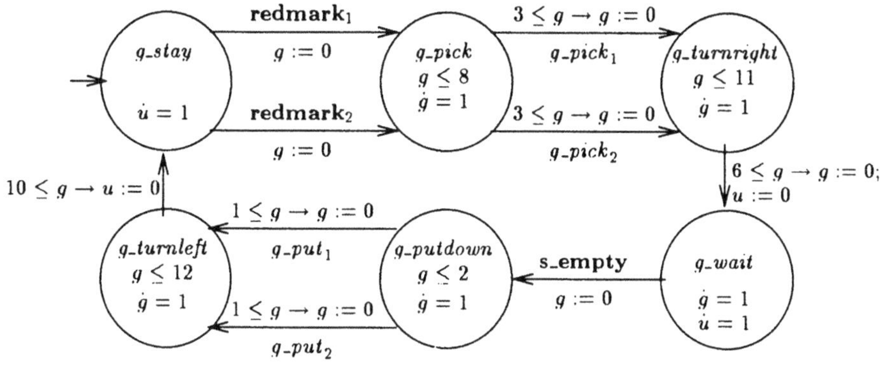

Fig. 15. Robot G

on the conveyor belt is processed by the manufacturing system. The conveyor belt is moving at a certain speed from one end to the other. The clock b_i measures the total time that box i spends on the conveyor belt, and thus determines the position of box i on the belt. A box requires 133 to 134 seconds to reach the red mark after it is placed on the belt by robot D. If a box is not picked up by robot G before the end of the belt, then the box falls off the belt 166 to 167 seconds after it is placed on the belt.

Robot G at the end of the conveyor belt is modeled by the automaton in Figure 15. The clock g measures the time needed to perform the actions of robot G. Initially robot G is looking at the red mark next to the conveyor belt (location *g_stay*). When it sees a processed box moving beyond the red mark, it picks up that box from the belt in 3 to 8 seconds (location *g_pick*), makes a right turn in 6 to 11 seconds (location *g_turnright*), waits for the service station to be empty (location *g_wait*), puts the box into the service station in 6 to 11 seconds (location *g_putdown*), makes a left turn back to the conveyor belt in 1 to 2 seconds (location *g_turnleft*), and stays there watching the red mark (location *g_stay*).

The service station is modeled by the automaton in Figure 16. Whenever the service station receives a processed box, it pops up an unprocessed box for robot D to pick up. The service station takes 8 to 10 seconds to switch the processed and unprocessed boxes, which is measured by the clock s. Initially both boxes are on the conveyor belt before the red mark. There are at most two boxes on the belt at any time, because the service station pops up a new box only when it receives a processed box from robot G.

According to Puri and Varaiya's specification, the transitions with the synchronization letters *s_ready*, **redmark**$_1$, **redmark**$_2$, and **s_empty**, are urgent; that is, robot D picks up a box from the service station as soon as it is ready and sees a box in the service station, etc. We treat the *s_ready* transitions as ordinary transitions, because this assumption will not affect our analysis. We use the clock u and the following global invariants to model the urgent transitions:

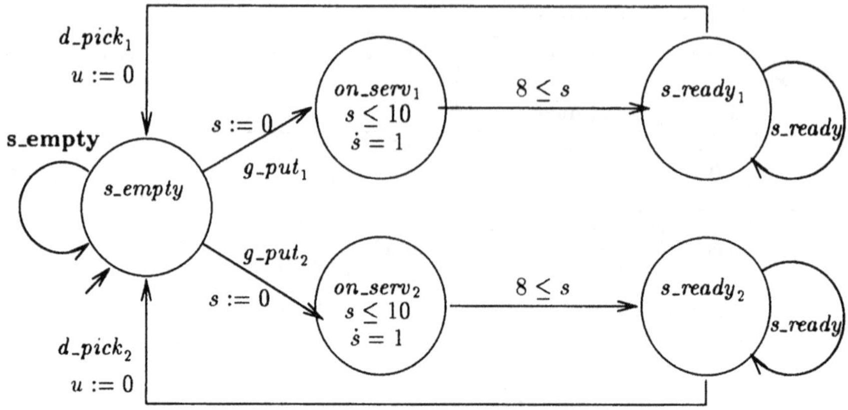

Fig. 16. The service station

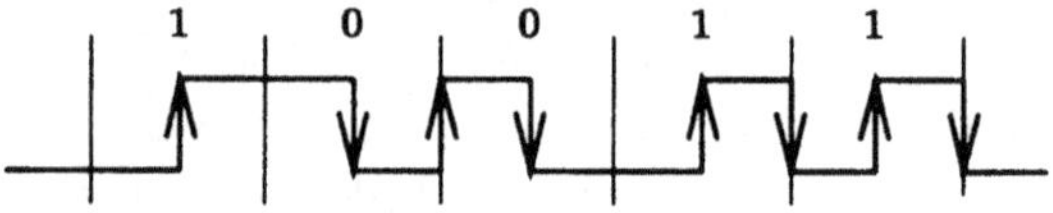

Fig. 17. The Manchester encoding

```
GlobalInvar = {{l[grobot]==gstay && l[box1]==movf, 0==u},
    {l[grobot]==gstay && l[box2]==movf, 0==u},
    {l[grobot]==gwait && l[station]==sempty, 0==u}}
```

We want to check the safety requirement that no box will ever fall off the conveyor belt. This requirement clearly depends on the initial positions of the two boxes on the belt. Then we use HyTech to analyze the reachability problem $(A, \varphi_I, \varphi_F)$, where A is the product of all five automata and

$$\varphi_I = (l[box_1] = mov\_m \wedge l[box_2] = mov\_m \wedge l[robot_G] = g\_stay \wedge l[robot_D] = d\_stay \wedge l[servicestation] = s\_empty \wedge u = 0),$$
$$\varphi_F = (l[box_1] = fall \vee l[box_2] = fall).$$

After we simplified the product automaton by eliminating unreachable locations and identifying locations in which a box is fallen, HyTech is able to return, in 163.41 minutes of CPU time, the following target region:

```
-1 <= -b1+b2 && -9 <= b1-b2 || -1 <= b1-b2 && -9 <= -b1+b2
```

using backward computation without approximation. It follows that

$$b2 - b1 > 9 \vee b1 - b2 > 9$$

is a necessary and sufficient condition on the initial condition of the system so that neither box will fall off the conveyor belt; that is, $|b_1 - b_2| > 9$.

5.3 The Philips Audio Control Protocol

In [BPV94], the timing-based Philips audio control protocol is modeled by an extension of the timed I/O automata model [LV92, LV93], and verified mathematically without computer support. We model the same protocol using linear hybrid automata, and verify its correctness for input strings up to length 8 using HyTech.

The protocol consists of a sender and a receiver. The sender uses the Manchester encoding to encode an input string of bits into a continuous signal (see Figure 17 for the encoding of 10011). The voltage on the communication bus is either high or low. A 0 bit is sent as a *down* signal from high to low voltage; a 1 bit is sent as a *up* signal from low to high voltage. The time line is divided into time slots of equal length, and the signals are sent in the middle of each time slot. The receiver decodes the continuous signal into an output string of bits. The protocol is correct iff the input and output strings match.

The time slots are measured by local clocks of the sender and receiver. These local clocks, however, may not be accurate, and their derivatives may vary within the rate interval $[\frac{19}{20}, \frac{21}{20}]$. Besides this potential 5% (or 1/20) timing error, the protocol faces also the following complications:

- The receiver does not know when the first time slot begins. The sender and the receiver can synchronize at the beginning of the transmission by knowing that (1) before the transmission, the voltage is low, and (2) the transmitted string starts with the bit 1.
- The receiver does not know the length of the bit string that is transmitted.
- The receiver sees only up signals and no down signals (because down signals are difficult to detect).

Using HyTech, we verify that whenever the sender encodes and sends a string of up to 8 bits, the receiver correctly decodes all bits in string. HyTech also shows that the protocol is incorrect in the case that the local clocks are subject to timing errors up to 1/15.

We use four linear hybrid automata to model the input, the sender, the receiver, and the output. The input automaton of Figure 18 generates all the possible bit strings up to a certain length. The length of the input string is decided by the initial value of the integer variable k. Whenever a bit is nondeterministically generated by the input automaton, the value of k is decremented by 1. When k becomes 0, the input automaton nondeterministically generates a suffix of one or two bits. So the input automaton generates all possible input strings of $k+1$ or $k+2$ bits. The integer variable c stores the message that is sent. If the bit 0 is sent, then c is updated to $2c$; if the bit 1 is sent, then c is updated to $2c+1$. The input automaton synchronizes with the sender through the synchronization letters $head_i$ and $input_i$, which correspond to looking at the next bit of the input string and sending the next bit of the input string, respectively.

The sender is modeled by the automaton of Figure 19. The variable x represents the drifting local clock of the sender, and its rate interval is $[\frac{19}{20}, \frac{21}{20}]$ for

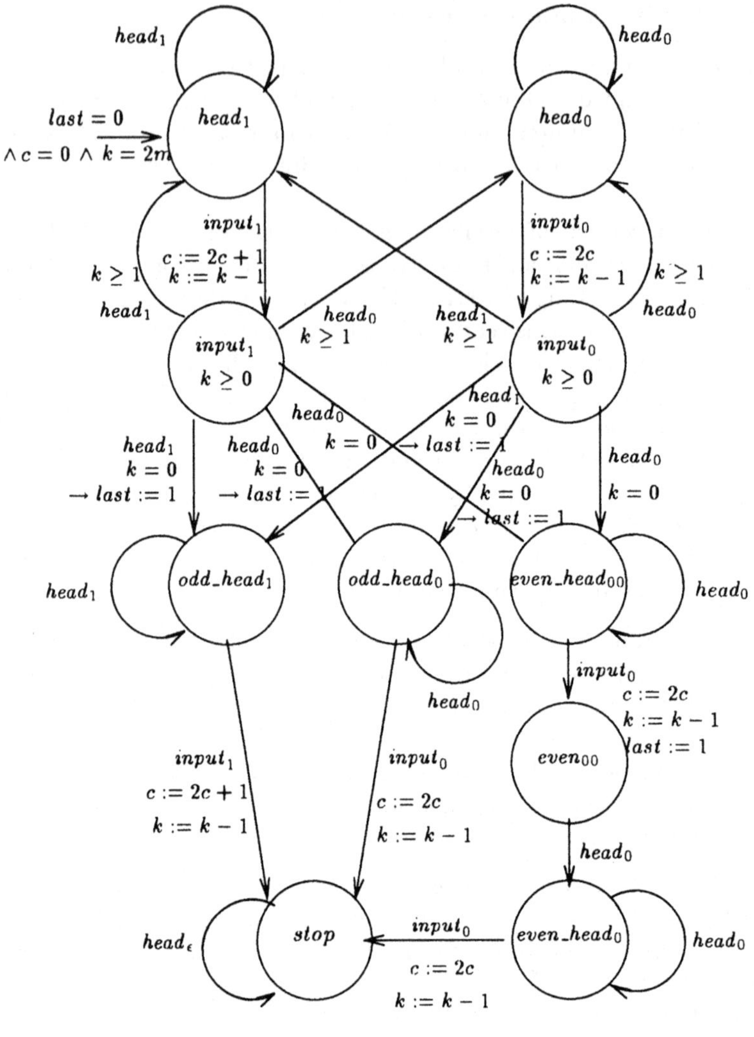

Fig. 18. The input automaton

all locations of the sender automaton. The sender synchronizes with the receiver through the synchronization letter *up*, which represents up signals.

The receiver is modeled by the automaton of Figure 20. The variable y represents the drifting local clock of the receiver, and its rate interval is $[\frac{19}{20}, \frac{21}{20}]$ for all locations of the receiver automaton. The receiver sees each up signal of the sender and decides if it encodes a 0 or a 1. If no up signal is received for a certain amount of time, the receiver times out and concludes that the transmission is completed. The receiver synchronizes with the output automaton through the synchronization letters $output_i$, which represent the bits of the decoded string.

The output automaton of Figure 21 stores the decoded bit string in the integer variable d. Then $c = d$ signals a correct decoding, and $c \neq d$ signals an

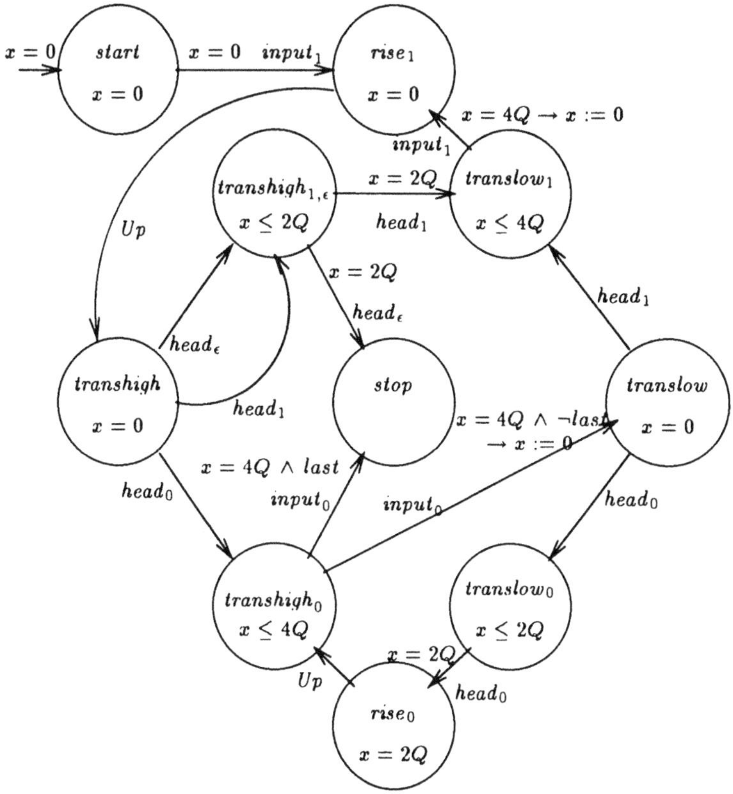

Fig. 19. The sender automaton

error in the protocol. We check if it is possible to reach a state that satisfies $c \neq d$ when the transmission is completed. A forward reachability analysis with HYTECH successfully verifies the protocol for $k = 6$. The performance of HYTECH is summarized in Table 22.

Remark. Since the submission of this paper, we have reimplemented HYTECH independent of MATHEMATICA [HHWT95]. That implementation is significantly more efficient, and achieves speedup of one to three orders of magnitude. For example, the distributed control system of Section 5.1 can be checked using 12 seconds of CPU time, and the manufacturing system of Section 5.2 can be checked using 353 seconds of CPU time.

Acknowledgement. We thank Howard Wong-Toi for the performance figures of the new implementation of HYTECH and a careful reading of this paper.

References

[ACH+95] R. Alur, C. Coucoubetis, N. Halbwachs, T.A. Henzinger, P.-H. Ho, X. Nicollin, A. Olivero, J. Sifakis, and S. Yovine. The algorithmic analysis of hybrid systems. *Theoretical Computer Science*, 138:3–34, 1995.

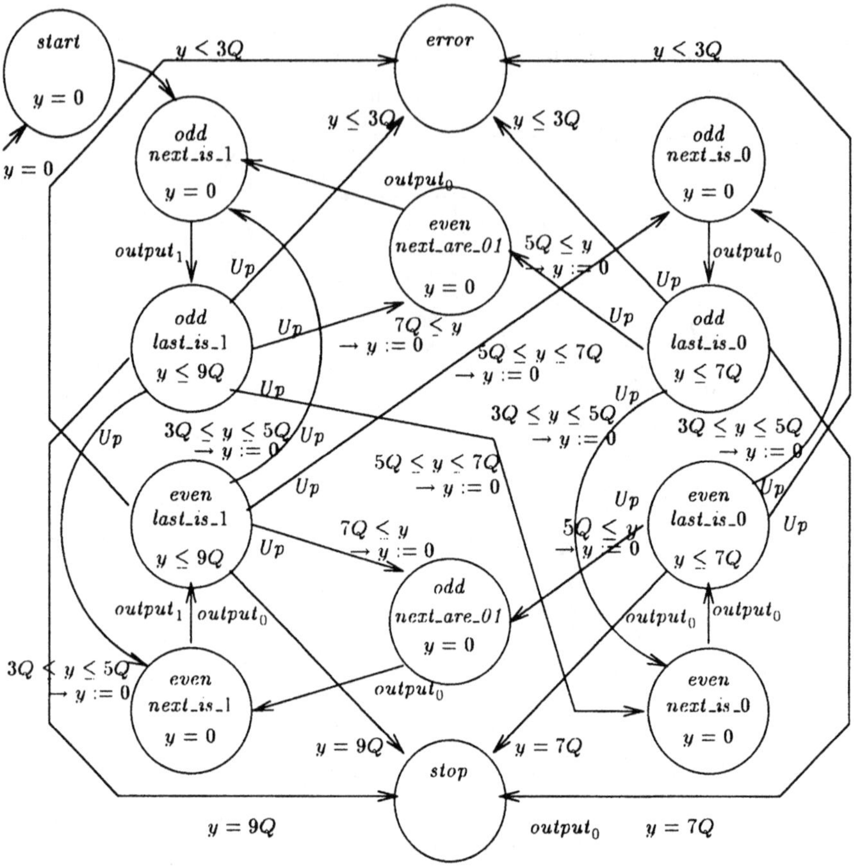

Fig. 20. The receiver automaton

[ACHH93] R. Alur, C. Courcoubetis, T.A. Henzinger, and P.-H. Ho. Hybrid automata: an algorithmic approach to the specification and verification of hybrid systems. In R.L. Grossman, A. Nerode, A.P. Ravn, and H. Rischel, editors, *Hybrid Systems*, Lecture Notes in Computer Science 736, pages 209–229. Springer-Verlag, 1993.

[AHH93] R. Alur, T.A. Henzinger, and P.-H. Ho. Automatic symbolic verification of embedded systems. In *Proceedings of the 14th Annual Real-time Systems Symposium*, pages 2–11. IEEE Computer Society Press, 1993.

[AHV93] R. Alur, T.A. Henzinger, and M.Y. Vardi. Parametric real-time reasoning. In *Proceedings of the 25th Annual Symposium on Theory of Computing*, pages 592–601. ACM Press, 1993.

[BPV94] D. Bosscher, I. Polak, and F. Vaandrager. Verification of an audio-control protocol. In H. Langmaack, W.-P. de Roever, and J. Vytopil, editors, *FTRTFT 94: Formal Techniques in Real-time and Fault-tolerant Systems*, Lecture Notes in Computer Science 863, pages 170–192. Springer-Verlag, 1994.

[CC77] P. Cousot and R. Cousot. Abstract interpretation: a unified lattice model

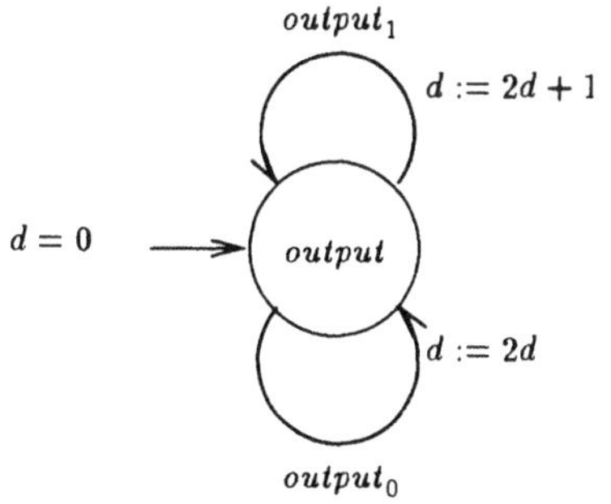

Fig. 21. The output automaton

Clock error	Input length	Location number	Transition number	CPU time	
1/20	5 or 6	1300	2795	681.8 sec.	Proved
	7 or 8			4275 sec.	
1/15	5 or 6			2018 sec.	Disproved

Fig. 22. Verification of the audio control protocol

for the static analysis of programs by construction or approximation of fixpoints. In *Proceedings of the Fourth Annual Symposium on Principles of Programming Languages*, pages 238–252, Los Angeles, California, 1977. ACM Press.

[CC92] P. Cousot and R. Cousot. Comparing the Galois connection and widening/narrowing approaches to abstract interpretation. In *PLILP*, Lecture Notes in Computer Science 631, pages 269–295. Springer-Verlag, 1992.

[CES86] E.M. Clarke, E.A. Emerson, and A.P. Sistla. Automatic verification of finite-state concurrent systems using temporal-logic specifications. *ACM Transactions on Programming Languages and Systems*, 8(2):244–263, 1986.

[CH78] P. Cousot and N. Halbwachs. Automatic discovery of linear restraints among variables of a program. In *Proceedings of the Fifth Annual Symposium on Principles of Programming Languages*. ACM Press, 1978.

[Cor94] J.C. Corbett. Modeling and analysis of real-time Ada tasking programs. In *Proceedings of the 15th Annual Real-time Systems Symposium*. IEEE Computer Society Press, 1994.

[DW95] D.L. Dill and H. Wong-Toi. Verification of real-time systems by successive over- and underapproximation. In *CAV 95: Computer-aided Verification*, Lecture Notes in Computer Science. Springer-Verlag, 1995.

[DY95] C. Daws and S. Yovine. Verification of multirate timed automata with KRONOS: two examples. Technical Report Spectre-95-06, VERIMAG, April 1995.

[Hal93] N. Halbwachs. Delay analysis in synchronous programs. In C. Courcoubetis, editor, *CAV 93: Computer-aided Verification*, Lecture Notes in Computer Science 697, pages 333–346. Springer-Verlag, 1993.

[HH95a] T.A. Henzinger and P.-H. Ho. Algorithmic analysis of nonlinear hybrid systems. In *CAV 95: Computer-aided Verification*, Lecture Notes in Computer Science. Springer-Verlag, 1995.

[HH95b] T.A. Henzinger and P.-H. Ho. A note on abstract-interpretation strategies for hybrid automata. This volume, 1995.

[HHWT95] T.A. Henzinger, P.-H. Ho, and H. Wong-Toi. HYTECH: The next generation. Submitted, 1995.

[HRP94] N. Halbwachs, P. Raymond, and Y.-E. Proy. Verification of linear hybrid systems by means of convex approximation. In B. LeCharlier, editor, *International Symposium on Static Analysis, SAS'94*, Lecture Notes in Computer Science 864, Namur (belgium), September 1994. Springer-Verlag.

[HW95] P.-H. Ho and H. Wong-Toi. Automated analysis of an audio control protocol. In *CAV 95: Computer-aided Verification*, Lecture Notes in Computer Science. Springer-Verlag, 1995.

[LV92] N.A. Lynch and F. Vaandrager. Action transducers and timed automata. In R.J. Cleaveland, editor, *CONCUR 92: Theories of Concurrency*, Lecture Notes in Computer Science 630, pages 436–455. Springer-Verlag, 1992.

[LV93] N.A. Lynch and F. Vaandrager. Forward and backward simulations, part ii: timing-based systems. Technical Report CS-R9314, CWI, Amsterdam, 1993.

[NOSY93] X. Nicollin, A. Olivero, J. Sifakis, and S. Yovine. An approach to the description and analysis of hybrid systems. In R.L. Grossman, A. Nerode, A.P. Ravn, and H. Rischel, editors, *Hybrid Systems*, Lecture Notes in Computer Science 736, pages 149–178. Springer-Verlag, 1993.

[PV95] A. Puri and P. Varaiya. Verification of hybrid systems using abstractions. This volume, 1995.

Hybrid Systems as Finsler Manifolds: Finite State Control as Approximation to Connections

Wolf Kohn*[1] and Anil Nerode**[2] and Jeffrey B. Remmel***[3]

[1] Sagent Corporation
Bellevue, Washington
e-mail: Wolf_Kohn%SAGENT@notes.worldcom.com
[2] Mathematical Sciences Institute
Cornell University, Ithaca, NY 14853
e-mail: anil@math.cornell.edu
[3] Department of Mathematics and Department of Computer Science
Universityof California at San Diego San Diego, CA 92093
e-mail; jremmel@ucsd.edu

Abstract. Hybrid systems are networks of interacting digital devices and continuous plants reacting to a changing environment. Our multiple agent hybrid control architecture ([KN93b], [KN93c]) is based on the notion of hybrid system state. The latter incorporates evolution models using differential or difference equations, logic constraints, and geometric constraints. The set of hybrid states of a hybrid system can be construed in a variety of ways as a differentiable (or a C^{∞}) manifold which we have called the carrier manifold ([KNRG95]). We have suggested that for control problems the coordinates of points of the carrier manifolds should be selected to incorporate all information about system state, control state, and environment needed to choose new values of control parameters.

This paper provides an outline of our ongoing work which expresses a wide class of control problems as relaxed, or convexified, calculus of variations problems on carrier manifolds. Weierstrass pointed out that many calculus of variations problems with time as an independent variable can be rewritten as parametric problems, that is, as problems where time is not an explicit variable, by moving to a larger manifold where time is an extra variable, $x_{n+1} = t$, with the extra constraint $\dot{x}_{n+1} = 1$. Then

* sponsored by Army Research Office contract DAAL03-91-C-0027, DARPA-US ARMY AMCCOM (Picatinny Arsenal, N. J.) contract DAAA21-92-C-0013 to ORA Corp., SDIO contract DAAH04-93-C-O113

** sponsored by Army Research Office contract DAAL03-91-C-0027, DARPA-US ARMY AMCCOM (Picatinny Arsenal, N. J.) contract DAAA21-92-C-0013 to ORA Corp., SDIO contract DAAH04-93-C-O113

*** sponsored by Army Research Office contract DAAL03-91-C-0027, DARPA-US ARMY AMCCOM (Picatinny Arsenal, N. J.) contract DAAA21-92-C-0013 to ORA Corp., SDIO contract DAAH04-93-C-O113 and NSF grant DMS-9306427

Finsler (1918) pointed out that each such parametric calculus of variations problem induces a metric ground form ds^2 associated with the manifold. We reduce a wide class of optimal control problems to calculating the Cartan affine connection associated with the Finsler metric ground form induced by the corresponding convexified calculus of variations problem. For any prescribed ϵ of deviation from the variational minimum, we then use chatterings of control flows on the manifold to approximate to a connection and use this to compute a finite state control program which enforces ϵ-optimal behavior. Thus we find that Finsler manifolds and their associated differential geometry and control flows form a computational basis for extraction and verification of hybrid systems and intelligent control systems in general. We remark that other connections, not arising from metrics, can be usefully employed in control too.

Here we discuss only the simplest problem of the calculus of variations and optimal control: find an "optimal" path $x(t)$ among admissible paths on the manifold M extending from initial point $x(t_0) = x_0$ to endpoint $x(t_1) = x_1$ with a given end direction $\dot{x}(t_1) = y_1$. The theory of Finsler manifolds is not familiar to most electrical engineers and certainly not to computer scientists. The goal of this paper is to give some historical background and a brief description of how one interprets connections as optimal controls and extraction of finite state control programs as extraction of approximations to connections. One can also unwind and use connections similarly for non-parametric problems. A book length mathematical treatment is in preparation.

1 Introduction

Recently we have dramatically improved the extraction algorithms to compute controls and control automata for hybrid systems by computing symbolic formulas for desired controls from either the Euler-Lagrange equations or the Hamilton-Jacobi equations ([KNRG95]). We observed that these formulas are essentially the Christoffel symbol expressions for connections ([LC26]), of the sort every student of Einstein's general relativity has learned. A literature search revealed that one proper mathematical context to explain this phenomenon is Finsler manifolds. The symbolic formulas for controls which we derived can be viewed as formulas for the coefficients of a Cartan connection on an appropriate Finsler manifold associated with a parametric calculus of variations formulation of the optimal control problem. Elie Cartan ([C34]) was one of the first persons to associate connections on Finsler manifolds with parametric variational problems. In fact, the parametric calculus of variations was itself the source of Finsler manifolds. The later development of the theory of Finsler manifolds moved away from its variational origins to concentrate on problems of formulating and proving generalizations for Finsler manifolds of results already known for Riemannian manifolds and to raising and answering global questions about Finsler spaces. We hope that our work will focus attention on the value of concrete Finsler manifold computations to elucidate and expedite computation of optimal controls and extraction of intelligent controls in general. Of course, control theory did not exist when Finsler invented the subject in 1918 ([F18]).

We think that our work represents the first systematic application of connections to extract optimal controls. If not, this work is certainly the first to use chattering vector fields to approximate to connections and to use this to extract finite control automata guaranteeing ϵ-optimal control. The latter is what makes the methodology an engineering tool. Isolated cases of connections describing controls have turned up before:, see, for instance, the Berry phenomenon in physics ([M93], [MRWZ93]). But apparently they have not previously been identified as a reflection of the Finsler structure of manifolds arising from variational problems.

2 Hybrid Systems and their Carrier Manifolds

Hybrid systems are interacting networks of continuous, generally nonlinear, plants and discrete automata. The discrete automata are control automata, generally controlling the actuators which are used to enforce desired plant behavior. Hybrid systems have emerged as a powerful methodology for modeling large complex systems, and the theory of hybrid systems and its application is a field of active research, see ([GNRR93]).

Simple Hybrid Systems.

Here is a brief description of a simple hybrid system ([KN93a]). It consists of a continuous plant interacting with a control automaton at times $n\Delta$. The plant runs open loop inside $[n\Delta, (n+1)\Delta]$ based on the control $c_n = c_n(t)$ supplied at time $n\Delta$. The control automaton receives as input at time $n\Delta$, a measurement of the state of the plant at time $n\Delta$, $x(n\Delta)$. The plant runs open loop with no further inputs till time $(n+1)\Delta$. Then based on the control automaton state at time $(n+1)\Delta$ and a measurement of the plant state $x((n+1)\Delta)$ at time $(n+1)\Delta$, it transmits a new control $c_{n+1}(t)$ to the plant to be used for the interval $[(n+1)\Delta, (n+2)\Delta]$, etc. We usually assume that the continuous plant is described by an ordinary vector differential equation

$$\dot{x} = f(x, c, d),$$

with x, c, d being points on specified (state) manifolds. It is assumed that for any time t_0, for any time interval $[t_0, t_0+\Delta]$, for any initial state $x(t_0)$, and for any admissible control and disturbance functions $c(t), d(t)$ defined on $[t_0, t_0+\Delta]$, there is a unique admissable $x(t)$ defined on $[t_0, t_0+\Delta]$ satisfying the differential equation. The function $x(t)$ is called the resulting plant state trajectory.

Continualization and the Carrier Manifold.

A hybrid system has a hybrid state, the simultaneous dynamical state of all plants and digital control devices. Properly construed, the hybrid states will form a differentiable manifold which we call the *carrier manifold* of the system. To incorporate the digital states as certain coordinates of points of the carrier manifold, we "continualize" the digital states. That is, we view the digital states as finite real valued piecewise constant funtions of continuous time and then we take smooth approximations to them. This also allows us to consider logical and differential or variational constraints on the same footing, each restricting the points allowed on the carrier manifold. In fact, all logical or discontinuous features can be "continualized" without practical side-effects. That is, any logical discrete states on successive time intervals can be represented first as step functions with values zero or one on these intervals and then smoothed. This is physically correct. That is, for any semiconductor chip used in an analog device, the zeros or ones are really just abbreviations for sensor readings of the continuous state of the chip.

Every constraint on the system, digital or continuous, is incorporated into the definition of what points are on the carrier manifold. Lagrange constraints are regarded as part of the definition of the manifold as well, being restrictions on what points are in the manifold. Following Caratheodory's lead ([C35]), we never eliminate constraints by Lagrange multipliers, but construe the hybrid system evolution of state as taking place on a differentiable manifold defined by the constraints. We then look for feedback functions $y = F(x)$ of points x on the manifold. For the purpose of this paper, we shall regard the choice of the value $F(x) = y$ as telling us in what direction $\dot{x} = y$ to steer when in state x to get to a goal. The goal, for the didactic purpose of the paper, is then simply to end in state x_1, steering at that time in a prespecified direction $\dot{y} = y_1$. If we use such a feedback control function to enforce the behavior of the system, the hybrid system trajectory that results is dependent on the feedback control function used.

The Fundamental Problem of Hybrid Systems.

Find algorithms which, given continuous plant differential equations and plant performance specifications (which may include logical constraints), extract digital control programs (mode switching programs) that force the state trajectories of the system to obey their performance specifications.

3 Optimal Control and the Variational Direct Method

In this section, we shall briefly describe our approach to solving the fundamental problem of hybrid systems. Our methods stem from Hilbert's direct method for proving the existence of solutions of variational problems, based on semicontinuity and compactness of function spaces. This is the origin of the relaxed controls of L. C. Young ([Y80]) and the convexified Lagrangians of Ekeland ([E83]) that we use. In contrast, Pontryagin's optimal control stems from the earlier classical paradigm of Weierstrass's variational calculus. In that paradigm, one first looks for relative minima by using necessary conditions (maximum principle, transversality, etc.). Then, among the solutions to the necessary condition, one hopes there is an optimal one. If such an optimal solution exists, then one tries to find it and tries to verify that it is optimal using second variations. Thus Pontryagin's method is based on necessary conditions. Our methods are based on constructivizing the existence proofs of Hilbert and Young for calculus of variations problems. We compute physically realizable feedback control functions that approximate non-physically realizable optimal controls that can be proved to exist non-constructively by Hilbert's method. The realization is by a finite control automaton and gives rise to a hybrid system as described above.

Here is Hilbert's inspirational dictum ([H05]) on variational problems.

> Eine jede reguläre Aufgabe der Variationsrechnung besitzt eine Lösung sobald hinsichtlich der Natur der gegebenen Grenzbedingungen geeignete einschränkende Annahmen erfüllt sind und nötigenfalls der Begriff der Lösung eine sinngemä*ß*e Erweiterung erfährt.

To paraphrase this in English, every regular calculus of variations problem has a solution provided the obvious boundary value conditions are satisfied and the notion of solution is taken in an appropriate sense. Here "regular" refers to the positive definiteness of an associated quadratic form everywhere, a notion going back to Legendre, Jacobi, etc.

For us, what is the appropriate "extended sense" for solutions? To apply Hilbert's dictum, we need to say what we mean by an "approximate" solution, and to specify a metric saying what it means for two such approximate solutions to be close. Then we can complete the metric space of approximate solutions. This completion will consist of the intended "extended sense" solutions, which are in no sense intended to be constructive or physically realizable. Our original method was to adopt the Young-Warga ([Y80], [W72]) definition of measure valued feedback functions F. In the Young-Warga theory, the value of $F(x)$ is a probability measure on the space of control parameter values, which is the completion of the finite support probability measures with respect to an appropriate metric. More recently, we have introduced a formulation ([GKNR95]) of this theory based on convex analysis ([E83], [ET76]) which is more transparently extendible from Euclidean space to manifolds.

Kohn discovered that the non-physical measure valued controls which the Young theory assigns as optimal solutions in the extended sense for calculus of variations problems can be approximately implemented by computing chattering controls, issued as orders to the plant by a finite control automaton. This led to what is now termed the Kohn-Nerode multiple agent hybrid control architecture (MAHCA).

The Kohn-Nerode Methodology.

We shall start with a basic outline of our method and we shall fill in some details of the outline in subsequent sections. We have previously developed a methodology and algorithms for a very general formulation of the fundamental problem of hybrid systems based on the relaxed calculus of variations on manifolds. (We also have a stability theory for hybrid systems ([KN93a], [KNRY95]). The algorithms guarantee that control programs extracted enforce suboptimal stable behavior. This we cannot discuss here.

First, corresponding to an autonomous problem, we formulate a non-negative

Lagrangian (or cost function) $L(x, \dot{x})$ on plant state trajectories which result from applying possible control functions $c(t)$ to produce the trajectory in such a way that the smaller the value of the integral of $L(x(t), \dot{x}(t))$ along a trajectory from origin to goal, the better the performance. The plant state trajectories lie on a carrier manifold.

Second, the performance specification for the plant is reformulated as the requirement that the plant state trajectory has an integral within a user prescribed ϵ of its minimum. A control function of time or state that gives such a trajectory is called an ϵ-optimal control function. ϵ-optimal control is all that is required in actual applications.

Third, an existence theorem is applied that says that the optimal control problem has a solution in an extended sense. The extended sense we now prefer is that, in the minimization above, $L(x, u)$ is replaced by its convexification $L^*(x, u)$ with respect to u . What does this mean? $L(x, u)$ is defined over the disjoint fibers of the tangent bundle of the carrier manifold. Convexification is done with respect to the restriction of $L(x, u)$ to the fiber defined by x. Since a fiber is a vector space, i.e. the tangent space corresponding to the point x of the manifold, we can convexify the function $g_x(u) = L(x, u)$ which maps that vector space to the set of reals. The result is a convex function over the fiber. The union of these functions g_x constitute the convexification $L^*$of L. (In addition, we sometimes need to convexify the carrier manifold under an embedding into Euclidean space. We omit that discussion here.) We make no attempt to give an iterative scheme to compute the optimum arbitrarily closely, but content ourselves with ϵ-optimal solutions, because we have proved that the general existence proof for optimal solutions is inherently non-constructive ([NG95]).

Fourth, for any prespecified ϵ, we compute an ϵ-optimal control for the original problem ([GKNR95]) and implement the ϵ-optimal control as a finite state, physically realizable, control automaton (appendix [KN93a]). The ϵ-optimal control is obtained as an approximation to the optimal control feedback law using algorithms from convex sets and automata theory. The actual control law issued by the finite control automaton when the system is in state x is a chattering control ([GKNR95]). The chatter is between approximations to some local extrema of $L(x, u)$. The use of this finite state automaton as controller assures ϵ-optimal control ([GKNR95]). (If we convexified the original constraint manifold, there is an additional step of approximating points on the convexified manifold by convex combinations of points on the original manifold to get an ϵ-optimal control that works on the original manifold.)

Fifth, we repeat this derivation of the control automaton on line, if due to unmodelled dynamics, the performance specification for the system is not met. Then we substitute a new control automaton for the old as required. This type of substitution represents the basic adaptaive capability of our architecture.

The new idea of this paper is that the symbolic computation of parallel transport using the Finsler metric as described below can provide significantly more efficient algorithms for extracting controls.

Remark. In the classical calculus of variations, going back to Euler and Lagrange, one reduces local minima problems with constraints to free problems with no constraints in a region of Euclidean space by introducing Lagrange multipliers. Computing Lagrange multipliers is computationally prohibitive except in very simple cases, and once done eliminates the geometry of the constraints. We follow Hilbert's illustrious student Caratheodory in sticking with manifold geometry as the vehicle for expressing constraints.

4 Parametric Problems of the Calculus of Variations

Weierstrass originated the parametric, or homogeneous, form of the calculus of variations. He pointed out that many calculus of variations problems with time as an independent variable can be rewritten as parametric problems, that is problems with time not an explicit variable, by moving to a larger manifold where time is an extra variable, $x_{n+1} = t$, with the extra constraint $\dot{x}_{n+1} = 1$ ([W30], [S42], [R66]). With mild smoothness assumptions, a wide variety of calculus of variations problems can be based on Lagrangians $L(x, \dot{x})$ defined on the line elements $(x, \dot{x})$ of the manifold.

Remarks.

1. We observe for those not familiar with the quaint notation of the calculus of variations that $\dot{x}$ is simply another variable with the same number of coordinates as x which are free, not the derivative of x. One should really write $L(x, u)$, substituting $x(t)$ for x and the time derivative $\dot{x}(t)$ for u when evaluating the Lagrangian on a curve $x(t)$, but that is not the historical convention.

2. We note that one should probably identify these line elements as members of the tangent bundle, not the contact bundle, because what is involved is calculations in the Lie algebra of derivations of the scalars, the real valued smooth functions on the manifold.

The basic hypotheses about $L(x, \dot{x})$ are the following.

***L* is homogeneous of degree 1 in $\dot{x}$.**

$$L(x, \lambda\dot{x}) = \lambda L(x, \dot{x}) \text{ for } \lambda \geq 0, \tag{1}$$

and

$L(x,\dot{x})$ is positive definite in $\dot{x}$.

$$(g_{ij}(x,\dot{x})) = \left(\frac{\partial^2(L^2(x,\dot{x}))}{\partial \dot{x}_i \partial \dot{x}_j}\right) \tag{2}$$

is a positive definite matrix for each fixed x.

The latter expresses regularity in Hilbert's sense.

The classic parametric calculus of variations problem takes the following form: Find an admissible curve on the constraint manifold satisfying the endpoint conditions and minimizing

$$\int_{t_0}^{t_1} L(x(t),\dot{x}(t))dt. \tag{3}$$

We call the value of this integral the *total cost* of the trajectory based on cost function L.

Remark. If one wishes at the end of a trajectory to enter a closed region, a natural attack is to try to enter normal to a tangent space to the region. One takes as the final point x_1 the point of contact of the tangent space and one takes as the final direction y_1 the inward directed normal to the surface at the contact point.

Caratheordory, who was a pupil of Hilbert, very much emphasized the geometric point of view in which constraints define a hypersurface in Euclidean space and variational solutions satisfying constraints are regarded as curves on hypersurfaces which possess a minimum property. This is the opposite of using Lagrange multipliers to reduce constrained problems to "free" problems without constraints in Euclidean space.

As judged by his thesis content ([F18]), the intent of Caratheodory's student Finsler was to generalize Gauss's theory of surfaces and, more generally, Riemannian geometry to the case when the surfaces have a Finsler metric ground form as their element of distance. Thus, Finsler studied angles, curvatures, etc, but put little emphasis on the implications for parametric variational calculus except to motivate the definition of his manifolds. A vast literature on Finsler geometry has since developed, most of it far removed from the needs of problem solving in the calculus of variations.

Finsler discovered that, with suitable smoothness assumptions, the Hessian g_{ij} of a regular parametric calculus of variations problem on a manifold induces a corresponding element of arclength or metric ground form, ds^2. This infinitesimal arclength for the Finsler manifold is defined by

$$ds^2 = \sum_{ij} g_{ij}(x,dx)dx_i dx_j \tag{4}$$

The homogeneity of degree 2 in $\dot{x}$ of $L^2(x,\dot{x})$ implies that

$$L^2(x,\dot{x}) = \sum_{ij} g_{ij}(x,\dot{x})\dot{x}_i\dot{x}_j \tag{5}$$

Thus another expression for the same thing is

$$ds^2 = L^2(x, dx). \tag{6}$$

This leads to defining $\dot{x}$ to be of *Finsler* length 1 if $L(x, \dot{x}) = 1$. This unit sphere defines a norm on the tangent space to the manifold at x. This unit sphere was called by Caratheodory the tangent indicatrix. Many calculus of variations problems can be reduced to determining its surface and how that surface changes as x moves over the manifold.

A special case of a Finsler metric ground form is a Riemannian metric ground form

$$ds^2 = \sum_{ij} g_{ij}(x) dx_i dx_j \tag{7}$$

where g_{ij} depends on the location x alone and not on the direction $\dot{x}$.

Remark. The reason that the Finsler metric involves $\dot{x}$ as well as x is simply that $\dot{x}$ is introduced when one transforms a non-parametric problem in the calculus of variations into a parametric form by introducing the time parameter t as an additional variable for the manifold. What is truly remarkable is how much of Riemannian geometry Finsler and his followers generalized with $\dot{x}$ present in the metric ground form, or expressed differently, how much differential geometry can be used in the parametric calculus of variations. One can conjecture that Weierstrass had some prescience about this when he studied the transformation, since much earlier Riemann (1854), a colleague of Weierstrass in Berlin, suggested one could develop the metric differential geometry of at least one non-Riemannian metric ground form, the positive fourth root of a certain quartic differential form ([R59]).

Finsler discovered the key fact that, generally, the geodesics under the metric ground form induced by L are the extremals of the corresponding parametric variational problem. That is, they are the solutions to the Euler-Lagrange equations.

$$\frac{d}{ds}\frac{\partial L(x, \dot{x}_i)}{\partial \dot{x}_i} - \frac{\partial L(x, \dot{x})}{\partial x_i} = 0 \tag{8}$$

Since this is the usual variational necessary condition for a minimum, the solutions to the parametric variational problem are then among the geodesics with respect to the Finsler metric ground form.

Remark. On the surface it may appear that the language of geodesics relative to ds^2 has eliminated use of the Lagrangian L. But in fact it only eliminates the mention of L, the derivatives of L enter into the definition of the g_{ij} and into every formula about geodesics, including the formulas to be discussed about connections. All that ds^2 does is to concentrate attention on differential geometric aspects of parametric variational problems and point toward using differential geometry based computation.

5 Connections and Control I

How do you compute controls with connections? We apologize in advance to differential geometers, relativists, and others for whom connections are daily bread and butter. Up to now, this group has not included control engineers, but we believe it will.

In this section, we describe links between the geodesics on Finsler manifolds described in the previous section and control policies (strategies). This description is based on the geodesics relative to the Finsler metric on the manifold. For the sake of comprehensibility we will not discuss here the consequences of a manifold having a complicated structure of geodesics. To see what is being avoided, define the set of goals (x_1, y_1) reachable from from x_0 as the set of all goals (x_1, y_1) such that there exists a direction y_0 and a geodesic on the manifold with slope y_0 starting at x_0, ending at that goal. These reachability sets can be quite complicated. We confine ourselves here to the illustrative case in which the goal (x_1, y_1) is reachable from every point x_0 by exactly one geodesic. Then all geodesics which end at that goal are defined globally and are uniquely determined by the endpoint conditions.

Illustrative Assumption for goal (x_1, y_1).

For any point x_0, on the manifold, there is a unique geodesic starting at x_0 and ending at x_1 with tangent vector y_1 at x_1.

In general, there is no escaping the phenomena of variational geometry such as focal points, conjugate points, transversality conditions, corner conditions, etc., which take us out of the illustrative case. On the other hand, in practice the illustrative case is quite common, because one can often choose the carrier manifold and Lagrangian to make it so.

When geodesics with respect to the Finsler ground form and extremals of the parametric variational problem coincide, to achieve the goal of reaching x_1 with tangent vector y_1, we should follow the appropriate geodesic starting at initial point x_0. The appropriate geodesic begins with *some* tangent vector y_0. The question then becomes: how do we determine the required y_0 to get to the goal? Clearly, to continue along the appropriate geodesic, at any intermediate point x, we always have to head along the tangent y to that geodesic at the point x. In summary, when at point x, the problem is to compute the tangent vector y of the required geodesic which leads to goal (x_1, y_1) in the tangent bundle. We can write $y = P(x, x_1, y_1)$, that is, the desired direction y to steer when at x is a function of the present position x and the tangent bundle goal (x_1, y_1). For a fixed goal (x_1, y_1), we can write $y = P(x)$. This is a vector field on the manifold associated with the goal in the illustrative case.

We can thus regard $P(x)$ as an optimal control policy relative to the goal (x_1, y_1), since following the geodesic is following the desired extremal path for the variational problem.

How do we compute policy $y = P(x, x_1, y_1)$ from the manifold description and the metric ground form? The oldest approach followed by Finsler and Cartan is to regard the manifold as embedded in a sufficently high dimensional Euclidean space. Following Cartan ([C34]), introduce an orthogonal basis, or moving frame, with origin at x on a geodesic with the principal normal to the geodesic as first member of the basis, so the rest is an orthogonal basis for the tangent space to the manifold. To see what to do, consider as Cartan did, and Levi Civita before him, a pair of infinitely close tangent planes, one at x, one at $x+dx$, along the geodesic. They intersect on a line. Think of the tangent plane at x as infinitesimally rotated over the line of intersection as axis to the tangent plane at $x + dx$, and write down the differential (Lie Algebra) relation obeyed by the infinitesimal rotation of the first moving frame to the second. This relation expresses the infinitesimal rotations of each axis of the moving frame around all the axes. Integrating this, as did Cartan ([C34]), we get the "equations of parallel transport" which compute how the moving frame moves down the geodesic starting at x_1 to get an image frame at x. The image of y_1 in the image frame is $y = P(x)$. This shows that, with this primitive interpretation, the steering direction needed at any point x is given symbolically by the Levi Civita parallel transport formulas, with the goal (x_1, y_1) as parameters.

These are the formulas we rediscovered in solving control problems on carrier manifolds, before finding by literature search that the correct context is Finsler spaces.

One of the most important features of Riemannian Geometry is covariant differentiation, introduced by M. G. Ricci in 1887-1896, and parallel transport, introduced by Levi-Civita in 1917 ([LC17]). This work was used extensively by Einstein and subsequent physicists in General Relativity to compute what happens with angles, measuring rods, and physical systems in general, as one moves through space-time. Connections are sometimes described as making sense in physics of the notion of a constant field in a neighborhood of a point. Affine connections were defined by Cartan in the 1920's and fully axiomatized by Kozul much later. Computing parallel transport is computing connection coefficients for a moving frame along trajectories, namely Christoffel coefficients, and integrating.

For us, the significance of covariant differentiation thus far has been that it gives a symbolic calculus for parallel transport, and parallel transport computes control policies.

For analytic details of the above process, see Cartan ([C34]), or Rund ([R59]). For the case when the manifold is a region of Euclidean space (but with a Finsler metric ground form), see the theses of Wren ([W30]) and Stokes ([S42]). The Wren and Stokes papers were written in the Bliss tradition of the University of Chicago School of the Calculus of Variations to squeeze out whatever is of use for calculus of variations from Cartan's book. That school died out at the end of World War II. Their work is less appreciated than it should be for supplying tools for solving variational problems. Perhaps this is because mathematics moved on to classification problems and global problems after World War II and away from helping scientists and engineers solve their variational problems. Of course, when they wrote, control theory did not exist in any form. This may also account for the neglect.

6 Geometric Preliminaries

The relevant notions are vector fields, derivations, and flows on manifolds M ([N67]). We act as if all manifolds are C^∞ to avoid special hypotheses. Let F_x be the space of real valued smooth functions f defined on a neighborhood of a point x in M. Let f and g be functions in F_x. A derivation v of F_x is a map:

$$v : F_x \to F_x$$

that satisfies the following two properties:

$$\begin{array}{ll} v(f+g)(x) = (v(f)+v(g))(x) & \text{(Linearity)} \\ v(f \cdot g)(x) = (v(f)\cdot g + f \cdot v(g))(x) & \text{(Leibniz's Rule)} \end{array}$$

Derivations define vector fields on M and also associated integral curves ([KN93c]). Suppose that C is a smooth curve on M parameterized by $\alpha : [t_0, t_1] \to M$ with $[t_0, t_1]$ a subinterval of R. Suppose that M is an n-dimensional manifold so that in local coordinates, $x = (x_1, \ldots, x_n)$ and C is given by n smooth functions

$$\alpha(t) = (\alpha_1(t), \ldots, \alpha_n(t))$$

whose derivative with respect to t is denoted by

$$\dot\alpha(t) = (\dot\alpha_1(t), \ldots, \dot\alpha_n(t)).$$

Introduce an equivalence relation on curves in M as the basis of the definition of tangent vectors at a point in M ([K90]). Two curves $\alpha(t)$ and $\beta(t)$ passing through x are said to be equivalent at x (notation: $\alpha \sim \beta$), if they satisfy the following condition:

There exists t and τ in $[t_0, t_1] \subset R$ such that

$$\alpha(t) = \beta(\tau) = x \text{ and } \dot\alpha(t) = \dot\beta(\tau). \tag{9}$$

Then $\sim$ is an equivalence relation on the class of curves in M passing through x. Let $[\alpha]$ be the equivalence class containing α. A tangent vector to $[\alpha]$ is a derivation $v\,|_x$, defined in the local coordinates[4] $(x_1, \ldots, x_n)$ by

Let $f : M \to R$ be a smooth function. Then

$$v\,|_x\,(f)(x) = \sum_{j=1}^{k} \dot\alpha_j(t) \frac{\partial f(x)}{\partial x_j} \quad \text{with } x = \alpha(t). \tag{10}$$

[4] Although we have expressed the tangent vector in terms of the chosen local coordinates, this can of course be done independent of local coordinates.

The tangent vector at x is an element of the tangent space at x, denoted by TM_x. The set of tangent spaces associated with the points of M form the tangent space (or tangent bundle manifold), denoted by TM:

$$TM = \cup_{x \in M} TM_x.$$

A vector field on M is an assignment of a derivation v to each point of M, $v \mid_x \in TM_x$, with $v \mid_x$ varying smoothly from point to point. In our modeling, we will always express vector fields in local coordinates. Thus if $x = (x_1, \ldots, x_n)$ in local coordinates, then

$$v \mid_x = \sum_j \lambda_j(x) \frac{\partial}{\partial x_j}. \tag{11}$$

Comparing (10) and (11) we see that if $\alpha(t)$ is a parameterized curve in M whose tangent vector at any point coincides with the value of v at the same point, then

$$\dot{\alpha}(t) = v \mid_{\alpha(t)}$$

for all t. In the local coordinates, we have that $x = (\alpha_1(t), \ldots, \alpha_n(t))$ must be a solution to the autonomous system of ordinary differential equations:

$$\frac{dx_j}{dt} = \lambda_j(x) \text{ for } j = 1, \ldots, n. \tag{12}$$

If v is a vector field, any parameterized curve $\alpha(t)$ passing through a point x in M is called an integral curve associated with v if $x = \alpha(t)$ and in local coordinates $v \mid_x$ satisfies (12). An integral curve associated with a field v, denoted by $\Psi(t, x)$, is termed the *flow generated by* v if it satisfies the following properties:

$$\Psi(t, \Psi(\tau, x)) = \Psi(t + \tau, p) \text{ (semigroup property)} \tag{13}$$
$$\Psi(0, x) = x \text{ (initial condition)}$$
and
$$\frac{d}{dt}\Psi(t, x) = v \mid_{\Psi(t,x)} \text{ (flow generation)}$$

To avoid details about maximal trajectories, we assume that flows yield unique global trajectories on M for any initial condition.

7 Connections and Control II

How does one compute with the connection along an integral curve on M? Suppose we are given x_0, x_1 on M and a curve

$$\alpha : [t_0, t_1] \to M. \tag{14}$$

Suppose that x_0 is the initial point and x_1 is the endpoint of the curve, i.e assume that $x_0 = \alpha(t_0)$ and $x_1 = \alpha(t_1)$. Suppose that curve lies entirely within some chart (U, ψ) of the manifold, where U is an open set of the manifold and ψ is a homeomorphism of U to a region of Euclidean space. Assume that the manifold is n-dimensional and that in the local coordinates of the chart, the curve α is given by

$$\psi(\alpha(t)) = (\alpha_1(t), \ldots, \alpha_n(t)).$$

Let TM_x denote the tangent space of M at the point x. Now suppose that we are also give a tangent vector $y_1 \in TM_{x_1}$.

We want to do a parallel transport of the tangent vector $y_1 \in TM_{x_1}$ at x_1 along α to x_0 to get a tangent vector y_0. This requires transporting y_1 to a vector $g(t)$ tangent to the surface at $\alpha(t)$ for all intermediate t. But then $g(t)$ is in the tangent space of the manifold at $\alpha(t)$ and will be of the form

$$g(t) = \sum g_i(t) \frac{\partial}{\partial x_i}. \tag{15}$$

The condition that $g(t)$ be parallel along α is that the covariant derivative be zero along that path. Omitting reference to α, this says that $g(t)$ satisfies :

$$0 = \nabla g = \sum_i \left[\frac{dg_i}{dt} \frac{\partial}{\partial x_i} + g_i \nabla \frac{\partial}{\partial x_i} \right] = \sum_{ijk} \left[\frac{dg_k}{dt} + g_i \frac{d\alpha_j}{dt} \Gamma_{ij}^k \right] \frac{\partial}{\partial x_k}. \tag{16}$$

Hence $g(t)$ is parallel along $\alpha(t)$ if and only if the coordinates functions $g_i(t)$ satisfy the following system of ordinary differential equations.

$$\frac{dg_k}{dt} + \sum_{ik} g_i \frac{d\alpha_j}{dt} \Gamma_{ij}^k = 0, \ k = 1 \ldots n. \tag{17}$$

Here the Γ_{ij}^k are the Christoffel symbols. These are computed by formulas from the chart for the manifold and, for the Finsler manifolds arising from parametric variational problems, from L and its derivatives which define the metric ground form.

When we impose the boundary condition that $g(t_1) = y_1$, the solution to the system of differential equations gives the desired y_0, namely, $g(t_0)$. Assume a global Lipschitz condition for the system of differential equations on the manifold. We can then apply the Picard method of successive approximations to compute g_i on the chart (U_i, ψ_i) from boundary condition $g(t_1) = y_1$. Thus we get a formula for computing $g(t)$ in general. If the curve extends over several charts, the uniqueness of solutions of Lipschitz condition ordinary differential equations plus the smoothness of the transition overlaps between charts implies that we can compute a formula for $g(t)$ from chart to chart and hence compute $y_0 = g(t_0)$.

Fields of Extremals.

Next we shall briefly discuss the relation between Euler-Lagrange equations and Hamilton-Jacobi equations in the non-parametric context with t present.

Given smooth f_i such that

$$\ddot{x}_i(t) = f_i(t, x_1(t), \ldots, x_n(t), \dot{x}_1(t), \ldots, \dot{x}_n(t)), \ i = 1, \ldots, n, \tag{18}$$

the vector field

$$\dot{x}_i(t) = \phi_i(t, x_1(t), \ldots, x_n(t)), \; i = 1, \ldots, n \tag{19}$$

necessarily satisfies the system

$$\frac{\partial \phi_i}{\partial t} + \sum_{j=1}^{n} \frac{\partial \phi_i}{\partial x} \cdot \phi_j = f_i(t, x_1(t), \ldots, x_n(t), \dot{x}_{1,}(t), \ldots, \dot{x}_n(t)), \; i = 1, \ldots, n. \tag{20}$$

To see this, differentiate (19) with respect to t,

$$\ddot{x}_i(t) = \frac{\partial \phi_i}{\partial t} + \sum_{j=1}^{n} \frac{\partial \phi_i}{\partial x_j} \cdot \frac{dx_j}{dt} \tag{21}$$

and use (18) to conclude that

$$f_i(t, x_1(t), \ldots, x_n(t)) = \frac{\partial \phi_i}{\partial t} + \sum_{j=1}^{n} \frac{\partial \phi_i}{\partial x} \cdot \phi_j. \tag{22}$$

Generalized Momentum and the Hamiltonian.

We give a computational discussion of what are essentially the Euler-Lagrange and Hamilton-Jacobi formulations in the non-parametric case, that is, with time as an explicit variable, because this fits with most textbook treatments. We return to the parametric Finsler case at the end of the secion.

The Euler-Lagrange conditions satisfied by a local extremum for $L(x, \dot{x})$ then takes the form:

$$\frac{d}{dt} L_{\dot{x}_i} - L_{x_i} = 0, \; i = 1, \ldots, n. \tag{23}$$

Usually, instead of n second order Euler-Lagrange equations, an equivalent canonical system of $2n$ first order equations is introduced. This system is written with respect to a variable x, a new variable p representing generalized momenta, and a new function $H(t, x, p)$ of these variables. This function is called Hamiltonian. The solutions of the canonical system are also called extremals, although values of solutions to this system are $2n$-dimensional vectors rather than the n-dimensional vectors of the Euler-Lagrange solutions. The Euler-Lagrange solutions are are the projections of the canonical solutions onto x-coordinates only.

The canonical solutions can be described by means of a particular partial differential equation which is called the Hamilton-Jacobi equation in mechanics and variational calculus and the Hamilton-Jacobi-Bellman equation in operations research and dynamic programming. The Hamilton-Jacobi equation is constructed using the Hamiltonian. This is the approach used by Kohn in his original declarative control for a variety of practical problems. For us the Hamiltonian together with the Hamilton-Jacobi equation will be important since we will use it later to describe how to extract chattering controls which approximate to the connection induced by the metric ground form.

These same generalized momenta and Hamiltonian can be used for another purpose. They are a means of formulating conditions ensuring that a given system of $2n$ first order differential equations yields the field of extremals for a given system of n (second order) Euler-Lagrange equations. Fields of extremals are used to formulate a sufficient condition for an Euler-Lagrange extremal to be a local minimum of the corresponding functional. We refrain from formulating the latter conditions due to space constraints.

The generalized momenta are

$$p_i = L_{\dot{x}_i}(t, x_1, \ldots, x_n, \dot{x}_1, \ldots, \dot{x}_n). \tag{24}$$

The Hamiltonian is

$$H = -L(t, x_1, \ldots, x_n, \dot{x}_1, \ldots, \dot{x}_n) + \sum_{i=1}^{n} p_i(t, x_1, \ldots, x_n, \dot{x}_1, \ldots \dot{x}_n)\dot{x}_i. \tag{25}$$

If we let

$$x = (x_1, \ldots, x_n) \text{ and} \tag{26}$$

$$\phi(t, x) = (\phi_1(t, x), \ldots, \phi_n(t, x)), \tag{27}$$

then the usual necessary and sufficient condition that (19) be a field of extremals for the variational problem corresponding to L is that

$$\frac{\partial p_i}{\partial x_j}(t, x, \phi(t, x)) = \frac{\partial p_j}{\partial x_i}(t, x, \phi(t, x)). \tag{28}$$

$$-\frac{\partial p_i}{\partial t}(t, x, \phi(t, x)) = \frac{\partial}{\partial x_i} H(t, x, \phi(t, x)). \tag{29}$$

Substitute (28) and (29) into (18) and simplify to get

$$L_{x_i} - \frac{d}{dt} L_{\dot{x}_i} = 0, \; i = 1, \ldots, n. \tag{30}$$

where p_i are the generalized momenta as defined by (24) and H is the Hamiltonian as defined by (25).

These are the Euler-Lagrange equations. Conversely, by starting with the Euler-Lagrange equations, we can get (28) and (29).

Remark. The Hamilton-Jacobi-Bellman equation for L is.

$$\frac{\partial S}{\partial t} + H(t, x, \frac{\partial S}{\partial x_1}, \ldots, \frac{\partial S}{\partial x_n}) = 0. \tag{31}$$

The general solution of this equation depends on n parameters. The partial derivatives of the solution with respect to the parameters gives the first integral of the canonical Euler equations referred to earlier. Note that the canonical equations are written in terms of generalized momenta and the Hamiltonian as

$$\dot{x}_i = \frac{\partial H}{\partial p_i} \tag{32}$$

$$\dot{p}_i = -\frac{\partial H}{\partial x_i} \tag{33}$$

For parametric problems and Finsler manifolds, the appropriate definition of generalized momenta and the Hamiltonion are respectively

$$p_i = \sum_j g_{ij}(x, \dot{x})\dot{x}_j, \tag{34}$$

$$H^2(x,p) = \sum_{ij} g^{ij}(x,p)p_i p_j. \tag{35}$$

where the matrix (g^{ij}) is the inverse transpose of (g_{ij}).

Then the generalized canonical equations take the form

$$\frac{\partial L(x,\dot{x})}{\partial x_i} = -\frac{\partial H(x,p)}{\partial x_i}, \tag{36}$$

$$p_i = \frac{\partial L}{\partial \dot{x}_i}, \text{ and} \tag{37}$$

$$\dot{x}_i = \frac{\partial H}{\partial p_i}. \tag{38}$$

For fixed x on the manifold, the p such that $H(x,p) = 1$ are called the vectors of *Hamiltonian* length 1, and define a norm on the tangent space at x based on coordinates (x,p). This unit sphere was named the figuratrix by Hadamard ([H10]). In the Hamilton-Jacobi formulation of the calculus of variations, many problems reduce to determining the surface of this sphere and how that surface changes as x varies over the manifold.

The beauty of Finsler spaces is that there is a duality between the indicatrix and the figuratrix. This duality, stemming from one of Minkowski, says that either sphere can be interpreted as the space of supporting hyperplanes to the other. We have made no effort to use the appropriate tensorial subscripts and superscripts to exhibit this, prefering to keep to elementary notation, see ([R59]).

8 Chattering

In our architecture, each infinitesimal control action ordered is implemented as a vector field on M. A control order is a sequence of such actions, resulting in a sequence of flows, each used in sequence for a prescribed time. We outline how we can approximate flows using a finite basis of infinitesimal control actions. A schedule of applying three successive flows in three successive subintervals which partition an intervals of length Δ interval looks like this.

$$\Psi(\tau,x) = \begin{cases} \Psi_{v_{n_1}}(\tau,x) \\ \quad \text{if } t \le \tau < t + \Delta_{n_1} \\ \Psi_{v_{n_2}}(\tau, \Psi_{v_{n_1}}(t+\Delta_{n_1}, x)) \\ \quad \text{if } t + \Delta_{n_1} \le \tau < t + \Delta_{n_1} + \Delta_{n_2} \\ \Psi_{v_{n_3}}(\tau, \Psi_{v_{n_2}}(t+\Delta_{n_1}+\Delta_{n_2}, \Psi_{v_{n_1}}(t+\Delta_{n_1}, x))) \\ \quad \text{if } t + \Delta_{n_1} + \Delta_{n_2} \le \tau < t + \Delta_{n_1} + \Delta_{n_2} + \Delta_{n_3} \end{cases} \tag{39}$$

with $\Delta_{n_1} + \Delta_{n_2} + \Delta_{n_1} = \Delta$.

We note that the flow Ψ given by (39) characterizes the evolution of the process. The vector field $v\mid_x$ associated with the flow Ψ is obtained by differentiation and the third identity in (13). This vector field applied at x is proportional to

$$v\mid_x = [v_{n_1}, [v_{n_2}, v_{n_3}]] \tag{40}$$

with second order error in the Lie-Taylor series. Here $[.,.]$ is Lie bracket. (The Lie bracket is defined as follows. Let v, w, be derivations on M and let f be any real valued smooth function $f : M \rightarrow R$. The Lie bracket of v and w is the derivation defined by $[v,w](f) = v(w(f)) - w(v(f))$).

Thus the control order $v\mid_x$ generated to control the process is a sequence of primitive actions of the form of (40). Moreover, from an appropriate version of the chattering theorem, ([K93]), we can show that we can choose appropriate primitive actions so that this action is essentially a *linear combination* of the primitive actions:

$$[v_{n_1}, [v_{n_2}, v_{n_3}]] = \sum_j \gamma_j(\alpha) v_j \tag{41}$$

$$\sum_j \gamma_j(\alpha) = 1$$

with the coefficients γ_j determined by the fraction of time $\beta_j(t,x)$ that each primitive action v_j is used. The composite action $v\mid_x$ and the vector of fraction functions β are inferred from an approximate solution of an optimization problem.

The effect of the field defined by the composite action $v\mid_x$ on any smooth function (equivalence function class) is computed by expanding the right hand side of (39) in a Lie-Taylor series ([K88]).

As outlined above, our optimization methods are for a Lagrangian L. This is a continuous, or in many variational problems only upper semicontinuous, function

$$L : T \times M \rightarrow R^+ \tag{42}$$

where T is time and R^+ is the positive real line.

The intention is that the Lagrangian is chosen to reflect by its values how well the agent is performing at time t. The smaller the non-negative value of L, the better the performance. This is characteristic of our use of Lagrangian optimization methods.

Specifically, we need to have two items: a *generator* for the Lagrangian $L(t,x)$ at time t, and the action issued. In summary, primitive control actions are implemented by infinitesimal generators of flows on M. Control orders are finite sequences of primitive control actions, each to be executed for a prescribed fraction of an interval.

The general structure of the Lagrangian in (42) at time t is given by

$$L(t,x) = F(C^u, \alpha)(t,x) \tag{43}$$

where F is smooth, C^u is a behavior modification function, and α is the command action issued.

We then express the change in the behavior modification function C^u due to the flow Ψ over the interval Δ in terms of $v\mid_x$. The evolution of the modified behavior function C^u over the interval $[t, t+\Delta)$ starting at point x is given by

$$C^u(t+\Delta, x) = C^u(t, \Psi(t+\Delta, x)) \tag{44}$$

Expanding the right hand side of (44) in a Lie-Taylor series around (t, x), we obtain

$$C^u(t+\Delta, x) = \sum_j \frac{(v\mid_x (C^u(t,x)))^j \cdot \Delta^j}{j!} \tag{45}$$

with

$$(v\mid_x (\cdot))^j = v\mid_x ((v\mid_x)^{j-1})$$

and

$$(v\mid_x)(\cdot)^0 = \text{ Identity operator.}$$

In general, the series on the righthand side of (45) will have infinitely many non-zero terms. However in the MACHA architecture, we work realtive to a finite subtopology (the small topology) of the topology of M generated by the knowledge base and certain logic clauses which is very crude and maintains behavior. Becuase of this we need only compute a finite number of non-zero terms. That is, we compute those powers of derivations which are required to distinguish between points which the small topology can distinguish. This is important because it allows the series on the right hand side of (45) to be generated by a locally finite automaton.

Remarks.

1) Since actuators are fixed physical devices, we should have the *same* set of primitive infinitesimal control actions ϕ^i to govern the actuators at all times and all points on the manifold. We formalize this by taking the set of primitive infinitesimal control actions available to be a set of global vector fields on the manifold, that is, flows. The term "infinitesimal control action" is derived from usage in dynamics and differential equations. The computations of flows to be used for control are generally infinitesimal in the traditional sense that they are Lie Algebra computations in the algebra of derivations corresponding to the flows. They are "local" computations uniform for all points and tangent spaces at those points. For instance, as described above, a common computation is to approximate to linear combinations of actions by iterated Lie brackets:

$$v = \alpha_1 v_{j_1} + \alpha_2 v_{j_2} \ldots \simeq [\alpha_1 v_{j_1}, [\alpha_2 v_{j_2}, \ldots] \ldots]$$

2) A (finite) control action ϕ is a finite sequence $(\phi^{i_1}, \delta_1), ..., (\phi^{i_m}, \delta_m)$ of primitive control actions ϕ^{i_k} applied for time δ_k in that order, each acting for a specified time before being replaced by the next, for a total elapsed time of

$\Delta = \sum_{k=1}^{m} \delta_k$. In terms of flows, this means that in control action ϕ, flow ϕ^{i_k} is allowed to flow for time δ_k after flow $\phi^{i_{k-1}}$ and before flow $\phi^{i_{k+1}}$.

3) If we desire to build a hybrid system which uses a finite state controller with finite output alphabet and all real time digital control programs have this property, then the set of primitive infinitesimal control actions available should be a finite set of vector fields on the manifold.

4) It is common to assume that the implementable flows are a Lie cone, closed under positive multiples and sums. But the flows chosen as primitive control actions and the length of times the flows will be assigned to evolve must be physically realizable by the system. Thus for a car, the primitive infinitesimal actions may be "steer" and "drive", but for physical realizability, limits on rates and magnitudes of these flows must be observed, which means one allows in practice rather less than a Lie Cone of controls and one has to check that controls derived in fact are within the performance limits of the physical system.

In a variational problem why should there be *any* finite set of control actions and finite control automata that guarantee ϵ-optimality? We shall see that they exist when there are constructive approximations to optimal solutions. Optimal solutions themselves often exist only in an extended sense and are generally not at all constructive.

We can arrange the construction of ([GKNR95],[KNRG95]) so that the extrema (local minima plus other extreme points) and convex combination coefficients for the convexification L^* mentioned earlier are continuous in the parameters involved. Then when we do parallel transport of a tangent vector y_1 at x_1 to get a tangent vector y_0 at x_0, we have y_0 continuous in x_0. Since, in using the connection to get the tangent direction of the appropriate geodesic at every point, we are simply solving a second order differential equation with a boundary condition on one end, this is easy to guarantee. Note that we also generally get continuity in the goal, since the goal is merely the initial condition for the differential equation. This will result in a policy $P(x, x_1, y_1)$ which is continuous and hence has a closed graph.

The construction in Appendix II of ([KN93a]), also in ([KNRY95]), will convert a suitable finite open cover of that closed graph into a finite state control automaton, with associated analog to digital and digital to analog converters, which approximates the control policy and enforces ϵ-optimality. In this model, x is converted by an analog to digital converter to an automaton state, x_1, y_1 are converted by an analog to digital converter to input symbols for the automaton. The automaton output is an output symbol converted by a digital to analog converter to an appropriate tangent vector y_0. The finite automaton is thus an "approximate" connection, at least along geodesics.

We note that such an automaton is generally not a discrete object, even though its input alphabet, output alphabet, and states are finite sets. These automata come equipped with finite, usually non-discrete, topologies on input, state, and output alphabets, which we need to verify controllability and observability of the resulting hybrid system.

Chattering Approximation to Parallel Transport.

We need only one more thing in order to compute the parallel transport required to guarantee ϵ-optimality. Given a convexified $\int L^*(t,x,\dot{x})dt$ with generalized momenta

$$p(t,x,\dot{x}) = F_{x_i}(t,x,\dot{x}), \tag{46}$$

let $g(t,x))$ be the parallel transport of goal tangent vector y_1 at x along an extremal. Then using the Hamilton-Bellman-Jacobi equation,

$$p_i(t,x,\phi) = g_{x_i}(t,x) \tag{47}$$

for $i = 1,\dots n$. Then

$$\frac{\partial g}{\partial t} = -H(t,x,\frac{\partial g}{\partial x_1},\dots,\frac{\partial g}{\partial x_n}). \tag{48}$$

is the Hamilton-Jacobi-Bellman equation for the goal g. We can then use the convex analysis of ([KNRG95], [GKNR95]) to represent an absolute minimum extremum for the convexified problem L^* as a convex combination of $\sum_{i=1}^m \alpha_i\phi^{(i)}$ of local extrema $\phi^{(i)}$ for the original problem L. Due to (29), these local extrema satisfy

$$-\frac{\partial p_i}{\partial t} = \min_{\underline{\alpha}}\frac{\partial H}{\partial x_i}(t,x,\sum_{k=1}^m \alpha_k\phi^{(k)}(t,x)) \tag{49}$$

for $i = 1,\dots,n$. These local extrema $\phi^{(i)}$ can be thought of as directions $\dot{x}$ in the tangent space. Hence they define infinitesimal control actions, or vector fields $\phi^{(1)},\dots,\phi^{(m)}$. The chattering theorem implies that an ϵ-optimal control field can be obtained from the convex combination $\sum_{i=1}^m \alpha_i\phi^{(i)}$ by chattering the $\phi^{(i)}$ in an appropriate fashion. To be a little more precise, an application of the chattering theorem shows that, for any ϵ, there is a Δ such that if we divide time into successive intervals of length Δ, we can bring $\int L(t,x,\dot{x})dt$ within ϵ of its absolute extremum by performing over each such interval Δ the following list of control actions (in any order) for the relative times indicated.

$$\phi(t,x) = \begin{pmatrix} \phi^{(1)} & \text{for time} & \delta_1 = \alpha_1\Delta \\ \phi^{(2)} & \text{for time} & \delta_2 = \alpha_2\Delta \\ \vdots & \vdots & \vdots \\ \vdots & \vdots & \vdots \\ \phi^{(m)} & \text{for time} & \delta_m = \alpha_m\Delta \end{pmatrix}$$

Here are some problems which arise in applications.

1. If, due to unmodelled disturbances, we get "off course", what policy do we follow? We follow the same policy. This is like Bellman's "Principle of Optimality".

2. What if we discover, en route, that due to sensed discrepencies, either the definition of the manifold or of the Lagrangian or of the goal is inappropriate. Then we use any such new information to formulate a new manifold, a new Lagrangian, and a new goal, and compute and use a new control policy. This is the form of adaptation needed, which we do in our architecture on line. We would hope we had done in advance adequate simulation and analysis of the dynamics of the environment and plant so that control policies exist and can be extracted to restore performance specifications, perhaps with lowered goal expectations. But there is no guarantee that modelling has covered all factors adequately, and no guarantee that adaptation will work in unforseen circumstances.

3. What happens if the goal direction $y_1 = 0$? A reasonable answer is that in practice this is an uninteresting case. Engineers are only interested in ϵ-optimal control. We can assume we have continuity of the total cost along geodesics as a function of the goal, else the problem is not worth trying to solve as formulated. Thus we need only look at a goal with y_1 small and non-zero. Since y_1 is non-zero, we can parallel transport to get y_0.

4. This point of view also shows how to handle the Bellman principle of optimality in the parametric calculus of variations on manifolds. That is, Finsler spaces have a duality theory ([R66], p. 19). One can then use the Hamiltonian and Hamilton-Jacobi equations there.

9 The Modern View on Connections

We should also touch on the contemporary treatment of affine connections, originally due to J. L. Kozul. It defines a connection on M as a map $\nabla_X Y$ from vector fields X, Y on the manifold to a vector field $Z = \nabla_X Y$ which is bilinear and covariant. That is, for any vector fields X, Y, X_1, X_2, Y_1, Y_2, any real valued scalar function f on the manifold, we assume that ∇ satisfies the following two properties.

Bilinearity

$$\nabla_{(X_1+X_2)}(Y_1 + Y_2) = \nabla_{X_1} Y_1 + \nabla_{X_2} Y_2 + \nabla_{X_1} Y_2 + \nabla_{X_2} Y_1$$

$$\nabla_{fX} Y = f \nabla_X Y$$

Covariance

$$\nabla_X (fY) = X(f) Y + f \nabla_X Y$$

where f is any scalar, that is, a real valued function on the manifold M.

What is the scalar $X(f)$ in the definition of covariance? A vector field X on the manifold is the direction field of an ordinary differential equation $\dot{u}(t) = X(u(t))$ which integrates to a flow (integral curve) $\Phi(t) : M \to M$. Thus

$(\Phi(t))(x)$ is the value $u(t)$ of the solution to the differential equation at time t with initial condition $u(0) = x$. The action of a vector field X on a scalar f (real valued functions on M) is then given by defining the scalar $X(f)$ for x on M by

$$X(f)(x) = \lim_{t \to 0} \frac{f(\Phi(t)(x)) - f(x)}{t}. \tag{50}$$

This is the directional derivative of f along X. The covariance formula is called a law of covariant differentiation. $\nabla_X Y$ is called covariant derivative of Y by X.

Relative to coordinates on a chart, the bilinearity makes each $\nabla_{\frac{\partial}{\partial x_i}}(\frac{\partial}{\partial x_j})$ a linear combination $\sum_{k=1}^n \Gamma_{ijk} \frac{\partial}{\partial x_k}$ of the basis $\frac{\partial}{\partial x_1}, \ldots, \frac{\partial}{\partial x_n}$ at any given point. Here the linear combination coefficients Γ_{ijk} are Christoffel symbols.

The covariance property implies that if we are given sums

$$X = \sum_i A_i \frac{\partial}{\partial x_i} \text{ and } Y = \sum_j B_j \frac{\partial}{\partial x_j},$$

then

$$\nabla_X Y = \sum_{ijk} (A_i \frac{\partial B_k}{\partial x_i} + A_i B_j \Gamma_{ijk}) \frac{\partial}{\partial x_k}$$

This shows how Christoffel symbols determine the connection. The action of connections on curves, which is what we have discussed with respect to control, is the easily described ([H77]).

Let $\alpha(t) : [t_0, t_1] \to M$ be a curve. A vector field V along curve α assigns a tangent vector $V(t)$ at each point $\alpha(t)$. Then an affine connection corresponds to each vector field V on α, a new vector field ∇V on α such that

1. For a smooth real valued f on $[t_0, t_1]$,

$$\nabla(f(t)V(t)) = \frac{df}{dt} V(t) + f(t) \nabla V(t)$$

2. If V_1, V_2 are vector fields along α, then

$$\nabla V_1 + \nabla V_2 = \nabla(V_1 + V_2)$$

3. If y_1 is a tangent vector at $\alpha(t_1) = x_1$, then there is a unique vector field V along α with $V(t_1) = y_1$ and $\nabla V = 0$. For each $t \in [t_0, t_1]$, $y_1 \to y_t = V(t)$, as y_1 ranges over the tangent space at $\alpha(t_1)$, is a vector space isomorphism from the tangent space at $\alpha(t_1) = x_1$ to the tangent space at $\alpha(t)$. The map $y_1 \to y_t$ is called the parallel transport of tangent vector y_1 to y_t along α.

The relation between this ∇ and the connection $\nabla_X Y$ is that if $\beta(t)$ is a curve and X, Y are vector fields and $\dot{x} = X(\beta(t))$, $V(t) = Y(\beta(t))$, then $\nabla V(t) = \nabla_X Y(\beta(t))$.

We note that for a given α, the existence of vector field $V(t)$ along α follows from the existence theorem for solutions to a system of homogeneous linear first order differential equations. To see this, let $\alpha(t) = (\alpha_1(t), \ldots, \alpha_n(t))$ in local coordinates and write $V(t)$ in local coordinates as

$$V(t) = \sum_{i=1}^{n} \frac{\partial}{\partial \alpha_i}(\alpha(t)), \tag{51}$$

Then the condition that $\nabla V(t) = 0$, when written in local coordinates, becomes

$$\frac{d}{dt}V_k(t) + \frac{d\alpha_i}{dt} \sum_{i,j=1}^{n} V_j(t) \Gamma_{ijk}(\alpha(t)) = 0. \tag{52}$$

It follows that if α is to be self-parallel, i.e if V is to be the direction field $\frac{d\alpha}{dt}$ along α, then we must get $\nabla \frac{d\alpha}{dt} = 0$. This makes the equations of autoparallelism non-linear second order ordinary differential equations for α. Without additional assumptions, these second order differential equations only have local solutions. Of course geodesics satisfy these equations when a metric is present.

10 Appendix I: Applications

As mentioned in section 3, we have developed a multiple agent distributed hybrid control architecture (MAHCA) for extracting control programs for systems of cooperating agents controlling distributed hybrid systems based on distributed Lagrangians and relaxed variational calculus. See ([KN92], [NY92], [NY93a], [NY93b], [NY93c], [KN93a], [KN93b], [KN93c], [KNR93], [NRY93a], [NRY93b], [KNRY95], [GKNR95], and [KNRG95].) We have not been able to incorporate the multiple agent model carrier manifolds of ([KNRG95]) in this paper.

We have shown how to formulate multiple cooperating autonomous software agents controlling a hybrid system and meeting dynamical and logical constraints and goals as a single distributed relaxed variational problem. We have introduced algorithms for solving the latter problem for the required controls. These algorithms stem from the measure valued calculus of variations, Bellman's dynamic programming, the chattering lemma, the Eilenberg-Schutzenberger theory of linear automaton equations, and the Lie theory of cones of controls in Lie algebras.

Specific applications are being developed for a variety of areas such as national distributed multi-media systems for interactive video-audio-text on demand, for manufacturing processes, for virtual enterprises, for distributed interactive simulation, for traffic control, for wireless-broadband communications networks, for battlefield command and control, for medical information systems, for medical diagnosis and health care delivery systems, for computer aided control system design, for distributed cost-benefit analysis, and very recently for compression-decompression algorithms for high definition video and video conferencing. See ([NJK94], [NJK94a], [NJK94b], [NJKD94], [NJKD94b], [NJKHA94], [LGKNC94], [CJKNRS95]).

Models have been implemented and simulated for traffic control, multiple cell factory floors, producer consumer games, and by the Kohn-Nerode group in cooperation with the U. S. Army Picatinny Arsenal for a simulation of multiple agent control of cooperating tanks.

11 Appendix II: Background

For those who would like to read more about the links between control theory, Finsler Manifolds, and connections, we provide a brief list of references that we have found useful. Here is suggested background reading.

A) Calculus of variations ([A62], [H77], [E83], [GF63], [KS80]).

B) Differential geometry and connections ([LC26], [S65] [BG88], [B87], [T83]).

C) Convexity in functional analysis ([ET76], [E90]).

D) Finsler spaces ([B90], [M86], [S42]).

E) Lie algebras and ordinary differential equations ([N67], [O86], [BK89]).

Acknowledgements.
We would like to acknowledge many valuable conversations with Dr. Alexander Yakhnis and Dr. John James.

References

[A62] Akhiezer, N. I.: *The calculus of variations.* Blaisdell. (1962)
[B90] Bejancu, A.: *Finsler geometry and applications.* Harwood. (1990)
[BG88] Berger M., Gostiaux, B.: *Differential geometry : manifolds, curves, surfaces.* Springer Verlag. (1988)
[BK89] Bluman, G. W., Kumei, S.: *Symmetries and differential equations.* Springer-Verlag. (1989)
[BMS83] Brockett, R., Millman, R.S., Sussman, H.J. (eds): *Differential geometric control theory.* Birkhäuser. (1983)
[B87] Burke, W.: *Applied differential geometry.* Cambridge University Press. (1987)
[C04] Caratheodory, C.: Uber die discontinuierlichen Lösungen in der Variationsrechnung, Diss., Göttingen, 71pp. (1904)
[C35] Caratheodory, C.: *Variationsrechnung und Partialle Differentialgleichcnngenerster Ordnung.* Teubner, Berlin. (1935) (Dover 2nd English edition, 1982)
[C34] Cartan, E.: *Les Espaces de Finsler.* Actualities scientifiques et industrielle 79, Exposes de geometrie II (1934)
[CP86] Crampin, M., Pirini, F.A.E.: *Applicable Differential geometry.* Cambridge University Press. (1986)
[E83] Ekeland, I.: *Infinite dimensional optimization and convexity.* University of Chicago Press. (1983)
[E90] Ekeland, I.: *Convexity methods in Hamiltonian mechanics.* Springer-Verlag. (1990)
[ET76] Ekeland, I., Temam, R.: *Convex analysis and variational problems.* North-Holland. (1976)

[F18] Finsler, P.: *Über Kurven und Flächen in allegmeinen Räumen*. Diss. Göttingen (1918) (republished by Verlag Birkhäuser Basel, 1951)

[F75] Friedman, A.: *Stochastic Differential Equations I, II*. Academic Press. (1975)

[CJKNRS95] Cummings, B., James, J., Kohn, W., Nerode, A., Remmel, J.B., Shell, K.: Distributed MAHCA Cost-Benefit Analysis. MSI Tech. Report, Cornell University. (1995)

[GKNR95] Ge, X., Kohn, W., Nerode, A., Remmel, J.B.: Algorithms for Chattering Approximations to Relaxed Optimal Control. MSI Tech. Report 95-1, Cornell University. (1995)

[GF63] Gelfand, I. M., Fomin, S. V.: *Calculus of variations*. Prentice-Hall. 1963.

[G83] Griffiths, P. A.: *Exterior differential systems and the calculus of variations*. Birkhäuser. (1983)

[G92] Grossman, R.L.: Managing persistent object stores of predicates. Oak Park Research Technical Report, Number 92-R1, May, 1992. (1992)

[GNRR93] Grossman, R.L., Nerode, A., Ravn, A., Rischel, H. (eds.): *Hybrid Systems*. Springer Lecture Notes in Computer Science, Springer-Verlag, Bonn. (1993)

[GVQ93] Grossman, R.L., Valsamis, D., Qin, X.: Persistent stores and hybrid systems. Proceedings of the 32st IEEE Conference on Decision and Control, IEEE Press. (1993) 2298–2302

[GL95] Grossman, R.L., Larson, R.G.: An algebraic approach to hybrid systems. Journal of Theoretical Computer Science. (to appear)

[GN95] Grossman, R.L., Nerode, A.: Quantum automata and hybrid systems. (in preparation)

[H10] Hadamard, J., Leçons sur le calcul des variations. Hermann et Fils, Paris. (1910)

[H73] Hermann, R.: *Geometry, Physics, and Systems*. Marcel Dekker. (1973)

[H75] Hermann, R.: *Ricci-Levi Civita's paper on Tensor Analysis*. Math. Sciences Press, Brookline, Mass. (1975)

[H77] Hermann, R.: *Differential geometry and the calculus of variations*. Math. Sci. Press. (1977)

[H05] Hilbert, D.: Über das Dirichletsche Prinzip, J. reine angew. Math. Bd. **129** (1905) 63–67

[KS80] Kinderlehrer, D., Stampacchia, G. : *An introduction to variational inequalities*. Academic Press. (1983)

[K88] Kohn, W.: A declarative theory for rational controllers. Proceedings of the 27th IEEE CDC, Vol. 1. (1988) 131–136.

[K90] Kohn, W.: Declarative hierarchical controllers. Proc. of the Workshop on Software Tools for Distributed Intelligent Control Systems, Pacifica, CA, July 17-19 (1990) 141–163.

[K93] Kohn, W.: Multiple agent inference in equational domains via Infinitesimal operators. Proc. of Application Specific Symbolic Techniques in High Performance Computing Environments, The Fields Institute, Oct. 17-20 (1993)

[KN92] Kohn, W., Nerode, A.: An autonomous control theory: an overview. IEEE Symposium on Computer Aided Control System Design (March 17–19, 1992, Napa Valley, CA) (1992) 204–210

[KN93a] Kohn, W., Nerode, A.: Models for hybrid systems: automata, topologies, controllability and observability. in [GNRR93] (1993)

[KN93b] Kohn, W., Nerode, A.: Multiple agent autonomous control. Proceedings of the 31st IEEE CDC. (1993) 2956–2966

[KN93c] Kohn, W., Nerode, A.: Multiple agent autonomous hybrid control systems. *Logical Methods* (Crossley,J., Remmel, J.B., Shore, R., Sweeder, M. eds.), Birkhauser. (1993)

[KNR93] Kohn, W., Nerode, A., Remmel, J.B.: Agents in hybrid control. MSI Technical Report 93-101, Cornell University. (1993)

[KNRG95] Kohn, W., Nerode, A., Remmel, J.B., Ge, X.: Multiple agent hybrid control: carrier manifolds and chattering approximations to optimal control. CDC94 (1994)

[KNRY95] Kohn, W., Nerode, A., Remmel, J.B., Yakhnis, A.: Viability in hybrid systems. J. Theoretical Computer Science. **138** (1995) 141-168

[KNS95] Kohn, W., Nerode, A., Subrhamanian, V.S.: Constraint logic programming: hybrid control and logic as linear programming. MSI Technical Report 93-80, Cornell University. (1993) (to appear in CLP93 1995)

[LC17] Levi-Civita, T.: Rendiconti del Circolo Matematico di Palermo, fascicolo XLII. (1917)

[LC26] Levi-Civita, T.: *The Absolute Differential Calculus.* Blackie and Sons. (1926) (Dover Reprint. 1977)

[LNRS93] Lu, J., Nerode, A., Remmel, J.B., Subrahmanian, V.S.: Toward a theory of hybrid knowledge bases. MSI Technical Report 93-14, Cornell University. (1993)

[LGKNC94] Lu, J., Ge, X., Kohn, W., Nerode, A., Coleman, N.: A Semi-autonomous multiagent decision model for a battlefield environment. MSI Technical Report, Oct. 1994, Cornell University. (1994)

[M93] Marsden, J. E.: Some Remarks on Geometric Mechanics. Report, Department of Mathematics, the University of California, Berkeley. (1993)

[MRWZ93] Marsden, J. E., O'Reilly, O.M., Wicklin, F.J., Zombro, B.W.: Symmetry, Stability, Geometric Phases, and Mechanical Integrators. Part I: Non-Linear Science Today, Vol. 1, no. 1, (1991) 4-11: Part II: Non-Linear Science Today, Vol. 1, no. 2, (1991) 14-21

[M86] Matsumoto, M.: *Foundations of Finsler geometry and special Finsler spaces.* Kaiseisha Press, Shigkan. (1986)

[N67] Nelson, E.: *Tensor Analysis* Princeton University Press. (1967)

[NJK92] Nerode, A., James, J., Kohn, W.: Multiple agent declarative control architecture: A knowledge based system for reactive planning, scheduling and control in manufacturing systems. Intermetrics Report, Intermetrics, Bellevue, Wash., Nov. (1992)

[NJK94] Nerode, A., James, J., Kohn, W.: Multiple Agent Hybrid Control Architecture: A generic open architecture for incremental construction of reactive planning and scheduling. Intermetrics Report, Intermetrics, Bellevue, Wash., June (1994)

[NJK94a] Nerode, A., James, J., Kohn, W.: Multiple agent reactive control of distributed interactive simulations. Proc. Army Workshop on Hybrid Systems and Distributed Simulation, Feb. 28-March 1, (1994)

[NJK94b] Nerode, A., James, J., Kohn, W.: Multiple agent reactive control of wireless distributed multimedia communications networks for the digital battlefield. Intermetrics Report, Intermetrics, Bellevue, Wash., June (1994)

[NJKD94] Nerode, A., James, J., Kohn, W., DeClaris, N.: Intelligent integration of medical models. Proc. IEEE Conference on Systems, Man, and Cybernetics, San Antonio, 1-6 Oct. (1994)

[NJKD94b] Nerode, A., James, J., Kohn, W., DeClaris, N.: Medical information systems via high performance computing and communications. Proc. IEEE Biomedical Engineering Symposium, Baltimore, MD, Nov. (1994)

[NJKHA94] Nerode, A., James, J., Kohn, W., Harbison, K., Agrawala, A.: A hybrid systems approach to computer aided control system design. Proc. Joint Symposium on Computer Aided Control System Design. Tucson AZ 7-9 March (1994)

[NLS95] Nerode, A., Lu, J., Subrahmanian, V.S.: Hybrid knowledge bases. IEEE Trans. on Knowledge and Data Engineering. (to appear)

[NRY93a] Nerode, A., Remmel, J.B., Yakhnis, A.: Hybrid system games: Extraction of control automata and small topologies. MSI Technical Report 93-102, Cornell University. (1993)

[NRY93b] Nerode, A., Remmel, J.B., Yakhnis, A.: Hybrid systems and continuous Sensing Games. 9th IEEE Conference on Intelligent Control, August 25-27, (1993)

[NG95] Nerode, A.,Ge, X.: Effective Content of Relaxed Calculus of Variations: Semicontinuity, MSI Tech. Report, Feb. 1995, Cornell University. (1995)

[NY92] Nerode, A., Yakhnis, A.: Modelling hybrid systems as games. CDC92. (1992) 2947-2952

[NY93a] Nerode, A., Yakhnis, A.: Control automata and fixed points of set-valued operators for discrete sensing hybrid systems. MSI Technical Report 93-105, Cornell University. (1993)

[NY93b] Nerode, A., Yakhnis, A.: An example of extraction of a finite control automaton and A. Nerode's AD-converter for a discrete sensing hybrid system. MSI Technical Report 93-104, Cornell University. (1993)

[NY93c] Nerode, A., Yakhnis, A.: Hybrid games and hybrid systems. MSI Technical Report 93-77, Cornell University. (1993)

[O86] Olver, P. J.: *Applications of Lie groups to differential equations.* Springer-Verlag. (1986)

[O66] O'Neill, Barrett: *Elementary Differential Geometry.* Academic Press. (1966)

[R1854] Riemann, B.: *Über die Hypothesen, welche der Geometriezugrunde liegen.* (1854)

[R59] Rund, H.: *The Differential Geometry of Finsler Spaces.* Springer. (1959)

[R66] Rund, H.: *The Hamilton Jacobi Theory.* Van Nostrand. (1966)

[S65] Spivak, M.: Calculus on Manifolds. Benjamin. (1965)

[S42] Stokes, E.C.: Applications of the Covariant Derivative of Cartan in the Calculus of Variations. in *Contributions to the Calculus of Variations*, 1938-1941 (G A Bliss, L M Graves , M. R. Hestenes, W. T. Reid, eds.). The University of Chicago Press. (1942)

[T83] Torretti, R.: *Relativity and Geometry.* Pergamon. (1983)

[W72] Warga, J.: *Optimal Control of Differential and Functional Equations.* Academic Press. (1972)

[W30] Wren, F.L.: A New Theory of Parametric Problems in the Calculus of Variations. in *Contributions to the Calculus of Variations.* The University of Chicago Press. (1930)

[Y80] Young, L.C.: *Optimal Control Theory.* Chelsea Pub. Co. N.Y. (1980)

Constructing Hybrid Control Systems from Robust Linear Control Agents

Michael Lemmon *, Christopher Bett,
Peter Szymanski, and Panos Antsaklis

Department of Electrical Eng.
University of Notre Dame
Notre Dame, IN 46556

Abstract. Hybrid control systems (HCS) arise when the plant, a continuous state system (CSS), is controlled by a discrete event system (DES) controller. It is assumed that there exists a specification on the plant's desired symbolic behaviour. This specification describes how various operational modes of the controlled plant should fit together. One design problem involves determining an HCS that "realizes" the specified behaviour in a "stable" manner. In this paper, a solution is posed which involves finding a sequence of subgoals and controllers achieving the specified behaviour in a "logically" stable manner. The solution can therefore be viewed as a "logically" stable extension of the previously given symbolic specification on the plant's behaviour. This paper shows how such an extension can be constructed using multiple H^∞ control agents exhibiting robust stability. The problem formulation leads to a design problem which can be solved using existing H^∞ control techniques.

1 Introduction

Hybrid control systems (HCS) arise when a continuous-state (CSS) plant is controlled by a discrete event system (DES). While "hybrid control" has recently generated considerable interest in the research community [1] [2] [3] [4] [5] [6] [7] [8], it should be noted that these systems are common in practice. Examples of these systems are found in such applications as flexible manufacturing, chemical process control, electric power distribution, computer communication networks, and a host of other applications characterized by a high degree of complexity. Due to this complexity, the plants in these applications are often required to operate in a variety of modes whose underlying dynamics are structurally distinct. Hybrid controllers are then used to coordinate these various operational modes.

A good example of such systems is found in the aerospace industry. Spacecraft, such as commercial communication satellites, represent enormously complex systems having several operational modes. These satellites have their normal station keeping modes, modes associated with momentum unloading, and acquisition modes which are invoked when normal station keeping has been lost. No single feedback control system is capable of achieving these diverse control objectives, so a mechanism for

* The authors gratefully acknowledge the partial financial support of the National Science Foundation (MSS92-16559) and the Electric Power Research Institute (RP8030-06).

automatically switching between these different control modes needs to be implemented by the on-board computer. Switching between operational modes is usually initiated by the occurrence of specific events. For example, the loss of Earth pointing will force the satellite to initiate a sequence of actions leading to the reacquisition of the Earth limb. The logic coordinating these actions is a computer program which is conveniently modeled as a discrete event system. Integrating this DES supervisor with the satellite's control systems yields a hybrid control system.

As can be seen from the preceding example, a systems engineer usually has some prior knowledge of the system's various operational modes, their associated set-points, and the legal (or illegal) transitions between these various modes. This prior information represents a symbolic specification on the controlled plant's desired behaviour. The problem faced by the designer involves finding a "stable realization" of this symbolic specification. Turning to the spacecraft application described above, a standard behavioural specification would be to require normal stationkeeping after the vehicle loses Earth pointing. A *stable* realization of the specified behaviour would reliably reacquire the Earth limb regardless of the satellite's initial attitude or body rates. The problem faced by the designer would be to find a "strategy" for achieving this specified goal. This can be accomplished by decomposing the original specification into a set of subtasks whose combined solutions realize the specified transition. In the spacecraft application, for example, this might involve a sequence of steps in which body rates are first controlled and then the vehicle begins a search for the Earth limb.

The preceding discussion provides a clear picture of one type of HCS design problem. The HCS design problem involves finding an "extension" of a previously given behavioural specification which can be realized in a "stable" manner by the HCS. The precise meaning of "stable" will be discussed below and is found in [8]. In practice, this problem is solved by systems engineers whose prior experience provides considerable guidance in identifying realizable specifications. This approach, however, means that HCS design is highly specific to the application area. The recent interest in HCS design can be traced to a desire to automate the design process. This paper describes a systematic method to HCS design based on H^∞ control concepts. The method constructs a stable extension to the original behavioural specification by adding control agents to the HCS which exhibit robust stability to bounded perturbations of the plant structure.

The remainder of this paper is organized as follows. A brief description of the assumed HCS modeling framework [11] establishes some of the notational conventions used in this paper (section 2). The HCS design problem is formally stated in section 3. This section also provides a formal definition for the stability of specified behavioural transitions. Section 4 describes the design method proposed in this paper. This paper's results and formalisms are then discussed (section 5) in the context of previous HCS design frameworks [12].

2 Hybrid Control Systems

This section describes a hybrid control system (HCS) modeling framework which has been extensively used by our group [11]. The hybrid control system consists of three subsystems; the *plant*, the *controller*, and the *interface*. Figure 1 shows how these subsystems are interconnected. Each of these subsystems is discussed in more detail below.

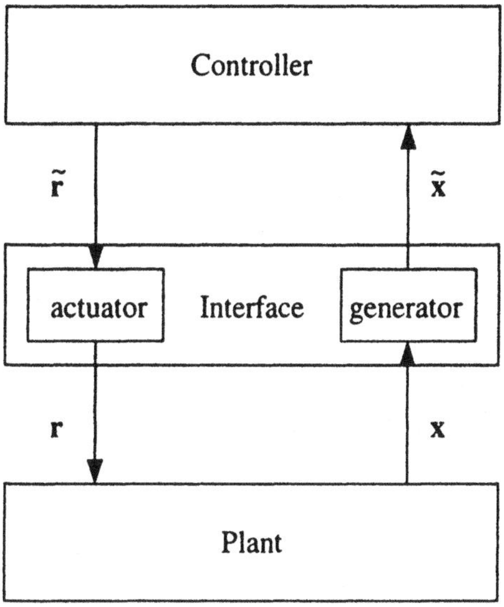

Fig. 1. Hybrid Control System

Plant: The plant is modeled by a non-autonomous vector differential equation,

$$\frac{d\bar{x}}{dt} = f(\bar{x}(t), \bar{r}(t)), \tag{1}$$

where $\bar{x} \in \Re^n$, $\bar{r} \in \Re^m$, and $f : \Re^n \times \Re^m \rightarrow \Re^n$ is Lipschitz continuous. The plant is a continuous-state system (CSS) which evolves over time.

Controller: The controller is a discrete event system (DES) represented by the following automaton

$$C = \left\{\tilde{S}_0, \tilde{S}, \tilde{X}, \tilde{R}, \xi, \psi\right\}, \tag{2}$$

where $\tilde{S}$ is an alphabet of controller states and $\tilde{S}_0 \subset \tilde{S}$ is a set of initial controller states. $\tilde{X}$ is an alphabet of *plant symbols*. These symbols are inputs to the DES controller. $\tilde{R}$ is an alphabet of *control directives*. These symbols are outputs from the controller. $\xi : \tilde{S} \times \tilde{X} \rightarrow \tilde{S}$ is the state transition mapping. $\psi : \tilde{S} \rightarrow \tilde{R}$ is the output mapping taking DES controller states onto specific control directives.

Interface: Note that the CSS plant is a continuous-state dynamical system while the controller is a DES evolving over a set of symbols. Because both of these systems are different, an *interface* is needed to facilitate communication between the plant and controller. Because both of these systems evolve over distinctly different types of state spaces, the combined system shown in figure 1 will be called a hybrid system.

The interface consists of two subsystems, the *generator* and the *actuator*. The generator transforms plant state trajectories, $\bar{x}(t)$, into a sequence of symbols, $\tilde{x}[n]$ drawn from the alphabet, $\tilde{X}$, of plant symbols. The actuator transforms a sequence, $\tilde{r}[n]$, of control directives issued by the DES controller into a continuous-time reference input, $\bar{r}(t)$, to the plant.

Generator: As noted above the generator transforms state trajectories into a sequence of symbols. For the class of hybrid systems treated in this paper, the transformation is done with respect to a finite collection of subsets of the state space. The resulting *goal set collection* is denoted

$$\mathbf{G} = \{\mathbf{g}_1, \mathbf{g}_2, \ldots, \mathbf{g}_N\}. \tag{3}$$

The generator issues a symbol $\tilde{x}_i$ whenever the plant state trajectory crosses (in a transverse manner) the boundary, $\partial\mathbf{g}_i$, of the ith goal set. The sequence, $\tilde{x}[n]$, issued by the generator marks the entrance of the plant's state into different goal sets in $\mathbf{G}$.

Actuator: As noted above the actuator transforms control directives into a continuous reference input. This is accomplished by a mapping, $\alpha : \tilde{R} \rightarrow \Re^m$, called the *actuator mapping*. α maps the ith control directive $\tilde{r}_i$ onto a single constant vector $\bar{r}_i$ in $\Re^m$. If $\tilde{r}[n]$ represents the sequence of control directives and $\tau[n]$ represents a sequence of times (measured with respect to the plant's clock) when these directives were issued by the controller, then the reference trajectory can be written as

$$\bar{r}(t) = \sum_{n=1}^{\infty} \alpha(\tilde{r}[n]) I(\tau_n, \tau_{n+1}), \tag{4}$$

where $I(t_0, t_1)$ is an indicator function taking the value of 1 on the interval $(t_0, t_1]$, and zero elsewhere.

Remark: The HCS modeling framework described above was first stated in [13]. The symbolic behaviour of dynamical systems over such state space partitions was discussed in [7]. A generalization [11] of this framework defined the generation of plant events with respect to an open cover of the plant's state space. This use of an open cover is very similar to a modeling framework discussed in [14]. Both models were developed independently and tended to emphasize different aspects of the hybrid control system. In particular, our model emphasizes the use of a simple and general interface between the plant and controller.

Remark: The preceding HCS modeling frameworks [11] [14] assume that the plant state evolves in a continuous-manner over the state space. Many hybrid systems are also well modeled as jump dynamical systems. In this case, the plant state can jump discontinuously between compact subsets of the state space. Examples of such models will be found in [15]. A good overview of all existing HCS modeling frameworks will be found in [16].

3 HCS Design Problem

The HCS design problems considered in this paper assume that the systems engineer has already formulated a specification on the controlled plant's desired logical behaviour. As was discussed in the introduction, the formal specification consists of a set of operational modes and allowed (legal) transitions between these modes. The HCS design problem involves finding the hybrid system (i.e., the actuator, generator, and controller) that realizes the specified behaviour in some appropriate manner. By "appropriate", this paper will mean "stable". In particular, this means that legal transitions within the specification possess a form of "transition" or T-stability that is to be defined below.

The remainder of this section discusses the HCS problem statment in more detail. Subsection 3.1 itemizes the designer's assumptions. Subsection 3.2 discusses the notion of T-stability. Subsection 3.3 formally states the HCS problem treated in this paper.

3.1 Assumptions

The original specification represents a set of initial assumptions about the hybrid system. These assumptions are listed below.

- **Control Policies:**
  The plant representation provided by equation 1 will be assumed to be affine in the command reference vector, $\bar{r}$. This means that the system equations are of the form
  $$\frac{d\bar{x}}{dt} = f_0(\bar{x}) + \sum_{i=1}^{m} r_i f_i(\bar{x}), \tag{5}$$
  where $f_0 : \Re^n \rightarrow \Re^n$ generates the plant's uncontrolled dynamics and $f_i : \Re^n \rightarrow \Re^n$ $(i = 1, \ldots, m)$ defines a set of vector fields over the state space associated with a specific controller. Each of these vector fields will be referred to as a *control policy.* The command reference vector $\bar{r} = (r_1 r_2 \ldots r_m)^T$ is a Boolean vector selecting out a single vector field f_i $(i = 1, \ldots, m)$ to regulate the uncontrolled plant. The plant, therefore, is controlled by switching between a finite set of available control policies. This set of policies will be called a *distribution* of vector fields,
  $$\Delta_c = \left\{ f_1 \, f_2 \cdots f_m \right\}. \tag{6}$$
  The designer has a preliminary distribution of control policies at his disposal. This knowledge represents information about the controllers and actuators which are available for regulating the plant. For technical reasons, the distributions will also need to be nonsingular and involutive so that the distribution has the maximal integral manifold property. With this assumption, we can use $(n-1)$-dimensional manifolds to form the basis of a local coordinate frame for the state space.
- **Goal Sets:**
  The designer is assumed to have a set of "distinguished" regions in the state space. These subsets are open bounded subsets of the state space called *goal sets.* The collection of initial goal sets is denoted
  $$\mathbf{G} = \left\{ \mathbf{g}_1 \, \mathbf{g}_2 \cdots \mathbf{g}_N \right\}. \tag{7}$$

Associated with the ith goal set is a *goal functional*, $g_i : \Re^n \rightarrow \Re$, $(i = 1, \ldots, N)$ such that

$$\mathfrak{g}_i = \{\bar{x} : g_i(\bar{x}) < 0\}\,. \tag{8}$$

The goal functionals, of course, are determined by the sensors available to monitor the process. The goal sets identify legal (or illegal) regions of the plant's state space.

- **Specification Automaton:**
  In addition to knowing the goal sets, the designer knows how the system should transition between these goal sets. This assumption represents prior information about legal and forbidden goal set transitions within the plant. This information can be conveniently represented by a non-deterministic finite automaton, $\mathcal{K}$. The resulting *specification automaton* is represented as

$$\mathcal{K} = \left(\tilde{S}_0, \tilde{S}, \tilde{X}, \tilde{R}, \psi, \xi\right) \tag{9}$$

  where $\tilde{S}$ is the automaton's state space, $\tilde{S}_0$ is an initial set of states, $\tilde{X}$ is an alphabet of input symbols, and $\tilde{R}$ is an alphabet of output symbols. The state transition mapping is given by $\psi : \tilde{S} \times \tilde{X} \rightarrow P(\tilde{S})$ where $P(\tilde{S})$ is the power set of $\tilde{S}$. Note that the specification automaton may be non-deterministic. The output mapping is $\xi : \tilde{S} \rightarrow \tilde{R}$. The language, $L(\mathcal{K})$, generated by this automaton represents the desired symbolic behaviour of the controlled plant.

The above assumptions specify what parts of the hybrid system are initially known by the designer. Knowing the control policy distribution implies knowledge of the plant's dynamics and controllers. Knowing the goal sets implies prior knowledge of the available sensors and prior knowledge of significant or forbidden regions within the plant's state space. Finally, knowing the specification automaton provides high-level requirements on the controlled plant's symbolic behaviour.

3.2 *T*-stability

Each of these assumptions therefore specifies a part of the HCS structure shown in figure 1. The goal sets specify the HCS generator, the distribution, Δ_c specifies the plant, and the specification automaton determines the DES controller. From this list it is apparent that the only part of the HCS which has not been initially specified by the designer is the interface actuator mapping, α. This observation suggests that HCS design simply involves determining an actuator to ensure that the specification automaton's language is "realized" by the HCS. As it turns out, this problem statement is overly simplistic and a more sophisticated formulation will be introduced below.

For the moment, however, let's consider this "simplistic" problem statement. To solve this problem, we first need to be precise about the sense in which the actuator "realizes" the specification automaton. In particular, we will say that the specified language, $L(\mathcal{K})$, is *realizable* if all symbolic transitions occur in a "stable" manner. By "stable", we mean that the symbol sequences issued by the generator are "invariant" to small perturbations of the CSS plant's initial state. The following definition [8] formally states this notion of transition or T-stability.

Definition 1. Let $\Phi_T(\bar{x}, \bar{r})$ be the state of the plant assuming initial condition $\bar{x}$ under the action of control directive $\bar{r}$. A transition from goal set $\mathbf{g}_0$ to $\mathbf{g}_T$ under control directive, $\bar{r}$, is said to be stable if and only if for all $\bar{x}_0 \in \mathbf{g}_0$ there exists an open neighborhood, $N(\bar{x}_0, \epsilon)$, and a finite time $T < \infty$ such that

- $\Phi_T(\bar{x}, \bar{r})$ is an open set in $\mathbf{g}_T$
- and the plant symbols, $\tilde{x}$, issued by the transition are identical

for all $\bar{x} \in N(\bar{x}_0, \epsilon)$.

Figure 2 illustrates an unstable transition. This figure shows a state trajectory which transfers the initial plant state, $\bar{x}_0$, between goal sets $\mathbf{g}_i$ and $\mathbf{g}_j$. As shown in the figure, the resulting state trajectory $\bar{x}(t)$ is tangential to the boundary of another goal set $\mathbf{g}_k$. If this transition were to occur in a stable manner, then the transfer must occur so that the time between successive events, τ, is bounded and so that the symbols issued by the interface are unchanged by small changes in initial state. Stability therefore requires that the "bundle" of state trajectories originating from an open neighborhood, $N(\bar{x}_0, \epsilon)$, not intersect the set $\bar{g}_k$. The trajectories in the figure do not possess this property, so this transition is not T-stable.

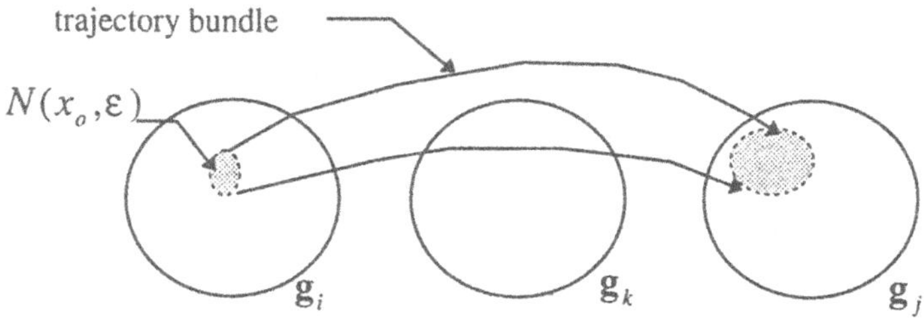

Fig. 2. Example of an unstable (unrealizable) symbolic transition

The definition of T-stability applies to a single transition. Clearly, this property will also extend to finite sequences of T-stable transitions. This fact is stated without proof in the following proposition.

Proposition 2. *Any finite sequence of connected T-stable transitions will also be T-stable.*

In the event that the control policies are linearly mixed as is done in equation 5, then it has been shown [8] that a sufficient condition for T-stable transitions can be cast as a set of inequality constraints. This sufficient system of inequalities are linear in the reference vector $\bar{r}$. The following proposition, stating these sufficient conditions, was proven in [8].

Proposition 3. *Consider a hybrid system whose plant state equations (eq. 5) are affine in the available control policies. A specified transition in $\mathcal{K}$ will be said to be T-stable if there exists a set of continuously differentiable positive definite functionals*

$V_i : \Re^n \rightarrow \Re$ $(i = 1,\ldots,N)$ which are zero on a closed proper subset of $\mathbf{g}_i \in \mathbf{G}$ $(i = 1,\ldots,N)$ such that for all $\bar{x} \notin \mathbf{g}_k$

$$\left(L_{f_0} V_k \; L_{f_1} V_k \cdots L_{f_m} V_k \right) \begin{pmatrix} 1 \\ r_1 \\ \vdots \\ r_m \end{pmatrix} < 0, \tag{10}$$

and for all $\bar{x} \in \mathbf{g}_i$ for $i = 1,\ldots,n$ and $i \neq k$,

$$\left(L_{f_0} V_i \; L_{f_1} V_i \cdots L_{f_m} V_i \right) \begin{pmatrix} 1 \\ r_1 \\ \vdots \\ r_m \end{pmatrix} > 0, \tag{11}$$

where $L_f V$ is the Lie (directional) derivative of V with respect to vector field f.

An algorithm searching for feasible points of the above inequality system will search for a command vector $\bar{r}$ which T-stabilizes the specified transition. By associating this vector with the control directive, $\bar{r}$, one obtains a characterization of a T-stabilizing actuator mapping, α, for the specified transition. The ellipsoid algorithm [18] provides just such an algorithm. An on-line version of this algorithm [8] was used to identify T-stable actuator mappings for simple hybrid systems. The ellipsoid method is well-known to converge after a finite number of updates if a feasible point exists. If such a point does not exist, then a stopping criterion exists which is known to scale in a polynomial manner with problem size. These results imply that we have an efficient algorithm for *deciding* whether or not a T-stabilizing actuator exists for a specified transition in $\mathcal{K}$.

The algorithm given in [8] converges after a finite number of updates *provided a T-stabilizing actuator exists.* If no such mapping exists, then the procedure terminates after a finite number of iterations, thereby implying that the specified transition cannot be realized. In general, it may be very difficult to ensure the existence of a T-stabilizing actuator. This fact is easily illustrated by the graphical example shown in figure 3.

Figure 3 shows an initial goal set $\mathbf{g}_i$ and a target goal set $\mathbf{g}_j$. It is assumed that the specification automaton allows a transition between these two goal sets under the action of a control directive, $\bar{r}_{ij}$. Now, assume there are only two control policies, $\Delta_c = \{f_1, f_2\}$ available to realize the specified transition. As seen in figure 3, it is highly unlikely for a single state trajectory to connect $\mathbf{g}_i$ and $\mathbf{g}_j$. In this case, therefore, there is no actuator mapping, α, which can ensure T-stable transitions between goal sets. Under certain conditions, however, it may be possible to find a chattering control policy between the two sets. The existence of this chattering or extremal control is well known from the theory of relaxed optimal control [17]. To implement such a switched control strategy as indicated in figure 3 would require expanding the original specification automaton to include the symbolic behaviour indicated by the chattering control. In other words, the existence results on extremal optimal controls appear to guarantee the existence of an extension of the original specification automaton which will be T-stable.

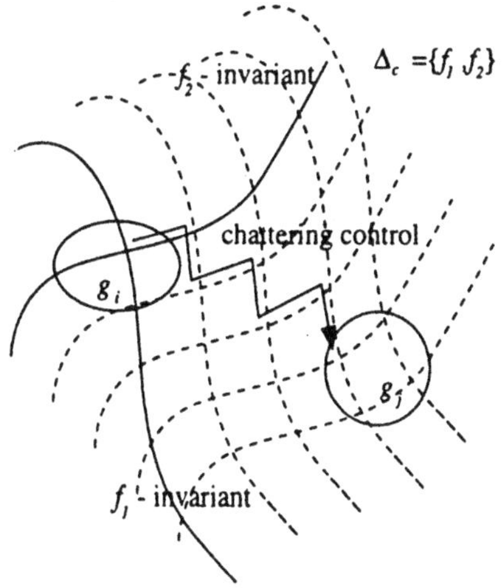

Fig. 3. Existence of Extremal Control between 2 goal sets

In light of the preceding discussion, it should be apparent that the underlying problem faced in designing hybrid control systems is much more complicated than just finding a T-stable actuator for the system. In general, an arbitrary specification language will not be realizable. A more complete problem statement, therefore, involves finding an extension of the original design specification which ensures T-stable transitions.

3.3 Problem Statement

As noted in the preceding section, the HCS design problem treated in this paper involves finding a "realizable" extension of an initial hybrid system structure. This subsection states that problem more precisely.

As noted in subsection 3.1, the designer's assumptions provide an initial hybrid control system. We will therefore represent this "initial" hybrid system by the following 4-tuple,

$$\mathcal{H} = (\mathcal{K}, \mathbf{G}, \Delta_c, \alpha), \tag{12}$$

where $\mathcal{K}$ is the specification automaton, Δ_c is the control policy distribution, $\mathbf{G}$ is the goal set collection, and α is an actuator mapping. The specification language, $L(\mathcal{K})$ will be said to be *realizable* if and only if all of its transitions are T-stabilized by the hybrid control system.

The algorithm cited above can be used to decide whether or not the specification language is realizable. In the event that it is not realizable, then an *extension* of the hybrid system needs to be determined. We therefore introduce an extension of the hybrid system that is denoted as follows.

$$\mathcal{H}^e = (\mathcal{K}^e, \mathbf{G}^e, \Delta_c^e, \alpha^e). \tag{13}$$

The extended collections, $\mathbf{G}^e$ and Δ_c^e, contain the original HCS collections, $\mathbf{G}$ and Δ_c. The extension's automaton, $\mathcal{K}^e$, is such that all strings $\sigma \in L(\mathcal{K})$ are contained in a string of σ^e in $L(\mathcal{K}^e)$. The extension's actuator mapping is determined so that all specified transitions are T-stable.

The HCS design therefore involves adding additional behaviours, goal sets, and controllers to the original hybrid control system, $\mathcal{H}$. The original hybrid system is therefore extended until all transitions within the extended specification language are T-stable. One recently proposed HCS design framework [12] tackles the HCS design problem by only adding intermediate goal sets to $\mathbf{G}$ and by extending the resulting specification language in the appropriate manner. A more general formulation of the HCS design problem would allow extensions to both $\mathbf{G}$ and Δ_c. This is precisely the problem formulation considered in this paper. A solution for this problem is described below.

4 Stable HCS Extensions

This section presents a method for constructing T-stable (realizable) extensions of a given behavioural specification. The method examines each transition in the original specification automaton, $\mathcal{K}$, and identifies those transitions which cannot be T-stabilized. We then determine those additional goal sets and control policies that must be added to $\mathbf{G}$ and Δ_c, respectively, to ensure T-stable transitions between these additional goal sets. The method adopted in this paper constructs the HCS extension out of linear control agents exhibiting robust stability to bounded plant perturbations. This method represents a significant departure from previous HCS design methods such as the Kohn-Nerode (KN) method [12] [25] which are based on the theory of relaxed optimal control [17].

Figure 4 illustrates how this extension can be obtained. The original transition specified by $L(\mathcal{K})$ is replaced by a sequence of T-stable transitions. Figure 4 shows the symbolic extension of the transition alongside of the actual situation in the plant's state space. Assume that there exists an ideal reference trajectory, $\bar{x}_r(t)$, between an initial goal set $\mathbf{g}_i$ and the terminal goal set $\mathbf{g}_j$. Define a sequence of goal sets $\mathbf{g}_m$ ($m = 1, \ldots, N_g$) which are all bounded subsets of the state space. As shown in figure 4, the goal sets are chosen to cover the reference trajectory. Associated with the mth goal set will be the mth controller. This controller will be selected to ensure T-stable transitions between intermediate goal sets. A specific approach for identifying T-stabilizing controllers and their associated goal sets will be discussed below. These new control policies are then added to Δ_c. The new "intermediate" goal sets are added to $\mathbf{G}$.

The approach outlined in figure 4 can be implemented once we are willing to adopt certain restrictions on the type of control policies to be added to Δ_c. The use of linear agents is motivated by the bounded nature of the goal set. Since $\mathbf{g}_m$ is bounded, it is possible to treat the plant as an uncertain linear system when its state is confined to the goal set. In this situation, we will use H^∞ control methods to determine linear control agents assuring T-stable transitions.

Linear control agents can only be used if the plant's perturbation over the reference trajectory is relatively smooth [19]. This notion of smoothness can be formalized as follows. Rewrite the plant's state equations in a linear state variable form,

$$\dot{\bar{x}} = A(\bar{x})\bar{x} + B(\bar{x})\bar{r} \tag{14}$$

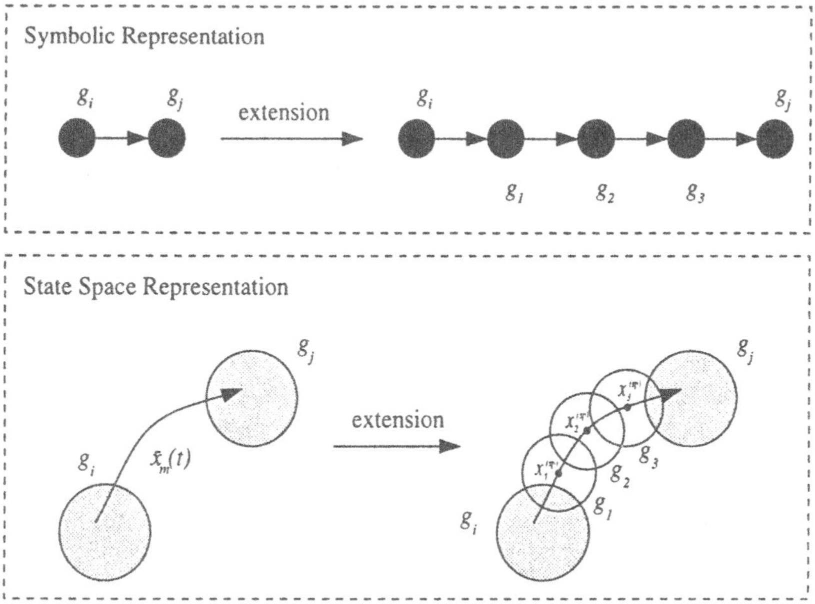

Fig. 4. A realizable extension of a specified transition is obtained by adding additional goal sets and control policies to the hybrid system.

$$\bar{y} = C(\bar{x})\bar{x}. \tag{15}$$

in which the system matrices are functions of the plant state, $\bar{x}$. At a point $\bar{x}_0$ within the plant's state space, the plant's nominal transfer function matrix is defined as

$$P(s|\bar{x}_0) = C(\bar{x}_0)\left(sI - A(\bar{x}_0)\right)^{-1} B(\bar{x}_0) \tag{16}$$

Now, consider points $\bar{x}$ near $\bar{x}_0$. We will assume that any local plant, $P(s|\bar{x})$, associated with a point $\bar{x}$ is related to $P(s|\bar{x}_0)$ through a bounded multiplicative perturbation yielding

$$P(s|\bar{x}) = (I + \Delta(s|\bar{x}, \bar{x}_0))P(s|\bar{x}_0) \tag{17}$$

with

$$\bar{\sigma}(\Delta(j\omega|\bar{x}, \bar{x}_0)) \leq \delta(\omega)||\bar{x} - \bar{x}_0||_2 \tag{18}$$

for all real ω. Here, $||\bar{x}||_2$ is a Euclidean 2-norm, $\Delta(s|\bar{x}, \bar{x}_0)$ is a stable, rational transfer function matrix, $\bar{\sigma}$ represents the maximum singular value, and $\delta(\omega)$ is a non-negative real function of ω. To model the plant in this fashion, we will assume that the multiplicative uncertainty does not change the number of closed right half plane (CRHP) poles of $P(s|\bar{x}_0)$. It will be further assumed that the plant state changes in a continuous manner with the output; i.e. $||\bar{x} - \bar{x}_0|| < K||\bar{y} - \bar{y}_0||$ and $K < \infty$. In the remainder of this paper we will assume that $K = 1$. The preceding bounds indicate that the local linear plant changes in a bounded manner as we move about within some neighborhood of $\bar{x}_0$.

The control agent architecture to be used is shown in figure 5. The controller is a model following state feedback control system. It uses an internal reference model which is represented by a stable minimum phase transfer function, $\tilde{P}(s)$. This reference represents an "ideal" trajectory, denoted $\bar{x}_r(t)$, between the two goal sets corresponding to a T-stable transition. It can be obtained by solving the Riccati equation associated with minimizing the performance functional,

$$J = \int_{t_0}^{t_f} (\bar{x}_r^T Q \bar{x}_r + \bar{u}^T R \bar{u}) dt, \tag{19}$$

subject to the differential constraint, $\dot{\bar{x}}_r = \bar{u}$, where Q and R are real symmetric positive definite matrices. As the reference model state moves along $\bar{x}_r(t)$, the controller that is used switches sequentially between N_g control agents. The index $m(= 1, \ldots, N_g)$ will be used to specify which control agent is "active". The mth control agent uses output feedback through the linear system $F_m(s)$.

Care must be taken in the design of the reference model so that the reference trajectory does not pass too close to regions in the state space associated with events other than initial event $\mathbf{g}_i$ and terminal event $\mathbf{g}_j$. For the remainder of the paper, we will assume that the reference trajectory remains at least a distance ϵ from goal sets other than $\mathbf{g}_i$ and $\mathbf{g}_j$. This condition is expressed as follows: *if* $\bar{x}_q \in \mathbf{g}_q$ *for* $q \neq i, j$ *then* $\|\bar{x}_q - \bar{x}_r(t)\|_2 > \epsilon$ *for all* $t \in (t_0, t_f)$. Proposition 3 determines one way in which this condition might be satisfied.

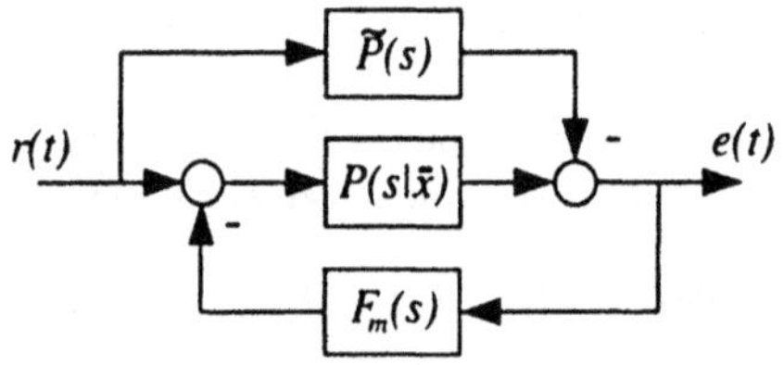

Fig. 5. HCS Control Agent

The objective of HCS design is to find intermediate goal sets and associated controllers that ensure T-stable transitions between intermediate goal sets. To accomplish this objective, we first generate the reference trajectory, $\bar{x}_r(t)$, according to the guidelines discussed above. Denote $\bar{x}_r^i = \bar{x}_r(t_i)$ for $t_0 \leq t_i < t_{i+1} \leq t_f$. The reference trajectory is sampled, yielding a sequence of *setpoints*,

$$S = \{\bar{x}_r^j\}_{j=1,\ldots,M} \,. \tag{20}$$

We will assume that the sample points are spaced closely enough so that the reference trajectory between consecutive setpoints will be interior to the union of two open spheres of radius γ centered at the setpoints. This will ensure that the reference trajectory is covered by the sequence of spheres centered at the setpoints. With the mth control agent, we will associate a subsequence of N_m contiguous setpoints denoted

$$\mathcal{S}_m = \{ \bar{x}_r^{m_i} | \bar{x}_r^{m_i} \in \mathcal{S}, \\ m_{i+1} = m_i + 1 \text{ for } i = 1, \ldots, N_m - 1 \} \tag{21}$$

For each m, there will be a unique m_1 and N_m which identifies the subsequence. The corresponding subset of the reference trajectory will be denoted

$$\mathcal{X}_m = \{\bar{x}_r(t) | t \in [t_{m_1}, t_{m_{N_m}}]\} \tag{22}$$

For each subsequence $\mathcal{S}_m$, a linear *aggregated plant*, $\hat{P}_m(s)$ will be constructed to satisfy

$$P(s|\bar{x}_r^{m_i}) = (I + \hat{\Delta}_m(s|\bar{x}_r^{m_i}))\hat{P}_m(s) \tag{23}$$

with

$$\bar{\sigma}(\hat{\Delta}_m(j\omega|\bar{x}_r^{m_i})) \leq \gamma\delta(\omega) \tag{24}$$

for all $\bar{x}_r^j \in \mathcal{S}_m$. $\hat{\Delta}_m(j\omega|\bar{x}_r^{m_i})$ is assumed to be a stable rational transfer function representing a multiplicative perturbation such that $P(s|\bar{x}_r^{m_i})$ and $\hat{P}_m(s)$ have the same number of CRHP poles for all $\bar{x}_r^j \in \mathcal{S}_m$. For each $\mathcal{S}_m$, the quantity N_m will be defined as the maximum length of $\mathcal{S}_m$ beginning at $\bar{x}_r^{m_1}$ such that an aggregated plant as defined in 23 and 24 may be constructed. Actual construction of the aggregated plant will be discussed later. For now, we will assume that $\hat{P}_m(s)$ is known.

With the mth control agent, the mth intermediate goal set will be defined as

$$\mathbf{g}_m = \bigcup_{\bar{x}_r \in \mathcal{X}_m} \mathbf{n}(\bar{x}_r, \gamma) \tag{25}$$

where $\mathbf{n}(\bar{x}_r, \gamma)$ is the special neighborhood set

$$\mathbf{n}(\bar{x}_r, \gamma) = \{\bar{x} : ||\bar{x} - \bar{x}_r||_2 < \gamma\}. \tag{26}$$

The goal set $\mathbf{g}_m$ is the set of all points within a distance γ of some point in $\mathcal{X}_m$. An example of the geometry described above is shown in Figure 6. We now state a proposition that relates all local plants in a goal set $\mathbf{g}_m$ to the aggregated plant $\hat{P}_m(s)$.

Proposition 4. *Let $\mathcal{S}$ be a sequence of setpoints drawn from the reference trajectory generated by $\tilde{P}(s)$. Let $\mathcal{S}_m$ be the mth local subsequence of $\mathcal{S}$ with aggregated plant $\hat{P}_m(s)$ as defined in 23 and 24 and associated goal set $\mathbf{g}_m$ as defined in equation 25. Then, for any $\bar{x} \in \mathbf{g}_m$,*

$$P(s|\bar{x}) = (I + \Delta_m(s|\bar{x}))\hat{P}_m(s) \tag{27}$$

with

$$\bar{\sigma}(\Delta_m(j\omega|\bar{x})) \leq \delta_1(\omega) \tag{28}$$

where

$$\delta_1(\omega) = (1 + \gamma\delta(\omega))^3 - 1 \tag{29}$$

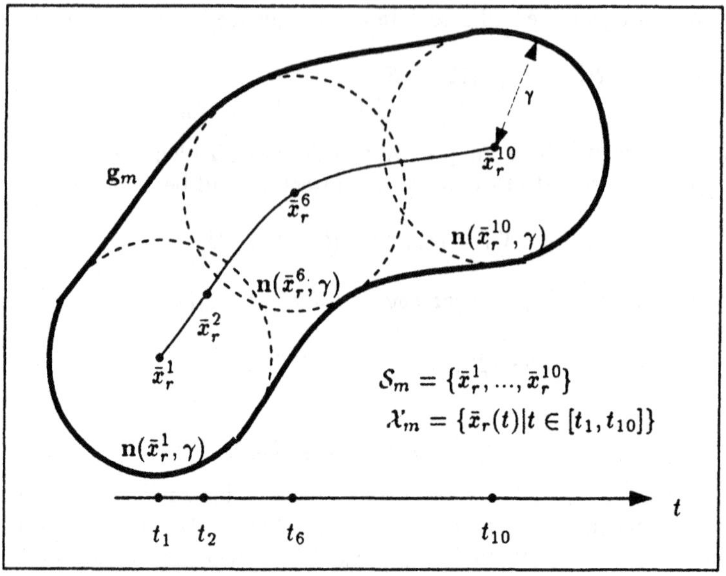

Fig. 6. Example of problem geometry

Proof: Consider any $\bar{x} \in \mathbf{g}_m$. From equations 17 and 18, there exists $\bar{x}_r \in \mathcal{X}_m$ such that

$$P(s|\bar{x}) = (I + \Delta(s|\bar{x}, \bar{x}_r))P(s|\bar{x}_r)$$

with $\bar{\sigma}(\Delta(j\omega|\bar{x}, \bar{x}_r)) \leq \gamma\delta(\omega)$. By our sampling assumption, for any $\bar{x}_r \in \mathcal{X}_m$, there exists a $\bar{x}_r^{m_i} \in \mathcal{S}_m$ such that

$$P(s|\bar{x}_r) = (I + \Delta(s|\bar{x}_r, \bar{x}_r^{m_i}))P(s|\bar{x}_r^{m_i})$$

with $\bar{\sigma}(\Delta(j\omega|\bar{x}_r, \bar{x}_r^{m_i})) \leq \gamma\delta(\omega)$. Finally, by construction of the aggregated plant,

$$P(s|\bar{x}_r^{m_i}) = (I + \hat{\Delta}_m(s|\bar{x}_r^{m_i}))\hat{P}_m(s)$$

with $\bar{\sigma}(\hat{\Delta}_m(j\omega|\bar{x}_r^{m_i})) \leq \gamma\delta(\omega)$ for any $\bar{x}_r^{m_i} \in \mathcal{S}_m$. Combining these relations, we obtain

$$\begin{aligned} P(s|\bar{x}) &= \Big[(I + \Delta(s|\bar{x}, \bar{x}_r))(I + \Delta(s|\bar{x}_r, \bar{x}_r^{m_i})) \\ &\qquad (I + \hat{\Delta}_m(s|\bar{x}_r^{m_i}))\Big] \hat{P}_m(s) & (30) \\ &= [I + \Delta_m(s|\bar{x})]\hat{P}_m(s) & (31) \end{aligned}$$

where

$$\begin{aligned} \bar{\sigma}(\Delta_m(j\omega)) &= \bar{\sigma}\Big((I + \Delta(j\omega|\bar{x}, \bar{x}_r))(I + \Delta(j\omega|\bar{x}_r, \bar{x}_r^{m_i})) \\ &\qquad (I + \hat{\Delta}_m(j\omega|\bar{x}_r^{m_i})) - I\Big) & (32) \\ &\leq (1 + \gamma\delta(\omega))^3 - 1 & (33) \end{aligned}$$

which completes the proof. □

The implication of the above result is that by designing a controller for a single uncertain linear plant $\hat{P}_m(s)$ that meets robust stability and peformance requirements, we simultaneously satisfy the criteria for all local plants $\{P(s|\bar{x})|\bar{x} \in \mathbf{g}_m\}$. In this case, the performance objective is to keep the plant state confined to the goal set $\mathbf{g}_m$ whenever $\bar{x}_r \in \mathcal{X}_m$. We now state the following proposition which provides a sufficient condition for the plant state to be confined to the goal set defined in equation 25.

Proposition 5. *Let $\mathcal{S}$ be a sequence of setpoints drawn from the reference trajectory generated by $\tilde{P}(s)$ with associated sets $\mathcal{S}_m$, $\mathcal{X}_m$, and $\mathbf{g}_m$ as defined in equations 21, 22, and 25. Let $\hat{P}_m(s)$ be the aggregated plant for $\mathcal{S}_m$ satisfying the conditions of proposition 4 so that*

$$P(s|\bar{x}) = (I + \Delta_m(s|\bar{x}))\hat{P}_m(s) \tag{34}$$

with $\bar{\sigma}(\Delta_m(j\omega|\bar{x})) \leq \delta_1(\omega)$ for all $\bar{x} \in \mathbf{g}_m$ where $\delta_1(\omega)$ is defined in proposition 4. If the plant state $\bar{x}(t_{m_1}) \in \mathbf{g}_m$, then the linear control agent $F_m(s)$ robustly stabilizes the local linear error dynamics and $\bar{x}(t) \in \mathbf{g}_m$ for all $t \in [t_{m_1}, t_{m_{N_m}}]$ provided the nominal closed loop characteristic function, $I + \hat{P}_m(s)F_m(s)$, has no zeros in the closed right half plane and

$$||W_1(\omega)S_m(j\omega)||_\infty + ||W_2(\omega)T_m(j\omega)||_\infty < 1 \tag{35}$$

where

$$S_m(s) = (I + \hat{P}_m(s)F_m(s))^{-1} = I - T_m(s) \tag{36}$$

and $W_1(\omega)$ and $W_2(\omega)$ satisfy

$$W_1(\omega) = \frac{1}{\gamma}[\bar{\sigma}(\hat{P}_m(j\omega) - \tilde{P}(j\omega)) + \delta_1(\omega)\bar{\sigma}(\hat{P}_m(j\omega))] \tag{37}$$

$$W_2(\omega) = \delta_1(\omega) \tag{38}$$

■

Proof: For $\bar{x} \in \mathbf{g}_m$, the local transfer function between the reference input $\bar{r}$ and the output $e(t)$ is given by

$$H_m(s|\bar{x}) = (I + P(s|\bar{x})F_m(s))^{-1}(P(s|\bar{x}) - \tilde{P}(s))$$

which may be rewritten as

$$H_m(s|\bar{x}) = S_m(s)(I + \Delta_m(s)T_m(s))^{-1}(P(s|\bar{x}) - \tilde{P}(s)) \tag{39}$$

using equations 34 and 36. Without loss of generality, assume that $||r||_2 = 1$ so that $||e||_2 = ||H(j\omega)||_\infty$. Robust stability of the local linear error dynamics is guaranteed if

- $I + \hat{P}_m(s)F_m(s)$ has no CRHP zeros
- $P(s|\bar{x})$ and $\hat{P}_m(s)$ have the same number of CRHP poles
- $\bar{\sigma}(T_m(j\omega))\delta_1(\omega) < 1$ for all ω

The first two conditions are satisfied by assumption. To see that the last condition is implied by equation 35, assume $||W_1(\omega)S_m(j\omega)||_\infty + ||W_2(\omega)T_m(j\omega)||_\infty < 1$. This implies

$$\bar{\sigma}(S_m(j\omega))\bar{\sigma}((P(s|\bar{x}) - \tilde{P}(s))/\gamma) + \bar{\sigma}(T_m(j\omega))\delta_1(\omega) < 1 \quad (40)$$

for all ω and for all $\bar{x} \in \mathbf{g}_m$ which guarantees $\bar{\sigma}(T_m(j\omega))\delta_1(\omega) < 1$ for all ω.

Since $||e(t)||_2 = ||\bar{x}(t) - \bar{x}_r(t)||_2$, $\bar{x}(t) \in \mathbf{g}_m$ for all $t \in [t_{m_1}, t_{m_{N_m}}]$ if $||H_m(j\omega|\bar{x})||_\infty < \gamma$ for all $\bar{x} \in \mathbf{g}_m$. Equation 40 implies

$$\bar{\sigma}(S_m(j\omega))\frac{\bar{\sigma}((P(s|\bar{x}) - \tilde{P}(s))/\gamma)}{\underline{\sigma}(I + \Delta_m(j\omega|\bar{x})T_m(j\omega))} < 1$$

for all ω and for all $\bar{x} \in \mathbf{g}_m$. This implies

$$\bar{\sigma}(S_m(j\omega))\bar{\sigma}((I + \Delta_m(j\omega|\bar{x})T_m(j\omega))^{-1})\bar{\sigma}((P(s|\bar{x}) - \tilde{P}(s))) < \gamma$$

for all ω and for all $\bar{x} \in \mathbf{g}_m$ which guarantees

$$\bar{\sigma}(S_m(j\omega))(I + \Delta_m(j\omega|\bar{x})T_m(j\omega))^{-1}(P(s|\bar{x}) - \tilde{P}(s)) < \gamma$$

for all ω and for all $\bar{x} \in \mathbf{g}_m$. Using equation 39, this reduces to $||H_m(j\omega|\bar{x})||_\infty < \gamma$ for all $\bar{x} \in \mathbf{g}_m$ which completes the proof. □

The sufficient condition in equation 35 may be realized in a number of ways. One possible method would be to find $F_m(s)$ which solves the mixed performance-sensitivity problem

$$\left\| \begin{array}{c} 2W_1(\omega)S_m(j\omega) \\ 2W_2(\omega)T_m(j\omega) \end{array} \right\|_\infty < 1 \quad (41)$$

which would guarantee equation 35. Another possible method substitutes $I - T_m(s)$ for $S_m(s)$ in equation 35, yielding a sufficient condition

$$||W_3(\omega)S_m(j\omega)||_\infty < 1 \quad (42)$$

where

$$W_3(\omega) = \frac{W_1(\omega) + W_2(\omega)}{1 - W_2(\omega)}$$

The choice will depend on the structure of the problem and which of the formulations in equations 41 and 42 is more convenient.

A collection of goal sets $\mathbf{g}_m$ will be said to be connected if their associated subsequences have non-null intersections. For connected collections of goal sets, $\mathbf{g}_m$, the following corollary results from the preceding proposition.

Corollary 6. *Consider a sequence of connected intermediate goal sets* $\{\mathbf{g}_m\}$ *(*$m = 1, \ldots, N_g$*) between initial goal set* $\mathbf{g}_I$ *and terminal goal set* $\mathbf{g}_F$ *with the non-null intersection* $(\mathbf{g}_1 \cap \mathbf{g}_I) \subset \mathbf{n}(\bar{x}_1^0, \gamma)$ *and* $\mathbf{n}(\bar{x}_m^M, \gamma) \subset \mathbf{g}_F$. *Let the symbol* $\tilde{x}_m$ *denote the entry of the reference model state into* $\mathcal{X}_m$ *and let* $\tilde{x}_{N_g+1}$ *indicate that the reference model state has reached* $\bar{x}_r^M \in \mathcal{S}$. *If the connected intermediate goal sets and their associated controllers satisfy the hypotheses of proposition 5, then the the sequence of transitions* $\tilde{x}_m \rightarrow \tilde{x}_{m+1}$ *for* $m = 1, \ldots, N_g$ *represents a T-stable transition* $\tilde{x}_I \rightarrow \tilde{x}_F$.

Proof: First, consider a single intermediate transition $\tilde{x}_m \rightarrow \tilde{x}_{m+1}$. Assume that the event $\tilde{x}_m$ occurs at $t = 0$. Let $\Phi_t(\bar{x}_0, m)$ represent the closed loop plant state at time $t \in [0, T]$ with controller $F_m(s)$ and initial state $\bar{x}_0$. Denote the first and last elements of $\mathcal{S}_m$ as $\bar{x}_m^0$ and $\bar{x}_m^T$ and note that $\bar{x}_m^T = \bar{x}_{m+1}^0$ since the goal sets are connected. Then $\mathbf{n}(\bar{x}_m^T, \gamma) = \mathbf{n}(\bar{x}_{m+1}^0, \gamma) \subset \mathbf{g}_{m+1}$. Consider any initial state $\bar{x}_0 \in \mathbf{n}(\bar{x}_m^0, \gamma) \subset \mathbf{g}_m$. If $F_m(s)$ satisfies the conditions of proposition 5, then $\Phi_t(\bar{x}_0, m) \in \{\bar{x} : \|\bar{x} - \bar{x}_r(t)\|_2 < \gamma\}$. This implies that $\Phi_T(\bar{x}_0, m) \in \{\bar{x} : \|\bar{x} - \bar{x}_r(T)\|_2 < \gamma\} \equiv \mathbf{n}(\bar{x}_m^T, \gamma) \subset \mathbf{g}_{m+1}$ for any initial state $\bar{x}_0 \in \mathbf{n}(\bar{x}_m^0, \gamma)$. But at $t = T$, $\tilde{x}_{m+1}$ occurs. Thus, for any transition $\tilde{x}_m \rightarrow \tilde{x}_{m+1}$,

$$\bar{x}_0 \in \mathbf{n}(\bar{x}_m^0, \gamma) \Rightarrow \Phi_T(\bar{x}_0, m) \in \mathbf{n}(\bar{x}_{m+1}^0, \gamma)$$

which is an open set of $\mathbf{g}_{m+1}$. Since we did not specify m, this property is true for all $m = 1, \ldots, N_g$. Additionally, note that $\{\bar{x} : \|\bar{x} - \bar{x}_m(t)\|_2 < \gamma\} \cap \mathbf{g}_q = \emptyset$ for $q \neq i, j$ if $\gamma < \epsilon$ which guarantees that there are no plant symbols issued during the transition $\tilde{x}_m \rightarrow \tilde{x}_{m+1}$ for $m = 1, \ldots, N_g$.

Now, consider the event $\tilde{x}_1$ with the plant state $\bar{x}_0$ interior to $(\mathbf{g}_1 \cap \mathbf{g}_I) \subset \mathbf{n}(\bar{x}_1^0, \gamma)$. By the above arguments, after N_g transitions $\tilde{x}_1 \rightarrow \tilde{x}_2 \rightarrow \ldots \rightarrow \tilde{x}_{N_g+1}$, the plant state $\bar{x} \in \mathbf{n}(\bar{x}_{N_g}^T, \gamma) \subset \mathbf{g}_F$. So, we have an open set of $\mathbf{g}_I$ mapping to an open set of $\mathbf{g}_F$. Furthermore, during all of the intermediate transitions, no other plant symbols are issued, so the transition is T-stable. □

We now discuss the derivation of the aggregated plant model. Derivation of an aggregated plant for a subsequence $\mathcal{S}_m$ is an important part of the design methodology presented here. One way of approaching the problem is to find a $\hat{P}_m(s)$ of the same order as the local plants which simultaneously solves a set of relative error model reduction problems (see [27]). Before presenting the next result, we review some notation. The set $\mathcal{RL}_\infty^{p\times q}$ represents the set of all rational transfer function matrices having finite infinity norm. Suppose $P_{11} \in \mathcal{RL}_\infty^{m\times n}$, $P_{12} \in \mathcal{RL}_\infty^{m\times p}$, $P_{21} \in \mathcal{RL}_\infty^{q\times n}$, $P_{22} \in \mathcal{RL}_\infty^{q\times p}$ and $K \in \mathcal{RL}_\infty^{p\times q}$. Then the linear fractional transformation (LFT) (see [26])

$$P_{11} + P_{12}K(I - P_{22}K)^{-1}P_{21}$$

is denoted

$$\mathcal{F}_\ell\left(\begin{bmatrix} P_{11} & P_{12} \\ P_{21} & P_{22} \end{bmatrix}, K\right).$$

The following proposition relates one way we might find an aggregated plant $\hat{P}_m(s)$ which satisfies proposition 4. The proof applies results from [27].

Proposition 7. *Suppose we are given a subsequence $\mathcal{S}_m \subset \mathcal{S}$. Let $P_i(s) = P(s|\bar{x}_r^{m_i}) \in \mathcal{RL}_\infty^{p\times q}$ for each $\bar{x}_r^{m_i} \in \mathcal{S}_m$. Let*

$$W_4(\omega) = \frac{1 - \delta_1(\omega)}{\delta_1(\omega)}$$

where $\delta_1(\omega)$ was defined in proposition 4. Let $V_i(s)$ represent a square left spectral factor of $P_i(s)$ when $q \leq p$ so that $P_i^*(s)P_i(s) = V_i(s)V_i^*(s)$ (where $M^*(s) = M^T(-s)$); let $U_i(s)$ represent a square right spectral factor of $P_i(s)$ when $p < q$ so that $P_i(s)P_i^*(s) = U_i^*(s)U_i(s)$. Suppose $\hat{P}_m(s) \in \mathcal{RL}_\infty^{p\times q}$. If

(a) $q \le p$ and

$$\left\| \mathcal{F}_\ell \left(\begin{bmatrix} Z_{11} & W_4 I_{p\times p} \\ Z_{21} & 0_{q\times p} \end{bmatrix}, \hat{P}_m \right) \right\|_\infty < 1. \tag{43}$$

where

$$Z_{11} = \left[-W_4 P_1 V_1^{*-1} \cdots -W_4 P_{N_m} V_{N_m}^{*-1} \right]$$

and

$$Z_{21} = \left[V_1^{*-1} \cdots V_{N_m}^{*-1} \right].$$

or if

(b) $p < q$ and

$$\left\| \mathcal{F}_\ell \left(\begin{bmatrix} Y_{11} & Y_{12} \\ I_{q\times q} & 0_{q\times p} \end{bmatrix}, \hat{P}_m \right) \right\|_\infty < 1. \tag{44}$$

where

$$Y_{11} = \begin{bmatrix} -W_4 U_1^{*-1} P_1 \\ \vdots \\ -W_4 U_{N_m}^{*-1} P_{N_m} \end{bmatrix}$$

and

$$Y_{12} = W_4 \begin{bmatrix} U_1^{*-1} \\ \vdots \\ U_{N_m}^{*-1} \end{bmatrix}$$

then for all $i = 1, \ldots, N_m$,

$$P(s|\bar{x}_r^{m_i}) = (I + \hat{\Delta}_m(s|\bar{x}_r^{m_i}))\hat{P}_m(s)$$

$$\bar{\sigma}(\hat{\Delta}_m(j\omega|\bar{x}_r^{m_i})) \le \delta_1(\omega)$$

Proof: Suppose that for all $i = 1, \ldots, N_m$,

$$\begin{aligned} P(s|\bar{x}_r^{m_i}) &= (I + \hat{\Delta}_m(s|\bar{x}_r^{m_i}))\hat{P}_m(s) \\ \bar{\sigma}(\hat{\Delta}_m(j\omega|\bar{x}_r^{m_i})) &\le \delta_1(\omega) \end{aligned} \tag{45}$$

where $\delta_1(\omega)$ is defined in proposition 4. Equation 45 is implied by

$$\begin{aligned} \hat{P}_m(s) &= (I + \tilde{\Delta}_m(s|\bar{x}_r^{m_i}))P(s|\bar{x}_r^{m_i}) \\ \bar{\sigma}(\tilde{\Delta}_m(j\omega|\bar{x}_r^{m_i})) &\le W_4^{-1}(\omega) = \tfrac{\delta_1(\omega)}{1-\delta_1(\omega)} \end{aligned} \tag{46}$$

(a) Suppose $q \le p$ and $P_i^*(s)P_i(s) = V_i(s)V_i^*(s)$ where $V_i(s)$ was defined in the proposition. Then equation 46 is guaranteed by

$$\|W_4(\omega)(\hat{P}_m(j\omega) - P_i(j\omega))V_i^{*-1}(j\omega)\|_\infty < 1 \tag{47}$$

for all $i = 1, \ldots, N_m$. We may solve this for all $i = 1, \ldots, N_m$ by finding a $\hat{P}_m(s) \in \mathcal{RL}_\infty^{p\times q}$, if one exists, which satisfies

$$\| W_4(\hat{P}_m - P_1)V_1^{*-1} \cdots W_4(\hat{P}_m - P_{N_m})V_{N_m}^{*-1} \|_\infty < 1 \tag{48}$$

where the dependance on ω has been omitted. Equation 48 may be rewritten as equation 43.

(b) Suppose $p < q$ and $P_i(s)P_i^*(s) = U_i^*(s)U_i(s)$ where $U_i(s)$ was defined in the proposition. Then equation 46 is guaranteed by

$$\|W_4(\omega)U_i^{*-1}(j\omega)(\hat{P}_m(j\omega) - P_i(j\omega))\|_\infty < 1 \tag{49}$$

for all $i = 1, \ldots, N_m$. We may solve this for all $i = 1, \ldots, N_m$ by finding a $\hat{P}_m(s) \in \mathcal{RL}_\infty^{p\times q}$, if one exists, which satisfies

$$\left\| \begin{array}{c} W_4 U_1^{*-1}(\hat{P}_m - P_1) \\ \vdots \\ W_4 U_{N_m}^{*-1}(\hat{P}_m - P_{N_m}) \end{array} \right\|_\infty < 1 \tag{50}$$

where the dependance on ω has been omitted. Equation 50 may be rewritten as equation 44. □.

The preceding results provide a very simple method for generating T-stable extensions of a given specification. This procedure is outlined below.

Step 1: Generate a reference trajectory, $\bar{x}_r(t)$, and obtain a sequence of M setpoints $\mathcal{S} = \{\bar{x}_r^j\}$ $(j = 1, \ldots, M)$.
Step 2: For each setpoint in $\mathcal{S}$ determine the local linear transfer function $P(s|\bar{x}_r^j)$.
Step 3: Initialize the iteration with i=1, N=0, and $m = 1$.
Step 4: Compute an aggregated plant, $\hat{P}_m(s)$ for all $j \in [i, i+N]$.
Step 5: Determine the H^∞ norm controller, $F_m(s)$. If no such controller exists then set $i = i + N - 1$, N=0, and $m = m + 1$. Otherwise set $N = N + 1$.
Step 6: Go to step 4 if $i + N < M$.
Step 7: If no controller was found in the last iteration of Step 5, then declare a failure. The procedure was unable to T-stablize the transition. Otherwise declare success.

5 Related Methods

The HCS design method described above is not the only way of solving the HCS design problem. A related approach known as the Kohn/Nerode (KN) method [12] was developed earlier. It will be instructive to briefly compare these two approaches.

Robust Stability: The KN-method relies heavily on formalisms associated with the relaxed optimal controllers [23] [17]. These approaches require that the plant models be known. As noted in the introduction, however, such knowledge is often imprecise. Consequently, we expect that the KN-method may exhibit sensitivity to modeling uncertainty. One way of handling this is to use an adaptor as proposed in [12]. Our approach, however, handles this uncertainty in a different manner. In particular, we deal with uncertainty using very mature formalisms that have recently been developed for controlling uncertain systems.

Open loop versus closed loop controllers: The KN-method symbolically solves a linear equation [24] to obtain a representation for those policy sequences reaching the final goal set. The method then determines the duration of time each policy is used by solving an associated linear programming problem [25]. The result can therefore be viewed as a sequence of controls which are switched on the basis of these previously determined "times". In this sense, the KN method is an open loop strategy. Our method, on the other hand, yields closed loop strategies which switch on the basis of sensed outputs (events) from the plant.

HCS Problem Statement: The KN-method expands the original behavioural specification by expanding the goal sets. It does not expand the distribution of available control policies. Our method finds expansions of the original specified behaviour by expanding both the control policy distribution and the goal sets.

Design Procedures: By fixing the control policy distribution, the KN-method allows conventional linear programming (LP) methods to be used in solving the HCS design problem. Our method allows variation of the control policy distribution which may be accomplished using existing H^∞ control techniques. Our research has shown that other uncertainty representations may be used as well [22]. For example, in certain cases where the model uncertainty manifests itself as interval matrices, the robust stability criteria may be formulated as a set of linear matrix inequalities (LMIs) [20]. In this case, our method leads to a nonconvex optimization problem which we can solve using fast alternating minimization techniques [9] [10]. In particular, related algorithms have been shown to have sublinear complexities comparable to that of interior-point([21]) LP algorithms.

Chattering vs. Non-chattering Controllers: The KN-method determines a chattering controller. Such control schemes are "optimal" in a well-defined mathematical sense. For certain applications which have a fixed set of controllers to work from, these control schemes are extremely appropriate. A good example might be found in spacecraft controls where different actuators (reaction wheels, thrusters, CMG's) are used depending on the current operational mode. In other applications, however, chattering controls cannot be tolerated. This is particularly evident in applications which need to minimize some measure of stress on mechanical components. For these systems, discontinuous changes in control direction can stress system components and cause premature failure. A good example of these systems will be found in the control of steam turbines used in electric power generation. Command profiles for these systems must be smooth enough to prevent excessive stress on turbine blades. In these applications, approaches such as the one being described here provided a valuable alternative to the KN-method.

6 Summary

This chapter has presented a method for HCS design which relies on modern concepts in H^∞ control. The method attempts to construct a T-stable extension of a formal specification on the plant desired symbolic behaviour. The extension involves finding additional goal sets and controllers which take the plant's state in a reliable (stable) manner to a desired goal state. The proposed procedure provides an alternative to the Kohn-Nerode hybrid control design method [12] which relies on results from the theory of relaxed optimal control [17]. The KN-method requires full knowledge of the plant and relies on adaptive mechanisms to ensure robust performance. Our method, on the other hand, uses an H^∞ controller to construct the hybrid system and therefore directly addresses the issue of plant uncertainty in a rigorous fashion. In addition to this, it is expected that our method can also be implemented using experimentally gathered data representing the plant's behaviour. This method might therefore be directly applicable to the design of complex systems for which analytic dynamical models are ill-defined or non-existent. A great deal of additional work is needed to more clearly understand the potential advantages and limitations of this method.

References

1. R.L. Grossman, A. Nerode, A.P. Ravn, and H. Rischel (eds.), *Hybrid Systems*, Lecture Notes in Computer Science 736, Springer Verlag, 1993.
2. A. Benveniste and P. Le Guernic, "Hybrid dynamical systems and the signal language", *IEEE Transactions on Automatic Control*, Vol. 35, No. 5, pp 535-546, May 1990.
3. A. Gollu and P. Varaiya, "Hybrid dynamical systems", In *Proceedings of the 28th Conference on Decision and Control*, pp. 2708-2712, Tampa, FL, Dec. 1989.
4. L.Holloway and B. Krogh, "Properties of behavioural models for a class of hybrid dynamical systems", In *Proceedings of the 31st Conference on Decision and Control*, pp. 3752-2757, Tuscon, AZ, Dec. 1992.
5. K.M. Passino and U. Ozguner, "Modeling and analysis of hybrid systems: examples", In *Proceedings of the 1991 IEEE International Symposium on Intelligent Control*, pp. 251-256, Arlington, VA, Aug. 1991.
6. P. Peleties and R. DeCarlo, "A modeling strategy with event structures for hybrid systems", In *Proceedings of the 28th Conference on Decision and Control*, pp. 1308-1313, Tampa, FL, Dec. 1989.
7. P.J. Ramadge, "On the periodicity of symbolic observations of piecewise smooth discrete-time systems", *IEEE Transactions on Automatic Control*, Vol. 35, No. 7, pp. 807-812, July 1990.
8. M.D. Lemmon and P.J. Antsaklis, "Inductively Inferring Valid Logical Models of Continuous-State Dynamical Systems", *Theoretical Computer Science*, 138 (1995) 201-210.
9. M.D. Lemmon and P.T. Szymanski, "Interior point implementations of alternating minimization training", to appear in *Advances in Neural Information Processing Systems* 7, pp. 574-582, San Mateo, California: Morgan Kaufmann publishers, 1995.
10. P.T. Szymanski, and M.D. Lemmon, "A modified interior point method for supervisory controller design", in *Proceedings of the 33rd IEEE Conference on Decision and Control*, Orlando Florida, pp. 1381-1386, December, 1994.
11. P. Antsaklis, J. Stiver, and M. Lemmon, "Hybrid system modeling and autonomous control systems", in [1] , pp. 366-392. Springer-Verlag, 1993.
12. A. Nerode and W. Kohn, "Multiple agent hybrid control architecture", in [1], pp. 297-316, Springer-Verlag, 1993.
13. J.A. Stiver and P.J. Antsaklis, "Modeling and analysis of hybrid control systems", In *Proceedings of the 31st Conference on Decision and Control*, pp. 3748-3751, Tuscon, AZ, Dec. 1992.
14. A. Nerode and W. Kohn, "Models for hybrid systems: automata, topologies, controllability, observability", in [1],pp. 317-356, Springer Verlag, 1993.
15. R.W. Brockett, "Hybrid models for motion control systems", Technical Report CICS-P-364, Center for Intelligent Control Systems, Massachusetts Institute of Technology, March 1993.

16. M.S. Branicky, V.S. Borkar, and S.K. Mitter, "A unified framework for hybrid control: background, model, and theory", Technical Report LIDS-P-2239, Laboratory for Information and Decision Systems, Massachusetts Institute of Technology, April 1994.
17. W.H. Fleming and R.W. Rishel, *Deterministic and Stochastic Optimal Control*, Springer Verlag, 1975.
18. M. Groetschel, L. Lovasz and A. Schrijver, *Geometric Algorithms and Combinatorial Optimization*. Springer-Verlag, Berlin, 1988.
19. J.S. Shamma, and M. Athans, "Analysis of gain scheduled control for nonlinear plants", *IEEE Trans. on Automatic Control*, Vol. AC-35, No. 8, pp. 898-907, August 1990.
20. S. Boyd, L. El Ghaoui, E. Feron, and V. Balakrishnan, *Linear matrix inequalities in system and control theory*, SIAM Studies in Applied Mathematics vol 15, 1994.
21. C.C. Gonzaga (1992), "Path-Following Methods for Linear Programming", *SIAM Review*, Vol 34(2), pp. 167-224, June 1992.
22. M.D. Lemmon, P.T. Szymanski, and C.Bett, "Interior point competitive learning of control agents in colony-style systems", to appear in *Proceedings of SPIE Orlando Symposium*, Volume 2492, April 1995.
23. L.C. Young, *Optimal Control Theory*, Chelsea Publishing Co. NY 1980.
24. W. Kuich and A. Salomaa, *Semirings, automata, and languages*, Springer Verlag, 1985.
25. X. Ge, W.Kohn, and A. Nerode, "Algorithms for chattering approximations to relaxed optimal controls", Technical Report 94-23, Mathematical Sciences Institute, Cornell University, April 1994.
26. J. Doyle, A. Packard, and K. Zhou, "Review of LFTs, LMIs and μ", *Proceedings of the 30th Conference on Decision and Control*, Brighton, England, Dec 1991.
27. K. Glover, "Multiplicative approximation of linear multivariable systems with L^∞ error bounds", in *Proceedings of the American Control Conference*, pp. 1705-1709, 1986.

Controllers as Fixed Points of Set-Valued Operators

Anil Nerode*[1] and Jeffrey B. Remmel**[2] and Alexander Yakhnis***[3]

[1] Mathematical Sciences Institute
Cornell University, Ithaca, NY 14853
e-mail anil@math.cornell.edu
[2] Department of Mathematics and Computer Science
University of California at San Diego, San Diego, CA 92093
e-mail:jremmel@ucsd.edu
[3] Mathematical Sciences Institute
Cornell University, Ithaca, NY 14853
e-mail: ayakh@math.cornell.edu

Abstract. We consider the problem of constructing a "controller" for a hybrid system which will solve the viability problem that all points of plant trajectories stay inside a given "viability set". Here, a "controller" is a network of three successive devices, a digital to analog converter, a digital program (a computer together with its control software), and an analog to digital converter. We model a controller as an input-output automaton, which we call a "control automaton". We give a necessary and sufficient condition that must be satisfied in order that a finite state control automaton solves a viability problem. Our results apply to plants modelled by vector differential equations with control and disturbance parameters, or to plants modelled by differential inclusions with a control parameter. We represent imprecise sensing of plant state. The main restrictions on the range of applicability of our results are that the set of admissible control laws is finite, and that we can neglect delays when control laws are reset. The latter restriction seems to be inessential.

1 Introduction

Classical control, despite many mathematical papers on more mathematically advanced techniques, still deals in practice with controlling a continuous plant by employing a single linear feedback control law implemented by an analog control device. Usually the linear feedback control law is obtained by using linear algebra (Laplace transforms) applied to a linear approximation of the plant (PID controllers, for instance). If the plant is highly non-linear, such an approximation can be expected to lead to poor performance.

* supported in part by Army Research Office contract DAAL03-91-C-0027
** supported in part by Army Research Office contract DAAL03-91-C-0027 and NSF grant DMS-9306427
*** supported by DARPA-US ARMY AMCCOM (Picatinny Arsenal, N. J.) contract DAAA21-92-C-0013 to ORA Corp.

The idea of a hierarchical finite state control automaton for mode switching, that is, switching between linear control laws, was developed to remedy this problem by employing a finite automaton which switches between different linear feedback laws as a function of what has been sensed about plant state. Such automata have traditionally been designed via bottom-up approaches. These are often based on finding, by trial and error, a decomposition of the state space of the plant into regions (modes) in each of which a different linear control law is applied. This bottom-up approach can cause serious problems. How do we verify that performance specifications are met? How do we verify stability, controllability, observability? The usual theorems about linear control do not apply when there is mode switching between linear laws.

We call interacting networks of finite state control automata and continuous plants "hybrid systems". Those in which a single plant interacts with a single control automaton are called "simple hybrid systems" [13].

In contrast to the bottom-up mode switching approach, Kohn-Nerode, in [14], [8], [15] and many other papers, have developed a top-down approach for extracting the required finite state control automata from performance specifications and simulation models based on the relaxed calculus of variations. The basic outline of this approach is the following.

First corresponding to an autonomous problem, we formulate a non-negative Lagrangian on plant state trajectories which result from applying possible control functions c of time to produce the trajectory, in such a way that the smaller the value of its integral along a trajectory, the better the performance will be. The original Lagrangian is usually non-convex. The performance specification for the plant is reformulated as the requirement that the plant state trajectory has an integral within a prescribed ϵ of its minimum. A control function of time or state that does this from a given initial state is called an ϵ-optimal control function. That is all that is required in actual applications.

Second the non-linear optimal control problem is solved for using the theory of relaxed control. This optimal control is usually a physically non-realizable measure valued feedback control law.

Third a finite state, physically realizable control automaton for mode switching, approximating to that optimal control feedback law, is extracted which assures control optimal within a prespecified ϵ.

This top-down approach has focused attention on investigating methods for finding finite control automata for mode switching from performance specifications. Part of every performance specification are the various constraints on the plant trajectory. One of the simplest and yet still important examples of such constraints are viability constraints. A viability constraint restricts the plant state to stay perpetually in a specified set VS of plant states. The set VS is often called a viability set. See Aubin [4] for a thorough treatment of viability for continuous systems.

Nerode and Yakhnis began the study of viability for hybrid systems using fixed points of set valued maps in a widely circulated but unpublished report [20]. This paper is an extension of that report. In [16], Kohn, Nerode, Remmel and Yakhnis gave a very general treatment of viability using graph kernels. But that general treatment gives somewhat weaker results in specific cases than the treatment here, which stems from [20]. See [5] for a recent follow up on [20] in yet a different direction.

Our results apply to nonlinear plants with disturbances and imprecise measurements of plant states. We will focus on plants for which delays in the imposition of new control laws can be neglected.

2 A Viability Problem

A Plant Model. Our plant model is that of [20] and [16], but we allow differential inclusions.

Following [13], here is a brief description of a simple hybrid system. It consists of a continuous plant interacting with a control automaton at times $n\Delta$. The plant runs open loop inside $[n\Delta, (n+1)\Delta]$ based on the control $c_n = c_n(t)$ supplied at time $n\Delta$. The control automaton receives as input at time $n\Delta$, a measurement of the state of the plant at time $n\Delta$, $x(n\Delta)$. The plant runs open loop with no further inputs till time $(n+1)\Delta$. Then based on the control automaton state at time $(n+1)\Delta t$ and a measurement of the plant state $x((n+1)\Delta)$ at time $(n+1)\Delta$, it transmits a new control $c_{n+1}(t)$ to the plant to be used for the interval $[(n+1)\Delta, (n+2)\Delta]$, etc. For simplicity, we assume that the continuous plant is described by an ordinary vector differential equation

$$\dot{x} = f(x, c, d),$$

with x, c, d being finite dimensional vectors belonging to some Euclidean spaces. It is assumed that for any time t_0, for any time interval $[t_0, t_0+\Delta]$, for any initial state $x(t_0)$, and for any admissible control and disturbance functions $c(t), d(t)$ defined on $[t_0, t_0+\Delta]$, there is a unique $x(t)$ defined on $[t_0, t_0+\Delta]$ satisfying the differential equation. The function $x(t)$ will be called the plant state trajectory which starts at $x(t_0)$ and which is determined by $c(t)$ and $d(t)$ or which is guided by $c(t)$ and $d(t)$.

We can give an equivalent description of the plant by means of a differential inclusion if we wish to eliminate the explicit mention of disturbances. Let D be the admissible set of values of the disturbance parameter d, let $F(x,c) = \{f(x,c,d) : d \in D\}$. Then, by a well known lemma of Filippov ([6]), an equivalent description of the plant is

$$\dot{x} \in F(x, c).$$

For any time interval and control function $c(t)$, there is a plant trajectory satisfying the differential inclusion over the interval if and only if there is a measurable disturbance function $d(t)$ such that the plant trajectory satisfies the ordinary differential equation modelling the plant with functions $c(t), d(t)$ inserted respectively for c and d. The premises of Filippov lemma are only mildly restrictive.

However, independent of what the premises are, one may wish to change to the viewpoint of differential inclusions because, whatever the function $f(x, c, d)$, its trajectories are also trajectories of the differential inclusion $\dot{x}(t) \in F(x, c(t))$. Thus, the inclusion preserves information. In addition, putting the plant model in a form of a differential inclusion indicates that we do not know exactly how disturbances effect the plant velocities in state space, and in practice this is usually the case. So we will assume that continuous plants are modelled by differential inclusions. Plants modelled by vector differential equations with control and disturbance parameters are widely used. So we state some of our results and proofs in this setting too.

We remind the reader that a solution to a differential inclusion is an absolutely continuous function $x(t)$ satisfying the inclusion almost everywhere over a time interval. All our assumptions about the plant will be phrased in terms of conditions on the set-valued right hand side of the inclusion. Our assumptions on F serve to insure one or more of the following.

(I) Existence of a plant trajectory for any point of VS and for any admissible control law inserted into the differential inclusion.

(II) Continuous dependence of plant trajectories on initial plant state. That is, suppose we are given a fixed control law c and a sequence x_n of plant states converging to x. Then, for any plant trajectory $y(t)$ which begins at x and is guided by c, there exists a sequence of plant trajectories $y_n(t)$ which begin at x_n and are guided by c and which converges to $y(t)$ uniformly over the sampling interval. (All plant trajectories take their values from the viability set VS.)

For instance, to insure (I), typical assumptions about F might be ([3]):

1. F is upper or lower semicontinuous, and
2. the values of F are compact convex sets.

Conditions assuring (II) are given in [7]. For the rest of this paper, we prefer to state the premises of our theorems in terms of properties (I) and (II).

As in [16], we may restrict the control laws to those defined over the interval $[0, \Delta]$. The actual control laws arise by linear translation to the current time interval. Thus control strategies need output a control law only over interval $[0, \Delta]$.

Viability Sets. Suppose that we are given a subset VS of plant state space. The viability problem is to find a subset VS' of VS and a control strategy which will ensure that if a plant state is initially in VS', then the plant trajectories guided by that strategy will stay in VS forever.

Ensuring viability is a quite common problem in control engineering. The following are either viability problems or include viability as a requirement.

By a "positioning problem" we mean the problem of positioning a plant in a prescribed neighborhood of a given plant state. The prescribed neighborhood of that state is then the set VS. This problem usually divides into two stages. The first stage is to force the plant state into VS. The second stage is to force the plant to stay in VS forever, or until some other prescribed event occurs. The latter part we refer to as "stabilizing the plant in VS". A commonplace example is stabilizing house temperature in the vicinity of a value specified for a thermostat. Another example is to stabilize the output of a chemical reactor in the vicinity of a desired value.

Suppose we are given plant dynamics F and a set VS of plant states. Suppose VS' is a subset of VS and suppose we have a control strategy which will insure that if a plant state is initially in VS', then the plant trajectories guided by the strategy will stay in VS forever. The set VS is then called **viable** with respect to the plant dynamics F.

We compare our set-up with Aubin's [4]. We allow a narrower class of control strategies. Those we allow are those that maintain a constant control law during intervals between sensor readings of plant state. A control law is a function of time. When we say "constant control law during an interval of time", we mean that only one control function of time is used to determine control parameters for the differential equation during that time interval. In contrast, Aubin allows control strategies which may change the control law applied at any time. We are modelling digital control strategies, which do maintain constant control laws during successive intervals of time. We note that there are Aubin control strategies which cannot be implemented exactly by using finite state control automata, similar to what we have discussed elsewhere for measure valued relaxed optimal controls. Generally only approximations to such control strategies can be implemented by digital programs and this is exactly our concern.

3 Control Automaton

We identify the concepts of controller and control strategy described above with the concept of an input-output automaton. Recall that we are assuming that the set of plant states is a a subset of a Euclidean space. We denote by $B(X, a)$ the set of states whose distance the set X is less than a. We interpret $e > 0$ as the bound on the size of deviations of the measurements of a plant state from the true state. The definitions below follow the style of [13], except that our definitions cover the case of imprecise measurements and that we use differential inclusions to model plants. If A and B are sets, then we let 2^B denote the set of all subsets of B. If $G : A \to 2^B$ is a set valued map, then the graph of G is the subset of $A \times B$ which consists of the set of all $(a, b) \in A \times B$ such that $b \in G(a)$.

Definition 1. Suppose we are given a plant with viability set VS, and a set of admissible control laws C over time interval $[0, \Delta]$. A control automaton for such a plant is a nondeterministic input-output (Mealy) automaton $(Q, C, TO, M, q_{in}, VS')$ consisting of the following.

1. Its input alphabet is the set of measurements of plant states $M = B(VS, e)$.
2. Its output alphabet is a subset of C.
3. Its set of states is a discrete set Q.
4. Its transition-output function is a set-valued function TO whose graph is a subset of $Q \times M \times Q \times C$.
5. Its initial internal state q_{in}.
6. A set $VS' \subseteq VS$ of admissible initial states of the plant.

We require that the automaton produces a new state and new output pair (q,c) simultaneously. The automaton can simultaneously switch to the new state q and use the output c for a new control law to be used starting at the control automaton state transition time.

Definition 2. An infinite plant trajectory *guided* by a control automaton is a concatenation of a sequence of plant trajectories $x_n(t)$ over intervals $[n\Delta, (n+1)\Delta]$ with the following properties. (These are referred to as "parts" of the infinite trajectory.)

1. The initial trajectory x_0 begins at some admissible plant initial state x_{in} from the VS' and either it satifies the differential inclusion $\dot{x}_0 \in F(x_0, c_0)$ if the plant is modeled by the differential inclusion $\dot{x} \in F(x, c)$ or there is an admissible disturbance d_0 such that it is the trajectory determined by $\dot{x_0} = f(x_0, c_0, d_0)$ if the plant is modeled plant is modeled by a vector differential equation $\dot{x} = f(x, c, d)$ where c_0 is an initial control law such that $(q_{in}, m_{in}, q', c_0)$ is in the graph of TO for some measurement m_{in} of x_{in} and state q'.
2. The part of the trajectory numbered $n+1$ begins at the end of the part of the trajectory numbered n.
3. The $(n+1)$-th part of the trajectory x_{n+1} either satifies the differential inclusion $\dot{x}_{n+1} \in F(x_{n+1}, c_{n+1})$ if the plant is modeled by the differential inclusion $\dot{x} \in F(x, c)$ or there is an admissible disturbance d_{n+1} such that it is the trajectory determined by $\dot{x}_{n+1} = f(x_{n+1}, c_{n+1}, d_{n+1})$ if the plant is modeled plant is modeled by a vector differential equation $\dot{x} = f(x, c, d)$ where c_{n+1} is the control law output by the control automaton for a measurement the plant state at time $(n+1)\Delta$ and control automaton state q_n of the endpoint of the n-th part of the trajectory.

A control automaton is **correct** with respect to the viability set VS and the plant model if every plant trajectory guided by the automaton lies in the set VS.

4 Controllability Operators

We define a controllability operator over subsets of the viability set VS. This operator is defined relative to the choice of a real number Δ. This number represents the length of a constant sampling and control interval throughout the plant control process. We say that a measurement m corresponds to a subset Z of VS if $m \in B(Z, e)$.

We define the value of the controllability operator on a set $Z \subseteq VS$ in two steps.

Definition 3. The Controllability Relation. Consider a set $Z \subseteq VS$. Call a subset Z' of Z *controllable with target* Z if for any measurement m corresponding to the subset Z', there exists a nonempty subset C' of the set of admissible control laws C such that for any point x of Z' whose measurement is m, there is a control law $c \in C'$ insuring the following.

1. Any plant trajectory which starts at x and is guided by the control law c will end in the set Z at the end of the sampling interval.
2. Any plant trajectory which starts at x and is guided by the control law c remains in the set VS throughout the interval.

Note that if U, V are two subsets of Z controllable with target Z, then $U \bigcup V$ is also controllable with target Z. Thus there exists a largest subset Z' of Z which is controllable with target Z.

Definition 4. Controllability Operators. The value of the controllability operator on a set $Z \subseteq VS$ is the largest subset of Z which is controllable with target Z.

If H is a controllability operator, the definition above requires H to be monotone decreasing. That is, for every subset Z of VS,

$$H(Z) \subseteq Z.$$

Moreover if $U \subseteq V \subseteq VS$ and H is a controllability operator, then $H(U) \subseteq H(V)$. That is, $H(U) \subseteq U$ is also controllable with target V, and since $H(V)$ is the largest subset of V which is controllable with target V, $H(U) \subseteq H(V)$.

5 Fixed Points and Correctness

Recall that a simple hybrid system is a pair consisting of a plant and a finite state control automaton whose input is a plant state measurement at the end of a sampling interval and whose output is a control which is used to control the plant for the next sampling interval. The question we address is how to find a correct control automaton for the viability problem. This means that we are constructing a control automaton which, together with the given differential inclusion plant model and viability set VS, constitutes a simple hybrid system whose trajectories are viable in the sense that the plant state is always in VS if the initial state is an element of a suitable subset VS' of VS. In fact, our results can sometimes be used to verify that a hybrid system handed to us produces viable trajectories.

For our nest result, we only need the existence of plant trajectories beginning at any point of VS for any admissible control law. Any property of a plant differential inclusion ensuring such can be used as a premise for the results. For example, we can assume that the plant differential inclusion satisfies conditions (1) and (2) of section 2.

Theorem 5. *Suppose that*

1. *the plant is given by a differential inclusion has only a finite set of control laws C and satisfies condition (I) of section 2 and that*
2. *the plant state measurements consist of all values which deviate from the actual plant states in the viability set VS by less than $e > 0$.*

Assume that a viability set VS and the length of the sampling interval Δ are given. Then there exists a correct finite state control automaton if and only if all of the following conditions hold.

(A) *The corresponding controllability operator has a non-empty fixed point V.*
(B) *There exists a sequence of subsets of V, $(V(1), \ldots, V(n))$ and a sequence of subsets of C, $(C(1), \ldots, C(n))$ such that $V = \bigcup_{i=1}^{N} V(k)$ and for any k with $1 \leq k \leq n$, if m is a measurement for $V(k)$, then there is a control law $c \in C(k)$ corresponding to m and an index j such that any any plant trajectory which starts at an $x \in V(k)$ whose measurement is m and which is guided by c ends in $V(j)$ at the end of the sampling interval and stays entirely in VS.*

Proof. Suppose that the set C of admissible control laws is finite. Suppose that $A = (Q, C, TO, M, q_{in}, VS')$ where $M = B(VS, e)$ is a correct finite state control automaton for the plant and the viability set VS. We will construct a nonempty fixed point of the controllability operator representing the plant, the set VS, and the sampling interval Δ.

Let VS' be the set of admissible initial plant states for the automaton. For any set $X \subseteq VS$, let $A(X, n\Delta)$ be the set of end points of plant trajectories which begin at points of X and guided by the control automaton A over the interval $[0, n\Delta]$.

Put

$$V_0 = VS',$$
$$V_{n+1} = A(VS', n\Delta),$$
$$V = \bigcup \{V_n : n \geq 0\}.$$

For any automaton we call the corresponding set V as defined above, the set *swept out* by the automaton.

Claim. The set swept out by the automaton A is a nonempty fixed point of the controllability operator H.

Note that since $H(V) \subseteq V$, to prove our claim it is sufficient to show that $V \subseteq H(V)$. Let m be a measurement corresponding to V. Let x be a point of V with the measurement m. Thus there is some n such that $x \in V_n$. Since our automaton is correct our definitions ensure that there $z_0 \in VS'$ such the plant trajectory which starts at z_0 which is determined by A ends at x at time $n\Delta$. Suppose the state of A at time $n\Delta$ is q. Then since A is correct, $TO(q, m)$ is a nonempty subset of pairs $(q', c') \in Q \times C$ such that the plant trajectory which starts at x and is guided by c' will stay in VS during the next time interval of length Δ and will clearly end in V_{n+1}. Thus $x \in H(V)$ and hence $V \subseteq H(V)$.

Our claim shows that condition (A) of the conclusion of the theorem holds. To prove condition (B), suppose that the set of states of the A is $\{q_1 = q_{in}, q_2, \ldots, q_n\}$. Then let $V(i)$ the set of $x \in V$ such that there exists an $a_0 \in VS'$ and an $n \geq 0$ with the property that there is a plant trajectory guided by A which starts at a_0 and ends in x at time $n\Delta$ and the corresponding state of A in the interval $[(n-1)\Delta, n\Delta]$ is q_i. Let $C(i)$ be the union of the set of all c such that there is a state q with $(q, c) \in TO(q_i, m)$ where m is a measurment for some $x \in V(i)$. It follows that if x is in $V(i)$, m is its measurement for x, and $(q_r, c) \in TO(q_i, m)$, then c will guide the plant into $V(r)$. This proves (B).

Conversely, suppose (A), (B) hold. Let $Q = \{1, 2, ..., n\}$, where n is as in (B). We define a deterministic control automaton with the set of states Q. Let $TO(k, m)$ be the set of all pairs (j, c) such that $c \in C(k)$ and for any $x \in V(k)$ whose measurement is m, the plant trajectory which starts at x and is guided by c ends in $V(j)$ at the end of the sampling interval. Now choose any k in Q and put $VS' = V(k)$. Then consider control automaton A defined by $Q, C, TO, M, q_{in} = k, VS' = V(k))$. It is easy easy to see that every plant trajectory guided by this automaton at the ends of sampling intervals is in some $V(\ell_m)$ at time $m\Delta$ and always stays in VS during the interval $[(m-1)\Delta, m\Delta]$ for all $m \geq 1$. Thus A will be correct for VS.

For next theorem we assume that the plant is modelled by a vector differential equation $\dot{x} = f(x, c, d)$ where c is the control paramater and d is a disturbance parameter which has the following properties which are analogous to conditions (I) and (II) for differential inclusions described in section 2.

(I*) For every point in the set VS and any pair consisting of an admissible control law and an admissible disturbance function, there is a unique plant trajectory beginning from that point and guided by the control law and the disturbance function which exists over entire sampling interval and

(II*) The above plant trajectories depend continuously on their starting point and are contained in VS.

Theorem Closures of fixed points *Suppose that*

1. *the plant is modeled by a vector differential equation* $\dot{x} = f(x, c, d)$ *satisfying conditions (I*) and (I**) and*
2. *the plant state measurements consist of all values which deviate from the actual plant states in the viability set* VS *by less than* $e > 0$.

If the viability set VS is closed, then the closure of any fixed point of the controllability operator is itself a fixed point of that same operator.

Proof. Let H be a controllability operator relative to VS and sampling interval length Δ and let $V \subseteq VS$ be a fixed point of H. Let U be the closure of V with respect to Euclidean topology of the plant state space. Since VS is closed, $U \subseteq VS$. Furthermore, $H(V) \subseteq H(U)$. Thus $V \subseteq H(U) \subseteq U$. We will show that all limit points of V are in $H(U)$ so that $U = H(U)$ as desired.

Let u be a limit of a sequence elements in V, $v_0, v_1, \ldots$. Note that the set of measurements for U is the same as that for V. Let m be any measurement for u. Then for some $k \geq 0$, m is a measurement of v_n for all $n \geq k$. Since V is controllable with target V, there is a nonempty set of control laws $C_{m,n} \subseteq C$ corresponding to m which guide any plant tractory which starts at v_n into V at the end of a sampling interval, staying all the time in VS. Since c is finite, there must exist a $c \in C$ such that $c \in C_{m,n}$ for infinitely many n. Let N_c denote the set of all n such that $c \in C_{m,n}$. Now fix any disturbance function d. Then a plant trajectory $x_u(t)$ determined by u, c, d is a limit of the plant trajectories $x_{v_n}(t)$ determined by v_n, c, d for $n \in N_c$. Then for any $0 \leq t \leq \Delta$,

$$lim_{n \to \infty, n \in N_c} x_{v_n}(t) = x_u(t).$$

Thus since it is the case that $x_{v_n}(t) \in VS$ for all $n \geq 0$ and all $0 \leq t \leq \Delta$ and VS is closed, it follows that $x_u(t) \in VS$ for all $0 \leq t \leq \Delta$. Moreover since since we are assuming that $x_{v_n}(\Delta) \in V$ for all $n \geq 0$ and all $0 \leq t \leq \Delta$ and U is the closure of V, $x_u(\Delta) \in U$. Thus c witnesses that $u \in H(U)$ and we have proved that $U \subseteq H(U)$. Hence U is a fixed point of the controllability operator.

The theorem above holds for plants modeled by differential inclusions satisfying the conditions for existence of plant trajectories and their continuous dependence on initial plants states, that is, for plants modeled by differential inclusions satisfying conditions (I) and (II) of section 2. The proof is similar to proof of Theorem 5.2.

A slightly more general theorem holds under the assumptions of the preceding theorem on a plant model. The proof is similar to that of the preceding theorem.

Theorem Closure of controllable sets *Suppose that*

1. *the plant is modeled by a vector differential equation $\dot{x} = f(x, c, d)$ satisfying conditions (I*) and (I**) and*
2. *the plant state measurements consist of all values which deviate from the actual plant states by less than $e > 0$.*

If a set Z is closed, then the closure of any subset of Z which controllable with target Z is also a subset of Z which is controllable with target Z.

6 Variations

The transition-output function of a control automaton is often given in practice by inequalities separating measurements at which transitions occur. We would prefer it if the correctness of the control automaton does not depend on whether a strict inequality is replaced by non-strict one, since exact values cannot be sensed in practice. This preference is often mathematically justified. We would like to explain why.

Consider a control automaton $A = (Q, C, TO, M, q_{in}, VS')$ with a finite set of output control laws C. Equip C and Q, the set of states for A, with the discrete topology. Let $M = B(VS, e)$, where VS is the viability set, have its usual topology inherited from Euclidean space. Put the product topology on the Cartesian products $Q \times M \times Q \times C$. Let the graph of $\overline{TO}$ be the closure of the graph TO in $Q \times M \times Q \times C$. Then define the *closure of the control automaton* A to be the automaton $\overline{A} = (Q, C, \overline{TO}, M, q_{in}, VS')$.

We note that in a wide variety of applications the only information we need to determine the control law required for the next interval is a measurement of the current plant state and the control law used in previous interval since the control law often contains the information about the settings of the various parameters of the actuators used to physically control the plant. In such a situation, we want to let the set of states Q of the automaton be identical with the finite set of control laws C for the output alphabet of the control automaton. Moreover the only transitions that we want is when the new control law is the same as the new state. Thus we want to act with a control automaton $A = (Q, C, TO, M, q_{in}, VS')$ where $Q = C$ and the graph of TO is a set tuples of $C \times M \times C \times C$ of the form (c, m, c', c').

Theorem 6. *Suppose that the plant is either modeled by a differential inclusion $\dot{x} \in F(x, c)$ satisfying condition (I) or is modeled by a vector differential equation $\dot{x} = f(x, c, d)$ satisfying condition (I*). Moroever assume that the set of plant state measurements M consist of all values which deviate from the actual plant states in the viability set VS by less than $e > 0$. Then if $A = (Q, C, TO, M, q_{in}, VS')$ is a control automaton with a finite set of control laws C as it output alphabet which is correct for the viability set VS, then the closure of A, $\overline{A}$, is also a correct for the viability set VS.*

Proof. Let $(x_0(t), x_1(t), \ldots)$ be a sequence of plant trajectories which are guided by the control automation $\overline{A}$. Thus there is a sequence of control laws $c_0, c_1, \ldots$, a sequence $q_0, q_1, \ldots$ of states of $\overline{A}$, and a sequence of measurements $m_0, m_1, \ldots$ such that

(i) $x_i(t)$ is defined on $[i\Delta, (i+1)\Delta]$ and takes values in VS for all $i \geq 0$,
(ii) $x_0(0) \in VS'$ and $x_i((i+1)\Delta) = x_{i+1}((i+1)\Delta)$ for all $i \geq 0$,
(iii) m_i is a measurement for $x_i(i\Delta)$ for all $i \geq 0$,
(iv) $q_0 = q_{in}$ and (q_0, m_0, q', c_0) is in the graph of $\overline{TO}$ for some q', and
(v) for all $i \geq 0$, $(q_i, m_{i+1}, q_{i+1}, c_{i+1})$ is in the graph of $\overline{TO}$.

Clearly to show that $\overline{A}$ is correct, it is enough to show that $(x_0(t), x_1(t), \ldots)$ is also a sequence of plant trajectories which are guided by the control automation A. Our idea is quite simple. That is, we shall show by induction that there is a sequence of measurements $m_0', m_1', \ldots$ such that (i)-(v) hold for the same sequence of control laws $c_0, c_1, \ldots$ and sequence of states $q_0, q_1, \ldots$ with A replacing $\overline{A}$ and $m_0', m_1', \ldots$ replacing $m_0, m_1, \ldots$. That is, suppose by induction that there is a sequence of measurements $m_0', m_1', \ldots, m_n'$ such that

(i) $x_i(t)$ is defined on $[i\Delta, (i+1)\Delta]$ and takes values in VS for all $0 \le i \le n$,
(ii) $x_0(0) \in VS'$ and $x_i((i+1)\Delta) = x_{i+1}((i+1)\Delta)$ for all $0 \le i < n$,
(iii) m_i' is a measurement for $x_i(i\Delta)$ for all $0 \le i \le n$,
(iv) $q_0 = q_{in}$ and (q_0, m_0', q', c_0) is in the graph of TO for some q', and,
(v) for all $n > i \ge 0$, $(q_i, m_{i+1}', q_{i+1}, c_{i+1})$ is in the graph of TO.

Now since $(q_n, m_{n+1}, , q_{n+1}, c_{n+1})$ is in the graph of $\overline{TO}$, there is a sequence

$$(q_{n,0}, m_{n+1,0}, q_{n+1,0}, c_{n+1,0}), (q_{n,1}, m_{n+1,1}, q_{n+1,1}, c_{n+1,1}), \ldots$$

of elements of the graph of TO which converges to $(q_n, m_{n+1}, q_{n+1}, c_{n+1})$ in the product topology. Thus there must exist a $k > 0$ such that $q_{n,\ell} = q_n$, $q_{n+1,\ell} = q_{n+1}$, and $c_{n+1,\ell} = c_{n+1}$ for $\ell \ge k$. Moreover for all $\delta > 0$, there exists a k_δ such that $|m_{n+1} - m_{n+1,\ell}| < \delta$ for all $\ell > k_\delta$. Now since m_{n+1} is a measurement for $x_n((n+1)\Delta)$, it must be the case that $|x_n((n+1)\Delta) - m_{n+1}| < e$. Then let $\delta' = \frac{e - |x_n((n+1)\Delta) - m_{n+1}|}{2}$. Then it is easy to see that if $\ell > p = max(k, k_{\delta'})$, then $q_{n,\ell} = q_n$, $q_{n+1,\ell} = q_{n+1}$, $c_{n+1,\ell} = c_{n+1}$, and $|x_n((n+1)\Delta) - m_{n+1,\ell}| < e$. Thus $m_{n+1}' = m_{n+1,p+1}$ is an measurment for $x_n((n+1)\Delta)$ and $(q_n, m_{n+1}', q_{n+1}, c_{n+1})$ is in the graph of TO so that we can extend our sequence of plant trajectories guided by A from $x_0(t), \ldots, x_n(t)$ to $x_0(t), \ldots, x_n(t), x_{n+1}(t)$ while keeping the same set of states and control laws as the corresponding plant trajectories guided by $\overline{A}$. Thus A is correct for the viability set VS if and only if $\overline{A}$ is correct for the viability set VS.

Example 1. Suppose that a plant is controlled by a control automaton $A = (Q, C, TO, M, q_{in}, VS')$ where $Q = C$ and the graph of TO is a set tuples form $C \times M \times C \times C$ of the form (c, m, c', c'). Suppose that A is correct with respect to a viability set VS relative to a set of initial plant states VS'. Suppose that the graph of TO is defined by

1. (x, c, c, c) is in the graph of TO if $x < g$
2. (x, c, c', c') is in the graph of TO, otherwise.

Then we can replace $<$ in this definition by $\le$ and the automaton is still correct.

7 Construction of a Fixed Point for the Controllability Operator

A controllability operator, according to its definition, is monotone decreasing. By transfinitely iterating the operator over the ordinals beginning from the set VS, we get a fixed point. Of course, it may turn out to be empty. Even it is nonempty, there is no guarantee that we will reach a fixed point after only finitely many or even countably many interations of the controllability operator. However we can show that if the viability set VS is compact, then it suffices to iterate the operator at most ω times.

Theorem 7. *Suppose that the plant is either modeled by a differential inclusion $\dot{x} \in F(x,c)$ satisfying conditions (I) and (II) or is modeled by a vector differential equation $\dot{x} = f(x,c,d)$ satisfying conditions (I*) and (II*). Assume that the set of plant state measurements M consist of all values which deviate from the actual plant states in the viability set VS by less than $e > 0$. Let VS be a viability set which compact and let H denote the controllability operator for the plant relative to VS and a some fixed length of a sampling interval Δ. Suppose that the admissible control laws maintain a constant control laws in any interval and that up to translation, there are only a finite number of control laws C. Let $Z_0, Z_1, \ldots$ be defined inductively by*

$$Z_0 = VS,$$
$$Z_{n+1} = H(Z_n), n \geq 0,$$

and let $V = \bigcap_{n \geq 0} Z_n$. Then

1. *If Z_n is empty for some n, the controllability operator H of the plant has no nonempty fixed points*
2. *If Z_n is nonempty for all n, then V is a maximal fixed point of the controllability operator H which is nonempty.*

Proof. Because the controllability operator is monotone, i.e. $U_1 \subseteq U_2 \subseteq VS$ implies $H(U_1) \subseteq H(U_2)$, it is easy to prove by induction that if $H(U) = U$ for $U \subseteq VS$, then $U \subseteq Z_n$ for all n. Thus if $Z_n = \emptyset$ for some n, then clearly H has no nonempty fixed points. This establishes part (1) of the theorem.

To prove part (2) of the theorem, note that if U is fixed point of H, then since $U \subseteq Z_n$ for all n, $U \subseteq V = \bigcap_{n>0} Z_n$. Thus to prove V is a maximal fixed point of H, we need only show $H(V) = V$. Since we automatically have that $H(V) \subseteq V$, we, in fact, need only show $V \subseteq H(V)$. Moreover since VS is closed, the controllability operator takes closed sets into closed sets. Thus it easily follows by induction that Z_n is compact for all $n \geq 0$. Hence by Cantor's intersection theorem, V is non-empty and compact. So let m be a measurement of some $x \in V$. Since $x \in Z_n$ and $H(Z_n) = Z_{n+1}$ for all $n \geq 0$, then for each n, there exists a finite nonempty set of admissible control laws $C_{m,n}$ such that any plant trajectory which starts at x and which is determined by a control law $c \in C_{m,n}$ and any admissible disturbance ends in Z_{n+1} and stays entirely in VS. Since C is finite, it follows that there must be some $c \in C$ such that $c \in C_{m,n}$ for infinitely many n. But then any plant trajectory which starts at x which is determined by a control law c and any admissible disturbance ends in Z_{n+1} for infinitely many n and stays entirely in VS. But since $\{Z_n\}_{n \geq 0}$ is a decreasing sequence, it follows that any plant trajectory which starts at x which is determined by a control law c and any admissible disturbance ends in V and stays entirely in VS. Thus c witnesses that $x \in H(V)$. Thus $V \subseteq H(V)$ and hence we can conclude that V is the maximal fixed point of H.

Remarks.

1) In many examples the sequence of sets above becomes constant at a finite index. Why this happens requires further investigation. The problem is investigated in [5].

2) If Z_0 is a set of initial plant states, and a control automaton is constructed using the initial segment of the sequence up to the n-th set (as indicated in the statement of Theorem 7, then the guided plant trajectories stay within the set VS during the time interval $[0, n\Delta]$. This means that the sequence $\{Z_k : k = 0, \ldots, n-1\}$ will yield a correct control automaton in case the control process terminates at $n\Delta$. Thus, approximations to fixed points, as well as the fixed points themselves, may yield useful software. Finally, we note that the process of translating an abstract finite state control automaton into AD and DA converters and a finite digital controller has been described in [13] and [16] and is implementable with off the shelf hardware.

References

1. Alur, R., Courcoubetis, C., Henzinger, T., Ho, P-H.: Hybrid Automata: An Algorithmic Approach to the Specification and Verification of Hybrid Systems. in [10] (1993) 209–229
2. Aubin, J.P.: *Set Valued Analysis*. Birkhauser. (1990)
3. Aubin, J.P., Cellina, A.: *Differential Inclusions*. Birkhauser. (1990)
4. Aubin, J.P.: *Viability Theory*. Birkhauser. (1991)
5. Deshpande, A.,Varaiya, P.: Viable Control of Hybrid Systems, this volume.
6. Filippov, A.F.: Classical Solutions of Differential Equationswith multivalued right hand side. SIAM J. Control and Optimization **5** (1967) 609–621
7. Filippov, A.F.: *Differential Equations with Discontinuous Right Hand Sides*. Kluwer Academic Publishers. (1988)
8. Ge, X., Kohn, W., Nerode, A., Remmel, J.B.: Algorithms for chattering approximations to relaxed controls. Tech. Report 94–23, Mathematical Sciences Institute, Cornell University. (1994)
9. Ge, X., Nerode, A.: Effective Content of Relaxed Calculus of Variations: semicontinuity. Tech Report, Feb. 1995, Mathematical Sciences Institute, Cornell University. (1995)

10. Grossman, R., Nerode, A., Rischel, H., Ravn, A., eds.: *Hybrid Systems.* Springer Lecture Notes in Computer Science. (1993)
11. Kohn, W., Nerode, A.: An autonomous control theory: an overview. IEEE Symposium on Computer Aided Control System Design (March 17–19, 1992, Napa Valley, CA), (1992) 204–210
12. Kohn, W., Nerode, A.: Multiple Agent Autonomous Control Systems. Proc. 31st IEEE CDC Tucson, Ar. 2956-2962 (1993)
13. Kohn, W., Nerode, A.: Models for Hybrid Systems: Automata, Topologies, Controllability, Observability. in [10], (1993) 317–356
14. Kohn, W., Nerode, A.: Multiple agent autonomous hybrid control systems. *Logical Methods: A Symposium in honor of Anil Nerode's 60th Birthday,* (J. N. Crossley, Jeffrey B. Remmel, Richard A., Shore, Moss E. Sweedler, eds), Birkhauser. (1993)
15. Kohn, W., Nerode, A., Remmel, J.B., Ge, X.: Multiple Agent Hybrid Control: Carrier Manifolds and Chattering Approximations to Optimal Control. CDC94 (1994)
16. Kohn, W., Nerode, A., Remmel, J.B., Yakhnis, A.: Viability in Hybrid Systems. Theoretical Comp. Sc. **138** (1995) 141–168
17. Nerode, A., Remmel, J.B., Yakhnis, A.: Hybrid Systems Games: Extraction of Control Automata with Small Topologies. Tech Report No. 93-102, December 1993, Mathematical Sciences Institute, Cornell University. (1993)
18. Nerode, A., Yakhnis, A.: Modeling Hybrid Systems as Games. CDC92 (1992) 2947-2952
19. Nerode, A., Yakhnis, A.: Hybrid Games and Hybrid Systems. Tech Report No. 93-77, October 1993, Mathematical Sciences Institute, Cornell University. (1993)
20. Nerode, A., Yakhnis, A.: Control Automata and Fixed Points of Set-Valued Operators for Discrete Sensing Hybrid Systems, Tech. Report 93-105, Mathematical Sciences Institute, Cornell University. (1993)
21. Roxin, E.O.: The Attainable set in Control Systems, in *Mathematical Theory of Control,* Lecture Notes in Pure and Applied Mathematics, volume 142, M. Joshi and A. Balakrishnan editors, Marcel Dekker, inc. (1993)
22. Warga, J.: *Optimal Control of Differential and Functional Equations.* Academic Press. (1972)

Verification of Hybrid Systems Using Abstractions*

Anuj Puri and Pravin Varaiya

Department of Electrical Engineering and Computer Science,
University of California, Berkeley, CA 94720

Abstract. A hybrid system models both discrete event and continuous dynamics. We present a modeling formalism and a verification methodology for hybrid systems. The verification methodology is based on abstracting the continuous dynamics in the hybrid system by simpler continuous dynamics. We present two methods for doing this: in the first method, a differential inclusion is replaced with a simpler differential inclusion; in the second method, we look at the timing information that is relevant to the verification problem, and construct an abstraction of the hybrid system with a timed automaton. We illustrate our methodology by applying it to the train-gate-controller example.

1 Introduction

Traditional control theory and system theory have been successfully used in design of a large class of systems. But in the design of large, complex distributed systems such as Air-Traffic Control Systems, Automated Factories, and Intelligent Vehicle Highway Systems [10] etc., issues such as communication, coordination, planning and logic also have a prominent role to play. Systems which combine discrete event dynamics to model the coordination and the logical behavior, and continuous time dynamics for modeling the continuous behavior are called hybrid systems. In this paper, we present a modeling formalism and a verification methodology for hybrid systems. For a survey of other work on hybrid systems, see [5].

A hybrid system consists of coordinating components. Each component is modeled by a hybrid automaton [1, 6]. Our methodology for verification is based on a conservative abstraction of the dynamics of a hybrid automaton with simpler dynamics. We provide two methods for doing this. With the first method, a differential equation is abstracted by a differential inclusion. In the second method, we look at the timing information in the system that is relevant to the verification problem, and abstract the system with a timed automaton. We illustrate how real-time verification tools can be used to verify hybrid systems once they are abstracted. We illustrate both the modeling formalism and the verification methodology by looking at the train-gate-controller example.

* Research supported by the California PATH program and by the National Science Foundation under grant ECS9417370. Email: anuj@eecs.berkeley.edu

In Section 2, we introduce hybrid automata and model the train-gate-controller example with them. In Section 3, we introduce the abstraction mechanism and our verification methodology, and illustrate it by applying it to the train-gate-controller example.

2 Hybrid Systems

2.1 Prelimineries

$\mathcal{R}$ are the reals, and $\mathcal{R}^+$ are the nonnegative reals. A differential equation is $\dot{x} = f(x)$ where $x \in \mathcal{R}^n$ and $f : \mathcal{R}^n \to \mathcal{R}^n$. A differential inclusion is written as $\dot{x} \in f(x)$ where $x \in \mathcal{R}^n$ and f is a set-valued map from $\mathcal{R}^n$ to $\mathcal{R}^n$ (i.e., $f(x) \subset \mathcal{R}^n$). A solution to the differential inclusion with initial condition $x_0 \in \mathcal{R}^n$ is any differentiable function $\phi(t)$, where $\phi : \mathcal{R}^+ \to \mathcal{R}^n$ such that $\phi(0) = x_0$ and $\dot{\phi}(t) \in f(\phi(t))$. A differential inclusion has a family of solutions starting with the same initial condition.

Definition 2.1 *For a differential inclusion $\dot{x} \in f(x)$, $\phi(x_0, t)$ is a solution of the differential inclusion with initial condition x_0 provided $\dot{\phi}(x_0, t) \in f(\phi(x_0, t))$. We denote the piece of the trajectory on the interval $[0, t]$ as $\phi(x_0, [0, t])$. For $X_0 \subset \mathcal{R}^n$, $\phi(X_0, t) = \{\phi(x_0, t) | x_0 \in X_0$ and $\phi(x_0, t)$ is a solution of the differential inclusion $\}$.*

During a transition in the hybrid automaton, the state may be initialized to a new value. This is formalized by associating an initialization relation $\lambda \subset \mathcal{R}^n \times \mathcal{R}^n$ with an edge, where for $(x, y) \in \lambda$, x is initialized to y. Define Γ to be the set of all initialization relations.

An enabling condition is $\delta \subset \mathcal{R}^n$. The set of all such conditions is Φ.

2.2 Hybrid Automata

A hybrid automaton consists of a finite set of control locations [1, 6]. At each location, the continuous state evolves according to some inclusion. The edges between control locations are labeled with enabling conditions. A jump can be made from one control location to another provided the enabling condition is satisfied.

Formally, a hybrid automaton is $H = (L, D, I, \mu, \Delta)$ where

- L is a finite set of control locations
- D associates the differential inclusion $D(l)$ with location l
- $I \subset L \times \mathcal{R}^n$ is the set of initial states
- μ associates the invariance condition $\mu(l) \subset \mathcal{R}^n$ with location l
- $\Delta \subset L \times \Phi \times \Gamma \times L$ is the set of edges

— an edge $e = (l, \delta, \lambda, l') \in \Delta$ is an edge from location l to location l' labeled with the enabling condition δ, and initialization relation λ.

The state of the hybrid automaton H is $(l, x) \in L \times \mathcal{R}^n$. Figure 1 illustrates a hybrid automaton. The arrow indicates the initial state. In this case the initial state is $(B, x = 0)$.

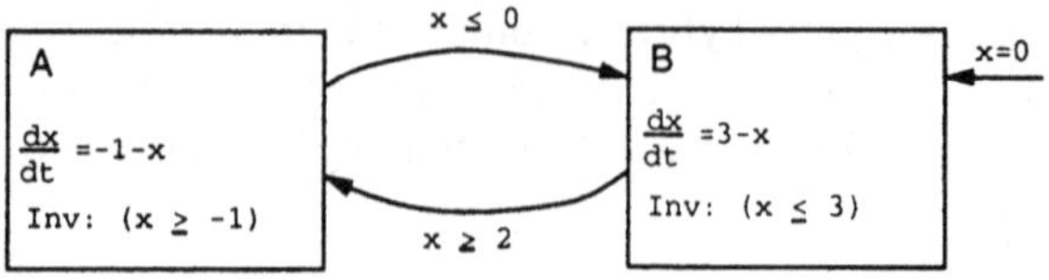

Fig. 1. Example Hybrid Automaton

2.3 Semantics

The state (l, x) of the hybrid automata changes in two different ways. Either the location l remains fixed and x evolves according to the differential inclusion $D(l)$, or a discrete transition is made from location l to location l'. Definition 2.2 defines the *time* transition, and definition 2.3 defines the *discrete* transition.

Definition 2.2 *For* $t = 0$, *we define* $(l, x) \xrightarrow{0} (l, x)$ *For* $t > 0$, *we say* $(l, x_0) \xrightarrow{t} (l, x)$ *provided* $x = \phi(x_0, t)$ *and* $\phi(x_0, [0, t]) \subset \mu(l)$, *where* $\phi(x_0, t)$ *is a solution to the differential inclusion* $\dot{x} \in D(l)(x)$.

Definition 2.3 $(l, x_0) \xrightarrow{(\delta,\lambda)} (l', x)$ *provided* $x_0 \in \delta$, $x \in \lambda[x]$, *and* $(l, \delta, \lambda, l') \in \Delta$.

A trajectory of the hybrid automaton is $\pi = (l_0, \phi_0, I_0)(l_1, \phi_1, I_1)\ldots$ where $l_i \in L$, I_i is an interval and $\phi_i : I_i \to \mathcal{R}^n$ is a continuous trajectory. The interval $I_i = [T_i, T_i']$ where $T_0 = 0$ and $T_{i+1} = T_i'$. For example $[0, 7][7, 7][7, 7][7, 10.3]\ldots$ is a valid interval sequence in which time progresses during the interval $[0, 7]$, there are three instantaneous transitions at time 7 and time again progresses during $[7, 10.3]$, etc.

2.4 Reach Set

We next extend definition 2.2 and definition 2.3 to sets of states. For a set of states R, we define $\mathcal{T}[R]$ to be all the states reached from R by a time transition, and $\mathcal{D}[R]$ to be all the states reached from R by a discrete transition.

Definition 2.4 $\mathcal{T}[R] = \{(l, x) | (l, x_0) \in R$, *and for some* $t \geq 0, (l, x_0) \xrightarrow{t} (l, x)\}$.

Note that $R \subset \mathcal{T}[R]$, since $(l, x) \xrightarrow{0} (l, x)$.

Definition 2.5 $\mathcal{D}[R] = \{(l', x) | (l, x_0) \in R$ *and* $(l, x_0) \xrightarrow{(\delta,\lambda)} (l', x)\}$

The Reach Set of a hybrid automaton is all the states that can be reached starting from an initial state. A state is reached by a sequence of *time* and *discrete* transitions.

We compute the Reach Set iteratively as follows :

$$R_0 = I$$
$$R_{k+1} = \mathcal{T}[R_k] \bigcup \mathcal{D}[\mathcal{T}[R_k]]$$

Definition 2.6 *We define the Reach Set, $R[H]$, of hybrid automaton H to be $R[H] = \bigcup_k R_k$*

As an example, we compute the Reach Set of the hybrid automaton of Figure 1.

Reach Set Computation:

$$R_0 = (B, x = 0)$$
$$R_1 = (A, 2 \leq x \leq 3) \bigcup (B, 0 \leq x \leq 3)$$
$$R_2 = (A, -1 \leq x \leq 3) \bigcup (B, -1 \leq x \leq 3)$$
$$R_3 = R_2 = R[H]$$

2.5 Coordination

A hybrid system will consist of several components with communication and coordination between them. We model each component by a hybrid automaton and define the coordination between the components to be the product hybrid automaton.

Given M components, we define the alphabet or the event set of the hybrid system to be $\Sigma = L_1 \times \cdots \times L_M$ where L_i is the set of control locations for component i. The continuous state of the system is $x = (x_1, \ldots, x_M) \in \mathcal{R}^n$, where $x_i \in \mathcal{R}^{k_i}$, and $\sum_{i=1}^{M} k_i = n$. Each component is modeled by a hybrid automaton. The continuous state of component i is $x_i \in \mathcal{R}^{k_i}$. The enabling condition on the edges are functions of the global state — combination of control location and continuous state of the other components.

Given component hybrid automata $H_i, i = 1, \ldots, M$, we define the coordination of the component hybrid automata to be the product hybrid automaton $H = H_1 \bigotimes \cdots \bigotimes H_M$. The product hybrid automaton represents coordination between the component automata. The control locations in the product hybrid automaton are the Cartesian product of the locations of the component hybrid automata. At location $l = (l_1, \ldots, l_M)$, the x_i component of the continuous state evolves with the inclusion of component i at location l_i. A transition can be made from one location to another provided the transition is permitted by each of the components.

2.6 Train-Gate-Controller Example

In this section, we illustrate the preceding concepts by an example. We model a system consisting of three components: the train, the gate, and the gate controller. The setup is shown in Figure 2. The train runs on a circular track. The train track passes across a road on which there is a gate which must be lowered to stop the traffic when the train approaches, and raised after the train has passed the road. The gate controller gets information from sensors located on the track and lowers or raises the gate. The length of the track is 25, the train location is indicated by the state variable $y \in [0,25)$, $y = 0$ is the position of the gate, the gate height is represented by state variable x and the sensors are located at $y = 5$ and $y = 15$.

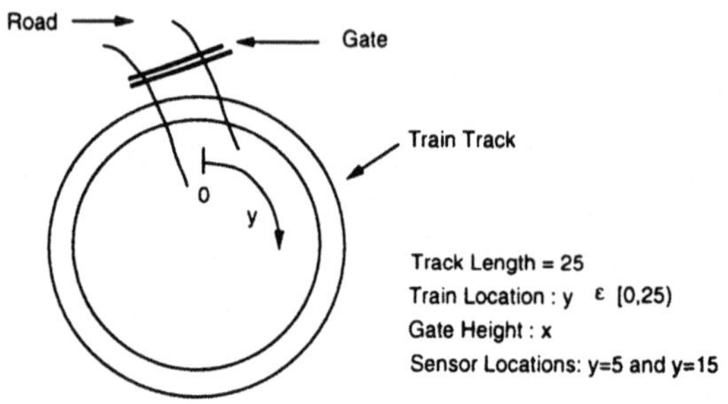

Fig. 2. Train-Gate Controller Example

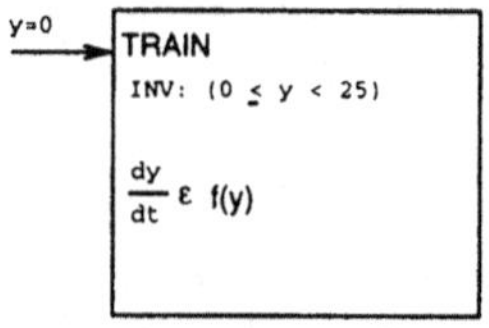

Fig. 3. Hybrid Automaton for a Train

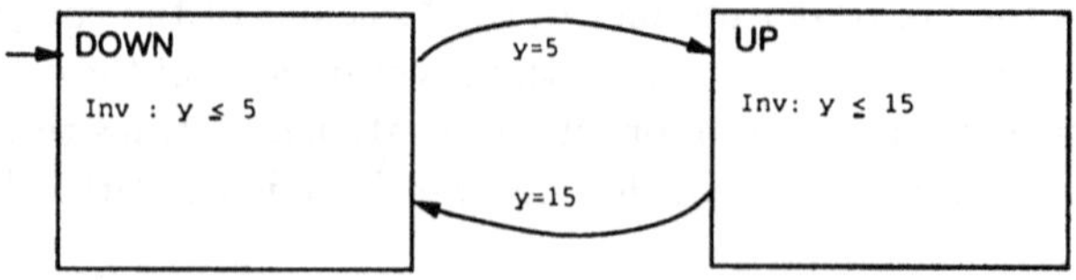

Fig. 4. Hybrid Automaton for the Gate Controller

The train automaton T, the gate controller automaton C and the gate automaton G are shown in Figures 3, 4, and 5. Figure 6 shows the product hybrid automaton $H = T \bigotimes C \bigotimes G$.

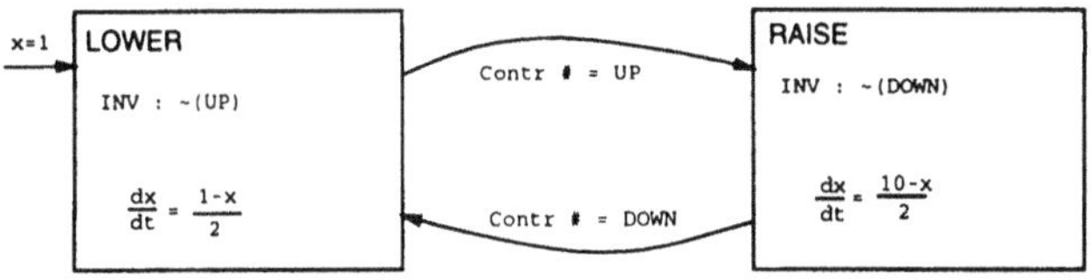

Fig. 5. Hybrid Automaton for the Gate

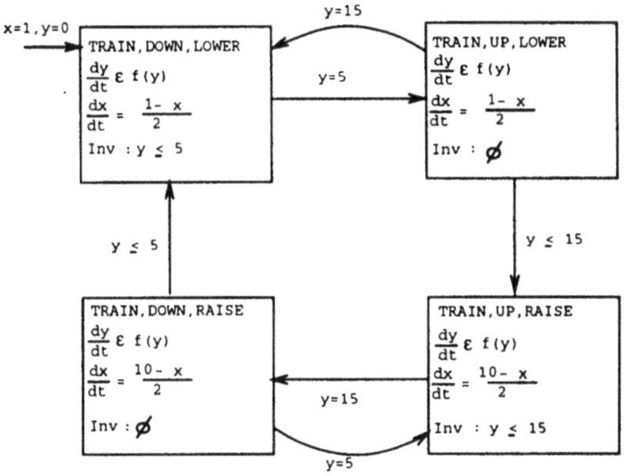

Fig. 6. Hybrid Automaton for the Train-Gate-Controller System

2.7 Specification

Given a hybrid automata with state space $L \times \mathcal{R}^n$, a specification is $S \subset L \times \mathcal{R}^n$.

Definition 2.7 *For a hybrid automaton $H = (L, D, I, \mu, \Delta)$ with specification $S \subset L \times \mathcal{R}^n$, we say H satisfies the specification provided $R[H] \subset S$.*

Example:
For the Train-Gate-Controllere example, we want to prove that the gate is always down when the train reaches the gate. We choose as our specification $S = L \times (\mathcal{R}^2 - (y = 0) \wedge (x > 4))$.

We will check if $R[H] \subset S$. If this is satisfied, then the gate is always down when the train reaches the gate.

3 Verification using Abstractions

To verify the hybrid system, we reason on two levels. At a logical level, the details of the differential equation are ignored, and we reason about sequence of events. Time constraints may be taken into account as well. In reasoning with the differential equation, we ignore other aspects of the system. Instead, we focus on a simple specification the differential equation has to satisfy. This specification becomes a conservative abstraction of the differential equation, and replaces it when we reason at the logical level.

The reasoning at the two levels has to be consistent. We should not arrive at a conclusion that is not valid. We do this by conservatively abstracting the differential equation, and using the abstraction to analyze the system.

We propose two methods to abstract a differential equation. With the first method, a differential equation is abstracted by a simpler differential inclusion. Instead of analyzing the system with the inclusion $\dot{x} \in f(x)$, we analyze the simpler inclusion $\dot{x} \in g(x)$ where $f(x) \subset g(x)$. The second inclusion is a conservative abstraction of the first inclusion, because the solutions of the first inclusion are also solutions of the second inclusion. The second inclusion $\dot{x} \in g(x)$ may be simpler; for example, it may be a piecewise constant inclusion.

In the second approach, we replace a differential equation with a timed automaton. The idea is to look at the timing information in the system that is relevant to the verification problem, and abstracting the system with a timed automaton which contains this information. We do this by getting bounds on the evolution of trajectories in a differential equation. These bounds are functions of time. We use a timed automaton to "encode" these bounds by forming a mapping from the states of the timed automaton to the space $\mathcal{R}^n$ of the differential equation.

3.1 Abstracting with Differential Inclusions

Suppose $\dot{x} \in f(x)$ is a differential inclusion at a control location of the hybrid automaton. The differential inclusion may be difficult to analyze. For this reason, we abstract it with something simpler, for example $\dot{x} \in g(x)$ where $f(x) \subset g(x)$. The inclusion $\dot{x} \in g(x)$ may be simpler; for example, it may be a piecewise constant inclusion [9].

Example:
The train $\dot{y} \in f(y)$ is abstracted with the simple inclusion $\dot{y} \in [v_{min}, v_{max}]$ where v_{min} is the minimum speed at which the train travels, and v_{max} is the maximum speed. Since v_{min} is the minimum speed of the train, and v_{max} is the maximum speed, $\dot{y} \in [v_{min}, v_{max}]$ is a conservative abstraction. We choose $v_{min} = 0$, and $v_{max} = 3$, the maximum speed of the train.

The hybrid automaton obtained by abstracting a differential inclusion is an abstraction.

Theorem 1. *If $H = (L, D, I, \mu, \Delta)$ is a hybrid automaton, and $H_A = (L, D_A, I, \mu, \Delta)$ where $D(l)(x) \subset D_A(l)(x)$ for every control location $l \in L$, then H_A is an abstraction of H and $R[H] \subset R[H_A]$.*

The reach set of the abstracted automaton is a conservative or upper bound on the reach set of the original hybrid automaton. Therefore, if we show H_A satisfies the specification, then clearly H does as well. For an application of this approach to an Automated Vehicle Highway System, see [8].

3.2 Abstracting with Timed Automata

A timed automaton [2] is a hybrid automaton in which all the continuous variables are clocks. A clock x moves with rate "1" (i.e., $\dot{x} = 1$). On a transition, the clock may get reset to "0".

To conservatively abstract a differential inclusion, we look at its evolution. For example, suppose we want to abstract the differential inclusion $\dot{x} \in f(x)$, which starts from an initial condition $x(0) \in X_0$. To abstract the system, we look at $\phi(X_0, t)$ — the set of trajectories starting from some initial condition in X_0. We obtain bounds on the evolution of the differential inclusion, and "encode" this information with a clock. The relationship between the differential inclusion and the clock is then made by a mapping.

Example:
Consider the differential equation $\dot{x} = \frac{(1-x)}{2}$ *in the gate automaton, whose solution is* $x(t) = (x(0) - 1)e^{-\frac{t}{2}} + 1$. *Suppose* $x(0) \in X_0 = [1, 10]$. *Then* $\phi(X_0, t) = [1, 9e^{-\frac{t}{2}} + 1]$. *Define the set-valued map* h *where* $h(t) = \phi(X_0, t)$. *If we think of "t" as a clock recording time, then* $h(t)$ *provides a bound on the region in which* $x(t)$ *lies, provided* $x(0) \in X_0$. *Notice, in general, it may be difficult to compute* $\phi(X_0, t)$. *In this case,* $h(t)$ *can be used to bound* $\phi(X_0, t)$, *i.e.,* $\phi(X_0, t) \subset h(t)$.

Each control location l can be abstracted in this manner provided we associate an initial set $X_0(l) \subset \mathcal{R}^n$ with it. This has the affect of forgetting the value of x when a transition is made into the location l, and instead bounding it with the set $X_0(l)$. Of course, to be a conservative bound, the state must actually be within $X_0(l)$ when a transition is made into location l. This assumption needs to be checked. That is, whenever a transition is made into control location l, the state must actually be within $X_0(l)$.

We illustrate this procedure by abstracting the gate automaton of figure 5. The gate automaton has two control locations, $LOWER$ and $RAISE$, and continuous state x which represents the height of the gate. The abstracted automaton is shown in figure 9. It has control locations LOW and $HIGH$, and a clock T. The initial sets associated with LOW and $HIGH$ are $X_0(LOW) = [1, 10]$ and $X_0(HIGH) = [1, 10]$. This assumption will be checked later on. There is a map from the states of the abstracted hybrid automaton to the original hybrid automaton. The map is a set valued map given by

$$h(T, HIGH) = RAISE \times [10 - 9e^{-\frac{t}{2}}, 10]$$

$$h(T, LOW) = LOWER \times [1, 9e^{-\frac{t}{2}} + 1]$$

Essentially, the map bounds the trajectories at time t in location l when $x(0) \in X_0(l)$. For a trajectory π of the hybrid automaton, there is a trajectory ψ in the abstracted hybrid automaton such that $\pi(t) \in h(\psi(t))$. More properly, we say the abstracted automaton simulates the original hybrid automaton. This terminology is justified by the following interpretation of the mapping h. In control location $HIGH$, the abstracted automaton simulates the original automaton when it is in control location $RAISE$ and initial condition $x(0) \in [0,10]$. For every trajectory $\pi = (RAISE, \phi)$ with $\phi(0) \in [0,10]$, there is a trajectory $\psi = (UP, T)$ with $T(0) = 0$ in the abstracted automaton such that $\pi(t) \in h(\psi(t))$. In control location LOW, the abstracted automaton simulates the original automaton when it is in control location $LOWER$, and initial condition $x(0) \in [1,10]$.

An enabling condition δ in the hybrid automaton depends on the states of the hybrid automaton. We replace it with $h^{-1}(\delta)$ so that it depends on the states of the abstracted hybrid automaton. The invariance conditions are similarly replaced.

The abstracted automata for the gate ($Gate_A$), train ($Train_A$), and controller ($Contr_A$) are shown in figure 7 - 9. Notice, that $H_A = Gate_A \bigotimes Train_A \bigotimes Contr_A$ is a closed system. We analyze the abstracted hybrid automaton rather than the original hybrid automaton. The relationship between the reach set of the original hybrid automaton, and the reach set of the abstracted hybrid automaton is $R[H] \subset h(R[H_A])$. If we show that the abstracted hybrid automaton satisfies the specification, then clearly, so does the original hybrid automaton since $R[H] \subset h(R[H_A]) \subset S$. The inclusion of hybrid automaton in figure 7 is a rectangular inclusion [7], and using the method in [4], it can be translated into a timed automaton. The continuous variables in the abstracted hybrid automaton are clocks. Hence, the abstracted hybrid automaton is a timed automaton for which it is possible to construct the reach set $R[H_A]$, and therefore $h(R[H_A])$.

We also need to check that our assumption about the initial set $X_0(l)$ was correct. We do this by looking at $h(R[H_A])$ and checking whether a state $w \notin X_0(l)$ is reachable by a transition into location l. If this is not the case, then our assumption about the initial sets is correct. When the assumptions are correct and $h(R[H_A]) \subset S$, then the system satisfies the specification. We used the timed verification tools [3, 6] to show that the specification is satisfied in the train-gate-controller example. That is, the gate is down when the train is at the gate.

We note that the abstractions that are created depend on the property to be proved. A different specification may require a different abstraction. For example, the specification that the gate is eventually raised to let the traffic through cannot be proved using our abstraction. A different abstraction would be needed to prove this property.

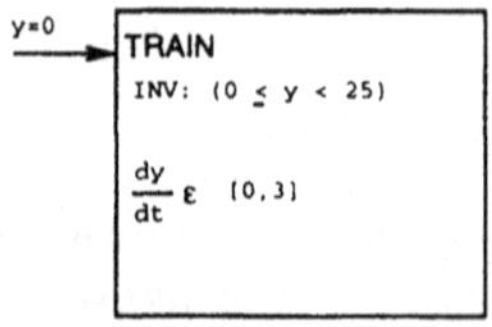

Fig. 7. Abstracted Automaton for the Train

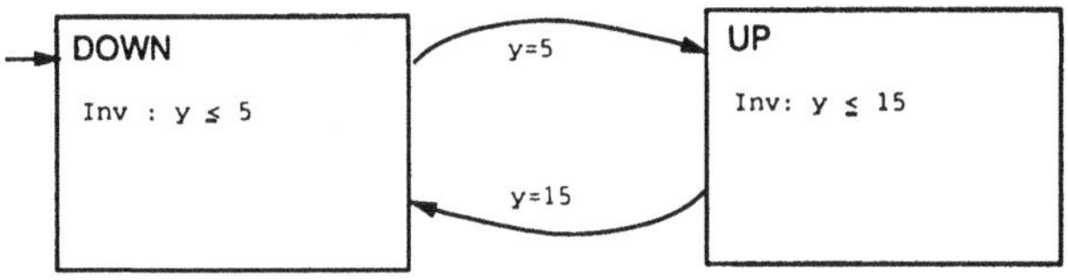

Fig. 8. Abstracted Automaton for the Gate Controller

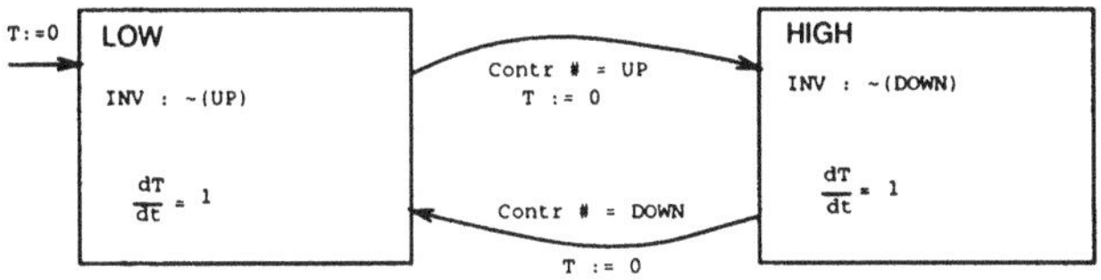

Fig. 9. Abstracted Automaton for the Gate

4 Conclusion

In this paper, we presented a modeling formalism and a verification methodology for hybrid systems. The hybrid system, H, consists of coordinating components, where each component is modeled with a hybrid automaton. The specification S is given as a predicate on the reach set. The verification problem is to show that $R[H] \subset S$.

A hybrid automaton with complicated dynamics may be difficult to analyze. Instead, we look at the component hybrid automata, and abstract them with automata which have simpler dynamics. We provided two methods for doing this. With the first method, a differential inclusion is abstracted with a simpler differential inclusion. In the second method, we work with the timing information that is relevant to verifying the system. Computer tools can then be used to analyze the abstracted hybrid automata that have simpler dynamics.

References

1. R.Alur, C.Courcoubetis, T.A. Henzinger and P.-H. Ho, Hybrid automata: an algorithmic approach to the specification and verification of hybrid systems, *Hybrid Systems*, LNCS 736, Springer-Verlag.
2. R. Alur and D. Dill, Automata for modeling real-time systems, *Proc. 17th ICALP*, LNCS 443, Springer-Verlag, 1990.
3. R.Alur, A.Itai, R.Kurshan and M.Yannakakis, Timing Verification by Successive Approximation, *Proc. 4th Workshop Computer-Aided Verification*, Lecture Notes in Computer Science 663, Springer-Verlag, 1992.
4. P.Kopke, T. Henzinger, A. Puri and P. Varaiya, What's Decidable About Hybrid Automata, *STOCS 1995*.
5. R.L. Grossman et al., eds., *Hybrid Systems*, LNCS 736, Springer-Verlag, 1993.
6. X.Nicollin, A. Olivero, J. Sifakis, and S.Yovine, An Approach to the Description and Analysis of Hybrid Systems, *Hybrid Systems*, LNCS 736, Springer-Verlag, 1993.

7. A. Puri and P. Varaiya, Decidability of Hybrid Systems with Rectangular Differential Inclusions, *Computer-Aided Verification 1994*, LNCS 818, Springer-Verlag, 1994.
8. A. Puri and P. Varaiya, Driving Safely in Smart Cars, *American Control Conference 1995.*
9. A. Puri, V. Borkar, and P. Varaiya, ϵ-Approximation of Differential Inclusions, Draft.
10. P.Varaiya. Smart Cars on Smart Roads: Problems of Control. *IEEE Transactions on Automatic Control*, 38(2):195-207, February 1993.

Control of Continuous Plants by Symbolic Output Feedback

Jörg Raisch*

Institut für Systemdynamik und Regelungstechnik
Universität Stuttgart
Pfaffenwaldring 9
D-70550 Stuttgart
Fed. Rep. of Germany
email: raisch@isr.uni-stuttgart.de

Abstract. This contribution addresses the following hybrid control problem: A continuous plant is to be controlled via symbolic, or quantized, measurement and control signals. Quantization levels may be arbitrarily coarse (e.g. "temperature is too high", "ok" or "too low", "valve open" or "closed"). Sensor quantization is assumed to partition the plant output space into a finite number of rectilinear cells, each of which is associated with a unique measurement symbol. Measurement symbols are processed by a control algorithm, which generates a control signal with finite range, i.e. a symbolic signal. The latter determines the evolution of the plant state in time. Interaction between plant and controller is modeled by an interrupt structure, where information exchange is represented by sequences of timed (measurement and control) events. It is shown that, if certain observability conditions hold, the continuous plant state can be reconstructed from a sequence of measurement events. Likewise, if certain controllability/reachability conditions hold, the plant state can be transferred to any desired point in $\mathbb{R}^n$ by applying a suitable sequence of control events. These results are used in a heuristic Certainty Equivalence feedback scheme reminiscent of "receding horizon" strategies.

Keywords: Symbolic Output Feedback, Hybrid Control Systems.

1 Introduction

We consider the following problem: A physical plant P (assumed to be adequately represented by a set of ordinary differential equations) with real-valued input and output signals is to be controlled by a device K ("the controller") which "sees" only a quantized, or symbolic, version of the plant output (e.g. "temperature is too high", "ok" or "too low"). It generates a piecewise constant

* The author wishes to acknowledge support by "Deutsche Forschungsgemeinschaft" through Grant Ra 516/2-1. Part of this work was done while the author was visiting the Department of Electrical and Computer Engineering at the University of Toronto.

signal with finite range (e.g. "valve open" or "closed"). This symbolic signal determines how the plant state evolves in time. Quantization levels in both input and output signal may be arbitrarily coarse. Sensor quantization is assumed to partition the plant output space into rectilinear boxes, where each box is associated with a unique measurement symbol.

Thus, our feedback configuration is characterized by the interaction of continuous and symbolic signals, and it can be interpreted as a special case of a hybrid dynamical system (e.g. [4]).

Such problems can be found in many areas of application. They are especially common in chemical process control, where P typically represents plant *and* lower level (continuous) control loops (dealing, for example, with set point regulation). K is a higher level, or supervisory, controller. It handles start-up and shut-down procedures (by switching between control symbols) and takes appropriate action in case of irregularities (represented by discrete-valued "alarm" signals).

A common approach when dealing with such control configurations is to convert the continuous (numerical, or quantitative,) plant description P to a discrete (or qualitative) model – in the simplest case by lumping all states of the original model that correspond to a specific measurement symbol (e.g. [10, 11, 19, 1]). In all practically relevant cases, the resulting discrete model will be nondeterministic – this feature reflecting the fact that, for a given input, not all of the underlying continuous system trajectories originating in a certain quantization box will "move" to the same quantization box from there. Refinements, where partitioning of the continuous plant state space is based on present *and* past measurement (and control) information, have been reported in [15, 16].

In contrast to this, we propose a framework where the continuous plant model is retained and used for on-line calculation of an appropriate control signal: We model interaction between the discrete-valued world of the controller and the continuous-valued world of the plant by an *interrupt structure*: The controller, instead of "looking permanently" at the incoming measurement symbols, is "notified" only if a measurement symbol changes. It only sends information to the plant if it wants to change a control symbol. It is assumed that both events can happen at *any* time $t \in \mathbb{R}^+$, i.e. time is considered to be continuous. This assumption is well justified whenever plant time constants are large compared to processing times in the control computer – a common situation, for example, in chemical engineering control problems. Hence, information exchange between plant and controller is represented by sequences of timed (measurement and control) events, each event consisting of a real number (time) and a symbol. Within this framework, a measurement event contains *precise* (albeit incomplete) information on the continuous plant state x: roughly speaking, it implies that at an "event time" $t = t^y$, $x(t)$ "crosses" or "hits" a well-defined hyperplane in the plant state space $\mathbb{R}^n$. If certain observability conditions hold, the plant state can be completely reconstructed from a sequence of measurement events. Likewise, if certain controllability/reachability conditions hold, the plant state can be transferred between any two points in (a subset of) $\mathbb{R}^n$ by applying a suitable sequence of control events. These conditions for controllability, reachability, and observability are the main results in this paper. Based on them, a heuristic feedback strategy reminiscent of "Receding Horizon Control" is suggested: the controller is supplied with a continuous plant model, it "learns" the plant state from the recorded sequence of measurement events and computes a sequence of future control events that will (within a given time T) drive the plant state to a desired (equilibrium) point. The state estimate is updated, and the sequence of future control events revised, when a new measurement event becomes available. The advantages of such an approach are obvious if specifications are "much finer" than measurement quantization levels, and are therefore easier to formulate in terms of continuous-valued plant variables. As an example, consider the case where we have only measurement symbols "liquid level in tank 1 is above value A", "below value B", or "between A and B", but want to keep the liquid level, on average, as close as possible to some value C between A and B.

This approach was proposed in [13, 14] for a particularly simple class of problems. In this paper, it is extended to a broader (and more realistic) class of models.

In the following section, we describe our set-up: plant, controller and interface between continuous- and discrete-valued signals. Specifically, we show how both controller inputs and outputs can be interpreted as sequences of timed (measurement and control) events. In section 3, we discuss which part of the continuous plant state space can be controlled to, or reached from, the origin by an admissible sequence of control events. In section 4, we address the problem of observing the plant state from a sequence of measurement events. In section 5, a heuristic control strategy is suggested. It is characterized by the fact that the event-processing controller contains a continuous model of the plant. Finally, we look at two simple examples where reachability and observability conditions are satisfied and the proposed control strategy makes intuitive sense.

This paper has been written from a control engineering point of view (with potential chemical engineering applications in mind). Hence, "standard control" terminology and notation are used throughout.

2 Control Configuration

We consider the control system shown in Fig. 1. Symbolic signals are represented by dashed lines and continuous-valued signals are drawn as solid lines. The dynamic behaviour of the plant is determined by the symbolic controller output u_d – details will be discussed in section 2.1. z and y are referred to as "external" and "internal" plant outputs. z is a suitably defined "artificial" signal which is used to formulate design specifications. A quantized, or symbolic, version y_d of the internal plant output y can be "seen" by the controller.

2.1 The Plant

We restrict ourselves to a plant model which is simple enough to be – to a certain extent – analytically tractable, but general enough to capture characteristic

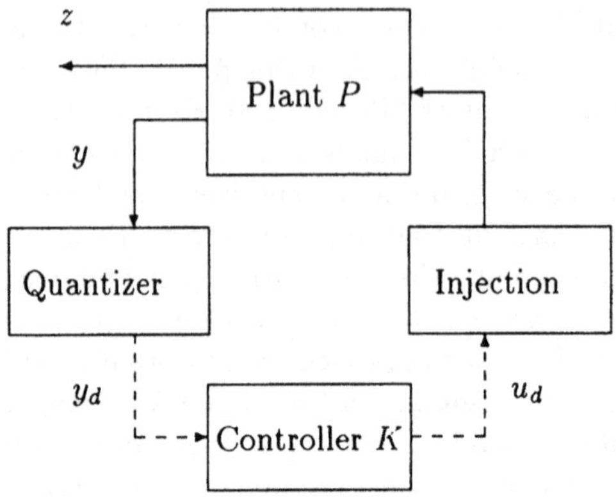

Fig. 1. Symbolic control of continuous plant

features of many practical problems: The plant state is assumed to "live" in $\mathbb{R}^n$ and to evolve according to

$$\dot{x}(t) = f(u_d(t))\, x(t) + g(u_d(t)) \;, \tag{1}$$

where u_d is a piecewise constant signal with finite range U_d:

$$u_d(t) \in U_d \;,$$
$$U_d = \left\{u_d^{(1)}, \ldots, u_d^{(N)}\right\} \;;$$

the functions f and g assign a real $n \times n$-matrix (a "dynamics matrix") and a real n-vector (a "forcing term") to every $u_d \in U_d$:

$$f : U_d \to \mathbb{R}^{n \times n} \;,$$
$$g : U_d \to \mathbb{R}^n \;.$$

The function $(f, g) : U_d \to \mathbb{R}^{n \times n} \times \mathbb{R}^n$ is injective (for every control symbol, we have a different combination of dynamics matrix and forcing term), whereas no such assumption is made for the "individual" maps f and g. Introduce the following notation:

$$\{A_1, \ldots, A_P\} := \text{image}(f) \;,$$
$$U_{d_i} := \left\{u_d^{(k)} \in U_d \mid f(u_d^{(k)}) = A_i\right\} \;,$$
$$\{b_{i_1}, \ldots, b_{i_{N_i}}\} := \text{image}(g_{|U_{d_i}}) \;,$$

where $g_{|U_{d_i}}$ is the restriction of g to U_{d_i} (Fig. 2). Obviously, $\sum_{i=1}^{P} N_i = N$.

Denote the rank of the real $n \times N_i$-matrix $[b_{i_1}, \ldots, b_{i_{N_i}}]$ by q_i. Find a basis B_i for the q_i-dimensional subspace of $\mathbb{R}^n$ spanned by the b_{i_k}, $k = 1, \ldots, N_i$, and write

$$[b_{i_1}, \ldots, b_{i_{N_1}}] = B_i\, [u_{i_1}, \ldots, u_{i_{N_i}}].$$

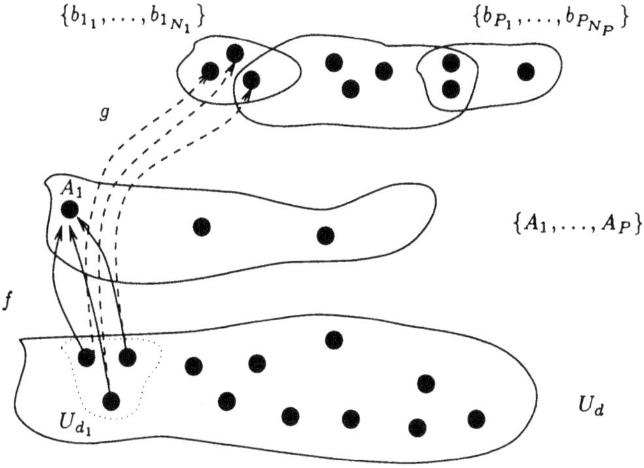

Fig. 2. Maps f and g

Obviously, the u_{i_k}, $k = 1, \ldots, N_i$, represent N_i different points in IR^{q_i}.

Now, we can re-interprete the situation in a more familiar way: If $u_d(t)$ is restricted to U_{d_i}, state equation (1) becomes

$$\dot{x}(t) = A_i x(t) + B_i u_i(t) \ , \tag{2}$$

where u_i is a piecewise constant signal "living" on $\{u_{i_1}, \ldots, u_{i_{N_i}}\} \subset \mathrm{IR}^{q_i}$. (2) is an old friend from linear control theory, the only difference to the familiar case being that the plant input is restricted to a finite number of points in IR^{q_i}. Switching to a control symbol $u_d(t)$ from a different subset U_{d_j}, changes the "model parameters" from (A_i, B_i) to (A_j, B_j).

Both external and internal output are assumed to depend linearly on the current plant state:

$$z(t) = C_z x(t) \ , \tag{3}$$

$$y(t) = C_y x(t) \ . \tag{4}$$

More general output equations (the right hand sides depending on the control symbol u_d, but still affine in x) wouldn't pose any additional problems. They would, however, be of little use in most practical problems, and cause notation to be even more cumbersome.

Example 1. Consider the three-tank system in Fig. 3. The states x_1, x_2 and x_3 describe the deviation of the water levels in tank A, B, C from a common reference value. Valve 1 and 2 are either "open" or "closed". The pump can perform five different actions: it can pump water from a common resource *into* either tank A or C (at a constant rate r), it can pump water *from* either tank A

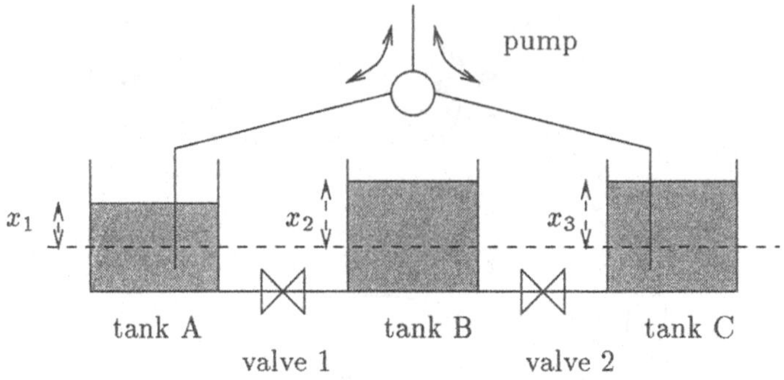

Fig. 3. Example 1

Table 1. Control symbols in Example 1

$u_d^{(1)}$:=	"valve 1 open" & "valve 2 open" & "pumping water into tank A"
$u_d^{(2)}$:=	"valve 1 open" & "valve 2 open" & "pumping water from tank A"
$u_d^{(3)}$:=	"valve 1 open" & "valve 2 open" & "pumping water into tank C"
$u_d^{(4)}$:=	"valve 1 open" & "valve 2 open" & "pumping water from tank C"
$u_d^{(5)}$:=	"valve 1 open" & "valve 2 open" & "pump switched off"
$u_d^{(6)}$:=	"valve 1 open" & "valve 2 closed" & "pumping water into tank A"
$u_d^{(7)}$:=	"valve 1 open" & "valve 2 closed" & "pumping water from tank A"
$u_d^{(8)}$:=	"valve 1 open" & "valve 2 closed" & "pumping water into tank C"
$u_d^{(9)}$:=	"valve 1 open" & "valve 2 closed" & "pumping water from tank C"
$u_d^{(10)}$:=	"valve 1 open" & "valve 2 closed" & "pump switched off"
⋮	⋮
$u_d^{(16)}$:=	"valve 1 closed" & "valve 2 closed" & "pumping water into tank A"
$u_d^{(17)}$:=	"valve 1 closed" & "valve 2 closed" & "pumping water from tank A"
$u_d^{(18)}$:=	"valve 1 closed" & "valve 2 closed" & "pumping water into tank C"
$u_d^{(19)}$:=	"valve 1 closed" & "valve 2 closed" & "pumping water from tank C"
$u_d^{(20)}$:=	"valve 1 closed" & "valve 2 closed" & "pump switched off"

or C (at rate r), or it can be switched off. Hence, in this example, U_d consists of $2 \cdot 2 \cdot 5$ symbols (Table 1).

Under the assumption that the flow rates between adjacent tanks are proportional to the difference in their water levels, we get:

$$A_1 := f(u_d^{(k)}) = \begin{bmatrix} -a & a & 0 \\ a & -(a+b) & b \\ 0 & b & -b \end{bmatrix}, \quad k = 1, \ldots, 5,$$

$$b_{1_1} := g(u_d^{(1)}) = \begin{bmatrix} r \\ 0 \\ 0 \end{bmatrix}, \; b_{1_2} := g(u_d^{(2)}) = \begin{bmatrix} -r \\ 0 \\ 0 \end{bmatrix},$$

$$b_{1_3} := g(u_d^{(3)}) = \begin{bmatrix} 0 \\ 0 \\ r \end{bmatrix}, \; b_{1_4} := g(u_d^{(4)}) = \begin{bmatrix} 0 \\ 0 \\ -r \end{bmatrix},$$

$$b_{1_5} := g(u_d^{(5)}) = \begin{bmatrix} 0 \\ 0 \\ 0 \end{bmatrix},$$

$$\vdots$$

$$A_4 := f(u_d^{(k)}) = \begin{bmatrix} 0\,0\,0 \\ 0\,0\,0 \\ 0\,0\,0 \end{bmatrix}, \quad k = 16, \ldots, 20,$$

$$b_{4_1} := g(u_d^{(16)}) = g(u_d^{(1)}), \; b_{4_2} := g(u_d^{(17)}) = g(u_d^{(2)}),$$
$$b_{4_3} := g(u_d^{(18)}) = g(u_d^{(3)}), \; b_{4_4} := g(u_d^{(19)}) = g(u_d^{(4)}),$$
$$b_{4_5} := g(u_d^{(20)}) = g(u_d^{(5)}).$$

Clearly, the rank of $[b_{i_1}, \ldots, b_{i_5}]$ is 2 (for $i = 1, \ldots, 4$), and we can write

$$[b_{i_1}, \ldots, b_{i_5}] = \underbrace{\begin{bmatrix} r & 0 \\ 0 & 0 \\ 0 & r \end{bmatrix}}_{B_i} \underbrace{\begin{bmatrix} 1 & -1 & 0 & 0 & 0 \\ 0 & 0 & 1 & -1 & 0 \end{bmatrix}}_{[u_{i_1} \ldots u_{i_5}]}, \quad i = 1, \ldots, 4.$$

Suppose, control specifications are in terms of all three state variables, and there is only one sensor available (placed in tank A). Then, the output equations (3), (4) become

$$z(t) = x(t) \; ,$$
$$y(t) = \begin{bmatrix} 1\,0\,0 \end{bmatrix} x(t) \; .$$

2.2 Quantization, Measurement Events, and Control Events

For every $t \in \mathbb{R}^+$, the quantizer Q assigns a symbol $y_d(t) \in \{y_d^{(1)}, \ldots, y_d^{(M)}\}$ to the current value $y(t)$ of the internal plant output. Hence, the function $Q : \mathbb{R}^p \to \{y_d^{(1)}, \ldots, y_d^{(M)}\}$ partitions the space $\mathbb{R}^p$ into M disjoint sets, or quantization cells. Each cell is assumed to be a rectilinear "box" with edges parallel to the coordinate axes (Fig. 4). The $y_i^{(i_k)}$, $i, i_k = 1, 2$, in Fig. 4 are referred to as quantization thresholds.

Notation in the following sections becomes simpler if we replace the "scalar symbol" $y_d(t)$ by a "vector symbol" $\begin{bmatrix} y_{d_1}(t) \ldots y_{d_p}(t) \end{bmatrix}$. This is equivalent to

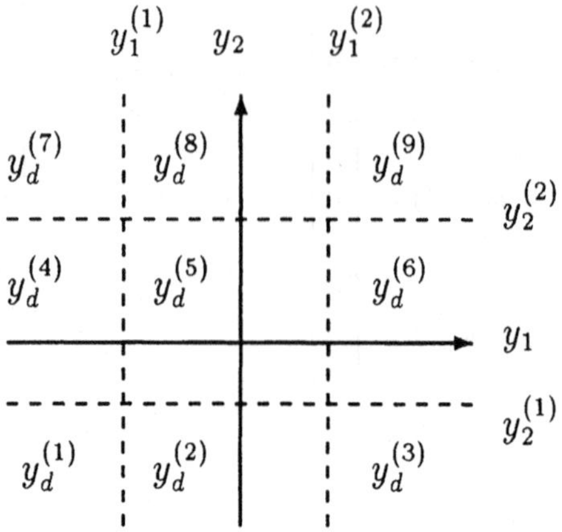

Fig. 4. Output quantization

replacing the quantizer Q for the vector $y(t)$ by independent quantizers for the elements $y_i(t)$ of $y(t)$, $i = 1, \ldots, p$:

$$Q_i : \mathbb{R} \rightarrow \{y_{d_i}^{(1)}, \ldots, y_{d_i}^{(M_i)}\}, \qquad \prod_{i=1}^{p}(M_i) = M, \tag{5}$$

assigns the symbol $y_{d_i}^{(r)}$ to every y_i within a given interval (Fig. 5):

$$Q_i(y_i) = \begin{cases} y_{d_i}^{(1)} & \text{for } -\infty < y_i \leq y_i^{(1)} \\ y_{d_i}^{(r)} & \text{for } y_i^{(r-1)} < y_i \leq y_i^{(r)},\ r = 2, \ldots, M_i - 1 \\ y_{d_i}^{(M_i)} & \text{for } y_i^{(M_i-1)} < y_i < \infty. \end{cases} \tag{6}$$

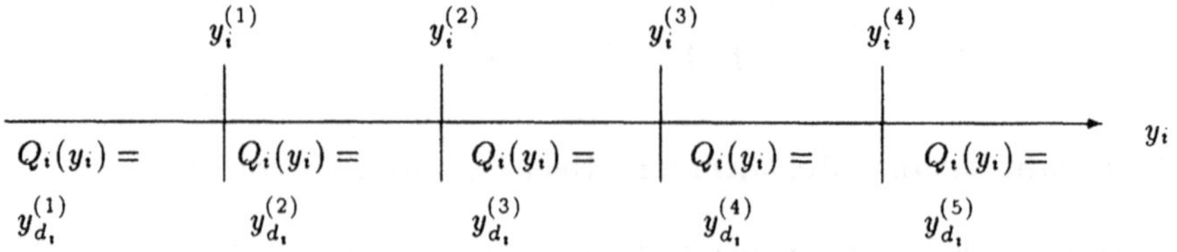

Fig. 5. Elementwise quantization of the plant output y

As time is continuous, knowledge of the symbolic measurement signals $Q_i(y_i)$, $i = 1, \ldots, p$, implies *precise* (albeit incomplete) information on the underlying continuous variable y. Suppose, for example, that $Q_i(y_i)$ changes from $y_{d_i}^{(r)}$ to

$y_{d_i}^{(r+1)}$ (or from $y_{d_i}^{(r+1)}$ to $y_{d_i}^{(r)}$) at time t_k^y. Then, at time t_k^y, y_i takes its threshold value $y_i^{(r)}$:

$$Q_i(y_i(t_k^y)) = y_{d_i}^{(r)} \quad \& \quad \lim_{(\epsilon>0)\to 0} Q_i(y_i(t_k^y \pm \epsilon)) = y_{d_i}^{(r+1)} \qquad \Longrightarrow \qquad y_i(t_k^y) = y_i^{(r)} \ .$$

Each such change, i.e. each new piece of precise information, can be interpreted as a *timed measurement event* consisting of a real-valued number (the time when the change occurs) and an element from the finite set of threshold values,

$$\langle t = t_k^y,\ y_i(t_k^y) = y_i^{(r)} \rangle, \quad k = 1, 2, \ldots \tag{7}$$

(or $\langle t_k^y,\ y_i^{(r)} \rangle$, for short). Conversely, from the initial information $y_d(0)$ and the sequence of measurement events (7), one can immediately tell the quantization box the signal $y(t)$ is currently in and therefore reconstruct its discrete-valued version $y_d(t)$. Hence, given $y_d(0)$, all incoming measurement information is captured in the sequence of events (7).

The controller output $u_d(t)$ – a piecewise constant symbolic signal – can also be interpreted as a sequence of events: Denote time instants where the value of u_d changes by t_k^u, $k = 1, 2, \ldots$, and let the "new" symbol be $u_d^{(s)}$:

$$\lim_{(\epsilon>0)\to 0} u_d(t_k^u - \epsilon) \neq u_d^{(s)} \quad \& \quad u_d(t_k^u) = u_d^{(s)}.$$

Each control signal is then simply a sequence of timed control events

$$\langle t = t_k^u,\ u_d(t_k^u) = u_d^{(s)} \rangle, \quad k = 1, 2, \ldots \tag{8}$$

(or $\langle t_k^u,\ u_d^{(s)} \rangle$, for short).

Quantization levels in both the controller input and output can be arbitrarily coarse; our set-up therefore covers extreme cases involving binary signals.

2.3 The Controller

The controller is a dynamical system which generates the sequence of control events (8) in response to the sequence of measurement events (7), the only restriction being causality.

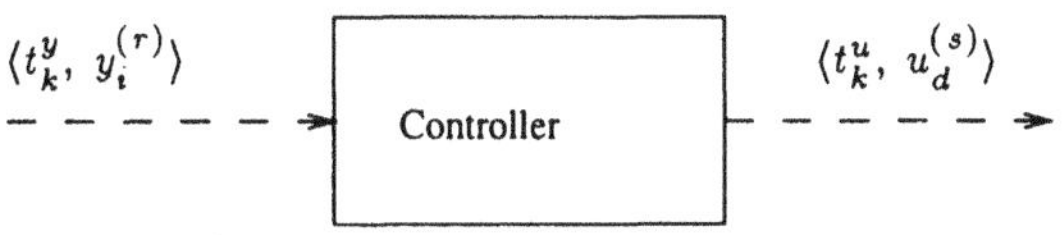

Fig. 6. Controller

3 Controllability and Reachability

In this section, we investigate controllability and reachability issues. We distinguish two cases: the general case described in section 2.1, and a special case characterized by the fact that the image of f is a singleton. The latter case is not too difficult: By invoking Pontryagin's maximum principle, the problem can be reduced to one with continuous-valued, but constrained, control inputs. Then, standard results from control theory can be applied to give necessary and sufficient conditions under which the whole $\mathbb{R}^n$ is controllable to, and reachable from, the origin using a piecewise constant input signal u_d with $u_d(t) \in U_d$. These conditions, however, are pretty restrictive (although they hold for an example which is introduced in section 6). Hence, if they fail, we also have to ask which *part* of the state space is controllable to, and reachable from, the origin using symbolic inputs.

In the general case, we expect control to be "more powerful", as the input signal u_d can change the dynamics of the plant model. Thus, we only ask under which conditions the whole $\mathbb{R}^n$ is controllable to, and reachable from, the origin using symbolic inputs. Not surprisingly, sufficient conditions are found to hold for example 1.

3.1 A Special Case

If the image of f is a singleton, we can rewrite (2) as

$$\dot{x}(t) = Ax(t) + Bu(t) \ , \tag{9}$$

where u is a piecewise constant signal, $u(t) \in \{u_1, \ldots, u_N\} \subset \mathbb{R}^q$, and the map from U_d, the set of symbols, to $\{u_1, \ldots, u_N\}$ is bijective. We denote this class of discrete input signals by $\mathcal{U}_d$ and call it D-admissible. Over any finite time interval, any D-admissible signal can be interpreted as a finite sequence of control events (8). Recall that, by construction, B has full column rank.

Definition 1 (D-controllability and D-reachability). Consider eqn. (9). The set of all states which can be controlled to the origin (reached from the origin) in given time T by applying input signals from $\mathcal{U}_d$ is denoted by $\mathcal{C}_d^T$ and $\mathcal{R}_d^T$, respectively. $\mathcal{C}_d$ and $\mathcal{R}_d$ are the sets of states that can be controlled to the origin (reached from the origin) by D-admissible inputs in arbitrary (but finite) time.

Now, consider the (hypothetical) case where the input signals are continuous-valued: An input signal u is called *C-admissible* if it is piecewise continuous and lives in the convex hull H of $\{u_1, \ldots, u_N\}$. Clearly, H is a convex polyhedron in $\mathbb{R}^q$ with dimension q. The set of C-admissible inputs is denoted by $\mathcal{U}_c$.

Definition 2 (C-controllability and C-reachability). Consider eqn. (9). The set of all states that can be controlled to the origin (reached from the origin) in given time T by applying C-admissible inputs is denoted by $\mathcal{C}_c^T$ and $\mathcal{R}_c^T$, respectively. $\mathcal{C}_c$ and $\mathcal{R}_c$ are the sets of states that can be controlled to the origin (reached from the origin) by C-admissible inputs in arbitrary (but finite) time.

Theorem 3. *Assume that*

1. $0 \in \{u_1, \dots, u_N\}$ *(i.e. $u(t) \equiv 0$ is both D- and C-admissible),*
2. *0 is an interior point of H, and*
3. *rank $[Bw, ABw, \dots, A^{n-1}Bw] = n$ for every w parallel to an edge of H. This is "almost always" true if rank$[B, AB, \dots, A^{n-1}B] = n$ and A is cyclic [6].*

Then,

$$x \in \mathcal{C}_c^T \text{ iff } x \in \mathcal{C}_d^T \tag{10}$$

$$x \in \mathcal{R}_c^T \text{ iff } x \in \mathcal{R}_d^T \;, \tag{11}$$

i.e. C-controllability (C-reachability) in time T implies and is implied by D-controllability (D-reachability) in time T.

Proof. The "if" part in (10) is trivial, as $\mathcal{U}_d^T \subset \mathcal{U}_c^T$. The "only if" part follows from an application of Pontryagin's maximum principle (e.g. [2]). Details of the proof can be found in [14].

The "if" part in (11) is also trivial. To prove the "only if" part of (11), define "controllability operators"

$$C_c^T : \mathcal{U}_c^T \to \mathbb{R}^n$$
$$C_d^T : \mathcal{U}_d^T \to \mathbb{R}^n$$

with action

$$C_c^T(u(t)) := -\int_0^T e^{-At}Bu(t)dt$$
$$C_d^T(u(t)) := -\int_0^T e^{-At}Bu(t)dt \;.$$

The ranges of these operators are just the sets of states that can be controlled to the origin in time T by C-admissible and D-admissible inputs, respectively. Because of (10), it is obvious that

$$\text{Range}C_c^T = \mathcal{C}_c^T = \mathcal{C}_d^T = \text{Range}C_d^T \;. \tag{12}$$

Analogously, we define "reachability operators"

$$R_c^T : \mathcal{U}_c^T \to \mathbb{R}^n$$
$$R_d^T : \mathcal{U}_d^T \to \mathbb{R}^n$$

with action

$$R_c^T(u(t)) := \int_0^T e^{A(T-t)}Bu(t)dt = -e^{AT}C_c^T(u(t)) \tag{13}$$

$$R_d^T(u(t)) := \int_0^T e^{A(T-t)}Bu(t)dt = -e^{AT}C_d^T(u(t)) \;. \tag{14}$$

Clearly, the range of R_c^T (R_d^T) determines the set of states that can be reached from the origin in time T by applying C-admissible (D-admissible) control inputs. From (12)-(14), it follows immediately that the range of R_c^T and the range of R_d^T coincide, and therefore

$$\mathcal{R}_c^T = \text{Range} R_c^T = \text{Range} R_d^T = \mathcal{R}_d^T \ . \tag{15}$$

□

Now, necessary and sufficient conditions for global reachability and controllability by D-admissible inputs follow as a corollary from their well known counterparts for C-admissible inputs in [3, 17] and from Theorem 3.

Corollary 4. *Consider eqn. (9). Assume that the conditions in Theorem 3 hold. Then,* $\mathcal{R}_d = \mathbb{R}^n$ *if and only if no eigenvalue of* A *has negative real part.* $\mathcal{C}_d = \mathbb{R}^n$ *if and only if no eigenvalue of* A *has positive real part.* $\mathcal{C}_d = \mathcal{R}_d = \mathbb{R}^n$ *if and only if all eigenvalues of* A *lie on the imaginary axis.*

Remark. Note that the last two conditions do not imply stability: Existence of non-trivial Jordan-blocks and hence instability of the plant model (9) are allowed.

Clearly, the eigenvalue conditions in Corollary 4 are pretty restrictive. If they are not satisfied, we'd still like to know which *part* of the plant state space is controllable (reachable) by applying D-admissible control signals (i.e. sequences of control events (8)). For this, we can make use of other results from control theory: How to determine the discrete-time version of $\mathcal{R}_c^T$ is discussed, for example, in [8] and [9]. Their results can be used to approximate $\mathcal{R}_c^T$ and, because of (14) and (15), $\mathcal{R}_d^T$ and $\mathcal{C}_d^T$.

It is particularly easy to find the maximal *subspaces* of $\mathbb{R}^n$ which are contained in $\mathcal{R}_d$ and $\mathcal{C}_d$: Suppose assumptions 1 and 2 in Theorem 3 hold. Then, it follows from Corollary 4 that $\tilde{\mathcal{R}}_d$, the maximal subspace contained in $\mathcal{R}_d$, and $\tilde{\mathcal{C}}_d$, the maximal subspace contained in $\mathcal{C}_d$, are given by:

$$\tilde{\mathcal{R}}_d = \text{Span}\,[B\ AB\ \dots\ A^{n-1}B] \cap (E_c(A) + E_u(A)) \ , \tag{16}$$

$$\tilde{\mathcal{C}}_d = \text{Span}\,[B\ AB\ \dots\ A^{n-1}B] \cap (E_c(A) + E_s(A)) \ , \tag{17}$$

where $E_s(A)$, $E_c(A)$, and $E_u(A)$ denote the stable, central, and unstable subspaces of $\dot{x} = Ax$, respectively[2].

[2] Recall that $E_s(A)$, $E_c(A)$, and $E_u(A)$ are spanned by the real and imaginary parts of all generalized eigenvectors of A corresponding to eigenvalues with negative, zero and positive real parts respectively.

3.2 The General Case

Now, we address the issue of global controllability and reachability for the general model described in section 2.1. It is not surprising that in this case verifiable necessary and sufficient conditions are not known. However, a variety of sufficient conditions can be stated easily. The following, for example, is intuitively clear:

Suppose there exists a subset U_{d_i} of control symbols, where the corresponding matrices A_i, B_i and the associated set of input vectors $\{u_{i_1}, \ldots, u_{i_{N_i}}\}$ satisfy the assumptions in Theorem 3. Suppose the same is true for A_j, B_j, and $\{u_{j_1}, \ldots, u_{j_{N_j}}\}$, which are associated with a subset U_{d_j}. Then, $\mathbb{R}^n$ is controllable to, and reachable from, the origin with D-admissible control signals if all eigenvalues of A_i are in the closed left halfplane (global controllability) and all eigenvalues of A_j lie in the closed right halfplane (global reachability).

The following is another simple result: As before, consider two subsets U_{d_i} and U_{d_j} of U_d. Suppose that assumptions 1 and 2 in Theorem 3 hold for A_i, B_i, $\{u_{i_1}, \ldots, u_{i_{N_i}}\}$ and A_j, B_j, $\{u_{j_1}, \ldots, u_{j_{N_j}}\}$. Denote by $\widetilde{\mathcal{R}}_d(i)$ the maximal reachable subspace when u_d is restricted to U_{d_i}. Then, $\widetilde{\mathcal{R}}_d(i)+\widetilde{\mathcal{R}}_d(j)$ is reachable with $u_d \in U_{d_i} \cup U_{d_j}$, if $\widetilde{\mathcal{R}}_d(i) - (\widetilde{\mathcal{R}}_d(i) \cap \widetilde{\mathcal{R}}_d(j))$ is A_j-invariant.

The proof of this statement is straightforward: Let $\widetilde{x} = \widetilde{x}_i + \widetilde{x}_j$ be an arbitrary point in $\widetilde{\mathcal{R}}_d(i) + \widetilde{\mathcal{R}}_d(j)$, with $\widetilde{x}_i \in \widetilde{\mathcal{R}}_d(i) - (\widetilde{\mathcal{R}}_d(i) \cap \widetilde{\mathcal{R}}_d(j))$ and $\widetilde{x}_j \in \widetilde{\mathcal{R}}_d(j)$. To drive the state from the origin to $\widetilde{x}$, first restrict u_d to U_{d_i}. Then, in finite time T_i, we can drive the state to any point in $\widetilde{\mathcal{R}}_d(i)$ and hence in $\widetilde{\mathcal{R}}_d(i) - (\widetilde{\mathcal{R}}_d(i) \cap \widetilde{\mathcal{R}}_d(j))$. Denote this point by $\bar{x}_i$. Now, restrict u_d to U_{d_j}. Then, in finite time $T_j - T_i$, we can steer the state to

$$x(T_j) = e^{A_j(T_j - T_i)}\bar{x}_i + \widetilde{x}_j,$$

where $\widetilde{x}_j$ is any desired point in $\widetilde{\mathcal{R}}_d(j)$. Now it remains to show that $\bar{x}_i \in \widetilde{\mathcal{R}}_d(i) - (\widetilde{\mathcal{R}}_d(i) \cap \widetilde{\mathcal{R}}_d(j))$ can be chosen such that $\widetilde{x}_i = e^{A_j(T_j - T_i)}\bar{x}_i$ is indeed the desired point in $\widetilde{\mathcal{R}}_d(i) - (\widetilde{\mathcal{R}}_d(i) \cap \widetilde{\mathcal{R}}_d(j))$. But this is obvious as $\widetilde{\mathcal{R}}_d(i) - (\widetilde{\mathcal{R}}_d(i) \cap \widetilde{\mathcal{R}}_d(j))$ is A_j- and hence $e^{A_j(T_j - T_i)}$- and $e^{-A_j(T_j - T_i)}$-invariant.

Example 1 (revisited). In example 1, the conditions for global controllability hold for $i = 1$. Consequently, the state can be driven to the origin from any initial condition with $u_d \in U_{d_1} = \{u_d^{(1)}, \ldots, u_d^{(5)}\}$, i.e. with valve 1 and 2 constantly open. Global reachability, on the other hand, cannot be achieved by restricting the control symbols to one of the subsets $U_{d_1}, \ldots, U_{d_4}$: for any $i \in \{1, \ldots, 4\}$, $\widetilde{\mathcal{R}}_d(i)$ is either a one-dimensional or a two-dimensional subspace of $\mathbb{R}^3$: If we choose the parameters as $a = 3$, $b = 2$, $r = 1$, we get:

$$\widetilde{\mathcal{R}}_d(1) = \text{Span}\left\{\begin{bmatrix}1\\1\\1\end{bmatrix}\right\},$$

$$\widetilde{\mathcal{R}}_d(2) = \text{Span}\left\{\begin{bmatrix}1\\1\\0\end{bmatrix}\begin{bmatrix}0\\0\\1\end{bmatrix}\right\},$$

$$\widetilde{\mathcal{R}}_d(3) = \text{Span}\left\{\begin{bmatrix}-\sqrt{2}\\1\\1\end{bmatrix}\begin{bmatrix}\sqrt{2}\\1\\1\end{bmatrix}\right\},$$

$$\widetilde{\mathcal{R}}_d(4) = \text{Span}\left\{\begin{bmatrix}1\\0\\0\end{bmatrix}\begin{bmatrix}0\\0\\1\end{bmatrix}\right\}.$$

With $u_d \in U_{d_1} \cup U_{d_4}$, however, the whole state space is reachable from the origin: $\widetilde{\mathcal{R}}_d(1) - (\widetilde{\mathcal{R}}_d(1) \cap \widetilde{\mathcal{R}}_d(4))$ is (trivially) A_4-invariant (recall that A_4 is a zero-matrix); hence, $\widetilde{\mathcal{R}}_d(1) + \widetilde{\mathcal{R}}_d(4) = \mathbb{R}^3$ is reachable. This result is intuitively clear: For $i = 1$ (i.e. valve 1 and 2 are open), the system state can be shifted from the origin along the direction of $\begin{bmatrix}1 & 1 & 1\end{bmatrix}'$ until we get a desired x_2-value. Then, we switch to $i = 4$ (i.e. we close both valve 1 and 2) and move the state anywhere on the plane $x_2 = \text{const}$.

4 Observability

Adapting standard terminology (e.g. [18]), we say that a given D-admissible control input $u(t)$ (i.e. a given sequence of control events $\langle t_k^u, u_d^{(s)}\rangle$, $k = 1, 2, \ldots$) *distinguishes* between two initial states $x(0)$ and $\tilde{x}(0)$, if it produces different sequences of measurement events for both initial conditions. $x(0)$ and $\tilde{x}(0)$ are called *distinguishable*, if there exists at least one sequence $\langle t_k^u, u_d^{(s)}\rangle$, $k = 1, 2, \ldots$ that distinguishes between them. Finally, the hybrid system consisting of plant (1), (4) and quantizer (5)-(6) is said to be *observable* on a set $\mathcal{O}_d$, if every pair of initial states in $\mathcal{O}_d$ is distinguishable.

Theorem 5. *Let T be any finite positive real number and denote the m-th row of C_y by c_m. The hybrid system (1), (4), (5), (6) is observable on $\mathcal{C}_d^T$ if for at least one $i \in \{1, \ldots, P\}$ and one $m \in \{1, \ldots, p\}$*

1. *there exists an $r \in \{1, \ldots, M_m - 1\}$ such that $c_m x = y_m^{(r)}$ intersects $\mathcal{C}_d^T(i) \cap \mathcal{R}_d^T(i)$ and*
2. *$\text{rank}[c_m', A_i'c_m', \ldots, (A_i')^{n-1}c_m'] = n$.*

($\mathcal{C}_d^T(i)$ and $\mathcal{R}_d^T(i)$ denote the sets of states of (1) which are controllable to (reachable from) the origin in time T when the control symbols are restricted to U_{d_i}.)

Proof. (Adapted from [13].) Suppose first that we have been able to find a sequence of control events that generates n measurement events $\langle t_k^y, y_m^{(r)}\rangle$, $k = 1, \ldots, n$. Then

$$\begin{bmatrix} y_m^{(r)} \\ \vdots \\ y_m^{(r)} \end{bmatrix} = \underbrace{\begin{bmatrix} c_m e^{A_i t_1^y} \\ \vdots \\ c_m e^{A_i t_n^y} \end{bmatrix}}_{:=O_{i,m}} x(0) + h(u_i(t)),$$

where $h(u_i(t))$ is known. $x(0)$ is the *unique* initial condition that results in this particular sequence of measurement events if and only if $O_{i,m}$ has rank n. For $O_{i,m}$ to be singular it is necessary (and sufficient) that the homogeneous equation

$$0 = O_{i,m}x(0) \tag{18}$$

has a non-trivial solution (i.e. we can start the continuous system (1),(4) with zero-input in $x(0) \neq 0$ and measure $y_m(t) = c_m x(t) = 0$ at n different instants of time $t_1^y, \ldots, t_n^y$). Because of the rank condition in the theorem, $y_m(t) \equiv 0 \;\; \forall t \geq 0$ is not possible. Therefore, the only way how (18) can occur is that the event times t_k^y are "pathological", i.e. they coincide with the points where the m-th component of the autonomous solution of (1),(4) intersects the t-axis. Clearly, these points form a subset of measure zero of $\mathbb{R}^+$. Now, it remains to show that there exists a control signal that generates measurement events with "non-pathological" timing. This is indeed the case for all initial states in $\mathcal{C}_d^T$: As the "event-generating" hyperplane $c_m x = y_m^{(r)}$ intersects $\mathcal{C}_d^T(i) \cap \mathcal{R}_d^T(i)$ (and therefore $\mathcal{C}_d^T$), there certainly exists a sequence of control events that drives $x(0)$ onto the hyperplane in any desired time $t_a \geq 2T$ (for example by forcing x_0 into the origin first, see Fig. 7). This can be repeated to generate other measurement events at any time t_b with $t_b - t_a \geq 2T$, t_c with $t_c - t_b \geq 2T$ and so on. In fact, any randomly chosen control signal that makes the state-trajectory cross the "event-generating" hyperplane often enough, will do this in a "nonpathological way", enabling us to distinguish between initial conditions and thus to reconstruct $x(0)$, with probability 1. □

Remark. In general, the "pathological" distances between measurement event times depend on *all* the eigenvalues of the matrix A_i. For asymptotically stable A_i and time going to infinity, they approach the inverse of the well known pathological sampling frequencies $(\lambda - \mu)/(2l\pi j)$, $\;l = \pm1, \pm2, \ldots$, where λ and μ are any two eigenvalues of A_i.

Remark. If hyperplanes $c_m x = y_m^{(r)}$ intersect $\mathcal{C}_d^T(i) \cap \mathcal{R}_d^T(i)$ for more than one m, the rank condition in the theorem can obviously be relaxed. If this is true for all $m \in \{1, \ldots, p\}$, then the rank condition reduces to

$$\operatorname{rank}\,[C', A_i'C', \ldots, (A_i')^{n-1}C'] = n.$$

Remark. The conditions in Theorem 5 are easy to check. Note, however, that only the *existence* of control sequences which distinguish between any two initial states is guaranteed by the above notion of observability (and hence the conditions in the theorem). Whether we will be able to find such a sequence depends on the system under consideration.

Example 1(revisited). Assume that the quantizer in our three tank problem gives the following information:

$$y_d(t) = \begin{cases} \text{"water level high" if } y(t) = \begin{bmatrix} 1 & 0 & 0 \end{bmatrix} x(t) > 1.5 \\ \text{"water level low" if } y(t) = \begin{bmatrix} 1 & 0 & 0 \end{bmatrix} x(t) \leq 1.5\,. \end{cases}$$

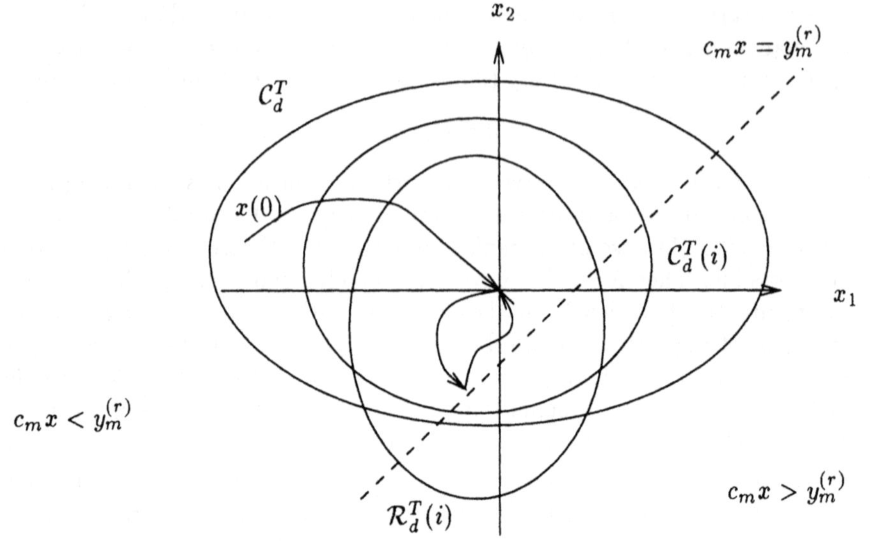

Fig. 7. Example for generating "nonpathologically" spaced measurement events

Then, it is easy to check that the (sufficient) observability conditions in Theorem 5 hold: We restrict control symbols to U_{d_1}, i.e. we "choose" (A_1, B_1) as system parameters. From the previous section, we know that $\mathcal{C}_d(1) = \mathbb{R}^n$ (and hence $\mathcal{C}_d = \mathbb{R}^n$); moreover – this has also been pointed out in the previous section – $\mathcal{R}_d(1)$ contains the subspace Span$\{[1\ 1\ 1]'\}$. Thus, the "event generating" hyperplane $x_1 = 1.5$ intersects $\mathcal{R}_d(1) \cap \mathcal{C}_d(1)$ and, for a large enough T, $\mathcal{R}_d^T(1) \cap \mathcal{C}_d^T(1)$: condition 1 is satisfied. Condition 2 also holds as

$$[C'\ A_1'C'\ (A_1')^2C'] = \begin{bmatrix} 1 & -a & a^2 \\ 0 & a & -2a^2 - ab \\ 0 & 0 & ab \end{bmatrix}$$

has rank 3. Hence, there exists a sequence of control events which allows reconstruction of the continuous plant state, and it is indeed trivial to come up with a suitable input signal.

5 "Stability" and a Heuristic Control Algorithm

Stability in the traditional (Lyapunov) sense[3] cannot be achieved for the hybrid control system in Fig. 1 unless – 1 – at least one control symbol $u_d^{(k)} \in U_d$ is

[3] Recall that an equilibrium point x_e is stable in the sense of Lyapunov, if for each $\epsilon > 0$ and each $t_0 \in \mathbb{R}^+$, there exists a $\delta(\epsilon, t_0) > 0$ such that $\|x(t_0) - x_e\| < \delta(\epsilon, t_0)$ implies $\|x(t) - x_e\| < \epsilon\ \forall t \geq t_0$ (e.g. [5]).

mapped to a stable dynamics matrix A_i or – 2 – the desired equilibrium point lies on an (intersection of) event generating hyperplane(s) and all trajectories can be forced into this equilibrium on the intersection ("chattering control"). The general impossibility of Lyapunov stability can be seen from a simple example: Suppose, $f(u_d^{(k)}) = A = \text{constant}\ \forall k \in \{1, \ldots, N\}$ and at least one eigenvalue of the matrix A lies in the open right half plane. Assume that 0 is the desired equilibrium, and $y = C_y x = 0$ is an inner point of a quantization box. Now, if $C_y x(0)$ lies inside this quantization box, and the orthogonal projection of $x(0)$ on the unstable subspace of A is nonzero, the state will "drift away" from the origin. The controller will not know about this until $C_y x(t)$ crosses into another quantization box, hereby generating a new measurement event.

Thus, stability (in the sense of Lyapunov) has to be replaced by a weaker notion. One might try to design for a "sufficiently small" attractive set (around a desired equilibrium point) which can be made globally attractive by suitable controller construction.

5.1 A Heuristic Model Based Controller

The idea underlying our heuristic approach is to use measurement events to reconstruct the (continuous) plant state and then to drive the plant state into a desired equilibrium point according to some optimality criterion. For state reconstruction and trajectory planning purposes, the controller (which is discrete in the sense that it accepts only (timed) discrete events and generates only (timed) discrete events) incorporates a continuous plant model P (eqns. (1), (3) - (4)). Clearly, this approach constitutes a dual control problem: Long-term benefits of "probing" (creating new measurement events for plant state reconstruction) and its possible negative short-term effects on the cost function somehow have to be traded off against each other. For this, we could use the following naive two-stage approach: a) Pure probing until enough measurement events have been generated to allow (in principle) reconstruction of the plant state. This is followed by b) a Certainty equivalence control policy reminiscent of "Receding horizon" approaches (e.g. [7] [12]): Based on the current plant state estimate, an open loop strategy is calculated and applied until a new measurement event becomes available. This is then used to update the plant state estimate and revise the future control strategy.

In this approach, the controller state "lives" in $\mathbb{R}^n \times \mathbb{N} \times \{P, CE\}$, its projection on $\mathbb{R}^n$ is the controller's estimate of the plant state, the projection on the binary set $\{P, CE\}$ tells whether the controller is in probing or Certainty Equivalence mode, and the projection on $\mathbb{N}$ "counts" the available measurement events during probing (and is held constant otherwise).

6 A Very Simple Example

The following simple example is meant to complement example 1. Whereas the latter is more general (the control input can change the characteristic features of the plant model), some of the concepts discussed in the previous sections can be understood more easily by looking at the simple mechanical system presented in the following.

Example 2. Assume that a cart of mass 1 moves without friction on a plane (Fig. 8).

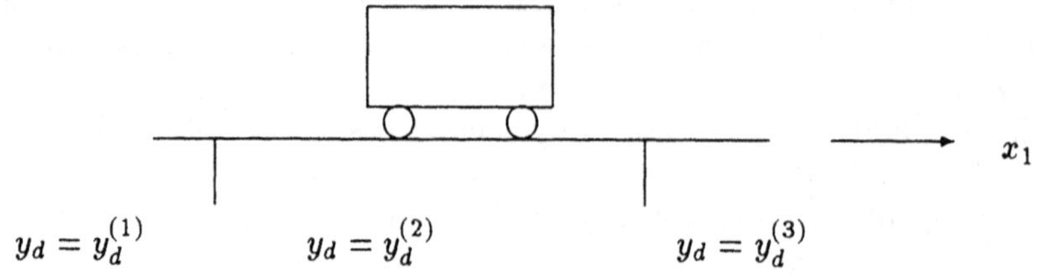

Fig. 8. Example: Cart on a plane

The states are defined as position (x_1) and velocity (x_2). An idealized system model is given by

$$\dot{x} = \underbrace{\begin{bmatrix} 0 & 1 \\ 0 & 0 \end{bmatrix}}_{A} x + g(u_d).$$

The forcing term depends on the control signal u_d, which can take three symbols:

$$\begin{aligned} U_d = \{u_d^{(1)} &= \text{“apply constant force to the right”}, \\ u_d^{(2)} &= \text{“don't apply any force”}, \\ u_d^{(3)} &= \text{“apply constant force to the left”} \ \} \ ; \end{aligned}$$

$$g(u_d^{(1)}) = \begin{bmatrix} 0 \\ 1 \end{bmatrix}, \quad g(u_d^{(2)}) = \begin{bmatrix} 0 \\ 0 \end{bmatrix}, \quad g(u_d^{(3)}) = \begin{bmatrix} 0 \\ -1 \end{bmatrix}.$$

Obviously, all forcing vectors are lineraly dependent; hence, we can write

$$\begin{aligned} [g(u_d^{(1)})\ g(u_d^{(2)})\ g(u_d^{(3)})] &= \begin{bmatrix} 0 \\ 1 \end{bmatrix} [1\ 0\ {-1}] \\ &:= B[u_1\ u_2\ u_3] \ . \end{aligned}$$

Suppose that velocity cannot be measured and position only in a highly quantized manner:

$$y_d(t) = Q(\underbrace{[1\ 0]\ x(t)}_{y(t)}) = \begin{cases} y_d^{(1)} \text{ if } y(t) \le y^{(1)} \ , \\ y_d^{(2)} \text{ if } y^{(1)} < y(t) \le y^{(2)} \ , \\ y_d^{(3)} \text{ if } y(t) > y^{(2)} \ , \end{cases}$$

where $y^{(1)} < 0$ and $y^{(2)} > 0$ are given threshold values. Thus, we have a continuous system ($x \in \mathbb{R}^2$) and symbolic control and measurement sets. From Corollary 4, it follows that

$$\mathcal{R}_d = \mathcal{C}_d = \mathbb{R}^2 \ ,$$

i.e. the whole plant state space is controllable to, and reachable from, the origin. This is also obvious from the top part of Fig. 9, which shows the state trajectories for the three possible control symbols: For $u_d = u_d^{(1)}$, the state trajectories are depicted as solid lines, for $u_d = u_d^{(2)}$ as dotted and for $u_d = u_d^{(3)}$ as dashed lines. For $u_d = u_d^{(2)}$, all points on the x_1 axis (x-marked) are stationary.

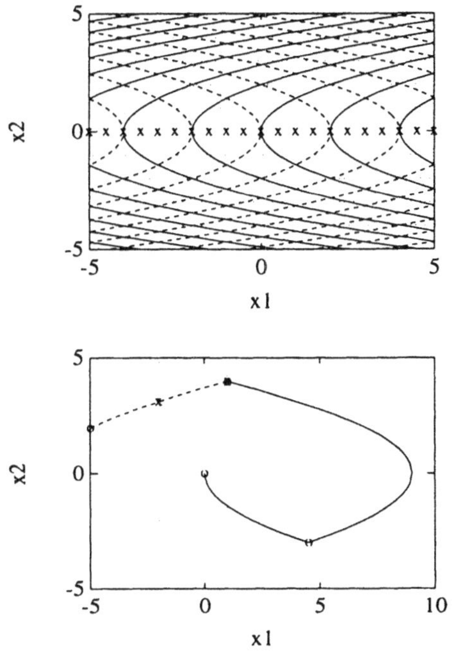

Fig. 9. All possible state trajectories (top), state trajectory for a candidate control strategy (bottom)

From Theorem 5, one deduces immediately that our system is observable on $\mathbb{R}^2$.

Suppose we want to drive the system state into the origin. Then, we can use the strategy shown in the bottom half of Fig. 9 (there, $y^{(1)} = -2$, $y^{(2)} = 1$, an "x" marks the occurence of a measurement event and an "o" the occurrence of a control event): Use the initial information $y_d(0)$ (i.e. the knowledge that, at time $t = 0$, the system is in the left/middle/right "quantization box") to generate a sequence of control events (a control signal) that creates two measurement events: For $y_d(0) = y_d^{(1)}$, we select $\langle t_1^u = 0,\ u_d^{(1)}\rangle$. This will eventually generate two measurement events $\langle t_1^y, y(t_1^y) = y^{(1)}\rangle$ – the state crosses into the middle quantization box – and $\langle t_2^y, y(t_2^y) = y^{(2)}\rangle$ – the state leaves the middle quantization box. Reconstruct the current state and drive it into the origin – e.g. in the time-optimal way shown in Fig. 9 by applying the sequence of control events $\langle t_2^u = t_2^y,\ u_d^{(3)}\rangle$ and $\langle t_3^u,\ u_d^{(1)}\rangle$; then switch the control to zero: $\langle t_4^u,\ u_d^{(2)}\rangle$. t_3^u and t_4^u can be computed from $x(t_2^y)$.

According to the strategy outlined in the previous section, we would have to update our state estimate with the new measurement event that becomes available once the plant state crosses again into the middle quantization box; then, we would also have to revise our control policy. As this example is only a thought experiment (no modelling errors, no unknown disturbances), there is no need for such an update – the plant state is perfectly known for $t \geq t_2^y$.

Under more realistic assumptions, we would also have to be aware that we will not be *exactly* in the origin (nor in any other point on the x_1-axis), once we switch the control to zero. This will cause the state to drift slowly away from its desired position (the origin). As this does not show in the quantized measurement until $y(t) = y^{(1)}$ or $y(t) = y^{(2)}$ (i.e. a new measurement event occurs), it will be hidden from the "outside world" and there is nothing we can do about it – the closed loop cannot be made stable in the sense of Lyapunov. Once a new measurement event occurs, we can update our state estimate and apply the same policy again.

7 Conclusions

In this contribution, a special hybrid problem has been investigated – the control of a continuous plant by symbolic output feedback. The common approach when dealing with such control configurations is to replace the continuous plant model by a discrete model, hence converting the mixed (continuous-discrete) problem into a purely discrete one. In contrast to this, we have proposed a framework which emphasizes the continuous aspect of the hybrid problem. This is more adequate whenever specifications are "much finer" than measurement quantization levels or are "naturally" expressed in terms of continuous variables. In this approach, interaction between plant and controller is modeled by an interrupt structure, and information exchange is represented by sequences of timed control and measurement events. The key questions within this framework are the following: Can the plant state be transferred between arbitrary points in $\mathbb{R}^n$ (or a subset thereof) by applying a suitable sequence of control events? Is it possible to reconstruct the continuous plant state from a sequence of measurement events? These questions regarding controllability/reachability and observability have been answered for the class of plant and interface models given by equations (1), (4), and (6). Based on these results, a heuristic Certainty-Equivalence-type controller has been suggested. The main idea and the qualitative effects of model errors and unmeasured disturbances have been illustrated by means of a (very) simple example. Quantitative robustness issues are currently being explored.

References

1. P. J. Antsaklis, J. A. Stiver, and M. Lemmon. Hybrid system modelling and autonomous control systems. In R. L. Grossman, A. Nerode, A. P. Ravn, and H. Rischel, editors. *Hybrid Systems*, Lecture Notes in Computer Science 736, pages 366–392. Springer, Berlin, 1993.
2. V. Boltyanski. *Mathematical Methods of Optimal Control*. Holt, Rinehart and Winston, New York, 1971.
3. R. F. Brammer. Controllability in linear autonomous systems with positive controllers. *SIAM J. Control*, 10:339–353, 1972.
4. R. L. Grossman, A. Nerode, A. P. Ravn, and H. Rischel, editors. *Hybrid Systems*, Lecture Notes in Computer Science 736, Springer, Berlin, 1993.
5. W. Hahn. *Stability of Motion*, Berlin, 1967. Springer.
6. T. Kailath. *Linear Systems*. Prentice-Hall, Englewood Cliffs, N. J., 1980.
7. S. S. Keerthi and E. G. Gilbert. Moving-horizon approximations for a general class of optimal nonlinear infinite-horizon discrete-time systems. In *Proc. 20th Annual Conf. Inform. Sci. Syst.*, pages 301–306, 1986.
8. J. B. Lasserre. Reachable, controllable sets and stabilizing control of constrained linear systems. *Automatica*, 29:531–536, 1993.
9. J. N. Lin. Determination of reachable set for a linear discrete system. *IEEE Transactions on Automatic Control*, 15:339–342, 1970.
10. J. Lunze. Ein Ansatz zur qualitativen Modellierung und Regelung dynamischer Systeme. *at - Automatisierungstechnik*, 41:451–460, 1993.
11. J. Lunze. Qualitative modelling of linear dynamical systems with quantized state measurements. *Automatica*, 30:417–431, 1994.
12. D. Q. Mayne and H. Michalska. Receding horizon control of nonlinear systems. *IEEE Transactions on Automatic Control*, 35:814–824, 1990.
13. J. Raisch. Simple hybrid control systems - continuous FDLTI plants with symbolic measurements and quantized control inputs. In G. Cohen and J.-P. Quadrat, editors, *11th International Conference on Analysis and Optimization of Systems*, Lecture Notes in Control and Information Sciences 199, pages 369–376. Springer, 1994.
14. J. Raisch. Simple hybrid control systems - continuous FDLTI plants with quantized control inputs and measurements. Systems Control Group Report 9305, Dept. of Electrical and Computer Engineering, University of Toronto, May 1993.
15. J. Raisch and S. D. O'Young. A discrete-time framework for control of hybrid systems. In *Proc. 1994 Hong Kong International Workshop on New Directions of Control and Manufacturing*, pages 34–40, 1994.
16. J. Raisch and S. D. O'Young. Symbolic control of uncertain hybrid systems. In *Proc. European Control Conference ECC95*, 1995.
17. W. E. Schmitendorf and B. R. Barmish. Null controllability of linear systems with constrained controls. *SIAM J. Control and Optimization*, 18:327–345, 1980.
18. E. D. Sontag. *Mathematical Control Theory*. Springer-Verlag, Berlin, 1990.
19. J. A. Stiver and P. J. Antsaklis. On the controllability of hybrid control systems. In *Proc. 32th IEEE Conference on Decision and Control*, 1993.

Hybrid control of a Robot - a case study*

Anders P. Ravn, Hans Rischel,
Michael Holdgaard, Thomas J. Eriksen
Department of Computer Science, bldg. 344

Finn Conrad, Torben O. Andersen
Institute of Control Engineering, bldg. 424
Technical University of Denmark
DK 2800 Lyngby, Denmark
Email: apr@id.dtu.dk

Abstract. An experiment with a distributed architecture to support a hybrid controller for a robot is described. For a desired trajectory, the controller plans a schedule for switching between a fixed set of control functions. Initial results indicate that the proposed architecture is better at achieving a desired trajectory than conventional control algorithms. The experiment also illustrates a division of concerns between software engineering and control engineering. Development of controlling real-time state machines and their mapping to processors and network is the task of software engineering, while the control engineer must identify plant phases, switching conditions and relevant control laws. These define algorithms for a planner and for the control functions. The format of schedules produced by the planner and the algorithms constitute the interface to software developers.

1 Introduction

The experiment reported here is motivated by questions that arise when w consider the growing interest in theories for hybrid systems within both the control theory and computer science communities.

- Do the ideas on hybrid control presented by, e.g., Nerode and Kohn [13] and Antsaklis et al. [1] have any role to play in the engineering of usual control systems, or are they just good for important, but highly specialized areas?
- How can one implement a distributed system based on the phase transition model exemplified in the work by Pnueli et al. [9, 10, 16], corresponding logics [17, 21] and design techniques [18, 6, 20] be used to develop hybrid architectures, e.g., the proposal by Najdm-Teherani in [12, 11]?

* Supported by the Danish Technical Research Council under the Co-design and IMCIA projects. This paper was written while Anders P. Ravn was visiting professor at Institut für Informatik und Praktische Mathematik, Christian-Albrechts-Universität zu Kiel, Germany, supproted by the German Research Council.

Such questions cannot be answered by theoretical investigations, they require experience with the techniques. An initial experiment was started at the end of 1993 as joint work by the Control Engineering Institute and the Computer Science Department of the Technical University. The main work force was two part-time MSc students, so from the start it was obvious that to reach some experimental results by mid 1994, we had to aim for a simple architecture and simple controllers which would shed some light on the issues, instead of going all out for a general, flexible, and intelligent architecture.

The initial experiment was so promising that we are continuing the collaboration to investigate the hybrid control paradigm in practice. The main part of this paper is an outline of the experiment, where we hope to give sufficient details such that others may be tempted to do similar experiments that confirms or questions our conclusions.

An important point in the experiment is that the hybrid control paradigm has a theoretical foundation, because one could probably develop a better performing controller for the specific system using a mixture of diverse paradigms (neural nets, expert systems, etc.) and ad hoc optimizations, but this would leave no directions for other experiments.

Overview: Section 2 describes the plant: a hydraulically controlled robot (manipulator). It also summarizes the issues in controlling the system, and gives a brief introduction to the hardware platform. Section 3 outlines the architectural principles and illustrates specification of requirements (performance constraints) and derivation of a top level design. Section 4 describes the lower level design and the distribution of processes over the network. Section 5 touches upon the concrete experiments with the developed system, and promising extensions. In the concluding section 6, we discuss a useful work division between control and software engineering in building such systems. A question that will arise if the hybrid system paradigm is successful.

2 The plant

The Robotic Institute of America (RIA) defines a robot as follows:

> 'A Robot is a re-programmable, multifunctional manipulator designed to move material, parts, tools or specialized devices through variable programmed motions for the performance of a variety of tasks'.

The robot we consider is a continuous path manipulator, where points along the path of motion matters. It is used where it is important to follow a complete path, possibly with high speed. Applications include spray painting, polishing, grinding and arc welding. When a robot is required to move high loads fast and precisely along a path, a hydraulic actuation device is often used.

Thus, the controlled plant in this experiment is a multi-axes, hydraulically powered, digitally controlled machine. It is a test facility for investigating the development and implementation problems arising in the digital control of oil

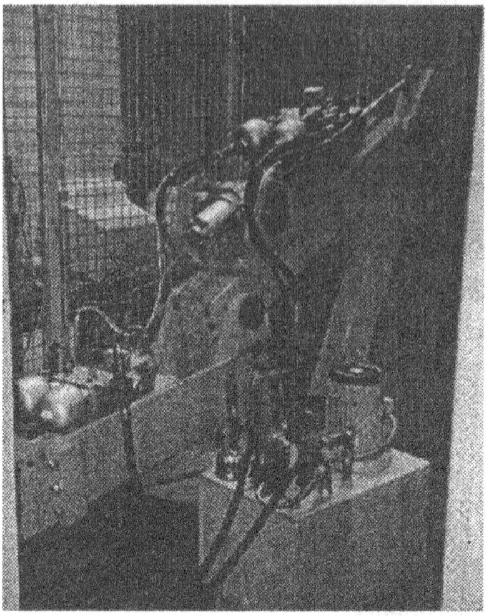

Fig. 1. The physical plant with perimeter fence.

hydraulic systems [3, 4]. It is a two link rotary arm manipulator with a high-frequency servo valve controlled hydraulic cylinder driving each link. The system is equipped with a measuring- and instrumentation system including sensors for the measurement of various dynamic variables. For safety reasons the system is connected to a hardwired shut down of oil flow (which stops the robot) in the case that a perimeter fence door is opened or if an emergency button is pressed. A photo of the equipment is shown in Figure 1.

Each arm can be regarded as stiff in the working range with a pay-load up to 50 kg. The joints are supported in pre-loaded rotary angular contact ball bearings to avoid stiction and give stiff support. The cylinders are specially designed and manufactured for high performance. The rods of the cylinders are supported by hydrostatic bearings and low friction seals for minimum friction. With membrane accumulators on both pressure- and tank ports, an almost constant supply pressure of 210 bar can be maintained. This adds to the linearity of the cylinders.

Each cylinder can yield a static force up to 20 kN. With maximum pay-load there is still adequate power to obtain tool center point velocities around 3.5 m/s. In this situation the centrifugal and Coriolis couplings as well as the gravity force and change of inertia moments have a significant influence on the dynamics, which implies a non-linear and coupled dynamic relation. Furthermore, the flow pressure relations in the valves are non-linear. That is, the same control input signal gives a different response in different arm positions. This will for instance be the case in a reference path shown in Figure 3, when speeds are close to maximal.

2.1 Hardware Platform

The robot is controlled through a network of 4 transputers. Two T225 without floating point directly connected to interface electronics for each of the two separate joints, and two T805 processors with floating point capability, where one is placed in a PC-based development system.

2.2 Mode switch control

In the absence of friction and other disturbances the dynamics of a n-link rigid manipulator can be written as:

$$H(q)\ddot{q} + C(q,\dot{q})\dot{q} + G(q) = \tau$$

where q is the $n \times 1$ vector of joint displacements, τ is the $n \times 1$ vector of applied joint torques (or forces), $H(q)$ is the $n \times n$ symmetric positive definite manipulator inertia matrix, $C(q,\dot{q})\dot{q}$ is the $n \times 1$ vector of centripetal and Coriolis torques and $G(q)$ is the $n \times 1$ vector of gravitational torques.

It is recognized that fixed linear control schemes are non-effective means of robot control because they are inable to cope with the highly non-linear coupled and time-varying characteristics of the robot. This is specially true in our case of a direct drive robot where inertia changes and gravity effects are significant. For appropriate control an adaptive controller could be used, however, the time needed for adaptation has turned out to be too long compared to the frequent change of operating conditions in demanding paths.

We deal with the problem by dividing the workspace of the robot into a number of modes, each characterized by a model and a controller. This implies that each time the robot enters a new mode of operation, a related controller is dispatched. Ideally this requires a "mode detector" which dynamically detects whether the mode has changed; but in the case study the characteristics of the workspace are known, and allows us to pre-plan the mode switches for a given reference path.

In the experiments, decentralized control schemes are used for controlling each joint of the robot, because they are less complex and require far less computation than centralized control schemes. This enables a higher sampling rate.

The control strategy transforms the modes in workspace into actuator space - reducing the problem of controlling the robot to that of controlling the actuators hereby regarding the manipulator as an external disturbance. The model structure therefore remains invariant, but the parameters depend heavily on the position of the manipulator in the workspace: The manipulator disturbs the actuators with an inertia load and a force, both varying with the position of the actuators.

The modes are selected by making the actuator workspace operate around a limited number of nominal operating points. For an operating point each actuator dynamics is approximated by a linear model such that a fixed linear controller - based on this model - performs well.

The control design objective was:

1. To guarantee a gain margin of 15 dB (the gain margin indicates how much further the loop gain can be increased without causing instability.
2. To make the response characteristic insensitive to changing operating points, non-linearities, unmodeled higher order dynamics etc.
3. To make the robot track the reference trajectory fast and smooth and make the overshoot as small as possible.

The linear open-loop model for the hydraulic position servos have the form:

$$G(s) = (K_q \omega_n^2)/(s(s^2 + 2\zeta\omega_n s + \omega_n^2))$$

where K_q, ω_n and ζ are positive time-varying parameters (depending on the actuator positions).

To satisfy design objectives 1. and 2, the parameter ranges were divided into four smaller ranges, each describing a mode. The linear robust feedback controllers for each mode were designed using a frequency domained optimization method based on H_∞ theory [15] resulting in a controller transfer function of the form:

$$F(s) = (a_3 s^3 + a_2 s^2 + a_1 s + a_0)/(s^4 + b_3 s^3 + b_2 s^2 + b_1 s + b_0)$$

where the coefficients a_i and b_i are changed when a new mode is entered. To satisfy design objective 3, a velocity reference feed-forward was added to the controller. The feedforward was chosen as a simple gain multiplied to the reference velocity and remains constant during operation. The overall structure of the controller is shown in Figure 2

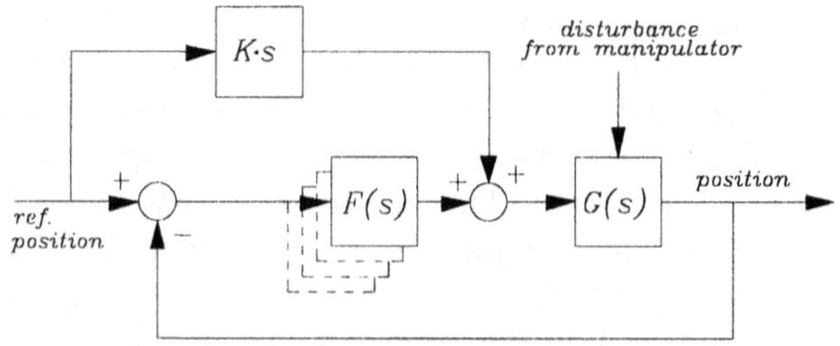

Fig. 2. Control structure.

To avoid unwanted "bumps" in the control signal at the instant when the parameters in the feedback controller are adapted, a filter is added where the time-constants determine how fast the transition from the old control signal to the new control signal will be.

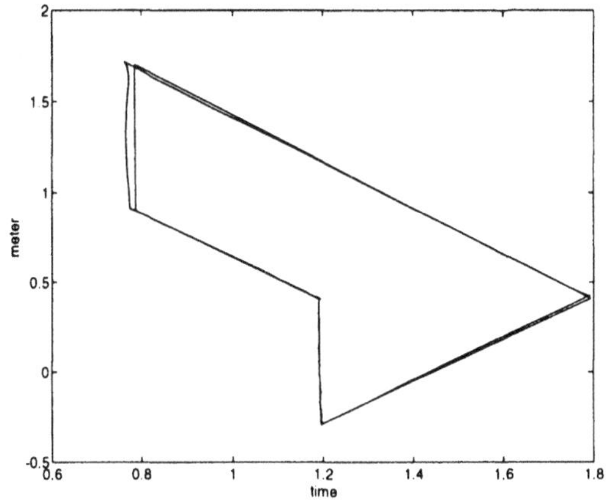

Fig. 3. The reference trajectory compared with the performance of the MSCS.

The performance of the mode-switching control scheme (MSCS) are illustrated in Figure 3, Figure 4 and Figure 5. In Figure 3 the reference trajectory is shown against the MSCS with the robot carrying a payload of 20 kg.

In Figure 4 is shown the error for the MSCS, following the reference trajectory, when the robot is carrying a payload of 0 kg and 20 kg. It is seen that the MSCS is insensitive to changing payload of the robot.

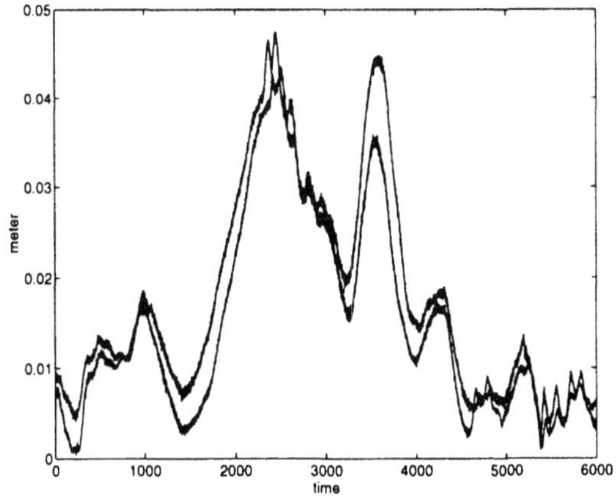

Fig. 4. The error for the MSCS with a payload of 0 kg and 20 kg.

Figure 5 is a verification of the design objective 3 regarding performance, in that it shows the error for the MSCS compared to a single robust control scheme. The comparing robust control scheme was designed to fullfil the design objectives 1 and 2.

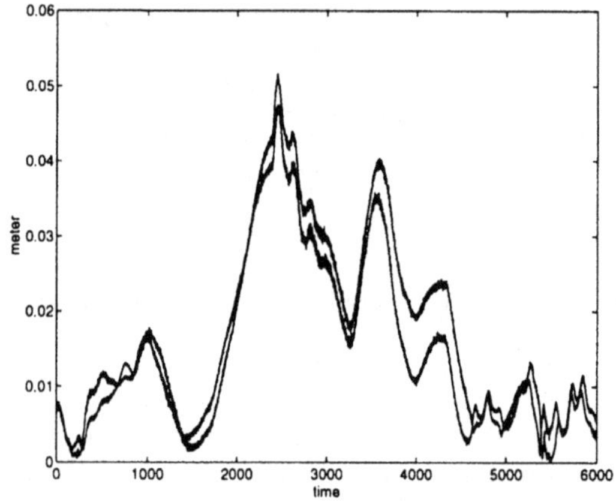

Fig. 5. The error for the MSCS compared with the error for a single robust control scheme.

3 Architecture and global design

The guiding principle in the following architecture is to model the system as a collection of state machines that coordinate their actions through shared states. The state machines may be composed in parallel or a given state machine may be serially [6] decomposed. A distinguished control state in each state machine is the *phase* of that machine. This forms the link to phase transition systems.

The main requirement or objective for the robot controller is to follow a path in the plane. A complication is that the robot is safety-critical because of the heavy forces used during movements. This warrants isolating some supervisory functions in a separate, trusted state machine. The main requirement is specified by means of a generalized trajectory function, describing a band within which the tool center point shall move during operation. Such a band B defines a tolerated set of positions $B(t)$ for each point of time t where the robot is active. A band is specified by a function from $\mathbf{R}^+$ (the non-negative reals) to subsets of the coordinate-space, *Coord*.

$$B : \mathbf{R}^+ \rightarrow \mathbf{P}(Coord)$$

The coordinates *Coord* can either be the ordinary $X - Y$ coordinates of the plane or the transformed positions for the actuator. The latter choice avoids the complexities of motion planning and coordinate transformations. For simplicity, it was therefore chosen in the experiment.

The band could for instance be the output from a CAD-CAM tool used to design items to be processed by the robot. A band must have a reasonable shape, it should at least contain some smooth trajectories that corresponds to the identified system. We shall call such a band *feasible*. In this outline, we assume a single feasible B which does not change during operation of the robot.

The required operation of the robot can then be given in terms of a Boolean state function *halt* which stops or starts the robot; the current position of the tool center point *tcp*; and a Boolean state *trace* which specifies when the robot is following a given band. These are states of a dynamic system, and are in Extended Duration Calculus given by total functions of *Time* (the non-negative reals)

$$\begin{aligned} &halt : Time \rightarrow \mathbf{Bool} \\ &tcp : Time \rightarrow Coord \\ &trace : Time \rightarrow \mathbf{Bool} \end{aligned}$$

The *halt* state is the phase of a state machine that *monitors* several conditions that should halt the system. These include an infeasible band, a *tcp* position outside a predefined safe working area, etc.

The *trace* state is the corresponding phase of the state machine that *controls* tracing of a path.

In the actual design there are further plant states that model discrete components like emergency stop. We shall omit those, because they only lead to a design of a more complex monitor.

The requirements are constraints on the phases and the plant state *tcp*. We elide the monitor constraints and concentrate on the constraints on the path control. They either define the interface between the monitor and the path control, or the performance objective of the path control.

One of the interface constraints is

$$R1 : \lceil \neg halt \rceil^{T} \longrightarrow \lceil trace \rceil$$

which uses a progress operator [18] to specify that if $\neg halt$ holds throughout any time interval of length T then this interval is followed by an interval where *trace* holds. The other interface constraint is the dual for *halt* and $\neg trace$.

The formula $R1$ requires the robot to enter a tracing state after an initial period of T where it can move to the start of the band.

The remaining formulas define the objective during *trace*: the tool center point shall stay within the band.

$$R2 : \forall r > 0 \bullet \lceil \neg trace \rceil \; ; \; \lceil trace \rceil^{r} \longrightarrow \mathbf{b}.tcp \in B(r)$$

This formula also uses the progress operator. Any interval which contains a transition from $\neg trace$ to *trace* (the subformula $\lceil \neg trace \rceil \; ; \; \lceil trace \rceil$) and where the duration of *trace* is the real value r is followed by an interval where the initial value of *tcp* is within $B(r)$.

There is an analogous requirement for stop of the robot when $\neg$*trace* is stable for some time.

In summary, the top level design consists of two parallel machines. One is the *Monitor* with phases *halt* and $\neg$*halt*, and the other is *Control* with phases *trace* and $\neg$*trace*.

3.1 Decomposition of Control

The path control is in the actual design decomposed by splitting the *trace* phase into subphases *plan*, *move* and *notfeas*. The latter is entered when a band during planning turns out to be infeasible. Renaming $\neg$*trace* to *stop* we have the design given in Figure 6.

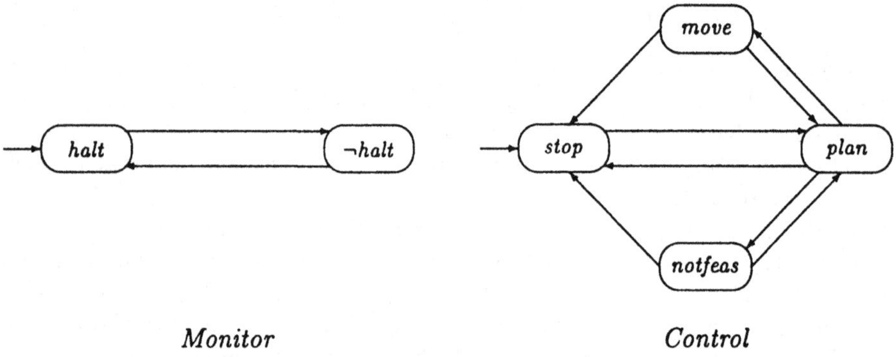

Fig. 6. Controlling and Monitoring Automatons

The constraints on the subphases take the forms already seen. The planning phase has for example the constraints:

Progress: The *plan* phase will be left within $T/2$ time units

$$\lceil plan \rceil^{T/2} \longrightarrow \lceil \neg plan \rceil$$

This constraint ensures that the planner under all circumstances moves within a bounded time.

Select-halt: A stable *halt* can only lead to the *stop* phase

$$(\lceil \neg plan \rceil \ ; \ \lceil plan \wedge halt \rceil) \longrightarrow \lceil plan \vee stop \rceil$$

This selection formula ensures that the **Progress** formula cannot be satisfied by moving to an arbitrary other phase when *halt* is stable.

Select-feas: And for a feasible band the only possibility is the *move* phase

$$(\lceil \neg plan \rceil \ ; \ \lceil plan \wedge \neg halt \wedge Feas(b) \rceil) \longrightarrow \lceil plan \vee move \rceil$$

Select-notfeas: And for an infeasible band, *notfeas* is the only possible next phase

$$(\lceil \neg plan \rceil \ ; \ \lceil plan \wedge \neg halt \wedge \neg Feas(b) \rceil) \longrightarrow \lceil plan \vee notfeas \rceil$$

Provided, that is, that *halt* is not asserted.

Remark: The division of the 'inputs' into: *halt*, $\neg halt \wedge Feas(b)$, or $\neg halt \wedge \neg Feas(b)$ $T/2$ is an implementation of a measurement function in the terminology of Nerode [13]. If a change takes place during the measurement, a continue value will give an arbitrary choice, but always in a bounded time.

The other phases have similar requirements at this level. The decomposition should allow a proof that it is a correct refinement, but this has actually not been checked, because we found it more interesting to press on to a more detailed design for *move*.

4 Detailed design and distribution

The monitor and the controller as outlined above is on the given platform implemented by an occam process (a thread) for the monitor and separate processes for each of the subphases of the controller. The serialisation of the subphases is ensured by a coordinator process. The decision to implement subphases as separate processes is given by the structure of the *move* subphases. This point is discussed briefly below, before a presentation of the implementation.

4.1 Schedules and motion planners

As a side effect of checking feasibility, the planning phase computes a sequence of activation records for subphases of *move*. These subphases are identified with a motion planner and controller for a segment of the band. An activation record contains:

1. A subphase (motion planner) identifier.
2. Time for activation (relative to start of *trace*).
3. Parameters.

The sequence of activation records is called a *schedule* because the activation times defines a unique motion planner for each subphase of *move*. It is analogous to the control script of Nerode [13].

Each motion planner is implemented so it may have several regulators that it can select one at a time. It is thus potentially a subplanner with a subschedule (now the switching time may be in the range below 1 millisecond). In order to allow smooth transitions, the motion planners may be receiving sensor inputs as they are sampled even when they are not activated. This is a technique that reduces the amount of parameter data that has to be transferred at a switch between motion planners.

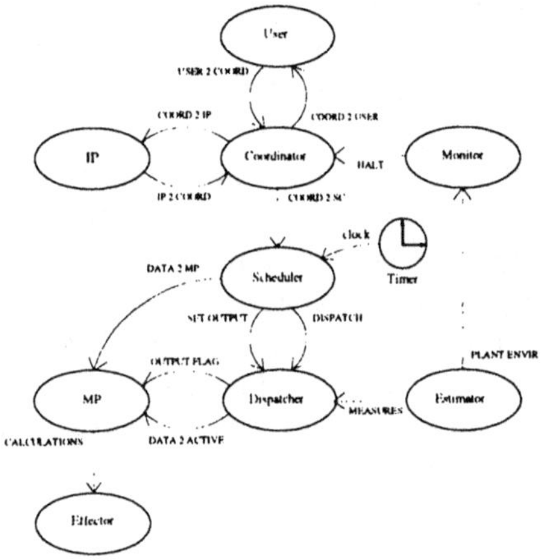

Fig. 7. Occam main process components.

4.2 Components of the implementation

The arguments above has lead to the process structure in Figure 7. The individual processes have the following functions:

User is initialisation and parameter (band) set-up.
Monitor is the *Monitor* described above. It receives input from the plant through the estimator (sensors).
Coordinator coordinates the subphases of *Control*.
IP is the *plan* subphase.
Scheduler distributes the activation records to motion planners and gives the activation times to the dispatcher.
Dispatcher starts and stops the individual motion planners according to the schedule.

MP is a set of motion planners.
Effector is the actuators.
Estimator is the sensors.

The split between *Scheduler* and *Dispatcher* is dictated by a desire to ensure prompt activation. This is also reflected in the physical placement, where the effector and estimator are placed on the two T225 processors. The rate of sampling and update can thus be in the range of 10 microseconds. The dispatcher and the motion planners occupy one T805 processor, while the remaining computation intensive processes are placed on the remaining T805.

The reasoning about timings and the resulting placement has been very informal [7], and it would be nice to have a compiler check that our timing assumptions actually hold, cf. [6].

5 The Planner and regulator

After having spent some time in developing a reasonably general architecture, one could dream of having sophisticated planning algorithms, but for the initial experiments we have settled for a fixed schedule for the chosen reference path.

5.1 Next steps

The experiment shows that a hybrid control scheme can improve performance. However, tailoring a schedule for each reference trajectory of a system is certainly time consuming, and also infeasible in a flexible control system. How can we then make better use of the architecture outlined above? At the moment, we are considering three ways of generalising the system:

1. A system identification may include a partitioning of the coordinate-space into regions with associated models. This could then form the basis for an efficient planner. Similar ideas are behind the switched Bond-graph paradigm discussed by Strömberg [19]. It might also have links to a simulation facility for switching systems as described by Back, Guckenheimer and Myers [2].
2. Another direction is to change the architecture such that there is feed-back from motion planners to the global planner. The idea would then be to implement a learning algorithm for a given reference trajectory, cf. [8].
3. A final direction would be to keep a very approximate global planner that creates an approximate schedule and then try to implement a dynamic switching between schemes in the motion planners. This would have to use a local architecture for motion planners analogous to Kohn's declarative control [14]. One may of course ask, why keep the global planner? Only experiments will tell whether it is safe to abandon it.

Experiments with these extensions are really in the domain of control engineering and may perhaps inspire new developments in control theory. However, the basic architecture and the implemented components of the system seem to be able to accommodate such changes.

6 Results and conclusion

We have described a modest experiment in implementing a hybrid controller architecture, and demonstrated that switching between control schemes improve performance for at least one rather demanding case of a motion path for the robot.

The experiment indicates that hybrid control is promising, not only in the context of highly complex control problems, but also as a theoretically sound approach that gives resonably simple solutions to practical problems.

The preceding section has indicated some of the possible avenues to be explored to get a more flexible controller within the architecture. The architecture accommodates such control experiments because the interface to control engineering is given by a few items: a planner algorithm, producing a schedule in a specific format, and a set of motion planners implementing local control schemes.

The components of the architecture have been systematically derived as parallel composition of state machines with serial decomposition of the individual machines, cf. Statecharts [5]. However, we are still at work on checking that the decompositions are correct refinement steps (and to change those that are not). It seems feasible to actually verify these decompositions using the Extended Duration Calculus.

The transformation to a network of distributed processes has been less systematic, and it looks a significant challenge to the techniques for real-time systems design.

Acknowledgement Help and assistance of Jens Ulrik Skakkebæk was instrumental in getting this paper ready. Comments from reviewers are gratefully acknowledged.

References

1. P. J. Antsaklis, J. A. Stiver, and M. Lemmon. Hybrid systems modeling and autonomous control systems. In R. L. Grossman, A. Nerode, A. P. Ravn, and H. Rischel, editors, *Hybrid Systems*, volume 736 of *LNCS*, pages 366–392, 1993.
2. A. Back, J. Guckenheimer, and M. Myers. A dynamical simulation facility for hybrid systems. In R. L. Grossman, A. Nerode, A. P. Ravn, and H. Rischel, editors, *Hybrid Systems*, volume 736 of *LNCS*, pages 255–267, 1993.
3. F. Conrad et al. On mechanical design and digital adaptive control of the fast tud-hydralic test robot manipulator. In *ASME WAM'91*, volume 91-WA-FPST-9. American Society of Mechanical Engineering, 1991.
4. F. Conrad et al. Transputer control of hydralic actuators and robots. *IEEE Trans. on Industrial Electronics*, October 1995.
5. D. Harel. Statecharts: A visual formalism for complex systems. *Science of Computer Programming*, 8:231–274, 1987.
6. Jifeng He, C. A. R. Hoare, M. Fränzle, M. Müller-Olm, E.-R. Olderog, M. Schenke, M. R. Hansen, A. P. Ravn, and H. Rischel. Provably correct systems. In H. Langmaack, W.-P. de Roever, and J. Vytopil, editors, *Formal Techniques in Real-Time and Fault-Tolerant Systems*, volume 863 of *LNCS*, pages 288–335, 1994.

7. M. Holdgaard, T. J. Eriksen, and A. P. Ravn. A distributed implementation of a mode switching control program. In *Proceedings of 7th Euromicro Workshop on Real-Time Systems*. IEEE Computer Society Press, 1995.
8. M. Lemmon, J. A. Stiver, and P. J. Antsaklis. Event identification and intelligent hybrid control. In R. L. Grossman, A. Nerode, A. P. Ravn, and H. Rischel, editors, *Hybrid Systems*, volume 736 of *LNCS*, pages 268–296, 1993.
9. O. Maler, Z. Manna, and A. Pnueli. From timed to hybrid systems. In J. W. de Bakker, C. Huizing, W.-P. de Roever, and G. Rozenberg, editors, *Real-Time: Theory in Practice, REX Workshop*, volume 600 of *LNCS*, pages 447–484, 1992.
10. Z Manna and A. Pnueli. Verifying hybrid systems. In R. L. Grossman, A. Nerode, A. P. Ravn, and H. Rischel, editors, *Hybrid Systems*, volume 736 of *LNCS*, pages 4–35, 1993.
11. S. Nadjm-Tehrani and J.-E. Strömberg. From physical modelling to compositional models of hybrid systems. In H. Langmaack, W.-P. de Roever, and J. Vytopil, editors, *Formal Techniques in Real-Time and Fault-Tolerant Systems*, volume 863 of *LNCS*, pages 583–604, 1994.
12. Simin Nadjm-Tehrani. *Reactive Systems in Physical Environments*. PhD thesis, Dept. Comp. and Inf. Science, Linköping University, Sweden, May 1994. Linköping Studies in Science and Technology, Dissertation no. 338.
13. A. Nerode and W. Kohn. Models for hybrid systems: Automata, topologies, controllability, observability. In R. L. Grossman, A. Nerode, A. P. Ravn, and H. Rischel, editors, *Hybrid Systems*, volume 736 of *LNCS*, pages 317–356, 1993.
14. A. Nerode and W. Kohn. Multiple agent hybrid control architectures. In R. L. Grossman, A. Nerode, A. P. Ravn, and H. Rischel, editors, *Hybrid Systems*, volume 736 of *LNCS*, pages 297–316, 1993.
15. R. Piché, S. Pohjolainen, and T. Virvalo. Design of robust controllers for position servos using H_∞ theory. *Journal of Systems and Control Engineering*, 205, 1991.
16. A. Pnueli. Development of hybrid systems (extended abstract). In H. Langmaack, W.-P. de Roever, and J. Vytopil, editors, *Formal Techniques in Real-Time and Fault-Tolerant Systems*, volume 863 of *LNCS*, pages 77–85, 1994.
17. A. P. Ravn and H. Rischel. Requirements capture for embedded real-time systems. In *Proc. IMACS-MCTS'91 Symp. on Modelling and Control of Technological Systems*, volume 2, pages 147–152. IMACS, May 1991.
18. A. P. Ravn, H. Rischel, and K. M. Hansen. Specifying and verifying requirements of real-time systems. *IEEE Trans. Software Engineering*, 19(1):41–55, January 1993.
19. Jan-Erik Strömberg. *A mode switching modelling philosophy*. PhD thesis, Dept. El. Eng., Automatic Control Group, Linköping University, Sweden, October 1994. Linköping Studies in Science and Technology, Dissertation no. 353.
20. Xinyao Yu, Ji Wang, Chaochen Zhou, and P. K. Pandya. Formal design of hybrid systems. In H. Langmaack, W.-P. de Roever, and J. Vytopil, editors, *Formal Techniques in Real-Time and Fault-Tolerant Systems*, volume 863 of *LNCS*, pages 738–755, 1994.
21. Zhou Chaochen, A. P. Ravn, and M. R. Hansen. An extended Duration Calculus for hybrid real-time systems. In R. L. Grossman, A. Nerode, A. P. Ravn, and H. Rischel, editors, *Hybrid Systems*, volume 736 of *LNCS*, pages 36–59, 1993.

Verifying Time-bounded Properties for ELECTRE Reactive Programs with Stopwatch Automata

Olivier Roux and Vlad Rusu*

Abstract. We present the automatic verification of time-bounded properties of programs written in the reactive language ELECTRE. For this, ELECTRE programs are translated into so-called *stopwatch automata*, automata with chronometers to measure time. Properties are expressed in the logic TCTL and model-checking algorithms are used to verify those properties on ELECTRE stopwatch automata. We argue that time-bounded TCTL is decidable on stopwatch automata.

Keywords: ELECTRE, TCTL, backward/forward analysis, stopwatch automata, decidability.

1 Introduction

Hard real-time systems are subject to externally defined timing constraints. These quantitative properties are so crucial that they must be checked for upon the expected behaviours of the application's tasks. *Reactive languages* intend to describe these behaviours through the possible executions of the actions with respect to the event occurrences.

The reactive languages may be splitted into two classes. The first class is that of the *synchronous* reactive languages [BB91] which are based on the strong synchrony hypothesis (ESTEREL [BD91], LUSTRE [HCRP91], SIGNAL [LLGL91], STATECHARTS [HP85] are examples of such imperative or data-flow or graphical synchronous languages). The second class is that of the *asynchronous* reactive languages, which deal with lengthing actions and non-simultaneous event occurrences in a continuous time domain. Among these, the ELECTRE language has a compiling technique which generates finite automata [PRH92], thus making it accessible to model-checking techniques for temporal properties verification.

More precisely, verification implies translating ELECTRE programs into the so-called *stopwatch automata*, finite state automata with chronometers that can be on or off. Stopwatch automata turn to be a class of *integrator systems* [KPSY93, ACHH93], with one further characteristic that provides decidability for *time-bounded properties*. But we have designed them specifically for ELECTRE, since simpler automata like *timed automata* [ACD90, NSY92] were not

* LAN (CNRS, Univ.Nantes, Ecole Centrale de Nantes) 44072 NANTES Cedex 03, France. E-mail: {rouxrusu}@lan.ec-nantes.fr

expressive enough to model the long-lasting actions which may be *preempted* by events and possibly later resumed.

Time-bounded properties are expressed in real-time temporal logic TCTL [ACD90, HNSY92]. We outline the arguments for decidability of time-bounded properties on stopwatch automata, and give the effective decision procedures. We extend this to linear hybrid systems [NOSY93, ACHH93]. A survey of decidability/undecidability results for hybrid systems can be found in [ACHH93]. Recent results are in [PV94, HKPV95].

2 ELECTRE and stopwatch automata

2.1 A foretaste of ELECTRE

ELECTRE is an asynchronous reactive language. Its basic constructs are the *module* (non zero and finite-time executing code) and the *event* (which can be imagined as either a hardware or software interrupt). A very simple example of ELECTRE program comes from the following: an operating system must iteratively execute tasks `A` and `B`. But the tasks can can be executed either in *calm* mode — from beginning to end, one after the other — or in *disturbed* mode — when some events `x` and `y` occur and preempt the tasks. When a task has been preempted, it will be later resumed at its preemption point.

This behaviour is described in figure 1 and is expressed by the ELECTRE program `[A↑@x:B↑@y]*`.

Here, `A` and `B` are *modules* (tasks), and `x`, `y` are events. Symbol ↑ denotes *preemption*, : denotes *launching* a module, and `*` is *iteration*. The `@` symbol in `@x` or `@y` means that events `x`, `y` are *fleeting*, i.e. they will be taken into account only if they are waited for, otherwise — since they are not memorized — they will be lost.

But events can have other properties (single or multiple memorizing) and modules also have properties (non-preemptible, or preemptible with restarting or resuming). Also, other operators describe parallel or exclusive composition of modules and events, leading to increasingly complex behaviours.

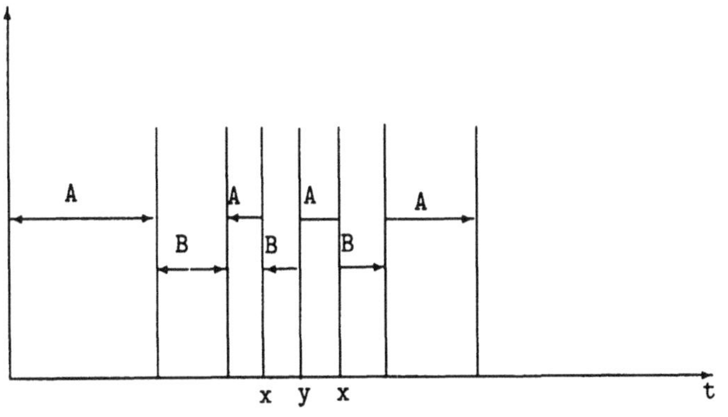

Fig. 1 *Behaviour of the* ELECTRE *program* `[A↑@x:B↑@y]*`

The operational semantics of ELECTRE has been defined in [PRH92, CR95]: ELECTRE programs are compiled into FIFO-automata, finite-state automata with FIFO queues to memorize a theoretically unbounded number of (memorizable) events. By making the realistic hypothesis that *only a bounded number of events should be memorized*, the FIFO automata can be transformed into finite state automata. For example, program `[A↑@x:B↑@y]*` gives the 4-state automaton in figure 2 — just ignore now the inequalities and the differential equations around the automaton.

States contain informations about modules: active, interrupted, idle, and transitions are labeled with events or module completions which cause some change in program execution.

2.2 Stopwatch automata

In order to include some information about real-time, we transform the finite-state automaton above into a hybrid system. On the given example, suppose that **A** lasts any time in $\mathbb{R}$ between 7 and 8 time units; **B** lasts any time in $\mathbb{R}$ between 4 and 5 time units; two consecutive **x** are separated by any time in $\mathbb{R}$ greater than 1 time unit; and two consecutive **y** are separated by any time in $\mathbb{R}$ greater than 2 time units. We obtain the stopwatch automaton in figure 2.

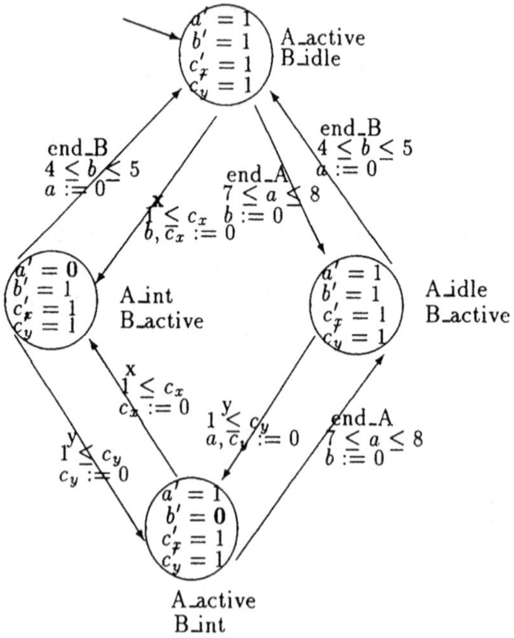

Fig. 2 *Stopwatch automaton for* `[A↑@x:B↑@y]*` *with timed assumptions*

This is done by adding to our finite state automaton 4 variables: a, b, c_x, c_y, used to measure the times of modules **A**, **B** and events **x**, **y**. We shall call the states *locations*, the transitions *edges*, and the variables *stopwatches*. Classically, the stopwatches evolve at locations following simple differential equations — when the equation of stopwatch a at location s is $a' = 1$, we say that the stopwatch is *on* — and edges can be taken provided conditions on stopwatches are satisfied, then maybe resetting some of the stopwatches.

Hereafter, the locations of the hybrid system will be in set S, and for each location $s \in S$ we have an *invariant* δ_s which must be *true* for the control to remain at that location; the edges are in set $E \subseteq S \times S$, and for each edge $< s, s' >$, we have a *label* $a^{<s,s'>}$, a *condition* $\phi^{<s,s'>}$ for the edge to be taken and $H^{<s,s'>}$ a *set of stopwatches to be reset* when taking the edge.

We obtain true *integrator systems* [KPSY93, ACHH93], because at some locations some stopwatches may remain constant, for expressing preemption of modules and resuming precisely at preemption point. More precisely, we obtain from ELECTRE programs so-called *stopwatch automata*.

Definition 1. A *stopwatch automaton* is an integrator system with the property that each of its cycles (in the sense of graphs) contains — for some stopwatch a — a resetting edge $a := 0$ for that stopwatch and a testing edge for the same stopwatch by some condition $k \leq a$ with $k \in \mathbb{N} - \{0\}$. □

The consequence is that **any cycle**, when run *twice* following the semantics of hybrid sytems, **will last at least some number of time units** - for instance, the upper-left cycle in figure 2 when run twice will last at least 4 time units.

This low-bounded duration characteristic provides decidability for time--bounded properties (sections 3.1, 3.2).

3 Time-bounded properties of stopwatch automata

We use real-time temporal logic TCTL [ACD90, HNSY92], which is syntactically defined by the grammar rule:

$$\psi ::= p|x \prec c|\neg\psi|\psi \wedge \psi|z.\psi|\psi\exists U\psi|\psi\forall U\psi$$

where $p \in P$ are *propositions*, z is a *new* stopwatch and $\prec \in \{<, \leq\}$.

The first four variants in the right of the above rule define the so called *state formulas*, that is, boolean combinations of propositions and inequalities on the stopwatches.

We focus on formulas of the form $\psi_1\exists U_{\leq c}\psi_2$ and $\psi_1\forall U_{\leq c}\psi_2$ with ψ_1 and ψ_2 two *state* formulas and $c \in \mathbb{N}$ (they are known to be abreviations for $z.(\psi_1\exists U_{\leq c}(\psi_2 \wedge z \leq c))$ and $z.(\psi_1\forall U_{\leq c}(\psi_2 \wedge z \leq c))$ respectively). We call $\psi_1\exists U_{\leq c}\psi_2$ *bounded reachability* formulas and $\psi_1\forall U_{\leq c}\psi_2$, *bounded response* formulas. We use for both the name of *time-bounded* formulas.

TCTL formulas are interpreted over the transition system which gives the semantics of a hybrid system. It is known [NOSY93, ACHH93] that the semantics of hybrid systems (in our context — of stopwatch automata) are infinite labeled transition systems. A *state* of such a transition system is a pair (s, v) where s is a state of the stopwatch automaton and v is a function that gives values to all the stopwatches. The *transitions* are of two kind: *continuous* transitions — that represent the *passing of time* at some location of the stopwatch automaton — and *discrete* transitions — that represent *taking some edge* of the stopwatch automaton.

A run of a stopwatch automaton, starting from location s with stopwatches valued by v, is a path in the transition system that represents its semantics, starting from state (s, v). Such a run will alternate letting time pass at locations and taking edges:

Definition 2. A *run* of a stopwatch automaton is an alternation of *continuous* and *discrete* transitions:

$$\rho = (s_0, v_0) \xrightarrow{t_0} (s_0, v_0') \xrightarrow{<s_0,s_1>} (s_1, v_1) \xrightarrow{t_1} (s_1, v_1') \xrightarrow{<s_1,s_2>} (s_2, v_2) \xrightarrow{t_2} \ldots$$

where the s_i are locations, the v_i and v_i' are values for the stopwatches such that:

- for all $i \in \mathbb{N}$, $v_i' = v_i +_{s_i} t_i$ that is, v_i' is obtained from v_i by adding t_i to the stopwatches which are *on* at location s_i, and letting the other stopwatches unchanged; moreover, invariant δ_{s_i} should be *true* while staying at s_i: $\forall t \in [0, t_i].\delta_{s_i}(t) = true$;
- for all $i \in \mathbb{N}$, $< s_i, s_{i+1} >$ is an edge of the stopwatch automaton; since it is taken at state (s_i, v_i'), we have $\phi^{<s_i,s_{i+1}>}(v_i') = true$ that is, the condition of edge $< s_i, s_{i+1} >$ is satisfied by valuation v_i';
- for all $i \in \mathbb{N}$, $v_{i+1} = v_i'[H^{<s_i,s_{i+1}>} := 0]$ — when edge $< s_i, s_{i+1} >$ is taken, the corresponding stopwatches are reset. □

By *duration* of a run we mean the total time spent at locations:

Definition 3. Let
$\rho = (s_0, v_0) \xrightarrow{t_0} (s_0, v_0') \xrightarrow{<s_0,s_1>} (s_1, v_1) \xrightarrow{t_1} (s_1, v_1') \xrightarrow{<s_1,s_2>} (s_2, v_2) \xrightarrow{t_2} \ldots$
$(s_{k-1}, v_{k-1}) \xrightarrow{t_{k-1}} (s_{k-1}, v_{k-1}') \xrightarrow{<s_{k-1},s_k>} (s_k, v_k)$ be a run that takes a finite number of edges. The *duration* of the run is $d(\rho) = \sum_{i=\overline{0,k-1}} t_i$. □

The truth of a TCTL formula involving operators $\exists U$ and $\forall U$ depends on states (s, v), and on runs starting from those states. For the *time-bounded* formulas, we can limit ourselves to time-bounded runs (definition 5). But let us first emphasize that runs are *piecewise dense* sets of states:

Definition 4. Given a run ρ (definition 2) of a stopwatch automaton, the *states of the run* are all the states $(s_i, v_i +_{s_i} t)$ for any $t \in [0, t_i]$. □

The states of a run are naturally ordered, which is used in the following definition:

Definition 5 (satisfaction). Let (s, v) be a state of (the semantics of) a stopwatch automaton and ψ_1, ψ_2 two state formulas.

- $(s, v) \models \psi_1 \exists U_{\leq c} \psi_2$ if there exists a **run** of **duration** $\leq$ c starting from (s, v) on which ψ_2 holds at some state and $\psi_1 \vee \psi_2$ holds at all the previous states;
- $(s, v) \models \psi_1 \forall U_{\leq c} \psi_2$ if for all **run** of **duration** = c starting from (s, v), ψ_2 holds at some state and $\psi_1 \vee \psi_2$ holds at all the previous states. □

3.1 Bounded reachability is decidable

The bounded rechability problem is: given a bounded reachability formula $\psi_1 \exists U_{\leq c} \psi_2$ and a stopwatch automaton, find all states of the stopwatch automaton that satisfy the formula.

The arguments for decidability It has been shown in general for reachability formulas that these states can be computed by 'backward analysis' [NOSY93], which has the intuitive meaning: **'go back in time' from states satisfying ψ_2, along runs where ψ_1 or ψ_2 hold, and gather all the encountered states.**

In particular for bounded reachability formula $\psi_1 \exists U_{\leq c} \psi_2$, to find all states satisfying the formula we have to 'go back in time' at most c time units from states satisfying ψ_2 (definition 5), and **because of the low-bounded duration of cycles** in stopwatch automata, 'going back' c time units means 'going back' less than $K = 2M \times \left(\left[\frac{c}{\delta}\right] + 1\right)$ edges[2], where M is the number of edges of the stopwatch automaton, and δ is the smallest duration of *twice running* some cycle[3].

Going back at most K edges is done precisely by executing at most K steps of the 'backward analysis' procedure in section 3.1. Since the number of steps to be executed is **finite**, our problem is **decidable**.

Decision procedure: backward analysis We now give the decision procedure. We first need some way to express in a compact form the set of states satisfying a TCTL formula; this is done with the so-called TCTL domains (a version of *unions of symbolic states* [Yov93]): n-uples of predicates on the stopwatches $(\psi^s)_{s \in S}$ — for each location $s \in S$, one predicate ψ^s to hold at that location. It turns out that the characteristic set of every TCTL formula can be written in that way; and the ψ^s turn out to be boolean combinations of *linear inequalities* on the stopwatches.

---

[2] [ ] is the *integer part*

[3] explanation: consider a run of K edges or more. This means running at least $\left[\frac{c}{\delta}\right] + 1$ times *twice* some cycle, and since running twice some cycle lasts at least δ, the whole run will take more that c time units.

Formulas *without temporal operators* are represented by TCTL domains in an obvious way:

- for propositions p, the representation is $(\psi^s)_{s\in S}$ where

$$\psi^s = \begin{cases} \delta_s & \text{if } p \text{ is true at } s \\ false & \text{otherwise} \end{cases}$$

- for $x \prec c$, the representation is $(\delta_s \wedge x \prec c)_{s\in S}$
- for $\neg\psi$, the representation is $(\neg\psi^s \wedge \delta_s)_{s\in S}$
- for $\psi_1 \wedge \psi_2$, the representation is $(\psi_1^s \wedge \psi_2^s)_{s\in S}$
- for the $z.\psi$ construction, the representation is $(\psi^s[z := 0])_{s\in S}$ — where $\psi^s[z := 0]$ is ψ^s in which z is replaced with 0.

Observation 1 Let s a location of a stopwatch automaton, and ψ^s be a boolean combination of linear inequalities on the stopwatches. **In the sequel, we shall identify ψ^s with the set of states** $\{(s,v)|\psi^s(v) = true\}$ and we shall allow us to mix set and boolean operators. □

The 'go-back-in-time operator' $\triangleright_s$. As in observation 1, let s be a location of a stopwatch automaton, and ψ_1^s, ψ_2^s be boolean combinations of inequalities (sets of states). Then $\psi_1^s \triangleright_s \psi_2^s$ is a boolean combination of inequalities (set of states) obtained **by going back in time at location s from set ψ_2^s by staying at all the intermediary in the union $\psi_1^s \cup \psi_2^s$:**

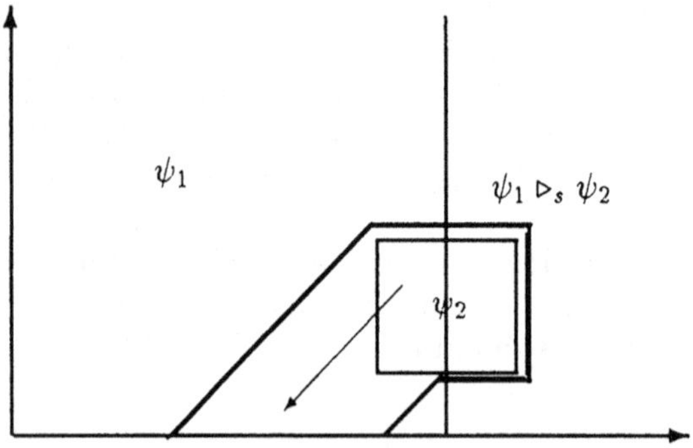

Fig. 3 *Geometrical interpretation of* $\psi_1^s \triangleright_s \psi_2^s$

or, alternatively, **the set of states from which ψ_2^s can be reached by letting time pass at a given location s, remaining at all the intermediary in the union $\psi_1^s \cup \psi_2^s$.** So $\psi_1^s \triangleright_s \psi_2^s$ has the following analytical form (see now ψ_1^s and ψ_2^s as boolean combination of inequalities):

$$\psi_1^s \triangleright_s \psi_2^s \;=\; \exists t \in \mathbf{R}^+.([\psi_2^s +_s t] \;\wedge\; \left[\forall\, t' \in [0,t].(\psi_1^s \;\vee\; \psi_2^s) +_s t'\right]) \qquad (1)$$

where notation $\psi^s +_s t$ means replacing in ψ^s any stopwatch a which is *on* at location s by $a + t$.

It is proven (see a very close result in [Yov93]) that quantifiers can be eliminated in formula (1), then giving a boolean combinations of inequalities.

Backward analysis. Theorem 6 describes how to compute the representation of $\exists U$ formulas. It consists in an algorithm that iterates the computing of operator $\triangleright_s$ above. Anything which is relative relative to step k of the algorithm is designed by symbol $\rfloor_k$.

Theorem 6 (backward analysis). If ψ_1 and ψ_2 are two TCTL formulas represented by TCTL domains $(\psi_1^s)_{s\in S}$ and $(\psi_2^s)_{s\in S}$, then $\psi_1 \exists U \psi_2$ is represented by the TCTL domain $(\psi^s)_{s\in S}$ such that $\forall s \in S$:

$$\psi^s = \bigvee_{k\in N} (\Delta^s\rfloor_k) \tag{2}$$

where the $\Delta^s\rfloor_k$ are boolean combinations of inequalities computed by: $\forall s \in S$,

$$\Delta^s\rfloor_0 = \psi_1^s \triangleright_s \psi_2^s$$

and $\forall s \in S, \forall k \in N$,

$$\Delta^s\rfloor_{k+1} = \bigvee_{<s,s'>\in E} (\psi_1^s \triangleright_s (\psi_1^s \wedge \phi^{<s,s'>} \wedge (\Delta^{s'}\rfloor_k)[H^{<s,s'>} := 0])) \qquad \square$$

Recall that for each edge $< s, s' > \in E$ of a stopwatch automaton we have $\phi^{<s,s'>}$ the condition to be satisfied for taking the edge, and $H^{<s,s'>}$ the set of stopwatches to be reset when taking the edge. Notation $(\Delta^{s'}\rfloor_k)[H^{<s,s'>} := 0]$ means precisely replacing all stopwatches of set $H^{<s,s'>}$ by 0 in boolean combination of inequalities $\Delta^{s'}\rfloor_k$. Note that in general, backward analysis needs to compute an infinite number of items $\Delta^s\rfloor_k$ and take their union (formula (2)). By the arguments of section 3.1, backward analysis for $\exists U_{\leq c}$ formulas needs to compute only a finite number of the $\Delta^s\rfloor_k$: [4]

Theorem 7. Backward analysis for bounded reachability formulas $\psi_1 \exists U_{\leq c} \psi_2$ on stopwatch automata **terminates** after less than $K = 2M \times \left(\left[\frac{c}{\delta}\right] + 1\right)$ steps — where M is the number of edges of the stopwatch automaton, and δ is the smallest duration of *twice running* some cycle. $\square$

The proof is available from the authors, and it is extendable to more general linear hybrid systems.

---

[4] In fact, we can show that in this case backward analysis *terminates by itself* in a finite number of steps, because in the infinite union (formula (2)), for $k \geq K$, all the $\Delta^s\rfloor_k$ are empty sets.

3.2 Bounded response is decidable

We formulate the bounded response problem: given two state formulas ψ_1 and ψ_2, decide whether for all states (s,v) in a set ψ^s, $(s,v) \models \psi_1 \forall U_{\leq c} \psi_2$. Location s is fixed, and the set ψ^s is identified with a boolean combination of linear inequalities on the stopwatches (observation 1).

This problem is simpler than the previous one; we only ask a question about the given $(s,v) \in \psi^s$ satisfying the formula, we do not ask for the whole set of states that satisfy the formula.

The arguments for decidability. The idea is similar to that for bounded reachability, but the other way around: **'go forth in time' from set of states ψ^s for c time units so at most $K = 2M \times \left(\left[\frac{c}{\delta}\right] + 1\right)$ edges, to see if on any run of duration c time units ψ_2 eventually holds and ψ_1 or ψ_2 hold until that**.

To recall the individual runs, we organize them in a tree starting from the set of states ψ^s; every path from the root to the leaves will represent runs of duration c from ψ^s; so we can check all those paths/runs to see if ψ_2 holds eventually and that ψ_1 or ψ_2 hold until that.

The tree is of **finite** depth (at most K) so our problem is **decidable**.

Decision procedure: tree forward analysis. We now give the decision procedure for the bounded response problem. First, we recall the representation in section 3.1 for state formulas ψ_1 and ψ_2 as TCTL domains $(\psi_1^s)_{s \in S}$ respectively $(\psi_2^s)_{s \in S}$; then, we need an operator te express the 'going forth in time' at locations.

The 'go-forth-in-time' operator $\lhd_s$. Recall that we are considering a set of states (boolean combination of inequalities) ψ^s. The set $\lhd_s(\psi^s)$ is defined to be **the set of states that can be obtained from ψ^s by letting time pass at location s**:

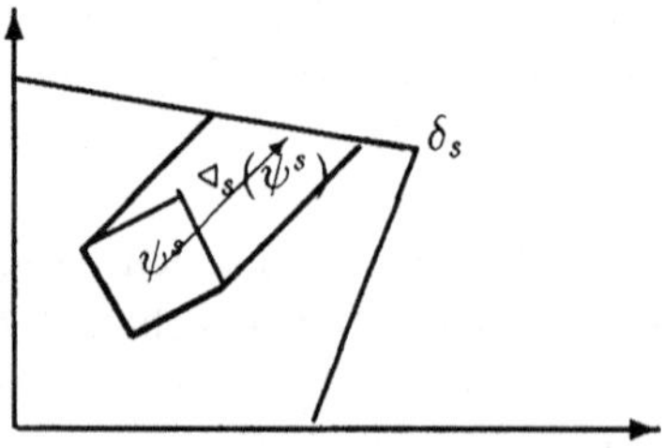

Fig. 4 *Geometrical interpretation of* $\lhd_s(\psi^s)$

Staying at location s means that invariant δ_s remains *true*, and since $\psi^s \subseteq \delta_s$, the analytical expression of $\lhd_s(\psi^s)$ is: $\delta_s \wedge \exists t \in \mathbb{R}^+.(\psi^s +_s (-t))$. Recall that notation $\psi^s +_s (-t))$ means adding $(-t)$ to the stopwatches that are *on* at location s, and letting the others unchanged. As in formula (1), the quantifier can be eliminated, giving a boolean combination of inequalities.

The decision procedure. Consider the resursive function in figure 5, with a PASCAL-like notation, in which above defined elements: ψ^s, $\triangleleft_s, \psi_1$, ψ_2, c are involved — the last three are supposed to be global. The procedure just builds the above-mentionned tree, and it stops when duration c is exceded. To check this, we add to the automaton a new stopwatch z to measure global time (it is always on and never reset) and we test the intersection with polyhedron $(z \leq c)$.

Function Response (s, ψ^s): **boolean**
    $\psi'^s \leftarrow \triangleleft_s(\psi^s) \cap (z \leq c)$
    if $\psi'^s \neq \emptyset$ **then**
        if $\psi'^s \not\subseteq \psi_1^s \cup \psi_2^s$ **then**
            return *false*
        else
            if $\psi'^s \subseteq \psi_2^s$ **then**
                return *true*
            else
                return
                    $\bigwedge_{<s,s'>\in E}$ **Response**$(s', (\psi'^s \wedge \neg\psi_2^s \wedge \phi^{<s,s'>})[H^{<s,s'>} \searrow 0])$
            endif
        endif
    else
        return *false*
    endif
End

Fig. 5 *Tree forward analysis*

Theorem 8 (Decidability of Bounded Response). Given two state formulas ψ_1, ψ_2, and function **Response** above, the call:
    satisfy $\leftarrow$ **Response** (s, ψ^s)
returns in boolean variable **satisfy** the answer to the bounded response problem: $\forall(s, v) \in \psi^s.(s, v) \models \psi_1 \forall U_{\leq c} \psi_2$. □

Proof. First, the set of states $\psi'^s = \triangleleft_s(\psi^s) \cap (z \leq c)$ that are reachable from current parameter ψ^s is computed, and only the states that do not exceed duration c are kept — if this set is empty, the answer is *false* because no state satisfying ψ_2 has been reached until c time units. Else, if this set of states is not included in $\psi_1^s \cup \psi_2^s$, then the definition of satisfying $\psi_1 \forall_{\leq c} U \psi_2$ (definition 5) is violated, so the answer is again *false*. If, on the contrary, ψ'^s is included in ψ_2^s this means all the runs so far end by satisfying ψ_2 before time c so the answer is *true*. If even this is not the case, then there are states satisying $\psi'^s \wedge \neg\psi_2^s$, so for these states the search should continue; the answer is the *and* of all boolean values returned by recursive calls at one more deeper level, taking edges $< s, s' >$ starting from current location s. When taking an edge, in polyhedron ψ'^s the stopwatches in $H^{<s,s'>}$ are set to 0, which we denote $\psi'^s[H^{<s,s'>} \searrow 0]$ [5].

[5] this is not the same as $\psi'^s[H^{<s,s'>} := 0]$, in fact it is the dual operation

4 Conclusion

Starting from ELECTRE programs and timed assumptions, we have defined Stopwatch automata, a kind of integrator systems for which bounded reachability and bounded response properties are decidable. These properties are known to be only semi-decidable in general; however, our decidability result and procedures hold for more general linear hybrid systems, provided they have the crucial *low-bounded duration* characteristic for cycles [RR95].

These more general linear hybrid systems are adapted to describe systems which mix continous evolutions and discrete changes, provided *the discrete changes are clearly separated in time*, a slightly stronger requirement than *not allowing a infinite number of discrete changes in finite time*. In this case, the problem of reachability/response in *bounded time* is solved.

Such systems in the litterature as the Water-level Monitor, Leaking Gas Burner, or Temperature Control System [NOSY93, KPSY93, ACHH93] are of this kind.

We are implementing the above decision procedures in program VALET (Vérificateur Adapté au Langage Electre avec Tctl). We use Polyhedral Library 'Chernikova' [Ver92, Wil93] for efficient computations on polyhedra. The final version of VALET should support more general linear hybrid systems.

References

[ACD90] R. Alur, C. Courcoubetis, and D. Dill. Model-checking for real-time systems. In *Proc. IEEE 5th Symp. Logic in Computer Science*, 1990.

[ACHH93] R. Alur, C. Courcoubetis, T.A Henzinger, and P. Ho. Hybrid automata : An algorithmic approach to the specification and verification of hybrid systems. In *Proc. Workshop on Theory of Hybrid Systems*, 1993.

[BB91] Albert Benveniste and Gérard Berry. The synchronous approach to reactive and real-time systems. *Proccedings of the IEEE*, 79(9):1270–1282, september 1991.

[BD91] Frédéric Boussinot and Robert De Simone. The ESTEREL language. *Proceedings of the IEEE*, 79(9):1293–1304, september 1991.

[CR95] F. Cassez and O. Roux. Compilation of the ELECTRE reactive language into finite transition systems. *Theoretical Computer Science*, 144, june 1995. to appear.

[HCRP91] Nicolas Halbwachs, Paul Caspi, Pascal Raymond, and Daniel Pilaud. The synchronous dataflow language LUSTRE. *Proceedings of the IEEE*, 79(9):1304–1320, september 1991.

[HKPV95] T.A. Henzinger, P.W. Kopke, A. Puri, and P. Varaiya. What's decidable about hybrid automata. To appear, 1995.

[HNSY92] T.A. Henzinger, X. Nicollin, J. Sifakis, and S. Yovine. Symbolic model-checking for real-time systems. In *Proc. IEEE 7th Symp. Logic in Computer Science*, 1992.

[HP85] David Harel and Amir Pnueli. On the development of reactive systems. In K.R. Apt, editor, *Logics and Models of Concurrent Systems*, volume 13, pages 477–498. NATO ASI Series, (Springer Verlag, New-York), 1985.

[KPSY93] Y. Kesten, A. Pnueli, J. Sifakis, and S. Yovine. Integration graphs : a class of decidable hybrid systems. In *Proc. Workshop on Theory of Hybrid Systems*, 1993.

[LLGL91] Paul Le Guernic, Michel Le Borgne, Thierry Gautier, and Claude Le Maire. Programming real-time applications with SIGNAL. *Proceedings of the IEEE*, 79(9):1321–1336, september 1991.

[NOSY93] X. Nicollin, A. Olivero, J. Sifakis, and S. Yovine. An approach to the description and analysis of hybrid systems. In *Proc. Workshop on Theory of Hybrid Systems*, 1993.

[NSY92] X. Nicollin, J. Sifakis, and S. Yovine. Compiling real-time specifications into extended automata. *IEEE Transactions on Software Engineering*, 18(9):794–804, 1992.

[PRH92] J. Perraud, O. Roux, and M. Huou. Operational semantics of a kernel of the language ELECTRE. *Theoretical Computer Science*, 97(1):83–104, april 1992.

[PV94] A. Puri and P. Varaiya. Decidability of hybrid systems with rectangular differential inclusions. In *Proc. Conference on Computer-Aided Verification*, 1994.

[RR95] O. Roux and V. Rusu. Verifying real-time systems with decidable hybrid automata. In *Second European Workshop on Real-Time and Hybrid Systems*, 1995.

[Ver92] H. Le Verge. A note on chernikova's algorithm. Technical report, 437, IRISA, Rennes, France, 1992.

[Wil93] D.K. Wilde. A library for doing polyhedral operations. Technical report, 2157, IRISA, Rennes, France, 1993.

[Yov93] S. Yovine. *Méthodes et outils pour la vérification symbolique de systèmes temporisés*. PhD thesis, Institut National Polytechnique de Grenoble, 1993.

Inductive Modeling: A Framework Marrying Systems Theory and Non-monotonic Reasoning

Hessam S. Sarjoughian
Bernard P. Zeigler

University of Arizona
Electrical and Computer Engr.
Tucson, AZ 85721
hessam@ece.arizona.edu
zeigler@ece.arizona.edu

Abstract

This article develops a framework for inductive modelling that works at the input/output level of system description. Rather than attempt to construct a state-space model from given observed data, an inductive modeler can employ non-monotonic logic to manage a data base of observed and hypothesized input/output time segments. Also, some basic criteria are established to guide the evaluation of the inductive modeler's performance.

1 Introduction

The theory of modeling will evolve as long as it continues to improve humankind's understanding of the world we live in. As this world becomes more complex, the techniques of modeling we employ are constantly challenged. Methodologies either prosper or become obsolete depending on their ability to meet the challenges facing them. If we want to continue exploring new problems — and studying old problems in new ways — we must find more powerful and increasingly flexible modeling methodologies.

A dynamical system may be modeled using differential equation, discrete-time, or discrete-event formalisms [5, 16, 38, 42, 44, 45]. Each of these formalisms is associated with a particular time base. Systems represented (modeled) using either a differential equation or a discrete-event formalism have a continuous time base, while those employing a discrete-time formalism have a discrete time base. Furthermore, dynamical systems can be viewed as either time-invariant or time-variant [5, 42, 44], deterministic or stochastic [42, 44], as well as causal or non-causal [21, 32, 38, 5, 40, 44]. The notion of time-invariance *vs.* time-variance, associated with a dynamical system, determines whether its behavior depends on a fixed time point or not. A deterministic model is not subject to uncertainty as is a stochastic model. The causal behavior of a model asserts that its previous behavior influences its present and future behavior. The past or present behavior of a noncausal model, however, may depend on its future behavior. In this work, we are concerned with a class of dynamical systems, which are time-invariant, deterministic, and causal.

Furthermore, a modeling methodology can be viewed as either deductive, abductive, or inductive depending on whether outputs, inputs, or structure of a system are the subject of inquiry, respectively [5, 12, 16, 44, 45, 42].

The inductive paradigm *attempts* to predict a system's behavior from input and output data sets without any any reference to the system's structure. In broad terms, inductive modeling consists of identifying input/output trajectories of interest, reasoning about the observed data (which would result in some postulated relationships among them), and gaining more and more confidence in the hypothesized relationships by unsuccessfully trying to refute them through additional experiments.

A very useful concept in modeling is *abstraction.* The flexibility provided by *multiple abstraction levels* has proven indispensable in modeling activities. It provides the means to study a system from different view points [15, 16, 18, 42, 44, 45, 46, 41].

The proposed inductive modeling approach requires partitioning of IO trajectories. It also requires reasoning about the use of input/output segments in predicting output trajectories as necessary. We use two classes of abstractions. In particular, a trajectory can be partitioned into smaller parts (segments) based on abstractions on the time base as well as the range of the input/output variables. These abstraction levels support varying degrees of data granularity. The other class of abstractions provides the basic means for reasoning about input/output segments. For instance, we may abstract the length of an input/output segment pair, or abstract a segment's initial state. The general concept for examining the correctness of an abstraction is *homomorphism* [39, 44, 42].

The ability to change the abstraction level based on explicit assumptions (justifying the use of a particular abstraction level) underlies the development of a non-standard modeling methodology. Obviously, a reasoning mechanism that is able to operate at varying abstraction levels, necessitated due to a lack of knowledge, has a non-monotonic character.

We can categorize modeling methodologies into *standard* and *non-standard* [36]. A model generated using a standard methodology cannot in itself emulate the tasks of a modeler. Whereas these models may contain varying forms of decision making, they are, in general, not subject to revision [5, 10, 18, 44, 42, 36].

An ideal *non-standard modeling methodology*, on the other hand, would support creating models that would emulate a modeler's activities. While this approach has to possess all the capabilities afforded by the standard methodologies, it should, in addition, possess reasoning capabilities similar to our own. It is important to observe that, whereas non-standard methodologies are prone to committing (hopefully correctable) *mistakes*, standard methodologies are expected to always make the *correct* decision.

To develop a non-standard inductive modeling methodology (see Figure 1), shows the components of a non-standard modeling methodology. we draw from two basic disciplines: Mathematical Systems Theory [28, 18, 42, 34] and Artificial Intelligence [12, 13, 30]. The former provides a framework for representing and studying dynamical systems. The latter supports explicit types of reasoning using declarative knowledge representation.

Although Systems Theory forms the foundation for all system representations, it is, by itself, insufficient to support modeling of complex and interconnected systems [18, 42, 45]. As a result, several modeling methodologies have been developed [44, 45, 46, 18, 42] that build on Systems Theory, but add several essential components to it. In our work, we shall use a hierarchy of system specifications with abstraction levels, starting from purely observed input/output behavior and ending with the IO systems [44, 45].

As stated earlier, non-standard methodologies need to deal with changing abstraction levels expressible in declarative form. Traditionally, reasoning has been within the confines of *declarative* representations due to its rich semantics and powerful means for manipulation of descriptive knowledge [6, 12, 25, 31]. In representing abstraction levels for *explicit* reasoning, we take the *logical approach* to knowledge representation [12], as opposed to other approaches, such as semantic nets [13].

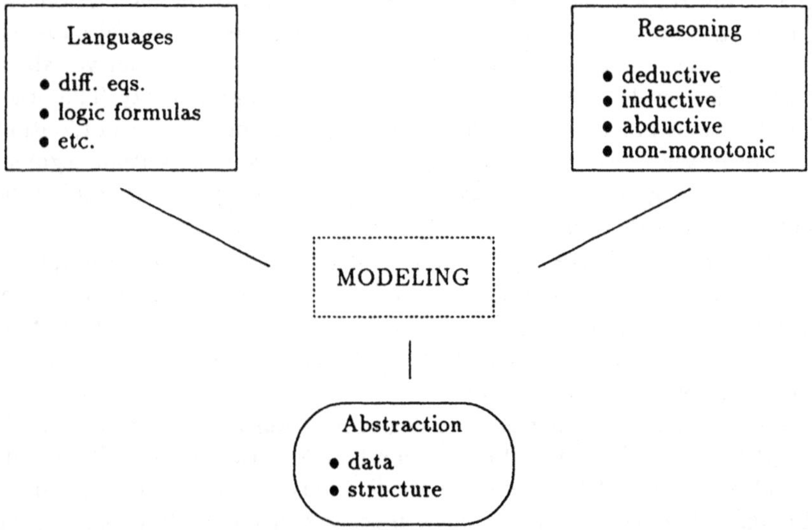

Fig. 1. Components of Modeling.

Non-standard modelling requires ***non-monotonic reasoning***, for which several approaches (formalized and unformalized) have been developed [4, 9, 29, 24, 25, 27, 22, 32, 33, 35, 38]. In this article, we use a variant of Truth Maintenance Systems [9] called Logic-based Truth Maintenance System [24, 11].

In this article, we emphasize the structure of the inductive reasoner in terms of systems theory. We describe the I/O Function Observation (IOFO) specification from a stratification of system specifications. We propose a way to dissect the IOFO specification in order to define its iterative specification. We discuss non-monotonic reasoning and how it can play a significant role in inductive modeling. However, we will not discuss how the iterative IOFO representation can be mapped into logic for reasoning purposes. A detailed discussion of the inductive reasoner along with examples are given in [36]. We give an outline of a non-monotonic inductive reasoner along with some guidelines specifying how it can be evaluated. We conclude with a summary.

2 Systems Concepts

Systems theory is based on basic formal concepts such as time, a system's behavioral properties (e.g., determinism and time-invariance), and the representation of input and output trajectories and how they may be manipulated. The notions that are needed for a comprehensive exposition of this and the following two sections are discussed in [44, 36]. We give a few of these in an attempt to maintain some continuity in our discussions.

A systems behavior can be represented as a set of histories

$$\omega : < t_i, t_f > \longrightarrow Z$$

where Z is a set, and t_i and t_f are the initial and final instants of time associated with the set. Two contiguous histories ω_i and ω_j can be concatenated as $\omega_i(t) \circ \omega_j(t)$.

Systems are often assumed to be time-invariant and deterministic, since systems that satisfy these properties can be treated with rigor and preciseness. The discussion of time-invariance and determinism is often given within the context of a system's structure as opposed to its observed behavior. Appropriate restrictions are put on a set of differential equations describing their underlying features. In our discussions, the behavior of a time-invariant and deterministic system (i.e., its histories) is a direct consequence of its structural representation, and hence time-invariance and determinism of histories follow from the same properties of the underlying system that generates these histories.

It is useful to define time-invariance and determinism directly as properties of input/output histories, rather than indirectly through the properties of a system's internal structure that accounts for the observed input/output histories.

Definition 1. A set of input/output histories:

$$IOspace = \{(\omega, \psi) \mid (\omega, \psi) \in (X,T) \times (Y,T), \qquad dom(\omega) = dom(\psi)\}$$

is called an *IOspace*.

A trajectory may be composed from its *generator segments*. A set of generator segments Ω_G for a set of histories Ω is a set of segments such that $\Omega_G^+ \supseteq \Omega$ [44]. Hence, given generator segments (Ω_G and Ψ_G), we have:

Definition 2. A set of input/output segments

$$\begin{aligned} IOspace_G = \{(\omega_G, \psi_G) \mid &(\omega_G, \psi_G) \in IOspace, \\ &\omega_G \in \Omega_G, \psi_G \in \Psi_G, \\ &IOspace_G \subseteq IOspace, \\ &dom(\omega_G) = dom(\psi_G)\} \end{aligned}$$

$\Omega_G = PJN(IOspace_G, 1)$ is the input generator segment set, and
$\Psi_G = PJN(IOspace_G, 2)$ is the output generator segment set.

Given IO generator segments, *Complete* IO segments which are causal can be defined using the $IOspace_G$ [36]. A Complete IO segment is one with initial state.

3 Abstract Stratification of Models

It is important to place our approach to inductive modelling with the stratification of system specifications [44, 18, 28, 42]:

Level 0:	Observation Frame (O)
Level 1:	Input/Output Relation Observation (IORO)
Level 2:	Input/Output Function Observation (IOFO)
Level 2.5:	Iterative Input/Output Function Observation (G_F)
Level 3:	Input/Output System (IOS)

As we proceed from the lowest level to higher levels of the hierarchy, specifications make more concrete claims about the structure of the "black box" system. Associations between two neighboring levels exist in both downward and upward directions. The former is in the direction of structure to behavior, while the latter attempts to hypothesize structure from behavior — the direction of inductive modelling. Indeed, we have introduced level 2.5 for inductive modeling and it will be discussed in Section 4.

3.1 Input/Output Function Observation

The Input/Output Function Observation (IOFO) is a structure [44]:

$IOFO = \langle T, X, \Omega, Y, F \rangle$ where:

T Time base
X Input Value Set
Y Output Value Set
Ω Input History Set
F I/O Function Set

F is a set of input/output functions with the constraint that:

$f \in F \qquad \text{implies} \qquad f \subseteq \Omega \times (Y,T) \qquad$ is a functional, causal IOspace.

Each $f \in F$ is associated with an initial state, thus ensuring a unique response, $\psi = f(\omega)$, for an input segment, ω. Each f is a grouping of those input/output segment pairs that are associated with the same initial state.

A discrete-event IOFO must operate on discrete-event histories is illustrated in Figure 2. Input/output events are `a`, `b`, `c` and `nil` where null event ϕ is represented as `nil`.

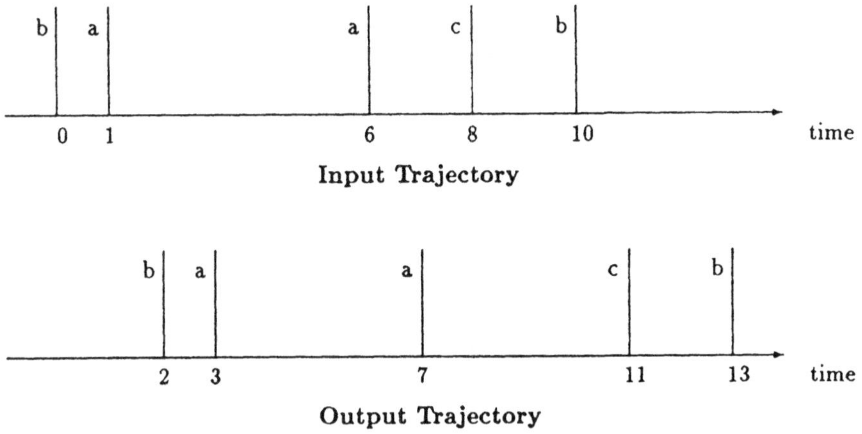

Fig. 2. A discrete-event input/output trajectory

Definition 3. A set of discrete-event input/output histories $DEVS(X,Y) = \{(\omega,\psi) \mid \omega \in DEVS(X), \quad \psi \in DEVS(Y)$, where $DEVS(X)$ and $DEVS(Y)$ are the semigroups generated by the two sets of generators, $G(X)$ and $G(Y)$ respectively, where $G(X)$ is the union of the following generator classes:

$\Omega_{\phi} = \{\phi_{\tau} \mid \tau \in \Re, \phi_{\tau} : <0,\tau> \longrightarrow \{\phi\}\}$[1],

$\Omega_{x} = \{x_{\tau} \mid \tau \in \Re, x_{\tau} : <0,\tau> \longrightarrow X^{\phi},\ x_{\tau}(0) = x,\ x_{\tau}(t) = \phi$ for $t \in (0,\tau)\}$

A discrete event IOFO is the obvious restriction of an IOFO to the input/output history spaces, $DEVS(X,Y)$ just described. From here on we restrict attention to discrete event IOFOs and omit mention of the restriction to the discrete event formalism.

[1] Due to closure under translation, restriction to segments starting at 0 is sufficient for complete specification.

4 Inductive Modeling and IOFO Specification

Earlier, inductive modeling was defined as a modeling paradigm that attempts to predict a system's behavior based on previously observed inputs and outputs. It considers a system as a *black box*, of which nothing is known but its inputs and outputs; in particular, no knowledge is claimed about the system's internal structure. Therefore, the IOFO specification provides a suitable formalism for inductive modeling.

A variant of IOFO specification $\langle T, X, S_i, Y, IOspace, F\rangle$ does contain knowledge about the initial states S_i that are associated with the respective input/output trajectories $IOspace$ [36]. In this representation, initial states of IO histories are represented explicitly and F is

$$F = \{f_i \mid f_i = (s_i, g_i), \quad \begin{array}{l} g_i : \{(\omega_{i,1}, \psi_{i,1}), \cdots, (\omega_{i,n}, \psi_{i,n})\} \\ s_i : \text{is an initial state associated with } g_i \end{array}$$
$$\omega_{i,j} \in \Omega, \quad \psi_{i,j} \in (Y,T) \qquad \text{for} \qquad i = 1, 2, \ldots, \quad j = 1, \ldots, n\}.$$

Given this IOFO specification, if we have a finite number of input/output trajectories and their initial states, all that we can expect is to match an initial state and its corresponding input trajectory with the repertoire of available data. That is, suppose we have a set F where $f = (s_i, \{(\omega_{i,j}, \psi_{i,j})\}) \in F$ for $i = 1, \ldots, n$, and $j = 1, \ldots, m$. Also, assume we have an initial state s' and input trajectory ω' for which we would like to determine the output trajectory. Then, if there exists f such that one of its input trajectories matches ω' and also its initial state matches s', then the output trajectory ψ is directly obtained from f. Obviously, the chances of finding a match for a given initial state s' and an input trajectory ω', given F, depends on the richness of F.

Iterative specifications of systems are the basis for computational manipulation [44, 36]. The above IOFO specification offers no iterative capability since it operates on input/output histories. However, if we can have an IOFO specification in terms of IO segments then it is possible to construct unobserved input/output histories by concatenating appropriate I/O segments. Hence we are interested in defining an *iterative IOFO structure*.

To this end, we are led to seek an input/output *segment generator* function, F_G that is similar in structure to F. The elements of F_G have the form $(s, \{(\omega_G, \psi_G)\})$ where $\omega_G \in PJN(IOspace_G, 1), \psi_G \in PJN(IOspace_G, 2)$. Constructing F_G requires partitioning of observed I/O trajectories (ω, ρ) into segments.

We envision two stages in generating F_G starting from a set of observed input/output histories F. In the first stage, the *partitioning* of trajectories into segments based on the segments' granularity, duration, and initial state assignment is dependent on a set of conjectures (assumptions). These assumptions constitute some belief set, which we denote as *assumption set-I*. This belief set, therefore, forms the basis upon which an iterative IOFO specification can be formalized.

In partitioning a trajectory into segments, if we consider two contiguous I/O segment pairs, we are assuming that the initial state of one pair immediately following another pair is the same as the final state of the previous pair. This is necessary since the systems that we are considering are assumed to be causal, and we are merely partitioning I/O trajectories.

If we are to *compose* trajectories from segments, another belief set is also needed. We denote this second belief set as *assumption set-II.* To compose two segments ω_1 and ω_2, they may be concatenated as $\omega_1 \circ \omega_2$. However, unless the final state of ω_1 is the same as the initial state of the ω_1, we are forced to make $\omega_1 \circ \omega_2$ a hypothesis. That is, forgiving a mismatch between the final state of ω_1 and the initial state of the ω_2 results in a hypothesized trajectory. This explains the requirement for assumption set-II.

At the IOFO specification level, our two transformations correspond to two sets of assumptions I and II. As motivated above, the former set facilitates the *partitioning* of trajectories into segments resulting in the *iterative IOFO structure.* The latter set allows the composition of incompatible I/O segments. Thereby construction of composite trajectories from segments leads to the concept of *Iterative IOFO structure.* We denote the iterative IOFO as G_F and the IOFO specification it elaborates to as $IOFO(G_F)$. Hence,

$$IOFO \longrightarrow G_F \longrightarrow IOFO(G_F).$$

Figure 3 depicts how each assumption set contributes to arriving at G_F and $IOFO(G_F)$.

The transformations from $IOFO$ to G_F and from G_F to $IOFO(G_F)$ can be considered to be independent. The interpretation of the independence of these transformations is that, once G_F is formed, the construction of $IOFO(G_F)$ is solely dependent on the assumption set-II.

There is, nevertheless, some interdependence between the transformations. This is due to the fact that, in constructing composite I/O trajectories, some segments may be hypothesized. The assumptions underlying the assignment of states of I/O segments for G_F and states of hypothesized I/O segments for $IOFO(G_F)$ might depend on each other. That is, the assumptions underlying the partitioning of a trajectory and the composition of I/O segments can be quite different.

First, we shall focus our attention on how one might partition an I/O trajectory given its initial state into I/O segments with assigned initial states.

Let us assume we have a function γ_G, which we call a *final state hypothesizer.* This function hypothesizes about a segment's final state based on assumption set-I.

$$\gamma_G : S_i \times R_G \longrightarrow S_f.$$

The outputs of γ_G are not necessarily the truly representative set of states of the system, but they may be the best that we are able to obtain given the input/output segments and their initial states only. We note that higher level knowledge such as the sets X and Y (elements of IOFO) can lead to different

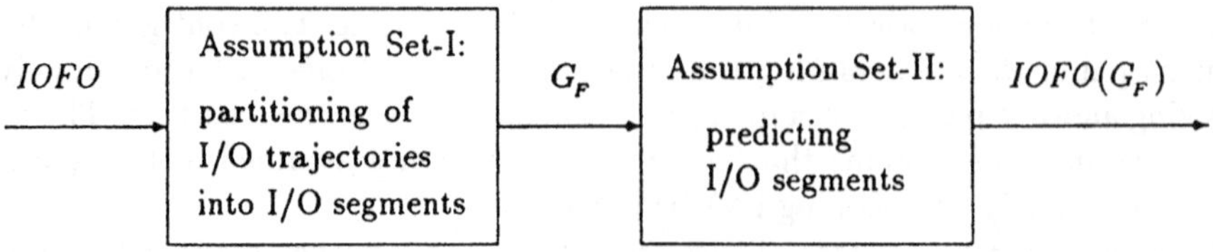

Fig. 3. Role of assumptions in constructing G_F and $IOFO(G_F)$.

sets of final states. That is, depending on how the number of levels (i.e., the quantization) is specified, the candidate final states may vary.

4.1 Iterative I/O Function Observation Structure

Using the alternative representation of an IOFO structure given above, we define an iterative specification for a causal, time-invariant I/O function observation structure as:

$$G_F(\gamma_G) = \langle T, X, S_i, Y, IOspace_G, F_G, \gamma_G \rangle \qquad where:$$

T	time base
X	input value set
Y	output value set
S_i	set of initial states
$IOspace_G$	causal, time-invariant input/output segment generator set
F_G	I/O function generator set
γ_G	final state hypothesizer

with the following constraints:
$F_G : S_i \longrightarrow$ partial $IOspace_G$,
$\gamma_G : S_i \times IOspace_G \longrightarrow S_i$.

The interpretation of the function γ_G is that it maps a given initial state and an input/output segment pair into another initial state.

To obtain an output segment $\psi_G^{i,j}$ for an input segment $\omega_G^{i,j} \in \Omega_G$, an initial state $s_i \in S_i$, and a given data set F_G (cf. Section 2.4.2), we define the output segment matching function as:

$$\eta_G : S_i \times \Omega_G \times F_G \longrightarrow \Psi_G,$$

and its *computational* counterpart as:

$$\eta_G(s_i, \omega_G, F_G) = F_G(s_i, \omega_G).$$

The elements of G_F are pairs of I/O generator segments. For example, Figure 4 shows four types of well-defined discrete-event generator segments for the I/O trajectory shown in Figure 2 [36].

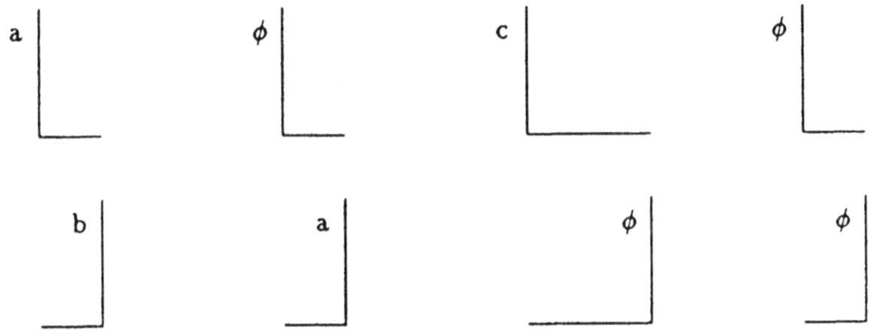

Fig. 4. Discrete-event input/output generator segments

The interpretation of the above function is that there exists an $f_i \in F_G$ such that $\eta_G : (s_i, \omega_G^{i,j}, f_i) \mapsto \psi_G^{i,j}$. For example, using the output segment matching function η_G and the special form of the final state hypothesizer function γ_G : $\Psi_G \rightarrow S_f$, we can determine the final state by consecutive application of γ_G followed by η_G. That is,

$$\gamma_G(\eta_G(s_i, \omega_{i,j}, f_i)) = \gamma_G(\psi_{i,j}) = s_f,$$

or using the general form of the final state hypothesizer, $\gamma_G : S_i \times IOspace_G^+ \rightarrow S_i$, we have the general form for combining η_G and γ_G:

$$\gamma_G(s_i, (\omega_{i,j}, \eta_G(s_i, \omega_{i,j}, f_i))) = \gamma_G(s_i, (\omega_{i,j}, \psi_{i,j})) = s_f.$$

Having defined the iterative IOFO structure G_F (level 2.5 in system specification hierarchy), we construct the IOFO structure $IOFO(G_F)$ generated from G_F. The details are supplied in [36].

4.2 Iterative Discrete-event IOFO Specification

We are interested in knowing how, if we have a time-invariant discrete-event input/output function observation specification, we can find its iterative counterpart. Our aim is to be able to represent and operate on I/O generator segments instead of I/O trajectories as discussed earlier. Note that the I/O trajectories and their initial states are in F. Hence we like to obtain their I/O generator segments and their initial states as F_G using assumption set-I.

We proceed to construct an iterative Discrete-event IOFO specification (DEVF).

Proposition 4. *Let* $D = \langle X_D, S_i, Y_D, IOspace_D, F_D \rangle$ *be a causal time-invariant discrete-event input/output function specification. Then, its iterative specification,* $G_F(D) = \langle T, X, S_i, Y, IOspace_G, F_G, \gamma_G \rangle$, *can be constructed provided that:*

1. *The mls decomposition of I/O trajectories* $\{(\omega, \psi) \mid \omega \in PJN(IOspace_D, 1)$ *and* $\psi \in PJN(IOspace_D, 2)\}$ *exists,*
2. *There exists a final state hypothesizer such that* $\gamma_G(s, (\omega_G, \psi_G)) = s_f \in S_i$,
3. *The composition of segments, using* η_G *as defined for the iterative I/O function observation structure, is satisfied within the discrete-event I/O function observation representation.*

If the above conditions are satisfied, the constructed discrete-event iterative I/O function observation has the following form:

$$G_F(D) = \langle T, X, S_i, Y, IOspace_G, F_G, \gamma_G \rangle \qquad \text{where:}$$

$T = \Re_0^+$
$X = X^{\phi}_{D,mls}$
$Y = Y^{\phi}_{D,mls}$
$IOspace_G \subseteq DEVS_G(X, Y)$
$F_G : S_i \longrightarrow$ causal time-invariant partial $IOspace_G$
$\gamma_G : S_i \times IOspace_G \rightarrow S_i$,

$X^{\phi}_{D,mls} = X_D \cup X_{mls} \cup \{\phi\}$,
X_{mls} is the set of input events obtained from segments generated from the *mls* decomposition of $PJN(IOspace_G, 1)$,

$Y^{\phi}_{D,mls} = Y_D \cup Y_{mls} \cup \{\phi\}$,
Y_{mls} is the set of output events obtained from segments generated from the *mls* decomposition of $PJN(IOspace_G, 2)$,

$$DEVS_G(X, Y) = \{(\omega_G, \psi_G) \mid \omega_G \in DEVS_G(X), \psi_G \in DEVS_G(Y)$$
$$DEVS_G(X) = \{\omega_G \mid \omega_G \in \Omega_G, \Omega_G = \Omega_x \cup \Omega_\phi\}$$
$$DEVS_G(Y) = \{\psi_G \mid \psi_G \in \Psi_G, \Psi_G = \Psi_y \cup \Omega_\phi\}\},$$

$\Omega_\phi = \{\phi_\tau \mid \tau \in \Re, \phi_\tau :< 0, \tau > \longrightarrow \{\phi\}\}$,
$\Omega_x = \{x_\tau \mid \tau \in \Re, x_\tau :< 0, \tau > \longrightarrow X^{\phi}, x_\tau(0) = x, x_\tau(t) = \phi \text{ for } t \in (0, \tau)\}$,
$\Omega_y = \{y_\tau \mid \tau \in \Re, y_\tau :< 0, \tau > \longrightarrow Y^{\phi}, y_\tau(0) = y, x_\tau(t) = \phi \text{ for } t \in (0, \tau)\}$.

Having constructed the iterative I/O function specification, we can proceed to construct the iterative I/O specification based on assumption set-II; i.e.,

$$IOFO(G_F(D)) = \langle X, \Omega_G^+, S_i, Y, \Psi_G^+, F_G^+ \rangle.$$

Based on the assumption that **all** input/output data and their corresponding initial states are available, a discrete-event IOFO specification can be canonically constructed [1]. However, data in the real world is never complete. Unlike a complete data set, a finite data set requires partitioning of trajectories and assignment of initial and final states in hope of predicting output histories that do not belong to the data set. The ability to revise our predictions in the light of new data is provided by non-monotonic reasoning, to which we now turn.

5 AI and Reasoning

We have already discussed the need for making tentative choices when defining the iterative IOFO specification due to lack of complete knowledge about the length, granularity, and possible state assignments to segments. Also, when there is no exact match for a given input segment, we proposed a set of assumptions to hypothesize about candidate segments.

System theory provided us with the basic means for representing input/output models (all variations of IOFO specifications). So it is fair to ask why we cannot use the same tools for reasoning. Unfortunately, the mathematics of system theory are too rigid to allow explicit forms of reasoning when knowledge cannot be assumed to be fixed and *apriori*.

Reasoning can be carried out by several schemes, each of which primarily follows one of two basic approaches: ***probabilistic reasoning*** [32] and ***logic-based reasoning*** [12]. The former can be split into extentional and intentional. Intentional approaches to probabilistic reasoning, which are also referred to as model-based reasoning approaches, compute conclusions within the framework of probability theory [19]. Extentional approaches (also referred to as production systems), however, use methods based on less stringent calculi such as the Dempster-Shafer certainty-factor calculus [8, 37], and fuzzy logic [43].

Logic-based approaches don't deal with the notion of certainty as it is defined for other approaches. They support reasoning based on the truth values TRUE, FALSE, and UNKNOWN assigned to entities to be reasoned with. They provide for a basic and simple type of reasoning, making tentative decisions that are subject to revision. Several ***non-monotonic reasoning*** (NMR) theories have been formalized based on this form of reasoning. The features of other formalisms — usually absent in logic-based approaches — have, however, been incorporated into several of its variants [12, 25, 17, 32].

Non-monotonic reasoning [25, 35, 27, 3, 2, 4, 22], which manifests itself in many types of problems, allows making tentative decisions, subject to future revocation. Just as NMR plays a vital role in our daily lives, any proposed intelligent machine needs to support NMR as well [26, 31, 12, 4, 20, 25].

Non-monotonic reasoning uses non-monotonic (***non-sound***) inference rules [12, 2, 4, 22] which is different than inductive inference rules. With such inference rules, an earlier conclusion may have to be retracted if another sentence is added to the database. That is, for a given database Δ, we allow non-sound inference rules to be used to infer *new* conclusions that do not logically follow from Δ when using sound inference rules only.

6 Application of NMR to Inductive Modeling

Default reasoning, inheritance, and reasoning about action are considered different variations of non-monotonic reasoning. In applications that exhibit non-monotonic characteristics, the primary reason for using NMR can eventually be traced back to the need for dealing with the qualification problem. What is more interesting is that, although NMR is itself inductive in nature, to our knowledge, NMR has not been used so far to deal with inductive forms of reasoning. Inductive inference rules, however, have been used for inductive modeling, (e.g., [12].)

For instance, NMR has been used in deductive qualitative modeling [16, 11, 13]. The inference engine for deductive qualitative modeling is built using deductive inference rules that essentially describe how implicit knowledge can be derived based on the axioms specifying how the system is supposed to work. These axioms, for instance, may describe how a motor or a thermostat is expected to operate under some well-defined conditions. Then, NMR is used to allow for properly handling situations that had not previously been expected, and for which consequently no provisions were explicitly specified. The role of NMR is to augment the existing theory, i.e., the set of axioms describing the behavior of the system, with additional knowledge such that the augmented theory is able to handle some unspecified situations as well. The inferencing mechanism of NMR is non-monotonic, and thus it is possible to withdraw decisions that are refutable given new axioms. Clearly both deductive and non-monotonic inference rules are involved in deductive qualitative modeling.

In contrast, our use of NMR relates to inductive reasoning modes. In inductive qualitative modeling, the situation is quite different. There are no theorems available that would describe the underlying behavior of the system. There exists no theory that can be subjected to deductive rules of inference to predict the system's behavior. Instead, the theory contains axioms expressing a system's observed behavior. Indeed in inductive modeling, we are constrained with some finite observed dataset. We use some inferencing mechanism to make decisions. These decisions are purely based on data rather than on theorems expressing how the system is supposed to operate. Hence, given a finite set of data and some assumptions, applicable inference rules have to be non-monotonic.

Therefore, the use of NMR in inductive qualitative modeling is quite different from its use in deductive qualitative modeling or any other application that is formulated based on either deductive or abductive paradigms. In inductive modeling as we have defined it, we need to rely on non-monotonic inference rules. Although there exists an extensive body of literature on inductive modeling and machine learning, to our knowledge, none of the previous research efforts in inductive modeling made use of non-monotonic reasoning.

For reasons that are discussed in [36], we have chosen Logic-based TMS [7, 23, 11] as the underlying machinery for reasoning.

7 Outline of a NMR Inductive Modeller

We now present a top-level description of the design of a NMR Inductive Modeller (NMRIM) working within the iterative IOFO framework for discrete event systems described earlier.

Given an input trajectory, initial state, and some stored IO segments (observed and previously hypothesized), the task of the inductive modeller is to generate an output trajectory for the given input trajectory. The inference engine begins by finding the longest input generator segment. It tries to find a match for this segment in its repertoire of ***observed*** IO segments. If it succeeds, it extracts the final state of the newly found output segment and uses it as the initial state for a new input trajectory with the previous candidate input segment chopped off from the previous input trajectory. Now, another candidate input segment can be selected and the same process repeated if applicable.

Most likely, there exists no observed IO segment which exactly matches the selected input segment and its initial state. In this case, the modeller first examines the hypothesized IO segments for a match. If it succeeds, it follows the previous steps with the exception that the match is with a hypothesized IO segment instead of an asserted IO segment.

If no exact match exists, the modeller hypothesizes an IO segment (using both hypothesized and asserted IO segments) and communicates it with its assigned truth value to the LTMS along with the abstraction assumptions that were used to bring about the match. Assumption set-II specifies different kinds of abstractions. Once again, given the hypothesized output segment, a truncated input trajectory with its initial state extracted from the previous hypothesized output segment is constructed as above. This process iterates until the concatenation of all the selected (partitioned) input segments is the same as the original input trajectory.

When the modeller receives new observed IO trajectories, it updates its databases as well as retracting any hypothesized IO segment which may be contradicted by the newly received data.

A critical test performed before accepting an hypothesized IO segment is for consistency with the current observed and believed segments. Consistency is based upon the ***causality*** property discussed earlier. Roughly, this requires that if two IO segments are associated with the same initial state, and one input is a left segment of the other, then the output segments also agree up to the segmentation point. The results of applying such a test are summarized in Figure 5. In the left column of Figure 5 — *Candidate IO Segment* — a predicted IO segment or an observed IO segment can be one of the followings only.

$$\{\text{asserted, hypothesized, rejected, ignored}, \star\}$$

Likewise, in the right column of Figure 5 — *Existing IO Segment* — a hypothesis can be one of:

$$\{\text{confirmed, rejected, unchanged}, \star\}.$$

Note that the elements of each of the above sets are mutually exclusive. For instance, when an IO segment cannot be asserted, hypothesized, confirmed, ignored, or left unchanged, it is denoted as $\star$. Besides assuring causally consistent databases, the consistency admission test can also serve to minimize the number of observed IO segments by maintaining only longest segments — a segment that is a causally consistent left segment of an existing segment may not be admitted into the database.

Candidate IO Segment	Existing IO Segment
Observed {asserted, ignored, $\star$}	**Assertion** {unchanged, $\star$}
Observed {asserted}	**Hypothesis** {confirmed, rejected, unchanged}
Predicted {hypothesized, rejected}	**Assertion** {unchanged}
Predicted {hypothesized, $\star$}	**Hypothesis** {unchanged, $\star$}

Fig. 5. How candidate and existing IO segments may fare w.r.t. to the consistency axioms.

8 Evaluation of NMR Inductive Modeller

How are we supposed to evaluate the performance of a non-monotonic inductive modeller? Figure 6 shows how some deductive (state-based) models can be constructed using some inductive approaches such as Neural Net [14]. In contrast, the non-monotonic inductive modeller does not work at level 3 of system specifications (Section 4). Instead, it supplements the asserted IO segments with some hypothesized IO segments from which IO trajectories are constructed (predicted.) It is based on level 2.5 of system specifications (Section 7).

The development of the inductive modeller is based on the language of Logic and on non-monotonic reasoning. Hence, an evaluation scheme for it has to rely on the theory of Logic and in particular the logic of non-monotonic reasoning.

A hypothesized IO segment can ultimately become a "valid" or a "invalid" IO segment. Those IO segments that are *valid* not only must be consistent with the available IO segments but also must be those that can be observed. An IO segment is *invalid* if it is inconsistent with the available IO segments. Those IO segments which are neither valid nor invalid are *possible*. The basis for evaluation of NMRIM should not be that it must generate valid IO segments only. This requirement is too strong because a segment which is "possible" w.r.t. the current asserted IO segments may become invalid after more observations. A minimum requirement for NMRIM is that it should generate only segments that are *valid* or *possible* w.r.t. the current observations. An account of when such a requirement can be satisfied is given in [36].

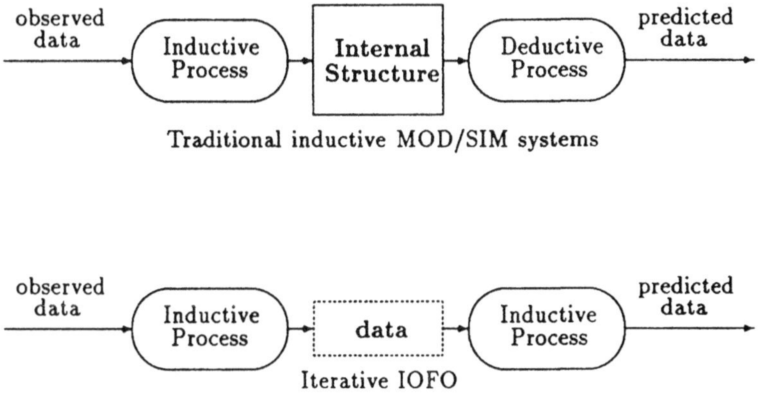

Fig. 6. A block diagram showing steps in predicting IO trajectories using deductive and inductive modeling approaches.

Once the minimum requirement is satisfied, NMRIM can be evaluated in both quantitative and qualitative settings [36]. In particular, we can assess (1) how accurate are the predicted IO trajectories, (2) how flexible is the modeller in making predictions; and (3) to what extent is the underlying reasoning explicit and accessible to the user.

Accuracy is a quantitative measure that can be used to rank alternative predictions, each supported by different abstractions (sets of assumptions.) *Flexibility* refers to NMRIM ability to make predictions on the basis of a few observed IO segments. Furthermore, flexibility includes incremental learning in that its predictions should improve as more data is accumulated. Finally, *accessibility to reasoning* enables the user to understand why and how a particular IO trajectory is predicted. Every hypothesized IO segment has a reason associated with it. The system is responsible for maintaining consistency among all IO segments and associating reasons for belief in them. These reasons should be explicitly available to the user. An inductive modeler satisfies these requirements [36].

9 Conclusions

We have taken some steps toward developing a theory of a non-monotonic inductive modeling methodology operating at the input/output level. A framework for such a methodology, utilizing the features of systems theory and non-monotonic reasoning, was presented. The proposed non-monotonic inductive modeling approach makes predictions which are supported with varying levels of abstractions. This approach is distinguishable from other inductive modeling approaches since the way it generates its outputs is comparable to what a human modeler might do. The inductive modeler takes an active role in understanding the data. It makes decisions (hypotheses) that are supported by the available data and is capable of making assumptions to enable predictions when the available data is insufficient. It is prepared to revise its predictions whenever warranted. In this way, some initial responsibilities of a human modeler can be handled by NMRIM.

Acknowledgment

We would like to thank Takao Asahi for his contribution at the earlier stages of this research. He and François Cellier have read parts of this manuscript which led to valuable improvements. Partial support of this research under NSF grant No. ASC-9318169 is acknowledged.

References

1. T. Asahi and B.P. Zeigler. "Behavioral characterization of discrete-event systems". In *AI, Simulation and Planning in High-Autonomy Systems*, pages 127–132, Tucson, AZ, Sept 1993. IEEE/CS Press.
2. P. Besnard. *An Introduction to Default Logic*. Springer-Verlag, New York, 1989.
3. D. Bobrow. Special Volume on Non-Monotonic Reasoning. *Artificial Intelligence*, 13(1–2), 1980.
4. G. Brewka. *Nonmonotonic Reasoning: Logical Foundations of Commonsense*. Cambridge University Press, Cambridge, 1991.
5. F. E. Cellier. *Continuous System Modeling*. Springer-Verlag, New York, 1991.
6. E. Davis. *Representation of Commonsense Knowledge*. Morgan Kaufmann, San Mateo, CA, 1990.
7. J. de Kleer. "An assumption based truth maintenance system". *Artificial Intelligence*, 28(2):127–162, 1986.
8. A.P. Dempster. "A generalization of bayesian inference". *Journal of the Royal Statistical Society, Series B*, 30(2):205–247, 1968.
9. J. Doyle. "A truth maintenance system". *Artificial Intelligence*, 12(1):231–272, 1979.
10. P.A. Fishwick. Simulation model design and execution. Department of Computer Science and Information Sciences, University of Florida, November 1993.
11. K.D. Forbus and J. de Kleer. *Building Problem Solvers*. MIT Press, Cambridge, 1993.

12. M. Genesereth and N.J. Nilsson. *Logical Foundations of Artificial Intelligence.* Morgan Kaufmann, 1987.
13. M.L. Ginsberg. *Essentials of Artificial Intelligence.* Morgan Kaufmann, San Mateo, CA, 1993.
14. S. Grossberg. *Neural Networks and Natural Intelligence.* MIT Press, 1988.
15. S.I. Hayakawa and A.R. Hayakawa. *Language in Thought and Action, Fifth Edition.* Harcourt Brace Jovanovich, Publishers, Orlando, Florida, 1990.
16. J.R. Hobbs and R.C. Moore, editors. *Formal Theories of the Commonsense World.* Ablex Publishing Corporation, Norwood, New Jersey, 1985.
17. D. Kirsh. Special Volume on Foundation of Artificial Intelligence. *Artificial Intelligence,* 47(1):1–346, 1991.
18. G.J. Klir. *Architecture of Systems Problem Solver.* Plenum Press, New York, 1985.
19. R.G. Laha. *Probability Theory.* John Wiley and Sons, New York, 1979.
20. M.L. Ginsberg M. Reinfrank, J. de Kleer and E. Sandewall, editors. *2nd International Workshop on Non-Monotonic Reasoning,* New York, June 13–15 1988. LNCS 346, Springer-Verlag.
21. J.L. Mackey. *The Cement of the Universe: A Study of Causation.* Oxford University Press, 1974.
22. V.W. Marek and M. Truszcynski. *Nonmonotonic Logic: Context-Dependent Reasoning.* Springer-Verlag, New York, 1993.
23. J.P. Martin and M. Reinfrank, editors. *Truth Maintenance Systems,* Stockholm, Sweden, August 6 1991. ECAI-90 Workshop, Springer-Verlag.
24. D.A. McAllester. An outlook on truth maintenance. Technical Report AIM-551, MIT, 1980.
25. J. McCarthy. *Formalizing Common Sense.* Ablex Publishing Corporation, Norwood, New Jersey, 1990. Collected Papers of John McCarthy on Commonsense Reasoning, edited by V. Lifschitz.
26. J. McCarthy and P. Hayes. "Some philosophical problems from the standpoint of artificial intelligence. In B. Meltzer and D. Michie, editors, *Machine Intelligence,* volume 4, pages 463–502. Edinburgh University Press, 1969.
27. D. McDermott and J. Doyle. "Non-monotonic logic I". *Artificial Intelligence,* 13(1–2):41–72, 1980.
28. M.D. Mesarovic and Y. Takahara. *Abstract System Theory.* Springer-Verlag, New York, 1989.
29. W. Marek A. Nerode and J. Remmel. "Nonmonotonic rule systems I". *Annals of Mathematics and Artificial Intelligence,* 1:241–273, 1990.
30. N.J. Nilsson. *Principles of Artificial Intelligence.* Tioga, Palo Alto, CA, 1980.
31. N.J. Nilsson. "Logic and artificial intelligence". *Artificial Intelligence,* 47(1–3):31–56, 1991.
32. J. Pearl. *Probabilistic Reasoning in Intelligent Systems: Networks of Plausible Inference.* Morgan Kaufmann, San Mateo, CA, 1988.
33. J.L. Pollock. *Knowledge and Justification.* Princeton University Press, Princeton, New Jersey, 1975.
34. P.L. Falk R.E. Kalman and M.A. Arbib. *Topics in Mathematical System Theory.* McGraw-Hill, New York, 1969.
35. R. Reiter. "A logic for default reasoning". *Artificial Intelligence,* 13(1–2):81–132, 1980.
36. H.S. Sarjoughian. *Inductive Modeling of Discrete-event Systems: A TMS-based Non-monotonic Reasoning Approach.* PhD thesis, University of Arizona, April 1995. Department of Electrical and Computer Engineering.

37. G.A. Shafer. *Mathematical Theory of Evidence.* Princeton University Press, Princeton, New Jersey, 1979.
38. Y. Shoham. *Reasoning About Change: Time and Causation from the Standpoint of Artificial Intelligence.* MIT Press, Cambridge, 1988.
39. H.S. Stone. *Discrete Mathematical Structures and Their Applications.* Science Research Associates, Inc., Chicago, 1973.
40. P. Suppes. *A Probabilistic Theory of Causation.* North-Holland, 1973.
41. D. Weld. "Reasoning about model accuracy. *Artificial Intelligence*, 56(2–3):255–300, 1992.
42. W.A. Wymore. *Model-based Systems Engineering: An Introduction to the Mathematical Theory of Discrete Systems and to the Tricotyledon Theory of System Design.* CRC, Boca Raton, 1993.
43. L.A. Zadeh. "Fuzzy logic and approximate reasoning". *Syntheses*, 3:407–428, 1975.
44. B.P. Zeigler. *Theory of Modeling and Simulation.* John Wiley and Sons, 1976.
45. B.P. Zeigler. *Multi-Facetted Modelling and Simulation.* Academic Press, 1984.
46. B.P. Zeigler. *Object-Oriented Simulation with Hierarchical, Modular Models: Intelligent Agents and Endomorphic Systems.* Academic Press, New York, 1990.

Semantics and Verification of Hierarchical CRP Programs*

R.K. Shyamasundar
Computer Science Group
Tata Institute of Fundamental Research
Bombay 400 005, India
e-mail: shyam@turing.tifr.res.in
S. Ramesh
Department of Computer Science and Engineering
Indian Institute of Technology
Bombay 400 076, India
e-mail: ramesh@cse.iitb.ernet.in

Abstract

Communicating Reactive Processes (CRP) paradigm unifies asynchronous and synchronous mechanisms of concurrent programming languages. As an example, CRP unifies ESTEREL and Hoare's Communicating Sequential Processes (CSP). It has been earlier shown that such a unification and in particular CRP can be used for the specification of hybrid systems and dynamic real-time systems. In this paper, we extend the CRP to support hierarchical refinement and describe a compositional semantics. Further, we show how verification can be done using the verification techniques and tools used for the verification of ESTEREL programs. We illustrate how a careful separation of ESTEREL and CSP mechanisms in CRP has enabled the use of the technique of *verification by reductions* for verification and illustrate the use of tools such as Auto/Autograph for the verification of CRP programs through the banker-teller example.

1 Introduction

Programming languages that have been used for real-time programming can be broadly categorized into:

1. *Asynchronous Languages*: Here, a program is treated as a set of loosely coupled independent execution units or processes, each process evolving at its own pace. Interprocess communication is done by mechanisms such as message passing. Communication as a whole is *asynchronous* in the sense that *an arbitrary amount of time can pass between the desire of communication and its actual completion.* This class includes languages such as Ada, Occam, CSP etc.

* The work was partially supported by IFCPAR (Indo-French Centre for the Promotion of Advanced Research), New Delhi, India.

2. *Perfectly Synchronous Languages*: In this class, programs react instantaneously to its inputs by producing the required outputs. Statements evolve in a tightly coupled input-driven way deterministically and communication is done by instantaneous broadcast where the receiver receives a message exactly at the time it is sent. That is, a perfectly synchronous program produces its outputs from its input with *no observable time delay*. Languages such as Esterel [2], Lustre, Signal, Statecharts belong to this category. These languages are often referred to as *reactive* languages. These languages use the multiform notion of time rather than clock time.

Programming languages based on the *perfect synchrony paradigm* have proven useful for programming reactive systems where determinism is a must and logical concurrency is required for good programming style. One of the main reasons for its success is due to the fact that it permits the programmer to focus on the logic of reactions and makes it possible to use several automata-based verification systems for correctness proofs. Further, the correctness proofs of programs follow their implementation very closely and hence, are more robust and reliable. On the other hand, asynchronous languages have proven useful in distributed processing and implementing algorithms on a network of computer systems where nondeterminism is an appropriate abstraction at the logical and physical levels. Presently, each class of languages is unable to handle problems to which the other class is tailored. Asynchronous languages are inappropriate for truly reactive systems that require deterministic synchronous (perfect) communication. On the other hand, existing synchronous languages lack support for asynchronous distributed algorithms. However, complex systems do require the abilities of both the languages. For instance, a robot driver must use a specific reactive program to control each articulation, but the global robot control may be necessarily asynchronous because of limitations of networking capabilities. General specification of complex reactive systems needs capabilities such as:

1. Relating asynchronous events (events that can happen arbitrarily close to each other),
2. Integration of discrete and continuous components (continuous components may cause continuous change in the values of some state variables according to some physical law), and
3. Referencing explicit clock times.

In [4], a paradigm referred to as *Communicating Reactive Processes* (CRP), has been proposed that provides a unification of perfect synchrony and asynchrony. CRP permits only top-level network parallelism. In [7] an extension of CRP, referred to as *Timed* CRP has been proposed having the capabilities to specify hybrid and asynchronous dense-time real-time systems and in [8], we have investigated the power of the CRP paradigm in providing a platform for programming dynamic real-time systems.

In this paper, we extend CRP to support hierarchical refinement. We show how the techniques for describing a compositional semantics of ESTEREL envisaged in [11], can be generalized for hierarchical CRP programs. The compositional behavioral semantics makes it possible to use techniques of ESTEREL verification for the verification of hierarchical CRP programs. Essentially, the compositional semantics allows us to split the proof into verification of ESTEREL nodes and the ESTEREL network. Thus, transformations such as behaviour abstraction, filtering and hiding can be applied for the nodes and for the network as well. We illustrate the verification of a CRP program using the above transformations using Auto/Autograph. In other words, we establish that proving CRP programs is no harder than proving ESTEREL programs. That is, the unification of perfect synchrony and asynchrony does not complicate the proof techniques. We illustrate the technique through the Banker-Teller example given in [4].

2 Hierarchical Communicating Sequential Processes

First, we give an informal introduction and rationale for CRP before describing the syntax and behavioral semantics.

2.1 Communicating Reactive Processes

A CRP program consists of independent locally reactive ESTEREL nodes that communicate with each other by rendezvous communication as in Hoare's Communicating Sequential Processes (CSP). CRP faithfully extends both ESTEREL and CSP and adds new possibilities such as precise local watchdogs on rendezvous.

A CRP program consists of a network M_1 // M_2 $\cdots$ // M_n of ESTEREL reactive programs or *nodes*, each having its own input/output reactive signals and its own notion of an instant. The network is asynchronous (in the sense discussed earlier) and each node M_i is locally reactive driving a part of a complex network that is handled globally by the network.

Basic Interface to Asynchrony

In full ESTEREL , external procedure calls are assumed to be instantaneous. Such a feature is not convenient for dealing with actions which are known to take time – for instance, even (long!) numerical computation required in many situations are known to take time. The new **exec** statement overcomes this draw back and provides a gateway to asynchrony. An asynchronous task is declared by "**task P**" (the precise syntax is given in the sequel) and task execution is controlled from ESTEREL by the statement

exec P

that waits for the task to terminate. Since there can be several occurrences of "**exec P**" in a module for the same task P, several simultaneously active tasks having the same name can coexist. In fact, to overcome the ambiguity, we permit the labelling of the **exec** statement (explicitly and implicitly).

Basic Rendezvous Statements

The central idea of establishing asynchronous communication between nodes lies in extending the basic **exec** primitive into a communication primitive. The usual **send** and **receive** asynchronous operations can be represented by particular tasks that handle the communication. Several **send/receive** interactions are possible according to various types of asynchronous communication. For instance, **send** can be non-blocking for full asynchrony, or **send** and **receive** can synchronize for CSP-like rendezvous communication. All choices can be implemented through ESTEREL . In this paper, we concern ourselves with the integration of ESTEREL with CSP and consider the rendezvous communication between nodes.

CRP nodes are linked by *channels.* We start by describing the pure synchronization case, where channels are symmetric. Pure channels are declared in CRP nodes by the declaration

```
channel C
```

A channel must be shared by at most two nodes. Communication over the channels is handled by the statement

```
rendezvous L:C
```

The **rendezvous** statement is a particular instance of **exec** statement and hence, the interface detailed above follows. A given channel can only perform one rendezvous at a time. Hence, in each node, the return signals of all the **rendezvous** statements on the same channel are *implicitly assumed (declared) to be incompatible*[2]. Values can be passed through unidirectional channels. In the following section, we describe the syntax of CRP and the declaration and calling conventions of asynchronous tasks.

2.2 Syntax of the Hierarchical CRP

In the following, we describe the abstract syntax of hierarchical CRP programs. We will not describe the syntax of ESTEREL statements for the sake of brevity; we will give the structure of kernel statements only.

```
< crp_prog >  ::=   < stat >  |  < network >
                   (* The usual ESTEREL kernel statements *)
< stat >   ::=   nothing
                 | halt
                 | emit S
                 | stat1; stat2
                 | loop stat end
                 | present S then stat1 else stat2 end
                 | do stat watching S
                 | stat1 || stat2
                 | trap T in stat end
                 | exit T
                 | signal S in stat end

                   (* Additional Statements in CRP *)
                         (* task declaration *)
                 | task task_id (par_lst)  return signal_nm (type)
                         (* channel declaration *)
                 | channel ch_name (type)
```

[2] This is needed for enforcing CSP restrictions only.

```
                    (* instantiating a task *)
        | exec task_id (par_lst) return signal_name

                    (* sending a value on the channel *)
        | rendezvous_send channel_id channel_name
            (expression) return signal_name (some_type)

                    (* sending a value on the channel *)
        | rendezvous_receive channel_id channel_name
            (variable) return signal_name (some_type)

                    (* Hiding Statement *)

        | [< network >]

< network > ::=    process1 || process2

< process > ::=    process begin < stat > end | < network >
```

The kernel statements are imperative in nature, and most of them are classical in appearance. The `trap-exit` constructs form an exception mechanism fully compatible with parallelism. Traps are lexically scoped. The local signal declaration "`signal` S `in` *stat* `end`" declares a lexically scoped signal S that can be used for internal broadcast communication within "*stat*". The `then` and `else` parts are optional in a present statement. If omitted, they are understood to be the same as `nothing`.

Task/channel declarations and the basic asynchronous and rendezvous executions are described below.

The Hierarchical Parallel Operator

In CRP [4], the network parallel operator can be used only at the top level. In the hierarchical CRP described in this paper, the network parallel operator can appear anywhere in the program text. This generality also permits us to introduce the analogue of "hiding" in CRP – thus, permitting hierarchical abstraction. We enforce the following discipline on the tasks with reference to sharing of signals and variables:

- Processes on different nodes do not share variables.
- Processes on different nodes can share only pure input signals.
- Any two processes denoting nodes can share common channel signals.

Task Declaration

An asynchronous task is declared as follows:

```
task task_id (par_lst)  return signal_nm (type);
```

where

- `par_lst` gives the list of parameters (reference or value);

- the signal returned by the task is given by the `signal_nm` with its type after the keyword **return**; the return signal is optional and it is possible to have multiple return signals.

A typical example of the declaration of a component is given by:

```
task MOUSE (y) return Mouse ({sanct,kill})
```

Channel Declaration

Channels are declared in CRP nodes by the declaration:

```
channel C (type)
```

A Channel is shared exactly by two nodes and the declarations should type-match.

Task Instantiation

Instantiation of the task is done through the primitive **exec**. For example, the above task can be instantiated from an ESTEREL program as follows:

```
exec MOUSE (y) return {win,loss,tie}
```

The execution of the above statement starts task MOUSE and waits for its termination; of course, the number and type of arguments, and the return signal type should match with the task declared. In other words, **exec** requests the environment to start the task and then waits for the return signal (which also indicates the termination of the task).

Since there can be several occurrences of **exec** T in a module for the same task T, several simultaneously active tasks having the same name can coexist. To avoid confusion among them one can assign an explicit label to each **exec** statement; a labeled **exec** statement is of the form **exec** L:T. The label name must be distinct from all other labels and input signal names. An implicit distinct label is given to **exec** statements.

Operational View of exec

The primitive **exec** provides an interface between ESTEREL and the asynchronous environment can be seen by the following interpretation. Given an **exec** statement labeled L, the asynchronous task execution is controlled from ESTEREL by three implicit signals **sL** ((output signal), L (input signal), and **kL** (output signal) corresponding to starting the task, completion and killing the execution of the task respectively. The output *start* signal **sL** is sent to the environment when the **exec** statement starts. It requests the start of an asynchronous incarnation of the task. The input *return signal* L, is sent by the environment when the task incarnation is terminated; it provokes instantaneous termination of the **exec** statement. the output *kill signal* kL is emitted by ESTEREL if the **exec** statement is preempted before termination, either by an enclosing watching statement or by concurrent exit from an enclosing trap. The return signals corresponding to the **exec** label can be used for declaring incompatibility with other input signals (this becomes handy in declaring channels).

Rendezvous Instantiation

The rendezvous in the sender is initiated through the command:

```
rendezvous_send channel_name (value_expression) return (some_type)
```

The rendezvous in the receiver is initiated through the command:

```
rendezvous_receive channel_name (variable) return (some_type)
```

The other explanations follow on the same lines as that of the task instantiation given above.

For convenience[3], we use the following construct for guard selection in CRP:

```
rend_case
    case  L1: rendezvous_receive channel1 (x) do stat_1
    case  L2: rendezvous_send channel2 (exp) do stat_2
end
```

The code corresponds to selecting a *receive* on `channel1` followed by the execution of the statement-sequence *stat1* or *send* on `channel2` followed by the execution of the statement-sequence *stat2*.

2.3 Behavioural Semantics

In the classical ESTEREL, given a module M and an input event I, the behavioral semantics determines a transition $M \underset{I}{\overset{O}{\longrightarrow}} M'$ where O is the generated output event and M', the derivative is another module suited to perform further reactions. In this reaction, the total event consists of all the input and output signals. The derivative M' has the same interface as M and differs by its body. The reaction to a sequence of input events is computed by chaining such elementary transitions. In [11], a compositional semantics of ESTEREL has been described with the purpose of adapting the technique of *verification by reductions* for verifying ESTEREL programs. The primary distinction of the transition systems generated lies in the structure of action labels. In the semantics, the action labels are triples consisting of: (i) necessarily present signals, (ii) necessarily absent signals and (iii) signals emitted during the reaction. In [11], a relation among the input and output signals during each instant is used for checking causality. However, this is no longer needed due to the causality schemes proposed recently by G. Berry [3] and Olivier Ploton [6]. In the context of CRP, we have to keep track of pending asynchronous tasking requests and do the bookkeeping of the successful completion of the task or preemption of the task. In the semantics of CRP described in [4], we used an action structure consisting of (i) the input signals present, (ii) the output signals emitted, (iii) the trap set or the termination levels and (iv) the set of pending asynchronous tasks. The bookkeeping for generating the appropriate kill signals was done through a syntactic decoration of the *present and trap* constructs. In this paper, we describe a compositional semantics of the hierarchical CRP without resorting to such ad-hoc decorations. It can be easily seen that "the syntactic decorations" are in a sense state-information. For this reason, in our approach we keep track of an explicit information with reference to preemptive tasks and signals separately as an environment. Thus, the transition system takes the form,

$$< stat, environment > \xrightarrow{<A>,L} < stat', environment' >$$

where:

[3] It may be noted that nondeterministic guard selection can be done in CRP without any additional operator; cf. [4].

Environment: Environment reflects the signals that can kill the rendezvous/asynchronous tasks. In other words, the environment, keeps track of the signals that can preempt (kill) asynchronous tasks or the rendezvous. Formally, the environment (usually denoted by η), is denoted by $\{< L_i, K_i >\}$ where L_i is an active rendezvous/asynchronous_task label which can be preempted by any of the set of signals in K_i. Note that K_i would have at least the signal kL_i which is the signal generated by the ESTEREL node for killing the task labelled L_i. Note that for killing a rendezvous/task labelled, L, it is necessary that a kill signal kL is generated in the ESTEREL node.

Action: The action label[4] is a tuple $< E >, L$ where

- E is the set of signals consisting of
  - The necessarily present signals, I^+,
  - The necessarily absent signals, J^-,
  - The output signals, O, emitted and
- L is the set of completed rendezvous labels/asynchronous task labels in the current instant.

Thus, an event is of the form $< I^+, J^-, O >, L$.

It may be noticed that , our action labels will not have the termination levels as we are omitting the traps to begin with. We will generalize the rules later to handle traps.

ESTEREL permits testing of absence of signals. Thus, we would like to capture the behaviour in terms of an event that is based on a set of signals present and a set of signals that are absent so that it will be possible to compose modules. Due to the synchrony hypothesis, there would be some cycles (or leads to equations which do not have solutions) which have to be eliminated for the computational semantics. Recently the problem of detecting cycles has been completely solved and the algorithms based on the electrical semantics and obtaining causal-free programs complete with respect to the behavioral semantics is reported in [3] and [6] respectively. We will not consider causality aspects in the sequel.

Notation:

1. We use a predefined signal "fin" to indicate sequential termination; it corresponds to a boolean flag or a general termination level used in the SOS rules in [1, 2]. Note that fin can only be emitted (similar to the $\_exit$ signal used in [11]).
2. We define operator # used in two ways, one in the context of environment, η, defined above and the other in the context of a set of pending asynchronous tasks, L.
   - Let $E =< I^+, J^-, O >$ be an event, η an environment, and S an input signal. Then, $< E, \eta > \#S = < I^+, J^-, O' >$ where
     $$O' = O \cup \{kL \mid < L, \{S\} > \in \eta, \text{ and } sL \notin O\}$$
     $$-\{sL \mid < L, \{S\} > \in \eta, \text{ and } sL \in O\}$$

     That is, $< E, \eta > \#S$ differs from E by the **rendezvous**'s/**exec**'s started before the current instant that are killed by S and that **rendezvous**'s started

[4] We ignore traps at this point.

at the current instant being watched under S are simply ignored or killed right away.

- Let $E = \langle I^+, J^-, O \rangle$ be an event, L be a set of rendezvous/exec labels. Then, $E\#L = \langle I^+, J^-, O' \rangle$ where

$$O' = O \cup \{k\ell | \ell \in L \text{ and } s\ell \notin O\} \\ -\{s\ell | \ell \in L, \text{ and } s\ell \in O\}$$

3. Let η be an environment and S a signal. Then,

$$\eta \uparrow S = \{\langle H, \alpha \cup \{S\}\rangle \mid \langle H, \alpha\rangle \in \eta\}$$

$$\eta \downarrow S = \{\langle H, \alpha \backslash \{S\}\rangle \mid \langle H, \alpha\rangle \in \eta\}$$

In the following, we describe the behavioural semantics using rules are of the form

$$\langle P, \eta \rangle \xrightarrow{E,L} \langle P', \eta' \rangle$$

For the sake of brevity, we consider pure ESTEREL rules and hence, in the rendezvous we do not distinguish between **rendezvous_send** and **rendezvous_receive**. Hence, for the sake of brevity, we use **rendezvous** for denoting either of them.

Rules for the basic constructs of ESTEREL are given in Table Ia. The interpretation of the rules (*halt*), (*nothing*), (*emit*), (*seq_1*), and (*seq_2*) follows on the lines of the rules for ESTEREL.

halt $\langle \mathbf{halt}, \emptyset \rangle \xrightarrow{\langle \emptyset, \emptyset, \emptyset \rangle, \emptyset} \langle \mathbf{halt}, \emptyset \rangle$
nothing $\langle \mathbf{nothing}, \emptyset \rangle \xrightarrow{\langle \emptyset, \emptyset, \{fin\} \rangle, \emptyset} \langle \mathbf{nothing}, \emptyset \rangle$
emit $\langle \mathbf{emit\ S}, \emptyset \rangle \xrightarrow{\langle \emptyset, \emptyset, \{S\} \cup \{fin\} \rangle, \emptyset} \langle \mathbf{nothing}, \emptyset \rangle$
Seq_1 $\dfrac{\langle stat_1, \eta_1 \rangle \xrightarrow{\langle I_1^+, J_1^-, O_1 \cup \{fin\} \rangle, \emptyset} \langle stat_1', \emptyset \rangle,\ fin \notin O_1 \quad \langle stat_2, \emptyset \rangle \xrightarrow{\langle I_2^+, J_2^-, O_2 \rangle, \emptyset} \langle stat_2', \eta_2' \rangle, \quad I_1^+ \cap J_2^- = I_2^+ \cap J_1^- = \emptyset,\ O_1 \cap J_2 = O_2 \cap J_1 = \emptyset}{\langle stat_1; stat_2, \eta_1 \rangle \xrightarrow{\langle I_1^+ \cup I_2^+, J_1^- \cup J_2^-, O_1 \cup O_2 \rangle, \emptyset} \langle stat_2', \eta_2' \rangle}$
Seq_2 $\dfrac{\langle stat_1, \eta_1 \rangle \xrightarrow{\langle I_1^+, J_1^-, O_1 \rangle, L_1} \langle stat_1', \eta_1' \rangle, fin \notin O_1}{\langle stat_1; stat_2, \eta_1 \rangle \xrightarrow{\langle I_1^+, J_1^-, O_1 \rangle, L_1} \langle stat_1'; stat_2, \eta_1' \rangle}$
Table Ia: Inductive Rules

The rule for the loop-statement is given below:

loop:

$$\frac{< stat, \eta > \xrightarrow{<I^+, J^-, O>, L} < stat', \eta' >}{< \texttt{loop}\ stat\ \texttt{end}, \eta > \xrightarrow{<I^+, J^-, O>, L} < stat';\ \texttt{loop}\ stat\ \texttt{end}, \eta' >}$$

The rules for the **exec** and **rendezvous** are given in Table Ib.

exec_st

$$< \texttt{exec}\ T, \emptyset > \xrightarrow{<\emptyset, \emptyset, \{sT\}>, \emptyset} < \texttt{exec\_wait}\ T,\ \{< T, \emptyset >\} >$$

exec_wt1

$$< \texttt{exec\_wait}\ T, \{< T, \emptyset >\} > \xrightarrow{<\emptyset, \emptyset, \emptyset>, \emptyset} < \texttt{exec\_wait}\ T,\ \{< T, \emptyset >\} >$$

exec_wt2

$$< \texttt{exce\_wait}\ T, \{< T, \emptyset >\} > \xrightarrow{<\emptyset, \emptyset, \{fin\}>, \{T\}} < \texttt{nothing}, \emptyset >$$

rend_st

$$\frac{H \text{ is a channel}}{rend < \texttt{rendezvous}\ H, \emptyset > \xrightarrow{<\emptyset, \emptyset, \{sH\}>, \emptyset} < \texttt{rend\_wait}\ H,\ \{< H, \emptyset >\} >}$$

rend_wt1

$$\frac{H \text{ is a channel}}{< \texttt{rend\_wait}\ H, \{< H, \emptyset >\} > \xrightarrow{<\emptyset, \emptyset, \emptyset>, \emptyset} < \texttt{rend\_wait}\ H,\ \{< H, \emptyset >\} >}$$

rend_wt2

$$\frac{H \text{ is a channel}}{< \texttt{rend\_wait}\ H, \{< H, \emptyset >\} > \xrightarrow{<\emptyset, \emptyset, \{fin\}>, \{H\}} < \texttt{nothing}, \emptyset >}$$

Table Ib: Inductive Rules

The rule *exec_st* is used to start a *asynchronous task* and the statement is rewritten into an auxiliary **exec_wait** statement that does not exist in the language proper but is used for convenience in defining the semantics. The environment on the left-side of the rule *exec_st* is empty as there cannot be any pending task at this point; on the right-side of this rule, the tuple $< T, \emptyset >$ is added reflecting the fact that the asynchronous task labelled T cannot be preempted by any other signal thus far. In rule (*exec_wt1*), nothing happens; that is, the ESTEREL node is waiting for the completion signals from the asynchronous task. In rule (*exec_wt2*), the task has been completed and hence, the tuple $< T, \emptyset >$ is removed from the environment. The rules *rend_st*, *rend_wt1* and rend_wt2 are the same as that of the corresponding *exec* rules except for checking for the channel declaration. In fact, in an isolated way, there is not much of a difference between the **exec** and the *rendezvous* locally; however, they differ in parallel composition (and hence, implementation). The rules *seq_1* and *seq_2* should be obvious except for the environment component. The environment component for $stat_1$ need not be empty due to the recursive application of the rule (note that there can be instantaneous transitions resulting in the addition of the environment); however, the environment for $stat_2$ remains empty as there cannot be any other pending tasks in it unless $stat_1$ terminates.

Rules for Watching Statement:

The rules are described in Table II. We express these rules in terms of the *present* and *watching_immediate* statements as the structure enables us to show the possible environment for the components and the corresponding updating. In the case of rule (*watching_stat*) the environment on the left is empty as there cannot be any pending task to start with and further, the watching-statement is rewritten as a present-statement. Note that the rule adds "S" as a possible preemptive signal for all the pending rendezvous/exec labels. It may be noted that we could have used the *watching* statement instead of the *watching-immediate* statement in this rule; the reason for using the latter is that we would like to add the preemptive signal only once; the addition of the preemptive signal is done only through the *watching* statement; note that *watching immediate* and *present* statements have similarities. *Watching-Immediate* and *Present* Statements: Rule (*watching_immed1*) reflects the behaviour when the signal S is not present in the current instant. In this case, the behaviour would be reflected by the transitions of *stat*; note that the environment need not be empty as it could have been obtained through a rewrite. Since the testing takes place immediately, there is no need to add in the environment the preemptive signal. The rule (*watching_immed2*) corresponds to preemption due to the presence of S in the current instant. It preempts all the pending rendezvous/asynchronous tasks.

Rules (*present1*) and (*present2*) correspond to situations where the transitions correspond to the then-clause and the else-part of the present statement. Since the execution corresponding to the then-clause can take place only after verifying the presence of S, the environment for $stat_1$ in rule (*present1*) is empty; in rule (*present2*) the environment for $stat_2$ need not be empty since instantaneous transitions after verifying the absence of S can lead to change in the environment. Note that (*present1*) and (*watching_immed2*) are alike. Actually, rule (*present2*) handles two situations: if the "present-statement" had arisen due to rewrite of the "watching-statement", then it implies that the embedded statement has inherited some environment; other wise, $\eta = \emptyset$, as the statement could not have been fired. In either case, the effect is given by the rule.

The rules for local signal declaration, the network parallel operator, and hiding are given in Table III. It may be noted, that the complete local signal-declarations is captured by rule *Signal_decl*. It is of interest to note this rule completely captures the behaviour since all operators maintain coherency of events, i.e., $S \in J^- \Rightarrow \not\exists(S \in O)$. The operators are explained below.

est_Par1

$$\frac{\begin{array}{l} < stat_1, \eta_1 > \xrightarrow{<I_1^+, J_1^-, O_1>, L_1} < stat_1', \eta_1' > \\ < stat_2, \eta_2 > \xrightarrow{<I_2^+, J_2^-, O_2>, L_2} < stat_2', \eta_2' >, \\ I_1^+ \cap J_2^- = I_2^+ \cap J_1^- = \emptyset, O_1 \cap J_2^- = O_2 \cap J_1^- = \emptyset,\ fin \notin O_1 \cap O_2 \end{array}}{\begin{array}{l} < stat_1 || stat_2, \eta_1 \cup \eta_2 > \xrightarrow{<I_1^+ \cup I_2^+, J_1^- \cup J_2^-, O_1 \cup O_2 - \{fin\}>, L_1 \cup L_2} \\ \qquad < stat_1' || stat_2', \eta_1' \cup \eta_2' > \end{array}}$$

est_Par2

$$\frac{\begin{array}{l} < stat_1, \eta_1 > \xrightarrow{<I_1^+, J_1^-, O_1>, L_1} < nothing, \emptyset >,\ fin \in O_1, \\ < stat_2, \eta_2 > \xrightarrow{<I_2^+, J_2^-, O_2>, L_2} < nothing, \emptyset >,\ fin \in O_2, \\ I_1^+ \cap J_2^- = I_2^+ \cap J_1^- = \emptyset,\ O_1 \cap J_2 = O_2 \cap J_1 = \emptyset, \end{array}}{< stat_1 || stat_2, \eta_1 \cup \eta_2 > \xrightarrow{<I_1^+ \cup I_2^+, J_1^- \cup J_2^-, O_1 \cup O_2 \cup \{fin\}>, L_1 \cup L_2} < nothing, \emptyset >}$$

watching_stat

$$\frac{< stat, \emptyset > \xrightarrow{<I_1^+, J_1^-, O_1>, L} < stat', \eta' >}{\begin{array}{l} < \texttt{do}\ stat\ \texttt{watching}\ S\ , \emptyset > \xrightarrow{<I_1^+, J_1^-, O_1>, L} \\ \qquad < \texttt{present}\ S\ \texttt{else do}\ stat'\ \texttt{watching immediate}\ S, \eta' \uparrow S > \end{array}}$$

where $\eta' \uparrow S$ is the operation that adds signal S as a possible kill signal for the pending rendezvous labels L.

watching_immed1

$$\frac{< stat, \eta > \xrightarrow{<I^+, J^-, O>, L} < stat', \eta' >,\ S \notin I^+ \cup O}{\begin{array}{l} < \texttt{do}\ stat\ \texttt{watching immediate}\ S, \eta > \xrightarrow{<I^+, J^-, O>, L} \\ \qquad < \texttt{do}\ stat'\ \texttt{watching immediate}\ S, \eta' > \end{array}}$$

watching_immed2

$$\frac{< stat, \eta > \xrightarrow{<I^+, J^-, O>, L} < stat', \eta' >,\ E \equiv < I^+, J^-, O >, S \notin J^- \cup O}{< \texttt{do}\ stat\ \texttt{watching immediate}\ S, \eta > \xrightarrow{<E[I/I \cup \{S\}], \eta> \# S} < \texttt{nothing}, \emptyset >}$$

where $E[\cdot/\cdot]$ denotes the usual substitution;

present1

$$\frac{S \notin J_1^- \cup O_1,\ < stat_1, \emptyset > \xrightarrow{E_1 \equiv <I_1^+, J_1^-, O_1>, L_1} < stat_1', \eta' >}{\begin{array}{l} < \texttt{present}\ S\ \texttt{then}\ stat_1\ \texttt{else}\ stat_2\ \texttt{end}, \eta > \xrightarrow{<E_1[I_1/I_1 \cup \{S\}], \eta'> \# S} \\ \qquad < stat_1', \eta' > \end{array}}$$

present2

$$\frac{S \notin I_2^+ \cup O_2,\ < stat_2, \eta > \xrightarrow{<I_2^+, J_2^-, O_2>, L_2} < stat_2', \eta' >}{\begin{array}{l} < \texttt{present}\ S\ \texttt{then}\ stat_1\ \texttt{else}\ stat_2\ \texttt{end}, \eta \uparrow S > \xrightarrow{<I_2^+, J_2^- \cup \{S\}, O_2>, L_2} \\ \qquad < stat_2', \eta' > \end{array}}$$

Table II: Inductive Rules

Signal_decl

$$\frac{< stat, \eta > \xrightarrow{E,L} < stat', \eta' >,\ S \in I^+ \Rightarrow S \in O}{< \mathbf{signal}\ S\ \mathbf{in}\ stat, \eta \downarrow S > \xrightarrow{E/\{S\}),L} < \mathbf{signal}\ S\ \mathbf{in}\ stat', \eta' \downarrow S >}$$

where $\eta \downarrow S$ corresponds to $\{< H, \alpha \backslash \{S\} > \mid < H, \alpha > \in \eta\}$

Net_Par_Syn

$$\frac{\begin{array}{l} < P_1, \eta_1 > \xrightarrow{E_1 \equiv < I_1^+, J_1^-, O_1 >, L_1} < P_1', \eta_1' > \\ < P_2, \eta_2 > \xrightarrow{E_2 \equiv < I_2^+, J_2^-, O_2 >, L_2} < P_2', \eta_2' >,\ L_1 \cap CC = L_2 \cap CC \end{array}}{< P_1 || P_2, \eta_1 \cup \eta_2 > \xrightarrow{< I_1^+ \cup I_2^+, J_1^- \cup J_2^-, O_1 - \{fin_1\} \cup O_2 - \{fin_2\} \cup fin' >, L_1 \cup L_2} < P_1' || P_2', \eta_1' \cup \eta_2' >}$$

where fin' is equal to $\{fin\}$ if $fin_1 \in O_1 \wedge fin_2 \in O_2$; otherwise, $fin' = \emptyset$.

Net_Par_Int

$$\frac{\begin{array}{l} < P_1, \eta_1 > \xrightarrow{E_1 \equiv < I_1^+, J_1^-, O_1 >, L_1} < P_1', \eta_1' > \\ \eta|_{P_1} = \eta_1,\ \eta|_{P2} = \eta'|_{P2}, \eta_1' = \eta'|_{P_1},\ L_1 \cap CC = \emptyset \end{array}}{< P_1 || P_2, \eta > \xrightarrow{E_1, L_1} < P_1' || P_2, \eta' >}$$

Hiding_initiate_1

$$< \mathbf{process\ begin}\ P\ \mathbf{end}, \emptyset > \rightarrow < P, \{< P, \emptyset >\} >$$

Hiding_ext1

$$\frac{< P, \eta > \xrightarrow{E \equiv < I^+, J^-, O >, L} < P', \eta' >,\ fin \notin O, kP \notin O}{< [P], \eta > \xrightarrow{< ext(I^+), ext(J^-), ext(O) >, ext(L)} < [P'], \eta' >}$$

Here ext denotes the projection of the signals to those that can be observed outside P; note that, it can share ESTEREL *signals. Note that there could be some some actions in L that are not observable.*

Hiding_ext2

$$\frac{< P, \eta > \xrightarrow{E \equiv < I^+, J^-, O >, L} < P', \eta' >,\ fin \in O, kP \notin E}{< [P], \eta > \xrightarrow{< ext(I^+), ext(J^-), ext(O) - \{fin\} >, ext(L)} < end_P, \eta' >}$$

Hiding_ext_ter_3

$$\frac{< P, \eta > \xrightarrow{E \equiv < I^+, J^-, O >, L} < P', \eta' >,\ kP \in E}{< [P_1], \eta_1 > \xrightarrow{E_1 \# kP_1, \emptyset} < end_P, \eta' >}$$

terminate_1

$$\frac{< P, kP > \in \eta}{< end_P, \eta > \xrightarrow{E,L} < nothing, \emptyset >}$$

Table III: Inductive Rules

Parallel Operator

Rules (*Net_Par_Syn*) and (*Net_Par_Int*) correspond to the rules corresponding to synchronization and interleaving actions. The synchronization is expressed with reference to the common set of channels, denoted CC, between the two processes. Note that it is here one can differentiate an *exec* and a *rendezvous*.

Nesting of Processes

The semantics of process essentially depends on the semantics of a "process", sharing of signals, variables and termination aspects. We impose the following conventions on the programs:

1. Variables are not shared as in the case of ESTEREL modules,
2. Nested processes can share only ESTEREL pure input signals.
3. When a process is preempted all its children get preempted.

Since processes can also be killed, we enrich the environment to consist of tuples of the form

- $<$ *process*, *set_of_kill_signals* $>$ in addition to the tuple
- $<$ *rendezvous labels*, *set_of_kill_signals* $>$

with the usual interpretation.

Rule (*hiding_initiate_1*) is similar to the semantics of a run-module of ESTEREL or the rendezvous of CRP. It rewrites into a process or module with the addition of possible process preemption signals (which are empty at this point). Rules (*hiding_ext1*), (*hiding_ext2*) and (*hiding_ext3*) correspond to the network that has neither finished nor preempted, properly finished and preempted respectively. In all these rules, if L is empty, then the action is treated as an internal action. In rule (*hiding_ext2*), we have used an auxiliary statement end_P_1 in place of `nothing`; this rewriting will be helpful if we want to synchronize with external clocks. Rule (*hiding_ext3*) corresponds to preemption of the network. As usual we have used priority for the `kill` signal.

Note:

1. The advantage of using end_P is that, the semantics can be easily modified for the case when the termination can be observed in the `tick` of the immediately enclosing module.
2. The semantics is consistent with the CRP semantics, in the sense, that termination of the embedded task also provides an input to the immediately enclosing module. This clearly models the asynchronous nature of the model.

3 Rules for Traps

In this section, we describe the behavioral rules of the `trap` and `exit` statements. First let us consider the priority definition among trap labels. Note that several traps may be exited simultaneously. The rule is simple: only the outermost trap matters, the other ones being discarded. For example in, *trap1*

$$\frac{\gamma \circ T \vdash < stat, \eta > \xrightarrow[T]{<I^+,J^-,O>,L} < stat', \eta' >,}{\gamma \vdash < \mathbf{trap}\ T\ \mathbf{in}\ stat, \eta > \xrightarrow[T]{<I^+,J^-,O_1 \cup \{fin\}>\#initiated(I^+),L} < nothing, \emptyset >}$$

where $O_1 = O_1 \cup \{k\ell \mid \ell \in pending(\eta)\}$; *i.e., kill* signals are added for the pending tasks and the rendezvous labels under the scope of trap T. The operator $\#initiated(I^+)$ removes all the tasks initiated in the current instant.

trap2

$$\frac{\gamma \circ T \vdash < stat, \eta > \xrightarrow[\ell]{<I^+,J^-,O\}>,L} < stat', \eta' >, \ell \neq T}{\gamma \vdash < \mathbf{trap}\ T\ \mathbf{in}\ stat, \eta > \xrightarrow[\ell]{<I^+,J^-,O>,L} < \mathbf{trap}\ T\ \mathbf{in}\ stat', \eta' >}$$

exit

$$< \mathtt{exit\ T}, \emptyset > \xrightarrow[T]{<\emptyset,\emptyset,\emptyset>,\emptyset} < halt, \emptyset >$$

est_par

$$\frac{\begin{array}{l} \gamma \vdash < stat_1, \eta_1 > \xrightarrow[\ell_1]{<I_1^+,J_1^-,O_1\}>,L_1} < stat_1', \eta_1' >, \\ \gamma \vdash < stat_2, \eta_2 > \xrightarrow[\ell_2]{<I_2^+,J_2^-,O_2\}>,L_2} < stat_2', \eta_2' >, \\ I_1^+ \cap J_2^- = I_2^+ \cap J_1^- = \emptyset \\ O_1 \cap J_2^- = O_2 \cap J_1^- = \emptyset \end{array}}{\begin{array}{l} \gamma \vdash < stat_1 || stat_2, \eta_1 \cup \eta_2 > \xrightarrow[nesting(\ell_1,\ell_2)]{<I_1^+ \cup I_2^+, J_1^- \cup J_2^-, O_1 \cup O_2>, L_1 \cup L_2} \\ \qquad < stat_1' || stat_2', \eta_1' \cup \eta_2' > \end{array}}$$

Table IV: Inductive Rules

```
trap T1 in
  trap T2 in
          exit T1
        ||
          exit T2
  end ; (* trap T1 *)
  emit O
end ; (* trap T2 *)
```

the traps T1 and T2 are exited simultaneously, the internal trap is discarded and O is not emitted.

Now, let us enrich the labels of the transitions by sets of trap labels. First, we use a trap-environment which consists of sets of sequences of trap-labels encountered. Thus, the transition $\gamma \vdash < stat, \eta > \xrightarrow[T]{E,L} < stat', \eta' >$ denotes the transitions with the trap set T assuming the trap-environment γ; T is a set of trap labels that contains the labels of the **exit** statements executes by *stat*.

The rules for the trap-statement, and the exit statement are given below. For giving the flavor of the tuple corresponding to the trap set, we also describe the rule for ESTEREL parallel operator. We use the following notation in the following:

- Let γ denote an element in the set of sequence of trap labels, Γ.
- Let *initiated*(I^+) = set of signals of the form $s\ell$ where ℓ is either an asynchronous task label or a rendezvous label. That is, *initiated*(I^+) yields the tasks initiated in the current instant.
- The function *pending* : *environment* $\rightarrow$ *task_labels* yields the set of pending tasks in the environment. That is, *pending*(η) collects all the pending tasks in η.
- We define an operator *nesting* taking two trap labels and outputs the label which nests the other label. The function *nesting* is realized by operations on the trap-environment γ.

In rule (*trap1*), the trap exited is labelled T (below the arrow). The trap essentially preempts all the pending asynchronous tasks and the embedded statement. The killing signals corresponding to the killed tasks are added to the output signals (which is achieved through the # operator). Note that the tasks completed can in fact does so; it is here, one can distinguish *weak preemption* (i.e., trap-construct) and strong preemption (i.e., do-watching construct). In (*trap2*), η' from *stat* is carried further; the addition of kill signals will be reflected when the actual trap is encountered (that is reflected in rule *trap1*). The halt-statement rewriting into a halt is reflected in rule (*exit*); note that in this rule the *trap* emitted is T. All other rules simply propagate the trap label.

In the full paper, we discuss abstract properties of the semantic equations such as compositionality, relationship with the CSP and `meije` semantics.

3.1 Illustrative Example

We illustrate the rules given earlier by describing the operational behaviour of a simple program containing two nested watching statements:

```
do
  do
     rendezvous H
  watching S1
watching S2.
```

Informally, the execution of this program consists of at least two steps: in the first step the rendezvous is initiated which in a latter step is either completed or preempted. The preemption can happen because of the occurrence of S1 or S2 with the latter having the priority over the former. Note that occurrence of S1 or S2 is ignored in the first step of the execution.

First Step: The behavior of the program in the first step is given by the last rewriting of the sequence of rewritings given below. This sequence describes the derivation of the last rewriting by applying the operational semantics. This derivation involves the application of '*rend_st*' rule once and '*watching_st*' rule twice.

$$\frac{<\mathit{rendezvousH},\emptyset> \xrightarrow{<\emptyset,\emptyset,\{sH\}>,\emptyset} <\mathit{rend\_waitH},\{<H,\emptyset>\}>}{\begin{array}{l}<\mathtt{do\ rendezvous}\ H\ \mathtt{watching}\ S1,\emptyset> \xrightarrow{<\emptyset,\emptyset,\{sH\}>,\emptyset} \\ \quad <\mathtt{present}\ S1\ \mathtt{else\ do\ rend\_wait}\ H\ \mathtt{watching\ immediate}S1, <H,\{S1\}>\end{array}}$$

$$\overline{\begin{array}{l}<\mathtt{do\ do\ rendezvous}\ H\ \mathtt{watching}\ S1\ \mathtt{watching}\ S2,\emptyset> \xrightarrow{<\emptyset,\emptyset,\{sH\}>,\emptyset} \\ \quad <\mathtt{present}\ S2\ \mathtt{else\ do\ present}\ S1\ \mathtt{else\ do\ rend\_wait}\ H \\ \quad \mathtt{watching\ immediate}\ S1\ \mathtt{watching\ immediate}\ S2, <H,\{S1,S2\}>\end{array}}$$

Next step: The behavior of the program in the next step depends upon whether S1 or S2 occurs or the rendezvous is completed. Four cases are distinguished and in each case, a derivation sequence leading to a possible behaviour is given.

Case 1: Neither S1 nor S2 occurs and the rendezvous does not get completed

$$\frac{<\mathtt{rend\_wait}\ H,<H,\emptyset>> \xrightarrow{<\emptyset,\emptyset,\emptyset>,\emptyset} <\mathtt{rend\_wait}\ H,\{<H,\emptyset>\}>}{\begin{array}{l}<\mathtt{do\ rend\_wait}\ H\ \mathtt{watching\ immediate}\ S1,<H,\emptyset>> \xrightarrow{<\emptyset,\emptyset,\emptyset>,\emptyset} \\ \qquad\qquad <\mathtt{do\ rend\_wait}\ H\ \mathtt{watching\ immediate}\ S1,<H,\emptyset>>\end{array}}$$

$$\overline{\begin{array}{l}<\mathtt{present}\ S1\ \mathtt{else\ do\ rend\_wait}\ H\ \mathtt{watching\ immediate}\ S1,<H,\{S1\}>> \xrightarrow{<\emptyset,\emptyset,\emptyset>,\emptyset} \\ \qquad\qquad <\mathtt{do\ rendezvous\_}\mathit{wait}\ H\ \mathtt{watching\ immediate}\ S1,<H,\emptyset>>\end{array}}$$

$$\overline{\begin{array}{l}<\mathtt{do\ present}\ S1\ \mathtt{else\ do\ rend\_wait}\ H\ \mathtt{watching\ immediate}\ S1\ \mathtt{watching\ immediate}\ S2, \\ \qquad\qquad <H,\{S1\}>> \\ \quad \xrightarrow{<\emptyset,\emptyset,\emptyset>,\emptyset} <\mathtt{do\ do\ rend\_wait}\ H\ \mathtt{watching\ immediate}\ S1\ \mathtt{watching\ immediate}S2, \\ \qquad\qquad <H,\emptyset>>\end{array}}$$

$$\overline{\begin{array}{l}<\mathtt{present}\ S2\ \mathtt{else\ do\ present}\ S1\ \mathtt{else\ do\ rend\_wait}\ H\ \mathtt{watching\ immediate} \\ \quad S1\ \mathtt{watching\ immediate}\ S2,\ <H,\{S1,S2\}>> \xrightarrow{<\emptyset,\emptyset,\emptyset>,\emptyset}\end{array}}$$
$$<\mathtt{do\ do\ rend\_wait}\ H\ \mathtt{watching\ immediate}\ S1\ \mathtt{watching\ immediate}\ S2,<H,\emptyset>>$$

Case 2: Neither S1 nor S2 occur but the rendezvous is completed.

$$\frac{<\mathtt{rend\_wait}\ H,<H,\emptyset>> \xrightarrow{<\emptyset,\emptyset,\{fin\}>,\{H\}} <\mathrm{nothing},\emptyset>>}{\begin{array}{l}<\mathtt{do\ rend\_wait}\ H\ \mathtt{watching\ immediate}\ S1,<H,\emptyset>> \\ \qquad\qquad \xrightarrow{<\emptyset,\emptyset,\{fin\}>,\{H\}} <\mathrm{nothing},\emptyset>\end{array}}$$

$$\overline{\begin{array}{l}<\mathtt{present}\ S1\ \mathtt{else\ do\ rend\_wait}\ H\ \mathtt{watching\ immediate}\ S1,<H,\{S1\}>> \\ \qquad\qquad \xrightarrow{<\emptyset,\emptyset,\{fin\}>,\{H\}} <\mathrm{nothing},\emptyset>\end{array}}$$

$$\overline{\begin{array}{l}<\mathtt{do\ present}\ S1\ \mathtt{else\ \ rend\_wait}\ H\ \mathtt{watching} \\ \qquad\qquad \mathtt{immediate}\ S1\ \mathtt{watching\ immediate}\ S2,<H,\{S1\}>> \\ \qquad\qquad\qquad \xrightarrow{<\emptyset,\emptyset,\{fin\}>,\{H\}} <\mathrm{nothing},\emptyset>\end{array}}$$

$$\overline{\begin{array}{l}<\mathtt{present}\ S2\ \mathtt{else\ do\ present}\ S1\ \mathtt{else\ do\ rend\_wait}\ H\ \mathtt{watching\ immediate}\ S1 \\ \quad \mathtt{watching\ immediate}\ S2,<H,\{S1,S2\}>> \xrightarrow{<\emptyset,\emptyset,\{fin\}>,\{H\}} <\mathrm{nothing},\emptyset>\end{array}}$$

Case 3: S1 occurs but S2 does not.

$$\dfrac{\dfrac{\dfrac{<\texttt{nothing}, \emptyset> \xrightarrow{<\emptyset,\emptyset,\{fin\}>,\emptyset} <\texttt{nothing}, \emptyset>}{<\texttt{present } S1 \texttt{ else do rend\_wait } H \texttt{ watching immediate } S1, < H, \{S1\} >> \xrightarrow{<\emptyset,\emptyset,\{fin\}>,\emptyset} <\texttt{nothing}, \emptyset>}}{<\texttt{do present } S1 \texttt{ else do rend\_wait } H \texttt{ watching immediate } S1 \texttt{ watching immediate } S2, < H, \{S1\} >> \xrightarrow{<\emptyset,\emptyset,\{fin\}>,\emptyset} <\texttt{nothing}, \emptyset>}}{<\texttt{present } S2 \texttt{ else do present } S1 \texttt{ else do rend\_wait } H \texttt{ watching immediate } S1 \texttt{ watching immediate } S2, < H, \{S1, S2\} >> \xrightarrow{<\emptyset,\emptyset,\{fin\}>,\emptyset} <\texttt{nothing}, \emptyset>}$$

Case 4: S2 occurs.

In this case, the derivation is similar to Case 3 but shorter having only the first and the last rewriting of the above sequence. Note that it does not matter whether $S1$ occurs or not.

Subsequent Steps: There would be subsequent computation steps provided Case 1 holds in the second step. Assuming this, we have the following rewritten program: '`do do rend_wait` H `watching immediate` S1 `watching immediate` S2' with the environment part being $\{H, \emptyset\}$. In subsequent steps this program is rewritten into itself with the environment remaining the same until either S1 or S2 occurs or the rendezvous is completed. When one of these events happen, the whole program terminates.

Remarks: In the above, it is assumed that S1 and S2 are distinct. If both are identical, then all the cases described above except the case that S1 occurs but S2 does not hold.

4 Verification

One of the distinct advantages of using synchronous languages for specifying reactive systems is that the description of the system analyzed/validated is very close to implementation. One of oldest languages in the family of synchronous languages ESTEREL has good developmental facilities such as efficient code generating compilers, verifiers etc. One of our motivation in the formulation of CRP [4] has been that we should preserve such an advantage as far as possible. In this section, we illustrate that it is indeed the case that the same verifying techniques/tools used on ESTEREL programs are also usable on CRP programs.

In the following, we describe briefly the compositional verification of ESTEREL before we show how the same techniques can be applied to hierarchical CRP and illustrate the verification on the banker-teller example using the system Auto/Autograph (or mauto/atg) described in [12].

4.1 Compositional Verification of ESTEREL

In [11], a compositional semantics of ESTEREL has been defined which enables compositional verification. The structure of the transitions has been discussed in the previous section that gives the power of compositionality for ESTEREL programs. The main idea behind the verification is that ESTEREL programs can be compiled into reactive automata ([11]) which can then be compared with specification automata. The power of "compositionality" for verification lies in applying transformations [11] such as:

1. Behaviour Abstraction: One of the most useful transformations of this class is *signal hiding*. In fact, we can define this transformation with reference to the set of signals say S on each of the syntactic constructs of the language. Essentially, the transformation removes the signals not in S from the transition events. Thus, the resultant automata may become nondeterministic keeping the number of states the same (hence, reachability properties would be preserved).
2. Symbolic Bisimulation: It is possible to define a bisimulation which will be congruent to the ESTEREL operators and commute with the hiding transformation.
3. Filtering: Better reductions can be achieved by guessing local contexts.

The above method will not be usable unless it is scalable. It is here the aid of tools comes to rescue. The verification system called Auto (the current version of this called Mauto) can be effectively used in checking the above transformations. A variety of automata operations with reference to the above transformations can be used to manipulate and compare automata. Some of the interesting operations are: bisimulation equivalence checking, deadlock detection, hiding and abstraction. An associated system Autograph/Atg ([12]) provides a user-friendly visual aid for viewing these automata. Using this system, Esterel programs can be verified to satisfy desired properties. In the context of verification, one major problem is the state-space explosion. A number of proposals are being offered in the literature. The above technique overcomes this problem as the transformations mentioned above lead to simplifications of the automata being verified.

4.2 Verifying CRP programs

The main issue lies in establishing that the compositional reductions can be achieved at the ESTEREL node level and the network level. The crux of verification lies in synthesizing *reactive automata* from the semantics. In the context of CRP, a transition depends on an event composed of (1) necessarily present signals, (2) necessarily absent signals, (3) output signals, and (4) the set of active/pending rendezvous actions which could be preempted. Since variables are not shared and the channels connect at most two processes, we can

1. Synthesize a local reactive automata at each node.
2. Synthesize a network reactive automata that reacts with reference to the rendezvous actions which can be locally preempted.

Let us consider the reactive automata at a node. It is obvious that each is locally reactive and the `ticks` are distinct at each node; they can synchronize only through the rendezvous actions. Thus, the reactive automata does not really depend upon the number of nodes but only on the number of nodes with which it interacts. The semantics outlined above enables to use the transformations described above both at the node and the network level.

As mentioned already, for a given CRP program there is an associated *reactive automata* which can be constructed using the operational semantics of CRP; the nodes in this automata corresponds to the CRP programs and its derivatives while the transitions capture the 'arrow' relation defined by the semantics. The transitions are more complex than the ESTEREL transitions as there is an additional component of completed rendezvous actions. This extra structure does not cause much problem in the verification process using Auto thanks to the underlying powerful meije calculus [10]. In fact, it contributes to a structuring of the proof process itself as follows:

- A CRP program can be abstracted out to be an ESTEREL program by hiding all the rendezvous actions and signals.
- A CRP program can be abstracted out to be a distributed program by hiding all the ESTEREL signals and retaining only the rendezvous actions.
- A CRP program can be viewed as a program consisting of processes interacting via either ESTEREL signals or rendezvous actions.

Depending upon the property under verification, appropriate abstractions can be carried out. This flexibility is possible because the semantics of CRP interprets rendezvous actions in terms of Esterel signals. As mentioned above, we can effectively use the above mentioned transformations for verifying CRP programs. An interesting and important observation of this paper is that these transformation rules are applicable to CRP programs as well and hence, they can help in containing the state explosion problem.

4.3 Illustrative Example of Verification: A Banker-Teller Program

As an illustration of the verification process, we shall take the Banker Teller example discussed in the original paper [4] and verify its correctness. This program consists of two CRP processes, Teller and Bank. Teller controls a machine through which the clients of the bank interact and carry out transactions. Teller processes controlling different machines interact with the Bank node which actually performs the money transfers. Each of these nodes, in addition to being locally reactive, interact with other nodes through rendezvous actions.

The teller machine reads a card, checks the code locally, and communicate by rendezvous with the bank to perform the actual transaction. During the whole process, the user can cancel the transaction by pressing a `Cancel` key.

The reactive interface is self-explanatory. The rendezvous interface consists of three valued channels. The output channels `CardToBank` and `AmountToBank` are used to send the card information and transaction amount. The input channel `Authorization` is used to receive a boolean telling whether the transaction is authorized. No rendezvous label is necessary here since there is exactly one `rendezvous` statement per channel. The output channel `CancelToBank` is used by the machine to inform Bank that the user has cancelled the current transaction.

The body repeatedly waits for a card and performs the transaction. The client-code checking section is purely local and reactive. The body after reading the valid code waits locally for the amount to be given (typed) by the client. Subsequently, the information about the card and amount are sent through the respective channels to Bank. Then the machine waits for authorization signal from the bank. This is a boolean valued signal and if it is true, one instantaneously delivers the money and returns the card exactly when the authorization rendezvous is completed, to ensure transaction atomicity. During the transaction, the user can press the **Cancel** key, in which case the transaction is gracefully abandoned, including the various rendezvous, and the card is returned. Since the bank considers the transaction to be performed when the **Authorization** rendezvous is completed, one must declare an exclusion relation between **Authorization** and **Cancel** to prevent these two events to happen simultaneously. The outer trap handles the abnormal cases of the transaction: the card is kept if the code is typed in incorrectly three times or if the transaction is not authorized. We use a full ESTEREL extension of the **trap** statement, in which one can write a trap-handler to be activated when the trap is exited. This extension is easily derivable from kernel statements.

In the code given below, we do not use explicit channel-declarations and further, we use the syntax of the prototype CRP compiler for the rendezvous statements. That is,

```
exec RdvSendCardToBank () (?Card) return CardToBank
```

stands for

```
rendezvous_send CardToBank (?Card) return  CardToBank
```

The statement denotes sending the value in the signal, Card, over the channel, *CardToBank*, and returns the pure signal CardToBank; note that we have not made distinctions between the channel name and the type of return signals. Similarly,

```
exec RdvRecvAuthorization()() return Authorization
```

stands for

```
exec RdvRecvAuthorization()() return Authorization
```

Here, the statement denotes receiving a value on the channel, Authorization, and the return signal is a pure signal Authorization. The other statements are similarly interpreted.

```
module BankerTeller :
type CardInfo; function CheckCode (CardInfo, integer) : boolean;
input Card : CardInfo;
input Code : integer, Amount : integer;
input Cancel;
output GiveCode, EnterAmount;
output DeliverMoney : integer;
output KeepCard, ReturnCard;
task RdvSendCardToBank () (CardInfo);
task RdvSendAmountToBank () (integer);
task RdvRecvAuthorization () () : boolean;
task RdvSendCancelToBank () ();
return CardToBank, AmountToBank;
return Authorization : boolean;
```

```
   return CancelToBank;
   relation Cancel # Authorization;
      loop
         trap ReturnCard, KeepCard in
            await Card;
                  do
               trap CodeOk in
                  emit GiveCode;
                  await Code;
                  if CheckCode(?Card, ?Code) then exit CodeOk end
               exit KeepCard
               end trap;
            watching Cancel timeout
               exit ReturnCard
            end;
            do (* watching Cancel again, but this time provokes *)
               (* cancellation rendezvous with the bank*)
               [
                  exec RdvSendCardToBank()(?Card) return CardToBank
               ||
                  emit EnterAmount;
                  await Amount
               ];

               exec RdvSendAmountToBank()(?Amount) return AmountToBank;

               exec RdvRecvAuthorization()() return Authorization;
               if ?Authorization then
                  emit DeliverMoney(?Amount);
                  exit ReturnCard
               else
                  exit KeepCard
               end if
            watching Cancel timeout
               exec RdvSendCancelToBank () () return CancelToBank;
               exit ReturnCard
            end;
         handle ReturnCard do
            emit ReturnCard
         handle KeepCard do
            emit KeepCard
         end trap
      end loop
   end module
```

The part of code of Bank that interacts with one Teller machine is given by the following segment:

```
module Bank:
type CardInfo;
output CARD_INFO: CardInfo;
output AMOUNT: integer;
```

```
input ACCEPT, REFUSE;
task RdvRecvCardToBank () () : CardInfo;
task RdvRecvAmountToBank () () : integer;
task RdvSendAuthorization () (boolean);
task RdvRecvCancelToBank () ();
return CardToBank: CardInfo;
return AmountToBank: integer;
return Authorization;
return CancelToBank;
loop
   [ do
         exec RdvRecvCardToBank()() return CardToBank;
         emit CARD_INFO(?CardToBank);
         exec RdvRecvAmountToBank()() return AmountToBank do
         emit AMOUNT(?AmountToBank);
         var Acceptation : boolean in
            await
                case ACCEPT do Acceptation := true
                case REFUSE do Acceptation := false
            end await
            exec RdvSendAuthorization()(Acceptation) return Authorization
         end var
     watching CancelToBannk
   ||
     do
        exec RdvRecvCancelToBank()() return CancelToBank do
     watching Authorization
   ]
end loop
end module
```

Using the operational semantics, the automata for the two processes can be obtained. The bank node can interact with a number of teller nodes (a fixed number) but for the purpose of verification, it is sufficient to look at the combined behavior of Bank with one Teller process. This is achieved by subjecting the Bank node to the behavioral abstraction transformation that hides all the signals of Bank other than those that are required for the interaction of one Teller process. After this abstraction, the Bank automata and the Teller automata are constructed. Now we can construct the automata describing the behavior of the Bank and the Teller; the automaton is shown in Figure 1 (for the sake of brevity, we have only shown the combined (merged) automata). This is rather large and can be abstracted out to a simple automata shown in Figure 2. The transitions in this automata are obtained by renaming and combining the transitions of the original automata. The exact relationship between the actions of the original automaton and that of the new one is given in the syntax of Auto as follows:

```
parse-criterion CR=
CARD = (/Card? and not /Cancel?),
RT-CODE = (/Code? and not /Cancel? and not /KeepCard!),
AMOUNT = (not /Cancel?):(/Amount? and not /Cancel?),
CANCEL = (/Cancel?):(not /CancelToBank?)*:(/CancelToBank?),
```

```
ACC = (not /Cancel?):(/ACCEPT? and not /Cancel?),
REF =(not /Cancel?):(/REFUSE? and not /Cancel?),
WRONG-CODE = (/Code? and /KeepCard!),
NO-MONEY = (/KeepCard! and /Authorization?),
SUCCESS = (/Authorization? and /DeliverMoney!);
```

The renamings given in the parse-criterion follow the usual regular expression syntax of Auto. For the sake of completeness, we briefly explain the parse-criterion:

The parse criterion is named CR and is a sequence of equations that specifies that any path in the original automata satisfying the right hand side of an equation should be replaced by an edge labeled by the corresponding left hand side in the new automaton. The right hand side of an equation is a regular expression over boolean expressions with ':' being the concatenation operator and '*' being the Kleene-closure operator. Any boolean expression on the right hand side is made up of basic propositions of the form $\backslash a$ where, a, is a signal. An edge satisfies the proposition $\backslash a$ provided the meije action labeling the edge contains signal a.

For example, the following equation

AMOUNT = (not /Cancel?):(/Amount? and not /Cancel?),

specifies that the new automaton has an edge labeled 'AMOUNT' in the place of a path in the old automaton, that consists of two edges, the label of the first edge not containing signal *Cancel* and the second edge containing *Amount* but not containing *Cancel*.

It is straightforward to see that the abstracted automaton with the parse criterion, CR, given above has the intended behavior of a correct Banker-Teller.

5 Discussion

In the previous sections, we have described a compositional behavioural semantics of CRP with nested network parallelism. The semantics given here differs from the semantics given in [4] is two aspects: (i) we no longer use "decoration of labels" and (ii) the transition labels consist of both *present* and *absent* information. We also do not use the termination levels as it leads to problems in compositional semantics; note that, the termination levels provide a convenient means of encoding traps. The nested parallelism permits hierarchical abstraction. We have also illustrated how the techniques of verification of ESTEREL programs can be adapted for CRP (with the hierarchy operator) programs. Further, through the example of banker-teller described in [4], we have shown how Auto/Autograph can be used for the verification of CRP programs. Since the verification is done at the reactive automata level we can also use it for timed CRP which enables specification of hybrid systems.

It may be noted that having given the semantics of hierarchical parallel operator, it is easy to provide a semantics for creation of objects. In other words, we can obtain a compositional semantics of object-oriented CRP. One of the reasons why we have not given the semantics of objects here is that it is not clear as to how to verify the resultant object CRP programs. It may be pointed out that we could effectively use the techniques of ESTEREL verification for the hierarchical CRP programs, as we could map the system into a set of interacting reactive automata expressible in terms of the powerful *meije* calculus [10] on which the verification tools have been built. However, with the additional notion of objects, we cannot use *meije* calculus

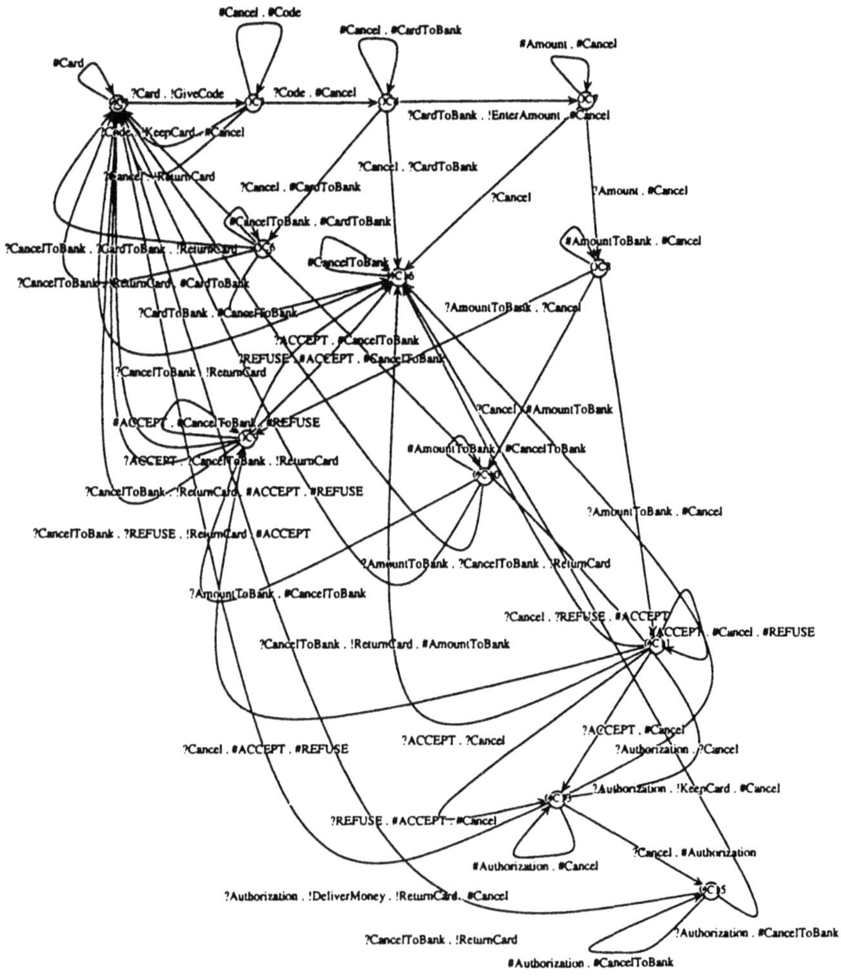

Fig. 1. Combined Automata

as the underlying calculus and hence, we have not been able to effectively use the verification techniques. We are also working[5] towards an abstraction of ESTEREL with multiple clocks with the idea that overcoming some of the drawbacks of the current module definitions of ESTEREL and also to provide a integrated formalism for perfect synchrony, asynchrony and objects.

Acknowledgments

It is great pleasure to thank Professors G. Berry, F. Boussinot and R. de Simone of CMA, Sophia Antipolis for invaluable discussions.

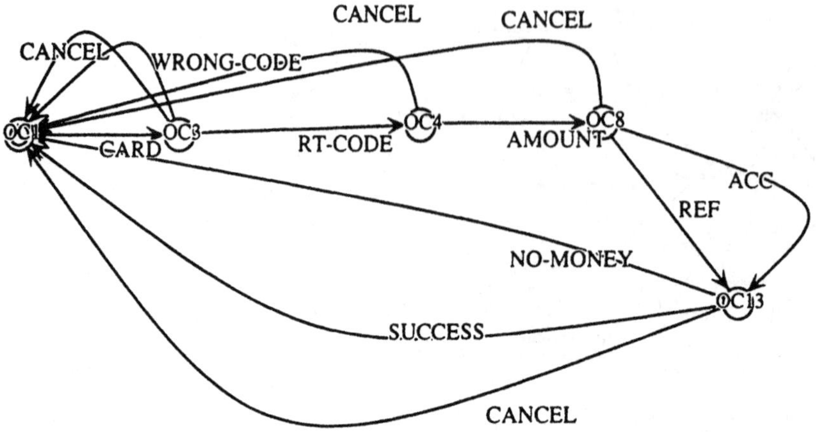

Fig. 2. Abstracted Automata

References

1. G. Berry (1992), *A hardware implementation of pure Esterel*, Sadhana: Academy Proceedings in Engineering Sciences, Indian Academy of Sciences, Special Issue on Real Time Systems, edited by RK Shyamasundar, 17 (1):95-139, 1992.
2. G. Berry and G. Gonthier, *The Esterel synchronous programming language: Design, semantics, Implementation*, SCP, Vol. 19, No.2, Nov. 92, pp. 87-152.
3. G. Berry, *Causality in* ESTEREL *Programs*, Dagstuhl Seminar on Synchronous Languages, 28 Nov - Dec 2'94.
4. G. Berry, S. Ramesh and R.K. Shyamasundar, *Communicating Reactive Processes*, 20th ACM POPL, South Carolina, Jan. 1993, pp. 85-99.
5. G. Berry, S. Ramesh and R.K. Shyamasundar, *Multi-Clock ESTEREL*, manuscript, December 1994.
6. O. Ploton, *Identifying Causality in* ESTEREL *Programs*, Dagstuhl Seminar on Synchronous Languages, 28 Nov - Dec 2' 94.
7. R.K. Shyamasundar, *Specification of Hybrid Systems in CRP*, Proc. of AMAST 93, Workshops in Computing Series from Springer-Verlag, Edited by M. Nivat, C. Rattray, T. Rus and G. Scollo, pp. 227-238, December 1993.
8. R.K. Shyamasundar, *Specification of Dynamic Real-Time Systems in CRP*, Proc. IFIP 94, Hamburg, Aug. 28 - Sept 2'94.
9. R.K. Shyamasundar and S. Ramesh, *Languages for Recative Specifications: Synchrony vs Asynchrony*, Proc. FTRTFT, Lubeck, LNCS, 863, Sept 1994.
10. R. de Simone, *Higher Level Synchronizing devices in Meije-SCCS*, TCS, 37, pp 245-267, 1985.
11. R. de Simone and A. Ressouche, *Compositional Semantics of ESTEREL and Verification by Compositional Reductions*, Proc. of Computer Aided Verification 94, June 94.
12. V. Roy and R. de Simone, *Auto and Autograph*, In R. Kurshan, editor, *proceedings of Workshop on Computer Aided Verification*, New-Brunswick, June 1990.

Interface and Controller Design for Hybrid Control Systems

James A. Stiver, Panos J. Antsaklis, and Michael D. Lemmon

Department of Electrical Engineering
University of Notre Dame, Notre Dame, IN 46556

Abstract. The hybrid control systems considered here consist of a continuous-time plant under the control of a discrete event system. Communication between the plant and controller is provided by an interface which can convert signals from the continuous domain of the plant to the discrete, symbolic domain of the controller, and vise-versa. When designing a controller for a hybrid system, the designer may or may not be free to design the interface as well. This paper examines these two cases. First, a methodology is presented for designing a controller when the interface and plant are given. This approach is based on the methodology for controller design in logical discrete event systems. Second, a method is presented to design both the interface and controller. This approach is based on the natural invariants of the system.

1 Introduction

The hybrid control systems considered in this paper consist of three chief components: a continuous-time plant, a discrete event system (DES) controller, and an interface. This work uses a modeling framework for hybrid control previously developed by the authors [1, 2, 3, 4, 5, 6, 7, 8, 9, 10]. Hybrid systems in general have attracted significant interest in recent years. Efforts include the work by Nerode, Kohn, et al. [11, 12, 13], Brockett [14], Ramadge, et al. [15, 16, 17], Varaiya, et al. [18, 19] and Tittus, et al. [20, 21]. Articles on these and other approaches can be found in [22].

In some of our previous work involving this framework, attention was focused on designing the controller, given a plant and interface [6, 8, 9]. In particular, a discrete event system, called the DES plant model, was developed to model the combined plant and interface. Then existing techniques for the design of discrete event system controllers were extended to design controllers for DES plant models.

Later work has focused on designing the interface as well as the controller [7, 10]. In this case the goal is to develop a method to design the interface and controller for a hybrid system when only the plant and control goals are given. The interface is designed to distinguish regions of the plant state space based on where the trajectories lead for a given control policy. Subsets of these regions, called *common flow regions*, are identified and then bounded using invariant manifolds. This provides a means for the system to determine when the state lies in such a region and to apply the appropriate control policy.

In this paper, we present both of these methods. First the modeling framework is described and the examples, which are used throughout, are presented. In Section

3, we describe controller design techniques for cases in which the plant and interface are both specified. This is refered to as the *logical approach*, because it is based on techniques developed for the control of logical discrete event systems. Then in the Section 4, the *invariant based approach* is presented. This is a method to design the interface using the invariants of the system, and the method leads directly to a controller design as well.

2 Hybrid Control System Modeling

A hybrid control system, can be divided into three parts, the plant, interface, and controller as shown in Figure 1. In this model, the plant represents the continuous-time components of the system, while the controller represents the discrete-event portions. The interface provides the necessary mechanism by which the former two communicate. The models used for each of these three parts, as well as the way they interact are now described.

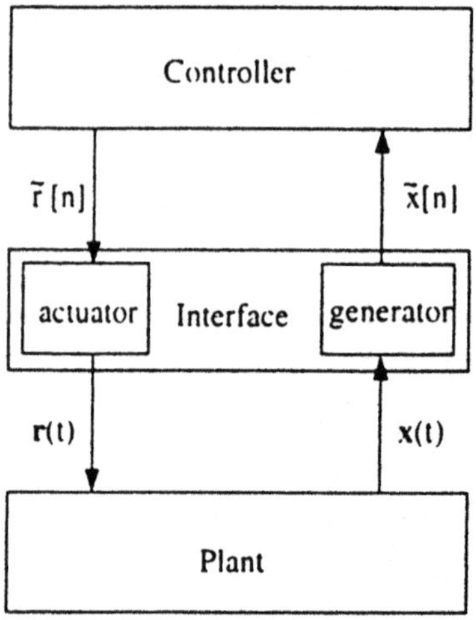

Fig. 1. Hybrid Control System

2.1 Plant

The plant is the part of the model which represents the entire continuous-time portion of the hybrid control system. The distinguishing feature of the plant is that it has a continuous state space, where the state takes on values that are real numbers, and evolves with time according to a set of differential equations. Motivated by tradition, this part of the model is referred to as the plant but since it contains all the continuous dynamics, it can also contain a conventional, continuous-time, controller.

Mathematically, the plant is represented by the equation

$$\dot{\mathbf{x}}(t) = f(\mathbf{x}(t), \mathbf{r}(t)) \tag{1}$$

where $\mathbf{x}(t) \in \Re^n$ and $\mathbf{r}(t) \in \Re^m$ are the state and input vectors respectively. $f : \Re^n \times \Re^m \to \Re^n$ is a continuous function which satisfies the Lipschitz conditions, thus guaranteeing the existence and uniqueness of its solutions. Note that the plant input and state are continuous-time vector valued signals. Boldface letters are used here to denote vectors and vector valued signals.

2.2 Controller

The controller is a discrete event system which is modeled as a deterministic automaton. This automaton is specified by a quintuple, $(\tilde{S}, \tilde{X}, \tilde{R}, \delta, \phi)$, where $\tilde{S}$ is the set of states, $\tilde{X}$ is the set of *plant symbols*, $\tilde{R}$ is the set of *controller symbols*, $\delta : \tilde{S} \times \tilde{X} \to \tilde{S}$ is the state transition function, and $\phi : \tilde{S} \to \tilde{R}$ is the output function. The symbols in set $\tilde{R}$ are called controller symbols because they are generated by the controller. Likewise, the symbols in set $\tilde{X}$ are called plant symbols and are generated based on events in the plant. The action of the controller is described by the equations

$$\tilde{s}[n] = \delta(\tilde{s}[n-1], \tilde{x}[n]) \tag{2}$$

$$\tilde{r}[n] = \phi(\tilde{s}[n]) \tag{3}$$

where $\tilde{s}[n] \in \tilde{S}$, $\tilde{x}[n] \in \tilde{X}$, and $\tilde{r}[n] \in \tilde{R}$. The index n is analogous to a time index in that it specifies the order of the symbols in the sequence. The input and output signals associated with the controller are sequences of symbols.

Tildes are used to indicate a symbol valued set or sequence. For example, $\tilde{X}$ is the set of plant symbols and $\tilde{x}[n]$ is the nth symbol of a sequence of plant symbols. Subscripts are also used, e.g. $\tilde{x}_i$ which denotes the ith member of the symbol alphabet $\tilde{X}$.

2.3 Interface

The controller and plant cannot communicate directly in a hybrid control system because each utilizes a different type of signal. Thus an interface is required which can convert continuous-time signals to sequences of symbols and vice versa. The way that this conversion is accomplished determines, to a great extent, the nature of the overall hybrid control system. The interface consists of two simple subsystems, the generator and actuator.

Plant Events and the Generator The *generator* is the subsystem of the interface which converts the continuous-time output (state) of the plant to an asynchronous, symbolic input for the controller. To perform this task, two processes must be in place. First, a triggering mechanism is required that will determine *when* a plant symbol should be generated, and second, a process is required to determine *which* particular plant symbol should be generated.

In the generator, the triggering mechanism is based on the idea of *plant events*. A plant event is simply an occurrence in the plant, an idea borrowed from the field of discrete event systems. For the hybrid control systems studied here, a plant event is defined by a hypersurface that separates the plant state space into two open regions. A plant event occurs whenever the plant state crosses its associated hypersurface in a given direction.

The set of plant events recognized by the generator is given by a set of smooth functionals, $\{h_i : \Re^n \to \Re, i \in I\}$, defined on the state space of the plant. Each functional must satisfy the condition,

$$\nabla_x h_i(\xi) \neq 0, \forall \xi \in \mathcal{N}(h_i), \tag{4}$$

where ∇_x denotes the gradient with respect to x. This condition ensures that the null space of the functional, $\mathcal{N}(h_i) = \{\xi \in \Re^n : h_i(\xi) = 0\}$, forms an $n-1$ dimensional smooth hypersurface separating the state space.

A plant event occurs whenever the state crosses a hypersurface, as given by the condition

$$\exists i \in I \text{ s.t. } h_i(\mathbf{x}(t)) = 0, \frac{d}{dt} h_i(\mathbf{x}(t)) \neq 0 \tag{5}$$

Notice that the condition stated above is true whenever a hypersurface is crossed, regardless of the direction.

The sequence of plant events is denoted as $\epsilon[n]$, where $\epsilon[n] = i$ indicates the nth event occurred by crossing the hypersurface, h_i. The sequence of plant event instants, that is the times at which the events occurred, is expressed as $\tau[n]$. These sequences are defined as follows,

$$\begin{aligned} \tau[0] &= 0 \\ \epsilon[0] &= 0 \end{aligned} \tag{6}$$

$$\begin{aligned} \tau[n] = \inf\{t \geq \tau[n-1] : \exists \epsilon[n] = \min\{i \in I : h_i(\mathbf{x}(t)) = 0 \\ \wedge \tfrac{d}{dt} h_i(\mathbf{x}(t)) \neq 0 \wedge (t > \tau[n-1] \vee i \neq \epsilon[n-1])\}\} \end{aligned}$$

Perhaps a bit of explanation is required for the above equation. The sequence of plant events, $\epsilon[n]$, is ordered according to the time at which the events occur, and for events occurring simultaneously, the order is determined by the value of i. So, for example, if h_2 and h_3 are crossed at the same time, then $e[n] = 2$ and $e[n+1] = 3$.

The generator must also determine which plant symbol will be generated when a plant event occurs. The plant symbol notifies the controller that a plant event has occurred. It can also reflect which particular plant event has occurred and provide information about the current value of the state. This is modeled by a set of *plant symbol generating functions*, one for each hypersurface, which maps the state, at the time of the event, to a plant symbol. In the case of a silent event, the plant symbol generating function maps to the *null symbol*, which the controller will not recognize as a plant symbol.

The sequence of plant symbols is defined as

$$\tilde{x}[n] = \begin{cases} \alpha_i(\mathbf{x}(\tau_e[n])) & \text{if } \quad \frac{d}{dt} h_{e[n]}(\mathbf{x}(\tau[n])) < 0 \\ \epsilon & \text{otherwise} \end{cases} \tag{7}$$

The ϵ indicates the null symbol which is issued when the hypersurface is crossed in the direction opposite of the way the plant event was defined. This is discussed further below.

Several interesting issues arise in considering this mechanism for defining plant events. One issue is the question of whether the plant events should be "one-sided" or "two-sided", that is, should a plant event occur when the hypersurface is crossed in only one, or in either direction. The advantage of one-sided events is that they make the model more general, the disadvantage is that they cause complications when the model is used for analysis (some of these complications will be noted in remarks later in this paper). The solution adopted here is to defined the events as two-sided, but to only recognize the event when the hypersurface is crossed in a defined direction. When the hypersurface is crossed in the other direction, the event is called a *silent event*. Thus, the model can handle one-sided events without the complications mentioned above.

The Actuator The actuator converts the sequence of controller symbols to a plant input signal, using the function $\gamma : \tilde{R} \to \Re^m$, as follows.

$$\mathbf{r}(t) = \sum_{n=0}^{\infty} \gamma(\tilde{r}[n]) I(t, \tau[n], \tau[n+1]) \tag{8}$$

where $I(t, \tau_1, \tau_2)$ is a characteristic function taking on the value of unity over the time interval $[\tau_1, \tau_2)$ and zero elsewhere. $\tau[n]$ is the time of the nth control symbol which is defined in equation 6.

The plant input, $\mathbf{r}(t)$, can only take on certain constant values, where each value is associated with a particular controller symbol. Thus the plant input is a piecewise constant signal which may change only when a controller symbol occurs. We refer to the various inputs as *control policies*. Each controller symbol initiates a particular control policy.

2.4 Example - Thermostat

Consider a system made up of a thermostat, room, and heater. If the thermostat is set at 70°F, and assuming it is colder outside, the system behaves as follows. If the room temperature falls below 70 degrees the heater starts and remains on until the room temperature exceeds 75 degrees at which point the heater shuts off. Note that the actual temperature settings in a real system may be different. For simplicity, we will assume that when the heater is on it produces heat at a constant rate.

The plant in this hybrid control system is made up of the heater and room, and it can be modeled with the following differential equation.

$$\dot{\mathbf{x}}(t) = .0025(T_O - \mathbf{x}(t)) + .02\mathbf{r}(t) \tag{9}$$

Here $\mathbf{x}(t)$ is the room temperature, T_O is the outside temperature, and $\mathbf{r}(t)$ is the voltage into the heater. Temperatures are in degrees Fahrenheit and time is in minutes.

The generator and controller are found in the thermostat. The generator partitions the state space with two hypersurfaces.

$$h_1(\mathbf{x}) = \mathbf{x} - 70 \tag{10}$$

$$h_2(\mathbf{x}) = -\mathbf{x} + 75 \tag{11}$$

The first hypersurface detects when the temperature falls below 70°F and the second detects when the temperature rises above 75°F. The events are represented symbolically to the controller.

$$\alpha_1(\xi) = cold \tag{12}$$

$$\alpha_2(\xi) = hot \tag{13}$$

It is common to see bimetallic strips performing this function in an actual thermostat, where the band is physically connected to the controller. The controller has two states (typically it is just a switch in the thermostat) as illustrated in Figure 2. The output function of the thermostat controller provides two controller symbols,

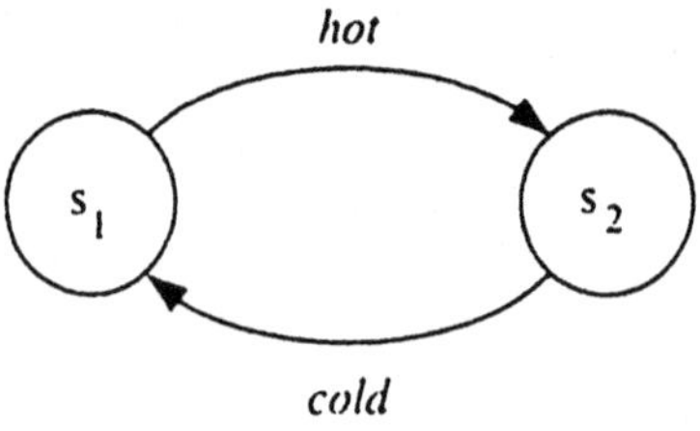

Fig. 2. Controller in Thermostat

on and *off*.

$$\phi(\tilde{s}_1) = on \qquad\qquad \phi(\tilde{s}_2) = off \tag{14}$$

Finally the actuator converts the symbolic output of the controller to a continuous input for the plant.

$$\gamma(on) = 110 \qquad\qquad \gamma(off) = 0 \tag{15}$$

In this case the plant input is the voltage supply to the heater. 0 or 110 volts. Physically, the symbolic output from the controller could be a low voltage signal, say 0 or 12 volts, or perhaps a pneumatic signal.

2.5 Example - Double Integrator

The following simple example will be used throughout the paper to illustrate the work. The system consists of a double integrator plant which is controlled by a discrete event system. The control goal is to drive the state of the plant to the region of the origin.

First, the plant and interface will be presented. The plant is given by the differential equation.

$$\dot{\mathbf{x}}(t) = \begin{bmatrix} 0 & 1 \\ 0 & 0 \end{bmatrix} \mathbf{x}(t) + \begin{bmatrix} 0 \\ 1 \end{bmatrix} \mathbf{r}(t) \tag{16}$$

The generator recognizes four plant events which occur when the following hypersurfaces are crossed.

$$h_1(\mathbf{x}) = x_1 \qquad h_2(\mathbf{x}) = -x_1 \tag{17}$$

$$h_3(\mathbf{x}) = x_1 + 10x_2 \; h_4(\mathbf{x}) = -x_1 - 10x_2 \tag{18}$$

Two of these hypersurfaces lie on the x_2 axis and the other two lie on a line of slope 0.1 passing through the origin. There are four events rather than two so that crossings can be detected in both directions for each hypersurface. Symbols are attached to the plant events as follows.

$$\alpha_1(\mathbf{x}) = \tilde{x}_1 \; \alpha_2(\mathbf{x}) = \tilde{x}_1 \tag{19}$$

$$\alpha_3(\mathbf{x}) = \tilde{x}_2 \; \alpha_4(\mathbf{x}) = \tilde{x}_2 \tag{20}$$

Notice that the same symbol can be used to label more than one plant event and that the value of the mapping α_i does not depend on the state, $\mathbf{x}(t)$, in this case. In this example the plant symbol only identifies which hypersurface was crossed. Figure 3 illustrates this.

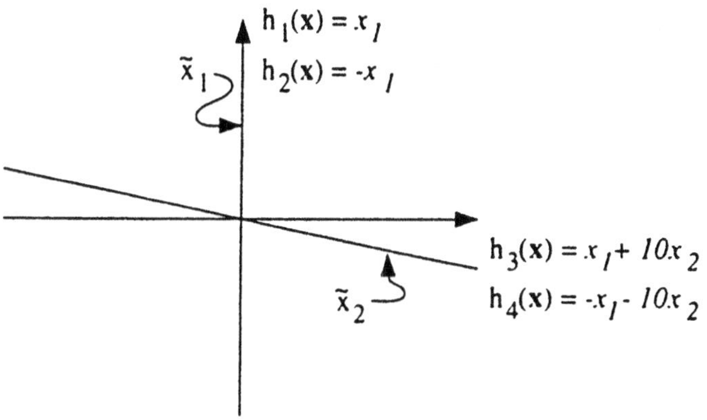

Fig. 3. Generator for Double Integrator Example

The actuator provides three possible inputs to the plant.

$$\gamma(\tilde{r}) = \begin{cases} -1 & \text{if } \tilde{r} = \tilde{r}_1 \\ 0 & \text{if } \tilde{r} = \tilde{r}_2 \\ 1 & \text{if } \tilde{r} = \tilde{r}_3 \end{cases} \tag{21}$$

These inputs were chosen so that the plant can be driven to the origin by applying them in the proper sequence.

Finally, the controller is the four state automaton pictured in Figure 4. The ouput function of the controller is the following.

$$\phi(\tilde{s}_1) = \tilde{r}_1 \; \phi(\tilde{s}_2) = \tilde{r}_2 \tag{22}$$

$$\phi(\tilde{s}_3) = \tilde{r}_2 \; \phi(\tilde{s}_4) = \tilde{r}_3 \tag{23}$$

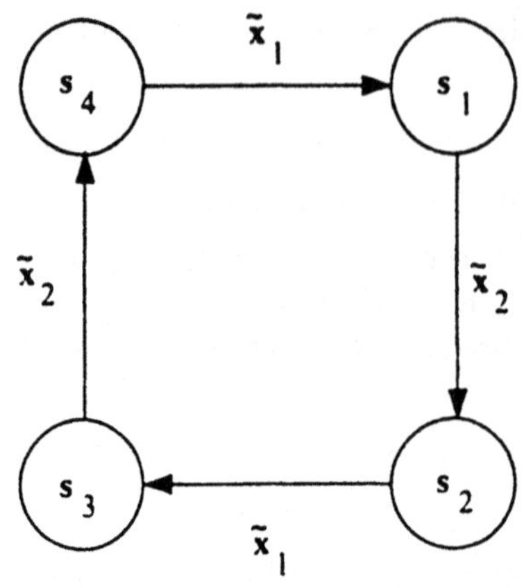

Fig. 4. Controller for Double Integrator Example

The controller in this example was designed to drive the state of the double integrator to the origin, and Figure 5 shows that this goal is indeed achieved. However, the design was adhoc. The following sections will show a more systematic method of designing an interface and controller. This example will be used again there.

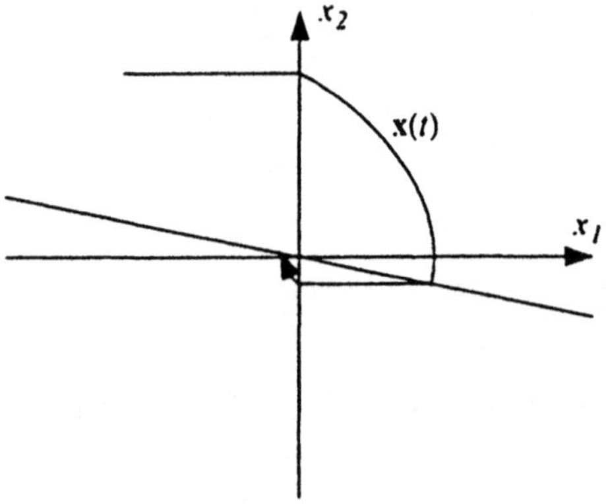

Fig. 5. State Space Trajectory for Double Integrator Example

3 Logical Approach

3.1 DES Plant Model

In a hybrid control system, the plant taken together with the actuator and generator, behaves like a discrete event system. It accepts symbolic inputs via the actuator and produces symbolic outputs via the generator. This situation is somewhat analogous to the way a continuous-time plant, equipped with a zero order hold and a sampler, "looks" like a discrete-time plant. In a hybrid control system, the DES which models the plant, actuator, and generator is called the *DES plant model*. From the DES controller's point of view, it is the DES plant model which is controlled.

It must be pointed out that the DES plant model is an approximation of the actual plant-actuator-generator combination. Since the DES plant model has a discrete state space, it cannot model the exact behavior of a system which has a continuous state space. The exact relationship between the two will be discussed after the description of the DES plant model.

The DES plant model is an automaton, represented mathematically by a quintuple, $(\tilde{P}, \tilde{X}, \tilde{R}, \psi, \lambda)$. $\tilde{P}$ is the set of states, $\tilde{X}$ is the set of plant symbols, and $\tilde{R}$ is the set of control symbols. $\psi : \tilde{P} \times \tilde{R} \rightarrow 2^{\tilde{P}}$ is the state transition function, for a given DES plant state and a given control symbol, it specifies which DES plant states are enabled. The output function, $\lambda : \tilde{P} \times \tilde{P} \rightarrow 2^{\tilde{X}}$, maps the previous and current state to a set of plant symbols.

The set of DES plant states, $\tilde{P}$, is based upon the set of hypersurfaces realized in the generator. Each open region in the state space of the plant, bounded by hypersurfaces, is associated with a state of the DES plant. Whenever a plant event occurs there is a state transition in the DES plant. Stating this more rigorously, an equivalence relation, $\equiv_P$, can be defined on the set $\{\xi \in \Re^n : h_i(\xi) \neq 0, i \in I\}$ as follows

$$\xi_1 \equiv_P \xi_2 \text{ iff } h_i(\xi_1) h_i(\xi_2) > 0, \forall i \in I. \tag{24}$$

Each of the equivalence classes of this relation is associated with a unique DES plant state. Thus it is convenient to index the set of states, $\tilde{P}$, with a binary vector, $b \in \mathbf{B}^I$, such that b_i is the ith element of b and $\tilde{p}_b$ is associated with the set $\{\xi \in \Re^n : b_i = 1 \Leftrightarrow h_i(\xi) < 0\}$. The equivalence relation is not defined for states which lie on the hypersurfaces. When the continuous state touches a hypersurface the DES plant model remains in its previous state until the hypersurface is crossed.

Formally, the set of DES plant states is defined as a set of equivalence classes on the state space of the plant.

Definition 1. The set of DES plant states, $\tilde{P}$, is defined as follows.

$$\tilde{P} = \{\xi \in \Re^n : h_i(\xi) \neq 0, i \in I\} / \equiv_P \tag{25}$$

So, for example, the state $\tilde{p}_b$ is defined as

$$\tilde{p}_b = \{\xi \in \Re^n : b_i = 0 \Rightarrow h_i(\xi) > 0 \text{ and } b_i = 1 \Rightarrow h_i(\xi) < 0\} \tag{26}$$

Now the DES plant state can be defined for a system.

Definition 2. The *DES plant state*, $\tilde{p}[n]$, is defined as follows.

$$\tilde{p}[n] = \tilde{p}_b \tag{27}$$

where

$$\lim_{\epsilon \to 0^+} \mathbf{x}(\tau[n] + \epsilon) \in \tilde{p}_b \tag{28}$$

So the current state of the DES corresponds to the most recently entered region of the plant state space. The limit must be used because at exactly $\tau[n]$ the continuous state will be on a boundary.

The reason for this definition of state for the DES plant model is that it represents how much can be known about the system by observing the plant symbols without actually calculating the trajectories. So after a plant symbol is generated nothing can be ascertained beyond the resulting region.

Now we are in a position to determine the state transition function, ψ, and the output function, λ. First we define adjacency for DES plant states.

Definition 3. Two DES plant states, $\tilde{p}_b, \tilde{p}_c$, are *adjacent* at $(i \in I, \xi \in \mathcal{N}(h_i))$ if for all $j \in I$,

$$\mathcal{N}(h_j) = \mathcal{N}(h_i) \Rightarrow b_j \neq c_j$$

$$\xi \in \overline{\tilde{p}_b} \cap \overline{\tilde{p}_c}.$$

where $\overline{\tilde{p}_b}$ represents the closure of $\tilde{p}_b$.

When two DES plant states are adjacent at (i, ξ) it means that the regions corresponding to these states are separated by the hypersurface $\mathcal{N}(h_i)$, and the point ξ lies on this hypersurface on the boundary of both regions. Thus ξ identifies a possible transition point between the regions.

The following proposition states that for a given DES plant state, $\tilde{p}_b$, and control symbol, $\tilde{r}_k$, a possible successor state is $\tilde{p}_c$ if the stated conditions are met.

Proposition 4. *Given a hybrid control system, described by (1) - (8), with f and h_i smooth, if $\exists i \in I$ and $\xi \in \mathcal{N}(h_i)$ such that following conditions are satisfied,*

- *$\tilde{p}_b$ and $\tilde{p}_c$ are adjacent at (i, ξ),*
- $b_i = 0 \Rightarrow \nabla_x h_i(\xi) \cdot f(\xi, \gamma(\tilde{r}_k)) < 0$
- $b_i = 1 \Rightarrow \nabla_x h_i(\xi) \cdot f(\xi, \gamma(\tilde{r}_k)) > 0$

then $\tilde{p}_c \in \psi(\tilde{p}_b, \tilde{r}_k)$.

Proof. Assume there exists $(i \in I, \xi \in \mathcal{N}(h_i))$ which satisfy the proposition for some $\tilde{p}_b, \tilde{p}_c$, and $\tilde{r}_k$. Consider a trajectory, $\mathbf{x}$, such that at time t, $\mathbf{x}(t) = \xi$ and $\dot{\mathbf{x}}(t) = f(\mathbf{x}(t), \gamma(\tilde{r}_k))$. By the adjacency assumption, we know that $\mathbf{x}(t) \in \overline{\tilde{p}_b}$ and along with the other two conditions of the proposition we know that $\mathbf{x}(t^-) \in \tilde{p}_b$. The adjacency assumption also means that $\mathbf{x}(t) \in \overline{\tilde{p}_c}$ and along with the other two conditions of the proposition, we know that $\mathbf{x}(t^+) \in \tilde{p}_c$. So therefore there is a state transition at time t from $\tilde{p}_b$ to $\tilde{p}_c$ with the control symbol $\tilde{r}_k$.

The usefulness of this proposition is that it allows the extraction of a DES automaton model of the continuous plant and interface. Note that in certain cases this is a rather straightforward task. For instance, it is known that if a particular region boundary is only crossed in one direction under a given command, then the conditions of the proposition need only be tested at a single point on the boundary. This condition is true for the double integrator example which follows. In general this may not be the case, but one can restrict the area of interest to an operating region of the plant state space thus reducing the computation required.

The output function, λ, can be found by a similar procedure described in the next proposition.

Proposition 5. *Given a hybrid control system described by (1) - (8), with f and h_i smooth, $\tilde{x}_\ell \in \lambda(\tilde{p}_b, \tilde{p}_c)$ if and only if $\exists (i, \xi)$ which satisfies Proposition 4 for some $\tilde{r}_k$ and such that $\alpha_i(\xi) = \tilde{x}_\ell$.*

Proof. This proposition follows immediately from the definition of the generator. In particular, the plant symbol generated by a plant event is defined as $\alpha_i(\xi)$ where ξ is the continuous-time plant state at the time of the plant event.

3.2 Example - Thermostat

The thermostat/heater example has a simple DES plant model which is useful to illustrate how these models work. Figure 6 shows the DES plant model for the heater/thermostat. The convention for labeling the arcs is to list the controller symbols which enable the transition followed by a "/" and then the plant symbols which can be generated by the transition. Notice that two of the transitions are labeled with null symbols, ϵ. This reflects the fact that nothing actually happens in the system at these transitions. When the controller receives a null symbol it remains in the same state and reissues the current controller symbol. This is equivalent to the controller doing nothing, but it serves to keep all the symbolic sequences, $\tilde{s}, \tilde{p}$, etc., in phase with each other.

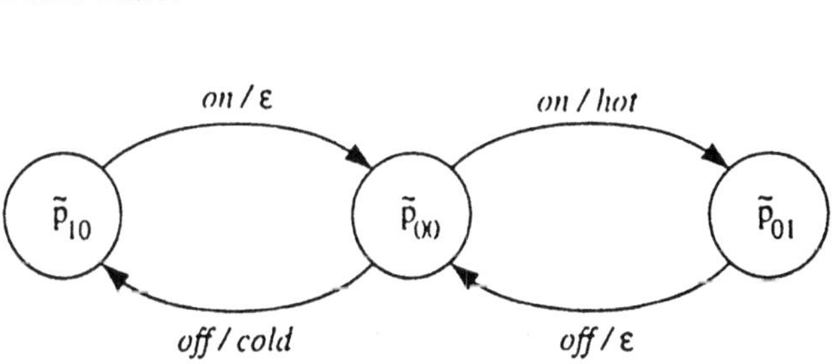

Fig. 6. DES Plant for Thermostat/Heater

3.3 Example - Double Integrator

Now we return to the double integrator example from Section 2. Using Proposition 4, we can extract the DES plant for this system. It is shown in Figure 7. To illustrate how the DES plant was extracted start with the DES plant state $\tilde{p}_9$ (i.e. $\tilde{p}_{1001}$) and consider whether $\tilde{p}_5 \in \psi(\tilde{p}_9, \tilde{r}_2)$. $i = 1$ and $\xi = [0\ 1]'$ satisfy the conditions of the proposition, showing that indeed $\tilde{p}_5 \in \psi(\tilde{p}_9, \tilde{r}_2)$. Proceeding in this way we extract the DES plant model. At the same time, Proposition 5 is used to find the plant symbols generated by the transitions. In the sample instance, $\lambda(\tilde{p}_9, \tilde{p}_5)$ there are two possible symbols, $\tilde{x}_1$ and ϵ. By convention the nonsilent symbol takes precedence so $\{\tilde{x}_1\} = \lambda(\tilde{p}_9, \tilde{p}_5)$.

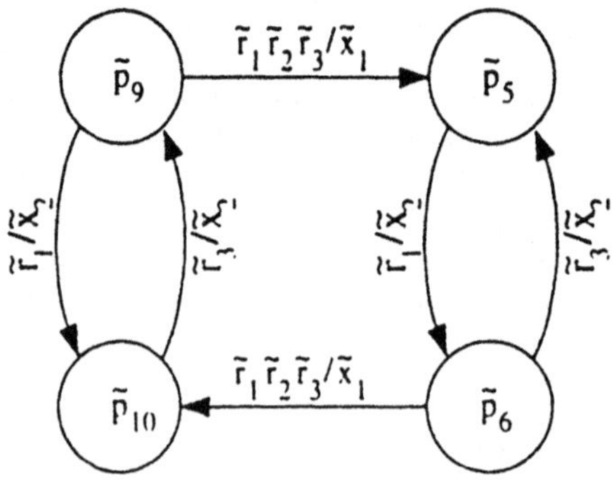

Fig. 7. DES Plant for Double Integrator

Now that the plant and interface have been converted to a discrete event system, techniques for controller design from that area can be applied.

3.4 Logical Approach to DES Control

In this section, we use the language generated by the DES plant to examine the controllability of the hybrid control system. This work builds upon the work done by Ramadge and Wonham on the controllability of discrete event systems in a logical framework [23, 24, 25, 26, 27]. Here we adapt several of those results and apply them to the DES plant model obtained from a hybrid control system.

Before existing techniques, developed in the logical DES framework can be extended, certain differences must be dealt with. The Ramadge-Wonham model (RWM) consists of two interacting DES's called here the *RWM generator* and *RWM supervisor*. The RWM generator is analogous to our DES plant and the RWM supervisor is analogous to the DES controller. The RWM generator shares its name with the generator found in the hybrid control system interface but the two should not be confused. In the RWM, the plant symbols are usually referred to as "events", but we will continue to call them plant symbols to avoid confusion. The plant symbols in the RWM are divided into two sets, those which are controllable and those which are uncontrollable: $\tilde{X} = \tilde{X}_c \cup \tilde{X}_u$. A plant symbol being controllable means that the supervisor can prevent it from being issued by the RWM generator. When the supervisor prevents a controllable plant symbol from being issued, the plant symbol is said to be *disabled*. The plant symbols in $\tilde{X}_c$ can be individually disabled, at any time and in any combination, by a command from the RWM supervisor, while the plant symbols in $\tilde{X}_u$ can never be disabled. This is in contrast to our DES plant where each command (controller symbol) from the DES controller disables a particular subset of $\tilde{X}$ determined by the complement of the set given by the transition function, ψ. Furthermore, this set of disabled plant symbols depends not only on the controller symbol but also the present state of the DES plant. In addition, there is no guarantee that any arbitrary subset of $\tilde{X}$ can be disabled while the other plant symbols remain enabled.

The general inability to disable plant symbols individually is what differentiates the DES plant model, in the hybrid system context, from the automata of earlier frameworks.

3.5 The DES Plant Language and Observability

The behavior of a DES can be characterized by the set of all finite sequences of symbols which it can generate. This set is referred to as the language of the DES, and is denoted L. Given the set of all plant symbols, $\tilde{X}$, the alphabet, $\tilde{X}^*$, refers to all finite sequences of symbols from the alphabet. The language, L, is a subset of $\tilde{X}^*$. The following defines which strings, $\tilde{x}$, are in the language of a given DES plant model.

Definition 6. Given a finite sequence of plant symbols, $\tilde{x} : \mathbf{N} \rightarrow \tilde{X}$, defined over the set $\mathbf{N} = \{1, ..., N\}$, then $\tilde{x} \in L$ if there exists $\tilde{p} \in \tilde{P}^*$ and $\tilde{r} \in \tilde{R}^*$, such that the following hold.

$$\tilde{p}[n+1] \in \psi(\tilde{p}[n], \tilde{r}[n]) \; \forall n \in \mathbf{N} \tag{29}$$

$$\tilde{x}[n] \in \lambda(\tilde{p}[n-1], \tilde{p}[n]) \; \forall n \in \mathbf{N} \tag{30}$$

The language of a DES plant model may or may not provide a useful feedback signal to the controller. For example, suppose there is only one plant symbol and it is associated with every plant event. The controller would not receive much useful information in such a case. On the other hand, if the language of the DES plant model is sufficiently rich that the current state of the DES plant can be ascertained from its initial state and past output, the output provides more useful feedback.

Definition 7. A DES plant model is *observable* if the current state can be determined uniquely from the previous state and plant symbol. That is, observability means that $\forall \tilde{p}_b, \tilde{p}_c, \tilde{p}_d \in \tilde{P}$ and $\tilde{x}_\ell \in \tilde{X}$, if

$$\tilde{x}_\ell \in \lambda(\tilde{p}_b, \tilde{p}_c)$$

and

$$\tilde{x}_\ell \in \lambda(\tilde{p}_b, \tilde{p}_d)$$

then

$$\tilde{p}_c = \tilde{p}_d.$$

The following proposition follows immediately from the above definition.

Proposition 8. *If a DES plant model is observable, then for any initial state, $\tilde{p}[0]$ and sequence of plant symbols, $\tilde{x} \in L$, produced by the DES, there exists a unique sequence of DES plant states, $\tilde{p}$, capable of producing the sequence, $\tilde{x}$.*

Proof. The definition of observability can be applied iteratively to prove that the each state of the sequence, $\tilde{p}$, is determined uniquely by the previous state and current plant symbol.

In cases where the DES plant model is observable, the above proposition implies the existence of a mapping, $obs : \tilde{P} \times L \to \tilde{P}^*$, which takes an initial state together with a string from the language and maps them to the corresponding sequence of states. The nth state in the sequence, $\tilde{p}[n]$, can also be written, $obs(q_0, \tilde{x})[n]$, where $q_0 \in \tilde{P}$ was the initial state.

3.6 Controllability and Supervisor Design

A DES is controlled by having various symbols disabled by the controller based upon the sequence of symbols which the DES has already generated. When a DES is controlled, it will generate a set of symbol sequences which lie in a subset of its language. If we denote this language of the DES under control as L_c then $L_c \subset L$.

It is possible to determine whether a given RWM generator can be controlled to a desired language [23]. That is, whether it is possible to design a controller such that the RWM generator will be restricted to some target language K. Such a controller can be designed if K is prefix closed and

$$\bar{K}\tilde{X}_u \cap L \subset \bar{K} \tag{31}$$

where $\bar{K}$ represents the set of all prefixes of K. A prefix of K is a sequence of symbols, to which another sequence can be concatenated to obtain a sequence found in K. A language is said to be prefix closed if all the prefixes of that language are also in the language.

When equation 31 is true for a given RWM generator, the desired language K is said to be controllable, and provided K is prefix closed, a controller can be designed which will restrict the generator to the language K. This condition requires that if an uncontrollable symbol occurs after the generator has produced a prefix of K, the resulting string must still be a prefix of K because the uncontrollable symbol cannot be prevented.

Since the DES plant model belongs to a slightly different class of automata than the RWM, we present another definition for controllable language which applies to the DES plant. We assume in this section that we are dealing with observable DES plant models, that all languages are prefix closed, and that q_0 is the initial state.

Definition 9. A language, K, is controllable with respect to a given DES plant if $\forall \tilde{x} \in K$, there exists $\rho \in \tilde{R}$ such that

$$\tilde{x}\lambda(q, \psi(q, \rho)) \subset K, \tag{32}$$

where $q = obs(q_0, \tilde{x})[N]$.

This definition requires that for every prefix of the desired language, K, there exists a control, ρ, which will enable only symbols which will cause string to remain in K.

Proposition 10. *If the language K is controllable according to (9), then a controller can be designed which will restrict the given DES plant to the language K.*

Proof. Let the controller be given by $con : \tilde{X}^* \rightarrow \tilde{R}$ where $con(\tilde{x}) \in \{\rho \in \tilde{R} : \tilde{x}\lambda(q, \psi(q, \rho)) \subset K, q = obs(q_0, \tilde{x})[N]\}$. $con(\tilde{x})$ is guaranteed to be non-empty by (32). We can now show by induction that $\tilde{x} \in L_{con} \Rightarrow \tilde{x} \in K$.

1. $\forall \tilde{x} \in L_f$ such that $|\tilde{x}| = 0$ we have $\tilde{x} \in K$. This is trivial because the only such $\tilde{x}$ is the null string ϵ and $\epsilon \in K$ because K is prefix closed.
2. Let $L_f{}^i = \{\tilde{x} : \tilde{x} \in L_f, |\tilde{x}| = i\}$, that is $L_f{}^i$ is the set of all sequences of length i found in L_f. Given $L_f{}^i$, $L_f{}^{i+1} = \{w \in \tilde{X}^* : w = \tilde{x}\lambda(q, \psi(q, con(\tilde{x})), \tilde{x} \in L_f{}^i\}$. Now with the definition of $con(\tilde{x})$ and (32) we have $L_f{}^i \subset K \Rightarrow L_f{}^{i+1} \subset K$.

So $\tilde{x} \in L_f \Rightarrow w \in K$.

Since the DES plant can be seen as a generalization of the original RWM, the conditions in (32) should reduce to those of (31) under the appropriate restrictions. This is indeed the case.

If the desired language is not attainable for a given DES, it may be possible to find a more restricted language which is. If so, the least restricted behavior is desirable. [23] and [26] describe and provide a method for finding this behavior which is referred to as the *supremal controllable sublanguage*, $K^\uparrow$, of the desired language. The supremal controllable sublanguage is the largest subset of K which can be attained by a controller. $K^\uparrow$ can be found via the following iterative procedure.

$$K_0 = K \tag{33}$$
$$K_{i+1} = \{w : w \in K, \bar{w}X_u \cap L \subset \overline{K_i}\} \tag{34}$$
$$K^\uparrow = \lim_{i \to \infty} K_i \tag{35}$$

Once again, this procedure applies to the RWM. For hybrid control systems, the supremal controllable sublanguage of the DES plant can be found by a similar iterative scheme.

$$K_0 = K \tag{36}$$
$$K_{i+1} = \{w \in K : \forall \tilde{x} \in \bar{w}\ \exists\, \rho \in \tilde{R} \text{ such that } \tilde{x}\lambda(q, \psi(q, \rho)) \subset K_i\} \tag{37}$$
$$K^\uparrow = \lim_{i \to \infty} K_i \tag{38}$$

This result yields the following proposition.

Proposition 11. *For a DES plant and language K, $K^\uparrow$ is controllable and contains all controllable sublanguages of K.*

Proof. From (37) we have

$$K^{\uparrow} = \{w \in K : \forall \tilde{x} \in \tilde{w}\ \exists\, \rho \in \tilde{R} \text{ such that } \tilde{x}\lambda(q, \psi(q, \rho)) \subset K^{\uparrow}\} \tag{39}$$

which implies

$$\tilde{x} \in K^{\uparrow} \Rightarrow \exists\, \rho \in \tilde{R} \text{ such that } \tilde{x}\lambda(q, \psi(q, \rho)) \subset K^{\uparrow} \tag{40}$$

From (40) it is clear that $K^{\uparrow}$ is controllable. We prove that every prefix closed, controllable subset of K is in $K^{\uparrow}$ by assuming there exists $M \subset K$ such that M is controllable but $M \not\subset K^{\uparrow}$ and showing this leads to a contradiction.

$$\exists M \subset K \text{ s.t. } M \not\subset K^{\uparrow} \tag{41}$$

$$\Rightarrow \quad \exists w \in M \text{ s.t. } w \notin K^{\uparrow} \tag{42}$$

$$\Rightarrow \quad \exists i \text{ s.t. } w \in K_i,\ w \notin K_{i+1} \tag{43}$$

$$\Rightarrow \quad \exists \tilde{x} \in \tilde{w} \text{ s.t. } \forall \rho \in \tilde{R}.\ \tilde{x}\lambda(q, \psi(q, \rho)) \not\subset K_i \tag{44}$$

$$\Rightarrow \quad \exists w' \in \tilde{x}\lambda(q, \psi(q, \rho)) \text{ s.t. } w' \in M,\ w' \notin K_i \tag{45}$$

$$\Rightarrow \quad \exists j < i \text{ s.t. } w' \in K_j,\ w' \notin K_{j+1} \tag{46}$$

If the sequence is repeated with $i = j$ and $w = w'$ we eventually arrive at the conclusion that $w' \in M$ but $w' \notin K_0$ which violates the assumption that $M \subset K$ and precludes the existence of such an M.

3.7 Example - Double Integrator

We use the double integrator example again because the DES plant was found earlier. This DES is represented by the automaton in Figure 8.

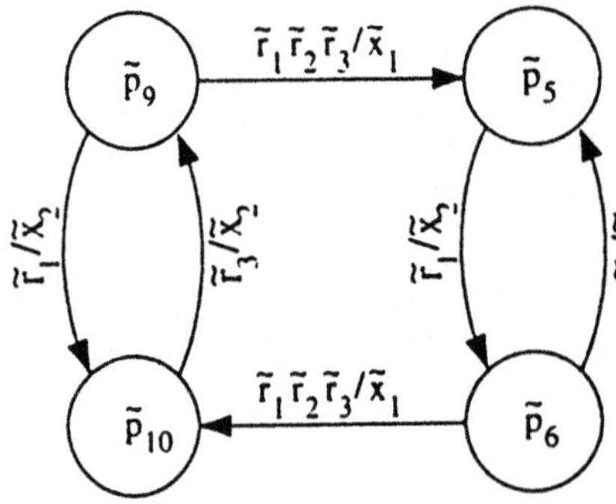

Fig. 8. DES Plant Model for Example 1

Let the initial state be $q_0 = \tilde{p}_5$. Then the language generated by this automaton is $L = \overline{(\tilde{x}_2(\tilde{x}_2\tilde{x}_2)^*\tilde{x}_1)^*}$. If we want to drive the plant in clockwise circles, then the desired language is $K = \overline{(\tilde{x}_2\tilde{x}_1)^*}$. It can be shown that this K is controllable because it satisfies Equation (32). Therefore according to Proposition 10, a controller can be designed to achieve the stated control goal.

3.8 A More Complex DES Plant Model

This example has a richer behavior and will illustrate the generation of a supremal controllable sublanguage as well as the design of a controller. We start immediately with the DES plant model shown in Figure 9.

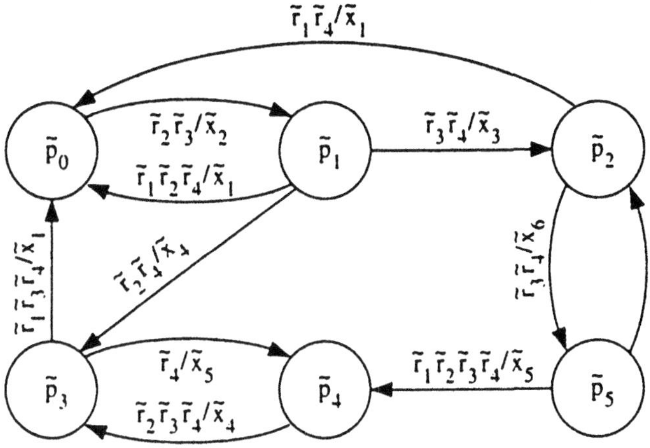

Fig. 9. DES Plant Model for Example 2

The language generated by this DES is $L = \overline{L_m}$ where

$$L_m = (\tilde{x}_2(\tilde{x}_1 + \tilde{x}_4(\tilde{x}_5\tilde{x}_4)^*\tilde{x}_1 + \tilde{x}_3(\tilde{x}_6\tilde{x}_3)^*(\tilde{x}_1 + \tilde{x}_6\tilde{x}_5\tilde{x}_4(\tilde{x}_5\tilde{x}_4)^*\tilde{x}_1)))^* \quad (47)$$

Suppose we want to control the DES so that it never enters state $\tilde{p}_4$. We simply remove the transitions to $\tilde{p}_4$ and then compute the resulting language. This desired language is therefore

$$K = \overline{(\tilde{x}_2(\tilde{x}_1 + \tilde{x}_4\tilde{x}_1 + \tilde{x}_3(\tilde{x}_6\tilde{x}_3)^*\tilde{x}_1))^*} \quad (48)$$

In this example, the language K is not controllable. This can be seen by considering the string $\tilde{x}_2\tilde{x}_3\tilde{x}_6 \in K$, for which there exists no $\rho \in \tilde{R}$ which will prevent the DES from deviating from K by generating $\tilde{x}_5$ and entering state $\tilde{p}_4$.

Since K is not controllable, we find the supremal controllable sublanguage of K as defined in equation (38). The supremal controllable sublanguage is

$$K^\uparrow = K_1 = \overline{(\tilde{x}_2(\tilde{x}_1 + \tilde{x}_4\tilde{x}_1 + \tilde{x}_3\tilde{x}_1))^*} \quad (49)$$

Obtaining a DES controller once the supremal controllable sublanguage has been found is straight forward. The controller is a DES whose language is given by $K^\uparrow$ and the output of the controller in each state, $\phi(\tilde{s})$, is the controller symbol which enables only transitions which are found in the controller. The existence of such a controller symbol is guaranteed by the fact that $K^{\uparrow}$ is controllable. For Example 2, the controller is shown in Figure 10 and its output function, ϕ, is as follows:

$$\phi(\tilde{s}_1) = \tilde{r}_2 \quad \phi(\tilde{s}_2) = \tilde{r}_4 \tag{50}$$

$$\phi(\tilde{s}_3) = \tilde{r}_1 \quad \phi(\tilde{s}_4) = \tilde{r}_1 \tag{51}$$

□

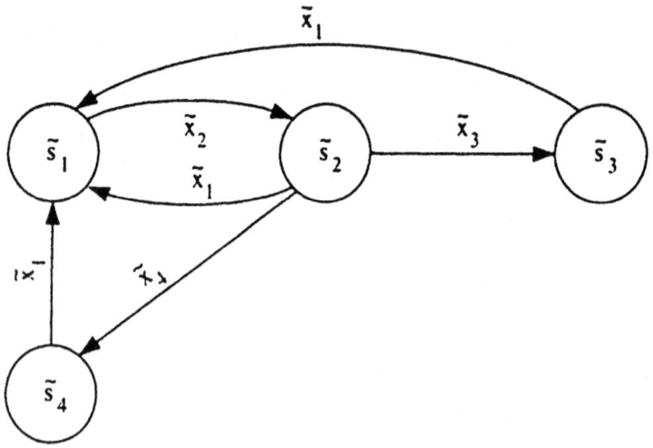

Fig. 10. DES Controller for Example 2

3.9 Remarks

The approach described above is also discussed in detail in [9]. There is, in addition, a discussion of determinism and quasideterminism. The hybrid control system is also extended to include systems with discrete time plants. An example with a nonlinear plant is presented.

For a detailed description of the derivation of a formula for computing the supremal controllable sublanguage of a given language see [28].

4 Invariant Based Approach

If the interface is not given the designer must design both the interface and the controller. One could, of course, design the interface using any technique and then use the logical approach to design the controller. Here a methodology is presented to design the controller and the interface together based on the natural invariants of a plant described by

$$\dot{\mathbf{x}}(t) = f(\mathbf{x}(t), \mathbf{r}(t)) \tag{52}$$

where certain smoothness assumptions apply.

In particular, this section discusses the design of the generator, which is part of the interface, and the design of the controller. We assume that the plant is given, the set of available control policies is given, and the control goals are specified as follows. Each control goal for the system is given as a starting set and a target set, each of which is an open subset of the plant state space. To realize the goal, the controller must be able to drive the plant state from anywhere in the starting set to somewhere in the target set using the available control policies. Generally, a system will have multiple control goals.

To successfully control the plant, the controller must know which control policy to apply and when to apply it. The controller receives all its information about the plant from the generator, and therefore the generator must be designed to provide that information which the controller requires.

We propose the following solution to this design problem. For a given target region, identify the states which can be driven to that region by the application of a single control policy. If the starting region is contained within this set of states, the control goal is achievable via a single control policy. If not, then this new set of states can be used as a target region and the process can be repeated. This will result in a set of states which can be driven to the original target region with no more than two control policies applied in sequence. This process can be repeated until the set of states, for which a sequence of control policies exists to drive them to the target region, includes the entire starting region (provided the set of control policies is adequate as mentioned below).

When the regions have been identified, the generator is designed to tell the controller, via plant symbols, which region the plant state is currently in. The controller will then call for the control policy which drives the states in that region to the target region.

4.1 Generator Design

To describe the regions mentioned above, we use the concept of the flow [29]. Let the flow for the plant (1) be given by $F_k : \mathbf{X} \times \Re \to \mathbf{X}$, where

$$\mathbf{x}(t) = F_k(\mathbf{x}(0), t). \tag{53}$$

The flow represents the state of the plant after an elapsed time of t, with an initial state of $\mathbf{x}(0)$, and with a constant input of $\gamma(\tilde{r}_k)$. Since the plant is time invariant, there is no loss of generality when the initial state is defined at $t = 0$. The flow is defined over both positive and negative values of time. The flow can be extended over time using the forward flow function, $F_k^+ : \mathbf{X} \to \mathbf{P}(\mathbf{X}^n)$, and the backward flow function, $F_k^- : \mathbf{X} \to \mathbf{P}(\mathbf{X}^n)$, which are defined as follows.

$$F_k^+(\xi) = \bigcup_{t \geq 0} \{F_k(\xi, t)\} \tag{54}$$

$$F_k^-(\xi) = \bigcup_{t \leq 0} \{F_k(\xi, t)\} \tag{55}$$

The backward and forward flow functions can be defined on an arbitrary set of states in the following natural way.

$$F_k^+(\mathbf{A}) = \bigcup_{\xi \in \mathbf{A}} \{F_k^+(\xi)\} \tag{56}$$

$$F_k^-(\mathbf{A}) = \bigcup_{\xi \in \mathbf{A}} \{F_k^-(\xi)\} \tag{57}$$

where $\mathbf{A} \subset \mathbf{X}$. For a target region. T. $F_k^-(T)$ is the set of initial states from which the plant can be driven to T with the input $\gamma(\tilde{r}_k)$. In addition. $F_k^+(T)$ is the set of states which can be reached with input $\gamma(\tilde{r}_k)$ and an initial state in T.

Now a generator design procedure can be described using the backward flow function. This is a preliminary procedure. upon which the final design method. developed subsequently. is based. For a given starting region. $S \subset \mathbf{X}$. and target region. $T \subset \mathbf{X}$. use the following algorithm.

1. If $S \subset T$, stop.
2. Identify the regions. $F_k^-(T). \forall \tilde{r}_k \in \tilde{R}$.
3. Let $T = \bigcup_{\tilde{r}_k \in \tilde{R}} F_k^-(T)$
4. Go to 1.

There are two problems associated with this algorithm as stated. First. it will not stop if there is no sequence of available control policies which will achieve the control goal. and second. actually identifying the regions given by the flow functions is quite involved. The first issue is related to the adequacy of the available control policies and will not be dealt with here. The second problem will be addressed. The difficulty in identifying a region given by a flow function is integrating over all the points in the target region. In the generator design procedure developed here. we will concentrate on finding a subset of the region $F_k^-(T)$. rather than the region itself. By definition. all the trajectories passing through $F_k^-(T)$ lead to the target region. T. and therefore all the trajectories found in a subset of $F_k^-(T)$ will also lead to the target.

Here. we will focus on identifying subsets of $F_k^-(T)$ which we call *common flow regions*. Common flow regions are bounded by invariant manifolds and an exit boundary. The invariant manifolds are used because the state trajectory can neither enter nor leave the common flow region through an invariant manifold. The exit boundary is chosen as the only boundary through which state trajectories leave the common flow region.

To design the generator. it is necessary to select the set of hypersurfaces. $\{h_i : \mathbf{X} \to \Re \mid i \in I\}$ and the associated functions. $\{\alpha_i : \mathcal{N}(h_i) \to \tilde{R} \mid i \in I\}$. described in Section 2.3. These hypersurfaces make up the invariant manifolds and exit boundaries mentioned above. as well as forming the boundary for the target region(s).

A target region. T. is specified as

$$T = \{\xi \in \mathbf{X} : \forall i \in I_T, h_i(\xi) < 0\}. \tag{58}$$

where I_T is the index set indicating which hypersurfaces bound the target region. A common flow region. B. is specified as

$$B = \{\xi \in \mathbf{X} : h_i(\xi) < 0, h_e(\xi) > 0, \forall i \in I_B\}. \tag{59}$$

where I_B is an index set indicating which hypersurfaces form the invariant manifolds bounding B and h_e defines the exit boundary for B.

The goal, of course, is that B should include only states whose trajectories lead to the target region. Figure 11 shows an example of this where $I_T = \{1\}$ and $I_B = \{2,3\}$. The target region, T, is surrounded by h_1, the common flow region lies between h_2 and h_3 above the exit boundary, h_e.

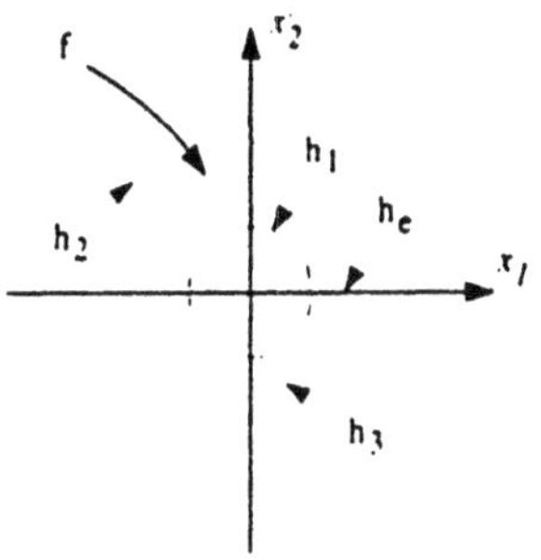

Fig. 11. Target Region and Invariants

We now present two propositions which can be used to determine the suitability of a set of hypersurfaces to achieve our goal of identifying a common flow region. In different situations, one of the propositions may be easier to apply than the other. The following propositions give sufficient conditions for the hypersurfaces bounding B and T to ensure that all state trajectories in B will reach the target region.

Proposition 12. *Given the following:*

1. *A flow generated by a smooth vector field,* f_k
2. *A target region,* $T \subset \mathbf{X}$
3. *A set of smooth hypersurfaces,* $h_i, i \in I_B \subset 2^I$
4. *A smooth hypersurface (exit boundary),* h_e

such that $B = \{\xi \in \mathbf{X} : h_i(\xi) < 0, h_e(\xi) > 0, \forall i \in I_B\} \neq \emptyset$. *For all* $\xi \in B$ *there is a finite time,* t, *such that* $F_k(\xi, t) \in T$, *if the following conditions are satisfied:*

1. $\nabla_\xi h_i(\xi) \cdot f(\xi) = 0, \forall i \in I_B$
2. $\exists \epsilon > 0, \nabla_\xi h_e(\xi) \cdot f(\xi) < -\epsilon, \forall \xi \in B$
3. $B \cap \mathcal{N}(h_e) \subset T$

Proof. The proof of this proposition is straightforward. The first condition of the proposition, which can be rewritten as

$$\frac{dh_i(\mathbf{x}(t))}{dt} = 0, \qquad (60)$$

precludes the state trajectory crossing any hypersurface indexed by the set I_B, thus ensuring no trajectory in B will leave B except through the remaining boundary. The second condition, which can be rewritten as

$$\frac{dh_e(\mathbf{x}(t))}{dt} < -\epsilon, \tag{61}$$

ensures that within a finite time,

$$t < \frac{h_e(\xi)}{\epsilon}, \tag{62}$$

the trajectory at $\xi \in B$ will cross the exit boundary. The final condition guarantees that any trajectory leaving B through the exit boundary will be in the target region when it does so. Together these conditions are sufficient to guarantee that any state in B will enter the target region in finite time.

The second proposition uses a slightly different way of specifying a common flow region. In addition to the invariant manifolds and the exit boundary, there is also a cap boundary. The cap boundary is used to obtain a common flow region which is bounded. So for this case

$$B = \{\xi \in \mathbf{X} : h_i(\xi) < 0, h_e(\xi) > 0, h_c(\xi) < 0, \forall i \in I_B\}. \tag{63}$$

Proposition 13. *Given the following:*

1. *A flow generated by a smooth vector field,* f_k
2. *A target region,* $T \subset \mathbf{X}$
3. *A set of smooth hypersurfaces,* $h_i, i \in I_B \subset 2^I$
4. *A smooth hypersurface (exit boundary),* h_e
5. *A smooth hypersurface (cap boundary),* h_c

such that $B = \{\xi \in \mathbf{X} : h_i(\xi) < 0, h_e(\xi) > 0, h_c(\xi) < 0, \forall i \in I_B\} \neq \emptyset$ *and* $\overline{B}$ *(closure of B) is compact. For all* $\xi \in B$ *there is a finite time,* t*, such that* $F_k(\xi, t) \in T$*, if the following conditions are satisfied:*

1. $\nabla_\xi h_i(\xi) \cdot f(\xi) = 0, \forall i \in I_B$
2. $\nabla_\xi h_c(\xi) \cdot f(\xi) < 0, \forall \xi \in B \cap \mathcal{N}(h_c)$
3. $B \cap \mathcal{N}(h_e) \subset T$
4. *There are no limit sets in* $\overline{B}$

Proof. As in Proposition 12, the first condition precludes the state trajectory crossing any hypersurface indexed by the set I_B, thus ensuring no trajectory in B will leave B except through one of the remaining boundaries. The second condition, which can be rewritten as

$$\frac{dh_c(\mathbf{x}(t))}{dt} < 0, \tag{64}$$

ensures that no trajectory can leave B through the cap boundary. Thus, the exit boundary provides the only available egress from B. The third condition guarantees that any trajectory leaving B through the exit boundary will be in the target region when it does so. The final condition permits the application of a previously known result [30], stating that any state within a compact set without limit sets will leave that compact set in finite time.

Consider the hypersurfaces defined by $\{h_i : i \in I_B\}$. These hypersurfaces must first be invariant under the vector field of the given control policy, f. This can be achieved by choosing them to be integral manifolds of an $n-1$ dimensional distribution which is invariant under f. An $n-1$ dimensional distribution, $\Delta(\mathbf{x})$, is invariant under f if it satisfies

$$[f(\mathbf{x}), \Delta(\mathbf{x})] \subset \Delta(\mathbf{x}), \tag{65}$$

where the $[f(\mathbf{x}), \Delta(\mathbf{x})]$ indicates the Lie bracket. Of the invariant distributions, those that have integral manifolds as we require, are exactly those which are involutive (according to Frobenius). This means

$$\delta_1(\mathbf{x}), \delta_2(\mathbf{x}) \in \Delta(\mathbf{x}) \Rightarrow [\delta_1(\mathbf{x}), \delta_2(\mathbf{x})] \in \Delta(\mathbf{x}). \tag{66}$$

Therefore by identifying the involutive distributions which are invariant under the vector field, f, we have indentified a set of candidate hypersurfaces. For details of these relationships between vector fields and invariant distributions, see [31].

Since an $n-1$ dimensional involutive distribution can be defined as the span of $n-1$ vector fields, over each of which it will then be invariant, and the control policy only gives one vector field, f, there will be more than one family of hypersurfaces which are all invariant under f. The set of all invariant hypersurfaces can be found in terms of $n-1$ functionally independent mappings which form the basis for the desired set of functionals, $\{h_i : i \in I_B\}$. This basis is obtained by solving the characteristic equation

$$\frac{dx_1}{f_1(\mathbf{x})} = \frac{dx_2}{f_2(\mathbf{x})} = \cdots = \frac{dx_n}{f_n(\mathbf{x})} \tag{67}$$

where $f_i(\mathbf{x})$ is the ith element of $f(\mathbf{x})$.

4.2 Controller Design

In previous work using this framework for hybrid control systems, the interface was assumed to be given and the controller was designed using the given plant and interface: see Section 3 and [6, 8, 9]. In those cases, the plant and interface were modeled as a discrete event system, called the *DES plant model*, and existing DES controller design techniques were adapted and used to obtain a controller. The drawback was that there was no guarantee that the desired behavior could be achieved with the given plant and interface.

Now, with the generator design technique described in Section 4.1, the controller design is anticipated by the design of the interface. This represents an improvement over the previous situation because now there is no question that the control goal can be achieved once the interface has been successfully designed, and furthermore the actual controller design has been largely determined by the interface design.

Once the interface has been designed as described in Section 4.1, the design of the controller involves two steps. The first step is to construct one subautomaton for each control goal. This is the step which is already determined by the interface design. The second step is the connection of these subautomata to create a single DES controller. This step will depend upon the order in which the simpler control goals are to be achieved. For example, if a chemical process is to produce a sequence of different products. then each subautomaton in the controller would be designed to produce one of the products, and these subautomata would be connected to produce the products in the desired sequence.

The hypersurfaces in the generator divide the state space of the plant into a number of *cells*. Two states are in the same cell exactly when they are both on the same side (positive or negative) with respect to each hypersurface. States which lie on a hypersurface are not in any cell.

The first step in creating the controller is the contruction of the subautomata, one for each individual control goal. Each subautomaton is constructed in the following way.

i. Create a controller state to represent each cell.
ii. Place transitions between states which represent adjacent cells.
iii. Label each transition with the plant symbol which is generated by the hypersurface separating the associated cells.

We now have a subautomaton which can follow the progress of the plant state as it moves from cell to cell. Next the controller output function must be designed for each subautomaton.

The controller symbol output by a given controller state depends on which common flow region contains the associated cell. Each common flow region was constructed using a specific control policy, and the control symbol which initiates that control policy should be output by controller states representing cells contained in that common flow region. However, in general, common flow regions will overlap, meaning a given cell can lie in more than one common flow region. In such cases treat the cell as lying within the common flow region which is closest to the target region. Distance, in this case, is the number additional control policies which must be used to reach the target region. If common flow regions are both the same distance, then the choice is arbitrary, though the common flow region which is favored in one case must then be favored in all such cases. States which represent cells not contained in any common flow region or target region will never be visited and can thus be deleted.

Once the individual subautomata have been constructed they must be connected to form a single controller. This can be accomplished by following these steps for each subautomaton.

i. Remove the state(s) which represent cells in the target region as well all transitions emanating from such states.
ii. Connect the dangling transitions to states in the subautomaton which achieves the next desired control goal. The connections will be to the states which represent the same cells as the states which were removed.

In this way, as soon as one control goal is achieved, the system will begin working on the next one. The actual order in which each control goals are pursued is up to the designer.

4.3 Example - Double Integrator

Consider the double integrator example from Section 2.5. Suppose we are given the plant,

$$\dot{\mathbf{x}}(t) = \begin{bmatrix} 0 & 1 \\ 0 & 0 \end{bmatrix} \mathbf{x}(t) + \begin{bmatrix} 0 \\ 1 \end{bmatrix} \mathbf{r}(t), \tag{68}$$

three available control policies,

$$\mathbf{r}(t) \in \{-1, 0, 1\}, \tag{69}$$

and the following control goal: drive the plant state to the interior of the unit circle from any initial point. So the starting set consists of the entire state space and the following control goal: drive the plant state to the interior of the unit circle from any initial point. So the starting set consists of the entire state space, and the target set is

$$T = \{\boldsymbol{\xi} \in \mathbf{X} : \xi_1^2 + \xi_2^2 < 1\}. \tag{70}$$

The target set is bounded by the hypersurface given by

$$h_T(\boldsymbol{\xi}) = \xi_1^2 + \xi_2^2 - 1 \tag{71}$$

The first step is to calculate the invariants which can be used to obtain hypersurfaces. There are three families of invariants, one for each of the three control policies.

$$\pm(\xi_1 + \frac{1}{2}\xi_2^2 + c_1) \tag{72}$$

$$\pm(\xi_1 - \frac{1}{2}\xi_2^2 + c_2) \tag{73}$$

$$\pm(\xi_2 + c_3) \tag{74}$$

The first hypersurface, h_1, is used to identify the target region.

$$h_1(\boldsymbol{\xi}) = h_T(\boldsymbol{\xi}) = \xi_1^2 + \xi_2^2 - 1 \tag{75}$$

A tube entering T under the first control policy, $\mathbf{r}(t) = -1$, is bounded by

$$h_2(\boldsymbol{\xi}) = -\xi_1 - \frac{1}{2}\xi_2^2 - .9 \tag{76}$$

$$h_3(\boldsymbol{\xi}) = \xi_1 + \frac{1}{2}\xi_2^2 - .9 \tag{77}$$

and

$$h_e(\boldsymbol{\xi}) = h_4(\boldsymbol{\xi}) = \xi_2 + .1 \tag{78}$$

These hypersurfaces satisfy Proposition 12. Identify this tube as B_1.

$$B_1 = \{\boldsymbol{\xi} : h_i(\boldsymbol{\xi}) < 0, h_4(\boldsymbol{\xi}) > 0, i \in \{2, 3\}\} \tag{79}$$

Likewise, a tube entering T under the third control policy is bounded by

$$h_5(\boldsymbol{\xi}) = -\xi_1 + \frac{1}{2}\xi_2^2 - .9 \tag{80}$$

$$h_6(\boldsymbol{\xi}) = \xi_1 - \frac{1}{2}\xi_2^2 - .9 \tag{81}$$

and

$$h_e(\xi) = h_7(\xi) = -\xi_2 + .1 \tag{82}$$

Identify this tube as B_2.

$$B_2 = \{\xi : h_i(\xi) < 0, h_7(\xi) > 0, i \in \{5,6\}\} \tag{83}$$

Figure 12 illustrates what we have so far. Now the target can be extended to include B_1 or B_2 and more tubes can be obtained. Let the new target be given by $T' = T \cup B_1$. A tube entering T' under the second control policy is bounded by choosing

$$I_{B_3} = \{7\} \tag{84}$$

and $e = 2$. A tube entering $T'' = T' \cup B_3$ under the third control policy is bounded by choosing

$$I_{B_5} = \{6\} \tag{85}$$

and $e = 7$. Figure 13 gives a final picture of the hypersurfaces and regions involved in this example.

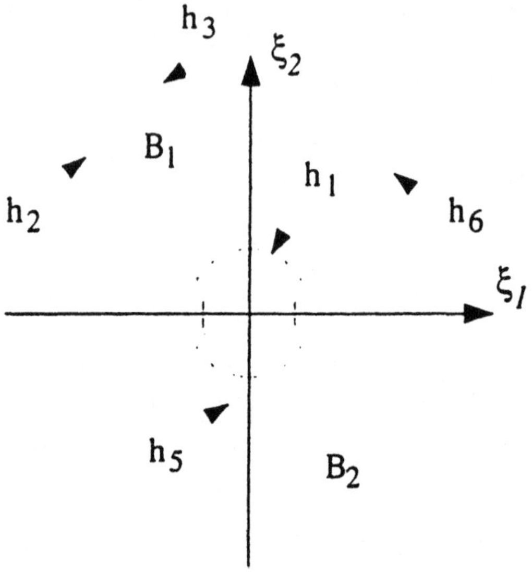

Fig. 12. Target Region and Invariants

There is only one control goal for this example and therefore the entire controller will consist of a single subautomata. Start by creating a controller state $\tilde{s}_T$ which is associated with the target region. Two tubes, labeled B_1 and B_2, were identified which lead to the target region. So create two more controller states, $\tilde{s}_1$ and $\tilde{s}_2$. B_1 consists of the trajectories which reach the target region under control policy $\tilde{r}_1$ and therefore $\phi(\tilde{s}_1) = \tilde{r}_1$, likewise $\phi(\tilde{s}_2) = \tilde{r}_3$. Connect $\tilde{s}_1$ to $\tilde{s}_T$ with a tranisition labeled

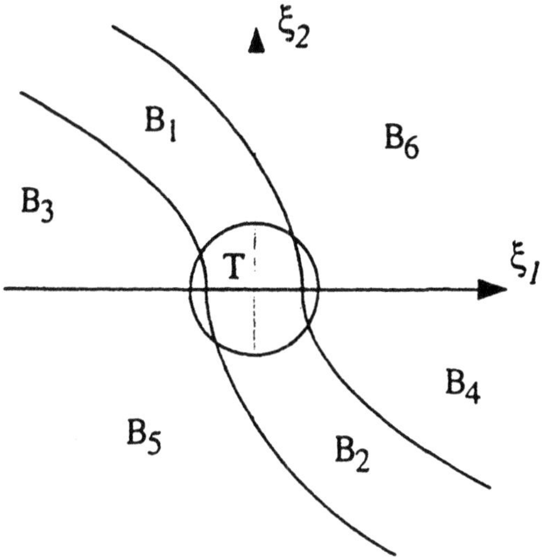

Fig. 13. Target Region and Invariants

$\tilde{x}_1$ which is generated when the plant state crosses h_1 to enter the target region. Do the same for $\tilde{s}_2$. Next, create $\tilde{s}_3$ to go with B_3, and add a transition to $\tilde{s}_1$ labeled $\tilde{x}_2$. When all the tubes have their associated states and transitions the controller shown in Figure 14 results.

4.4 Example - Triple Integrator

With the double integrator example, it is easy to see how the invariant surfaces are used. The technique can also be used in more complicated cases where it is not so intuitively obvious. Consider the triple integrator,

$$\dot{\mathbf{x}}(t) = \begin{bmatrix} 0 & 1 & 0 \\ 0 & 0 & 1 \\ 0 & 0 & 0 \end{bmatrix} \mathbf{x}(t) + \begin{bmatrix} 0 \\ 0 \\ 1 \end{bmatrix} \mathbf{r}(t), \tag{86}$$

with the same three available control policies,

$$\mathbf{r}(t) \in \{-1, 0, 1\}. \tag{87}$$

This time the control goal is to drive the state to the unit sphere from any initial state. First find a basis for the invariants by solving the characteristic equation

$$\frac{dx_1}{\mathbf{x}_2} = \frac{dx_2}{\mathbf{x}_3} = \frac{dx_3}{\mathbf{r}} \tag{88}$$

Two functions are obtained,

$$h_a(\xi) = r\xi_2 - \frac{1}{2}\xi_3^2 + c_1 \tag{89}$$

$$h_b(\xi) = r^2\xi_1 + \frac{1}{3}\xi_3^3 - r\xi_2\xi_3 - rc_2. \tag{90}$$

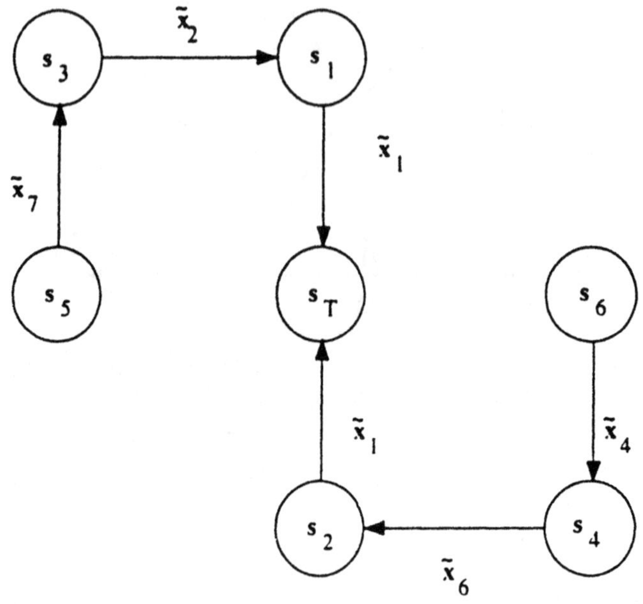

Fig. 14. Controller

where r is the input (control policy) and c_1 and c_2 are arbitrary constants. Example hypersurfaces for $r = 1$ and $c_1 = c_2 = 0$ are shown in Figures 15 and 16.

The target region is bounded by the hypersurface, h_1, i.e. $I_T = \{1\}$.

$$h_1(\xi) = \xi_1^2 + \xi_2^2 + \xi_3^2 - 1 \tag{91}$$

Now identify a "tube" of trajectories which enters the target region under the input $\mathbf{r}(t) = 1$. The following hypersurfaces are used.

$$h_2(\xi) = -\xi_2 + \frac{1}{2}\xi_3^2 - \sqrt{2} \tag{92}$$

$$h_3(\xi) = \xi_2 - \frac{1}{2}\xi_3^2 - \sqrt{2} \tag{93}$$

$$h_4(\xi) = \xi_1 + \frac{1}{3}\xi_3^3 - \xi_2\xi_3 - \sqrt{2} \tag{94}$$

$$h_5(\xi) = -\xi_1 - \frac{1}{3}\xi_3^3 + \xi_2\xi_3 - \sqrt{2} \tag{95}$$

The tube runs through the origin, where the target is centered, and the constant values, $\pm\sqrt{2}$, where choosen so that the tube passes through the target.

4.5 Remarks

Preliminary results were presented in [7], where they were discussed in the context of digital control. A more extensive presentation can be found in [32].

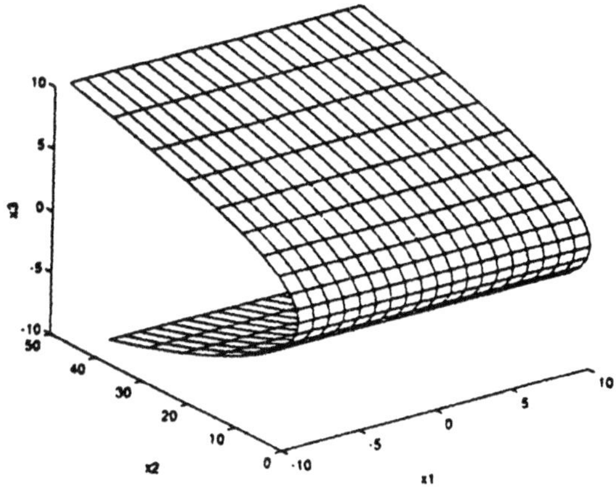

Fig. 15. Invariant for h_a

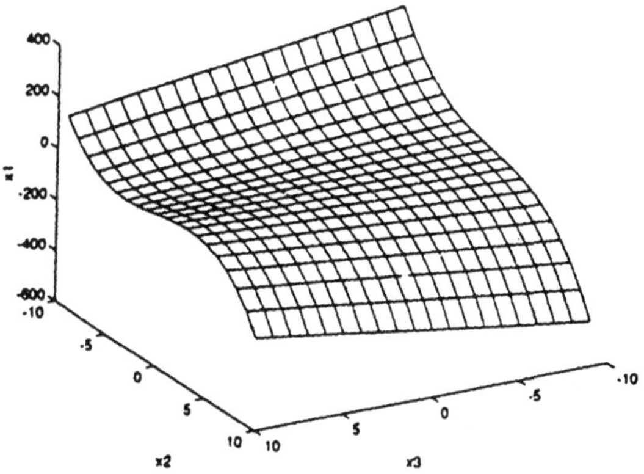

Fig. 16. Invariant for h_b

5 Conclusion

This paper presented two methods to design a hybrid control system. The first method can be used when the plant and interface of the system have been specified and a controller is required. The second method allows the design of the interface (the generator) as well as of the controller.

References

1. P. J. Antsaklis, J. A. Stiver, and M. D. Lemmon. "Hybrid system modeling and autonomous control systems", In *Hybrid Systems*, edited by R. L. Grossman, A. Nerode, A. P. Ravn and H. Rischel, vol. 736 of *Lecture Notes in Computer Science*, pp. 366–392. Springer-Verlag, 1993.

2. J. A. Stiver, "Modeling of hybrid control systems using discrete event system models", Master's thesis, Department of Electrical Engineering, University of Notre Dame, Notre Dame, IN, May 1991.
3. J. A. Stiver and P. J. Antsaklis, "A novel discrete event system approach to modeling and analysis of hybrid control systems", In *Proceedings of the Twenty-Ninth Annual Allerton Conference on Communication, Control, and Computing*, University of Illinois at Urbana-Champaign, Oct. 1991.
4. J. A. Stiver and P. J. Antsaklis, "Modeling and analysis of hybrid control systems", In *Proceedings of the 31st Conference on Decision and Control*, pp. 3748–3751, Tucson, AZ, Dec. 1992.
5. J. A. Stiver and P. J. Antsaklis, "State space partitioning for hybrid control systems", In *Proceedings of the American Control Conference*, pp. 2303–2304, San Francisco, California, June 1993.
6. J. A. Stiver and P. J. Antsaklis, "On the controllability of hybrid control systems", In *Proceedings of the 32nd Conference on Decision and Control*, pp. 3748–3751, San Antonio, TX, Dec. 1993.
7. J. A. Stiver, P. J. Antsaklis, and M. D. Lemmon, "Digital control from a hybrid perspective", In *Proceedings of the 33rd Conference on Decision and Control*, pp. 4241–4246, Lake Buena Vista, FL, Dec. 1994.
8. P. J. Antsaklis, M. D. Lemmon, and J. A. Stiver, "Learning to be autonomous: Intelligent supervisory control", Technical Report of the ISIS Group ISIS-93-003, University of Notre Dame, Notre Dame, IN, April 1993. To appear as a chapter in the IEEE Press book Intelligent Control: Theory and Applications.
9. J. A. Stiver, P. J. Antsaklis, and M. D. Lemmon, "A logical des approach to the design of hybrid systems", Technical Report of the ISIS Group (Interdisciplinary Studies of Intelligent Systems) ISIS-94-011, University of Notre Dame, October 1994.
10. J. A. Stiver, P. J. Antsaklis, and M. D. Lemmon, "Interface Design for Hybrid Control Systems", Technical Report of the ISIS Group (Interdisciplinary Studies of Intelligent Systems) ISIS-95-001, University of Notre Dame, January 1995.
11. A. Nerode and W. Kohn, "Models for Hybrid Systems: Automata, Topologies, Controllability, Observability", In *Hybrid Systems*, edited by R. L. Grossman, A. Nerode, A. P. Ravn and H. Rischel, pp. 317–356, Springer-Verlag, 1993.
12. W. Kohn, A. Nerode, J. Remmel, and X. Ge, "Multiple agent hybrid control: Carrier manifolds and chattering approximations to optimal control", In *Proceedings of the 33rd IEEE Conference on Decision and Control*, pp. 4221–4227, Lake Buena Vista, FL, Dec. 1994.
13. W. Kohn, J. James, A. Nerode, and N. DeClaris, "A hybrid systems approach to integration of medical models", In *Proceedings of the 33rd IEEE Conference on Decision and Control*, pp. 4247–4252, Lake Buena Vista, FL, Dec. 1994.
14. R. Brockett, "Language driven hybrid systems", In *Proceedings of the 33rd IEEE Conference on Decision and Control*, pp. 4210–4214, Lake Buena Vista, FL, Dec. 1994.
15. P. J. Ramadge, "On the periodicity of symbolic observations of piecewise smooth discrete-time systems", *IEEE Transactions on Automatic Control*, vol. 35, no. 7, pp. 807–812, July 1990.
16. C. Chase and P. J. Ramadge, "Dynamics of a switched n buffer system", In *Proceedings of the Twenty-Eighth Annual Allerton Conference on Communication, Control, and Computing*, pp. 455–464, University of Illinois at Urbana-Champaign, Oct. 1991.
17. S. Di Gennaro, C. Horn, S. Kulkarni, and P. Ramadge, "Reduction of timed hybrid systems", In *Proceedings of the 33rd IEEE Conference on Decision and Control*, pp. 4215–4220, Lake Buena Vista, FL, Dec. 1994.

18. A. Gollu and P. Varaiya. "Hybrid dynamical systems". In *Proceedings of the 28th Conference on Decision and Control*, pp. 2708–2712. Tampa. FL. Dec. 1989.
19. A. Deshpande and P. Varaiya. "Viable control of hybrid systems". In the Ph.D. Dissertation of the first author. June 1994. ftp: eclair.eecs.berkeley.edu.
20. M. Tittus and B. Egardt, "Control-law synthesis for linear hybrid systems". In *Proceedings of the 33rd IEEE Conference on Decision and Control*. pp. 961–966. Lake Buena Vista. FL. Dec. 1994.
21. B. Lennartson. B. Egardt. and M. Tittus. "Hybrid systems in process control". In *Proceedings of the 33rd IEEE Conference on Decision and Control*. pp. 3587–3595. Lake Buena Vista. FL. Dec. 1994.
22. R. L. Grossman. A. Nerode. A. P. Ravn. and H. Rischel. editors. *Hybrid Systems*. vol. 736 of *Lecture Notes in Computer Science*. Springer-Verlag. 1993.
23. P. J. Ramadge and W. M. Wonham. "Supervisory control of a class of discrete event processes". Systems Control Group Report 8515, University of Toronto. Toronto. Canada. Nov. 1985.
24. P. Ramadge and W. M. Wonham. "Supervisory control of a class of discrete event processes". *SIAM Journal of Control and Optimization*. vol. 25. no. 1. pp. 206–230. Jan. 1987.
25. P. Ramadge and W. M. Wonham. "The control of discrete event systems". *Proceedings of the IEEE*. vol. 77. no. 1. pp. 81–89. Jan. 1989.
26. W. M. Wonham and P. J. Ramadge. "On the supremal controllable sublanguage of a given language". Systems Control Group Report 8312, University of Toronto. Toronto. Canada. Nov. 1983.
27. W. M. Wonham and P. J. Wonham. "On the supremal controllable sublanguage of a given language". *SIAM Journal of Control and Optimization*. vol. 25. no. 3. pp. 637–659. May 1987.
28. X. Yang. P. J. Antsaklis. and M. D. Lemmon. "On the supremal controllable sublanguage in the discrete event model of nondeterministic hybrid control systems". Technical Report of the ISIS Group ISIS-94-004. University of Notre Dame. Notre Dame. IN. March 1994.
29. H. Nijmeijer and A. J. van der Schaft. *Nonlinear Dynamical Control Systems*. Springer-Verlag. New York. 1990.
30. R. Miller and A. Michel. *Ordinary Differential Equations*. Academic Press. New York. NY. 1982.
31. A. Isidori. *Nonlinear Control Systems*. Springer-Verlag, Berlin. 2 edition, 1989.
32. P. J. Antsaklis. "On Intelligent Control: Report of the IEEE CSS Task Force on Intelligent Control". Technical Report of the ISIS Group (Interdisciplinary Studies of Intelligent Systems) ISIS-94-001. University of Notre Dame. January 1994.

Hybrid Objects*

Michael Tittus and Bo Egardt

Control Eng. Lab., Chalmers University of Technology
S-412 96 Göteborg, Sweden
email: mt/be@control.chalmers.se

Abstract. Combining the modelling power of hybrid automata and the principles of object-orientation, the concept of hybrid objects is introduced. Hybrid objects are general, reusable and encapsulated models of physical systems. Furthermore, the principles of an algorithm, which automatically generates correct control code given the desired behavior of the object, are presented.

1 Introduction

The behavior of hybrid systems is described by means of continuous and discrete variables. A number of different philosophies have been pursued in order to establish a mathematical framework that is able to handle both continuous and discrete processes. These range from the construction of hybrid automata [ACHH93] and structures [NOSY92], over the definition of interfaces between the continuous and the discrete part [SA92] to more complex mathematical frameworks [Bro93]. Other approaches use Petri Nets [LBAD91], bond-graphs [STS93] or a synchronuous approach [BLG90], [BB91], just to name a few.

In this paper we shall combine the modelling power of hybrid automata with the principles of object-orientation. The goal is to model reality by reusable objects that can easily be adapted to different applications—*hybrid objects*. That way, models of hybrid systems can be designed and can be used as components when modelling larger and more complex processes.

The present work is based on the hybrid framework introduced in [TE94b], and the object model introduced in [TE92]. In our framework a hybrid system is interpreted as consisting of a discrete controller within an analog environment (process). Hybrid automata then are generalized finite state machines where discrete transitions transfer the system between a finite number of control states or locations. The global states of a system can change continuously with time according to the laws of physics and these changes are governed for each control location by a set of differential equations.

The object model consists of a general part, which describes the system's full functionality, and a specific part, which makes the object easily adaptable to different specifications and control strategies. The specific part contains exchangable recipes (the object's methods) which are functions of the given specifications.

* Research supported by the Swedish Research Councel for Engineering Sciences (TFR) under the project number 92-185 and by the Sewedish National Board for Industrial and Technical Development under the project number 90-01076.

We shall, in this paper, introduce a hybrid object-model together with an algorithm that allows us to automatically generate a set of correct methods for a restricted class of hybrid systems. This generation is done off-line and results in a deterministic control-law for each method. It is a function of the given specifications. In the following sections we shall discuss the hybrid framework applied, introduce the hybrid object-model, and show how correct methods can be generated.

2 The Hybrid Framework

The framework used here is identical to the one presented in [TE94b] and is based on the hybrid automata first introduced in [ACHH93].

2.1 Some Definitions

Given a set of continuous and discrete variables of a system, the system's *state* is an interpretation of these variables. In accordance with the two types of variables (discrete and continuous) we encounter two kinds of state changes—namely continuous and discrete ones.

In our model of hybrid systems the discrete state or control location is comprised of a number of discrete (often binary) control variables. All these control variables can be manipulated by discrete inputs, and different settings of the control variables correspond to different control locations ("mode-switching"). In order to identify the different control locations an index is defined, which ranges over the finite set of control locations and thus labels the discrete state of the system.

Usually, a transfer from one location to another not only causes a change of state but often also a change in the system's future behavior or the way the continuous variables behave.

Since changes in the discrete state happen at discrete points in time we divide the time axis into *phases*, with each change of the discrete state-variable separating two adjacent phases. The transition between two phases we shall refer to as a *phase-transition*. Within each phase the system's state-transformation can be described by a set of differential equations.

Hence, a *system state* is defined as a pair consisting of the *data-state* (i.e. an interpretation of all continuous or data variables) and the *control location*, encoded by its index.

2.2 Hybrid Automata

Adopting the terminology of [ACHH93], we define a hybrid automata as an eight-tuple $\mathcal{A} = (V_D, Q, \Sigma_c, \delta_c, \mu_1, \mu_2, \mu_3, Q_m)$.

- A finite set V_D of real-valued data variables $\{x_i\}$. The set of all data states is denoted by Σ_D. We shall write x for the data-variable vector.
- A finite set Q of control locations. Each location $l \in Q$ is encoded as a distinct integer value of the index i, which therefore ranges over the set Q. We write $Q = \{l_1, l_2, \ldots, l_n\}$, that is, $i = \{1, \ldots, n\}$. Thus, a system state is a pair (x, l) from the set $\Sigma = \Sigma_D \times Q$, as defined above. *Phase* is a synonym for *location*.

- A finite set Σ_c of control events causing a change in the discrete control location of the system.
- A partially defined discrete transition function $\delta_c : Q \times \Sigma_c \to Q$ that defines the dynamics of the discrete part of $\mathcal{A}$. Since δ_c is a partially defined function it also determines the connectivity between phases.
- A labeling function μ_1 that is used to assign to each phase a function describing the data-state transformation (activity) $\mu_1(l)$. $\mu_1(l)$ has the form of a system of first-order differential equations.
- A labeling function μ_2 that assigns to each location $l \in Q$ a phase-invariant $\mu_2(l) \subseteq \Sigma_D$. As long as the system resides in a phase l it has to satisfy the invariant $\mu_2(l)$, otherwise an exception is triggered. The notation $\mu_2(l)$ is also used to designate the ***sub-space*** in the system's state-space that is defined by the invariant $\mu_2(l)$. The correct interpretation is usually clear from the context.
- A labeling function μ_3 that assigns to each pair $e \in Q^2$ of locations, for which the transition function δ_c is defined, a phase-transition relation $\mu_3(e) \subseteq \Sigma_D^2$. A state (x', l') is called a successor of the state (x, l) iff $(x, x') \in \mu_3(l, l')$. In general, μ_3 is defined as
$$\mu_3(l_i, l_j): \quad x := f(x)$$
Unlike in [ACHH93] the phase-transition relation is not guarded here. The actual execution of a transition is enforced according to δ_c by the generated control-law.
- A finite set $Q_m \subseteq Q$ of marked locations. A marked location is a location in which the system's state does not change, that is, $\mu_1(l)$: $\dot{x} = 0$. Thus, they can be seen as singular points or sets of points in the system's state space, depending on the invariant $\mu_2(l)$. Seen that way, marked *locations* actually are marked *states*. They serve as starting and ending points of executable paths.

Restrictions Made in this Paper. In this treatment we restrict ourselves to *linear* hybrid systems which have the following features:

- The activities $\mu_1(l)$ within a phase l are restricted to linear functions defined by a set of first-order differential equations of the form $\dot{x} = k_l$ with k_l denoting a constant vector.
- The phase invariant $\mu_2(l)$ is defined by means of a set of linear inequalities over Σ_D and has the form $A \cdot x + b \leq 0$ with A and b being constant matrices and $x \in \Sigma_D$.
- The phase-transition relation $\mu_3(e)$ is defined by an assignment $x := \alpha_x$ for $x \in \Sigma_D$, where α_x is a linear term over Σ_D of the form $R \cdot x + c$, with R and c being again constant matrices.

3 The Hybrid Object-Model

In [TE92] an object model has been developed with the objective to completely model a part of the physical (hybrid) world. The system is described with respect to its functionality (What can be done?), its physical architecture (Who does it?), and its behavior (When and how is it done?). Mathematical equations, together with states and transitions are used to model system behavior.

The object-oriented paradigm tries to copy the way we think about reality and therefore is very intuitive.

An object is an encapsulated *abstraction* of reality that offers specific services to the surrounding world. These services are called *methods* and are the only way the object can be operated on. The methods are offered to prospective users by means of an *interface*, the actual implementation being hidden to the user. A method is initiated by sending a *message* to the respective object telling it to execute this specific method.

There are certain powerful tools and principles that make object-orientation suitable for the modelling of control systems.

- *Modularization*, that is, the decomposition of large and complex systems into smaller *modules* or objects that interact with each other is an inherent feature of object-orientation.
- *Classes* of objects can be defined in advance and stored in a library. Each object then is created as an *instance* of an already existing class and contains the same features as its class. This feature ensures reusability of models.
- The concept of reusability becomes even more powerful with the *inheritance* feature. A new class of objects can be easily created as specializations and/or extensions of already existing ones. A class created as an heir of another class inherits all of the parent's features plus adds its own. This allows quick prototyping.

The reason then why we use an object-oriented framework is not that the model could not do without it but because it results in a very intuitive approach with powerful features.

Our hybrid object model combines these features of the object-oriented paradigm with the modelling capabilities of hybrid automata and consists of two parts, a general and a specific one, as depicted in Fig. 1. All the object's methods are contained in the application-dependent specific part. Each method is executed as a result of an external message sent to the hybrid object and has as its purpose to transfer the object into a different state. The methods contain sequences of control events which are triggered depending on the continuous state of the object. These events cause a change in the current control location and thus in the continuous behavior of the system. In that way, the object is controlled into another state while satisfying all given specifications.

We shall in the following consider the general and specific part in more detail.

3.1 The General Part

The general part of the hybrid object models the physical system as a hybrid automaton and is the reusable part of the model. The full functionality of the system which physically is possible is modelled and no specifications—except the physically motivated ones—are included. This part models the system from the "What can be done" point of view and constitutes the reusable part of the hybrid object.

The hybrid automaton also contains all control actions that have to be taken within the object in order to switch between the different phases.

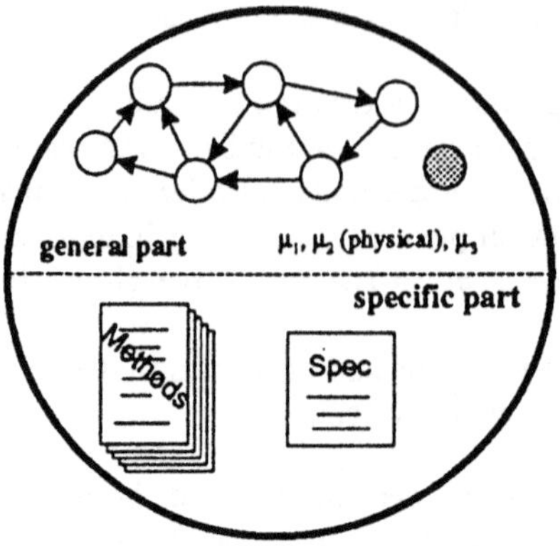

Fig. 1. Schematic of the hybrid object-model

As discussed in detail in [TE92], our approach suggests that an object may reside in a number of *object states*. *Object transitions*, or simply *transitions*, are the means of transferring the object in a controlled way from one state to another. These transitions obey continuous and/or discrete laws and can either happen instantaneously or over time. They are accomplished by the previously mentioned methods of the object.

States and Transitions. An object resides in a state if it is in a situation that is static in the sense that the object can stay in that state temporally unrestricted by, for example, physical laws. Continuous control functions are only allowed if they do not alter the system state. This includes for example control-loops that keep a constant set-point. The object's states coincide with the marked locations of the hybrid model of the system, which were defined as keeping a constant system-state. In the general part, the object has one generic marked state, which in itself is an instantiable class. Later on, in the specific part, instances of this marked state will be specified.

A transition (object transition) corresponds to an executable path in the hybrid automaton that connects two marked locations, i.e. it is accomplished by a sequence of phase-transitions. The general part thus implicitly contains the infinite set of all physically possible transitions, since it expresses the object's full functionality. Creating a finite number of instances of the marked state, specifies a subset of desired transitions in the specific part. Furthermore, the manner of executing a transition depends on the given specifications and is also generated in the specific part.

Programming by Contract. Each object offers the user a contract for each of its methods, which contains a require and an ensure option. In order to be able to execute a method the object is required to be in a certain marked state. If this precondition is satisfied then the object guarantees that the method will leave the object in the desired destination state. Otherwise an exception is triggered.

3.2 The Specific Part – Modelling Dynamics

The guiding factor in the design of control laws are specifications and restrictions imposed on the system by the laws of physics, and the intentions of system engineers. Both, specifications and restrictions are formulated as invariants. Only the "acceptable" is defined and thus everything that is not explicitly denoted as acceptable is by definition an exception.

The specific part contains a specification attribute, and the object's methods in executable form. The specification attribute, which is exchangable and application-dependent, contains conditions for correct system behavior. Each method is created as a function thereof. Therefore, unlike in the traditional object-oriented setting, the hybrid object's methods and specifications are seldom reusable, but are generated anew for each setting.

Specification Attribute. The desired and/or required behavior can be specified in five ways:

- *System-invariants* that are required to hold independent of the location or state the system is in. They apply automatically if not overwritten by stricter local specifications. Examples might be physical restrictions as for example maximum levels or pressures. Those of the system invariants that are physically motivated belong to the general part of the hybrid object.
- For each location l of the hybrid automaton a specific *phase invariant* $\mu_2(l)$ can be defined in the form of a matrix inequality, as described earlier. If there is no phase invariant defined for a certain variable then the respective system-invariant applies automatically.
- A set Q_m of marked states s are defined by means of phase invariants $\mu_2(s)$ and as instances of the generic marked state.
- A set Q_f of forbidden phases, marked as well as unmarked ones, can be specified. Forbidding marked phases corresponds to inhibiting certain object states.
- A set of forbidden phase transitions can be specified by declaring the corresponding phase-transition relation as undefined.

All this information is loaded as a kind of recipe into the specification attribute.

Methods. A *method* contains the control-law for a specific transition between two object states. It governs the system's dynamical behavior under certain specifications and can be seen as the object's recipes. Methods are generated by the object itself as a function of the specification attribute.

Hybrid systems are assumed to be controlled by discrete inputs that make them change their phase or control location. All discrete variables can be manipulated. Hence, the system can be controlled to run through any possible sequence of phases, the only restriction being the invariants of the different phases. In [TE94b] controllability with respect to a pair (s_1, s_2) of marked states has been defined as follows:

Definition 1. A hybrid system $\mathcal{A}$ is said to be *controllable with respect to a state-transition* $\Phi(s_1, s_2)$ with $s_1, s_2 \in Q_m$, and an invariant $\mu_2(l)$ for each location l iff there exists at least one sequence of discrete control-inputs that gives $\mathcal{A}$ an acceptable run $\pi(s_1, \ldots, s_2)$. That is, it *(1)* takes $\mathcal{A}$ from s_1 to s_2, and *(2)* all invariants applying to $\Phi(s_1, s_2)$ are continuously satisfied.

Hence, a necessary and sufficient condition for the existence of a correct method for a state transition is the controllability of this specific state transition. In the last section we shall show how this definition can be used to generate correct methods as deterministic sequences of control signals. Only controllable state transitions are offered as methods by the object.

We shall proceed with an example illustrating the role of methods. The details of method generation however, are deferred to the last section.

3.3 Example

Consider the tank-system of three identical tanks depicted in Fig. 2. All three tanks are identical and all flows cause the tank-levels to rise or decrease by 1 unit/time unit. The pump in the center has four different positions. It is either shut off, or pumps fluid from Tank 1 to Tank 3 (location l_1), from Tank 3 to Tank 2 (location l_2), or from Tank 2 to Tank 1 (location l_3). While the pump is shut off, fluid can be added to the system via Valve 1 (location l_5) or the system can be drained via Valve 2 (location l_4). The pump is assumed to have a time delay of 0.5 time units before rotating anti-clockwise to the next position.

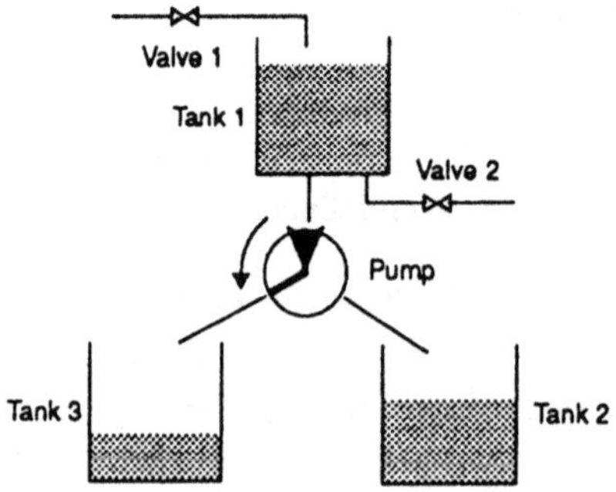

Fig. 2. Example Tanksystem

In Fig. 3 the system (hybrid object) with its specification is modelled as a hybrid automaton. Transitions to and from locations l_4 and l_5 are governed by special rules set in the specification in the specific part. As data variables, the three tank levels have been chosen. The time delay has been modelled as a discrete change of the levels when changing control location. The labeling functions μ_3 have been omitted

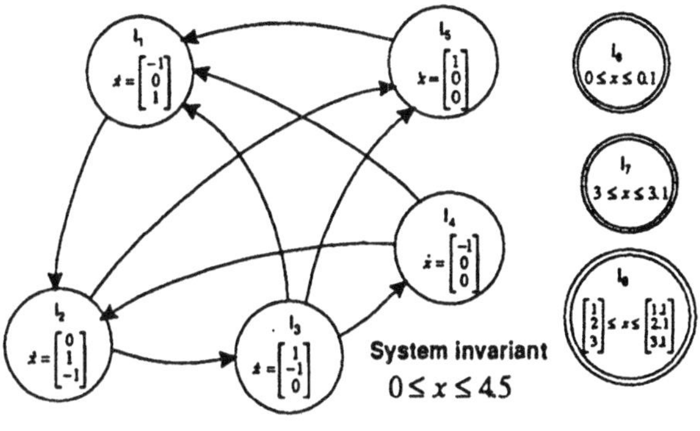

Fig. 3. Automata representation of the tank-system

to keep the figure simpler. Thus we assume for all transitions from l_1

$$\mu_3(l_1,\cdot): x := x + \begin{bmatrix} -0.5 \\ 0 \\ 0.5 \end{bmatrix}$$

and analogously for all transitions from l_2 and l_3. The hybrid automaton including the physically motivated system invariant, but without the marked states constitutes the reusable general part of the hybrid object *Tanksystem*.

The specification attribute contains the control objective to transfer the system between the three marked states l_6 to l_8. Each marked state is assumed to be bidirectionally connected with all non-marked locations.

The system is found to be controllable and the following correct paths together with sets of control laws have been computed:

$$\langle 6,5,1,2,5,1,2,5,7\rangle$$
$$\langle 6,5,1,2,5,1,2,5,8\rangle$$
$$\langle 7,3,4,8\rangle$$
$$\langle 7,4,2,3,4,2,3,4,6\rangle$$
$$\langle 8,4,2,3,4,2,3,4,6\rangle$$
$$\langle 8,5,1,2,7\rangle$$

Each of these control laws corresponds to a method in the specific part. A simulation of the hybrid system transferring between the marked locations l_7 and l_6 is shown in Fig. 4. In this case, a control strategy was chosen that forces the system to change control location as late as possible.

4 Implementation Model

With the methods generated we can take a look at a possible implementation of the now no longer general object. Together with the physical system the object forms

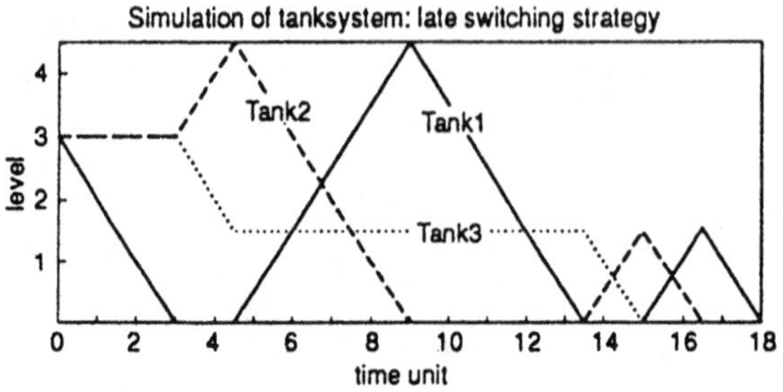

Fig. 4. Simulation of the three tank levels

a plant–supervisor structure with the object's implementation acting as discrete supervisor.

The plant–supervisor system is shown in Fig. 5. The physical plant is represented as the general hybrid automaton $\mathcal{A} = (V_D, Q, \Sigma_c, \delta_c, \mu_1, \mu_2, \mu_3, q_m)$, which is identical with the general part of the object model. q_m denotes the generic marked state and μ_2 in the general part only contains the physical restrictions. The supervisor, on the other hand, is modelled as a discrete Mealy machine that interprets different recipes. The supervisor consists of a general (application-independent) part and a specific (application-dependent) part.

4.1 The General Part

The general part consists of a Mealy machine and an interface. It interprets the recipes provided by the methods. The supervisor uses the continuous system state as an input and outputs control events to the plant.

Mealy machine. The Mealy machine is defined by the following six-tuple: $\mathcal{S} = (Q_S, \Sigma, \Delta, \delta, \lambda, q_0)$

- The set Q_S of locations is identical to the one used for the modelling of the plant $\mathcal{A}$, i.e. $Q_S = Q$.
- The set Σ denotes a finite input alphabet of events. We shall write $\sigma_{i,j}$ for the event that causes a transition from location q_i to q_j with $q_i, q_j \in Q_S$.
- The set Δ stands for the finite output alphabet of $\mathcal{S}$. This set contains the necessary control events that cause the plant to switch location and is identical to the set Σ_c of the plant $\mathcal{A}$.
- The transition function δ maps $Q \times \Sigma$ to Q. That is, $\delta(q_i, \sigma_{i,j}) = q_j$.
- In a Mealy machine, the function $\lambda(q, \sigma)$ defines the output associated with the transition from location q on event σ. In other words, λ maps $Q \times \Sigma$ to Δ.
- Location $q_0 \in Q$ denotes the initial state the system is in when it is started.

The main purpose of the Mealy machine is to keep track of the current location of the hybrid system and generate control events.

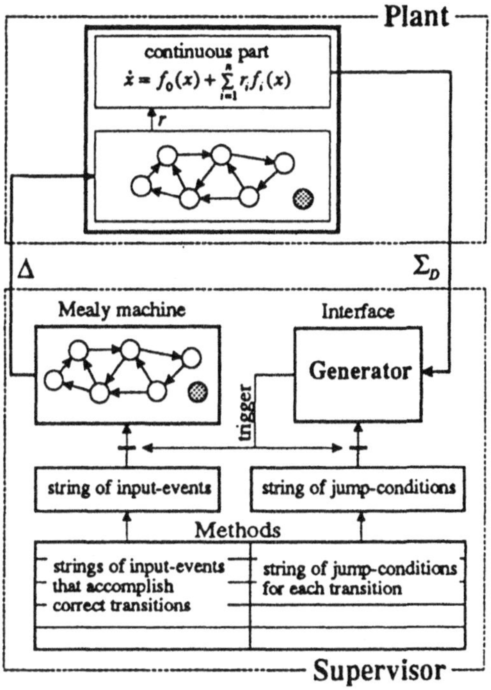

Fig. 5. Implementation of the hybrid object

Interface. The interface contains two buffers and a generator. When the object receives a message to execute a method transferring the system from a state s_1 to a state s_2, the corresponding string of input events $\langle \ldots \sigma_{i,j}, \sigma_{j,k}, \ldots \rangle$ which accomplishes this transition is loaded into a buffer. The other buffer contains a string with the corresponding jump-conditions. (Figure 5)

When starting the execution of the method the first input event is triggered causing a transition in the Mealy-machine and the generation of a control event that lead to a corresponding phase transition in the plant. At the same time the first jump-condition is loaded into the generator. The generator now compares the plant's data state with the jump-condition and generates a trigger signal as soon as the jump-condition is satisfied, which corresponds to a control strategy where transitions are executed as soon as possible. This trigger signal causes the next input event to be triggered and the loading of the next jump-condition. This continues until the two buffers are empty and the plant has reached its destination state (last jump-condition satisfied). The object then sends a message about the completion of the method.

4.2 Specific Part

The application-dependent specific part contains all the methods or recipes in the form of strings of input events together with the corresponding jump-conditions for each phase-transition. For each controllable transition the object offers a method specifying the two marked states that serve as initial and final location for this specific transition. A method can only be called if the object satisfies the pre-condition of the method, i.e. it is in the correct state. Otherwise an exception is triggered.

5 Generating Methods

In this last section we briefly discuss the principles needed to decide the controllability of an object's state transitions. Furthermore, we show how to automatically generate methods for controllable state transitions for the case of linear hybrid systems. A more thorough treatment about the concept of controllability for linear hybrid systems is given in [TE94a] and an algorithm to decide the controllability is presented in detail in [Tit93].

5.1 Some Definitions

An important requirement when it comes to a phase-transition between two phases, l_1 and l_2, is that such a discrete state change or "jump" will not lead the system into forbidden areas in location l_2, i.e. trigger an exception. This problem might occur if l_1 and l_2 do not have identical phase invariants or if data variables are changed by a phase-transition. Hence, we define for each possible transition from a location l_1 to a location l_2 a sub-set $\lambda(l_1 \rightarrow l_2) \subseteq \Sigma_D$ of the state space called the "*jump-set* from l_1 to l_2". $\lambda(l_1 \rightarrow l_2)$ even denotes the set of linear inequalities used as a mathematical description of this sub-set—the so-called *jump conditions.*

Definition 2. The *jump-set* $\lambda(l_1 \rightarrow l_2)$ is the set of all points of the system's data state-space Σ_D that satisfy $\mu_2(l_1)$ and from which $\mu_2(l_2)$ is reachable by means of the phase-transition relation $\mu_3(l_1, l_2)$.

The jump conditions thus make sure that the system will always "jump" into the invariant space of the destination phase. Since both $\mu_2(l)$ and $\mu_3(l_1, l_2)$ are given by linear expressions, the jump-conditions will of necessity be of the same form. The jump-conditions have to be satisfied in order for the system to be allowed to transfer to the next phase and can be compared to the transition guards used by [ACHH93]. At the same time, these jump-conditions are at the bottom of the design of correct control laws, as will be seen later.

Even with correctly defined jump-condition $\lambda(l_1 \rightarrow l_2)$ we cannot guarantee that *each* allowed data state in the destination phase l_2 (defined by $\mu_2(l_2)$) is reachable by the phase-transition $e = (l_1, l_2)$. Only a sub-set of $\mu_2(l_2)$—the *reachable sub-set* $\rho(l_2 \leftarrow l_1)$ from l_1 to l_2—can be reached from l_1.

Definition 3. The *reachable sub-set* $\rho(l_2 \leftarrow l_1)$ is the set of all points in the invariant $\mu_2(l_2)$ that is reachable from within the jump-set $\lambda(l_1 \rightarrow l_2)$ via the phase-transition relation $\mu_3(l_1, l_2)$.

One last definition:

Definition 4. The *extended jump-set* $\lambda_{\text{ext}}(l_1 \to l_2)$ is the union of $\lambda(l_1 \to l_2)$ and all states $x \in \mu_2(l_1)$ such that $\lambda(l_1 \to l_2)$ is reachable from x via the data-state transformation $\mu_1(l_1)$.

Thus, the extended jump-set $\lambda_{\text{ext}}(l_1 \to l_2)$ guarantees, when satisfied, that a correct transition $e = (l_1, l_2)$ sooner or later will be possible. That is, once the system has entered phase l_1 and is within $\lambda_{\text{ext}}(l_1 \to l_2)$ the system will "drift" into the jump-set $\lambda(l_1 \to l_2)$ because of the phase activity $\mu_1(l_1)$.

Figure 6 shows the three sub-sets of the state-space and their relationship to each other. We shall represent these sub-sets as two-dimensional areas within the nodes of the respective location. The area of the node l itself represents the phase invariant $\mu_2(l)$. The set $\lambda(l_1 \to l_2)$ is the intersection of the inverse $\mu_3(l_1, l_2)$-mapping of the phase invariant of location l_2, and the phase invariant of location l_1, while $\rho(l_2 \leftarrow l_1)$ is the $\mu_3(l_1, l_2)$-mapping of $\lambda(l_1 \to l_2)$ into the phase invariant of location l_2.

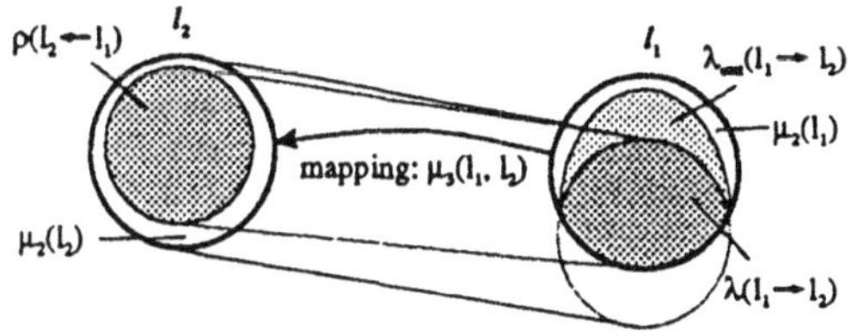

Fig. 6. The different sub-spaces occuring at a phase-transition

5.2 Deciding Controllability for State-Transitions

A transition from a location l_i to another location l_{i+1} is or becomes possible only if the system constantly satisfies $\lambda_{\text{ext}}(l_i \to l_{i+1})$ while in phase l_i. If we continue this line of thought, then, in order to have a controllable path $\pi(l_{i-1}, l_i, l_{i+1})$ we have to make sure that the transition $e = (l_{i-1}, l_i)$ transfers the system into $\lambda_{\text{ext}}(l_i \to l_{i+1})$, so as to guarantee a continuation of the path. Hence, for the path π the allowed space for phase l_i is $\lambda_{\text{ext}}(l_i \to l_{i+1})$. We say that the invariant state-space of location l_i is reduced to $\lambda_{\text{ext}}(l_i \to l_{i+1})$ for this specific path.

In Fig. 7 this principle is illustrated. In order to check if a certain path is controllable it is tested if the final state can be reached from the initial state. This is achieved by generating the extended jump-spaces for each step in the path starting with the final state and going backwards. The invariant $\mu_2(l)$ of each phase in the path is reduced to the extended jump-set before considering the next transition along the path. If $\lambda_{\text{ext}}(s_1 \to l_1)$ turns out not to be void, then the system is controllable along this specific path. Otherwise, a different path has to be tried.

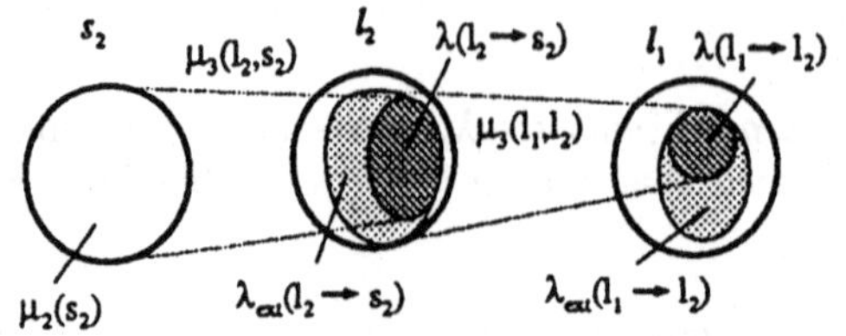

Fig. 7. Deciding controllability along a path

In the case of s_1 and s_2 consisting not only of a single point in the state-space, but a sub-set thereof we note that an acceptable run $\pi(s_1, l_1, \ldots, l_n, s_2)$ does not guarantee that each point in s_2 can be reached. Neither do we crave controllability from every point $x \in s_1$ for π to qualify as acceptable but we demand that there exists at least one state $x \in s_1$ from which some state $y \in s_2$ can be reached along π. The obvious reason for this is that $\rho(s_2 \leftarrow l_n) \subseteq \mu_2(s_2)$, and $\lambda_{\text{ext}}(s_1 \to l_1) \subseteq \mu_2(s_1)$.

Loops. A path is said to contain a possible loop if one and the same location can appear more than once in a path. Sometimes it is necessary to include a loop in a path in order to find an acceptable run of the system, since a loop may increase the extended jump-set of a phase. Figure 8 shows a part of a possible path that includes a loop. From location l_1 the path continues over location l_i towards the marked location s_2.

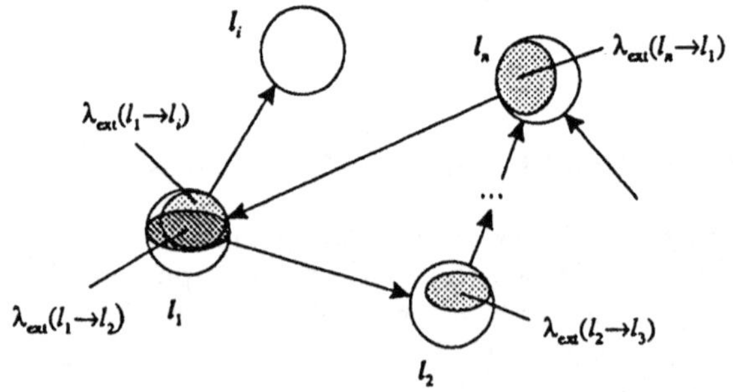

Fig. 8. Loop reduction

The iteration backwards from s_2 until the beginning of the loop (l_1) has resulted in $\lambda_{\text{ext}}(l_1 \to l_i)$, i.e. the extended jump-set from location l_1 to the next location in the path. With $\lambda_{\text{ext}}(l_1 \to l_i)$ as a starting point we iterate the sequence $\langle l_1, l_n, \ldots, l_2, l_1 \rangle$. This iteration ends with the computation of the set $\lambda_{\text{ext}}(l_1 \to l_2)$, which guarantees a successful run from any point within $\lambda_{\text{ext}}(l_1 \to l_2)$ to s_2 via the loop. In other words, the system is controllable from location l_1 to the final state s_2 if it starts from within either $\lambda_{\text{ext}}(l_1 \to l_2)$ or $\lambda_{\text{ext}}(l_1 \to l_i)$. In the first case, the acceptable path includes the sequence $\langle l_1, l_2, \ldots, l_n, l_1 \rangle$ while in the later case the loop is not needed. Consequently the new invariant for location l_1, considering both alternatives, is changed according to

$$\lambda_{\text{ext}}(l_1 \to l_i) := \lambda_{\text{ext}}(l_1 \to l_2) \cup \lambda_{\text{ext}}(l_1 \to l_i)$$

With a changed $\lambda_{\text{ext}}(l_1 \to l_i)$ even the other jump-sets in the loop can be updated (i.e. enlarged). We therefore can continue to circle the loop, each time with a new and enlarged $\lambda_{\text{ext}}(l_1 \to l_i)$ as a starting point, until the sequence of sets $\lambda_{\text{ext}}(l_1 \to l_i)$ reaches the invariant set

$$\lambda^*_{\text{ext}}(l_1 \to l_i) = \lambda_{\text{ext}}(l_1 \to l_2) \cup \lambda^*_{\text{ext}}(l_1 \to l_i)$$

where it cannot be enlarged anymore. This procedure has to be repeated for each possible loop.

Algorithm. An algorithm to decide the controllability of a hybrid system $\mathcal{A}$ has been implemented in MATLAB using the approach outlined above. The algorithm decides whether or not the system is controllable by calculating one correct path for each ordered pair of marked locations. Only paths with one loop at a time can be handled in the present version. Because of the nature of loops the problem is semi-decidable, i.e. the algorithm does not necessarily terminate. Besides correct paths it calculates also the corresponding sequence of jump-sets that act as transition guards for the different paths. A detailed description of the algorithm can be found in [Tit93].

5.3 Method Generation

The controllability of a state transition is checked by finding a sequence of phase transitions (path) with their corresponding jump-sets that guarantees a correct transfer of the system. If no requirements concerning the optimality of a method are specified, methods can be generated without further calculations. In the present implementation of the algorithm [Tit93] each generated method has the least possible number of phase transitions. The algorithm is semi-decidable.

Three control-strategies can be used. Either the object is transferred to the next phase along the generated path as soon as possible, that is, as soon as the system enters the respective jump-space, or as late as possible, that is, just before the jump-conditions are violated. In the second case, knowledge about the system's dynamic is used to predict the data-state, and to check if it is about to violate the jump-condition.

For a given sequence of phase transitions leading between marked states s_1 and s_2 it is also possible to find a time-optimal switching strategy. A clock is introduced into each phase and the algorithm is applied. When a correct path π between s_1 and s_2 is found, the jump-set $\lambda(s_1 \to l_1)$ will contain an intervall for the clock variable, giving the fastest and slowest transition time. By introducing the optimal (fastest) time as a constraint in s_2 the algorithm will after an additional execution come up with the time-optimal switching strategy for path π.

6 Conclusions

In this paper we have developed a model of a hybrid object which combines the advantages of the object-oriented paradigm with the concepts of hybrid automata. Our purpose was to create general, reusable, and encapsulated models of physical systems that, given a specification, automatically generate a correct control-code to shuttle the system between its states. The principles of an algorithm that generates control-code for linear hybrid systems were also presented.

Recent work includes the modelling and control of larger systems in the field of batch control with hybrid objects as components, see [TFL95].

Acknowledgement: We thank the anonymous reviewer for many helpful comments.

References

[ACHH93] R. Alur, C. Courcoubetis, T.A. Henzinger, and P.-H. Ho. Hybrid automata: An algorithmic approach to the specification and verification of hybrid systems. In R.L. Grossman, A. Nerode, A.P. Ravn, and H. Rischel, editors, *Theory of Hybrid Systems*, Lect. N. in Comp. Sci. 736, pages 209–229. Springer Verlag, 1993.

[BB91] A. Benveniste and G. Berry. The synchronous approach to reactive and real-time systems. *Proceedings of the IEEE*, 79(9):1270–1282, Sep. 1991.

[BLG90] A. Benveniste and P. Le Guernic. Hybrid dynamical systems theory and the SIGNAL language. *IEEE Transactions on Automatic Control*, 35(5):535–546, 1990.

[Bro93] R.W. Brockett. Hybrid models for motion control systems. In H.L. Trentelman and J.C. Willems, editors, *Essays on Control: Perspectives in the Theory and its Applications*, chapter 2, pages 29–53. Birkhäuser, 1993.

[LBAD91] J. Le Bail, H. Alla, and R. David. Hybrid petri nets. In *European Control Conference*, pages 1472–1477, Grenoble, France, 1991.

[NOSY92] X. Nicollin, A. Olivero, J. Sifakis, and S. Yovine. An approach to the description and analysis of hybrid systems. In *Workshop on Theory of Hybrid Systems*, Lyngby, Denmark, 1992.

[SA92] J.A. Stiver and P.J. Antsaklis. Modeling and analysis of hybrid control systems. In *Proc. of the 31st CDC*, pages 3748–3751, Tucson, Arizona, 1992.

[STS93] J.-E. Strömberg, J. Top, and U. Söderman. Modelling mode switching in dynamical systems. In J.W. Nieuwenhuis, C. Praagman, and H.L. Trentelman, editors, *Proc. of the 2nd ECC '93*, pages 848–853, Groningen, The Netherlands, 1993.

[TE92] M. Tittus and B. Egardt. An object-oriented approach for control-system design. In *1st Int. Conf. on Intelligent Systems Engineering*, pages 135–140, Edinburgh, UK, 1992.

[TE94a] M. Tittus and B. Egardt. Control design for linear hybrid systems. Technical Report CTH/RT/R-94/010, Contr. Eng. Lab, Chalmers Univ. of Techn., Gothenburg, Sweden, 1994. Submitted to IEEE TAC.

[TE94b] M. Tittus and B. Egardt. Controllability and control-law synthesis of linear hybrid systems. In G. Cohen and J.-P. Quadrat, editors, *11th Int. Conf. on Analysis and Optimization of Systems*, Lect. N. in Contr. and Inf. Sci. 199, pages 377–383. Springer Verlag, Sophia-Antipolis, France, 1994.

[TFL95] M. Tittus, M. Fabian, and B. Lennartson. Controlling and coordinating recipes in batch applications. Submitted to 34th CDC, New Orleans, USA, 1995.

[Tit93] M. Tittus. An algorithm for the generation of control laws for linear hybrid systems. Technical Report CTH/RT/R-93/0013, Contr. Eng. Lab, Chalmers Univ. of Techn., Gothenburg, Sweden, 1993.

Modelling of Hybrid Systems Based on Extended Coloured Petri Nets*

Y. Y. Yang, D. A. Linkens, and S. P. Banks

Department of Automatic Control and Syetems Egnineering
University of Sheffield
Sheffield S1 4DU, UK

Abstract. In this paper the modelling of hybrid systems, which are composed of both discrete event subsystems and continuous subsystems, is addressed. A general system structure for the hybrid system modelling is proposed along with their formulations. Based on these concepts, an extended coloured Petri net (ECPN) is proposed in order to implement the hybrid system modelling. Key ideas introduced in extended coloured Petri nets are dynamic colours, dynamic transitions, and dynamic places, which make ECPNs capable to model both variables governed by continuous dynamics and the ordinary Petri net tokens for discrete events. Thus, two aspects of hybrid system, continuous variables and discrete events, are unified in ECPNs by the number of tokens and the value of the tokens (defined as the colour state in this paper). A formal definition of ECPNs is also given. Finally, an illustrative example is given which shows how to use the developed approach for hybrid system modelling.
Keyward: Hybrid systems, coloured Prtri net, modelling and simulation.

1 Introduction

Systems modelling and simulation have to this time been classified mainly as either discrete event or continuous variable. It seems as if they were two totally different categories and little effort has been devoted to develop a unified methodology for hybrid systems, where discrete events coexist and interact closely with continuous dynamics. Yet, in reality, the need for a unified approach to deal with the modelling and simulation of hybrid systems is continuously increasing. This is because few practical processes can be considered in an entirely continuous or discrete manner when viewed at an integrated level. Even the majority of continuous processes experience significant discrete changes superimposed on their predominantly continuous behaviour. Such changes typically arise from equipment failures, planned operational shifts, start-up and shut-down operations, process maintenance, and so on. Similar situations exist in many discrete event systems, where some of the system activities (events) are influenced by the state of related continuous counterparts. For example, it is quite common for some of the discrete events to have certain preconditions related to continuous variables, such as the threshold values of a specific event. In many important industrial systems, such as paper mills, soaking pit/rolling mills, batch chemical processes, aerospace missions, nuclear plants, etc., both discrete events and continuous dynamics are not negligible. It is therefore natural to

* This research is supported by an EPSRC Grant GR/H/73585.

think that, in many situations, modelling, simulation and control of a practical system should be developed in a hybrid manner as opposed to a purely continuous or purely discrete event one. The proper combination of the continuous dynamics and discrete events is believed to lead to a superior system structure. As early as 1970, D. A. Fahrland [1] had proposed the concept of combined discrete-continuous system simulation. Pantelides, Barton and others have proposed a formulation of DAE (ordinary Differential and Algebraic Equations) for the modelling and simulation of combined discrete-continuous systems [2, 3]. Recently, there has been an increasing interest in the modelling and control of hybrid systems. Nerode and Kohn [4] proposed a hybrid system model, where the hybrid control system is defined as the control of a continuous plant by a digital sequential control automata. Antsaklis and his colleagues [5, 6] have studied the event identification problem for the purpose of intelligent hybrid control. Alur et al [7] developed a hybrid automata which can be used for the specification and verification of hybrid systems. Henzinger, Manna, and Pnueli [8] have studied hybrid control systems via a temporal logic approach.

The aim of this paper is to develop an efficient modelling methodology for hybrid systems. Model development is crucial to both computer simulation and system analysis for hybrid systems. Petri nets [9] are selected as the basis for this purpose because of the many outstanding properties they possess: the modelling power, the formal representation and analysis ability, and the visualisability. The paper is organised as follows. In Section 2, a formulation is given for the modelling of hybrid systems, and some structural characteristics and basic concepts about the model framework are addressed. Section 3 gives a brief review of Petri nets, and extensions to coloured Petri nets (ECPNs) are made in order to accommodate hybrid system modelling. Techniques for modelling hybrid systems based on ECPNs are also developed in this section. A case study is presented in Section 4 to illustrate the concepts and methods developed in Sections 2 and 3. Finally, conclusions are given in Section 5.

2 Problem Formulation for the Modelling of Hybrid Systems

Many industrial systems could be better described in the framework of a hybrid system model for the purpose of integration of control, optimisation and scheduling in a plant-wide level. In this section we first give some basic concepts of hybrid system modelling, and then formulations for modelling of hybrid systems are proposed.

No clear definition has yet been established for hybrid systems. While many people think that a hybrid system is a continuous plant plus a discrete (digital) controller [4, 7], others prefer a wider view that the plant itself can demonstrate both continuous and discrete behaviour [10, 11]. In this paper, a hybrid system is defined as a compound system with three primary components: the discrete event subsystems (DESs), the continuous variable subsystems (CVSs), and the mixed discrete-continuous subsystems (MDCSs). These components are closely related and interacting. Notice that here we suggest the MDCSs as a basic component of the hybrid system. The reason is that in some practical systems it is often impossible or very difficult to separate the discrete part and the continuous part in a subsystem because they are strongly interconnected both in the space domain and in the time domain.

Taking the soaking pit process in a steel mill as an example, the heating process of the ingot in a soaking pit is governed by the continuous thermal dynamics, but the ingots charging and discharging activities of the soaking pit governed by the demand of the rolling mill and the availability of the other resources, such as cranes, are typical discrete events. Both the discrete events and the continuous dynamics reside in the same physical process and overlap in the time domain and can hardly be separated. The introduction of MDCS will facilitate the modelling of such kind of sub-processes, which will become more clear in the later sections.

Traditionally, modelling and analysis of discrete event systems and continuous systems are developed independently with different methods and tools, due to the different characteristics encountered in DESs and in CVSs. In a CVS, the state trajectories are constantly changing, with the states taking continuous real values and being generally driven by some continuous inputs. But in a DES, the state trajectories are piecewice constants and are often event-driven. While the event duration is generally a continuous variable, the states usually take values from a finite discrete set. Compared with CVSs, which can be thought of as an abstraction of a physical process, DESs are a relatively recent phenomena mainly resulting from high technology man-made systems, such as flexible manufacturing systems, computer networks, communication systems, etc [12]. For CVSs, it is usually sufficient to employ differential equations or difference equations to describe the dynamic behaviour. There exists no such unique tool in the case of DESs. Recently, efforts have been made to develop systematic methodologies for DESs, and various formal or semi-formal methods have been proposed for DESs, such as Petri nets, minimax algebra, temporal logic, automata, etc. But none of these methods in DESs has reached the same development as differential equations in CVSs. Although the conventional way of treating DESs and CVSs separately and independently greatly simplifies the problem of modelling and analysis, it also introduces inefficiency and inaccuracy due to the lack of interaction in such models. Also, it is obvious that either the DES or the CVS model is only an approximation of a real process, so that a significant amount of knowledge is lost in such a single mode model due to the neglecting of the other part, resulting in a poor performance when the model is used for design and control.

In our hybrid system, the overall system is divided into three primary subsystems, which are themselves composed of many sub-subsystems of the same category. Suppose we have a hybrid system which has the topology of Fig. 1, i.e., we have N_c CVSs, N_d DESs, and N_m MDCSs. Now we will propose a semi-formal formulation for the hybrid system modelling.

2.1 Formulation of the CVSs

In our hybrid system, the CVS could be imagined as a continuous system with time variant model parameters. In other words, although the primary behaviour of CVS is governed by continuous dynamics, there exist possibly many time instants along the time trajectory when the system parameters may change abruptly due to the external interactions with the DESs where some events, such as start/stop operation of certain processes, sudden material and environment change, random pocess faults, etc. For such kind of piecewice continuous systems, we propose the following generalised differential algebraic equations (GDAE), which is an extended

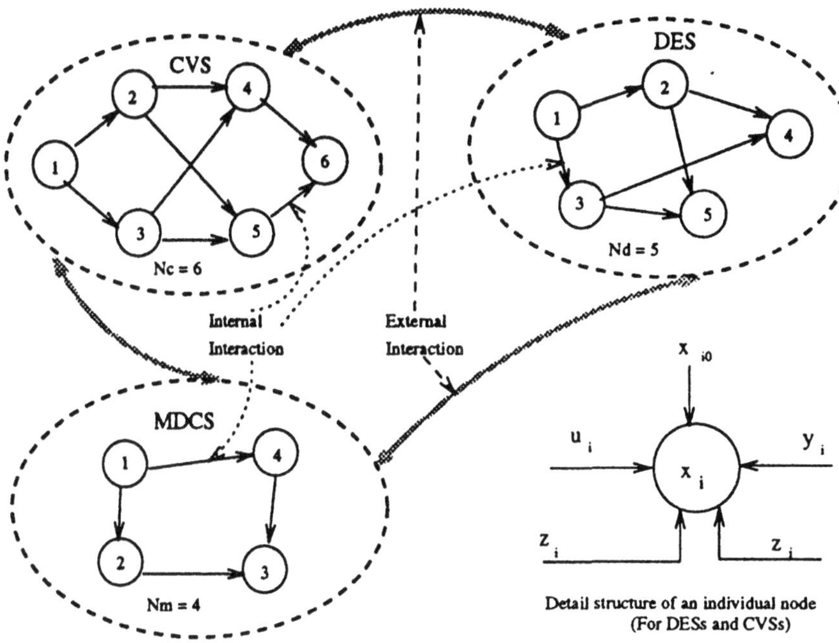

Fig. 1. Topology of a hybrid system

version of differential algebraic equations [13], as the modelling tool:

$$
\begin{aligned}
\frac{dx_c}{dt} &= f_c(x_c, w, u_c, \beta, t) \\
y_c = g_c(x_c, u_c, t), \quad \beta &= \varphi(z_{co}, t) \\
x_c(0) &= x_{c0}
\end{aligned}
\tag{1}
$$

where x_c is a differential variable, w is an algebraic variable, u_c and y_c are the input and output variables respectively, β is a dynamic model parameter, and z_{co} is an external interaction variable. All variables are in vector form with appropriate dimensions.

The key extension made in equation (1) is the introduction of dynamic model parameter β. The physical explanation is that in a CVS some of the model parameters are changing according to the events happening in DESs and MDCSs. A function, $\beta = \varphi(z_{co}, t)$, is introduced to map such changes in the outside world onto the parameter space within the CVS model. In this way the CVS is integrated into a unified environment of the hybrid system. It is assumed here that the CVS's model structure is not changed whatever outside event happens. This assumption will not affect the generality of the CVS model. If a CVS does not satisfy this assumption, we can easily move this subsystem into the category of MDCS, where facilities are provided to treat more strong interaction between DESs and CVSs.

For the CVSs shown in Fig. 1, we can write a more detailed GDAE:

$$\frac{dx_{ci}}{dt} = f_{ci}(x_{ci}, z_{cii}, w_{ci}, u_{ci}, \beta_i, t)$$

$$y_{ci} = g_{ci}(x_{ci}, u_{ci}, t), \qquad z_{cii} = h_{ci}(y_c, t) \tag{2}$$
$$\beta_i = \varphi(z_{coi}, t), \qquad x_{ci}(0) = x_{ci0} \quad i = 1, 2, \ldots, 6$$

(where subscript i represents the ith subsystem in the CVSs, z_{cii} and z_{coi} represent the internal and external interaction of subsystems i, respectively. Other symbols have the same meaning as described in equation (1). If we consider the simplest interaction of direct signal connection, the interaction equation $z_{cii} = h_{ci}(y_c, t)$ can be expressed in a matrix form:

$$z_c = H \times y_c \tag{3}$$

where z_c and y_c are the internal interaction vector and output vector, respectively, and $H = \{h_{ij}\}_{N_c \times N_c}$ is the interconnection matrix. If there is an interconnection from subsystem j to subsystem i, then $h_{ij} = 1$, otherwise $h_{ij} = 0$. Thus, for the CVSs shown in Fig. 1, we have:

$$H = \begin{vmatrix} 0\,0\,0\,0\,0\,0 \\ 1\,0\,0\,0\,0\,0 \\ 1\,0\,0\,0\,0\,0 \\ 0\,1\,1\,0\,0\,0 \\ 0\,1\,1\,0\,0\,0 \\ 0\,0\,0\,1\,1\,0 \end{vmatrix} \tag{4}$$

2.2 Formulation of the DESs

As has been discussed in Section 1, the DESs have a quite different behaviour compared to that of the CVSs. Usually, the evolution of a DES in the time domain depends on the complex interactions of the timing of various discrete events, such as the arrival or the departure of a job, and the initiation or the termination of a task. The state history of such dynamic systems is piece-wise constant and changes at discrete time instants. According to Ho [12], we can conceptually visualise such a DES as consisting of jobs (such as assemble, transfer, process) and resources (such as machines, robots, operators). In a practical system, there are usually a large number of jobs and resources in a DES and the interrelationships among jobs and resources are often complicated, resulting in an endless variety of queuing disciplines, priorities, service requirements, routings, scheduling, and general logical conditions that need to be met for an interaction to take place. Although in recent years much effort has been made to develop the modelling paradigms for DESs, it is still in its infancy and the general model structure has not yet been established. Some commonly used tools to tackle DES modelling are Markov chain theory, automata, Petri nets, minimax algebra, queuing network theory, and each of them deals only with some specific aspects of DES while being incapable of, or inefficient for, other aspects.

It is quite clear that the most important behaviour of DES can be captured by the state sequence $\{s_i\}$ and the event sequence $\{e_i\}$. Abstractly, the state sequence can be expressed by an automata:

$$A = \{I, O, S, ST, OT, s_0\} \tag{5}$$

where I is the input set, O is the output set, S is the state set, $ST : S \times I \longrightarrow S$ is the state transition map, $OT : S \longrightarrow O$ is the output transition map, and s_0

is the initial state. To serve the purpose of hybrid system modelling, we will focus our attention mainly on the state and event sequences, and equation (5) can be specialised, by adopting the conventional function notation, into the following form:

$$
\begin{aligned}
x_d(k+1) &= f_d(x_d(k), u_d(k), k) \\
y_d(k) &= g_d(x_d(k), k) \\
e(k) &= \varphi_d(x_d(k), u_d(k), z_o(t_k), k) \\
t_k &= t_{k-1} + \tau(e(k)) \\
x_d(0) &= x_{d0}
\end{aligned}
\tag{6}
$$

where $x_d(k) \in S$ is the kth state of the DESs, $u_d(k) \in I$ is the kth input of the DESs, $y_d(k) \in O$ is the kth output of the DESs, $e(k) \in E$ (E denotes the possible event set) is the kth event in the DES; t_k is the time when the kth event $e(k)$ happens; z_{do} is the external interaction variable, $\tau(e(k))$ is the time duration between the k-1th and the kth event, $f_d \subseteq ST$ is the functional representation of the state transition map, and $g_d \subseteq OT$ is the functional representation of the output transition map.

If we know the interrelationships in the DESs, we can also write the dynamic equations for each individual DES, as we have done in the CVS case. It should be noted that equation (6) is only meaningful symbolically. Because of the complicated interconnections encountered in DESs, a closed form of equation (6) is hardly possible. In Section 3 we will show how ECPNs can be used to implement this model structure.

2.3 Formulation of the MDCSs

The reason we introduce the MDCSs is that, often in a hybrid system, there are some sub-processes which involve a very strong connection of the discrete events and continuous dynamics such that they cannot be further separated either in the time domain or in the spatial domain. Now let us look at a single node of the MDCSs shown in Fig. 1, where the signal relationship for the node can be graphically represented by Fig. 2. Notice that all of the state, input, and output of a given node, say node i, are composed of two parts: the continuous one and the discrete one. Alternatively, we have $x_{mdci} = [x_{mdi}, x_{mci}]^T$, $u_{mdci} = [u_{mdi}, u_{mci}]^T$, and $y_{mdci} = [y_{mdi}, y_{mci}]^T$. Moreover there are essentially two interconnected dynamics inside a single MDCS node which drive the corresponding states and outputs in conjunction.

In our MDCSs, it is assumed that the discrete part is in active mode. This means that the associated continuous dynamics are activated initially by some discrete events, and then the continuous part will be governed by its dynamics within the time domain. The discrete parts can send signals in order to change the model parameters as well as the model structures of the continuous part, based on the evolution of the discrete activities and the detected/continuous state. Meanwhile, the evolution of the discrete events in the MDCSs could be affected by the continuous state, which is accessible in the discrete event part. These continuous states may be some preconditions for the discrete activity to happen, or the threshold values related to certain discrete events. Based on these active-passive interrelationships in the MDCS, the modelling method developed earlier for DESs and CVSs can be

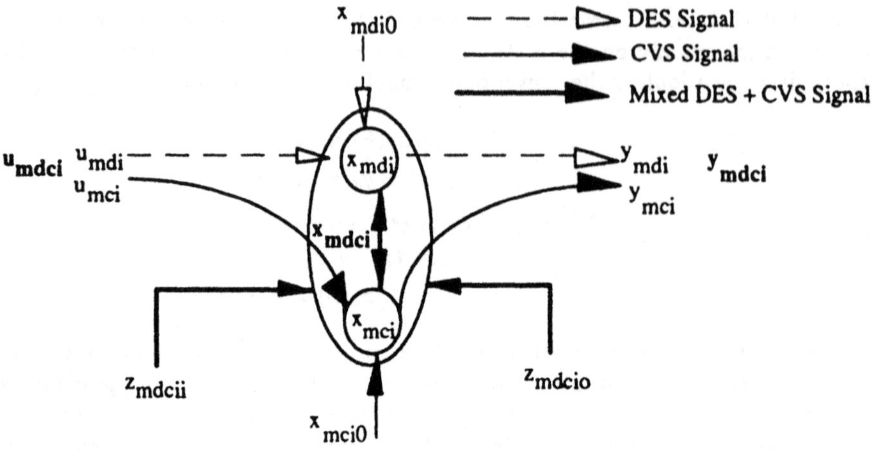

Fig. 2. Detailed signal flowchart of a MDCS

applied readily to the MDCSs, with some necessary extensions. The final model structure of the MDCSs are shown in equations (7-8):

$$\begin{aligned} x_{md}(k+1) &= f_{md}(x_{md}(k), u_{md}(k), k) \\ y_{md}(k) &= g_{md}(x_{md}(k), k) \\ e(k) &= \varphi_{md}(x_{md}(k), u_{md}(k), z_{mdo}(k), x_{mc}(t_k), k) \\ t_k &= t_{k-1} + \tau(e(k)) \\ x_{md}(0) &= x_{md0} \end{aligned} \tag{7}$$

$$\begin{aligned} f_{mc}^{(k)}(x_{mc}^{(k)}, \frac{dx_{mc}^{(k)}}{dt}, w_{mc}^{(k)}, u_{mc}^{(k)}, \beta^{(k)}, t) &= 0 \\ g_{mc}(y_{mc}^{(k)}, x_{mc}^{(k)}, u_{mc}^{(k)}, t) &= 0 \\ x_{mc}^{(k)}(0) &= x_{mc0}^{(k)} \\ t \in [t_k, t_{k+1}), \quad \forall k &= 0, 1, \ldots, n \end{aligned} \tag{8}$$

where the superscript k in equation (8) represents the entity related to the kth sub-domain of time, and other symbols have the similar meanings as defined in equations (1-6). The basic idea behind equation (8) is the partitioning of the whole time domain into n continuous time sub-domains $\{[t_k, t_{k+1}); \forall k = 0, 1, \ldots, n\}$ which are related to certain discrete events. The sub-domain time boundaries $(t_k, k = 1, 2, ..., n,)$ are determined by the evolution of the related discrete events governed by equation (7), representing the time instant at which the discontinuity occurs (either model parameter changes or model structure changes). The starting time boundary is $t_0 = 0$.

3 Modelling of Hybrid Systems Based on ECPNs

In this section we will develop the methodology of hybrid system modelling based on ECPNs. Basic model formulations proposed in Section 2 are adopted for the hybrid systems, and we will see how these model formulas can be implemented in an ECPN framework.

3.1 A brief review of Petri nets

The formal definition of a Petri net can be given as follows [14]:

Definition 1: An unmarked Petri net is a 4-tuple, $N = (P, T, Pre, Post)$, where $P = \{P_1, P_2, ..., P_n\}$ is a finite set of places, $T = \{T_1, T_2, ..., T_m\}$ is a finite set of transitions, $P \cap T = \emptyset$, $Pre : P \times T \longrightarrow \{0, 1, 2, 3, \ldots\}$ is the input incidence application, and $Post : P \times T \longrightarrow \{0, 1, 2, 3, \ldots\}$ is the output incidence application.

Graphically, a Petri net can be represented by a directed graph which consists of arcs and two kinds of nodes (places and transitions). $Pre(P_i, T_j)$ is the weight of the arc $P_i \longrightarrow T_j$, with $Pre(P_i, T_j) = 0$ meaning no arc from place P_i to transition T_j. Similarly, $Post(P_i, T_j)$ is the weight of the arc $T_j \longrightarrow P_i$.

Definition 2: A marked Petri net is a pair $M =< N, M_0 >$ in which N is the unmarked Petri net and M_0 is an initial marking.

A marking assigns to each place a non-negative integer, interpreted as the number of tokens in that place. The marking of the Petri net is denoted by M, an n-vector where n is the total number of places. The ith component of M, denoted by $M(i)$, is the number of tokens in place P_i. Fig. 3 is a graphical representation of a simple Petri net, where $Pre(1,1) = 2$, $Pre(2,1) = 1$, $Pre(3,1) = 0$, $Post(1,1) = 0$, $Post(2,1) = 0$, $Post(3,1) = 2$, and the initial marking is $M_0 = [2, 2, 0]^T$.

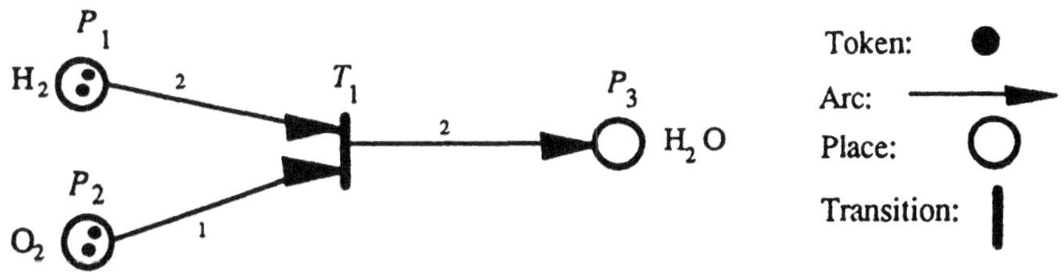

Fig. 3. A simple Petri net

Definition 3 (Enabled transition): In any marking M, a transition $T_j \in T$ is enabled iff

$$\forall P_i \in {}^\circ T_j, \; M(P_i) \geq Pre(P_i, T_j) \tag{9}$$

where ${}^\circ T_j$ denotes the set of input places for T_j, i.e., ${}^\circ T_j = \{P_i | Pre(P_i, T_j) \geq 1\}$.

Definition 4 (Firing transition): A transition T_j may be fired only if T_j is enabled, and when it is fired the Petri net will have the new marking M' determined by:

$$\begin{array}{c} \forall P_i \in {}^{\circ}T_j,\ M'(i) = M(i) - Pre(P_i, T_j); \\ \forall P_i \in T_j^{\circ},\ M'(i) = M(i) + Post(P_i, T_j); \\ \forall P_i \notin ({}^{\circ}T_j \cup T_j^{\circ}),\ M'(i) = M(i) \end{array} \tag{10}$$

where T_j° denotes the set of output places for T_j, i.e., $T_j^{\circ} = \{P_i | Post(P_i, T_j) \geq 1\}$, M is the marking before the firing of transition T_j. According to Definition 4, the marking of the Petri net in Fig. 3 after the firing of T_1 is $M' = [0, 1, 2]^T$.

A major strength of Petri nets is their support for analysing many properties and problems associated with discrete event systems. Two types of properties can be studied with a Petri net model: those which depend on the initial markings (also referred to as behavioural properties) such as reachability, liveness, safeness, boundedness, etc., and those which are independent of the initial marking (called structural properties) such as the structural conflict, place invariance, etc. [15, 9].

To cope with different modelling and analysis requirements, there are substantial developments in Petri net theory and concepts. Many extensions have been made to the original Petri nets, among them are timed Petri nets [16], coloured Petri nets (CPNs) [17, 18], stochastic Petri nets [19, 20], hybrid Petri nets [21, 22], etc.

CPNs [17] were developed based on the basic Petri nets, with the introduction of a new tuple of colour. The aim of introducing the colour is to simplify the Petri net structure by enfolding the original Petri net for these parts which have the same subnet-structure. Thus, with CPNs, we will be able to model more complicated systems with a better visualisation because of a relatively small network structure.

Definition 5: A CPN is a 6-tuple $CPN = (P, T, Pre, Post, M_0, C)$, where $P = \{P_1, P_2, ..., P_n\}$ is a finite set of places, $T = \{T_1, T_2, ..., T_m\}$ is a finite set of transitions, Pre and $Post$ are input and output incidence functions relating to the firing colours, $C = \{C_1, C_2, ...\}$ is the set of colours, and M_0 is an initial marking.

Letting $C(P_i)$ and $C(T_j)$ denote the set of colours in P_i and T_j, respectively, we have:

$$\begin{array}{l} C(P_i) = \{C_{i1}, C_{i2}, \ldots, C_{iu}\} = C_I,\ C_I \subseteq C,\ u = \|C(P_i)\|, \quad i = 1, \ldots, n; \\ C(T_j) = \{C_{j1}, C_{j2}, \ldots, C_{jv}\} = C_J,\ C_J \subseteq C,\ v = \|C(T_j)\|, \quad j = 1, \ldots, m. \end{array} \tag{11}$$

The behaviour of firing transition T_j is determined by the current firing colour $C_j^* \in C_J$. So in a CPN, the Pre and $Post$ incident functions are defined by:

$$Pre(P_i, T_j, C_j^*) = f_{ij}(C_j^*) = \sum_{s=1}^{u} N_{1s}(C_j^*) C_{is}$$

$$Post(P_i, T_j, C_j^*) = g_{ij}(C_j^*) = \sum_{s=1}^{u} N_{2s}(C_j^*) C_{is}; \quad C_j^* \in C_J \tag{12}$$

where N_{1s} and N_{2s} are non-negative integer functions.

Having defined the Pre and $Post$ functions in equation (12), the definitions of enabling and firing of CPNs are similar to those of ordinary Petri nets, with the extension of considering the current firing colour for the transition.

Definition 6 (Enabled transition): In any marking M, a transition $T_j \in T$ is enabled against the firing colour C_j^* ($C_j^* \in C_J$) iff

$$\forall P_i \in {}^\circ T_j,\ M(P_i) \geq Pre(P_i, T_j, C_j^*)$$

Definition 7 (Firing transition): A transition T_j may be fired only if T_j is enabled, and when it is fired with firing colour C_j^* the CPN will have the new marking M' determined by:

$$\begin{aligned} &\forall P_i \in {}^\circ T_j,\ M'(i) = M(i) - Pre(P_i, T_j, C_j^*); \\ &\forall P_i \in T_j^\circ,\ M'(i) = M(i) + Post(P_i, T_j, C_j^*); \\ &\forall P_i \notin ({}^\circ T_j \cup T_j^\circ),\ M'(i) = M(i) \end{aligned} \tag{13}$$

Fig. 4 is used to illustrate the basic ideas of CPNs. Here, $C = \{<a>, <b>, <c>, <e>, <h>, <x>, <>\}$, $C_I|_{I=1,2} = \{<a>, <b>, <c>, <e>\}$, $C_I|_{I=3} = \{<a>, <e>, <h>, <x>, <>\}$. The initial marking $M_0 = [4<a> +3<b> +6<e>, 5<b> +4<c> +2<e>, 0]^T$. By Definition 6, the enabling firing colours for T_1 are either $<b>$ or $<e>$, and the resulting marking after firing T_1 under the fringing colour $C_j^* = <e>$ is $M' = [4<a> +3<b> +3<e>, 5<b> +4<c> + <e>, 4<e>]^T$, according to Definition 7.

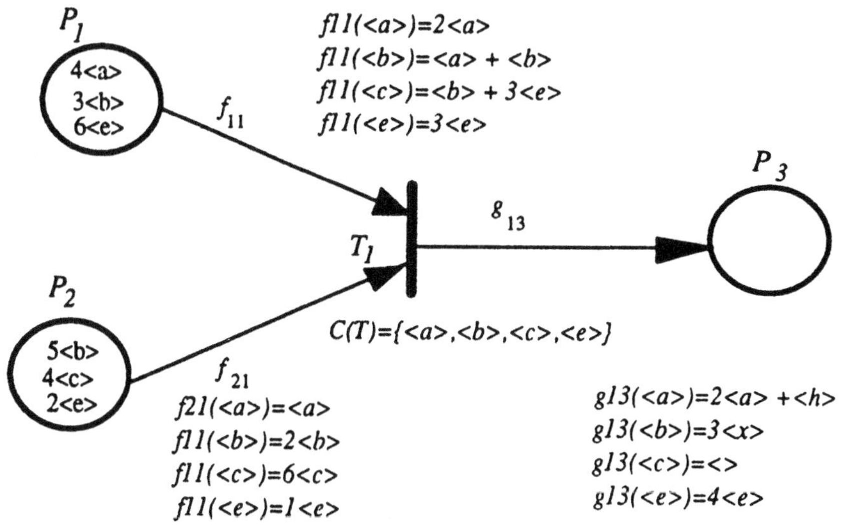

Fig. 4. A coloured Petri net

3.2 Extended coloured Petri nets

The CPN defined in Section 3.1 can be converted into an ordinary Petri net by means of unfolding [17]. Obviously, the introduction of colour is very useful for simplifying the net structure and makes modelling and analysis of larger problems possible, but it adds no extra power to the modelling ability. In other words, it is equivalent to the ordinary Petri net with respect to the modelling power. In the following, we will make further extensions to the CPN to provide capabilities of modelling the continuous dynamics without losing the existing modelling power.

First, we introduce the concepts of dynamic colours, which have a set of continuous attributes attached to the tokens. These continuous attributes (called color states in this paper) are defined as continuous variables associated with the dynamic colour. For example, suppose we have a dynamic colour of $< Ice >$ attached with a colour state of 'Temperature', then we can determine the difference between two tokens with the same colour $< Ice >$ by checking its colour state, and the colour state can then be used to represent the dynamic behaviour of the related colour of $< Ice >$.

Definition 8: An ECPN is a 7-tuple $ECPN = (P, T, Pre, Post, M_0, C, Dyn)$, where $P = \{P_1, P_2, ..., P_n\}$ is a finite set of places (including dynamic colors), $T = \{T_1, T_2, ..., T_m\}$ is a finite set of transitions, Pre and $Post$ are input and output incidence functions relating to the firing colours, $C = \{C_1, C_2, ...\}$ is the set of colours, M_0 is an initial marking, and $Dyn = \{DynP, DynT, DynC\}$ is the connecting map of places, transitions, and colors to their dynamic properties, respectively. Dyn can be expressed by:

$$\begin{aligned} DynP &: P \longrightarrow \{0,1\} \\ DynT &: T \longrightarrow \{0,1\} \\ DynC &: C \longrightarrow \{0,1\} \end{aligned} \qquad (14)$$

$DynC(C_k) = 0$ means colour C_k is an ordinary colour, while $DynC(C_k) = 1$ means colour C_k is a dynamic colour. The continuous attributes (colour state) of a dynamic colour C_k is denoted by x_k. Dynamic places and dynamic transitions are critical components in the ECPN. Dynamic places can hold tokens with dynamic colours, while dynamic transitions can have functions related with the dynamic colours of the tokens residing in its input (or output) places. The formal definitions of dynamic places and dynamic transitions are given in Definitions 9-10.

Definition 9: A place P_i is a dynamic place (D-place) $iff \exists k, C_k \in C_I \cap DynC(C_k) = 1$; In other words, a place P_i is a D-place if it can contain at least one dynamic colour. DP is the set of all D-places in the Petri net, i.e. $DP = \{P_i | C_k \in C_I \cap DynC(C_k) = 1\}$. X_i is the set of colour states contained in a D-place P_i, which is the collection of color state of the individual D-colours existing in P_i, i.e., $X_i = \{x_k | C_k \in C_I \cap DynC(C_k) = 1\}$. X is the set of all colour states of the Petri net, i.e., $X = \bigcup_{i=1,n} X_i$.

Definition 10: A transition T_j is a dynamic transition (it D-transition) $iff \exists i$, such that

$$\sum_i \Gamma_{ij} \cap \sum_k f_{jk} \neq \emptyset, \quad \forall P_i \in {}^{\circ}T_j, \forall P_k \in T_j^{\circ} \qquad (15)$$

where Γ_{ij} and f_{jk} are defined by equations (16-17). That is, a dynamic transition T_j will change some of the colour states of the output tokens or the firing will depend on some of the colour states of the input tokens. The set of D-transitions in the Petri net is denoted as DT.

The enabling rules for ECPNs are the same as those for ordinary CPNs except for these dynamic transitions. For a D-transition, an extra condition check, which is in the form of a Boolean expression related to the set of colour state X_i for each input place P_i, will be conducted before the transition T_j can be enabled. We can express this extra condition check as:

$$PreD(P_i, T_j, C_j^*) = \Gamma_{ij}(X_i, C_j^*) \tag{16}$$

where Γ_{ij} is the Boolean function whose value is determined by the set of colour states X_i and the firing colour C_j^* . Only when $PreD(P_i, T_j, C_j^*) = true$ can transition T_j be enabled against firing colour C_j^*, supposing the normal enabling conditions of $Pre(P_i, T_j, C_j^*) = true$ are satisfied.

Similarly, the firing rules are the same as those of ordinary CPNs except those of D-transitions. For a D-transition, extra functions are executed to update the value of each colour state x_k belonging to the set X_i for the related output place P_i. The colour state updation is usually governed by a equation derived from the behaviour of the continuous system dynamics:

$$X_l' = f_{jl}(X_{i1}, \ldots, X_{in}, C_j^*) \tag{17}$$

where X_l' is the set of colour state in the output place P_l after the transition T_j is fired, and X_{i1}, ..., X_{in} are the sets of colour states for each of the input places of transition T_j.

3.3 Modelling of hybrid systems based on ECPNs

Now we will propose a procedure for the modelling of hybrid systems based on ECPNs. For the modelling of the DES part, we can use the design approach developed in CPNs for ECPNs, since the components of D-colour and D-transition are not necessary for DES, and the ECPN reduce to an ordinary CPN.

We shall consider the CVSs as given by equation (1). A discrete time version of equation (1), which can easily be obtained by the finite difference approach, has the form:

$$\begin{aligned} x_c(k+1) &= f_c(x_c(k), u_c(k), \beta(k), k) \\ \beta(k) &= \varphi(z_{co}, k), \\ x_c(0) &= x_{c0} \end{aligned} \tag{18}$$

To model this CVS we introduce the D-colours of $< x >, < u >, < \beta >$, D-places of P_1, P_2, P_3, P_4, and D-transitions of T_1, T_2, T_3. The calculation of equation (18) can be implemented as a series of events at $k\Delta\tau$, $(k = 0, 1, 2, \ldots)$, with f_c embedded into T_1, and φ embedded into T_2. According to the concepts developed in the previous sections, the corresponding ECPN implementing this CVS subsystem is shown in Fig. 5.

In Fig. 5, new symbols introduced for D-place, D-transition, and I/O arc are shown. The I/O arc is a special arc that satisfies $Pre(P_i, T_j, C_t) = Post(P_i, T_j, C_t)$, where P_i and T_j are the place and transition which the I/O arc connects. The introduction of the I/O arc is to simplify the implementation for the CVS modelling,

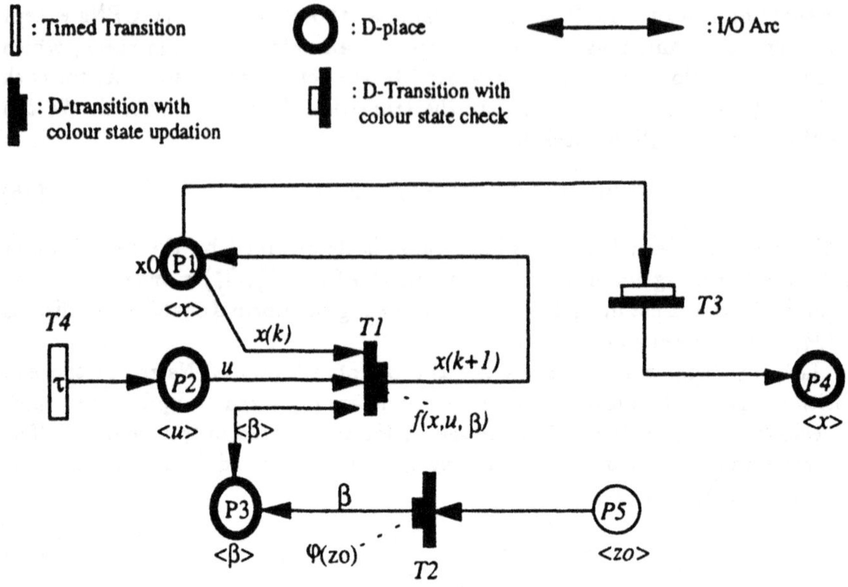

Fig. 5. Modelling of CVS by ECPNs

since in many continuous processes there is no token movement involved, only the colour state is passed to the required elements. Transition T_4 is introduced for the timing ($\Delta\tau$) of calculation for T_1 as well as to provide the input for the CVS, and T_3 is fired when the terminating condition for the continuous calculation is satisfied. Transition T_2 is to provide those model parameters which are determined by the evolution of the external DESs and MDCSs. Modelling of MDCSs can be carried out in a similar way with the incorporation of DESs modelling, and will be shown in Section 4 where a case study is given. Thus, all three subsystems of CVSs, DESs, and MDCSs involved in a hybrid system can be modelled in a unified way using ECPNs.

4 Case Study – Modelling of a Soaking Pit Process in a Steel Mill

The soaking pit/rolling mill complex is an important part in an iron and steel plant which may be divided into: (1) productive units, such as blast furnaces, steelworks, and rolling mills, etc., and (2) non-productive parts, such as the soaking pits, and the molten mixers, etc. The soaking pits, which act as an intermediate process between the steelworks and the rolling mills, have two main functions: to heat the ingots to a uniform temperature level in preparation for rolling, and to store the hot ingots until the mill is ready to roll them.

4.1 Description of the process

The practical soaking pit/rolling mill complex consists of fifteen soaking pits arranged in the soaking pit bay, and four mills with different capabilities to produce a variety of shapes and sizes for its products. Two overhead travelling cranes which span the pit-mill area (including all tracks) charge incoming ingots from the steelworks into the pits and draw the ingots out for rolling. Ingot transfer in the mill area is carried out by the track conveyor systems. The time required to transport hot ingots from the steelworks to the soaking pits is subject to stochastic variations caused by congestion and equipment failure. The number of ingots charged into a soaking pit depends on its size. The charging time depends on the destination pit, the source of input ingots, and the interaction between cranes. Fig. 6 depicts a schematic diagram of the ingot flow within the steel mill complex. In this paper we will focus our effort on the modelling of soaking pits process only. More detail information of this soaking pit/rolling mill process can be found in Yang et al [23].

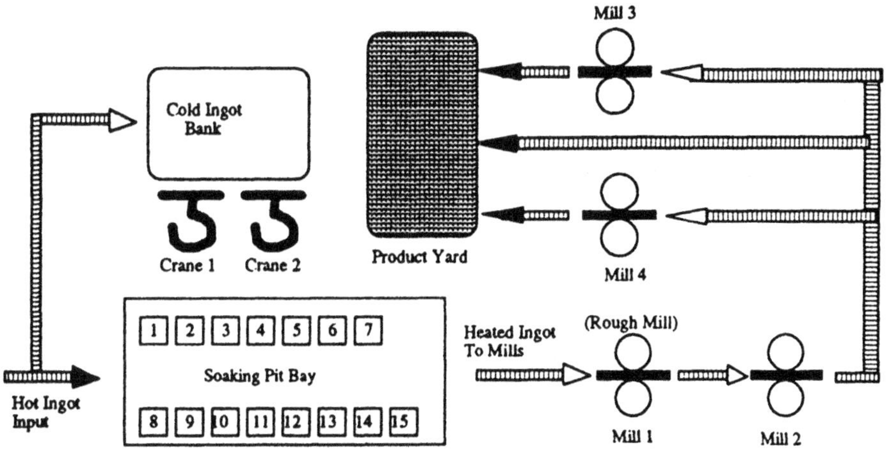

Fig. 6. The layout of a soaking pit/rolling mill complex

4.2 Modelling of the soaking pits

Based on the methodology developed in Section 2 and 3, the soaking pit process, which is a hybrid process, can be modelled by ECPNs. When a soaking pit is empty and ready for charging, a batch of ingots (15 ingots in this paper) is moved into the soaking pit one at a time. The preconditions of charging activity for a soaking pit are as follows: the pit is empty (or not full during the charging period), ingots are available for charging, and at least one of the cranes is available (two cranes are shared resources for both the charging and discharging operations). After the charging process is finished, the soaking process begins to drive the ingot temperature to its pre- specified level. For the discharging operation, the preconditions are: at least one of the cranes is available, the roughing mill is ready to roll, and the thermal state of the ingot in the soaking pit has met the requirement for rolling. Fig. 7 is the resulting ECPN model, where only two soaking pits are shown for the sake of graphical simplicity. It is easy to draw the overall ECPN model by duplicating a single soaking pit.

Notice that in Fig. 7 the Transitions T_{21} and T_{22} are shown in an abstracted level. Besides the normal token movements along the arcs of these transitions, the colour state (in this case the ingot temperature distribution) attached to the colour tokens must be updated whenever the transitions are fired. In order to implement these two transitions, we need to establish the dynamics of the ingot soaking process.

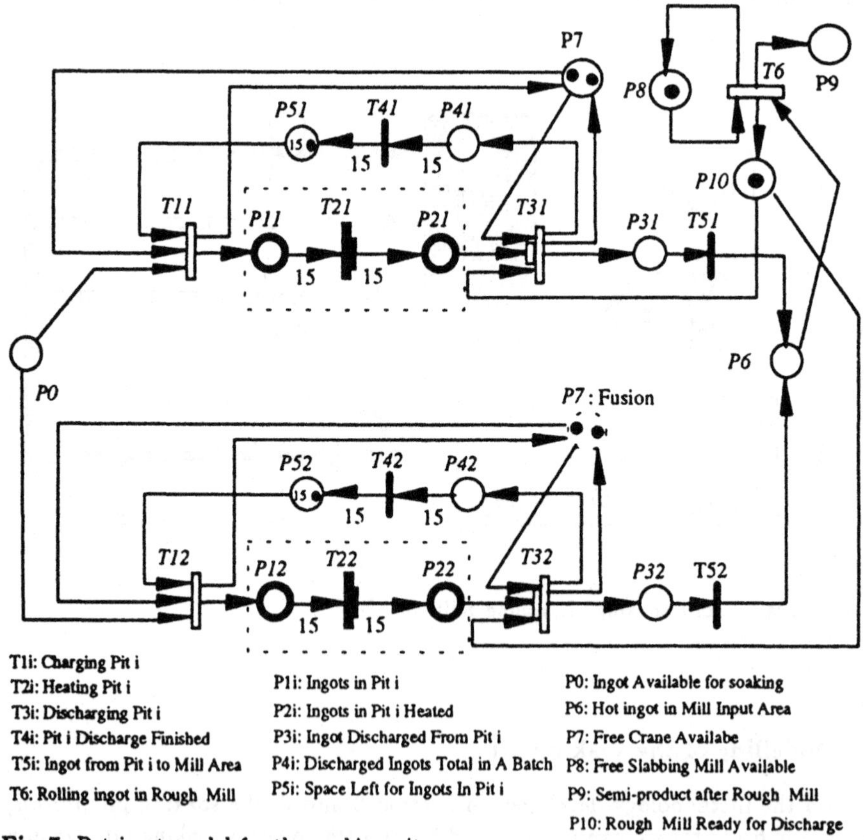

Fig. 7. Petri net model for the soaking pit area

According to Lu and Williams [24], the temperature distribution within the ingot is assumed to be one-dimensional under a cylindrical co-ordinate system. Based on the heat balance principle and some basic heat transfer equations, the temperature distribution within the ingots can be represented by a partial differential equation together with the related boundary conditions and initial conditions:

$$\begin{aligned}\frac{\partial T_s(r,t)}{\partial t} = \frac{1}{c\rho r}\frac{\partial}{\partial r}\left(kr\frac{\partial T_s(r,t)}{\partial r}\right), \quad r < R,\ t_0 \geq t \geq t_f \\ \frac{\partial T_s(r,t)}{\partial r}\Big|_{r=R} = -q(T_s, T_p) \\ T_s(r,t_0) = T_{s0}(r)\end{aligned} \tag{19}$$

where $T_s(r,t)$ is the temperature distribution of the ingot; $T_p(t)$ is the pit temperature; k, ρ, and c are the specific conductivity, density and specific heat coefficients of the ingot, respectively; R is the equivalent radius of the ingot; $q(T_s, T_p)$ is the heat flow from pits to the ingot surface; $T_{s0}(r)$ is the initial ingot temperature distribution, and t_f is the total soaking time.

Equation (19) can be converted further into a discrete state space model if we properly discretize the space and time domain. The resulting discrete state space model takes the form:

$$\begin{aligned}x(k+1) = f_d(x(k), u(k), \beta(k), k) \\ \beta(k) = f_\beta(z_o) \\ x(0) = x_0\end{aligned} \tag{20}$$

where $x(k) = [T_s(1,k), T_s(2,k), \ldots, T_s(N,k)]^T$ is the temperature distribution of the ingot, $u(k)$ is the input of the model, f_d is the discrete state transition function, and f_β is the model parameter vector. Adopting the modelling techniques developed in Section 3, the detailed Petri net structure for T_{21} and T_{22} is sketched in Fig. 8, which is obtained based on the scheme presented in Fig. 5.

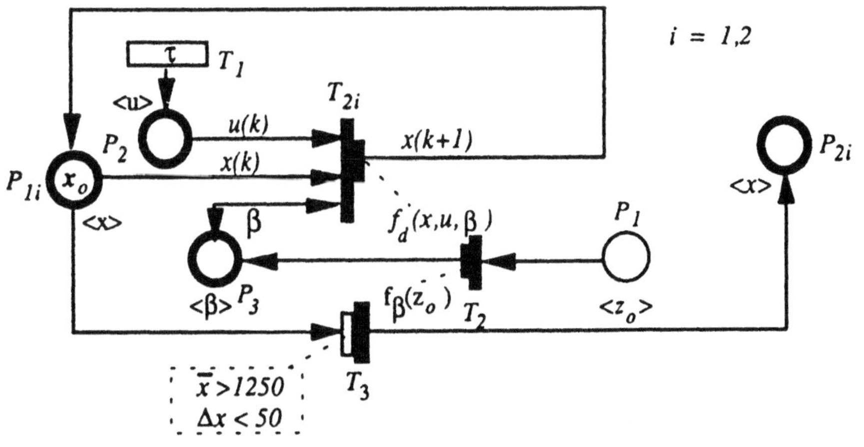

Fig. 8. Details for transition T11 and T12

In Fig. 8, the colour state $x(k)$ is calculated every τ time using equation (20), based on the current colour state, the input and the model parameters. Transition T_3 is designed to terminate the colour state updation if the colour state reaches a certain status (the ingot is hot enough (>1250C) and the ingot temperature difference in small enough (<50C) for rolling). As soon as T_3 is fired the tokens in place P_{1i} will be moved to P_{2i} and T_{2i} will be disabled, indicating the end of the soaking process and the ingots in the soaking pit being ready for discharge. Transition T_2 is used to provide the correct model parameters based on the current hybrid system environments. Typical simulation results for the soaking pit operations is shown in Fig 9 and Fig. 10 using the developed ECPN model. The temperature specifications for ingot rolling are $\bar{x}(K_f) \geq 1250°C$, and $\Delta x(K_f) \leq 50°C$. There are fourteen soaking pits online for this simulation, and the average rolling rate for Mill 1 is 1.5 minute/per ingot.

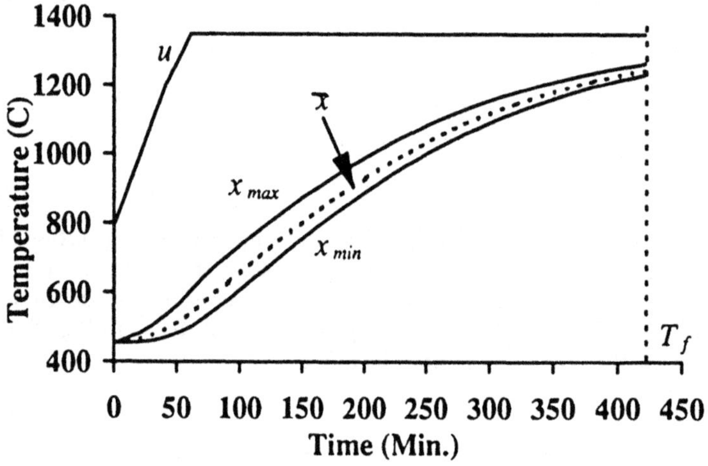

Fig. 9. Typical ingot temperature trajectory during soaking period

5 Conclusions

In this paper we have proposed a novel approach for the modelling of hybrid systems based on ECPNs. First we have proposed a general model structure for the hybrid system, which is composed of three basic components of CVSs, DESs, and MDCSs. The signal relationships between these subsystems are discussed, along with the model paradigm for each component. A generalised DAE is proposed in order to formulate the modelling of the CVSs in our hybrid system. Then, extensions to the CPNs are proposed for ECPNs. The key components of the extensions are the dynamic colours, dynamic places, and dynamic transitions. The continuous properties

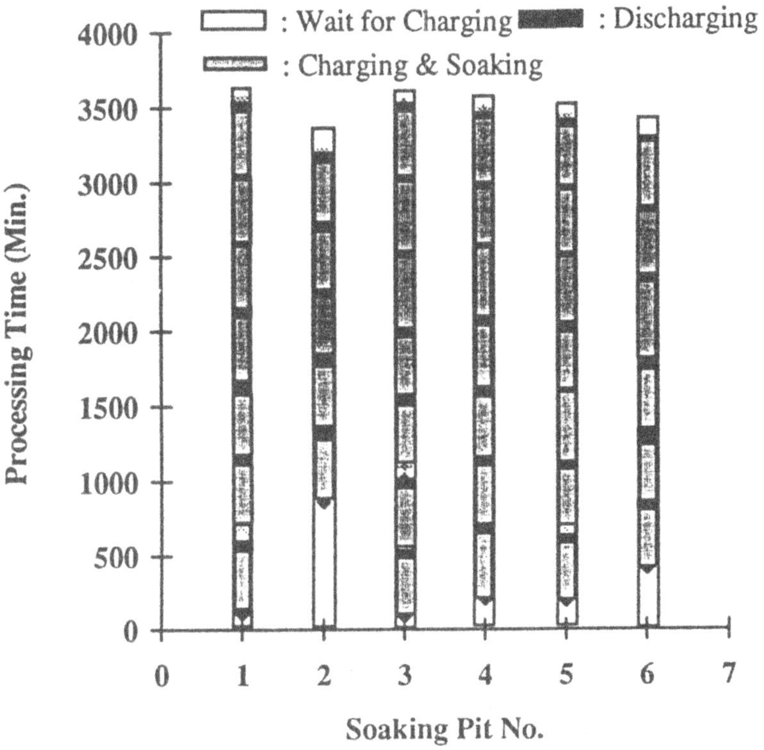

Fig. 10. Typical soaking pit production cycle sequences obtained through simulation

and the discrete event properties of hybrid systems can be unified by the concept of tokens with dynamic colour: the number of the tokens for discrete events and the value of the tokens (colour state) for continuous properties. This extension is a crucial step toward the hybrid system modelling using ECPNs. Two additional components for dynamic transitions are introduced: an extra conditional check for enablement, and the updating of the colour state during firing. These extensions make ECPNs suitable for modelling all of the three components involved in hybrid systems. An illustrative example is given to show the developed methodology. We believe the ideas developed in this paper are original and will benefit the modelling of hybrid systems.

As stated in the introduction, the study of modelling and control of hybrid systems is still in its infancy, and the efforts made in this paper are an attempt at a unified approach for hybrid system modelling. In order to verify such efforts, much more work remains to be done both theoretically and practically. Besides, the potential utilisation of the hybrid model should be considered. An obvious application will be control and optimisation for the hybrid system based on an ECPN model. This will be our future research topic.

References

1. Fahrland, D. A., Combined discrete event continuous system simulation, *Simulation*, 1970, Feb., 61-72.
2. Pantelides, C. C., and Barton, P. I., Equation-oriented dynamic simulation: current status and future perspectives, *Computers and Chemical Engineering*, 1992 17, 263-285.
3. Pantelides, C. C., Gritsis, D., Morison, K. R., and Sargent, R. W. H., The mathematical modelling of transient systems using differential-algebraic equations, *Computers and Chemical Engineering*, 1988, 12, 449-454.
4. Nerode, A., and Kohn, W., Models for hybrid systems: automata, topologies, controllability, observability, *Hybrid Systems, Lecture Notes in Computer Science 736*, Springer-Verlag, 1993, 317-346.
5. Antsaklis, P. J., Stiver, J. A., and Lemmon, M., Hybrid system modelling and autonomous control, *Hybrid Systems, Lecture Notes in Computer Science 736*, Springer-Verlag, 1993, 366-392.
6. Lemmon, M., Stiver, J. A., and Antsaklis, P. J., Event identification and intelligent hybrid control, *Hybrid Systems, Lecture Notes in Computer Science 736*, Springer-Verlag, 1993, 268-296.
7. Alur, R., Courcoubetis, C., Henzinger, T. A., and Ho, P. H., Hybrid automata: an algorithmic approach to the specification and verification of hybrid systems, *Hybrid Systems, Lecture Notes in Computer Science 736*, Springer-Verlag, 1993, 209-229.
8. Henzinger, T. A., Manna, Z., Pnueli, A., Towards refining temporal specifications into hybrid systems, *Hybrid Systems, Lecture Notes in Computer Science 736*, Springer-Verlag, 1993, 60-76.
9. Peterson, J. L., Petri Nets, *Computer Surveys*, 1977, 9(3), 223-253.
10. Yang, Y. Y., Linkens, D. A., and Banks, S. P., An unified approach for modelling of hybrid systems, *Proceedings of the European Simulation Multiconference*, Barcelona, Spain, 1-3 June, 1994, 240-244.
11. Deshpande, A. R., Control of Hybrid System, *PhD Thesis*, University of California, Berkeley, USA, 1994.
12. Ho, Y. C. (Ed.), *Discrete Event Dynamic Systems–Analysing Complexity and Performance in Modern World*, IEEE Press, 1992.
13. Barton, P. I., and Pantelides, C. C., The modelling and simulation of combined discrete/continuous processes, *The 1991 ICHEME Event*, 1991, 115-118.
14. David, R. and Alla, H., *Petri Nets and Grafcet: tools for modelling discrete event systems*, Printice Hall, 1992.
15. Murata, T., Petri nets: Properties, analysis and applications, *Proceedings of IEEE*, 1989, 77 (4), 541-580.
16. Zuberek, W. M., Timed Petri nets: definition, properties and applications, *Microelectron. Reliab.*, 1991, 31(4).
17. Jensen, K., *Coloured Petri Nets: Basic Concepts, Analysis Methods, and Practical Use*, Vol. 1, Springer - Verlag, 1992.
18. Christensen, S., and Petrucci, L., Towards a modular analysis of coloured Petri nets, in Jensen, K. (Ed.), *Application and Theory of Petri Nets 1992, Lecture Notes in Computer Science 616*, Springer - Verlag, 1992, 113-133.

19. Chiola, G., Ajmone, M., Balbo, G., and Conte, G.(1993), Generalised stochastic Petri nets: A definition at the net level and its implementations, *IEEE Trans. on Software Eng.*, 19(2), 89-107.
20. Hatono, I., Yamagata, K., and Tamura, H., Modelling and online scheduling of flexible manufacturing system using stochastic petri nets, *IEEE Trans. on Software Engineering*, 1991, 17(2).
21. Dubois, D., and Stecke, K. E., Dynamic analysis of repetitive decision-free discrete-event process: the algebra of timed marked graphs and algorithmic issues, *Annuls. of Operations Research*, 1990, 26, 151-193.
22. Dubois, E., and Alla, H., Hybrid Petri nets with a stochastic discrete part, *Proc. of European Control Conference 93*, Netherlands, June 1993, 144-149.
23. Yang, Y. Y., Linkens, D. A., and Mort, N., Modelling of a soaking pit/rolling mill process based on extended coloured Petri nets, *Proceedings of the IFAC/IFORS/IMACS Symposium of Large Scale Systems Theory and Applications*, London, UK, July 11-13, 1995.
24. Lu, Y. Z., and Williams, T. J., *Modelling, Estimation and Control of Soaking Pits*, ISA Publisher, 1983.

DEVS Framework for Modelling, Simulation, Analysis, and Design of Hybrid Systems

Bernard P. Zeigler
Electrical and Computer Engineering
The University of Arizona
Tucson, AZ 85721, U.S.A.

Hae Sang Song and Tag Gon Kim
Department of Electrical Engineering
Korea Advanced Institute of Science and Technology
Taejon 305-701, Korea

Herbert Praehofer
Systems Theory and Information Engineering
Johannes Kepler University of Linz
A-4040 Linz, Austria

Abstract

We make the case that Discrete Event System Specification (DEVS) is a universal formalism for discrete event dynamical systems (DEDS). DEVS offers an expressive framework for modelling, design, analysis and simulation of autonomous and hybrid systems. We review some known features of DEVS and its extensions. We then focus on the use of DEVS to formulate and synthesize supervisory level controllers.

1 Introduction

Formal treatment of discrete event dynamical systems is receiving ever more attention[5]. Work on a mathematical foundation of discrete event dynamic modeling and simulation began in the 70s [13, 14, 16] when DEVS (discrete event system specification) was introduced as an abstract formalism for discrete event modeling. Because of its system theoretic basis, DEVS is a universal formalism for discrete event dynamical systems (DEDS). Indeed, DEVS is properly viewed a short-hand to specify systems whose input, state and output trajectories are piecewise constant[17]. The step-like transitions in the trajectories are identified as discrete events.

Recently, interest in hybrid systems (mixed discrete-continuous) systems has been growing. Such systems are prominent in such areas as intelligent control [1, 7] and reactive system design [3]. This article proposes that DEVS-based systems theory provides a sound, general framework within which to address modelling, simulation, design, and analysis issues for hybrid systems.

Fig. 1 shows the basic approach to hybrid systems research proposed here. We briefly address the ideas raised in this figure.

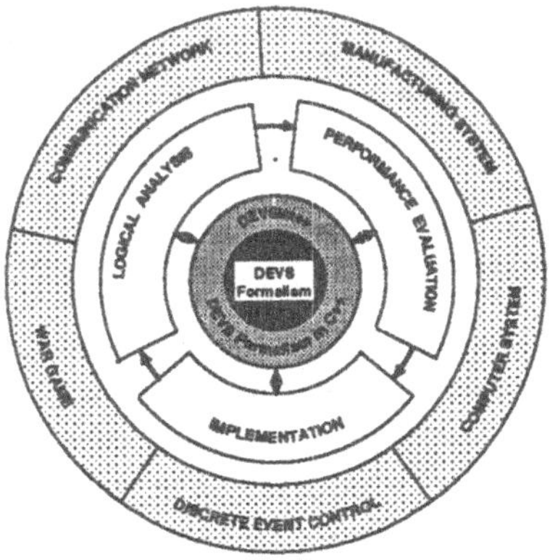

Fig. 1. DEVS-Centered Hybrid Systems Research

1.1 Modelling

Discrete event modelling has its origin in job shop scheduling simulations in operations research. More recently control theorists developed their own discrete event formalisms[5]. The universality claims of the DEVS just cited are addressed by characterizing the class of dynamical systems which can be represented by DEVS models. Prahofer and Zeigler (in press) showed that any causal dynamical system which has piecewise constant input and output segments can be represented by DEVS. We call this class of systems *DEVS-Representable*[17]. In particular, Differential Equation Specified Systems (DESS) are usually used to represent the plant (system under control) in hybrid systems and are controlled by high-level, symbolic, event-driven control schemes. Sensing and actuation in such systems are event-like and they have piecewise constant input and output trajectories when viewed within the frame of their sensing/actuator interface. Consequently, within such an interface, the plant is representable by a DEVS. Likewise, the controller has natural DEVS representation.

Closure under coupling[14, 9] is a desirable property for subclasses of dynamical systems since it guarantees that coupling of class instances results in a system in the same class. The class of DEVS-representable dynamical systems is closed under coupling. This justifies hierachical, modular construction of both DEVS models and the (continuous or discrete) counterpart systems they represent.

1.2 Simulation

DEV&DESS is an extension of the DEVS formalism for combined discrete/continuous modeling and simulation[9, 4]. DEV&DESS includes both the DEVS and DESS classes of systems. In addition it provides a means of specifying systems with interacting continuous and discrete trajectories. The formalism is fully expressive of hybrid systems in that it is shown to be closed under coupling. In particular, the resultants of coupled models that contain components of either class are expressibel as basic models in the formalism. Since it is also implemented in software, DEV&DESS offers a powerful modeling formalism for simulating hybrid systems.

1.3 Design and Analysis

Ramage and Wonham[11] were the first control theorists to develop an approach to designing discrete event controllers. They modelled the plant as an automaton and used language theory to design a controller that forces the plant to exhibit behavior consistent with given objectives. The goal of this paper is to demonstrate the applicability of DEVS as unifying framework for hybrid control. We do so by reformulating the Ramage-Wonham approach to controller design within the DEVS formalism.The resulting methodology is both more general and more intuitively appealing than the original. Since it is universal for general piecewise constant systems, DEVS, can represent timing information about the plant that eludes an automaton formulation. Such information can be employed to improve controller designs. Moreover, DEVS distinguishes between internal and external state transitions, a fact that enables us to derive a controller DEVS from a plant DEVS model in a rather direct manner.

Since the appeal of the proposed methodology rests upon the universality of DEVS representation, we first briefly review systems formalisms and their mapping into DEVS. Then we discuss the DEVS-based methodology for discrete event control of hybrid plants. We conclude with open questions for future research.

2 Review of General Dynamical Systems and DEVS

2.1 General Dynamical Systems

Based on[8, 13, 12] we define a general dynamical system as follows:

$$DS = (T, X, Y, \Omega, Q, \Delta, \Lambda)$$

with

- T is the time base,
- X is the set of input-values,
- Y is the set of output-values,
- Ω is the set of admissible input segments $\omega :< t_1, t_2 > \rightarrow X$ over T and X and Ω is closed under concatenation as well as under left segmentation [13],
- Q is the set of states;
- $\Delta : Q \times \Omega \rightarrow Q$ is the global state transition function,
- $\Lambda : Q \times X \rightarrow Y$ is the output function.

The global state transition function of the general dynamical system DS has to fulfill the following properties [8, 13]:

$$\text{Consistency: } \Delta(q, \omega_{(t,t>}) = q, \tag{1}$$

Semigroup property:

$$\forall \omega :< t_1, t_2 > \rightarrow X \in \Omega, t \in < t_1, t_2 >: \Delta(q, \omega_{< t_1, t_2 >}) = \Delta(\Delta(q, \omega_{< t_1, t >}), \omega_{< t, t_2 >}), \tag{2}$$

Causality:

$$\forall \omega, \bar{\omega} \in \Omega \text{ if } \forall t \in < t_1, t_2 >: \omega(t) = \bar{\omega}(t) \text{ then } \Delta(q, \omega_{< t_1, t_2 >}) = \Delta(q, \bar{\omega}_{< t_1, t_2 >}). \tag{3}$$

Causality, the semigroup property and closure of admissible segments under left segmentation justifies defining the state trajectory resulting from every initial state $q \in Q$ and input segment $\omega :< t_1, t_2 > \rightarrow X \in \Omega$ in the following way:

$$STRAJ_{q,\omega} :< t_1, t_2 > \rightarrow Q \text{ with } \forall t \in < t_1, t_2 > STRAJ_{q,\omega}(t) = \Delta(q, \omega_{< t_1, t >}). \tag{4}$$

Similarly we define the output trajectory $OTRAJ_{q,\omega} :< t_1, t_2 > \rightarrow Y$ by: for every initial state $q \in Q$, input segment $\omega :< t_1, t_2 > \rightarrow X \in \Omega$ and $t \in < t_1, t_2 >$

$$OTRAJ_{q,\omega}(t) = \Lambda(STRAJ_{q,\omega}(t), \omega(t)). \tag{5}$$

Now the input/output behavior R_{DS} of the dynamical system is given by

$$R_{DS} = \{(\omega, OTRAJ_{q,\omega}) : \omega \in \Omega, q \in Q\}. \tag{6}$$

2.2 The Discrete Event System Specification (DEVS) Formalism

An atomic DEVS is a structure [13, 14, 16]

$$DEVS = (X_M, Y_M, S, \delta_{ext}, \delta_{int}, \lambda, ta)$$

where

- X_M is the set of inputs,
- Y_M is the set of outputs,
- S is the set of sequential states,
- $\delta_{ext} : Q \times X \rightarrow S$ is the external state transition function,
- $\delta_{int} : S \rightarrow S$ is the internal state transition function,
- $\lambda : S \rightarrow Y$ is the output function,
- $ta : S \rightarrow \mathcal{R}_0^+ \cup \{\infty\}$ is the time advance function.

A DEVS specifies a dynamical system DS in the following way:

- the time base T is the real numbers $\mathcal{R}$;
- $X = X_M \cup \{\Phi\}$, i.e., the input set of the dynamical system is the input set of the DEVS together with a symbol $\Phi \notin X_M$ specifying the non-event,

- $Y = Y_M \cup \{\Phi\}$, i.e., the output set of the dynamical system is the output set of the DEVS together with Φ.
- $Q = \{(s,e) : s \in S, 0 \leq e \leq ta(s)\}$ the set of states of the dynamical system consists of the sequential states of the DEVS paired with a real number e giving the elapsed time since the last event,
- the admissible input segments is the set of all DEVS-segments over X and T (DEVS-segments [13] are characterized by the fact that for any $\omega :< t_1, t_2 > \rightarrow X \in \Omega$, there is only a finite number of event times $\{\tau_1, \tau_2, \ldots, \tau_n\}, \tau_i \in < t_1, t_2 >$ with $\omega(\tau_i) \neq \Phi$. As a special case, a so-called Φ segment has no events.
- for any DEVS input segment $\omega :< t_1, t_2 > \rightarrow X \in \Omega$ and state $q = (s,e)$ at time t_1 the global state transition function Δ is defined as follows: $\Delta(q, \omega_{<t_1,t_2>}) =$

$$(s, e + t_2 - t_1) \tag{7}$$

$$\text{if } e + t_2 - t_1 < ta(s) \wedge \omega \text{ is the } \Phi \text{ segment,}$$

$$\Delta((\delta_{int}(s), 0), \omega_{<t_1+ta(s)-e,t_2>}) \tag{8}$$

$$\text{if } e + t_2 - t_1 \geq ta(s) \wedge \neg\exists t \in < t_1, t_1 + ta(s) - e >: \omega(t) \neq \Phi,$$

$$\Delta((\delta_{ext}(s, e + t - t_1, \omega(t)), 0), \omega_{(t,t_2>}) \tag{9}$$

$$\text{if } \omega(t) \neq \Phi \wedge t \leq t_1 + ta(s) - e \wedge \omega \text{ restricted to } < t_1, t > \text{ is a } \Phi \text{ segment.}$$

- the output function Λ of the dynamical system is given by

$$\Lambda((s,e), x) = \lambda(s). \tag{10}$$

For a given input segment, ω three different cases can be identified:

- First, if ω is a non-event segment and an internal event does not occur in interval $< t_1, t_2 >$, then the sequential state is left unchanged but the elapsed time component e of the total state has to be updated (7).
- Second, if a time event scheduled by the time advance function does occur in the time interval $< t_1, t_2 >$ and there is no input event prior to the internal event time, then the internal transition function defines a new sequential state, the elapsed time component is set to zero, and the rest of the input segment is applied to this new state (8).
- Third, if the input segment ω defines an input event prior to the occurrence of an internal event, then the external state transition function is applied with the extenal input value and elapsed time and the elapsed time is set to zero. The rest of the input segment is applied to this new state (9).

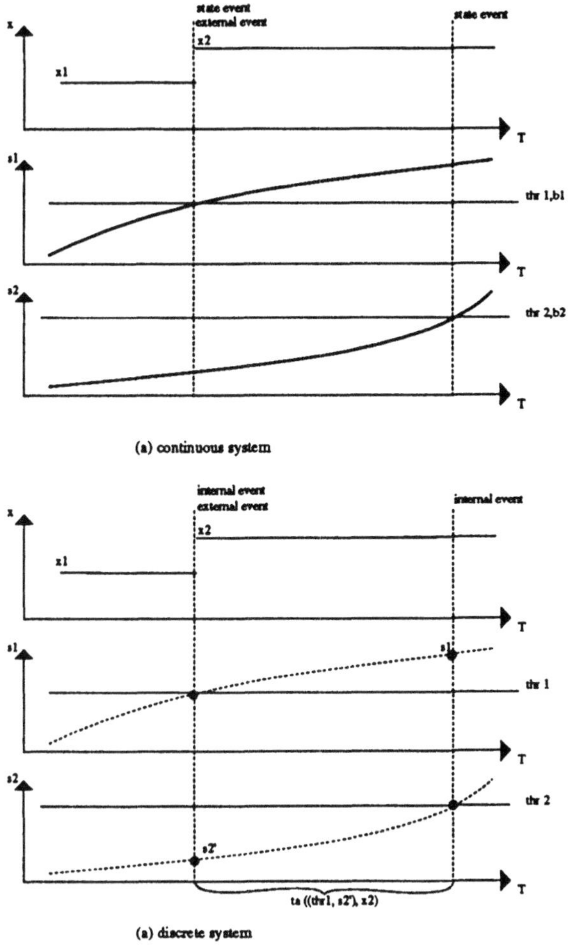

Fig. 2. Illustrating Discrete Event Trajectories

2.3 DEVS-Representation of Constant Input/Output Dynamical Systems

We define a *piecewise constant trajectory* $\omega :< t_1, t_2 > \rightarrow X$ in the following way: there is a finite (possibly empty) subset $\{\tau_1, \tau_2, \ldots, \tau_n\}, \tau_i \in < t_1, t_2 >$ for which $\omega(\tau_i) \neq \omega(\tau_i - \epsilon)$ and $\epsilon \in \mathcal{R}^+$ is arbitrarily small. Piecewise constant trajectories are equivalent to DEVS-segments in the sense that they can be transformed into each other. The transformations are as follows:

- *Piecewise constant trajectory* ω *to DEVS-segment* $\bar{\omega}$: For every event time $\tau_i \in \{\tau_1, \tau_2, \ldots, \tau_n\}$ of ω there is an value $\bar{\omega}(\tau_i) = \omega(\tau_i)$ and for all $\tau \neq \tau_i \Rightarrow \bar{\omega}(\tau) = \Phi$.

- *DEVS-segment $\bar{\omega}$ to piecewise constant trajectory ω*: Let $\{\tau_1, \tau_2, \ldots, \tau_n\}, \tau_i \in< t_1, t_2 >$ be the event times of the DEVS-segment $\bar{\omega}$ with $\bar{\omega}(\tau_i) \neq \Phi$. Then for every $\tau \in< t_1, t_2 >$ we define the respective value of the piecewise constant trajectory $\omega(\tau)$ by the value $\bar{\omega}(\tau_x)$ with τ_x being the largest time in $\{\tau_1, \tau_2, \ldots, \tau_n\}$ with $\tau_x \leq \tau$. If such a number τ_x does not exist then $\omega(\tau) = \Phi$.

For $x \in X$ let $x_{<t_1,t_2>}$ denote the segment $\omega :< t_1, t_2 >\longrightarrow X$ with constant value x, i.e., $\forall t \in< t_1, t_2 > \omega(t) = x$. If it is clear from the context, we write x for $x_{<t_1,t_2>}$.

We now define a *constant input/output dynamical system* as a dynamical system whose input trajectories $\omega \in \Omega$ and associated output trajectories $OTRAJ_{q,\omega}$ are piecewise constant only. Our main interest is to show how such a system can be represented in the DEVS formalism.

Given a constant input/output dynamical system $DS = (T, X, Y, \Omega, Q, \Delta, \Lambda)$, we define a
$DEVS_{DS} = (X_M, Y_M, S, \delta_{ext}, \delta_{int}, \lambda, ta)$

in the following way:

- $X_M = X$,
- $Y_M = Y$,
- $S = Q \times X$, and
- for every total state $((q, x), 0)$ and input segment $\omega :< t_1, t_2 >\longrightarrow X$ we define

$$ta((q, x)) = min\{e|OTRAJ_{q,x}(t_1) \neq OTRAJ_{q,x}(t_1 + e), \tag{11}$$

$$\delta_{int}((q, x)) = (STRAJ_{q,x}(t_1 + ta(s)), x), \tag{12}$$

$$\delta_{ext}((q, x), e, x') = (STRAJ_{q,x}(t_1 + e), x'), \tag{13}$$

$$\lambda((q, x)) = \Lambda(q, x). \tag{14}$$

Theorem 1: The $DEVS_{DS}$ as constructed above and the original DS are behaviorally equivalent, i.e., they have the same input/output behavior when started in the same state.

Proof: The main work is to show that for all input trajectories, ω and initial states, q the state in the constructed discrete event system $DEVS_{DS}$ is equal to the state of the original dynamical system DS at both the event times $\tau_e \in \{\tau_{e1}, \tau_{e2}, \ldots, \tau_{en}\}$ of ω and the event times $\tau_i \in \{\tau_{i1}, \tau_{i2}, \ldots, \tau_{in}\}$ of the associated output trajectory $OTRAJ_{q,\omega}$. This is done by examining the state and output trajectories of the dynamical system $\bar{DS} = (\bar{T}, \bar{X}, \bar{Y}, \bar{\Omega}, \bar{Q}, \bar{\Delta}, \bar{\Lambda})$ specified by $DEVS_{DS}$. From the definition of the output in (14), and from the fact that outputs do not change between event times it follows that the output behavior of the discrete event system $DEVS_{DS}$ is equal to that of the original dynamical system DS. The formal proof is presented in Prahofer and Zeigler (in press).

2.4 Coupled Models

Models constructed from components are formalized in the DEVS formalism as coupled models, DN

$$DN = < X, Y, M, EIC, EOC, IC, SELECT >$$

where

X : input events set;
Y : output events set;
M : DEVS components set;
$EIC \subseteq DN.IN \times M.IN$: external input coupling relation;
$EOC \subseteq M.OUT \times DN.OUT$: external output coupling relation;
$IC \subseteq M.OUT \times M.IN$: internal coupling relation;
$SELECT : 2^M - \emptyset \rightarrow M$: tie-breaking selector.

Here DN.IN, DN.OUT, M.IN, and M.OUT refer to the input and output ports of the newly constructed coupled model and component models, respectively. The three elements, EIC, EOC, IC specify the connections between the set of models M and input, output ports X, Y. $SELECT$ function acts as a tie-breaking selector.

Closure under coupling of DEVS-representable systems is stated as:

Theorem 2: A modular coupled system whose components are DEVS- representable and which does not contain algebraic loops is a DEVS-representable dynamical system.

The definitions required and formal proof are presented in Praehofer and Zeigler (in press).

2.5 DEVS-based Combined Discrete/Continuous Modeling of Hybrid Systems

In this section, we demonstrate how the DEVS representation applies to a special subclass of dynamical systems – those whose state dynamic is defined by a set of first order differential equations and whose input and output trajectories are piecewise constant. We introduce the ***output-partitioned*** *DEV&DESS* formalism for hybrid system modeling and show how a DEVS abstraction is constructed.

Let us consider a continuous system with an n-dimensional continuous state space $S_c = S_{c1} \times S_{c2} \times ... \times S_{cn} = R^n$ whose continuous behavior can be modeled by a set of first order differential equations $\phi = \{\phi_1, \phi_2, ..., \phi_n\}$. However, we assume that the system receives input stimulations which are piecewise constant time functions. Moreover, the system is outfitted with a set of threshold like sensors. Such a threshold sensor only changes its state at particular fixpoints f_{i,b_i} - *threshold values* - of continuous state variables s_i. For simplicity lets assume that only boolean valued sensors $thr_{f_{i,b_i}}$ are used which define a boolean value $x = thr_{f_{i,b}}$ for all $s_i < f_{i,b_i}$ and the opposite value $\neg x$ for all values $s_i > f_{i,b_i}$.

Then we define the output partitioned DEV&DESS in the following way:
Definition: An output-partitioned DEV&DESS is a structure:

$$op - DEV\&DESS = (X, Y, Q, \phi, \lambda, F) \tag{15}$$

where

- X is the set of input values which is an arbitrary set,
- $S = S_1 \times S_2 \times \ldots \times S_n = R^n$, i.e., is the continuous state space which is an n-dimensional vectorspace (it has n state variables),
- $\phi = \{\phi_1, \phi_2, ..., \phi_n\}$ is a set of rate of change function with $\phi_i : S \times X \rightarrow S_{ci}$,
- $F = \bigcup_{i \in \{1,...,n\}} F_i$ with $F_i = \{f_{i,1}, ..., f_{i,m_i}\}$ is the set of threshold values for all dimensions i,
- $Y = \mathcal{B}^F$ is the output value set which is the crossproduct of all boolean value sets of the threshold sensors $thr_{f_{i,b_i}}, f_{i,b_i} \in F$,
- $\lambda : S \rightarrow Y_M$ is the output function which is defined by

$$\lambda(s) = (thr_{f1,1}(s_1), thr_{f1,2}(s_2), ..., thr_{fi,b_i}(s_i), ..., thr_{fn,b_n}(s_n)) \tag{16}$$

  i.e., the output is defined by the values of the threshold sensors.

Such a specification defines a dynamical system DS in the following way:

- the time base T is the real numbers $\mathcal{R}$;
- input set X, output set Y, and state set $Q = S$ of the dynamical system are identical to those of the DEV&DESS,
- the admissible input segments Ω is the set of all piecewise constant segments over X and T,
- for any piecewise constant input segment $\omega :< t_1, t_2 > \rightarrow X \in \Omega$ and state s at time t_1 the global state transition function Δ is defined by the set of first order differential equations as follows:

$$\Delta(s, \omega_{<t_1,t_2>}) = s + \int_{t_1}^{t_2} f(STRAJ_{s,\omega}(\tau), \omega(\tau))d\tau, \tag{17}$$

- the output function Λ of the dynamical system is given by

$$\Lambda(s, x) = \lambda(s). \tag{18}$$

By the threshold values $f_{ib_i} \in F$ each dimension i of the continuous state space is partitioned into a set of ***output blocks*** [1]

$$OB_i = \{(f_{i,0}, f_{i,1} >, < f_{i,1}, f_{i,2} >, ..., < f_{i,b_i-1}, f_{i,b_i} >, ..., < f_{i,m_i}, f_{i,m_i+1})\} \tag{19}$$

with $f_{i,0} = -\infty$ and $f_{i,m_i+1} = +\infty$. The cross product $OB = \times_{i \in \{1,...,n\}} OB_i$ of the output form the so-called ***output partition*** of the state space. An output block

[1] For simplicity reasons, we do not distinguish between open or closed intervals in this work.

$(< f_{1,b_1-1}, f_{1,b_1} >, \ldots, < f_{i,b_i-1}, f_{i,b_i} >, \ldots, < f_{n,b_n-1}, f_{n,b_n} >)$ of the output partition is characterized by the fact that all states in the same output block have the same threshold values, i.e.,

$$\lambda_d(s) = \lambda_d(\bar{s}) \Leftrightarrow \forall i \exists f_{i,b_i} : s_{ci} \in < f_{i,b_i-1}, f_{i,b_i} > \wedge \bar{s}_{ci} \in < f_{i,b_i-1}, f_{i,b_i} > . \quad (20)$$

The output values do not change as long as the continuous states stay in one output block and, therefore, the output trajectory is a piecewise constant time function $OTRAJ_{s,\omega} :< t_1, t_2 > \rightarrow Y$ with $OTRAJ_{s,\omega}(t) = \lambda(STRAJ_{s,\omega}(t))$.

2.6 DEVS Abstraction of Output-Partitioned DEV&DESS Models

Since an output partitioned DEV&DESS is a piecewise constant input/output dynamical system if has a DEVS representation. The transition functions of a DEVS are derived from an DEV&DESS in the following way: for every total state $((s, x), 0)$ and input segment $\omega :< t_1, t_2 > \rightarrow X$ we define:

- the time advance is the time interval to the next output change, i.e., threshold crossing, provided a constant input x,

$$ta((s, x)) = min\{e | \lambda(s) \neq \lambda(s + \int_{t_1}^{t_1+e} f(STRAJ_{s,x}(\tau), x) d\tau\}, \quad (21)$$

- the internal transition function computes the state at the next threshold crossing occuring at time $t_1 + ta(s, x$

$$\delta_{int}((s, x)) = (s + \int_{t_1}^{t_1+ta(s,x)} f(STRAJ_{s,x}(\tau), x) d\tau, x), \quad (22)$$

- the external transition function updates the state at the input event

$$\delta_{ext}((s, x), e, x') = (s + \int_{t_1}^{t_1+e} f(STRAJ_{s,x}(\tau), x) d\tau, x'), \quad (23)$$

- the output function is inherited from the DEV&DESS

$$\lambda((s, x)) = \Lambda(s, x). \quad (24)$$

3 Supervisory control of discrete event systems

Supervisory control [2] is a kind of feedback control that supervises the behavior of a plant to achieve desired objectives. A design problem for a discrete event controller can be formulated as follows: given a plant with known dynamics and given control objectives, design a discrete event controller that satisfies the control objectives. To design a discrete event controller, we have to specify both the dynamics of the plant as well as the behavior of the plant that satisfies the control objectives. Given its universality, we model the plant behavior using the DEVS formalism. From the informally stated control objectives and the plant

DEVS model, we specify a desired state trajectory that satisfies the control objectives. A discrete event controller is then derived by transforming the desired state trajectory into a DEVS model. The plant and the designed controller are coupled through proper interfaces resulting in a controlled system that satisfies the required behavior. We describe the basic concepts and design steps in more detail in the following sections.

3.1 Behavior of Atomic and Coupled Models

This section makes the connection between DEVS atomic and coupled models on the one hand and concurrent system concepts on the other. We present a graphical notation that exposes the duality between external and internal transition that will play an important role in the design methodology to be discussed later.

Recall that the dynamic behavior of a DEVS is determined by a sequence of internal or external transitions. An internal transition happens autonomously after a state determined sojourn time has expired and an external transition is caused by an external event.

(a) $\delta_{ext}(s_i, p?m) = s_j$ (b) $\lambda(s_i) = p!m \rightarrow \delta_{int}(s_i) = s_j$

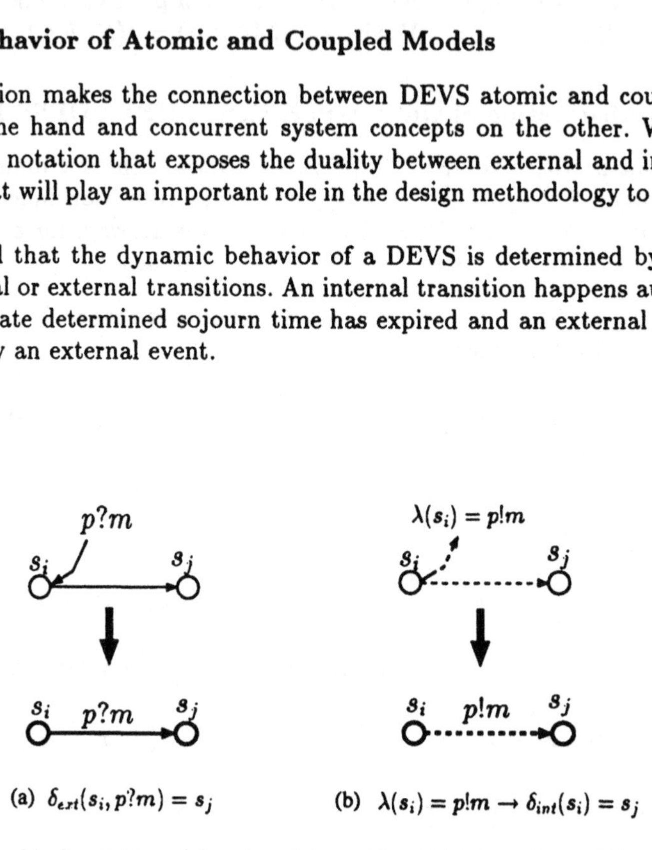

Fig. 3. graphical notation: (a) external transition (b) internal transition

In Fig. 3, (a) represents an external transition. An input event is specified using '?'. For example, an input event $in?m$ means that a message m is input at the input port in. Fig. 3(b) denotes an internal transition. An output event is specified using '!'. For example, an output event $out!m$ means that a message m is output at the port out. A dotted line represents an internal state transition specified by the internal transition function, δ_{int}. A sold line represents a state transition specified by the external transition function, δ_{ext}. This notation derives from CSP (Communicating Sequential Processes)[6] where a receiving process waits at a '?' step in its program until the sending process sends a matching '!' message. Although DEVS allows arbitrary communication between sender and receiver, we will adopt the CSP synchronous communication protocol restrictions for the following presentation. Thus, to communicate, we require that a component M_i sends an output event, $p_i!m_i$ and a receiving component M_j be waiting for it as an input event $p_j?m_j$, where the ports and messages must match (i.e., $p_i = p_j$ and $m_i = m_j$). In the case of a successful communication such as this, M_i undergoes an internal transition denoted as in Fig. 3(b) while concurrently, M_j undergoes an external transition denoted as in Fig. 3(a).

Since only the control portion of the state is typically shown graphically, we indicate transition conditions that depend on other state variables with a separator @. A time advance is attached to a state node to represent its sojourn time. An empty output event is denoted by $\emptyset$.

The concept of *controllability* will be formulated to ascertain whether or not a discrete event controller exists that can enforce a desired state trajectory. If a plant is controllable, the concept of an it inverse DEVS can be used to transform the desired state trajectory into the specification of a discrete event controller. In this section, we define these two concepts.

3.2 Controllable DEVS

Given a discrete event plant $M =< X, S, Y, \delta_{int}, \delta_{ext}, \lambda, ta >$, a *path* is a sequence of elements of S.

Definition 1 weak controllability. A path ST is weakly controllable if a series of internal and external transitions can take each state on the path to the next state on the path. More formally, a path ST is weakly controllable if:
for all s_i, s_{i+1} such that s_{i+1} follows s_i in ST, either:
a) $\delta_{int}(s_i)= s_{i+1}$, or
b) there is a pair (e, x), where $0 < e < ta(s_i)$. and $x = p?m$, such that $\delta_{ext}(s_i, e, x) = s_{i+1}$, or
c) the path $\delta_{int}(s_i)$, s_{i+1}, $path(i + 2)$ is weakly controllable, or
d) there is a pair (e, x), where $0 < e < ta(s_i)$. and $x = p?m$, such that the path $\delta_{ext}(s_i, e, x)$, s_{i+1}, $path(i + 2)$ is weakly controllable,
where $path(i)$ is the subsequence of the path following s_i.

Definition 2 strong controllability. A path is strongly controllable if either an internal or external transition can take each state on the path to the next state on the path. More formally, a path ST is strongly controllable if:
for all s_i, s_{i+1} such that s_{i+1} follows s_i in ST, either:
a) $\delta_{int}(s_i)= s_{i+1}$, or
b) there is a pair (e, x), where $0 < e < ta(s_i)$. and $x = p?m$,
such that $\delta_{ext}(s_i, e, x) = s_{i+1}$.

The states on a strongly controllable path never veer away from those specified in the path. We now strengthen the requirements further by removing the ability of the controller to find an input after some elapsed time to keep the plant on the desired path:

Definition 3 very strong controllability. A strongly controllable path ST is very strongly controllable if:
for all s_i, s_{i+1} such that s_{i+1} follows s_i in ST, either:
a) $\delta_{int}(s_i) = s_{i+1}$, or
b) there is an input $x = p?m$ such that $\delta_{ext}(s_i, e, x) = s_{i+1}$ for all $0 < e < ta(s_i)$.

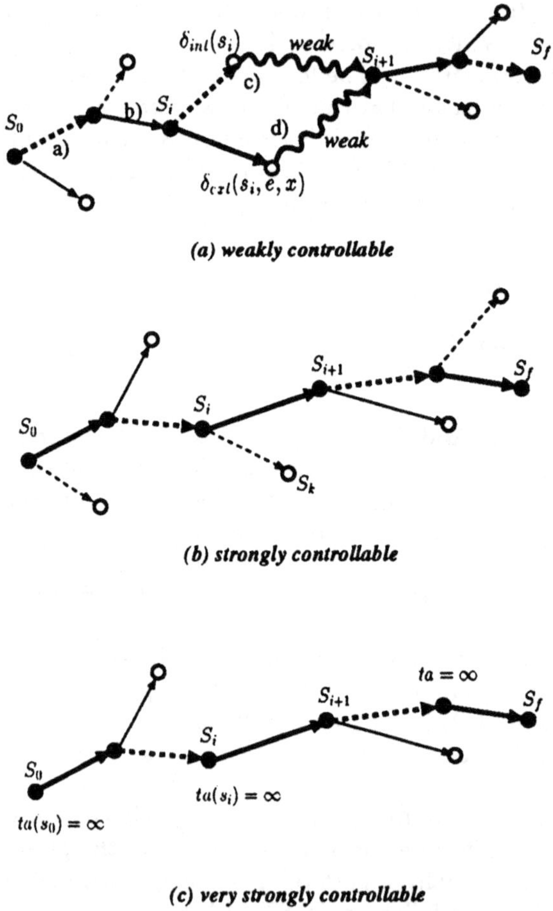

Fig. 4. Controllability Definition: (a) weakly controllable (b) strongly controllable (c) very strongly controllable path. Annotations a),b),c) and d) on Figure (a) shows four cases of the definition of weak controllability.

Figure4 shows cases for the three definitions of controllability. Filled circles represent states on the desired path, $ST = (s_0, s_1, \cdots, s_i, s_{i+1}, \cdots, s_f)$. Thick lines denote internal or external transitions that eventually lead one state on

ST to the next. The annotations a), b), c) and d) on Figure (a) correspond to four cases of the definition of weak controllability. The curly thick line on the path marked c) and d) illustrate weak controllability since there is at least one sequence of internal and external transitions that drive every state on the path to the next. In Figure (b), illustrating strong controllability, there are direct internal or external transitions from every state on the desired path to the next. Note that although there are internal transitions from states on the path to states not on the path (for instance, from s_i to s_k) , a controller can schedule a command within the plant's holding time, $ta(s_i)$ to force it onto the desired path, (i.e., to s_{i+1}). Figure (c) illustrates very strong controllability. Here, every state on the desired path with path-straying internal transitions is controllable by a command that does not depend on the time the plant has been in the state. As far as the controller is concerned, this is equivalent to such states being passive (i.e., $ta(s_i)$ is infinity). After entering a passive state, the plant merely waits for the proper command to drive it to the next desired state. In other words for very strong controllability, the plant behavior must be such that either it moves through the desired state path autonomously or if it needs an input to do so it passively waits for the command to arrive.

In the following, we focus on designing controllers for very strongly controllable plants. We remark that a weakly controllable plant may have to be enhanced with additional sensors and actuators to make it very strongly controllable. Thus there is a tradeoff between complicating the controller and complicating the plant.

3.3 Inverse DEVS

Definition 4 inverse DEVS. DEVS $M_i(M_j)$ is said to be an *inverse DEVS* of $M_j(M_i)$ iff the following properties hold:
(a) State set bijection
$S_i \rightarrow S_j$
(b) Dual I/O events set
$X_i = \{p?m | p!m \in Y_j\}$
$Y_i = \{p!m | p?m \in X_j\}$
(c) Dual transition relations
$T_{i,int} = \{(q, p!m, r) | (q, p?m, r) \in T_{j,ext}\}$
$T_{i,ext} = \{(q, p?m, r) | (q, p!m, r) \in T_{j,int}\}$

(a) states that there is one-to-one correspondence between states of two DEVSs. (b) indicates that input events of one correspond to the output events of the other, and vice versa. (c) denotes that internal transitions match external transitions and conversely.

More specifically, the transition relations for DEVS $M_k, k = i, j$ above are defined by:
$T_{k,int} = \{(s_i, p!m, s_j) \mid s_j = \delta_{k,int}(s_i),\ p!m = \lambda_k(s_i) \in Y_k,\ \ s_i, s_j \in S_k\}$,
$T_{k,ext} = \{(s_i, p?m, s_j) \mid s_j = \delta_{k,ext}(s_i, e, p?m),\ 0 < e < ta(s_i),\ p?m \in X_k,\ \ s_i, s_j \in S_k\}$.

When inversely related DEVS are started in corresponding states, they undergo corresponding transitions which leave them in corresponding states.

An operation on a DEVS M_i creating an inverse M_j is called an ***inverse DEVS transformation***. The inverse transformation is used to obtain a discrete event controller from a desired plant strongly controllable state trajectory.

4 Design methodology for DEC

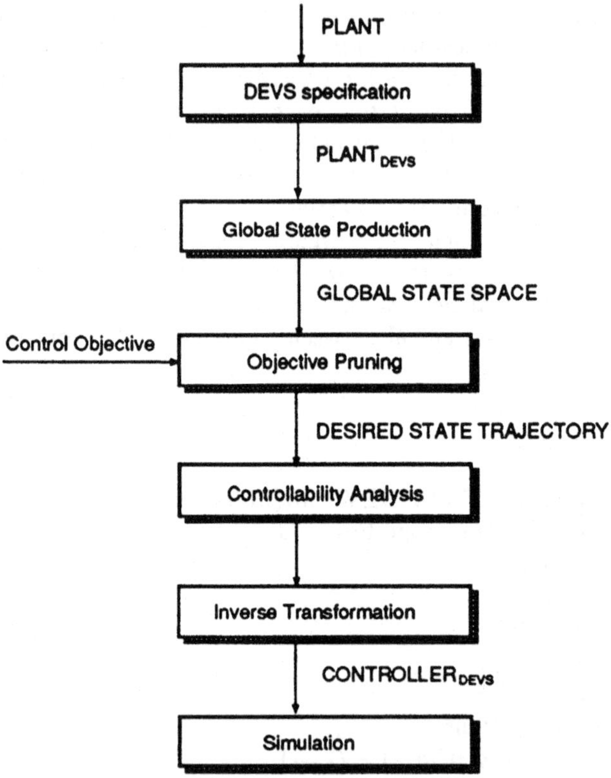

Fig. 5. Design steps for a discrete event controller

A methodology for design of a discrete event controller based on the DEVS formalism and the concepts of inverse DEVS is outlined in Fig. 5. First, a plant is represented in the DEVS formalism based on the theory discussed earlier. Typically such a representation is a coupled model consisting of several atomic models. We then obtain a global state space which is a crossproduct of the component state sets, from which we extract a desired state path satisfying given objectives. The next step is to check the controllability of the desired state path. If the plant is very strongly controllable, we map the desired state path to a DES controller using the inverse DEVS transformation. State reduction is then applied to produce a minimal controller. If the plant is not very strongly controllable, we consider augmenting it with further sensors and actuators. The performance of the controller may be checked by simulation with a realistic model of the plant.

5 A simple example: control of water supply system

We will clarify the methodology with a simple example of a water supply system.

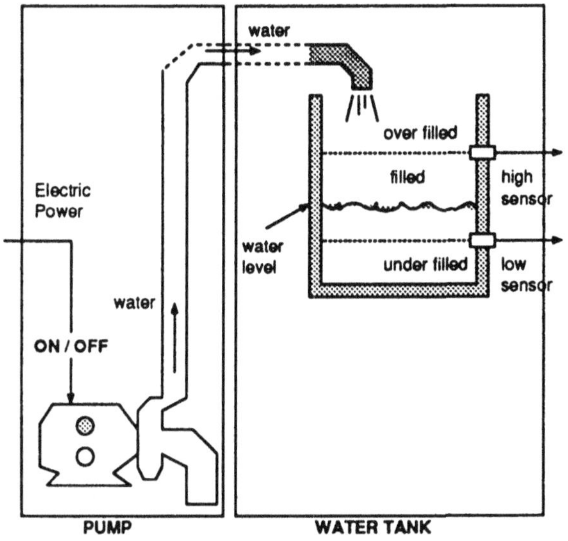

Fig. 6. The plant of a water supply system

5.1 Problem description

An informal problem description is a starting point for the design of a controller for the water supply system. In Fig. 6, a water supply plant consists of two subsystems, a water pump and a water tank. In the water tank, there are two water level sensors (low, high) and a pipe. The water pump is controlled by a ON/OFF switch. It fills the water tank through the water pipe.

5.2 DEVS model of the plant

Representation of the plant in the DEVS formalism is the first step in the design a discrete event controller whose objective is to keep the water level between the low and high sensors.

To specify a plant in the view of a discrete event system, we have to identify events occuring in the plant. These are:
1) the ON/OFF position changes of the motor switch.
2) the starting and stopping of water flow through the pipe.
3) low sensor switching between ON/OFF, the high sensor between ON/OFF

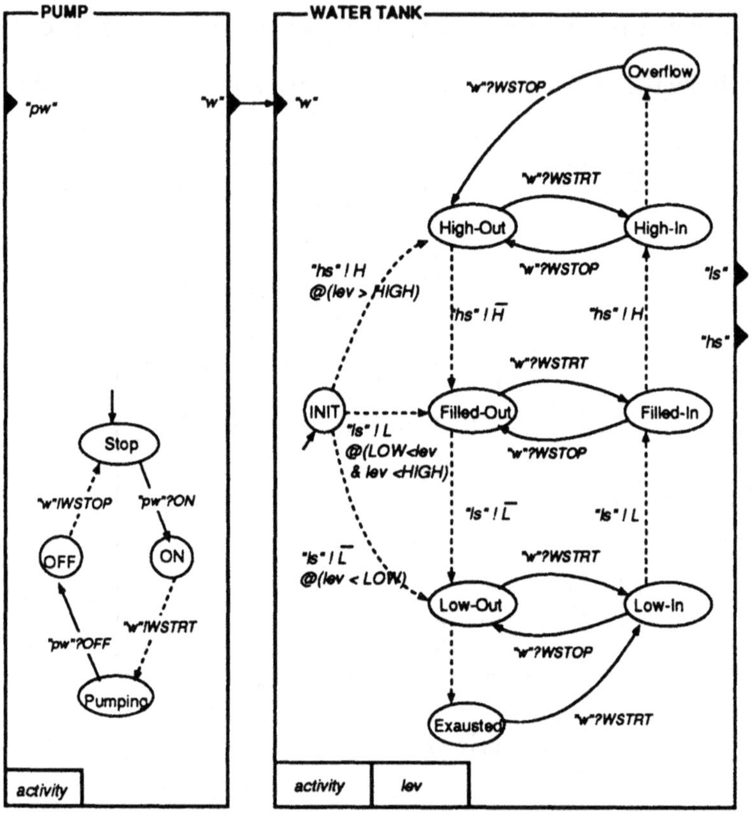

Fig. 7. DEVS Representation of the water supply plant

Fig. 7 shows a DEVS model of the plant. The initial state of the water tank is designated INIT. The level is represented by:

Low = under-filled,
Filled = filled,
High = over-filled

The input flow is represented by :

Out = stopped
In = flowing

With water flowing in, the state transitions from Low-In to Filled-In to High-In, generating proper sensor signals. If the in-flow is continued even though the water level exceeds the level of the high sensor, the water tank eventually overflows (Overflow). With the water pump stopped, the water level recedes as customers draw upon it. The state transitions from High-Out to Filled-Out to Low-Out. If the pump is not turned back on the water level reaches the bottom (Exhausted). As the water recedes, corresponding level sensor signals are generated.

The input port of the PUMP is "pw" and the output port is "w". If the water pump is turned on ("pw"?ON), it starts pumping water through the water pipe ("w"!WSTRT). If it is turned off ("pw"?OFF), it stops pumping and the flow of water will be stopped ("w"!STOP). The PUMP has four states

Stop = passive
ON = transient, issue a start output
Pumping = active,
OFF = transient, issue a stop output.

The two components, pump and tank, are coupled through port "w" (the the pipe). The coupled model, which is the plant representation, has one input port "pw" (power ON/OFF) and two output ports for the high and low sensors.

5.3 Controlled state trajectory

The next step is to get a global state transition diagram (GSTD) from the plant representation. This is obtained by a Cartesian product of state diagrams of each component. (Alternatively, we can use an approach called concurrency analysis that we have developed which significantly reduces the states that need to be examined.)

Let a state s in the GSTD be $(P.s, W.s)$, where $P.s$ denotes a state of the water pump and $W.s$ a state of the water tank. We formulate the desired state path, denoted K, by the constraints:

1. If $\exists$ `(`x`, Low-Stop)`, then eventually `(Stop, High-Out)`, where x means *don't care*.
2. $\exists$ no s such that `(`x`,Exhausted) or (`x`,Overflow)`.
3. hysteresis: $\exists$ no transition between `(`x`,Filled-In)` and `(`x`,Filled-Out)`

Now we can extract a controlled state trajectory from GSTD and the control objectives. Fig. 8 is the desired state trajectory. Initially, the state is changed into (Stop,Low-Out) after a few internal transitions. Then the motor should be ON, which is followed by a concurrent event "w"#WSTRT ("w"!WSTRT and "w"?WSTRT) If the water level reaches to the over-filled region, the motor should be OFF. Then, water stops flowing in. By internal transitions, the water level recedes, and then the motor should be turned back ON, and so on. This cyclic state path satisfies the three objectives.

The next step is to check if the controllability of the desired state path. Since every successive pair of states is connected by an internal or external transition with unconditional input, the plant is very strongly controllable.

However, consider Figure 9 in which the transition marked 1' occurs from

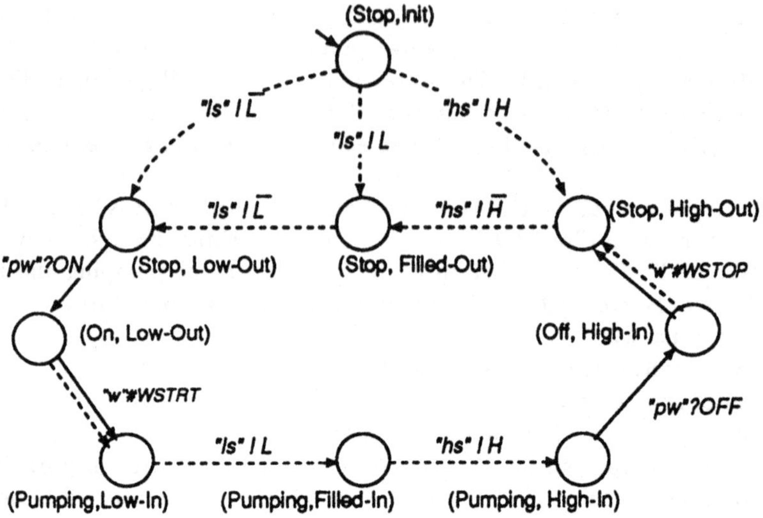

Fig. 8. desired state trajectory of the plant

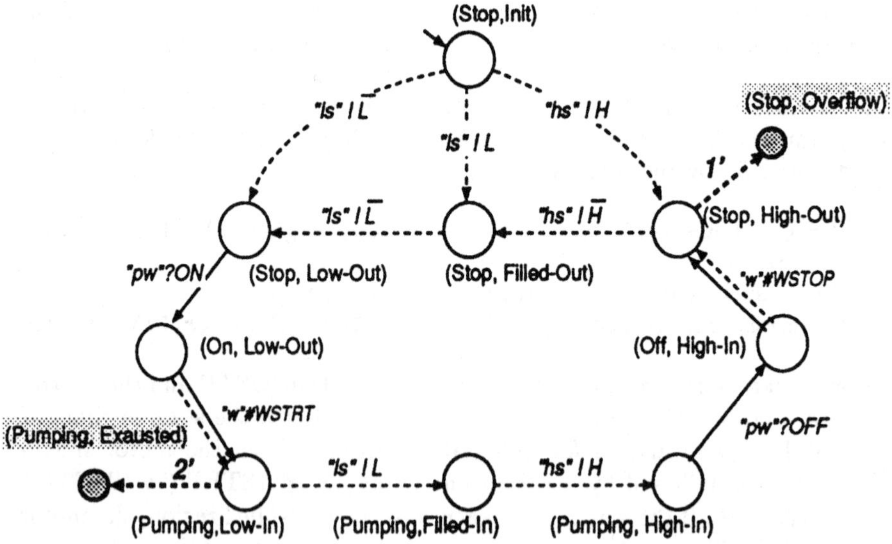

Fig. 9. Possible state trajectory with high sensor failure

(Stop, High-Out) to (Stop, Overflow). This is an internal transition which results from the remaining water in the pipe continuing to flow into the tank after the pump is turned off. This transition will occur if the high sensor is too close to the top of the tank. However, this internal transition is not on the desired path which requires (Stop,High-Stop) to be followed by (Stop, Filled-Out), i.e., for the water to start receding after the influx ceases **without** first overflowing. Moreover, there is **no** external transition that can rectify the situation since the pump has already been turned off. A similar internal transition, marked 2', occurs when the water reaches bottom despite the pump being turned back on due to the transport delay in the pipe.

We see that in this case, the plant is not weakly controllable. However, it can be made very strongly controllable by placing the high and low sensors sufficiently far away from the top and bottom of the tank, respectively.

5.4 Deriving a Discrete Event Controller

At this point we have represented a plant as a coupled DEVS, obtained the state transition behavior of its resultant, and interpreted the informally stated control objectives as a desired path in the state space of the resultant. This path is strongly controllable (assuming proper placement of the high/low sensors) so an associated state trajectory exists. Now, we derive a discrete event controller from the state trajectory using the inverse DEVS transformation.

Before the transformation, we can reduce the states of the desired state trajectory. The upper part of Fig. 10 is obtained by suitable state reduction from the trajectory in Fig. 8.

A discrete event controller is obtained by inverse DEVS transformation of the desired reduced state trajectory. The lower part of Fig. 10 shows the DEC after the transformation. Note that the transformation is straightforward and intuitive.

The two components, the controller, and the plant are to coupled through ports that have the same name to form a hierarchical coupled DEVS, shown in Figure 11.

The dynamics of the resultant follow the desired state trajectory since the controller and plant, once started in corresponding states, maintain this correspondence forever.

6 Discussion

More realistic DEVS models include states that generate null outputs. This can represent internal transitions in the plant which are not disclosed to the outside world via sensors. The inverse transformation may fail to produce a proper controller since the plant cannot generate an informative input to it.

An example of the above may occur when a sensor fails. To add more autonomy to the controller, it may be augmented with event-based control[15, 16, 10]. The controller then has a time-window in which it expects a sensor response. If

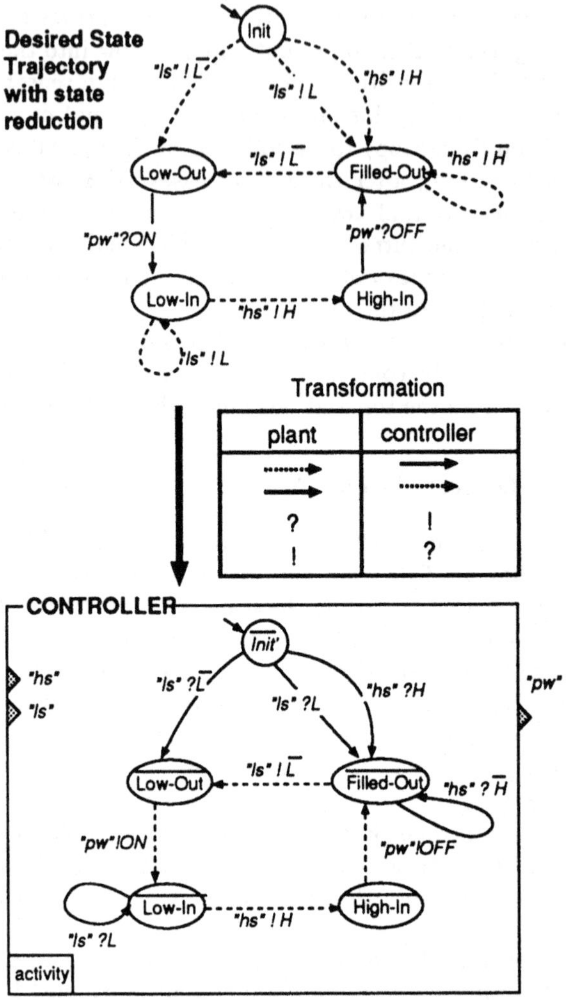

Fig. 10. Design of a DES controller

this response does not occur within the expected window, the controller knows a failure has occurred and can initiate a diagnosis while taking corrective action. This enables the DEVS event-based controller to be used for monitoring the plant for occurrence of malfunctions. The controller employs a DEVS model which is an abstraction of the plant and which takes into account normal variations in its response time. The error messages issued by the DEVS-based controller contain important information about possible causes for the systems' malfunctions and can be used by a diagnosing unit.

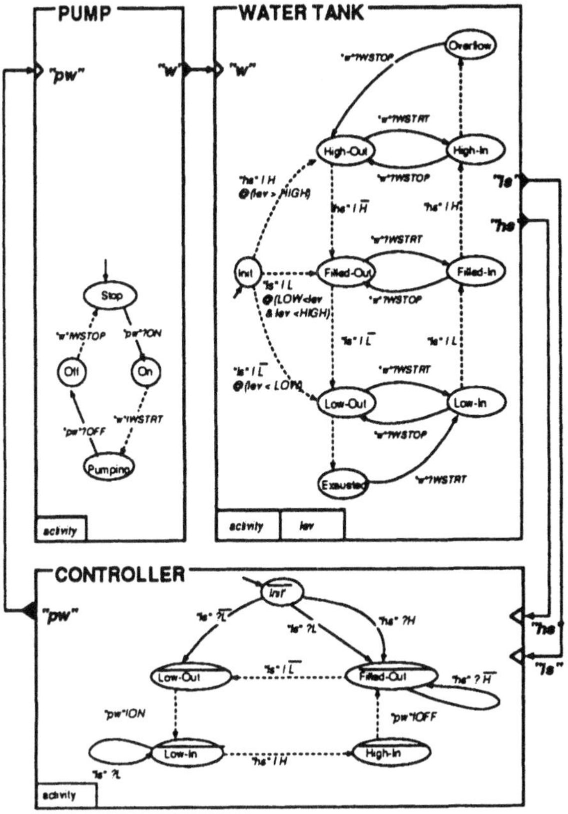

Fig. 11. Whole water supply system : coupling the plant and the controller designed

7 Conclusions

We have reviewed the DEVS formalism and provided examples of its application in support of our main theme: DEVS, and its parent system theory, provide a sound, general framework within which to address modelling, simulation, design, and analysis issues for hybrid systems. While special theories will continue to be main lines of research development, we believe that it is important for transmission to future generations that such theories also be encompassed within a general framework such as that offerred here. Embedding special theories into DEVS enables improved understanding of their strengths and limitations and can lead to stronger, more applicable, approaches. For example, much hybrid systems research employs discrete -time automata as the decision making components. Here, the limitation of a fixed step size may impose unnecessary limitations on the mapping of plant states to output symbols. This limitation is completely removed when the controller may be drawn from the class of DEVS models as illustrated in this paper. We therefore propose that a new important direction for future research in hybrid systems should be the integration of approaches within the DEVS-based general framework. The payoff for such integration includes the more direct transfer of theoretical results into simulation testing and ultimately, implementation in practice.

References

1. P.J. Anstaklis, M.D. Passino, and S.J. Wang. Towards intelligent autonomous control systems. *Journal of Intelligent and Robotic Systems*, 1(4):315–342, 1989.
2. Cristos G. Cassandras. *Discrete Event Systems*. IRWIN, 1993.
3. D. Harel et al. Statemate: A working environment for the specification of compex reactive system. *IEEE Trans. on SE*, 16(4):403–414, 1990.
4. Fr.Pichler and H. Schwaertzel, editors. *CAST Methods in Modelling*, chapter 3, pages 123–241. Springer-Verlag, 1992.
5. Y. C. Ho. Special issue on discret event dynamic systems. *Proceedings of the IEEE*, 77(1), 1989.
6. C.A.R. Hoare. *Communicating Sequential Processes*. Prentice-Hall, 1985.
7. A. Meystel. Intelligent control: A sketch of the theory. *Journal of Intelligent and Robotic Systems*, 2(2):97–107, 1989.
8. L. Padula and M. A. Arbib. *Systems Theory*. Saunders, Philadelphia, 1974.
9. H. Praehofer. *System Theoretic Foundations for Combined Discrete-Continuous System Simulation*. PhD thesis, Johannes Kepler University of Linz, Linz, Austria, 1991.
10. H. Praehofer and B.P. Zeigler. Automatic abstraction of event-based control models from continuous base models. *submitted to IEEE Trans. Systems, Man and Cybernetics*, 1995.
11. P.J.G. Ramadge and W.M. Wohnham. The control of discrete event systems. *Proceedings of the IEEE*, 77(1):81–98, 1989.
12. G. Wunsch, editor. *Handbuch der Systemtheorie*. Akademie-Verlag, Berlin, 1986.
13. B. P. Zeigler. *Theory of Modelling and Simulation*. John Wiley, New York, 1976.
14. B. P. Zeigler. *Multifacetted Modelling and Discrete Event Simulation*. Academic Press, London, 1984.
15. B. P. Zeigler. DEVS representation of dynamical systems: Event-based intelligent control. *Proceedings of the IEEE*, 77(1):72–80, 1989.
16. B. P. Zeigler. *Object-Oriented Simulation with Hierarchical, Modular Models*. Academic Press, London, 1990.
17. B. P. Zeigler and W. H. Sanders. Preface to special issue on environments for discrete event dynamic systems. *Discrete Event Dynamic Systems: Theory and Application*, 3(2):110–119, 1993.

Synthesis of Hybrid Constraint-Based Controllers

Ying Zhang* and Alan K. Mackworth**

Department of Computer Science
University of British Columbia
Vancouver, B.C.
Canada, V6T 1Z4
E-mail: zhang,mack@cs.ubc.ca

Abstract. A robot is an integrated system, with a controller embedded in its plant. We take a robotic system to be the coupling of a robot to its environment. Robotic systems are, in general, hybrid dynamic systems, consisting of continuous, discrete and event-driven components. We call the dynamic relationship of a robot and its environment the behavior of the robotic system. The problem of control synthesis is: given a requirements specification for the behavior, and given dynamic models of the plant and the environment, generate a controller so that the behavior of the robotic system satisfies the specification. We have developed a formal language, Timed Linear Temporal Logic (TLTL) [17], for requirements specification. We have also developed a semantic model, Constraint Nets [19], for modeling hybrid dynamic systems. In this paper, we study the problem of control synthesis using these representations. Control synthesis in general is difficult. We first focus on a special class of requirements specification, called constraint-based specification, in which constraints are associated with properties such as safety, reachability and persistence. Then we develop a systematic approach to synthesizing controllers using constraint methods, in which controllers are embedded constraint solvers that solve constraints in real-time. Finally, we consider hierarchical control structures, in which the higher levels embody digital/symbolic event-driven control derived from discrete constraint methods and the lower levels incorporate analog control based on continuous constraint methods. We illustrate these techniques using a robot soccer player as a running example.

1 Control Synthesis
Where does the problem fit in?

Robots are generally composed of electromechanical parts with multiple sensors and actuators. Robots should be reactive as well as purposive systems, closely coupled with their environments; they must deal with inconsistent, incomplete, unreliable and delayed information from various sources. Robots are usually complex, hierarchically organized and physically distributed; each component functions according to its own

* Current address: The Wilson Center for Technology, Xerox Corporation, 800-Phillips Road, M/S 128-51E, Webster, N.Y., U.S.A., 14580. E-mail: zhang@wrc.xerox.com
** Fellow, Canadian Institute for Advanced Research.

dynamics. The overall behavior of a robot emerges from coordination among its various parts and interaction with its environment. We call the coupling of a robot and its environment a robotic system, and the dynamic relationship of a robot and its environment the behavior of the robotic system.

A robot controller (or control system) is a subsystem of a robot, designed to regulate its behavior to meet certain requirements. In general, a robot controller is an integrated software/hardware system implemented on various digital/analog devices. Thus, from the systemic point of view, a robotic system consists of a controller, a plant and an environment as shown in Fig. 1, where X is a set of plant variables, U is the set of control variables, and Y is a set of environment variables. Some of the plant and environment variables may not be observable by the controller.

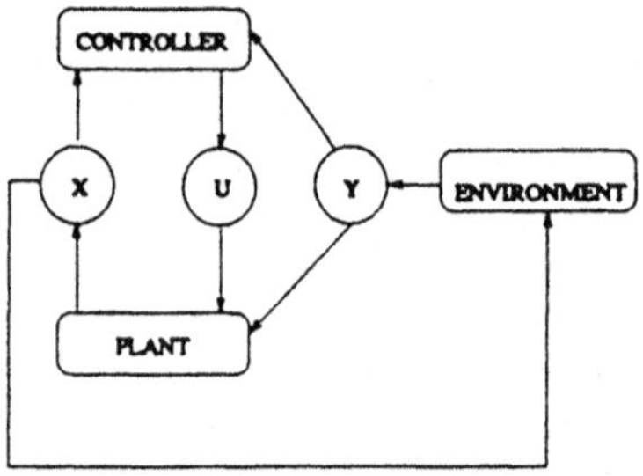

Fig. 1. A robotic system

Robotic systems are typical hybrid systems, consisting of continuous, discrete and event-driven components. In order to develop a robotic system, analyze its behavior and understand its underlying physics, we have developed a semantic model for hybrid dynamic systems, that we call Constraint Nets (CN). Using this model, we can characterize the components of a system and derive the behavior of the overall system.

Control systems are designed to meet certain requirements. Typical requirements include safety, reachability and persistence. Safety declares that a system should never be in a certain situation. Reachability declares that a system should reach a certain goal eventually. Persistence declares that a system should achieve a certain goal infinitely often. In order to specify requirements formally, we have developed two languages, Timed Linear Temporal Logic (TLTL) and timed ∀-automata. In this paper, we will introduce only TLTL.

The problem of control synthesis is stated as follows: given a requirements specification, and the dynamic models of the plant and the environment, produce (a dynamic model of) a controller such that the behavior of the robotic system meets the given requirements. Control synthesis in general is difficult. As a first step, we focus on a special class of requirements specification — constraint-based specification, in which constraints are associated with properties such as safety, reachability and persistence. We then develop a systematic approach to control synthesis from constraint-based specifications, in which controllers are constraint solvers that solve constraints in real-time.

Fig. 2 presents an overall framework of our approach to the design and analysis of robotic systems, indicating where the problem of control synthesis fits in. In this framework, control synthesis and behavior verification are coupled together to generate a "correct" system. In this paper, we will focus only on control synthesis.

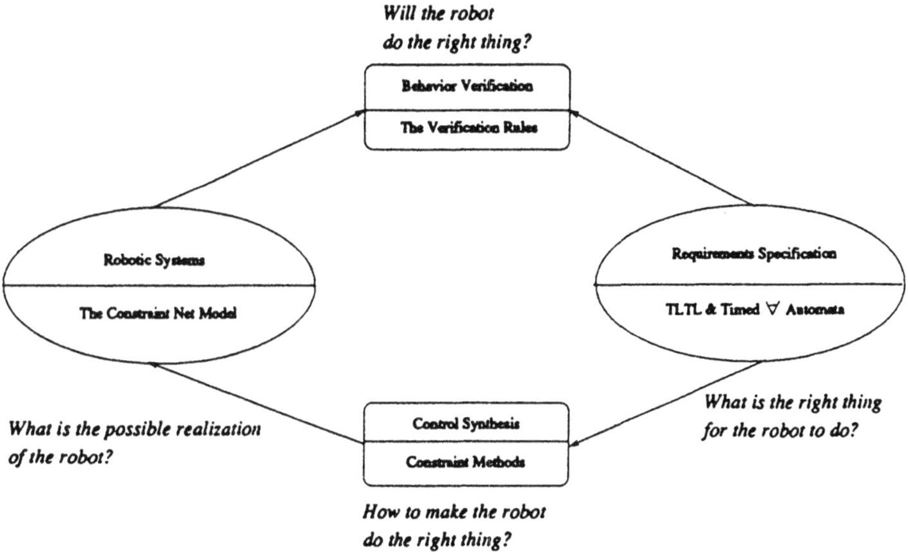

Fig. 2. Where does control synthesis fit in?

Let us introduce an example that will be used throughout this paper. In our Laboratory for Computational Intelligence, a testbed has been developed for radio-controlled cars playing soccer using visual feedback [15]. Each "soccer player" has a car-like mobile base. It can move forward and backward with a throttle setting, and can make turns by steering its two front wheels. However, it cannot move sideways and its turns are limited by mechanical stops in the steering gear.

Fig. 3 illustrates the configuration of a car. Let v be the velocity of the car and α be the current steering angle of the front wheels; v and α can be considered as the control inputs to the car. The dynamics of the car can be simply modeled by the following differential equations [7]:

$$\dot{x} = v\cos(\theta), \quad \dot{y} = v\sin(\theta), \quad \dot{\theta} = v/R \tag{1}$$

where (x, y) is the position of the tail of the car, θ is the heading and $R = L/\tan(\alpha)$ is the turning radius given the length of the car L. The environment of this system is a moving ball, a target (the goal) and a field with boundaries. A requirements specification for the robot is to kick or dribble the ball to the target repeatedly. The

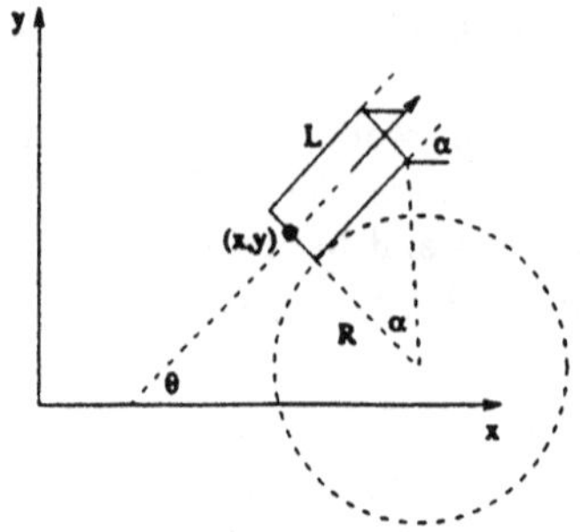

Fig. 3. The configuration of a car

controller of the car is equipped with both digital and analog devices [15]. How can we synthesize such a controller?

The rest of this paper is organized as follows. Section 2 briefly introduces the Constraint Net model. Section 3 presents TLTL. Section 4 focuses on constraint-based specification. Section 5 develops constraint-based control. Section 6 discusses general control structures. Finally, Section 7 concludes the paper and proposes further research on control synthesis.

2 Constraint Nets
What is synthesized?

Constraint Nets (CN) is a semantic model for hybrid dynamic systems. CN is an abstraction and generalization of dataflow networks [8, 2, 4, 3]. Any (causal) system with discrete/continuous time, discrete/continuous (state) variables, and asynchronous/synchronous event structures can be modeled. Furthermore, a system can be modeled hierarchically using aggregation operators; the dynamics of the environment as well as the dynamics of the plant and the controller can be modeled individually and then integrated.

Our approach to developing this model for hybrid systems is motivated by the following arguments. First, hybrid systems consist of interacting discrete and continuous components. Instead of fixing a model with particular time and domain structures, a model for hybrid systems should be developed on both abstract time structures and abstract data types. Second, hybrid systems are complex systems with multiple components. A model for hybrid systems should support hierarchy and modularity. Third, hybrid systems are generalizations of basic discrete or continuous systems. A model for hybrid systems should be at least as powerful as existing models of computation and control. In short, a model for hybrid systems should be unitary, modular, and powerful.

In the rest of this section, we briefly introduce some concepts of dynamic systems, and the syntax and semantics of CN. For a comprehensive introduction to CN, the reader is referred to [19] or [17].

2.1 Basic concepts

Let $\mathcal{R}$ be the set of real numbers and $\mathcal{R}^+$ be the set of nonnegative real numbers. A *metric* on a set X is a function $d : X \times X \rightarrow \mathcal{R}^+$ satisfying the following axioms for all $x, y, z \in X$:

1. $d(x, y) = d(y, x)$.
2. $d(x, y) + d(y, z) \geq d(x, z)$.
3. $d(x, y) = 0$ iff $x = y$.

A *metric space* is a pair $\langle X, d \rangle$ where X is a set and d is a metric on X. In a metric space $\langle X, d \rangle$, $d(x, y)$ is called "the distance between x and y." We will use X to denote the metric space $\langle X, d \rangle$ if no ambiguity arises.

Following are some basic concepts of dynamic systems developed in [17].

- *Time* can be abstracted as a linearly ordered set $\langle \mathcal{T}, \leq \rangle$ with a least element 0 as the start time point, and a metric d on set $\mathcal{T}$. Using this abstract notion, time can be either discrete, continuous or event-based.
- A *domain* can be either simple or composite. A simple domain denotes a simple data type, such as reals, integers, Booleans and characters; a composite domain denotes a structured data type, such as arrays, vectors, strings, objects, structures and records. Formally, a *simple domain* is a pair $\langle A \cup \{\perp_A\}, d_A \rangle$ where A is a set, $\perp_A \notin A$ means undefined in A, and d_A is a metric on A. A *composite domain* is a product of simple domains. Using this abstract notion, a domain can be numerical or symbolic, discrete or continuous.
- A trace characterizes the change of values of a variable over time. Formally, a *trace* is a mapping $v : \mathcal{T} \rightarrow A$ from time $\mathcal{T}$ to domain A. An *event trace* is a special type of trace whose domain is Boolean. An *event* in an event trace is a transition from 0 to 1 or from 1 to 0. An event trace characterizes some event-based time where the set of events in the trace is the time set.
- A *transduction* is a mapping from a tuple of input traces to an output trace that satisfies the causality constraint between the inputs and the output, i.e., the output value at any time depends only on its inputs up to that time. For instance, a state automaton with an initial state defines a transduction on discrete time; a temporal integration with a given initial value is a typical transduction on continuous time. Just as nullary functions represent constants, nullary transductions represent traces.
  We characterize two classes of atomic transductions: simple transductions and simple event-driven transductions.
- A *simple transduction* is a transliteration or a delay.
- A *transliteration* is a pointwise extension of a function. Intuitively, a transliteration is a transformational process without memory (internal state), such as a combinational circuit.
- A *delay* transduction is a sequential process where the output value at any time is the input value at a previous time. Discrete-time state transitions and continuous-time integrations can be modeled using delays.
- A simple *event-driven transduction* is a simple transduction augmented with an extra input which is an event trace; the transduction operates at every event and its output value holds between two events.

2.2 The Constraint Net model

A constraint net consists of a finite set of locations, a finite set of transductions and a finite set of connections. Formally, a ***constraint net*** is a triple $CN = \langle Lc, Td, Cn \rangle$, where Lc is a finite set of ***locations***, Td is a finite set of labels of ***transductions***, each with an ***output port*** and a set of ***input ports***, Cn is a set of ***connections*** between locations and ports, with the following restrictions: (1) there is at most one output port connected to each location, (2) each port of a transduction connects to a unique location and (3) no location is isolated. A location can be regarded as a wire, a channel, a variable, or a memory cell. Each transduction is a causal mapping from inputs to outputs over time, operating according to a certain reference time or activated by external events.

A location is an ***output*** of the constraint net if it is connected to the output of some transduction; it is otherwise an ***input***. A constraint net is ***open*** if there is an input location; it is otherwise ***closed***.

Semantically, a constraint net represents a set of equations, with locations as variables and transductions as functions. The ***semantics*** of the constraint net, with each location denoting a trace, is the least solution of the set of equations [19, 17].

A complex system is generally composed of multiple components. We define a ***module*** as a constraint net with a set of locations as its interface. A constraint net can be composed hierarchically using modular and aggregation operators on modules. The semantics of a system can be obtained hierarchically from the semantics of its subsystems and their connections.

Given CN, a constraint net model of a dynamic system, the abstract behavior of the system is the semantics of CN, denoted $[\![CN]\!]$, i.e., the set of input/output traces satisfying the model.

A constraint net is depicted by a bipartite graph where locations are depicted by circles, transductions by boxes and connections by arcs. A module is depicted by a box with rounded corners. For example, the dynamics of the car, characterized by Eq. (1), is denoted by an open constraint net, as shown in Fig. 4 where *sin*, *cos*, *tan* and $*$ are transliterations, and $\int$ is a temporal integrator. The car module is defined by choosing locations v, α, x, y, θ as the interface.

We can model a control system as a module that may be further decomposed into a hierarchy of interactive modules. The higher levels are composed of event-driven transductions and the lower levels are analog control components. Furthermore, the environment of the robot can be modeled as a module as well. A robotic system can be modeled as an integration of a plant, a controller and an environment (Fig. 1). The semantics (or behavior) of the system is the solution of the following equations:

$$X = PLANT(U, Y), \tag{2}$$
$$U = CONTROLLER(X, Y), \tag{3}$$
$$Y = ENVIRONMENT(X). \tag{4}$$

Note that $PLANT$, $CONTROLLER$ and $ENVIRONMENT$ are transductions (not simple functions), and the solution gives X, Y and U as tuples of traces (not values).

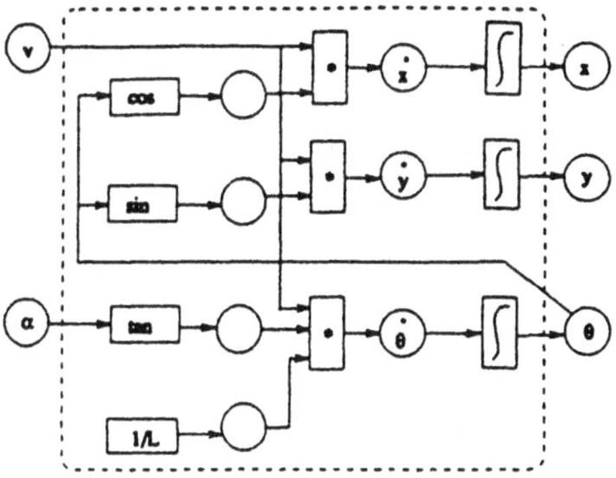

Fig. 4. The constraint net for the dynamics of the car, Eq. (1)

As we can see here, a robot, composed of a plant and a controller, is an open system, and a robotic system, composed of a robot and its environment, is a closed system.

3 Timed Linear Temporal Logic Specification
What is required?

A model of a dynamic system represents the whole system as a set of components and their connections. However, the behavior of the system is not explicitly represented, since most dynamic systems have no closed form solutions.

Most design requirements can be expressed by qualitative properties and can be satisfied by many models (implementations). A ***requirements specification*** $\mathcal{R}$ for a system CN is a set of allowable input/output traces of the system: the behavior of CN satisfies its requirements specification $\mathcal{R}$, written $[\![CN]\!] \models \mathcal{R}$, iff $[\![CN]\!] \subseteq \mathcal{R}$. For example, $CN : x = \int(x_0)(-x)$ is a model of dynamic system whose behavior, $x(t) = x_0e^{-t}$, satisfies the requirements specification $\mathcal{R} : \lim_{t\rightarrow\infty} x(t) = 0$.

The problem of control synthesis can be formalized as follows: given a requirements specification $\mathcal{R}$, the model of the plant $PLANT$ and the model of the environment $ENVIRONMENT$, synthesize a model of the controller $CONTROLLER$, such that $[\![CN]\!] \models \mathcal{R}$ where CN is modeled by Eqs. (2), (3) and (4).

We have developed Timed Linear Temporal Logic (TLTL) and timed ∀-automata, generalized from Linear Temporal Logics [11] and ∀-automata [10], for specifying requirements. In this paper, we introduce only TLTL.

The basic form of the propositional linear temporal logic (PLTL) is the classical propositional logic extended with temporal operators. Formally, the syntax of the logic is defined as follows. Let Φ be a set of propositions. The basic syntax is defined in BNF:

$$F ::= \mathit{false} \mid p \mid F_1 \rightarrow F_2 \mid F_1\,\mathcal{U}\,F_2 \mid F_1\,\mathcal{S}\,F_2$$

where $p \in \Phi$ is a proposition, $\rightarrow$ is a logical connective denoting "implication," $\mathcal{U}$ is a temporal operator denoting "until" and $\mathcal{S}$ is a temporal operator denoting "since."

Formally, the semantics of the logic is defined as follows. Let $\mathcal{T}$ and A be time and domain, respectively, and $v : \mathcal{T} \rightarrow A$ be a trace. Note that $\mathcal{T}$ and A can be either discrete or continuous. Let $V : \Phi \rightarrow 2^A$ be an interpretation that assigns to each proposition $p \in \Phi$ a subset $V(p)$ of A. We will use $a \models p$ to denote $a \in V(p)$. Then $v \models_t F$ denotes that trace v satisfies formula F at time t:

- $v \not\models_t false$.
- $v \models_t p$ for $p \in \Phi$ iff $v(t) \models p$.
- $v \models_t F_1 \rightarrow F_2$ iff $v \models_t F_1$ implies $v \models_t F_2$.
- $v \models_t F_1 \mathcal{U} F_2$ iff $\exists t' > t, v \models_{t'} F_2$ and $\forall t'', t < t'' < t', v \models_{t''} F_1$.
- $v \models_t F_1 \mathcal{S} F_2$ iff $\exists t' < t, v \models_{t'} F_2$ and $\forall t'', t' < t'' < t, v \models_{t''} F_1$.

We will use $v \models F$ to denote that v satisfies F initially, i.e., $v \models_0 F$.

More logical connectives and temporal operators can be defined using the basic logic connective $\rightarrow$ and the basic temporal operators $\mathcal{U}$ and $\mathcal{S}$.

Some commonly used logical connectives are defined as follows:

- *Negation*: $\neg F \equiv F \rightarrow false$.
- *True*: $true \equiv \neg false$.
- *Disjunction*: $F_1 \vee F_2 \equiv \neg F_1 \rightarrow F_2$.
- *Conjunction*: $F_1 \wedge F_2 \equiv \neg(F_1 \rightarrow \neg F_2)$.

Some commonly used temporal operators are defined as follows:

- *Eventually*: $\Diamond F \equiv F \vee true \mathcal{U} F$.
- *Always*: $\Box F \equiv \neg \Diamond \neg F$.
- *Next*: $\bigcirc F \equiv F \mathcal{U} F$.
- *Previous*: $\ominus F \equiv F \mathcal{S} F$.

Various stronger and weaker variations of these temporal operators [5] can also be defined.

We define some more abbreviations that are convenient to use in many situations.

- $final \equiv \neg \bigcirc true$.
- $initial \equiv \neg \ominus true$.
- $rise(p) \equiv (\neg p \wedge \bigcirc p) \vee (\ominus \neg p \wedge p)$.
- $change(p) \equiv rise(p) \vee rise(\neg p)$.
- $event(p) \equiv (\ominus \neg p \wedge p) \vee (\ominus p \wedge \neg p)$.

Some important properties of behaviors can be specified using PLTL.

- *Safety*: If B is a proposition denoting a bad situation, $\Box \neg B$.
- *Reachability*: If G is a proposition denoting a final goal, $\Diamond \Box G$.
- *Persistence*: If P is a proposition denoting a persistent condition, $\Box \Diamond P$.

In order to specify the metric properties of time, we develop Timed Linear Temporal Logic (TLTL). The basic syntax and semantics of TLTL are the same as those of PLTL. In addition, we augment the basic form of PLTL with two real-time operators. Let $\tau > 0$ be a positive real number, $T_{t+\tau} = \{t' | t < t', d(t, t') \leq \tau\}$ and $T_{t-\tau} = \{t' | t' < t, d(t', t) \leq \tau\}$. Two real-time operators are defined as follows:

- $v \models_t F_1 \mathcal{U}^\tau F_2$ iff $\exists t' \in T_{t+\tau}$, $v \models_{t'} F_2$ and $\forall t'', t < t'' < t', v \models_{t''} F_1$.
- $v \models_t F_1 \mathcal{S}^\tau F_2$ iff $\exists t' \in T_{t-\tau}$, $v \models_{t'} F_2$ and $\forall t'', t' < t'' < t, v \models_{t''} F_1$.

Other real-time and temporal operators can be defined using the two basic real-time operators.

- $\Diamond^\tau F \equiv true\, \mathcal{U}^\tau F$.
- $\Box^\tau F \equiv \neg(\Diamond^\tau \neg F)$.
- $\Diamond_\tau F \equiv true\, \mathcal{S}^\tau F$.
- $\Box_\tau F \equiv \neg(\Diamond_\tau \neg F)$.

With real-time operators, real-time properties can be specified, for example, real-time response can be specified as $\Box(E \rightarrow \Diamond^\tau R)$.

In our framework the control synthesis problem is stated as follows: given a set of TLTL formulas, and given constraint net models of the plant and the environment, produce a constraint net model of the controller so that the behavior of the overall system satisfies the given requirements. In its full generality this problem is too difficult. As a first step, we focus on a special class of requirements and develop a systematic, but not automatic, approach to control synthesis.

4 Constraint-Based Specification

What should be satisfied?

We consider constraints as relations on a set of state variables; the solution set of the constraints consists of the state variable tuples that satisfy all the constraints. We call the behavior of a dynamic system constraint-based if the system is asymptotically stable at the solution set of the given constraints, i.e., whenever the system diverges because of some disturbance, it will eventually return to the set satisfying the constraints.

Most robotic systems are constraint-based, where the constraints may include physical limitations, environmental restrictions, and safety and goal requirements. Most learning and adaptive dynamic systems exhibit some forms of constraint-based behaviors as well.

A constraint-based specification characterizes a desired constraint-based behavior. Formally, let C be a set of constraints and $sol(C)$ be the solution set of C, and let C^ϵ be a neighborhood of $sol(C)$; C^ϵ and $sol(C)$ can be taken as propositions whose interpretations are themselves. TLTL formulas $\Box C^\epsilon$, $\Diamond\Box C^\epsilon$ and $\Box\Diamond C^\epsilon$ are the specifications for safety, reachability and persistence, respectively, with respect to constraints C and error ϵ. Note that it is important to specify a desired property with some error, since many constraint methods cannot "solve" the constraint satisfaction problem in finite time (or with finite resource). A more formal notion of this kind of *open specification* is introduced in [12] and discussed in [17].

The strongest constraint-based requirement is the safety property which requires that the system never move away from the solution set of the constraints. The weakest constraint-based requirement is the persistence property which allows the system to be only asymptotically stable at the solution set of the constraints. Between these two extremes, there is the reachability property which ensures that the system approaches the solution set of the constraints eventually.

For example, the state variables of the robot soccer player (Fig. 3) are (x, y, θ). Let the desired goal be (x_d, y_d, θ_d). The set of constraints will be $x = x_d, y = y_d, \theta = \theta_d$. Let G denote an ϵ-neighborhood of (x_d, y_d, θ_d). A constraint-based specification for the soccer player is the persistence property $\Box\Diamond G$.

As another example, if the boundary of the soccer field is described by a manifold in the configuration space [9] of the car $f(x, y, \theta) = 0$, a safety property for the soccer player could be $\Box f(x, y, \theta) > 0$, that is, the car is always on the field. The corresponding constraint for this specification is the inequality $f(x, y, \theta) \geq \epsilon$ for some $\epsilon > 0$.

5 Constraint-Based Control

How does the controller satisfy the constraints?

Given a constraint-based requirements specification, the design of the controller becomes the synthesis of an embedded constraint solver which, together with the dynamics of the plant and the environment, solves the given constraints on-line.

5.1 Constraint methods

We have viewed constraint satisfaction as a dynamic process that approaches the solution set of the constraints [20]. We have also specified this constraint-based behavior using TLTL and timed $\forall$-automata [18]. Let C be a set of constraints, *sol*(C) be the solution set of C and C^ϵ be the ϵ-neighborhood of *sol*(C); a constraint net CN is a ***constraint solver*** for C, iff $\forall\epsilon > 0$, $[\![CN]\!] \models \Diamond\Box C^\epsilon$. For example, a constraint net corresponding to $\dot{x} = -x$ is a constraint solver for $x = 0$.

There are typically two kinds of constraint satisfaction problems, namely, global consistency and optimization [20]. The problem of ***global consistency*** is to find a solution tuple that satisfies all the given constraints. The problem of ***unconstrained optimization*** is to minimize a function $\mathcal{E} : \mathcal{R}^n \to \mathcal{R}$. Global consistency is for solving ***hard*** constraints and unconstrained optimization is for solving ***soft*** constraints. A problem of the first kind can be translated into one of the second by introducing an energy function representing the degree of global consistency. For example, given a set of equations $g(x) = 0$, let $\mathcal{E}_g(x) = \| g(x) \|_2$ where $\| \cdot \|_2$ is a quadratic norm. If a constraint solver CN solves $\min \mathcal{E}_g(x)$, CN solves $g(x) = 0$ if a solution exists.

Various constraint methods fit into our framework for constraint satisfaction [20]. There are two classes of constraint methods, ***discrete relaxation*** which can be implemented as state transition systems, and ***differential optimization*** which can be implemented as state integration systems. Both can be modeled as constraint nets [20]. We will illustrate these methods with two examples. For more comprehensive analysis of these methods, the reader is referred to [20].

Newton's method [16] is a typical discrete relaxation method that minimizes a second-order approximation of the given function at each iterative step. Let $\nabla\mathcal{E} = \frac{\partial\mathcal{E}}{\partial x}$ and J be the Jacobian of $\nabla\mathcal{E}$. At each step with current point $x^{(k)}$, Newton's method minimizes the function:

$$\mathcal{E}_a(x) = \mathcal{E}(x^{(k)}) + \nabla\mathcal{E}^T(x^{(k)})(x - x^{(k)}) + \frac{1}{2}(x - x^{(k)})^T J(x^{(k)})(x - x^{(k)}).$$

Let $\frac{\partial \mathcal{E}_k}{\partial x} = 0$, we have:

$$\nabla \mathcal{E}(x^{(k)}) + J(x^{(k)})(x - x^{(k)}) = 0.$$

The solution of the above equation becomes the next point, i.e.,

$$x^{(k+1)} = x^{(k)} - J^{-1}(x^{(k)}) \nabla \mathcal{E}(x^{(k)}).$$

Newton's method defines a state transition system, with the next state as a function of the current state.

The *gradient method* [14] is a typical differential optimization method, based on the gradient descent algorithm, where state variables slide downhill in the direction opposed to the gradient. Formally, if the function to be minimized is $\mathcal{E}$, the vector that points in the direction of maximum increase of $\mathcal{E}$ is the gradient of $\mathcal{E}$. Therefore, let $\nabla \mathcal{E} = \frac{\partial \mathcal{E}}{\partial x}$, the following gradient descent equation models the gradient method

$$\dot{x} = -k \nabla \mathcal{E}(x), \quad k > 0. \tag{5}$$

Various constraint methods [13, 20] can be applied to control synthesis under this framework. More importantly, as we have shown in [20], most constraint methods are associated with an energy function, which can be directly used as a Liapunov function in behavior verification [18].

5.2 Constraint-based controllers

Constraint solvers are closed nets, with all state variables controllable (outputs). Controllers are open systems, with both sensory inputs and controllable outputs. We call a controller an *embedded constraint solver* if the controller, together with the plant and the environment, satisfies the given constraint-based specification.

Most constraint methods can be applied to embedded constraint solvers as well. Embedded constraint solvers can be either discrete or continuous according to the constraint method used. Continuous solvers, based on the gradient method, generalize potential functions [6]. Discrete solvers, based on numerical or symbolic relaxation methods, are more flexible in many applications.

If Newton's method or the gradient method is used, the design of an energy function depends on the type of constraint-based specification. For reachability or persistence constraints, the energy function defines the degree of satisfaction of the constraints; for safety constraints, the energy function defines the degree of satisfaction of the constraints within C^ϵ while the energy outside of C^ϵ becomes infinity. For example, given a persistence specification $\Box\Diamond C^\epsilon$ with C defined as $f(x) = 0$, an energy function for this specification can be $\frac{1}{2}f^2(x)$. If $\Box(f(x) > 0)$ is required, an energy function can be $\max(-\ln \frac{f(x)}{\epsilon}, 0)$, i.e., if $f(x) \geq \epsilon$, then $\mathcal{E}(x) = 0$, if $0 < f(x) < \epsilon$, then $\mathcal{E} > 0$, and if $f(x) \to 0$, then $\mathcal{E}(x) \to \infty$.

Various existing controllers, from simple linear proportional derivative (PD) control to complex nonlinear adaptive control or potential field methods [6], can be derived and analyzed in this framework. Various learning algorithms and neural net computations [14] can be formalized in this framework as well.

In order to synthesize a controller for the robot soccer player, we must define a mapping from the plant state variables $\langle x, y, \theta \rangle$ to the control variables $\langle v, \alpha \rangle$, so that the system satisfies $\Box\Diamond C^{\epsilon}$ where C is the set of constraints $\{x = x_d, y = y_d, \theta = \theta_d\}$. We use the gradient method and define an energy function for the controller as

$$\mathcal{E} = \frac{k_p}{2}(x_d - x)^2 + \frac{k_p}{2}(y_d - y)^2 + \frac{k_t}{2}(\theta_d - \theta)^2.$$

The controller is designed to make $\mathcal{E}$ go to its minimum.

The derivation of the controller is as follows. Let p be the length of the path. We have $v = \dot{p}$. Using the gradient method on the energy function $\mathcal{E}$, we have

$$\dot{p} = -k_1 \frac{\partial \mathcal{E}}{\partial p}, \quad \dot{\theta} = -k_2 \frac{\partial \mathcal{E}}{\partial \theta}$$

where $\frac{\partial \mathcal{E}}{\partial p}$ and $\frac{\partial \mathcal{E}}{\partial \theta}$ can be computed as follows:

$$-\frac{\partial \mathcal{E}}{\partial p} = k_p(x_d - x)\frac{\partial x}{\partial p} + k_p(y_d - y)\frac{\partial y}{\partial p} + k_t(\theta_d - \theta)\frac{\partial \theta}{\partial p}$$

where

$$\frac{\partial x}{\partial p} = \frac{\dot{x}}{v} = \cos(\theta), \quad \frac{\partial y}{\partial p} = \frac{\dot{y}}{v} = \sin(\theta), \quad \frac{\partial \theta}{\partial p} = \frac{\dot{\theta}}{v} = \frac{\tan(\alpha)}{L}$$

and

$$-\frac{\partial \mathcal{E}}{\partial \theta} = k_p(x_d - x)\frac{\partial x}{\partial \theta} + k_p(y_d - y)\frac{\partial y}{\partial \theta} + k_t(\theta_d - \theta)$$

where

$$\frac{\partial x}{\partial \theta} \doteq -v\sin(\theta), \quad \frac{\partial y}{\partial \theta} \doteq v\cos(\theta).$$

Let $d = \sqrt{(x_d - x)^2 + (y_d - y)^2}$ and $\theta' = \tan^{-1}(y_d - y, x_d - x)$. According to the dynamics of the plant modeled by Equation (1), we have $\alpha = \tan^{-1}(\frac{L}{v}\dot{\theta})$. Therefore, the control law for the tracking problem is:

$$v = k_1[k_p d\cos(\theta' - \theta) + k_t(\theta_d - \theta)\frac{\tan(\alpha)}{L}]$$

$$\alpha = \tan^{-1}(\frac{L}{v}k_2(k_p v d\sin(\theta' - \theta) + k_t(\theta_d - \theta))).$$

Now we are able to analyze this control law using the energy function for the controller as the Liapunov function of the system. Since

$$\begin{aligned}\dot{\mathcal{E}} &= -[k_p(x_d - x)\dot{x} + k_p(y_d - y)\dot{y} + k_t(\theta_d - \theta)\dot{\theta}] \\ &= -[k_p(x_d - x)\cos(\theta) + k_p(y_d - y)\sin(\theta) + k_t(\theta_d - \theta)\frac{\tan(\alpha)}{L}]v \\ &= -\frac{1}{k_1}v^2 \leq 0,\end{aligned}$$

the Liapunov function is nonincreasing, therefore, the system is "stable." However, strictly speaking, the controller is not an embedded constraint solver since the system is not asymptotically stable at the solution: if $|\theta' - \theta| = \frac{\pi}{2}k$ and $\theta_d = \theta$ we get $v = 0$ even when $d \neq 0$. We prove that they are the only singularities of this control law.

Proposition 5.1 *This control law satisfies the condition $v = 0$ iff*

$$(d = 0 \vee |\theta' - \theta| = \frac{\pi}{2}k) \bigwedge (\theta_d = \theta).$$

Proof: According to the control law for α, $v = 0$ implies $\theta_d = \theta$. According to the control law for v, $v = 0$ implies $d\cos(\theta' - \theta) = 0$. □

Therefore, the car can stop at singularities without reaching the desired goal. In order to solve this problem, we use a hierarchical control structure.

6 Robotic Architecture

How should we compose a control system?

A control system, in general, is implemented in a vertical hierarchy [1] corresponding to a hierarchical abstraction of time and domains (Fig. 5). The bottom level sends

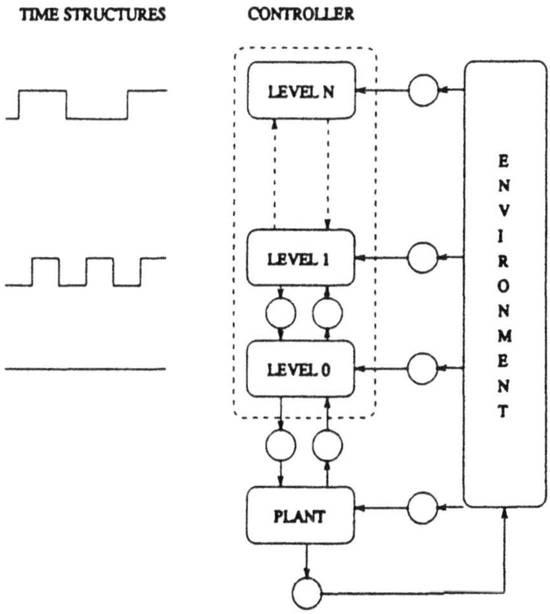

Fig. 5. Abstraction hierarchy

control signals to various actuators, and at the same time, senses the state of actuators. Control signals flow down and the sensing signals flow up. Sensing signals from the environment are distributed over levels. Each level is a black box that represents the causal relationship between the inputs and the outputs. The inputs consist of the control signals from the higher level, the sensing signals from the environment and the current states from the lower level. The outputs consist of the control signals to the lower level and the current states to the higher level. Usually, the bottom level is implemented by analog circuits that function with continuous dynamics and the higher levels are realized by distributed computing networks.

In our framework of control synthesis, constraints are specified at different levels on different domains, with the higher levels more abstract and the lower levels more plant-dependent.

A control system can be synthesized as a hierarchy of interactive embedded constraint solvers. Each abstraction level solves constraints on its state space and produces the input to the lower level. The higher levels are composed of digital/symbolic event-driven control derived from discrete constraint methods and the lower levels embody analog control based on continuous constraint methods.

For the robot soccer player, we have designed a two level controller. The low level tracks the given goal position (x_d, y_d, θ_d) in continuous time and the high level, driven by events from below, sets a new goal position according to the current ball position, the target position and the boundary of the field. Obstacle and singularity avoidance are considered at this level too. We will not describe this level in any detail here. There are various kinds of events to trigger the high level planner; the current version uses a simple event: the car coming to a stop (which implies that either the current goal is achieved or the car is stuck because of singularity or obstacles). Other events such as the ball moving significantly or an obstacle obstructing the path can be incorporated. Fig. 6 shows an overall CN control structure, with the models of the car (plant) and the ball (environment).

7 Developing Hybrid Control Systems
Where are we and where are we going?

We have considered control synthesis as one of the four key problems in developing hybrid control systems, together with modeling, specification and verification.

We have developed a unified framework for synthesizing continuous/discrete hybrid control systems based on a simple principle: on-line constraint satisfaction. However, we are not aiming to replace existing control theory, rather to formalize and generalize the underlying principles that are used in practice.

Local minima and/or singularities can be a major problem for this type of controller. In general, a complex robot control system should be developed and organized hierarchically. Normally singularities can be avoided if a higher level control strategy is used to detect singularities and to produce a sequence of intermediate configurations between the actual and the target configurations. Such a higher level event-triggered control strategy becomes more important when the robot is embedded in a complex environment.

We have developed a modeling and simulation environment, called ALERT (A Laboratory for Embedded Real-Time systems), in which the robot soccer player has been modeled and simulated. We are experimenting with an implementation on the real soccer players.

In the future, we also plan to characterize some restricted classes of hybrid control systems and investigate semi-automatic or automatic control synthesis methods for

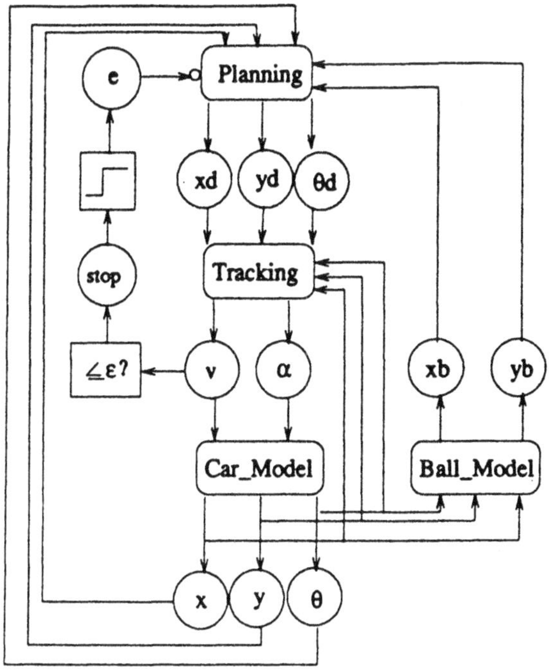

Fig. 6. A control structure for the robot soccer player

these classes. The fundamental goal is to provide foundations for the design and analysis of robotic systems and behaviors.

References

1. J. S. Albus. *Brains, Behavior, and Robotics.* BYTE Publications, 1981.
2. E. A. Ashcroft. Dataflow and eduction: Data-driven and demand-driven distributed computation. In J. W. deBakker, W.P. deRoever, and G. Rozenberg, editors, *Current Trends in Concurrency*, number 224 in Lecture Notes on Computer Science, pages 1 - 50. Springer-Verlag, 1986.
3. A. Benveniste and P. LeGuernic. Hybrid dynamical systems theory and the SIGNAL language. *IEEE Transactions on Automatic Control*, 35(5):535 - 546, May 1990.
4. P. Caspi, D. Pilaud, N. Halbwachs, and J. A. Plaice. LUSTRE: A declarative language for programming synchronous systems. In *ACM Proceedings on Principles of Programming Languages*, pages 178 - 188, 1987.
5. E. Emerson. Temporal and modal logic. In Jan Van Leeuwen, editor, *Handbook of Theoretical Computer Science*, volume B: Formal Models and Semantics. Elsevier, MIT Press, 1990.
6. D. E. Koditschek. Robot planning and control via potential functions. In J. Craig O. Khatib and T. Lozano-Perez, editors, *The Robotic Review 1.* MIT Press, 1989.
7. J. C. Latombe. *Robot Motion Planning.* Kluwer Academic Publishers, 1991.

8. J. Lavignon and Y. Shoham. Temporal automata. Technical Report STAN-CS-90-1325, Robotics Laboratory, Computer Science Department, Stanford University, Stanford, CA 94305, 1990.
9. T. Lozano-Perez. Spatial planning: A configuration space approach. *IEEE Transactions on Computers*, C-32(2), February 1983.
10. Z. Manna and A. Pnueli. Specification and verification of concurrent programs by ∀-automata. In *Proc. 14th Ann. ACM Symp. on Principles of Programming Languages*, pages 1–12, 1987.
11. Z. Manna and A. Pnueli. *The Temporal Logic of Reactive and Concurrent Systems*. Springer-Verlag, 1992.
12. A. Nerode and W. Kohn. Models for hybrid systems: Automata, topologies, controllability, observability. In R. L. Grossman, A. Nerode, A. P. Ravn, and H. Rischel, editors, *Hybrid Systems*, number 736 in Lecture Notes on Computer Science, pages 317 – 356. Springer-Verlag, 1993.
13. D. K. Pai. Least constraint: A framework for the control of complex mechanical systems. In *Proceedings of American Control Conference*, pages 426 – 432, Boston, 1991.
14. J. Platt. Constraint methods for neural networks and computer graphics. Technical Report Caltech-CS-TR-89-07, Department of Computer Science, California Institute of Technology, 1989.
15. M. Sahota and A. K. Mackworth. Can situated robots play soccer? In *1994 Canadian Artificial Intelligence, Banff, Alberta*, May 1994.
16. J. T. Sandfur. *Discrete Dynamical Systems: Theory and Applications*. Clarendon Press, 1990.
17. Y. Zhang. A foundation for the design and analysis of robotic systems and behaviors. Technical Report 94-26, Department of Computer Science, University of British Columbia, 1994. Ph.D. thesis.
18. Y. Zhang and A. K. Mackworth. Specification and verification of constraint-based dynamic systems. In A. Borning, editor, *Principles and Practice of Constraint Programming*, Lecture Notes in Computer Science 874, pages 229 – 242. Springer Verlag, 1994.
19. Y. Zhang and A. K. Mackworth. Constraint Nets: A semantic model for hybrid dynamic systems. *Theoretical Computer Science*, (138):211 – 239, 1995. Special Issue on Hybrid Systems.
20. Y. Zhang and A. K. Mackworth. Constraint programming in constraint nets. In V. Saraswat and P. Van Hentenryck, editors, *Principles and Practice of Constraint Programming*, pages 49 – 68. MIT Press, 1995.

Index

Andersen 391
Antsaklis 322, 462
Asarin 1
Banks 509
Benveniste 21
Bett 322
Birdwell 45
Bobrow 226
Bouajjani 61
Caines 86
Conrad 391
DeClaris 106
Deshpande 128
Dogruel 148
Echahed 61
Egardt 193
Eriksen 391
Fabre 21
Godbole 166
Grossman 191
Guckenheimer 202
Gupta 226
Henzinger 252,265
Ho 252, 265
Holdgaard, 391
Jagadeesan 226
Johnson 202
Kim 529
Kohn 291
Le Guernic 21
Lemmon 322, 462
Levy 21
Linkens 509
Lygeros 166
Mackworth 552
Maler 1
Moore 45
Nerode 291, 341
Özgüner 148
Pnueli 1
Praehofer 529
Puri 359
Raisch 370
Ramesh 136
Ravn 391
Remmel 291, 341
Rischel 391
Robbana 61
Roux 105
Rusu 105
Saraswat 226
Sarjoughian 117
Sastry 166
Shyamasundar 136
Song 529
Stiver 462
Su 106
Sweedler 191
Szymanski 322
Tittus 193
Varaiya 128, 359
Wei 86
Yakhnis 341
Yang 509
Zeigler 117, 529
Zhang 552

Lecture Notes in Computer Science

For information about Vols. 1–920

please contact your bookseller or Springer-Verlag

Vol. 921: L. C. Guillou, J.-J. Quisquater (Eds.), Advances in Cryptology – EUROCRYPT '95. Proceedings, 1995. XIV, 417 pages. 1995.

Vol. 922: H. Dörr, Efficient Graph Rewriting and Its Implementation. IX, 266 pages. 1995.

Vol. 923: M. Meyer (Ed.), Constraint Processing. IV, 289 pages. 1995.

Vol. 924: P. Ciancarini, O. Nierstrasz, A. Yonezawa (Eds.), Object-Based Models and Languages for Concurrent Systems. Proceedings, 1994. VII, 193 pages. 1995.

Vol. 925: J. Jeuring, E. Meijer (Eds.), Advanced Functional Programming. Proceedings, 1995. VII, 331 pages. 1995.

Vol. 926: P. Nesi (Ed.), Objective Software Quality. Proceedings, 1995. VIII, 249 pages. 1995.

Vol. 927: J. Dix, L. Moniz Pereira, T. C. Przymusinski (Eds.), Non-Monotonic Extensions of Logic Programming. Proceedings, 1994. IX, 229 pages. 1995. (Subseries LNAI).

Vol. 928: V.W. Marek, A. Nerode, M. Truszczynski (Eds.), Logic Programming and Nonmonotonic Reasoning. Proceedings, 1995. VIII, 417 pages. 1995. (Subseries LNAI).

Vol. 929: F. Morán, A. Moreno, J.J. Merelo, P. Chacón (Eds.), Advances in Artificial Life. Proceedings, 1995. XIII, 960 pages. 1995 (Subseries LNAI).

Vol. 930: J. Mira, F. Sandoval (Eds.), From Natural to Artificial Neural Computation. Proceedings, 1995. XVIII, 1150 pages. 1995.

Vol. 931: P.J. Braspenning, F. Thuijsman, A.J.M.M. Weijters (Eds.), Artificial Neural Networks. IX, 295 pages. 1995.

Vol. 932: J. Iivari, K. Lyytinen, M. Rossi (Eds.), Advanced Information Systems Engineering. Proceedings, 1995. XI, 388 pages. 1995.

Vol. 933: L. Pacholski, J. Tiuryn (Eds.), Computer Science Logic. Proceedings, 1994. IX, 543 pages. 1995.

Vol. 934: P. Barahona, M. Stefanelli, J. Wyatt (Eds.), Artificial Intelligence in Medicine. Proceedings, 1995. XI, 449 pages. 1995. (Subseries LNAI).

Vol. 935: G. De Michelis, M. Diaz (Eds.), Application and Theory of Petri Nets 1995. Proceedings, 1995. VIII, 511 pages. 1995.

Vol. 936: V.S. Alagar, M. Nivat (Eds.), Algebraic Methodology and Software Technology. Proceedings, 1995. XIV, 591 pages. 1995.

Vol. 937: Z. Galil, E. Ukkonen (Eds.), Combinatorial Pattern Matching. Proceedings, 1995. VIII, 409 pages. 1995.

Vol. 938: K.P. Birman, F. Mattern, A. Schiper (Eds.), Theory and Practice in Distributed Systems. Proceedings,1994. X, 263 pages. 1995.

Vol. 939: P. Wolper (Ed.), Computer Aided Verification. Proceedings, 1995. X, 451 pages. 1995.

Vol. 940: C. Goble, J. Keane (Eds.), Advances in Databases. Proceedings, 1995. X, 277 pages. 1995.

Vol. 941: M. Cadoli, Tractable Reasoning in Artificial Intelligence. XVII, 247 pages. 1995. (Subseries LNAI).

Vol. 942: G. Böckle, Exploitation of Fine-Grain Parallelism. IX, 188 pages. 1995.

Vol. 943: W. Klas, M. Schrefl, Metaclasses and Their Application. IX, 201 pages. 1995.

Vol. 944: Z. Fülöp, F. Gécseg (Eds.), Automata, Languages and Programming. Proceedings, 1995. XIII, 686 pages. 1995.

Vol. 945: B. Bouchon-Meunier, R.R. Yager, L.A. Zadeh (Eds.), Advances in Intelligent Computing - IPMU '94. Proceedings, 1994. XII, 628 pages.1995.

Vol. 946: C. Froidevaux, J. Kohlas (Eds.), Symbolic and Quantitative Approaches to Reasoning and Uncertainty. Proceedings, 1995. X, 420 pages. 1995. (Subseries LNAI).

Vol. 947: B. Möller (Ed.), Mathematics of Program Construction. Proceedings, 1995. VIII, 472 pages. 1995.

Vol. 948: G. Cohen, M. Giusti, T. Mora (Eds.), Applied Algebra, Algebraic Algorithms and Error-Correcting Codes. Proceedings, 1995. XI, 485 pages. 1995.

Vol. 949: D.G. Feitelson, L. Rudolph (Eds.), Job Scheduling Strategies for Parallel Processing. Proceedings, 1995. VIII, 361 pages. 1995.

Vol. 950: A. De Santis (Ed.), Advances in Cryptology - EUROCRYPT '94. Proceedings, 1994. XIII, 473 pages. 1995.

Vol. 951: M.J. Egenhofer, J.R. Herring (Eds.), Advances in Spatial Databases. Proceedings, 1995. XI, 405 pages. 1995.

Vol. 952: W. Olthoff (Ed.), ECOOP '95 - Object-Oriented Programming. Proceedings, 1995. XI, 471 pages. 1995.

Vol. 953: D. Pitt, D.E. Rydeheard, P. Johnstone (Eds.), Category Theory and Computer Science. Proceedings, 1995. VII, 252 pages. 1995.

Vol. 954: G. Ellis, R. Levinson, W. Rich. J.F. Sowa (Eds.), Conceptual Structures: Applications, Implementation and Theory. Proceedings, 1995. IX, 353 pages. 1995. (Subseries LNAI).

VOL. 955: S.G. Akl, F. Dehne, J.-R. Sack, N. Santoro (Eds.), Algorithms and Data Structures. Proceedings, 1995. IX, 519 pages. 1995.

Vol. 956: X. Yao (Ed.), Progress in Evolutionary Computation. Proceedings, 1993, 1994. VIII, 314 pages. 1995. (Subseries LNAI).

Vol. 957: C. Castelfranchi, J.-P. Müller (Eds.), From Reaction to Cognition. Proceedings, 1993. VI, 252 pages. 1995. (Subseries LNAI).

Vol. 958: J. Calmet, J.A. Campbell (Eds.), Integrating Symbolic Mathematical Computation and Artificial Intelligence. Proceedings, 1994. X, 275 pages. 1995.

Vol. 959: D.-Z. Du, M. Li (Eds.), Computing and Combinatorics. Proceedings, 1995. XIII, 654 pages. 1995.

Vol. 960: D. Leivant (Ed.), Logic and Computational Complexity. Proceedings, 1994. VIII, 514 pages. 1995.

Vol. 961: K.P. Jantke, S. Lange (Eds.), Algorithmic Learning for Knowledge-Based Systems. X, 511 pages. 1995. (Subseries LNAI).

Vol. 962: I. Lee, S.A. Smolka (Eds.), CONCUR '95: Concurrency Theory. Proceedings, 1995. X, 547 pages. 1995.

Vol. 963: D. Coppersmith (Ed.), Advances in Cryptology - CRYPTO '95. Proceedings, 1995. XII, 467 pages. 1995.

Vol. 964: V. Malyshkin (Ed.), Parallel Computing Technologies. Proceedings, 1995. XII, 497 pages. 1995.

Vol. 965: H. Reichel (Ed.), Fundamentals of Computation Theory. Proceedings, 1995. IX, 433 pages. 1995.

Vol. 966: S. Haridi, K. Ali, P. Magnusson (Eds.), EURO-PAR '95 Parallel Processing. Proceedings, 1995. XV, 734 pages. 1995.

Vol. 967: J.P. Bowen, M.G. Hinchey (Eds.), ZUM '95: The Z Formal Specification Notation. Proceedings, 1995. XI, 571 pages. 1995.

Vol. 968: N. Dershowitz, N. Lindenstrauss (Eds.), Conditional and Typed Rewriting Systems. Proceedings, 1994. VIII, 375 pages. 1995.

Vol. 969: J. Wiedermann, P. Hájek (Eds.), Mathematical Foundations of Computer Science 1995. Proceedings, 1995. XIII, 588 pages. 1995.

Vol. 970: V. Hlaváč, R. Šára (Eds.), Computer Analysis of Images and Patterns. Proceedings, 1995. XVIII, 960 pages. 1995.

Vol. 971: E.T. Schubert, P.J. Windley, J. Alves-Foss (Eds.), Higher Order Logic Theorem Proving and Its Applications. Proceedings, 1995. VIII, 400 pages. 1995.

Vol. 972: J.-M. Hélary, M. Raynal (Eds.), Distributed Algorithms. Proceedings, 1995. XI, 333 pages. 1995.

Vol. 973: H.H. Adelsberger, J. Lažanský, V. Mařík (Eds.), Information Management in Computer Integrated Manufacturing. IX, 665 pages. 1995.

Vol. 974: C. Braccini, L. DeFloriani, G. Vernazza (Eds.), Image Analysis and Processing. Proceedings, 1995. XIX, 757 pages. 1995.

Vol. 975: W. Moore, W. Luk (Eds.), Field-Programmable Logic and Applications. Proceedings, 1995. XI, 448 pages. 1995.

Vol. 976: U. Montanari, F. Rossi (Eds.), Principles and Practice of Constraint Programming — CP '95. Proceedings, 1995. XIII, 651 pages. 1995.

Vol. 977: H. Beilner, F. Bause (Eds.), Quantitative Evaluation of Computing and Communication Systems. Proceedings, 1995. X, 415 pages. 1995.

Vol. 978: N. Revell, A M. Tjoa (Eds.), Database and Expert Systems Applications. Proceedings, 1995. XV, 654 pages. 1995.

Vol. 979: P. Spirakis (Ed.), Algorithms — ESA '95. Proceedings, 1995. XII, 598 pages. 1995.

Vol. 980: A. Ferreira, J. Rolim (Eds.), Parallel Algorithms for Irregularly Structured Problems. Proceedings, 1995. IX, 409 pages. 1995.

Vol. 981: I. Wachsmuth, C.-R. Rollinger, W. Brauer (Eds.), KI-95: Advances in Artificial Intelligence. Proceedings, 1995. XII, 269 pages. (Subseries LNAI).

Vol. 982: S. Doaitse Swierstra, M. Hermenegildo (Eds.), Programming Languages: Implementations, Logics and Programs. Proceedings, 1995. XI, 467 pages. 1995.

Vol. 983: A. Mycroft (Ed.), Static Analysis. Proceedings, 1995. VIII, 423 pages. 1995.

Vol. 984: J.-M. Haton, M. Keane, M. Manago (Eds.), Advances in Case-Based Reasoning. Proceedings, 1994. VIII, 307 pages. 1995.

Vol. 985: T. Sellis (Ed.), Rules in Database Systems. Proceedings, 1995. VIII, 373 pages. 1995.

Vol. 986: Henry G. Baker (Ed.), Memory Management. Proceedings, 1995. XII, 417 pages. 1995.

Vol. 987: P.E. Camurati, H. Eveking (Eds.), Correct Hardware Design and Verification Methods. Proceedings, 1995. VIII, 342 pages. 1995.

Vol. 988: A.U. Frank, W. Kuhn (Eds.), Spatial Information Theory. Proceedings, 1995. XIII, 571 pages. 1995.

Vol. 989: W. Schäfer, P. Botella (Eds.), Software Engineering — ESEC '95. Proceedings, 1995. XII, 519 pages. 1995.

Vol. 990: C. Pinto-Ferreira, N.J. Mamede (Eds.), Progress in Artificial Intelligence. Proceedings, 1995. XIV, 487 pages. 1995. (Subseries LNAI).

Vol. 991: J. Wainer, A. Carvalho (Eds.), Advances in Artificial Intelligence. Proceedings, 1995. XII, 342 pages. 1995. (Subseries LNAI).

Vol. 992: M. Gori, G. Soda (Eds.), Topics in Artificial Intelligence. Proceedings, 1995. XII, 451 pages. 1995. (Subseries LNAI).

Vol. 993: T.C. Fogarty (Ed.), Evolutionary Computing. Proceedings, 1995. VIII, 264 pages. 1995.

Vol. 994: M. Hebert, J. Ponce, T. Boult, A. Gross (Eds.), Object Representation in Computer Vision. Proceedings, 1994. VIII, 359 pages. 1995.

Vol. 997: K.P. Jantke, T. Shinohara, T. Zeugmann (Eds.), Algorithmic Learning Theory. Proceedings, 1995. XV, 319 pages. 1995.

Vol. 998: A. Clarke, M. Campolargo, N. Karatzas (Eds.), Bringing Telecommunication Services to the People – IS&N '95. Proceedings, 1995. XII, 510 pages. 1995.

Vol. 999: P. Antsaklis, W. Kohn, A. Nerode, S. Sastry (Eds.), Hybrid Systems II. VIII, 569 pages. 1995.